# Principles of
# Digital Transmission
## With Wireless Applications

# Information Technology: Transmission, Processing, and Storage

Series Editor:  Jack Keil Wolf
*University of California at San Diego*
*La Jolla, California*

---

Principles of Digital Transmission: With Wireless Applications
Sergio Benedetto and Ezio Biglieri

# Principles of
# Digital Transmission

## With Wireless Applications

Sergio Benedetto and
Ezio Biglieri

*Politecnico di Torino*
*Torino, Italia*

**KLUWER ACADEMIC / PLENUM PUBLISHERS**
NEW YORK, BOSTON, DORDRECHT, LONDON, MOSCOW

Library of Congress Cataloging in Publication Data

Benedetto, Sergio.
    Principles of digital transmission: with wireless applications / Sergio Benedetto and Ezio
    Biglieri.
        p.     cm.—(Plenum series in telecommunications)
    Includes bibliographical references and index.
    ISBN 0-306-45753-9
    1. Digital communications. 2. Wireless communication systems. I. Biglieri, Ezio. II. Title.
III. Series.
TK5103.7.B464   1999                                                                              98-46066
621.382—dc21                                                                                              CIP

2

ISBN 0-306-45753-9

©1999 Kluwer Academic / Plenum Publishers, New York
233 Spring Street, New York, N.Y. 10013

10 9 8 7 6 5 4 3 2 1

A C.I.P. record for this book is available from the Library of Congress.

Printed in the United States of America

95.00                                ASH-5210                                          ·  10/11

# Preface

*Quelli che s'innamoran di pratica sanza scienzia,*
*son come 'l nocchieri ch'entra in navilio sanza timone o bussola,*
*che mai ha la certezza dove si vada.*
*Leonardo da Vinci, Codex G, Bibliothèque de l'Institut de France, Paris.*

This books stems from its ancestor *Digital Transmission Theory*, published by Prentice-Hall in 1987 and now out of print. Following the suggestion of several colleagues who complained about the unavailability of a textbook they liked and adopted in their courses, two out of its three former authors have deeply revised and updated the old book, laying a strong emphasis on wireless communications. We hope that those who liked the previous book will find again its flavor here, while new readers, untouched by nostalgia, will judge it favorably.

In keeping with the cliché "every edition is an addition," we started planning what new topics were needed in a textbook trying to provide a substantial covering of the discipline. However, we immediately became aware that an in-depth discussion of the many things we deemed appropriate for inclusion would quickly make this book twice the size of the previous one. It would certainly be nice to write, as in the Mahābhārata, "what is in this book, you can find somewhere else; but what is not in it, you cannot find anywhere." Yet such a book, like Borges' map drawn to 1:1 scale, would not hit the mark. For this reason we aimed at writing an entirely new book, whose focus was on (although not totally restricted to) wireless digital transmission, an area whose increasing relevance in these days need not be stressed. Even with this shift in focus, we are aware that many things were left out, so that the reader should not expect an encyclopedic coverage of the discipline, but rather a relatively thorough coverage of some important parts of it.

Some readers may note with dismay that in a book devoted, at least partially, to wireless communications, there is no description of wireless *systems*. If we were to choose an icon for this book, we would choose Carroll's Cheshire Cat of Wonderland. As Martin Gardner notes in his "Annotated Alice," the phrase "grin without a cat" is not a bad description of pure mathematics. Similarly, we think of this phrase as a good description of "communication theory" as contrasted to "communication systems." A book devoted to communication systems alone would be a cat without a grin: thus, due to the practical impossibility of delivering both, we opted for the grin. Another justification is that, as the Cheshire Cat is identified only by its smile, so we have characterized communications by its theoretical foundations.

Our goal is primarily to provide a textbook for senior or beginning-graduate students, although practicing engineers will probably find it useful. We agree with Plato, who in his *Seventh Letter* contrasts the dialectic method of teaching, exemplified by Socrates' personal, interactive mode of instruction, with that afforded by the written word. Words can only offer a shallow form of teaching: when questioned, they always provide the same answer, and cannot convey ultimate truths. Instruction can only take place within a dialogue, which a book can never offer. Yet, we hope that our treatment is reflective enough of our teaching experience so as to provide a useful tool for self-study.

We assume that the reader has a basic understanding of Fourier transform techniques, probability theory, random variables, random processes, signal transmission through linear systems, the sampling theorem, linear modulation methods, matrix algebra, vector spaces, and linear transformations. However, advanced knowledge of these topics is not required.

This book can serve as a text in either one-semester or two-semester courses in digital communications. We outline below some possible, although not exhaustive, roadmaps.

1. *A one-term basic course in digital communications:*
   Select review sections in Chapters 2, 3, 4, and 5, parts of Chapters 7 and 9.

2. *A one-term course in advanced digital communications:*
   Select review sections in Chapters 4 and 5, then Chapters 6, 7, 8, 9, and 13.

3. *A one-term course in information theory and coding:*
   Chapters 3, 9, 10, 11, 12, and parts of 13.

4. *A two-term course sequence in digital communications and coding:*
   (A) Select review sections in Chapters 2, 3, 4, 5, 6, and 7.
   (B) Chapters 9, 10, 11, 12, 13, and 14.

History tells us that Tolstoy's wife, Sonya, copied out "War and Peace" seven times. Since in these days wives are considerably less pliable than in 19th-century Russia, we produced the whole book by ourselves using LaTeX: this implies that we are solely responsible not only for technical inaccuracies, but also for typos. We would appreciate it if the readers who spot any of them would write to us at <benedetto,biglieri>@polito.it. An errata file will be kept and sent to anyone interested.

As this endeavor is partly the outcome of our teaching activity, it owes a great deal to our colleagues and students who volunteered to read parts of the book, correct mistakes, and provide criticism and suggestions for its improvement. We take this opportunity to acknowledge Giuseppe Caire, Andrea Carena, Vittorio Curri, G. David Forney, Jr., Roberto Garello, Roberto Gaudino, Jørn Justesen, Guido Montorsi, Giorgio Picchi, Pierluigi Poggiolini, S. Pas Pasupathy, Fabrizio Pollara, Bixio Rimoldi, Giorgio Taricco, Monica Visintin, Emanuele Viterbo, and Peter Willett. Participation of E.B. in symposia with Tony Ephremides, Ken Vastola, and Sergio Verdú, even when not strictly related to digital communications, was always conducive to scholarly productivity. Luciano Brino drew most of the figures with patience and skill.

*namo Gaṇeśāya vighnēśvarāya*

# Contents

# Contents

# Contents xvii

# Principles of
# Digital Transmission
## With Wireless Applications

# 1

# Introduction and motivation

In the information era we are experiencing, communication networks are a vital part of society's infrastructure. These networks accept information in electronic form and carry it to the final destination. Although a significant portion of the information to be transmitted is *analog* in nature (voice, TV signals, music), it is often converted into a *digital* signal before accessing the network. In the last two decades, the transmission of information in digital form has become predominant, and the analog-to-digital conversion has been pushed closer and closer to the source. Among the many reasons for this irreversible process, we can cite the availability of highly reliable, low-cost, small-sized digital circuitry, and the larger insensitivity of digital signals to channel impairments, the ease in enciphering the information, and the great variety of information signals to be transmitted (voice, images, alphanumeric characters, etc.), which can be made uniform after digitization.

From the point where analog-to-digital conversion takes place, the digital information is transmitted through the (possibly broadband) integrated services digital network (B-ISDN or simply ISDN). This scenario concerns then a large variety of digital signals of different formats and speeds that must be routed and switched to their proper destinations. Recently, the worldwide explosive success of the Internet, together with the advent of digital TV in encoded MPEG format and the great success of wireless telephony in digital form (GSM, IS54, IS95), have created a peak in the already steadily growing demand of digital transmission capacity.

The amount of data to be transmitted and the speed of transmission vary enormously from one application to another. When batches of data are sent for processing on a distant mainframe computer (an infrequent occurrence nowadays, with the advent of cheap and powerful desk-top personal computers), a

1

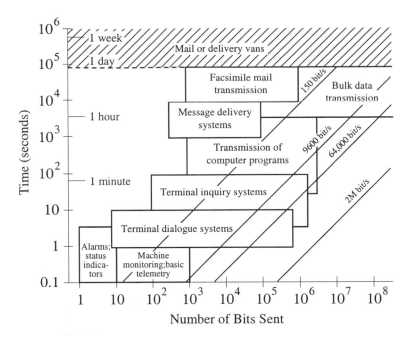

Figure 1.1: *Desirable delivery or response time and quantity of data for typical use of data transmission*

delivery time of the order of tens of minutes or longer may be acceptable. However, when a person-computer dialogue is taking place, the responses must be returned to the person quickly enough so as not to impede his or her train of thoughts. Response times between 100 milliseconds and 2 seconds are fine, but whoever has tried to download files using the Internet knows that the delays involved are often significantly longer. In real-time systems, where a machine or a process is being controlled, response times can vary from a few milliseconds to some seconds. Fig. 1.1 shows some of the common requirements for delivery times or response times and the amounts of data transmitted.

The block labeled "terminal dialogue systems," for example, indicates a response time from 1 to 10 seconds, and a message size ranging from 1 character (8 bits) to about 30,000 characters (around 240,000 bits), corresponding roughly to a JPEG picture. The transmission speed required by the communication link equals the number of transmitted bits (reported in the horizontal axis of Fig. 1.1) divided by the delivery time of one-way messages (reported in the vertical axis of Fig. 1.1). Straight lines on the figure correspond to a given speed, and some of them are also indicated. For most of the applications shown in the figure, the

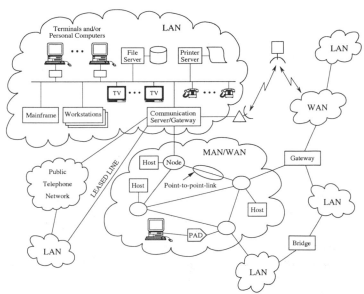

Figure 1.2: *Local area networks, metropolitan are networks (MAN), and wide area networks (WAN) interconnected through various telecommunication links.*

speeds allowed on telephone channels (up to, say, 64,000 bits per second) are sufficient. Of course, this concerns individual applications. We know, on the other hand, that the traffic incoming from several sources is often multiplexed to efficiently exploit the capacity of a single digital carrier. An impressive example of this is the up-to-date capacities of wave-division multiplexed fiber optic systems, which can convey more than 1 terabit per second in a single fiber!

A common form of digital communication nowadays consists of people sitting in front of a terminal (normally a personal computer) exchanging informations with other terminals (or a mainframe computer) or down-loading information from a provider. A community of users in a limited area is interconnected through a *local area network* (LAN) offering a variety of services like computing resources, voice and facsimile communication, teleconferencing, electronic mail, and access to distant Internet information providers.

Different LANs can exchange information over a packet-switched long-distance telecommunication network (e.g., through the Asynchronous Transfer Mode). Geographic networks of this kind are Metropolitan Area Networks (MAN) and Wide Area Networks (WAN), which can connect several nodes through high capacity links in a *ring* or *star* topology. This scenario can be represented as in Fig. 1.2.

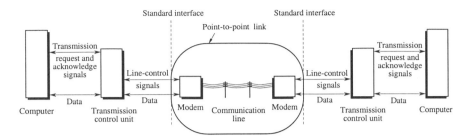

Figure 1.3: *Point-to-point communication link between two computers.*

In a geographic network, the communication engineer must solve a variety
of global problems, such as designing the topological structure of the network,
its link capacity allocation, and the routing and flow control procedures, as well
as local problems, such as the choice of the multiplexing scheme, the number of
message sources per concentration point, the access technique (polling, random
access, etc.), and the buffer size. The final system choices will be the result
of a trade-off between costs and performance, such as the (average, maximum)
response time and the specified reliability.

The exchange of information in a packet-switched network is governed by a
layered protocol architecture, such as that described, for example, in the ISO/OSI
reference model. Level 1 of this layered architecture concerns the point-to-point
communication between two nodes of the network. According to the physical
medium that connects the nodes, different problems are encountered to establish
a reliable link. Moreover, to access one of the nodes of the packet-switched
network, the user may have to exploit the dialed public telephone network or a
leased voice line.

Let us isolate and examine in greater detail one communication link in the
system of Fig. 1.2, for example, the one denoted as "point-to-point link," which
establishes a connection between two computers.[1] It is shown magnified in
Fig. 1.3. To be transmitted on the physical channel, the digital stream emitted by the transmitting computer must be converted into a sequence of waveforms suited to the channel. This operation is performed by a device known as
a *modem*, short for modulator/demodulator. The modem converts the data into a
signal whose range of frequencies matches the available bandwidth of the channel. Besides data, the terminal and the modem exchange various line-control
signals according to a standardized interface. At the other side, a modem converts the received waveforms into a digital stream that is sent to the receiving

---

[1] Very similar considerations could be applied to different forms of point-to-point connections,
like, for example, those regarding a mobile and base station in a wireless communication system.

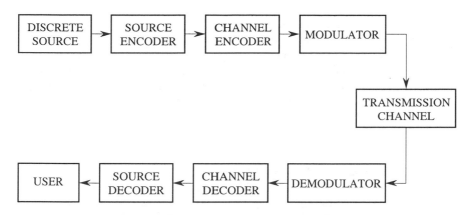

Figure 1.4: *Functional block diagram of a point-to-point digital communication system.*

computer through a transmission control unit that supervises the communication and implements the various layers of the ISO/OSI reference model.

The design of this *point-to-point* communication link is related to the choices made for the network in terms of available speeds, response times, multiplexing and access techniques. In particular, matching the sources of information to the channel speed may involve *source encoding* (like JPEG or MPEG for still and moving images, respectively), *channel bandwidth* and the choice of *modulation schemes*. The response time and the access techniques pose constraints on the modem setup time, that is, on the choice of the *synchronization* and *adaptive equalization* algorithms. The transmission quality is usually given in terms of *bit error probability*, which, in turn, depends on *channel encoding* (*error control*), the *transmitted power* and the modulation schemes.

This book is devoted to the theory of point-to-point digital communication. To resort to a more general and abstract context, let us expand the point-to-point connection of Fig. 1.3 into the functional block diagrams of Fig. 1.4. We only consider discrete information sources. When a source is analog in nature, such as a microphone activated by speech or a TV camera scanning a scene, we assume that a process of *sampling, quantizing* and *coding* takes place within the source, so that the output is a sequence of discrete symbols or letters. Discrete information sources are characterized by a *source alphabet*, a *source rate* (expressed in symbols per second), and a probability law governing the emission of sequences of symbols, or *messages*. From these parameters we can construct a probabilistic model of the information source and define a source information rate (denoted by $R_s$) in bits (binary digits) per second. The input to the second block of Fig. 1.4, the *source encoder*, is then a sequence of discrete symbols occurring at a certain rate. The source encoder converts the symbol sequence into a binary sequence by assigning *code words* to the symbols of the input sequence according to a specified rule. This encoding process has the goal of reducing the *redundancy* of the source (i.e., of obtaining an output data rate approaching $R_s$). At the receiver, the source decoder will convert the binary output of the channel decoder into a symbol sequence that is passed to the user.

Because the redundancy of the source information has been removed, the binary sequence at the output of the source encoder is highly vulnerable to errors occurring during the process of transmitting the information to its destination. The *channel encoder* introduces a controlled redundancy into the binary sequence so as to achieve highly reliable transmissions. At the receiver, the channel decoder recovers the information-bearing bits from the coded binary stream. Both the encoder and decoder can operate either in block mode or in a continuos sequential mode.

The communication channel provides the electrical connection between the source and the destination. The channel may be a pair of wires, a telephone link, an optical fiber, or free space over which the signal is radiated in the form of electromagnetic waves. In all cases, communication channels introduce various forms of impairments. Having finite bandwidths, they distort the signal in amplitude and phase. Moreover, the signal is attenuated and corrupted by unwanted additive and/or multiplicative random signals referred to as *noise* or *fading*. For these reasons, an exact replica of the transmitted signal cannot be obtained at the receiver input. The primary objective of a good communication system design is to counteract the effects of noise and distortion so as to achieve a faithful estimate of the transmitted signal.

The modulator converts the input binary stream into a waveform sequence suitable for transmission over the available channel. Being a powerful tool in the hands of the designer, modulation will receive considerable attention in this book. It involves a large number of choices, such as the number of waveforms, their shape, duration, and bandwidth, the power (average and/or peak), and more, allowing great flexibility in the system design. At the receiver, the demodulator extracts the binary sequence (*hard* demodulation) or suitably *sufficient statistics* (*soft* demodulation) from the received waveforms. Due to the impairment introduced by the channel, this process entails the possibility of errors between the input sequence to the modulator and the the output sequence from the demodulator (in the case of hard decoding), or a poor sufficient statistics (in the case of soft demodulation). A result of both types of degradation is a nonzero *bit error probability*. It is the goal of the *channel decoder* to exploit the redundancy introduced by the channel encoder to retrieve the transmitted information either by correcting the binary errors of the demodulator (hard decoding), or by suitably processing the sufficient statistics (soft decoding).

In practical point-to-point communication systems, other functional blocks exist, which for simplicity are not shown in Fig. 1.4. They are, for example, the *adaptive equalizer*, which reduces the channel distortions, the *carrier* and *clock synchronizers*, which allow coherent demodulation and proper sampling of the received signals, *scramblers* and *descramblers*, which are used to prevent unwanted strings of symbols at the channel input, and *enciphering* and *deciphering* devices, which ensure secure communication. Some of these blocks will be decribed in the book.

The book is organized as follows. Chapter 2 reviews the main results from the theory of random processes, spectral analysis, and detection theory, which can be considered as prerequisites to the remaining chapters. In Chapter 3 we look at probabilistic models for discrete information sources and communication channels. The main results from classical information theory are introduced as a conceptual background and framework for the successive material. Chapter 4 is devoted to memoryless waveform transmission over additive Gaussian noise channels. By using results from detection theory, optimum demodulator structures are derived, and the calculation of their error probabilities is presented. A distinction is made between coherent and noncoherent demodulation. In Chapter 5, the main modulation techniques employed for digital transmission are described, and their performances are compared in terms of error probability, energy, and bandwidth efficiency. Chapter 6 presents some modulation schemes specifically intended for transmission on wireless channels. In Chapter 7 we show how to evaluate the performance of systems affected by *intersymbol interference*, derive the optimization criteria for the overall system transfer function, and, finally, discuss the maximum-likelihood receiver structure. Chapter 8 is devoted to receivers for intersymbol-interference channels: adaptive receivers and channel equalization are covered. Chapter 9 deals with carrier and clock synchronization problems in modems. Chapter 10 describes linear block codes applied to improve the channel reliability, by error detection and correction. The most important classes of block codes and a few decoding techniques are described. The first part of Chapter 11 is devoted to linear convolutional codes. Their performance in terms of bit error probability is analyzed, and the maximum-likelihood decoding algorithm, the celebrated Viterbi algorithm, is described in detail (Appendix F is also devoted to it and to a maximum-a-posteriori decoding algorithm). The second part of Chapter 11 deals with concatenated codes, and particular attention is paid to the recently discovered, high-performance *turbo* codes. Chapter 12 covers the important topic of trellis-coded modulation, a technique to improve the channel reliability that merges modulation and channel coding in a very successful manner. Chapter 13 introduces models of fading channels and describes techniques for analysis and design of coding schemes operating on them.  Finally, Chapter 14 deals with digital transmission over nonlinear channels.

# A mathematical introduction

Signal theory, system theory, probability, and stochastic processes are the basic mathematical tools for the analysis and design of digital communication systems. Since a comprehensive treatment of all these topics requires several volumes, rather than attempting a comprehensive survey we devote this chapter to a selection of some points that are especially important in the developments that follow. The topics selected and the depth of their presentation were decided according to two criteria. First, where possible, laborious and sophisticated mathematical apparatuses have been omitted. This entails a certain loss of rigor, but it should improve the presentation of the subject matter. Second, those topics most likely to be familiar to the reader are reviewed very quickly, whereas more attention is devoted to certain specialized points of particular relevance for applications.

The topics covered in this chapter are deterministic and random signal theory for both discrete- and continuous-time models, linear and nonlinear system theory, and detection theory. Extensive bibliographical notes at the end of the chapter will guide the reader wishing to become more conversant with a specific topic.

## 2.1. Signals and systems

In this section we briefly present the basic concepts of the theory of linear and certain nonlinear systems. We begin with the time-discrete model for signals and systems and continue with the time-continuous model. To provide a higher level of generality to our presentation, we introduce and extensively employ complex time functions. The reasons for their use are explained in Section 2.4.

### 2.1.1.  Discrete signals and systems

A discrete-time signal is a sequence of real or complex numbers, denoted by $(x_n)_{n=n_1}^{n_2}$, defined for every integer index $n$ ranging in the interval $n_1 \le n \le n_2$. The index $n$ is usually referred to as the *discrete time*. Whenever $n_1 = -\infty$ and $n_2 = \infty$, or when the upper and lower indexes need not be specified, we shall simply write $(x_n)$. A *time-discrete system*, or for short a *discrete system*, is a mapping of a sequence $(x_n)$, called the *input* of the system, into another sequence $(y_n)$, called the *output* or *response*. We write

$$y_n = S[(x_n)] \tag{2.1}$$

for the general element of the sequence $(y_n)$.

A discrete system is *linear* if, for any pair of input signals $(x'_n)$, $(x''_n)$ and for any pair of complex numbers $A'$, $A''$, the following holds:

$$S[(A'x'_n + A''x''_n)] = A'S[(x'_n)] + A''S[(x''_n)] \tag{2.2}$$

Equation (2.2) means that if the system input is a linear combination of two signals, its output is the same linear combination of the two responses.

A discrete system is *time-invariant* if the rule by which an input sequence is transformed into an output sequence does not change with time. Mathematically, this is expressed by the condition

$$S[(x_{n-k})] = y_{n-k} \tag{2.3}$$

for all integers $k$. This is tantamount to saying that, if the input is delayed by $k$ time units, the output is delayed by the same quantity.

If $(\delta_n)$ denotes the "unit impulse" sequence

$$\delta_n = \begin{cases} 1, & n = 0, \\ 0, & n \neq 0 \end{cases} \tag{2.4}$$

and S is a linear, time-invariant discrete system, its response $(h_n)$ to the input $(\delta_n)$ is called the *(discrete) impulse response* of the system. Given a linear, time-invariant discrete system with impulse response $(h_n)$, its response to any arbitrary input $(x_n)$ can be computed via the *discrete convolution*

$$\begin{aligned} y_n &= \sum_{k=-\infty}^{\infty} x_k h_{n-k} \\ &= \sum_{k=-\infty}^{\infty} h_k x_{n-k} \end{aligned} \tag{2.5}$$

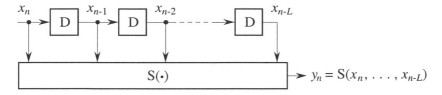

Figure 2.1: *Transversal-filter implementation of a time-invariant discrete system with memory L.*

It may happen that the system output at time $\ell$, say $y_\ell$, depends only on a certain subset of the input sequence. In particular, the system is said to be *causal* if $y_\ell$ depends only on $(x_n)_{n=-\infty}^\ell$. This means that the output at any given time depends only on the past and present values of the input, and not on its future values. In addition, the system is said to have a *finite memory L* if $y_\ell$ depends only on the finite segment $(x_n)_{n=\ell-L}^\ell$ of the past input. When $L = 0$, and hence $y_\ell$ depends only on $x_\ell$, the system is called *memoryless*. For a linear time-invariant system, causality implies $h_n = 0$ for all $n < 0$. A linear time-invariant system with finite memory $L$ has an impulse response sequence $(h_n)$ that may be nonzero only for $0 \leq n \leq L$. For this reason, a finite-memory system is often referred to also as a *finite impulse response* (FIR) system. A system with memory $L$ can be implemented as in Fig. 2.1. The blocks labeled D denote unit-delay elements (i.e., systems that respond to the input $x_n$ with the output $y_n = x_{n-1}$). A cascade of such unit-delay elements is called a *shift register*, and the resulting structure is called a *tapped delay line*, or *transversal*, filter. Here the function S( $\cdot$ ) defining the input-output relationship has $L + 1$ arguments. When the system is linear, S( $\cdot$ ) takes the form of a linear combination of its arguments:

$$S(x_n, x_{n-1}, \ldots, x_{n-L}) = \sum_{k=0}^{L} h_k x_{n-k} \tag{2.6}$$

In this case, the structure of Fig. 2.1 becomes the linear transversal filter of Fig. 2.2.

**Discrete Volterra systems**

Consider a time-invariant, nonlinear discrete system with memory $L$, and assume that the function S( $\cdot$ ) is sufficiently regular to be expanded in a Taylor series in a neighborhood of the origin $x_n = 0$, $x_{n-1} = 0$, ..., $x_{n-L} = 0$. We have the

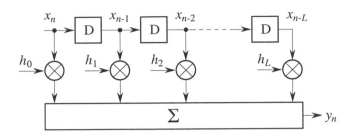

Figure 2.2: *Linear discrete transversal filter.*

representation

$$y_n = S(x_n, x_{n-1}, \ldots, x_{n-L}) = h^{(0)} + \sum_{i=0}^{L} h_i^{(1)} x_{n-i}$$

$$+ \sum_{i=0}^{L} \sum_{j=0}^{L} h_{ij}^{(2)} x_{n-i} x_{n-j} + \sum_{i=0}^{L} \sum_{j=0}^{L} \sum_{k=0}^{L} h_{ijk}^{(3)} x_{n-i} x_{n-j} x_{n-k} + \cdots \quad (2.7)$$

called a discrete *Volterra series*. It is seen that the system is completely characterized by the coefficients of the expansion, say

$$h^{(0)}, h_i^{(1)}, h_{ij}^{(2)}, h_{ijk}^{(3)}, \ldots \qquad i, j, k = 0, 1, 2, \cdots, L,$$

which are proportional to the partial derivatives of the function $S(\cdot)$ at the origin. These are called the system's *Volterra coefficients*. The expansion (2.7) can be generalized to systems with infinite memory, although in the computational practice only a finite number of terms will be retained. In general the Volterra system representation involves an infinite number of infinite summations. Thus, if a truncation of the series is not performed, we must associate with each series a suitable convergence condition to guarantee that the representation is meaningful (see, e.g., Rugh, 1981).

**Example 2.1**  Consider the discrete system shown in Fig. 2.3 and obtained by cascading a linear, time-invariant, causal system with impulse response $(h_n)$ to a memoryless nonlinear system with input-output relationship $y_n = g(w_n)$. Assume that $g(\cdot)$ is an analytic function, with a Taylor series expansion in the neighborhood of the origin

$$g(w) = \sum_{\ell=0}^{\infty} a_\ell w^\ell \qquad (2.8)$$

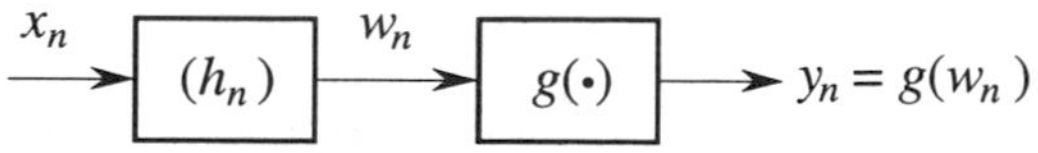

Figure 2.3: *A discrete nonlinear system.*

The input-output relationship for the system of Fig. 2.3 is then

$$
\begin{aligned}
y_n &= g\left(\sum_{i=0}^{\infty} h_i x_{n-i}\right) \\
&= a_0 + a_1 \sum_{i=0}^{\infty} h_i x_{n-i} + a_2 \sum_{i=0}^{\infty} \sum_{j=0}^{\infty} h_i h_j x_{n-i} x_{n-j} + \cdots
\end{aligned} \tag{2.9}
$$

so that the Volterra coefficients for the system are:

$$
\begin{aligned}
h^{(0)} &= 0 \\
h_i^{(1)} &= a_1 h_i \\
h_{ij}^{(2)} &= a_2 h_i h_j \\
&\cdots
\end{aligned}
$$

The following should be observed. First, if $g(\cdot)$ is a polynomial of degree $K$, the coefficients $a_{K+1}, a_{K+2}, \ldots$, in (2.8) are zero, so that only a finite number of summations will appear in (2.9). Second, if the impulse response sequence $(h_n)$ is finite (i.e., the linear system of Fig. (2.3) has a finite memory), then all the summations in (2.9) will include only a finite number of terms. □

**Discrete signals and systems in the transform domain**

Given a sequence $(x_n)$, we define its Fourier transform $\mathcal{F}[(x_n)]$ as the function of the frequency $f$ defined as

$$
X(f) \triangleq \sum_{n=-\infty}^{\infty} x_n e^{-jn2\pi f} \tag{2.10}
$$

where $j = \sqrt{-1}$. $X(f)$ is a periodic function of $f$ with period 1, so it is customary to consider it only in the interval $-1/2 \le f \le 1/2$. The *inverse Fourier transform* yields the elements of the sequence $(x_n)$ in terms of $X(f)$:

$$x_n = \int_{-1/2}^{1/2} X(f)e^{jn2\pi f}\, df \tag{2.11}$$

The Fourier transform $H(f)$ of the impulse response $(h_n)$ of a linear time-invariant system is called the *frequency response*, or *transfer function*, of the system. We call $|H(f)|$ the *amplitude* and $\arg[H(f)]$ the *phase* of the transfer function. The derivative of $\arg[H(f)]$ taken with respect to $f$ is called the *group delay* of the system. A basic property of the Fourier transform is that the response of a linear, time-invariant discrete system with transfer function $H(f)$ to a sequence with Fourier transform $X(f)$ has the Fourier transform $H(f)X(f)$.

### 2.1.2.  Continuous signals and systems

A *continuous-time signal* is a real or complex function $x(t)$ of the real variable $t$ (the *time*). Unless otherwise specified, the time is assumed to range from $-\infty$ to $\infty$. A *continuous-time system* is a mapping of a signal $x(t)$, the system *input*, into another signal $y(t)$, called the *output* or *response*. We write

$$y(t) = S[x(t)] \tag{2.12}$$

A continuous-time system is *linear* if for any pair of input signals $x'(t)$, $x''(t)$ and for any pair of complex numbers $A'$, $A''$, the following holds:

$$S[A'x'(t) + A''x''(t)] = A'S[x'(t)] + A''S[x''(t)] \tag{2.13}$$

A continuous-time system is *time-invariant* if (2.12) implies

$$S[x(t - \tau)] = y(t - \tau) \tag{2.14}$$

for all $\tau$. Let $\delta(t)$ denote the delta function, characterized by the *sifting property*

$$\int_{-\infty}^{\infty} \delta(t)\phi(t)\, dt = \phi(0) \tag{2.15}$$

valid for every function $\phi(t)$ continuous at the origin. The response $h(t)$ of a linear, time-invariant continuous system to the input $\delta(t)$ is called the *impulse response* of the system. For a system with a known impulse response $h(t)$, the response $y(t)$ to any input signal $x(t)$ can be computed via the convolution integral

$$
\begin{aligned}
y(t) &= \int_{-\infty}^{\infty} x(\tau)h(t - \tau)\, d\tau \\
&= \int_{-\infty}^{\infty} h(\tau)x(t - \tau)\, d\tau
\end{aligned}
\tag{2.16}
$$

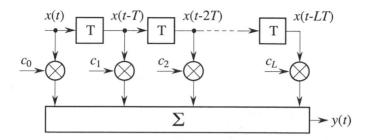

Figure 2.4: *Linear continuous transversal filter.*

It may happen that the system output $y(t)$ at time $t$ depends on the input $x(t)$ only through the values taken by $x(t)$ in the time interval I. If $I = (-\infty, t]$, the system is said to be *causal*. If $I = (t - t_0, t]$, $0 < t_0 < \infty$, the system is said to have a *finite memory* $t_0$. If $I = \{t\}$ (i.e., the output at any given time depends only on the input at the same time), the system is called *memoryless*. It is easily seen from (2.16) that, for a linear time-invariant system, causality is equivalent to having $h(t) = 0$ for all $t < 0$. A general time function $h(t)$ with the latter property is sometimes called *causal*.

A linear system is said to be *stable* if its response to any bounded input is bounded. A linear, time-invariant system is stable if and only if its impulse response is absolutely integrable.

**Example 2.2** Figure 2.4 represents a linear, time-invariant continuous system with finite memory. The blocks labeled T are delay elements, that is, systems with impulse response $\delta(t - T)$. A cascade of such elements is called a *(continuous) tapped delay line* and the structure of Fig. 2.4 a linear *transversal filter*. The system has an impulse response

$$h(t) = \sum_{\ell=0}^{L} c_\ell \delta(t - \ell T) \tag{2.17}$$

and a memory $LT$.      □

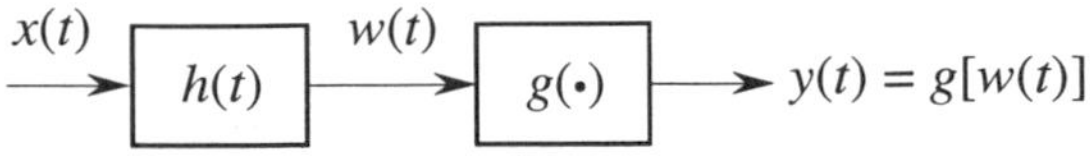

Figure 2.5: *A continuous nonlinear system.*

**Continuous Volterra systems**

To motivate our general discussion of Volterra series, consider as an example the time-invariant, nonlinear continuous system shown in Fig. 2.5. Assume that the first block represents a linear time-invariant system with impulse response $h(t)$ and that $g(\cdot)$ is a function as in Example 2.1, so (2.8) holds. The input-output relationship for this system can thus be expanded in the form

$$
\begin{aligned}
y(t) &= g\left[\int_{-\infty}^{\infty} h(\tau)x(t-\tau)\,d\tau\right] \\
&= a_0 + a_1 \int_{-\infty}^{\infty} h(\tau)x(t-\tau)\,d\tau \\
&\quad + a_2 \int_{-\infty}^{\infty}\int_{-\infty}^{\infty} h(\tau_1)h(\tau_2)x(t-\tau_1)x(t-\tau_2)\,d\tau_1\,d\tau_2 + \cdots (2.18)
\end{aligned}
$$

By defining

$$
\begin{aligned}
h_0 &= a_0 \\
h_1(t) &= a_1 h(t) \\
h_2(t_1,t_2) &= a_2 h(t_1)h(t_2) \\
&\vdots
\end{aligned}
\qquad (2.19)
$$

Eq. (2.18) can be rewritten as

$$
\begin{aligned}
y(t) &= h_0 + \int_{-\infty}^{\infty} h_1(\tau)x(t-\tau)\,d\tau \\
&\quad + \int_{-\infty}^{\infty}\int_{-\infty}^{\infty} h_2(\tau_1,\tau_2)x(t-\tau_1)x(t-\tau_2)\,d\tau_1\,d\tau_2 + \cdots \quad (2.20) \\
&\quad + \int_{-\infty}^{\infty}\cdots\int_{-\infty}^{\infty} h_k(\tau_1,\tau_2,\ldots,\tau_k)\left[\prod_{i=1}^{k} x(t-\tau_i)\,d\tau_i\right] + \cdots
\end{aligned}
$$

Equations (2.19) and (2.20) represent the input-output relationship of the system of Fig. 2.5. More generally, (2.20) without the definitions (2.19), that is, for a general set of functions $h_0$, $h_1(t)$, $h_2(t_1, t_2)$, ..., provides an input-output relationship for nonlinear time-invariant continuous systems. The RHS of (2.20) is called a *Volterra series*, and the functions $h_0$, $h_1(t)$, $h_2(t_1, t_2)$, ..., are called the *Volterra kernels* of the system. As a linear, time-invariant continuous system is completely characterized by its impulse response, so a nonlinear system whose input-output relationship can be expressed as a Volterra series is completely characterized by its Volterra kernels. It can be observed that the first-order kernel $h_1(t)$ is simply the impulse response of a linear system. The higher-order kernels can thus be viewed as higher-order impulse responses, which characterize the various orders of nonlinearity of the system. The zero-order term $h_0$ accounts for the response to a zero input.

It can be shown (see Problem 2.6) that a time-invariant system described by a Volterra series is causal if and only if, for all k,

$$h_k(t_1, t_2, \ldots, t_k) = 0 \qquad \text{for all } t_i < 0, \qquad i = 1, 2, \ldots, k \qquad (2.21)$$

A Volterra series expansion can be made simpler if it is assumed that the system kernels are symmetric functions of their arguments. That is, for every $k \geq 2$ any of the $k!$ possible permutations of the $k$ arguments of $h_k(t_1, t_2, \ldots, t_k)$ leaves the kernel unchanged. It can be proved (see Problem 2.5) that the assumption of symmetric kernels does not entail any loss of generality.

Volterra series can be viewed as "Taylor series with memory." As such they share with Taylor series some limitations, a major one being slow convergence. Moreover, the complexity in computation of the $k$th term of a Volterra series increases quickly with increasing $k$. Thus, it is expedient to use Volterra series only when the expansion (2.20) can be truncated to low-order terms, i.e., the system is "mildly nonlinear."

### Continuous signals and systems in the transform domain

With the notation $X(f) = \mathcal{F}[x(t)]$ we shall denote the Fourier transform of the signal $x(t)$; that is,

$$X(f) = \int_{-\infty}^{\infty} x(t) e^{-j2\pi ft} \, dt \qquad (2.22)$$

Given its Fourier transform $X(f)$, the signal $x(t)$ can be recovered by computing the *inverse Fourier transform* $\mathcal{F}^{-1}[X(f)]$:

$$x(t) = \int_{-\infty}^{\infty} X(f) e^{j2\pi ft} \, df \qquad (2.23)$$

The Fourier transform of a signal is also called the *amplitude spectrum* of the signal. If $h(t)$ denotes the impulse response of a linear, time-invariant system, its Fourier transform $H(f)$ is called the *frequency response*, or *transfer function*, of the system. We call $|H(f)|$ the *amplitude* and $\arg[H(f)]$ the *phase* of the transfer function. The derivative of $\arg[H(f)]$ taken with respect to $f$ is called the *group delay* of the system. It is seen from (2.22) that, when $x(t)$ is a real signal, the real part of $X(f)$ is an even function of $f$, and the imaginary part is an odd function of $f$. It follows that for a real $x(t)$ the function $|X(f)|$ is even, and $\arg[X(f)]$ is odd.

An important property of Fourier transform is that it relates products and convolutions of two signals $x(t)$, $y(t)$ with convolutions and products of their Fourier transforms $X(f)$ and $Y(f)$:

$$\mathcal{F}[x(t)y(t)] = \int_{-\infty}^{\infty} X(\alpha)Y(f - \alpha)\,d\alpha \tag{2.24}$$

and

$$\mathcal{F}\left[\int_{-\infty}^{\infty} x(\tau)y(t - \tau)\,d\tau\right] = X(f)Y(f) \tag{2.25}$$

In particular, (2.25) implies that the output $y(t)$ of a linear, time-invariant system with a transfer function $H(f)$ and an input signal $x(t)$ has the amplitude spectrum

$$Y(f) = H(f)X(f). \tag{2.26}$$

**Example 2.2 (continued)** The transfer function of the system shown in Fig. 2.4 is obtained by taking the Fourier transform of (2.17):

$$H(f) = \sum_{\ell=0}^{L} c_\ell e^{-j2\pi f \ell T}. \tag{2.27}$$

It is left as an exercise for the reader to derive the conditions for which this system exhibits a linear phase. $\square$

**Example 2.3** An important family of linear systems is provided by the Butterworth filters. The transfer function of the $n$th-order low-pass Butterworth filter with cutoff frequency $f_c$ is

$$H(f) = \frac{1}{D_n(jf/f_c)} \tag{2.28}$$

where

$$D_n(s) \stackrel{\triangle}{=} \prod_{i=1}^{n} \left[s - e^{j\pi(2i+n-1)/2n}\right] \tag{2.29}$$

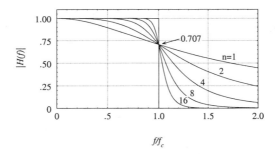

Figure 2.6: *Amplitude of the transfer function of low-pass Butterworth filters of various orders.*

is an $n$th degree polynomial. Expressions of these polynomials for some values of $n$ are

$$
\begin{aligned}
D_1(s) &= 1+s \\
D_2(s) &= 1+\sqrt{2}s + s^2 \\
D_3(s) &= 1+2s+2s^2+s^3
\end{aligned}
\tag{2.30}
$$

Figure 2.6 shows the amplitude $|H(f)|$ of the transfer function of the low-pass Butterworth filters for several values of their order $n$. It is seen that the curves of all orders pass through the 0.707 point at $f = f_c$. As $n \to \infty$, $|H(f)|$ approaches the ideal low-pass ("brickwall") characteristics:

$$
|H(f)| = \begin{cases} 1, & |f| < f_c \\ 0, & \text{elsewhere.} \end{cases}
\tag{2.31}
$$

$\square$

## 2.2.    Random processes

### 2.2.1.    Discrete-time processes

A *discrete-time random process*, or *random sequence*, is a sequence $(\xi_n)$ of real or complex random variables (RV) defined on some sample space. The index $n$ is usually referred to as the discrete time. A discrete-time process is completely characterized by providing the joint cumulative distribution functions (cdf) of the $N$-tuples $\xi_{i+1}, \xi_{i+2}, \ldots, \xi_{i+N}$ of RVs extracted from the sequence, for all integers $i$ and $N$, $N > 0$. If the process is complex, these joint distributions are the joint $2N$-dimensional distributions of the real and imaginary components of $\xi_{i+1}, \ldots, \xi_{i+N}$. The simplest possible case occurs when the RVs in the sequence are independent and identically distributed (iid). In this case the joint cdf of any $N$-tuple of RVs factors into the product of individual marginal cdfs. For a real process,

$$F_{\xi_{i+1}, \xi_{i+2}, \cdots, \xi_{i+N}}(x_{i+1}, x_{i+2}, \cdots, x_{i+N}) = \prod_{j=1}^{N} F_\xi(x_{i+j}) \qquad (2.32)$$

where $F_\xi(\,\cdot\,)$ is the common cdf of the RVs. Thus, a sequence of iid RVs is completely characterized by the single function $F_\xi(\,\cdot\,)$.

A random sequence is called *stationary* if for every $N$ the joint distribution function of $\xi_{i+1}, \xi_{i+2}, \ldots, \xi_{i+N}$ does not depend on $i$. In other words, a stationary random sequence is one whose probabilistic properties do not depend on the time origin, so that for any given integer $k$ the sequences $(\xi_n)$ and $(\xi_{n+k})$ are identically distributed. An iid sequence extending from $n = -\infty$ to $+\infty$ is an example of a stationary sequence.

The *mean* of a random sequence $(\xi_n)$ is the sequence $(\xi_n)$ of mean values

$$\mu_n \overset{\triangle}{=} \mathrm{E}[\xi_n] \qquad (2.33)$$

The *autocorrelation* of $(\xi_n)$ is the two-index sequence $(r_{n,m})$ such that

$$r_{n,m} \overset{\triangle}{=} \mathrm{E}[\xi_n \xi_m^*] \qquad (2.34)$$

For a stationary sequence,

**(a)** $\mu_n$ does not depend on $n$, and

**(b)** $r_{n,m}$ depends only on the difference $n - m$. Thus, the autocorrelation sequence has a single index.

Conditions (a) and (b), which are necessary for the stationarity of the sequence $(\xi_n)$, are generally not sufficient. If (a) and (b) hold true, we say that $(\xi_n)$ is *wide-sense* (WS) *stationary*. Notice that wide-sense stationarity is exceedingly simpler to check for than stationarity. Thus, it is always expedient to verify whether wide-sense stationarity is enough to prove the properties that are needed. In practice, although stationarity is usually invoked, wide-sense stationarity is often sufficient.

### Markov chains

For any real sequence $(\xi_n)_{n=0}^{\infty}$ of independent RVs, we have, for every $n$,

$$F_{\xi_n \mid \xi_{n-1}, \xi_{n-2}, \dots, \xi_0}(x_n \mid x_{n-1}, x_{n-2}, \dots, x_0) = F_{\xi_n}(x_n) \qquad (2.35)$$

where $F_{\xi_n \mid \xi_{n-1}, \xi_{n-2}, \dots, \xi_0}(\cdots)$ denotes the conditional cdf of the random variable $\xi_n$ given all the "past" RVs $\xi_{n-1}, \xi_{n-2}, \dots, \xi_0$. Equation (2.35) reflects the fact that $\xi_n$ is independent of the past of the sequence. A first-step generalization of (2.35) can be obtained by considering a situation in which, for any $n$,

$$F_{\xi_n \mid \xi_{n-1}, \xi_{n-2}, \dots, \xi_0}(x_n \mid x_{n-1}, x_{n-2}, \dots, x_0) = F_{\xi_n \mid \xi_{n-1}}(x_n \mid x_{n-1}) \qquad (2.36)$$

that is, $\xi_n$ depends on its past only through $\xi_{n-1}$.

When (2.36) holds, $(\xi_n)_{n=0}^{\infty}$ is called a *discrete-time (first-order) Markov process*. If in addition every $\xi_n$ can take only a finite number of possible values, say the integers $1, 2, \dots, q$, then $(\xi_n)$ is called a *(finite) Markov chain*, and the values of $\xi_n$ are referred to as the *states* of the chain. To specify a Markov chain, it suffices to give, for all times $n \geq 0$ and $j, k = 1, 2, \dots, q$, the probabilities $P\{\xi_n = j\}$ and $P\{\xi_{n+1} = k \mid \xi_n = j\}$. The latter quantity is the probability that the process will move to state $k$ at time $n + 1$ given that it was in state $j$ at time $n$. This probability is called the one-step *transition probability function* of the Markov chain.

A Markov chain is said to be *homogeneous* (or to have stationary transition probabilities) if the transition probabilities $P\{\xi_{\ell+m} = k \mid \xi_\ell = j\}$ depend only on the time difference $m$ and not on $\ell$. We then call

$$p_{jk}^{(m)} = P\{\xi_{\ell+m} = k \mid \xi_\ell = j\}, \quad \ell \geq 0, \quad m \geq 1, \quad j, k = 1, 2, \dots, q \quad (2.37)$$

the *m-step transition probability function* of the homogeneous Markov chain $(\xi_n)_{n=0}^{\infty}$. In other words, $p_{jk}^{(m)}$ is the conditional probability that the chain, being in state $j$ at time $\ell$, will move to state $k$ after $m$ time instants. The one-step transition probabilities $p_{jk}^{(1)}$ are simply written $p_{jk}$:

$$p_{jk} = P\{\xi_{\ell+1} = k \mid \xi_\ell = j\}, \qquad \ell \geq 0, \qquad j, k = 1, 2, \dots, q \qquad (2.38)$$

These transition probabilities can be arranged into a $q \times q$ *transition matrix* $\mathbf{P}$:

$$\mathbf{P} = \begin{bmatrix} p_{11} & p_{12} & \cdots & p_{1q} \\ p_{21} & p_{22} & \cdots & p_{2q} \\ & & \cdots & \\ p_{q1} & p_{q2} & \cdots & p_{qq} \end{bmatrix} \qquad (2.39)$$

The elements of **P** satisfy the conditions

$$p_{jk} \geq 0, \qquad j, k = 1, \ldots, q \tag{2.40}$$

and

$$\sum_{k=1}^{q} p_{jk} = 1, \qquad j = 1, 2, \ldots, q \tag{2.41}$$

(i.e., the sum of the entries in each row of **P** equals 1). Any square matrix that satisfies conditions (2.40) and (2.41) is called a *stochastic matrix* or a *Markov matrix*.

For a homogeneous Markov chain $(\xi_n)_{n=0}^{\infty}$, let $w_k^{(n)}$ denote the unconditional probability that state $k$ occurs at time $n$; that is,

$$w_k^{(n)} = P\{\xi_n = k\}, \qquad k = 1, 2, \ldots, q \tag{2.42}$$

The row $q$-vector of probabilities $w_k^{(n)}$,

$$\mathbf{w}^{(n)} = [w_1^{(n)} \ w_2^{(n)} \ \ldots w_q^{(n)}] \tag{2.43}$$

is called the *state distribution vector* at time $n$. With $\mathbf{w}^{(0)}$ denoting the initial state distribution vector, at time 1 we have

$$w_k^{(1)} = \sum_{j=1}^{q} w_j^{(0)} p_{jk}, \qquad k = 1, \ldots, q \tag{2.44}$$

or, in matrix notation,

$$\mathbf{w}^{(1)} = \mathbf{w}^{(0)} \mathbf{P} \tag{2.45}$$

Similarly, we obtain

$$\begin{aligned} \mathbf{w}^{(2)} &= \mathbf{w}^{(1)} \mathbf{P} \\ &= \mathbf{w}^{(0)} \mathbf{P}^2 \end{aligned} \tag{2.46}$$

and, iterating the process,

$$\begin{aligned} \mathbf{w}^{(m)} &= \mathbf{w}^{(m-1)} \mathbf{P} \\ &= \mathbf{w}^{(0)} \mathbf{P}^m \end{aligned} \tag{2.47}$$

More generally, we have

$$\mathbf{w}^{(\ell+m)} = \mathbf{w}^{(\ell)} \mathbf{P}^m \tag{2.48}$$

Equation (2.48) shows that the elements of $\mathbf{P}^m$ are the $m$-step transition probabilities defined in (2.37). This proves in particular that a homogeneous Markov chain $(\xi_n)_{n=0}^{\infty}$ is completely described by its initial state distribution vector $\mathbf{w}^{(0)}$ and its transition probability matrix $\mathbf{P}$. In fact, these are sufficient to evaluate $P\{\xi_n = j\}$ for every $n \geq 0$ and $j = 1, 2, \ldots, q$, which, in addition to the elements of $\mathbf{P}$, characterize a Markov chain.

Consider now the behavior of the state distribution vector $\mathbf{w}^{(n)}$ as $n \to \infty$. If the limit

$$\mathbf{w} = \lim_{n \to \infty} \mathbf{w}^{(n)} \tag{2.49}$$

exists, the vector $\mathbf{w}$ is called the *stationary distribution vector*. A homogeneous Markov chain such that $\mathbf{w}$ exists is called *regular*. It can be proved that a homogeneous Markov chain is regular if and only if all the eigenvalues of $\mathbf{P}$ with unit magnitude are identically 1. If, in addition, 1 is a simple eigenvalue of $\mathbf{P}$ (i.e., a simple root of the characteristic polynomial of $\mathbf{P}$), then the Markov chain is said to be *fully regular*. For a fully regular chain, the stationary state distribution vector is independent of the initial state distribution vector and can be evaluated by finding the unique solution of the system of homogeneous linear equations

$$\mathbf{w}\mathbf{P} = \mathbf{w} \tag{2.50}$$

subject to the constraints

$$\sum_{k=1}^{q} w_k = 1, \qquad w_k \geq 0, \qquad k = 1, 2, \ldots, q \tag{2.51}$$

Also, for a fully regular chain the limiting transition probability matrix

$$\mathbf{P}^{\infty} = \lim_{n \to \infty} \mathbf{P}^n \tag{2.52}$$

exists and has identical rows, each row being the stationary distribution vector $\mathbf{w}$:

$$\mathbf{P}^{\infty} = \begin{bmatrix} \mathbf{w} \\ \mathbf{w} \\ \vdots \\ \mathbf{w} \end{bmatrix} \tag{2.53}$$

The existence of $\mathbf{P}^{\infty}$ in the form (2.53) is a sufficient, as well as necessary, condition for a homogeneous Markov chain to be fully regular.

**Example 2.4**  Consider a digital communication system transmitting the symbols 0 and 1. Each symbol passes through several blocks. At each block there is a probability $1 - p$, $p < 1/2$, that the symbol at the output is equal to that at the input. Let $\xi_0$ denote the symbol entering the first block and $\xi_n$, $n \geq 1$, the symbol at the output of the $n$th block of the system. The sequence $\xi_0, \xi_1, \xi_2, \ldots$, is then a homogeneous Markov chain with transition probability matrix

$$\mathbf{P} = \begin{bmatrix} 1 - p & p \\ p & 1 - p \end{bmatrix}$$

The $n$-step transition probability matrix is

$$\mathbf{P}^n = \begin{bmatrix} \frac{1}{2} + \frac{1}{2}(1 - 2p)^n & \frac{1}{2} - \frac{1}{2}(1 - 2p)^n \\ \frac{1}{2} - \frac{1}{2}(1 - 2p)^n & \frac{1}{2} + \frac{1}{2}(1 - 2p)^n \end{bmatrix}$$

The eigenvalues of $\mathbf{P}$ are 1 and $1 - 2p$, so for $p \neq 0$ the chain is fully regular. Its stationary distribution vector is $\mathbf{w} = [\frac{1}{2} \ \frac{1}{2}]$, and

$$\mathbf{P}^\infty = \frac{1}{2} \begin{bmatrix} 1 & 1 \\ 1 & 1 \end{bmatrix}$$

which shows that as $n \to \infty$ a symbol entering the system has the same probability $1/2$ of being received correctly or incorrectly.                                        □

### Shift-register state sequences

An important special case of a Markov chain arises from the consideration of a stationary random sequence $(\alpha_n)$ of independent random variables, each taking on values in the set $\{a_1, a_2, \ldots, a_M\}$ with probabilities $p_k = P\{\alpha_n = a_k\}$, $k = 1, \ldots, M$, and of the sequence $(\sigma_n)_{n=0}^\infty$, with

$$\sigma_n = (\alpha_{n-1}, \ldots, \alpha_{n-L}) \tag{2.54}$$

If we consider an $L$-stage shift register fed with the sequence $(\alpha_n)$ (Fig. 2.7), $\sigma_n$ represents the content (the "state") of the shift register at time $n$ (i.e., when $\alpha_n$ is present at its input). For this reason, $(\sigma_n)$ is called a *shift-register state sequence*. Each $\sigma_n$ can take on $M^L$ values, and it can be verified that $(\sigma_n)$ forms a Markov chain. To derive its transition matrix, we shall first introduce a suitable ordering for the values of $\sigma_n$. This can be done in a natural way by first ordering the elements of the set $\{a_1, a_2, \ldots, a_M\}$ (a simple way to do this is to stipulate that $a_i$ precedes $a_j$ if and only if $i < j$) and then inducing the following "lexicographical" order among the $L$-tuples $a_{j_1}, a_{j_2}, \ldots, a_{j_L}$:

$$(a_{j_1}, a_{j_2}, \ldots, a_{j_L}) \text{ precedes } (a_{i_1}, a_{i_2}, \ldots, a_{i_L})$$

$$\text{if and only if} \begin{cases} j_1 < i_1, \text{ or} \\ j_1 = i_1 \text{ and } j_2 < i_2, \text{ or} \\ j_1 = i_1, \ j_2 = i_2, \text{ and } j_3 < i_3, \text{ etc.} \end{cases} \tag{2.55}$$

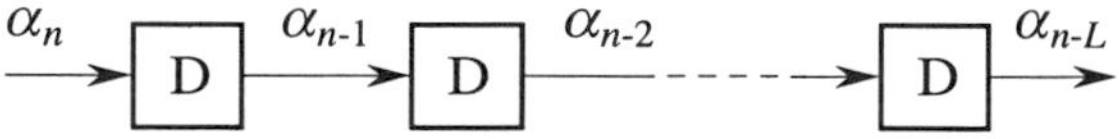

Figure 2.7: *Generating a shift-register sequence.*

Once the state set has been ordered according to the rule (2.55), each state can be represented by an integer number expressing its position in the ordered set. Thus, if $i$ represents the state $(a_{i_1}, a_{i_2}, \ldots, a_{i_L})$ and j represents the state $(a_{j_1}, a_{j_2}, \ldots, a_{j_L})$ the one-step transition probability $p_{ij}$ is given by

$$
\begin{aligned}
p_{ij} &\stackrel{\triangle}{=} P\{\sigma_n = (a_{j_1}, a_{j_2}, \ldots, a_{j_L}) \mid \sigma_{n-1} = (a_{i_1}, a_{i_2}, \ldots, a_{i_L})\} \\
&= P\{\alpha_{n-1} = a_{j_1}, \ldots, \alpha_{n-L} = a_{j_L} \mid \alpha_{n-2} = a_{i_1}, \ldots, \alpha_{n-L-1} = a_{i_L}\} \\
&= p_{j_1} \delta_{i_1 j_2} \delta_{i_2 j_3} \cdots \delta_{i_{L-1} j_L},
\end{aligned} \tag{2.56}
$$

where $\delta_{ij}$ denotes the Kronecker symbol ($\delta_{ii} = 1$ and $\delta_{ij} = 0$ for $i \neq j$).

**Example 2.5** Assume $M = 2$, $a_1 = 0$, $a_2 = 1$, and $L = 3$. The shift register has eight states, whose lexicographically ordered set is

$$\{(000), (001), (010), (011), (100), (101), (110), (111)\}.$$

The transition probability matrix of the corresponding Markov chain is

$$
\mathbf{P} = \begin{array}{c}
\begin{array}{cccccccc}
(000) & (001) & (010) & (011) & (100) & (101) & (110) & (111)
\end{array} \\
\left[ \begin{array}{cccccccc}
p_1 & 0 & 0 & 0 & p_2 & 0 & 0 & 0 \\
p_1 & 0 & 0 & 0 & p_2 & 0 & 0 & 0 \\
0 & p_1 & 0 & 0 & 0 & p_2 & 0 & 0 \\
0 & p_1 & 0 & 0 & 0 & p_2 & 0 & 0 \\
0 & 0 & p_1 & 0 & 0 & 0 & p_2 & 0 \\
0 & 0 & p_1 & 0 & 0 & 0 & p_2 & 0 \\
0 & 0 & 0 & p_1 & 0 & 0 & 0 & p_2 \\
0 & 0 & 0 & p_1 & 0 & 0 & 0 & p_2
\end{array} \right]
\begin{array}{l}
(000) \\
(001) \\
(010) \\
(011) \\
(100) \\
(101) \\
(110) \\
(111)
\end{array}
\end{array} \tag{2.57}
$$

As one can see, from state $(xyz)$ the shift register can move only to states $(wxy)$, with probability $p_1$ if $w = 0$ and $p_2$ if $w = 1$. □

Consider now the $m$-step transition probabilities. These are the elements of the matrix $\mathbf{P}^m$. Since the shift register has $L$ stages, its content after time $n + m$,

$m \geq L$, is independent of its content at time $n$. Consequently, the states $\sigma_{n+m}$, $m \geq L$, are independent of $\sigma_n$; so, for $m \geq L$,

$$P\{\sigma_{n+m} = (a_{j_1}, a_{j_2}, \ldots, a_{j_L}) \mid \sigma_n = (a_{i_1}, a_{i_2}, \ldots, a_{i_L})\} = \prod_{\ell=1}^{L} p_{j_\ell} \qquad (2.58)$$

Thus, $\mathbf{P}^L = \mathbf{P}^{L+1} = \cdots$, and $\mathbf{P}^L$ has identical rows. We can write

$$\mathbf{P}^L = \mathbf{P}^\infty \qquad (2.59)$$

which shows, in particular, that the shift-register state sequence defined in (2.54) is a fully regular Markov chain.

**Example 2.5 (continued)**  We have, by direct computation from (2.57) or using (2.58), that $\mathbf{P}^3$ has the structure (2.53), with $\mathbf{w}$, the stationary distribution vector, being equal to

$$\mathbf{w} = [p_1^3, \ p_1^2 p_2, \ p_1^2 p_2, \ p_1 p_2^2, \ p_1^2 p_2, \ p_1 p_2^2, \ p_1 p_2^2, \ p_2^3] \qquad (2.60)$$

□

### 2.2.2.  Continuous-time processes

A *continuous-time random process* (or *random continuous signal*) is a family of real or complex signals $\xi(t)$ defined on some probability space. At any $N$-tuple of times $t_1, t_2, \ldots, t_N$, the quantities $\xi(t_1), \xi(t_2), \ldots, \xi(t_N)$ are RVs. Consequently, a random process can be described by providing the joint distribution functions of the $N$ RVs $\xi(t_1), \xi(t_2), \ldots, \xi(t_N)$ for all integers $N$ and $N$-tuples of time instants.

A continuous-time random process is called *stationary* if for every $N$, for any $N$-tuple $(t_1, t_2, \ldots, t_N)$ and for every real $\tau$, the $N$-tuples of RVs $\xi(t_1), \xi(t_2)$, $\ldots, \xi(t_N)$ and $\xi(t_1 + \tau), \xi(t_2 + \tau), \ldots, \xi(t_N + \tau)$ are identically distributed. Stated in another way, a stationary random process is one whose probabilistic properties do not depend on the time origin. Thus, for any given $\tau$ the processes $\xi(t)$ and $\xi(t + \tau)$ are identically distributed.

The *mean* of the process $\xi(t)$ is the deterministic signal

$$\mu(t) \triangleq \mathrm{E}[\xi(t)] \qquad (2.61)$$

The *autocorrelation* of $\xi(t)$ is the function

$$R_\xi(t_1, t_2) \triangleq \mathrm{E}[\xi(t_1)\xi^*(t_2)] \qquad (2.62)$$

For a stationary process,

**(a)** $\mu(t)$ does not depend on time, and

**(b)** $R_\xi(t_1, t_2)$ depends only on the difference $t_1 - t_2$. Consequently, we can write

$$R_\xi(t_1 - t_2) \overset{\triangle}{=} \mathrm{E}[\xi(t_1)\xi^*(t_2)] \tag{2.63}$$

Conditions (a) and (b) are generally not sufficient for the stationarity of $\xi(t)$. If (a) and (b) hold true, we say that $\xi(t)$ is *wide-sense* (WS) stationary. A random process $\xi(t)$ is called *cyclostationary* with period $T$ if its probabilistic properties do not change when the time origin is shifted by a multiple of $T$; that is, we consider $\xi(t + kT)$, $k$ an integer, instead of $\xi(t)$. *Wide-sense cyclostationarity* can also be defined as follows: $\xi(t)$ is WS cyclostationary if

**(a)** $\mu(t)$ is a periodic function of time with period $T$, and

**(b)** the autocorrelation of the process has the property

$$R_\xi(t + \tau, t) = R_\xi(t + \tau + kT, t + kT) \tag{2.64}$$

$k$ any integer. Equation (2.64) can be interpreted by saying that $R_\xi(t+\tau, t)$, when considered as a function of $t$, is periodic with period $T$.

**Example 2.6** Consider the deterministic finite-energy signal $s(t)$ and a WS stationary sequence $(\alpha_n)$ of random variables with correlation $(r_n)$. The random signal

$$\xi(t) \overset{\triangle}{=} \sum_{\ell=-\infty}^{\infty} \alpha_\ell \, s(t - \ell T)$$

is WS cyclostationary with period $T$. In fact

$$\mu(t) = \mathrm{E}[\alpha_\ell] \sum_{\ell=-\infty}^{\infty} s(t - \ell T)$$

is periodic with period $T$. Moreover,

$$\begin{aligned} R_\xi(t_1, t_2) &= \sum_{\ell=-\infty}^{\infty} \sum_{m=-\infty}^{\infty} \mathrm{E}[\alpha_\ell \alpha_m^*] s(t_1 - \ell T) s^*(t_2 - mT) \\ &= \sum_{\ell=-\infty}^{\infty} \sum_{m=-\infty}^{\infty} r_{\ell-m} s(t_1 - \ell T) s^*(t_2 - mT), \end{aligned}$$

and it can be verified that (2.64) holds. □

Some important properties of stationary and cyclostationary processes are the following:

(a) If a stationary (cyclostationary) process is passed through a stable time-invariant system, it retains its stationarity (cyclostationarity).

(b) The sum of two stationary processes is a stationary process. The sum of a cyclostationary process and a stationary process is a cyclostationary process.

(c) Let $\xi(t)$ be a WS cyclostationary process with period $T$, and let $\eta(t)$ denote the randomly translated process

$$\eta(t) \stackrel{\triangle}{=} \xi(t + \theta), \tag{2.65}$$

where $\theta$ is a random variable statistically independent of $\xi(t)$ and uniformly distributed in the interval $(0, T)$. Then the process $\xi(t)$ is WS stationary.

**Gaussian processes**

A real random process $\xi(t)$ is called Gaussian if, for any given time instant $t$, $\xi(t)$ is a Gaussian random variable. Formally, $\xi(t)$ is a Gaussian process if for any $N$-tuple $t_1, t_2, \ldots, t_N$ of time instants, $N$ any integer $\geq 1$, the row $N$-vector of random variables $\boldsymbol{\xi} \stackrel{\triangle}{=} [\xi(t_1), \xi(t_2), \ldots, \xi(t_N)]$ has a Gaussian distribution, that is, a probability density function of the form

$$f_{\boldsymbol{\xi}}(\mathbf{x}) = \frac{1}{(2\pi)^{N/2}(\det \boldsymbol{\Lambda})^{1/2}} \exp\left[-\frac{1}{2}(\mathbf{x} - \boldsymbol{\mu})\boldsymbol{\Lambda}^{-1}(\mathbf{x} - \boldsymbol{\mu})'\right] \tag{2.66}$$

where $\boldsymbol{\mu}$ is the mean vector

$$\boldsymbol{\mu} \stackrel{\triangle}{=} \mathrm{E}[\boldsymbol{\xi}] = (\mathrm{E}[\xi(t_1)], \mathrm{E}[\xi(t_2)], \cdots, \mathrm{E}[\xi(t_N)]) \tag{2.67}$$

and $\boldsymbol{\Lambda}$ is the $N \times N$ *covariance matrix*

$$\boldsymbol{\Lambda} \stackrel{\triangle}{=} \mathrm{E}[(\boldsymbol{\xi} - \boldsymbol{\mu})'(\boldsymbol{\xi} - \boldsymbol{\mu})] \tag{2.68}$$

Now, let $\xi(t)$ be a complex random process, and let

$$\xi(t) = \xi_P(t) + j\xi_Q(t) \tag{2.69}$$

where $\xi_P(t)$, $\xi_Q(t)$ are real processes. The process $\xi(t)$ is called Gaussian if the joint distribution of $\xi_P(t_1)$, $\xi_P(t_2)$, ..., $\xi_P(t_N)$, $\xi_Q(t_1)$, $\xi_Q(t_2)$, ..., $\xi_Q(t_N)$ is $2N$-dimensional Gaussian for any $N$-tuple of time instants and for any integer $N \geq 1$.

Gaussian processes have the following properties:

**(a)** The output of any linear system whose input is a Gaussian process is still Gaussian.

**(b)** Let $\xi(t)$ be a WS stationary *real* Gaussian process. Then $\xi(t)$ is stationary.

**(c)** Let $\xi(t)$ be a WS stationary complex Gaussian process. Then $\xi(t)$ is stationary if and only if the average $E[\xi(t_1)\xi(t_2)]$ is a function only of the time difference $t_1 - t_2$.

Property (c) deserves some comments. Wide-sense stationarity of $\xi(t)$ implies that $E[\xi(t)\xi^*(s)]$ is a function of $t - s$, and $E[\xi(t)]$ is a constant. For the stationarity, one must show that $E[\xi_P(t)\xi_P(s)]$, $E[\xi_P(t)\xi_Q(s)]$, $E[\xi_Q(t)\xi_Q(s)]$ all depend only on the difference $t - s$. But this is equivalent to showing that $E[\xi(t)\xi^*(s)]$ and $E[\xi(t)\xi(s)]$ depend only on $t - s$. To verify the latter property, it is sometimes useful to apply *Grettenberg's theorem* (Grettenberg, 1965). It states that for a complex Gaussian process $\xi(t)$ with mean zero we have $E[\xi(t)\xi(s)] = 0$ if and only if, for all $0 \leq \theta \leq 2\pi$, the processes $\xi(t)$ and $e^{j\theta}\xi(t)$ are identically distributed; that is, $\xi(t)$ is invariant under phase rotations.

## 2.3. Spectral analysis of deterministic and random signals

In the representation of signals in the Fourier transform domain, one associates with each frequency $f$ a measure of its contribution to the signal. This representation is particularly useful when the signal is transformed by a linear time-invariant system, because in this case each of the frequency components of the signal is independently weighted by the system transfer function, according to the rule (2.26) (it holds for discrete and continuous signals). In this section we extend this concept to the spectral analysis of certain energetic quantities that one may want to associate with a given signal, such as its energy or its power (to be suitably defined). Specifically, assume that, for a given signal $\xi$, either discrete or continuous, deterministic or random, we have defined a nonnegative energetic quantity $\Pi_\xi$. The density spectrum of $\Pi_\xi$ is a frequency function, say $V_\xi(f)$, carrying information regarding how much of $\Pi_\xi$ is associated with each frequency $f$. The function $V_\xi(f)$ is nonnegative, and the two following properties hold:

**(a)** The integral of $V_\xi(f)$ gives $\Pi_\xi$:

$$\Pi_\xi = \int_I V_\xi(f)\, df. \tag{2.70}$$

**(b)** Let $\Pi_\eta$ be the same energetic quantity defined at the output of a linear, time-invariant system with transfer function $H(f)$ and input $\xi(t)$. Then

$$\Pi_\eta = \int_I |H(f)|^2 V_\xi(f)\, df \tag{2.71}$$

In (2.70) and (2.71), $I = (-\infty, \infty)$ if $\xi$ is a continuous-time signal, and $I = (-1/2, 1/2)$ if $\xi$ is a discrete-time signal.

Let us now specialize this general definition to some cases of practical interest.

**Energy density spectrum:** *Continuous deterministic signals*

Given a continuous deterministic signal $x(t)$, we define its energy as the quantity

$$\mathcal{E}_x \overset{\triangle}{=} \int_{-\infty}^{\infty} |x(t)|^2 \, dt \tag{2.72}$$

provided that the integral in (2.72) is finite. In the transform domain, the energy of a signal $x(t)$ whose Fourier transform is $X(f)$ can be expressed in the form

$$\mathcal{E}_x = \int_{-\infty}^{\infty} |X(f)|^2 \, df \tag{2.73}$$

Equality (2.73) is a special case of *Parseval's theorem*. This states that for two signals $x_1(t)$, $x_2(t)$ with Fourier transforms $X_1(f)$, $X_2(f)$, respectively, the following holds:

$$\int_{-\infty}^{\infty} x_1(t)x_2^*(t) \, dt = \int_{-\infty}^{\infty} X_1(f)X_2^*(f) \, df \tag{2.74}$$

The function

$$\mathcal{S}_x(f) \overset{\triangle}{=} |X(f)|^2 \tag{2.75}$$

is the *energy (density) spectrum* of $x(t)$. It is easily seen that with this definition both (2.70) and (2.71) hold.

**Power density spectrum:** *Continuous deterministic signals*

For a continuous *aperiodic* deterministic signal $x(t)$ whose energy is not finite, define its *average power* as the quantity

$$\mathcal{P}_x \overset{\triangle}{=} \lim_{a \to \infty} \frac{1}{a} \int_{-a/2}^{a/2} |x(t)|^2 \, dt \tag{2.76}$$

provided that this limit exists. If we define the truncated signal

$$x_a(t) \overset{\triangle}{=} \begin{cases} x(t), & -\dfrac{a}{2} < t < \dfrac{a}{2}, \\ 0, & \text{elsewhere} \end{cases} \tag{2.77}$$

the average power of $x(t)$ can be written

$$\mathcal{P}_x = \lim_{a \to \infty} \frac{1}{a} \mathcal{E}_a \tag{2.78}$$

where $\mathcal{E}_a$ denotes the energy of $x_a(t)$. Hence, for the signal $x(t)$ we define its power (density) spectrum as the function

$$\mathcal{G}_x(f) \triangleq \lim_{a \to \infty} \frac{1}{a} |X_a(f)|^2 \tag{2.79}$$

where $|X_a(f)|^2$ is the energy spectrum of the truncated signal (2.77).

For a *periodic signal* $x(t)$ with period $T$, its average power is defined as

$$P_x \triangleq \frac{1}{T} \int_{-T/2}^{T/2} |x(t)|^2 \, dt$$

Define its Fourier-series expansion

$$x(t) = \frac{1}{\sqrt{T}} \sum_{k=-\infty}^{\infty} c_k e^{jk2\pi t/T}$$

where

$$c_k \triangleq \frac{1}{\sqrt{T}} \int_{-T/2}^{T/2} e^{-jk2\pi t/T} \, dt$$

Its power spectrum is then given by

$$\mathcal{G}_x(f) \triangleq \frac{1}{T} \sum_{k=-\infty}^{\infty} |c_k|^2 \delta(f - n/T)$$

**Average power density spectrum:** *Discrete stationary random signals*

Consider a WS stationary random sequence $(\xi_n)$ with autocorrelation $(r_n)$. Its *average power* is defined as

$$P_\xi \triangleq \mathrm{E}\{|\xi_n|^2\} \tag{2.80}$$

The (average) power (density) spectrum $\mathcal{G}_\xi(f)$ of $(\xi_n)$ is the Fourier transform of the autocorrelation sequence $(r_n)$; that is,

$$\mathcal{G}_\xi(f) = \sum_{n=-\infty}^{\infty} r_n e^{-jn2\pi f}, \qquad |f| \le \frac{1}{2} \tag{2.81}$$

Let us show that with this definition (2.70) holds. We have

$$\int_{-1/2}^{1/2} \mathcal{G}_\xi(f) \, df = \sum_{n=-\infty}^{\infty} r_n \int_{-1/2}^{1/2} e^{-jn2\pi f} \, df = r_0 \tag{2.82}$$

and $r_0$ equals $\mathrm{E}\{|\xi_n|^2\}$ because of (2.34) and the assumption of WS stationarity. Property (2.71) can be proved similarly.

**Average power density spectrum:** *Continuous stationary random signals*

Let $\xi(t)$ be a WS stationary continuous random process with autocorrelation function $R_\xi(\tau)$. Its *average power* is defined as

$$\mathcal{P}_\xi \triangleq \mathrm{E}\{|\xi(t)|^2\} \tag{2.83}$$

The *(average) power (density) spectrum* $\mathcal{G}_\xi(f)$ of $\xi(t)$ is the Fourier transform of the autocorrelation function $R_\xi(\tau)$:

$$\mathcal{G}_\xi(f) = \int_{-\infty}^{\infty} R_\xi(\tau) e^{-j2\pi f\tau}\, d\tau \tag{2.84}$$

In this situation, (2.71) takes the form

$$\mathcal{P}_\eta = \int_{-\infty}^{\infty} |H(f)|^2 \mathcal{G}_\xi(f)\, df \tag{2.85}$$

where $\eta(t)$ is the response of a linear time-invariant system with transfer function $H(f)$ to the input $\xi(t)$.

**Example 2.7 (White noise)**   A process with autocorrelation function

$$R_\xi(\tau) = \frac{N_0}{2}\delta(\tau) \tag{2.86}$$

has a power spectrum

$$\mathcal{G}_\xi(f) = \frac{N_0}{2}, \qquad -\infty < f < \infty \tag{2.87}$$

Such a process is called a *white noise*. In practice, this process is not realizable, as its power $\mathcal{P}_\xi$ is not finite. However, this process can be very useful in instances where the actual process has an approximately constant spectral density over a frequency range wider than the bandwidth of the system under consideration. On the other hand, the observation of any process will be made through a measuring device whose bandwidth is finite: consequently, when we observe a constant spectral density it is mathematically convenient to assume that the underlying process (which we do not, and cannot, observe) is a white noise.

At the output of a linear time-invariant system with transfer function $H(f)$ we get the average power

$$\mathcal{P}_\eta = \frac{N_0}{2} \int_{-\infty}^{\infty} |H(f)|^2\, df \tag{2.88}$$

which is finite provided that the integral in the RHS converges. In this situation, it is customary to define the *equivalent noise bandwidth* of the system as

$$B_{\mathrm{eq}} \triangleq \frac{1}{2}\, \frac{\displaystyle\int_{-\infty}^{\infty} |H(f)|^2\, df}{\displaystyle\max_f |H(f)|^2} \tag{2.89}$$

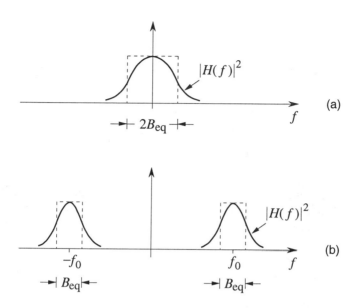

Figure 2.8: *Equivalent noise bandwidth for (a) low-pass systems, and (b) bandpass systems.*

Notice the presence of the factor $1/2$ in (2.89), which can be interpreted by saying that the bandwidth is only defined for positive frequencies. This convention is assumed throughout this book for every possible definition of the bandwidth of a signal or a system. For linear systems with a real impulse response, $|H(f)|$ is an even function. Hence, the factor $1/2$ can be omitted in the RHS of (2.89) and the integration carried out from 0 to $\infty$. With definition (2.89), the power at the output of a linear, time-invariant system with equivalent noise bandwidth $B_{\mathrm{eq}}$ and whose input is a white noise with power spectral density $N_0/2$ turns out to be

$$\mathcal{P}_\eta = N_0 \cdot B_{\mathrm{eq}} \cdot \max_f |H(f)|^2 \qquad (2.90)$$

Equation (2.90) shows that $B_{\mathrm{eq}}$ can be interpreted as the bandwidth of a system with a rectangular transfer function, whose amplitude squared is $\max |H(f)|^2$. Fig. 2.8 illustrates this fact for a low-pass and a bandpass system.

For example, the low-pass Butterworth filters defined in Example 2.3 have an equivalent noise bandwidth

$$B_{\mathrm{eq}} = f_c \frac{\pi/(2n)}{\sin[\pi/(2n)]} \qquad (2.91)$$

From (2.91) it is easily seen that, as $n \to \infty$, $B_{\text{eq}} \to f_c$, $f_c$ being the cutoff frequency of the filter. $\qquad\qquad\qquad\qquad\qquad\qquad\qquad\qquad\qquad\qquad\qquad\qquad\qquad\qquad\qquad\quad$ □

**Average power density spectrum:** *Continuous nonstationary random signals*

Consider now a nonstationary continuous random process $\xi(t)$. Clearly, definition (2.83) is not valid anymore because in general $\text{E}\{|\xi(t)|^2\}$ varies with time. In this situation, the definition of average power that should be used is

$$\mathcal{P}_\xi \stackrel{\triangle}{=} \lim_{a \to \infty} \frac{1}{a} \int_{-a/2}^{a/2} \text{E}\{|\xi(t)|^2\} \, dt \qquad\qquad (2.92)$$

that is, the time average of the mean value of the instantaneous power $|\xi(t)|^2$. With this definition, a spectral density function that satisfies properties (2.70) and (2.71) can also be defined for nonstationary processes, provided that we restrict our attention to an appropriate subclass of processes. This subclass is that of *harmonizable processes* (Loève, 1963, pp. 474–477). Roughly speaking, a process is harmonizable if we can define its Fourier transform:

$$\Xi(f) \stackrel{\triangle}{=} \int_{-\infty}^{\infty} \xi(t) \text{e}^{-j2\pi ft} \, dt \qquad\qquad (2.93)$$

Equation (2.93) defines a new random process in the variable $f$. In certain cases, a proper interpretation of (2.93) requires some care. In fact, (2.93) is an equality in the sense of distribution theory (i.e., it becomes an equality if a linear operator is applied to both sides and the order of integrations is reversed in the RHS). Incidentally, this is the correct way to interpret equalities like

$$\delta(t) = \int_{-\infty}^{\infty} \text{e}^{j2\pi ft} \, df$$

Harmonizable processes are a first-step generalization of WS stationary random processes. It has been shown (Cambanis and Liu, 1970) that, under some mild conditions, any random process obtained at the output of a linear system is harmonizable. The system may be randomly time variant and the input process need not be stationary, or even harmonizable.

For a harmonizable process $\xi(t)$, the power spectrum can be obtained as follows. Compute first the function

$$\Gamma_\xi(f_1, f_2) \stackrel{\triangle}{=} \text{E}[\Xi(f_1)\Xi^*(f_2)] \qquad\qquad (2.94)$$

Consider then the bisector $f_1 = f_2$ of the plane $(f_1, f_2)$ and the line masses of $\Gamma_\xi(f_1, f_2)$ located on it. The distribution of these line masses provides us with a function $\mathcal{G}_\xi(f)$, the power spectrum of $\xi(t)$. Specifically, if $\Gamma_\xi(f_1, f_2)$ can be written in the form

$$\Gamma_\xi(f_1, f_2) = \mathcal{G}_\xi(f_1)\delta(f_1 - f_2) + \Delta_\xi(f_1, f_2) \tag{2.95}$$

where $\Delta_\xi(f_1, f_2)$ has no line masses located on the bisector $f_1 = f_2$, then $\mathcal{G}_\xi(f)$ is the required spectrum. [It may happen that $\mathcal{G}_\xi(f)$ is identically zero; in this case the process has finite energy.] Using (2.93), it can easily be seen that $\Gamma_\xi(f_1, f_2)$ can be written in a form equivalent to (2.94):

$$\Gamma_\xi(f_1, f_2) = \int_{-\infty}^{\infty} \int_{-\infty}^{\infty} R_\xi(\tau_1, \tau_2) e^{-j2\pi(f_1\tau_1 - f_2\tau_2)} \, d\tau_1 \, d\tau_2 \tag{2.96}$$

Equation (2.96) shows that $\Gamma_\xi(f_1, f_2)$ is the two-dimensional Fourier transform of the autocorrelation function of the process $\xi(t)$. This is tantamount to saying that $R_\xi(\tau_1, \tau_2)$ is the inverse Fourier transform of $\Gamma_\xi(f_1, f_2)$:

$$R_\xi(\tau_1, \tau_2) = \int_{-\infty}^{\infty} \int_{-\infty}^{\infty} \Gamma_\xi(f_1, f_2) e^{j2\pi(f_1\tau_1 - f_2\tau_2)} \, df_1 \, df_2 \tag{2.97}$$

**Example 2.8** Let $\xi(t)$ be WS stationary. Its autocorrelation function depends only on $\tau_1 - \tau_2$. Thus, (2.96) yields

$$\begin{aligned}
\Gamma_\xi(f_1, f_2) &= \int_{-\infty}^{\infty} \int_{-\infty}^{\infty} R_\xi(\tau_1 - \tau_2) e^{-j2\pi[f_1(\tau_1 - \tau_2) + (f_1 - f_2)\tau_2]} \, d\tau_1 \, d\tau_2 \\
&= \int_{-\infty}^{\infty} R_\xi(\tau) e^{-j2\pi f_1 \tau} \, d\tau \cdot \delta(f_1 - f_2),
\end{aligned} \tag{2.98}$$

which is consistent with (2.84) (as it should be). Also notice that, using (2.97), one sees that $R_\xi(\tau_1, \tau_2)$ depends on the difference $\tau_1 - \tau_2$ only if $\Gamma_\xi(f_1, f_2)$ has the form

$$\Gamma_\xi(f_1, f_2) = \mathcal{G}_\xi(f_1)\delta(f_1 - f_2) \tag{2.99}$$

(see Fig. 2.9). $\qquad\qquad\qquad\qquad\qquad\qquad\qquad\qquad\qquad\qquad\qquad\qquad\qquad\quad$ $\square$

**Example 2.9** Let $\xi(t)$ be a WS cyclostationary process with period $T$. Using the property (2.64), it is seen that $R_\xi(\tau_1, \tau_2)$ can be expanded in the Fourier series

$$R_\xi(t + \tau, t) = \sum_{n=-\infty}^{\infty} g_n(\tau) e^{jn2\pi t/T} \tag{2.100}$$

where

$$g_n(\tau) \triangleq \frac{1}{T} \int_{-T/2}^{T/2} R_\xi(t + \tau, t) e^{-jn2\pi t/T} \, dt \tag{2.101}$$

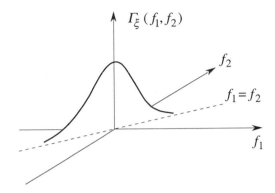

Figure 2.9: *The function $\Gamma_\xi(f_1, f_2)$ for a wide-sense stationary process.*

Using (2.96), we get

$$\Gamma_\xi(f_1, f_2) = \sum_{n=-\infty}^{\infty} \int_{-\infty}^{\infty} \int_{-\infty}^{\infty} g_n(\tau) e^{jn2\pi t/T} e^{-j2\pi[(f_1-f_2)t+f_1\tau]} \, d\tau \, dt$$

$$= \sum_{n=-\infty}^{\infty} G_n(f_1)\delta\left(f_1 - f_2 - \frac{n}{T}\right) \qquad (2.102)$$

where $G_n(\cdot)$ is the Fourier transform of $g_n(\cdot)$, $-\infty < n < \infty$. Equation (2.102) shows that $\Gamma_\xi(f_1, f_2)$ consists of line masses located on the lines $f_1 = f_2 + n/T$, $-\infty < n < \infty$, which are parallel to the bisector of the plane $(f_1, f_2)$. This situation is shown qualitatively in Fig. 2.10.

The power spectrum of $\xi(t)$ is then

$$\mathcal{G}_\xi(f) = G_0(f) \qquad (2.103)$$

It can also be shown that the power spectrum (2.103) can be obtained by considering the WS stationary process (2.65) and using (2.84).                                    □

### 2.3.1. Spectral analysis of random digital signals

In Chapter 4, devoted to the transmission of digital information using continuous signals, the following random process will be considered:

$$\xi(t) = \sum_{n=-\infty}^{\infty} s(t - nT; \alpha_n, \sigma_n) \qquad (2.104)$$

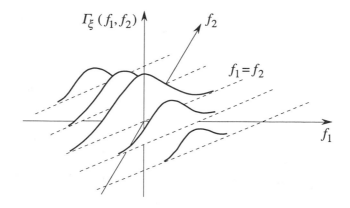

Figure 2.10: *The function* $\Gamma_\xi(f_1, f_2)$ *for a wide-sense cyclostationary process.*

This is called a *digitally modulated random signal*, or for short a *digital signal*. The sequence $(\alpha_n)$ of discrete RVs is WS stationary, and will be referred to as the sequence of *source symbols*. The sequence $(\sigma_n)$ is a stationary sequence of discrete random variables referred to as the *states of the modulator*. The random waveforms $s(t; \alpha_n, \sigma_n)$ take values in a set $\{s_i(t)\}_{i=1}^M$ of deterministic, finite-energy signals. They are output sequentially by the modulator, one every $T$ seconds, in accordance with the values of the source symbols and the modulator states.

Several special cases of (2.104) are of interest. If the modulator states $\sigma_n$ do not appear in (2.104), the modulator is called *memoryless*, and we have

$$\xi(t) = \sum_{n=-\infty}^{\infty} s(t - nT; \alpha_n) \tag{2.105}$$

If in addition

$$s(t; \alpha_n) = \alpha_n s(t), \tag{2.106}$$

that is, the waveforms of the set $\{s_i(t)\}_{i=1}^M$ are scalar multiples of one and the same signal $s(t)$, the modulator is called *linear*, and we have

$$\xi(t) = \sum_{n=-\infty}^{\infty} \alpha_n s(t - nT). \tag{2.107}$$

Here we evaluate the power density spectrum of the signal (2.104), which is generally nonstationary. The Fourier transform of $\xi(t)$ is given by

$$\Xi(f) = \sum_{n=-\infty}^{\infty} S(f; \alpha_n, \sigma_n) e^{-j2\pi fnT} \tag{2.108}$$

where $S(f; \alpha_n, \sigma_n)$, the Fourier transform of $s(t; \alpha_n, \sigma_n)$, takes values in the set $\{S_i(f)\}_{i=1}^{M}$, with $S_i(f) \triangleq \mathcal{F}[s_i(t)]$, $i = 1, 2, \ldots, M$. Thus, from (2.94) we get

$$\Gamma_\xi(f_1, f_2) = \sum_{m=-\infty}^{\infty} \sum_{n=-\infty}^{\infty} \mathrm{E}\{S(f_1; \alpha_m, \sigma_m) S^*(f_2; \alpha_n, \sigma_n)\} e^{-j2\pi(f_1 m - f_2 n)T}$$

$$= \sum_{\ell=-\infty}^{\infty} \sum_{n=-\infty}^{\infty} \mathrm{E}\{S(f_1; \alpha_{n+\ell}, \sigma_{n+\ell}) S^*(f_2; \alpha_n, \sigma_n)\} e^{-j2\pi f_1 \ell T} e^{-j2\pi(f_1 - f_2)nT}$$

As the sequences $(\alpha_n)$, $(\sigma_n)$ are stationary, the expectation in the last line of the previous equation depends only on $\ell$ and not on $n$. Thus, recalling the equality (see, e.g., Jones, 1966, p. 135)

$$\sum_{m=-\infty}^{\infty} e^{-j2\pi mwz} = \frac{1}{w} \sum_{m=-\infty}^{\infty} \delta\left(z - \frac{m}{w}\right) \tag{2.109}$$

we obtain

$$\Gamma_\xi(f_1, f_2) = \frac{1}{T} \sum_{\ell=-\infty}^{\infty} \mathrm{E}\{S(f_1; \alpha_{n+\ell}, \sigma_{n+\ell}) S^*(f_2; \alpha_n, \sigma_n)\} e^{-j2\pi f_1 \ell T}$$

$$\sum_{m=-\infty}^{\infty} \delta\left(f_1 - f_2 - \frac{m}{T}\right) \tag{2.110}$$

Compare now (2.110) with (2.95). It is apparent that the power spectrum of $\xi(t)$ is given by

$$\mathcal{G}_\xi(f) = \frac{1}{T} \sum_{\ell=-\infty}^{\infty} G_\ell(f) e^{-j2\pi f \ell T} \tag{2.111}$$

where

$$G_\ell(f) \triangleq \mathrm{E}\{S(f; \alpha_{n+\ell}, \sigma_{n+\ell}) S^*(f; \alpha_n, \sigma_n)\} \tag{2.112}$$

It is customary, in the computation of spectral densities, to separate their continuous part from their discrete part (line spectrum). This can be done in our situation by defining

$$G_\infty(f) \triangleq \lim_{\ell \to \infty} G_\ell(f)$$

$$= |\mathrm{E}\{S(f; \alpha_n, \sigma_n)\}|^2 \tag{2.113}$$

(this does not depend on $n$ because of stationarity) and rewriting (2.111) in the form

$$\mathcal{G}_\xi(f) = \frac{1}{T} \sum_{\ell=-\infty}^{\infty} [G_\ell(f) - G_\infty(f)] e^{-j2\pi f\ell T}$$
$$+ \frac{1}{T^2} G_\infty(f) \sum_{\ell=-\infty}^{\infty} \delta\left(f - \frac{\ell}{T}\right) \qquad (2.114)$$

where (2.109) was used again. The second term in the RHS of (2.114) is a line spectrum with lines spaced $1/T$ Hz apart. The first term is line-free if $G_\ell(f) - G_\infty(f)$ tends to zero fast enough as $\ell \to \infty$ for all $f$. We shall assume in the following that this is the case.

Equation (2.114) can be rewritten in a slightly different form by observing that, from definition (2.112), it follows that

$$G_{-\ell}(f) = G_\ell^*(f) \qquad (2.115)$$

Thus, denoting by $\mathcal{G}_\xi^{(c)}(f)$ and $\mathcal{G}_\xi^{(d)}(f)$ the continuous and the discrete part of the power spectrum, respectively, we finally get

$$\mathcal{G}_\xi(f) = \mathcal{G}_\xi^{(c)}(f) + \mathcal{G}_\xi^{(d)}(f) \qquad (2.116)$$

where

$$\mathcal{G}_\xi^{(c)}(f) = \frac{2}{T} \Re \left\{ \sum_{\ell=0}^{\infty} [G_\ell(f) - G_\infty(f)] e^{-j2\pi f\ell T} \right\} - \frac{1}{T} [G_0(f) - G_\infty(f)] \quad (2.117)$$

and

$$\mathcal{G}_\xi^{(d)}(f) = \frac{1}{T^2} G_\infty(f) \sum_{\ell=-\infty}^{\infty} \delta\left(f - \frac{\ell}{T}\right) \qquad (2.118)$$

We shall now proceed to specialize (2.116)–(2.118) to a number of cases of practical interest.

### Linearly modulated digital signals

When the modulator is linear, that is, (2.104) reduces to (2.116)–(2.118), from (2.112) we get, with $S(f)$ denoting the Fourier transform of $s(t)$,

$$G_\ell(f) = \mathrm{E}\{\alpha_{n+\ell}\alpha_n^*\}|S(f)|^2 \qquad (2.119)$$

If

$$\mathrm{E}\{\alpha_n\} = \mu \qquad (2.120)$$

and

$$E\{\alpha_\ell \alpha_m^*\} = \sigma_\alpha^2 \rho_{\ell-m} + |\mu|^2 \qquad (2.121)$$

with $\rho_0 = 1$ and $\rho_\infty = 0$, then the power spectrum of $\xi(t)$ is given by

$$\mathcal{G}_\xi(f) = \mathcal{G}_\xi^{(c)}(f) + \mathcal{G}_\xi^{(d)}(f) \qquad (2.122)$$

where

$$\mathcal{G}_\xi^{(c)}(f) = \frac{\sigma_\alpha^2}{T}|S(f)|^2 \left\{ 2\Re \sum_{\ell=0}^{\infty} \rho_\ell e^{-j2\pi f \ell T} - 1 \right\} \qquad (2.123)$$

and

$$\mathcal{G}_\xi^{(d)}(f) = \frac{|\mu|^2}{T^2}|S(f)|^2 \sum_{\ell=-\infty}^{\infty} \delta\left( f - \frac{\ell}{T} \right) \qquad (2.124)$$

It is seen from (2.124) that $\mu = 0$ is a sufficient condition for $\mathcal{G}_\xi(f)$ to have no lines in its spectrum.

When the random variables $\alpha_n$ are uncorrelated (i.e., $\rho_\ell = \delta_{0,\ell}$), we get from (2.123)

$$\mathcal{G}_\xi^{(c)}(f) = \frac{\sigma_\alpha^2}{T}|S(f)|^2 \qquad (2.125)$$

Notice from (2.123) the two factors that separately influence the shape of $\mathcal{G}_\xi^{(c)}(f)$. The first is the waveform $s(t)$ through its energy spectrum. The second is the correlation of the sequence $(\alpha_n)$, which appears in the bracketed factor of (2.123). If this factor is rewritten as

$$2\Re \sum_{\ell=0}^{\infty} \rho_\ell e^{-j2\pi f \ell T} - 1 = \sum_{\ell=-\infty}^{\infty} \rho_\ell e^{-j2\pi f \ell T}$$

it is seen that it turns out to be the Fourier transform of the sequence $(\rho_n)$. In practice, the fact that $\mathcal{G}_\xi(f)$ depends on two independent factors provides a degree of freedom that can be used to shape the signal spectrum. Indeed, a given spectrum can be obtained by choosing appropriately the waveform $s(t)$, or the correlation of $(\alpha_n)$, or both.

**Example 2.10**  Perhaps the simplest way to introduce correlation in a discrete sequence is to pass it through a linear system. Thus, let $(\beta_n)$ denote a sequence of iid RVs with $E\beta_n = 0$ and $E|\beta_n|^2 = 1$, and let $(\alpha_n)$ denote a new sequence with

$$\alpha_n = \sum_m h_m \beta_{n-m}, \qquad (2.126)$$

where $(h_n)$ is the impulse response of a linear, time-invariant system. In this situation a simple computation shows that

$$E\{\alpha_{n+\ell} \alpha_n^*\} = \sum_m h_{m+\ell} h_m^*. \qquad (2.127)$$

Thus, the power spectrum of (2.107), when $(\alpha_n)$ is as in (2.126), is

$$\mathcal{G}_\xi(f) = \frac{1}{T}|S(f)|^2|H(fT)|^2, \qquad (2.128)$$

where $H(f)$ is the transfer function of the discrete linear system:

$$H(f) \overset{\triangle}{=} \sum_m h_m e^{-j2\pi mf} \qquad (2.129)$$

It is immediately apparent from (2.128) that the same power spectrum for $\xi(t)$ could be obtained by using, instead of $(\alpha_n)$, the sequence $(\beta_n)$ and a signal whose Fourier transform is $S(f)H(fT)$.                                                                        □

### Nonlinearly modulated digital signals

We shall now consider the computation of the power spectrum of the digital signal $\xi(t)$ expressed by (2.104) when the sequence $(\sigma_n)$ is assumed to have a special structure. In particular, we assume that $(\alpha_n)$ is an iid sequence, and that $(\sigma_n)$ depends on $(\alpha_n)$ as follows:

$$\sigma_{n+1} = g(\alpha_n, \sigma_n), \qquad (2.130)$$

where $g(\,\cdot\,)$ is a completely known deterministic function. Equation (2.130) describes in which state the encoder is forced to move at time $n + 1$, when at time $n$ it was in state $\sigma_n$ and the source symbol is $\alpha_n$. The modulator uses the value of the pair $\alpha_n$, $\sigma_n$ to choose the waveform $s(t; \alpha_n, \sigma_n)$ from the set $\{s_i(t)\}_{i=1}^M$, which is then output sequentially.

For this model of a digital signal to be fully specified, it is sufficient to provide the function $g(\,\cdot\,)$ and the mapping between pairs $\alpha_n$, $\sigma_n$ and waveforms of the set $\{s_i(t)\}_{i=1}^M$. We assume, hereafter, that $\sigma_n$ takes on the $q$ values $\Sigma_1, \Sigma_2,$ $\ldots, \Sigma_q$, and $\alpha_n$ takes on the $L$ values $a_1, a_2, \ldots, a_L$ ($q$ and $L$ both finite). Thus, our description of $\xi(t)$ can be done through two $L \times q$ tables whose rows are labeled $a_1, a_2, \ldots, a_L$ and whose columns are labeled $\Sigma_1, \Sigma_2, \ldots, \Sigma_q$. In the first table we display the waveforms corresponding to the pairs $(a_i, \Sigma_j)$, and in the second the values of $g(a_i, \Sigma_j)$. An equivalent representation is in the form of a *state diagram*. This is a directed graph consisting of $q$ vertexes, each representing one state; an oriented branch is drawn from state $\Sigma_i$ to state $\Sigma_j$ if and only if there is a source symbol $a_k$ such that $g(a_k, \Sigma_i) = \Sigma_j$. The branch is then labeled by $a_k$ and by the waveform, say $s_\ell(t)$, corresponding to the pair $(a_k, \Sigma_i)$ (see Fig. 2.11). Before proceeding further, we provide some examples of nonlinearly modulated digital signals and their representations.

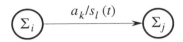

Figure 2.11: *Element of the state-diagram representation of a modulated digital signal.*

| $\sigma_n$ $\alpha_n$ | $\Sigma_+$ | $\Sigma_-$ |
|---|---|---|
| 0 | 0 | 0 |
| 1 | $s(t)$ | $-s(t)$ |

| $\sigma_n$ $\alpha_n$ | $\Sigma_+$ | $\Sigma_-$ |
|---|---|---|
| 0 | $\Sigma_+$ | $\Sigma_-$ |
| 1 | $\Sigma_-$ | $\Sigma_+$ |

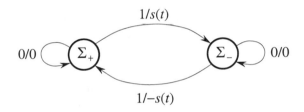

Figure 2.12: *Representation of the bipolar-encoded digital signal: Tabular form and state diagram.*

**Example 2.11 ("Bipolar-encoded" digital signal)**  The modulator has $q = 2$ states, say $\Sigma_+$ and $\Sigma_-$, and the source is binary; that is, $a \in \{0, 1\}$. The modulator responds to a source symbol 0 with a zero waveform and to a source symbol 1 with the waveform $s(t)$ or $-s(t)$, according to whether its state is $\Sigma_+$ or $\Sigma_-$, respectively. Source symbol 1 makes the modulator change its state. The tabular and state-diagram representations of this signal are provided in Fig. 2.12.                                                    □

**Example 2.12 ("Miller-encoded" digital signal)**  The modulator has $M = 4$ waveforms, $q = 4$ states, and the source is binary. Figure (2.13) describes this digital signal.
□

| $\sigma_n$ / $\alpha_n$ | $\Sigma_1$ | $\Sigma_2$ | $\Sigma_3$ | $\Sigma_4$ |
|---|---|---|---|---|
| 0 | $s_4(t)$ | $s_4(t)$ | $s_1(t)$ | $s_1(t)$ |
| 1 | $s_2(t)$ | $s_3(t)$ | $s_2(t)$ | $s_3(t)$ |

| $\sigma_n$ / $\alpha_n$ | $\Sigma_1$ | $\Sigma_2$ | $\Sigma_3$ | $\Sigma_4$ |
|---|---|---|---|---|
| 0 | $\Sigma_4$ | $\Sigma_4$ | $\Sigma_1$ | $\Sigma_1$ |
| 1 | $\Sigma_2$ | $\Sigma_3$ | $\Sigma_2$ | $\Sigma_3$ |

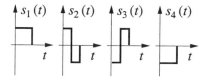

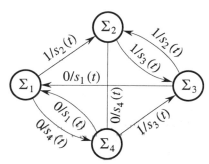

Figure 2.13: *Representation of the Miller-encoded digital signal: Tabular form, waveforms, and state diagram.*

**Example 2.13 ("TCM" digital signal)**   The modulator has $M = 8$, $q = 4$, and the source is quaternary. The available signals are

$$s_i(t) = \exp\left\{ j\left[2\pi f_0 t + (i-1)\frac{\pi}{4}\right]\right\}, \quad i = 1,\ldots,8, \quad 0 \le t < T, \quad f_0 T \gg 1$$

Fig. 2.14 describes the resulting digital signal.   □

For our future computations, the following quantities must be defined:

**(a)** The *state transition matrices* $\mathbf{E}_k$, $k = 1, 2, \ldots, L$, which are the $q \times q$

| $\sigma_n$ $\alpha_n$ | $\Sigma_1$ | $\Sigma_2$ | $\Sigma_3$ | $\Sigma_4$ |
|---|---|---|---|---|
| 0 | $s_1(t)$ | $s_2(t)$ | $s_3(t)$ | $s_4(t)$ |
| 1 | $s_5(t)$ | $s_6(t)$ | $s_7(t)$ | $s_8(t)$ |
| 2 | $s_3(t)$ | $s_4(t)$ | $s_1(t)$ | $s_2(t)$ |
| 3 | $s_7(t)$ | $s_8(t)$ | $s_5(t)$ | $s_6(t)$ |

$$s_i(t) = \exp\left\{ j\left[ 2\pi f_0 t + (i-1)\frac{\pi}{4} \right] \right\}$$

$$0 \le t < T$$

$$f_0 T \gg 1$$

| $\sigma_n$ $\alpha_n$ | $\Sigma_1$ | $\Sigma_2$ | $\Sigma_3$ | $\Sigma_4$ |
|---|---|---|---|---|
| 0 | $\Sigma_1$ | $\Sigma_3$ | $\Sigma_1$ | $\Sigma_3$ |
| 1 | $\Sigma_1$ | $\Sigma_3$ | $\Sigma_1$ | $\Sigma_3$ |
| 2 | $\Sigma_2$ | $\Sigma_4$ | $\Sigma_2$ | $\Sigma_4$ |
| 3 | $\Sigma_2$ | $\Sigma_4$ | $\Sigma_2$ | $\Sigma_4$ |

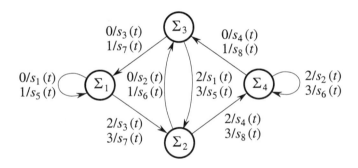

Figure 2.14: *Representation of the TCM digital signal: (a) Tabular form; (b) state diagram.*

matrices whose entry $[\mathbf{E}_k]_{ij}$ is equal to 1 if $g(a_k, \Sigma_i) = \Sigma_j$, and zero otherwise. In clarification, the matrix $\mathbf{E}_k$ has a 1 in row $i$ and column $j$ if the source symbol $a_k$ forces a transition of the modulator from state $\Sigma_i$ to state $\Sigma_j$. Otherwise, it has a zero.

**(b)** The row $q$-vectors $\mathbf{s}_k(f)$, $k = 1, 2, \ldots, L$, whose $q$ entries are the Fourier transforms of the waveforms of the set $\{s_i(t)\}_{i=1}^{M}$, according to the rule

$[\mathbf{s}_k(f)]_i = \mathcal{F}[s(t; a_k, \Sigma_i)]$. That is, $\mathbf{s}_k(f)$ includes the amplitude spectra of the modulator waveforms corresponding to the source symbol $a_k$ for the different modulator states.

**Example 2.11 (continued)** In this case, letting $a_1 = 0$ and $a_2 = 1$, we have

$$\mathbf{E}_1 = \begin{bmatrix} 1 & 0 \\ 0 & 1 \end{bmatrix}, \qquad \mathbf{E}_2 = \begin{bmatrix} 0 & 1 \\ 1 & 0 \end{bmatrix}$$

and

$$\mathbf{s}_1(f) = [0 \ 0], \qquad \mathbf{s}_2(f) = S(f) [1 \ -1]$$

where $S(f)$ is the Fourier transform of $s(t)$. □

**Example 2.12 (continued)** In this case, letting $a_1 = 0$, $a_2 = 1$, we have

$$\mathbf{E}_1 = \begin{bmatrix} 0 & 0 & 0 & 1 \\ 0 & 0 & 0 & 1 \\ 1 & 0 & 0 & 0 \\ 1 & 0 & 0 & 0 \end{bmatrix}, \qquad \mathbf{E}_2 = \begin{bmatrix} 0 & 1 & 0 & 0 \\ 0 & 0 & 1 & 0 \\ 0 & 1 & 0 & 0 \\ 0 & 0 & 1 & 0 \end{bmatrix},$$

and

$$\mathbf{s}_1(f) = S(f) [-1 - z \quad -1 - z \quad 1 + z \quad 1 + z]$$
$$\mathbf{s}_2(f) = S(f) [1 - z \quad -1 + z \quad 1 - z \quad -1 + z]$$

where

$$S(f) \triangleq \frac{T}{2} \frac{\sin \pi f T/2}{\pi f T/2}$$

and

$$z \triangleq e^{-j\pi f T}$$

□

**Example 2.13 (continued)** In this case, letting $a_i = i - 1$, $i = 1, 2, 3, 4$, we have

$$\mathbf{E}_1 = \mathbf{E}_2 = \begin{bmatrix} 1 & 0 & 0 & 0 \\ 0 & 0 & 1 & 0 \\ 1 & 0 & 0 & 0 \\ 0 & 0 & 1 & 0 \end{bmatrix} \qquad \mathbf{E}_3 = \mathbf{E}_4 = \begin{bmatrix} 0 & 1 & 0 & 0 \\ 0 & 0 & 0 & 1 \\ 0 & 1 & 0 & 0 \\ 0 & 0 & 0 & 1 \end{bmatrix}$$

and

$$\mathbf{s}_1(f) = S(f) [w^0 \quad w^1 \quad w^2 \quad w^3]$$

$$\mathbf{s}_2(f) = S(f) \begin{bmatrix} w^4 & w^5 & w^6 & w^7 \end{bmatrix}$$

$$\mathbf{s}_3(f) = S(f) \begin{bmatrix} w^2 & w^3 & w^0 & w^1 \end{bmatrix}$$

$$\mathbf{s}_4(f) = S(f) \begin{bmatrix} w^6 & w^7 & w^4 & w^5 \end{bmatrix}$$

where

$$S(f) \triangleq T \frac{\sin \pi (f - f_0)T}{\pi(f - f_0)T} \qquad w \triangleq e^{j\pi/4}$$

$\square$

We want now to evaluate the power spectrum of the digital signal (2.104). The assumption that $(\alpha_n)$ is an iid sequence, along with (2.130), implies that the state sequence $(\sigma_n)$ is a homogeneous Markov chain. In fact, the probability that the encoder is in a given state at time $n + 1$ depends only on the state $\sigma_n$ and on the symbol $\alpha_n$, and not on the preceding states $\sigma_{n-1}, \sigma_{n-2}, \ldots$. The transition matrix of this chain has entries

$$\begin{aligned}
[\mathbf{P}]_{ij} &\triangleq P\{\sigma_{n+1} = \Sigma_j \mid \sigma_n = \Sigma_i\} \\
&= P\{g(\alpha_n, \sigma_n) = \Sigma_j \mid \sigma_n = \Sigma_i\} \\
&= \sum_{k=1}^{L} P\{g(\alpha_n, \sigma_n) = \Sigma_j \mid \alpha_n = a_k, \sigma_n = \Sigma_i\} P\{\alpha_n = a_k\} \\
&= \sum_{k=1}^{L} p_k [\mathbf{E}_k]_{ij}
\end{aligned} \qquad (2.131)$$

where, as already defined,

$$p_k \triangleq P\{\alpha_n = a_k\}, \qquad k = 1, 2, \ldots, L \qquad (2.132)$$

Thus, the transition matrix $\mathbf{P}$ is a linear combination of the matrices $\mathbf{E}_k$:

$$\mathbf{P} = \sum_{k=1}^{L} p_k \mathbf{E}_k \qquad (2.133)$$

We assume that the Markov chain is fully regular, and that its starting time is $n = -\infty$. This implies that, for any finite $n$, $\mathbf{w}^{(n)} = \mathbf{w}$. Thus, the transition matrix $\mathbf{P}$ provides a complete characterization of the sequence of modulator states; in particular, the stationary state probabilities

$$w_i \triangleq P\{\sigma_n = \Sigma_i\} \qquad (2.134)$$

are obtained as the entries of vector $\mathbf{w}$ computed from (2.50) and (2.51).

Let us now define four quantities that play an important role in the expression of the power spectral density we are seeking. The first is the average value, taken over the source symbols, of the vectors $s_k(f)$:

$$c_2(f) \triangleq \sum_{i=1}^{L} p_k s_k(f) \tag{2.135}$$

The $i$th component of $c_2(f)$ is then the average amplitude spectrum of the waveforms available to the modulator when it is in state $\Sigma_i$.

The second is the $q$-vector $c_1(f)$ whose $j$th component is the average amplitude spectrum of the waveforms that, when output by the modulator, force it to state $\Sigma_j$.

This $j$th component of $c_1(f)$ is then given by

$$[c_1(f)]_j \triangleq \sum_{k=1}^{L} \sum_{i=1}^{q} p_k w_i [E_k]_{ij} \mathcal{F}[s(t; a_k, \Sigma_i)] \tag{2.136}$$

(recall from the definition of $E_k$ that $[E_k]_{ij} = 1$ only if the source symbol $a_k$ takes the modulator from state $\Sigma_i$ to state $\Sigma_j$). If we define the $q \times q$ diagonal matrix

$$D \triangleq \text{diag}(w_1, w_2, \ldots, w_q) \tag{2.137}$$

we have from (2.136)

$$c_1(f) = \sum_{k=1}^{L} p_k s_k(f) D E_k \tag{2.138}$$

Our third quantity is the average amplitude spectrum of the waveforms available from the modulator:

$$
\begin{aligned}
\mu(f) &\triangleq \text{E}\{\mathcal{F}[s(t; \alpha_n, \sigma_n)]\} \\
&= \sum_{k=1}^{L} \sum_{i=1}^{q} p_k w_i \mathcal{F}[s(t; a_k, \Sigma_i)] \\
&= \sum_{k=1}^{L} p_k w s'_k(f) \\
&= w c_2^*(f)
\end{aligned}
\tag{2.139}
$$

Finally, the fourth quantity of interest is the average energy spectrum of the waveforms available from the modulator:

$$
\begin{aligned}
c_0(f) &\triangleq \sum_{k=1}^{L} \sum_{i=1}^{q} p_k w_i |\mathcal{F}[s(t; a_k, \Sigma_i)]|^2 \\
&= \sum_{k=1}^{L} p_k s_k^*(f) D s'_k(f)
\end{aligned}
\tag{2.140}
$$

Before proceeding further, we evaluate these four quantities in a few examples.

**Example 2.11 (continued)**   Assuming that the source symbols 0 and 1 are equally likely, we have

$$\mathbf{P} = \mathbf{P}^\infty = \frac{1}{2}\begin{bmatrix} 1 & 1 \\ 1 & 1 \end{bmatrix}$$

so that

$$\mathbf{w} = [\frac{1}{2}\ \frac{1}{2}]$$

Moreover,

$$\begin{aligned}
\mathbf{c}_2(f) &= S(f)[\frac{1}{2}\ -\frac{1}{2}] \\
\mathbf{c}_1(f) &= S(f)[-\frac{1}{4}\ \frac{1}{4}] \\
\mu(f) &= 0
\end{aligned}$$

and

$$c_0(f) = \frac{1}{2}|S(f)|^2$$

$\square$

**Example 2.13 (continued)**   Assuming that the source symbols 0 and 1 are equally likely, we have

$$\mathbf{P} = \frac{1}{2}\begin{bmatrix} 0 & 1 & 0 & 1 \\ 0 & 0 & 1 & 1 \\ 1 & 1 & 0 & 0 \\ 1 & 0 & 1 & 0 \end{bmatrix} \qquad \mathbf{P}^\infty = \frac{1}{4}\begin{bmatrix} 1 & 1 & 1 & 1 \\ 1 & 1 & 1 & 1 \\ 1 & 1 & 1 & 1 \\ 1 & 1 & 1 & 1 \end{bmatrix}$$

Thus, $\mathbf{w} = [\frac{1}{4}\ \frac{1}{4}\ \frac{1}{4}\ \frac{1}{4}]$, and

$$\begin{aligned}
\mathbf{c}_2(f) &= S(f)[-z\ -1\ 1\ z], \\
\mathbf{c}_1(f) &= \frac{1}{4}[1+z\ 1-z\ -1+z\ -1-z], \\
\mu(f) &= 0 \\
c_0(f) &= 2|S(f)|^2.
\end{aligned}$$

$\square$

**Example 2.13 (continued)**   Assuming that the source symbols 0, 1, 2, 3, are equally likely, we have

$$\mathbf{P} = \frac{1}{2}\begin{bmatrix} 1 & 1 & 0 & 0 \\ 0 & 0 & 1 & 1 \\ 1 & 1 & 0 & 0 \\ 0 & 0 & 1 & 1 \end{bmatrix} \qquad \mathbf{P}^{\infty} = \mathbf{P}^2 = \frac{1}{4}\begin{bmatrix} 1 & 1 & 1 & 1 \\ 1 & 1 & 1 & 1 \\ 1 & 1 & 1 & 1 \\ 1 & 1 & 1 & 1 \end{bmatrix}$$

Moreover,

$$\begin{aligned}
\mathbf{c}_2(f) &= \mathbf{0} \\
\mathbf{c}_1(f) &= \mathbf{0} \\
\mu(f) &= 0,
\end{aligned}$$

and

$$c_0(f) = |S(f)|^2$$

$\square$

Consider now the computation of the power spectrum. This will be undertaken by applying (2.116)–(2.118). From (2.112) we have, for $\ell > 0$,

$$\begin{aligned}
G_\ell(f) &= \sum_{h=1}^{L}\sum_{k=1}^{L}\sum_{i=1}^{q}\sum_{j=1}^{q} S(f; a_h, \Sigma_j)S^*(f; a_k, \Sigma_i) \\
&\quad \cdot P\{\alpha_{n+\ell} = a_h, \alpha_n = a_k, \sigma_{n+\ell} = \Sigma_j, \sigma_n = \Sigma_i\} \quad (2.141)
\end{aligned}$$

The probabilities appearing in (2.141) can be put in the form

$$\begin{aligned}
&P\{\alpha_{n+\ell} = a_h, \alpha_n = a_k, \sigma_{n+\ell} = \Sigma_j, \sigma_n = \Sigma_i\} \\
&= P\{\alpha_{n+\ell} = a_h, \sigma_{n+\ell} = \Sigma_j \mid \alpha_n = a_k, \sigma_n = \Sigma_i\} \cdot p_k w_i \quad (2.142)
\end{aligned}$$

As the source symbols are independent, we have

$$\begin{aligned}
&P\{\alpha_{n+\ell} = a_h, \sigma_{n+\ell} = \Sigma_j \mid \alpha_n = a_k, \sigma_n = \Sigma_i\} \\
&= p_h\, P\{\sigma_{n+\ell} = \Sigma_j \mid \alpha_n = a_k, \sigma_n = \Sigma_i\} \\
&= p_h \sum_{m=1}^{q} P\{\sigma_{n+\ell} = \Sigma_j \mid \sigma_{n+1} = \Sigma_m, \alpha_n = a_k, \sigma_n = \Sigma_i\} \\
&\quad \cdot P\{\sigma_{n+1} = \Sigma_m \mid \alpha_n = a_k, \sigma_n = \Sigma_i\} \\
&= p_h \sum_{m=1}^{q} P\{\sigma_{n+\ell} = \Sigma_j \mid \sigma_{n+1} = \Sigma_m\}\,[\mathbf{E}_k]_{im} \\
&= p_h \sum_{m=1}^{q} [\mathbf{P}^{\ell-1}]_{mj}\,[\mathbf{E}_k]_{im} \quad (2.143)
\end{aligned}$$

For $\ell = 0$, we get instead

$$G_0(f) = \sum_{k=1}^{L} \sum_{i=1}^{q} |S(f; a_k, \Sigma_i)|^2 \, p_k \, w_i. \tag{2.144}$$

By combining together equations (2.141) to (2.144), we have

$$G_\ell(f) = \begin{cases} \displaystyle\sum_{h=1}^{L} \sum_{k=1}^{L} p_h p_k \mathbf{s}_k^*(f) \mathbf{D} \mathbf{E}_k \mathbf{P}^{\ell-1} \mathbf{s}_h'(f), & \ell > 0 \\[2mm] \displaystyle\sum_{h=1}^{L} p_h \mathbf{s}_h^*(f) \mathbf{D} \mathbf{s}_h'(f), & \ell = 0 \end{cases} \tag{2.145}$$

and, using definitions (2.135) to (2.140),

$$G_\ell(f) = \begin{cases} \mathbf{c}_1^*(f) \mathbf{P}^{\ell-1} \mathbf{c}_2'(f), & \ell > 0 \\ c_0(f), & \ell = 0 \end{cases} \tag{2.146}$$

Also, from (2.113) and the definition (2.139) of $\mu(f)$, we get

$$G_\infty(f) = |\mu(f)|^2 \tag{2.147}$$

or, equivalently, if (2.146) is used,

$$G_\infty(f) = \mathbf{c}_1^*(f) \mathbf{P}^\infty \mathbf{c}_2'(f) \tag{2.148}$$

In conclusion, the continuous and discrete parts of the power spectrum of our digital signal are given by

$$\mathcal{G}_\xi^{(c)}(f) = \frac{1}{T}[c_0(f) - |\mu(f)|^2] + \frac{2}{T}\Re[\mathbf{c}_1^*(f) \mathbf{\Lambda}(f) \mathbf{c}_2'(f)] \tag{2.149}$$

and

$$\mathcal{G}^{(d)}(f) = \frac{1}{T^2} |\mu(f)|^2 \sum_{\ell=-\infty}^{\infty} \delta\left(f - \frac{\ell}{T}\right) \tag{2.150}$$

where

$$\mathbf{\Lambda}(f) \triangleq \sum_{\ell=1}^{\infty} [\mathbf{P}^{\ell-1} - \mathbf{P}^\infty] e^{-j2\pi f \ell T} \tag{2.151}$$

Whenever there exists a finite $N$ such that $\mathbf{P}^N = \mathbf{P}^\infty$ [e.g., when $(\sigma_n)$ is a shift-register state sequence], $\mathbf{\Lambda}(f)$ involves a finite number of terms, and its computation is straightforward. If such an $N$ does not exist, we need a technique to evaluate the RHS of (2.151).

Observe that, from the equality $\mathbf{P}^k \mathbf{P}^\infty = \mathbf{P}^\infty$, we have

$$
\begin{aligned}
\mathbf{P}^k - \mathbf{P}^\infty &= (\mathbf{I} - \mathbf{P}^\infty)(\mathbf{P}^k - \mathbf{P}^\infty) \\
&= (\mathbf{I} - \mathbf{P}^\infty)(\mathbf{P} - \mathbf{P}^\infty)^k
\end{aligned}
\tag{2.152}
$$

for all $k > 0$. Thus

$$
\begin{aligned}
\mathbf{\Lambda}(f) &= e^{-j2\pi fT} \sum_{\ell=0}^{\infty} (\mathbf{P}^\ell - \mathbf{P}^\infty) e^{-j2\pi f\ell T} \\
&= e^{-j2\pi f\ell T}(\mathbf{I} - \mathbf{P}^\infty) \sum_{\ell=0}^{\infty} (\mathbf{P}^k - \mathbf{P}^\infty)^\ell e^{-j2\pi f\ell T} \\
&= (\mathbf{I} - \mathbf{P}^\infty)[e^{j2\pi fT}\mathbf{I} - (\mathbf{P}^k - \mathbf{P}^\infty)]^{-1}
\end{aligned}
\tag{2.153}
$$

where the last equality holds because the matrix $(\mathbf{P}^k - \mathbf{P}^\infty)$ has all its eigenvalues with magnitude less than 1 (see Cariolaro and Tronca, 1974, for a proof).

It is seen from (2.153) that the matrix $\mathbf{\Lambda}(f)$, necessary to evaluate the RHS of (2.149), can be computed for each value of $f$ by inverting a $q \times q$ matrix. This procedure is computationally inefficient because, if the spectrum value is needed for several $f$, many matrix inversions must be performed. For a more efficient technique, observe that $\mathbf{\Lambda}(f)$ is an analytic function of the matrix

$$
\mathbf{A} \triangleq \mathbf{P} - \mathbf{P}^\infty
\tag{2.154}
$$

so that $\mathbf{\Lambda}(f)$ can be written in the form of a polynomial in $\mathbf{A}$ whose coefficients depend on $f$, say,

$$
\mathbf{\Lambda}(f) = (\mathbf{I} - \mathbf{P}^\infty) \sum_{i=0}^{K-1} \beta_i(f)\mathbf{A}^i
\tag{2.155}
$$

The expansion (2.155) is not unique, unless we restrict $K$ to take on its minimum possible value (i.e., the degree of the minimal polynomial of $\mathbf{A}$). Here we assume that the reader is familiar with the basic results of matrix calculus, as summarized in Appendix B. In this situation, equating the RHS of (2.153) and (2.155), we get

$$
[e^{j2\pi fT}\mathbf{I} - \mathbf{A}] \sum_{i=0}^{K-1} \beta_i(f)\mathbf{A}^i - \mathbf{I} = 0
\tag{2.156}
$$

As the LHS of (2.156) is a polynomial in $\mathbf{A}$ having degree $K$, its coefficients must be proportional to those of the minimal polynomial of $\mathbf{A}$. Denoting this minimal polynomial by

$$
\Delta(\lambda) = \sum_{i=0}^{K} \delta_i \lambda^i, \qquad \delta_K = 1
\tag{2.157}
$$

and equating the coefficients of $\mathbf{A}_i$, $i = 0, \ldots, K$, in (2.156) and in the identity

$$\sum_{i=0}^{K} \delta_i \mathbf{A}^i = 0 \tag{2.158}$$

we get the coefficients $\beta_i(f)$, $i = 0, \ldots, K - 1$, needed to compute $\mathbf{\Lambda}(f)$ according to (2.155). This procedure allows one to express $\mathbf{\Lambda}(f)$ as a closed-form function of $f$, which can be computed for each value of $f$ with modest computational effort.

Although the use of the minimal polynomial of $\mathbf{A}$ to obtain the representation (2.155) leads to the most economical way to compute the spectrum, every polynomial $\Delta(\lambda)$ such that (2.158) holds can be used instead of the minimal polynomial. In particular, the use of the characteristic polynomial of $\mathbf{A}$ (which has degree $q$) leads to a relatively simple computational algorithm (due to Faddeev and first applied to this problem by Cariolaro and Tronca, 1974). According to this technique, $\mathbf{\Lambda}(f)$ can be given the form

$$\mathbf{\Lambda}(f) = (\mathbf{I} - \mathbf{P}^\infty) \frac{1}{\Delta(e^{j2\pi fT})} \mathbf{B}(e^{j2\pi fT}) \tag{2.159}$$

where $\Delta(\lambda)$ is now the characteristic polynomial of $\mathbf{A}$, and $\mathbf{B}(\cdot)$ is a $q \times q$ matrix polynomial:

$$\mathbf{B}(\lambda) = \lambda^{q-1}\mathbf{B}_0 + \lambda^{q-2}\mathbf{B}_1 + \cdots + \mathbf{B}_{q-1} \tag{2.160}$$

The polynomials $\mathbf{B}(\cdot)$ and $\Delta(\cdot)$ can be computed simultaneously by using the following recursive algorithm (Gantmacher, 1959). Starting with $\delta_q = 1$ and $\mathbf{B}_0 = \mathbf{I}$, let

$$\begin{aligned}
\mathbf{Q}_k &= \mathbf{A}\mathbf{B}_{k-1} \\
\delta_{q-k} &= -\frac{1}{k}\text{tr } \mathbf{Q}_k \\
\mathbf{B}_k &= \mathbf{Q}_k + \delta_{q-k}\mathbf{I}
\end{aligned} \tag{2.161}$$

for $k = 1, 2, \ldots, q$. At the final step, $\mathbf{B}_q$ must be equal to the null matrix, and $\delta_0 = 0$, because the matrix $\mathbf{A}$ has a zero eigenvalue.

**Example 2.11 (continued)**  In this case $\mathbf{P} = \mathbf{P}^\infty$; thus, from (2.151) we have

$$\mathbf{\Lambda}(f) = (\mathbf{I} - \mathbf{P}^\infty)e^{-j2\pi fT} = \frac{1}{2}\begin{bmatrix} 1 & -1 \\ -1 & 1 \end{bmatrix} e^{-j2\pi fT}$$

so that

$$\mathcal{G}_\xi(f) \equiv \mathcal{G}^{(c)}(f) = \frac{1}{2T}|S(f)|^2(1 - \cos 2\pi fT)$$

$\square$

**Example 2.12 (continued)**   We have

$$\mathbf{A} \overset{\triangle}{=} \mathbf{P} - \mathbf{P}^{\infty} = \frac{1}{4} \begin{bmatrix} -1 & 1 & -1 & 1 \\ -1 & -1 & 1 & 1 \\ 1 & 1 & -1 & -1 \\ 1 & -1 & 1 & -1 \end{bmatrix}$$

Application of the Faddeev algorithm gives

$$\delta_4 = \delta_3 = 1, \qquad \delta_2 = \frac{1}{2}, \qquad \delta_1 = \delta_0 = 0$$

$$\mathbf{B}_1 = \frac{1}{4} \begin{bmatrix} 3 & 1 & -1 & 1 \\ -1 & 3 & 1 & 1 \\ 1 & 1 & 3 & -1 \\ 1 & -1 & 1 & 3 \end{bmatrix} \qquad \mathbf{B}_2 = \frac{1}{4} \begin{bmatrix} 1 & 0 & 0 & 1 \\ 0 & 1 & 1 & 0 \\ 0 & 1 & 1 & 0 \\ 1 & 0 & 0 & 1 \end{bmatrix}$$

and

$$\mathbf{B}_3 = 0$$

Thus, using (2.149) and (2.159), we get

$$
\begin{aligned}
\mathcal{G}_\xi(f) &\equiv \mathcal{G}_\xi^{(c)}(f) \\
&= \frac{T}{2} \left( \frac{\sin \pi f T/2}{\pi f T/2} \right)^2 \frac{3 + \cos \pi f T + 2 \cos 2\pi f T - \cos 3\pi f T}{9 + 12 \cos 2\pi f T + 4 \cos 4\pi f T}
\end{aligned}
\tag{2.162}
$$

□

**Example 2.13 (continued)**   From (2.149) we get

$$\mathcal{G}_\xi(f) \equiv \mathcal{G}_\xi^{(c)}(f) = \frac{1}{T}|S(f)|^2 \tag{2.163}$$

□

**A special case**

We finally observe an important special case of the digital signal considered. If the modulator has only one state, or, equivalently, the waveform emitted at time $nT$ depends, in a one-to-one way, only on the source symbol at the same instant, we have, from (2.149) and (2.150) and after some computations,

$$\mathcal{G}_\xi^{(c)}(f) = \frac{1}{T} \left[ \sum_{i=1}^{M} p_i |S_i(f)|^2 - \left| \sum_{i=1}^{M} p_i S_i(f) \right|^2 \right] \tag{2.164}$$

and

$$G_\xi^{(d)}(f) = \frac{1}{T^2} \left| \sum_{i=1}^{M} p_i S_i(f) \right|^2 \sum_{\ell=-\infty}^{\infty} \delta\left(f - \frac{\ell}{T}\right) \qquad (2.165)$$

where $\{S_i(f)\}_{i=1}^{M}$ are the Fourier transforms of the waveforms available from the modulator.

## 2.4.  Narrowband signals and bandpass systems

When the signal $x(t)$ is real, its Fourier transform $X(f)$ shows certain symmetries around the zero frequency. In particular, the real part of $X(f)$ is an even function of $f$, and its imaginary part is odd. As a consequence, to be in a position to reconstruct $x(t)$, it is sufficient to specify $X(f)$ only for $f \geq 0$. Now suppose that $x(t)$ is passed through a linear, time-invariant system whose transfer function is the step function $\alpha u(f)$, $\alpha$ a constant. At the output of this system we observe a signal from which $x(t)$ can be recovered without information loss. The impulse response of this system is

$$\frac{\alpha}{2}\left[\delta(t) + j\frac{1}{\pi t}\right]$$

so its response to $x(t)$ is $\alpha/2 \cdot [x(t) + j\hat{x}(t)]$, where

$$\hat{x}(t) \triangleq \frac{1}{\pi} \int \frac{x(\theta)}{t - \theta}\, d\theta \qquad (2.166)$$

is called the *Hilbert transform* of $x(t)$. Notice that, because of the singularity in the integrand, the meaning of the RHS of (2.166) has to be made precise. Specifically, the integral is defined as the Cauchy principal value. The choice $\alpha = 2$ yields

$$x(t) = \Re[\mathring{x}(t)] \qquad (2.167)$$

where $x(t)$, the system output, is

$$\mathring{x}(t) \triangleq x(t) + j\hat{x}(t) \qquad (2.168)$$

Equation (2.167) shows that the original signal $x(t)$ can be recovered from the output of a system with transfer function $2u(f)$ by simply taking its real part. The complex signal $\mathring{x}(t)$ is called the *analytic signal* associated with $x(t)$.

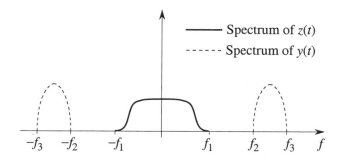

Figure 2.15: *Spectra of a baseband signal $z(t)$ and of a narrowband signal $y(t)$.*

**Example 2.14** Let $x(t) = \cos(2\pi f_0 t + \phi)$. Its Hilbert transform is $\hat{x}(t) = \sin(2\pi f_0 t + \phi)$, so the corresponding analytic signal turns out to be $\mathring{x}(t) = \exp\{j(2\pi f_0 t + \phi)\}$. We see from this simple example that the analytic signal representation is a generalization of the familiar complex representation of sinusoidal signals.  □

Among the properties of analytic signals, two are worth mentioning here.

**(a)** The operation transforming the real signal $x(t)$ into the analytic signal $\mathring{x}(t)$ is linear and time invariant. In particular, if $x(t)$ is a Gaussian random process, $\mathring{x}(t)$ is a Gaussian random process.

**(b)** Consider two real signals $z(t)$ and $y(t)$, and their product

$$x(t) \stackrel{\triangle}{=} z(t)y(t) \qquad (2.169)$$

Assume that $z(t)$ is a baseband signal, that is, its (amplitude or energy or power) spectrum is zero for $|f| > f_1$ and $y(t)$ is a narrowband signal, that is, its spectrum is nonzero only for $f_2 < |f| < f_3$, $f_2 > f_1$ (see Fig. 2.15). With these assumptions, from our definition of an analytic signal it follows that

$$\mathring{x}(t) = z(t)\mathring{y}(t) \qquad (2.170)$$

that is, $\mathring{x}(t)$ is the product of the real signal $z(t)$ and the analytic signal associated with $y(t)$.

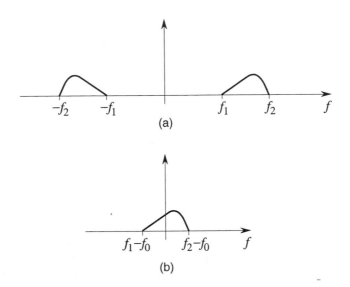

Figure 2.16: *(a) Spectrum of a narrowband signal; (b) spectrum of its complex envelope.* *(Figures not to scale.)*

**Example 2.15 (Amplitude modulation of a sinusoidal carrier)**    Let $y(t) = \cos 2\pi f_0 t$, and let $z(t)$ be a deterministic baseband signal whose Fourier transform $Z(f)$ is confined to the interval $(-f_1, f_1)$, $f_1 < f_0$. The analytic signal associated with their product is

$$\mathring{x}(t) = z(t)e^{j2\pi f_0 t} \tag{2.171}$$

which shows that the amplitude spectrum of $\mathring{x}(t)$ is $Z(f - f_0)$, that is, it is obtained by translating the amplitude spectrum of $z(t)$ around the frequency $f_0$. □

### 2.4.1. Narrowband signals: Complex envelopes

A *narrowband signal* is one whose spectrum is to a certain extent concentrated around a nonzero frequency. We define a real signal to be narrowband if its (amplitude or energy or power) spectrum is zero for $|f| \notin (f_1, f_2)$, where $(f_1, f_2)$ is a finite frequency interval not including the origin (see Fig. 2.16 (a)). On the other hand, a signal whose spectrum is concentrated around the origin of the frequency axis is referred to as a *baseband signal*. For a given narrowband signal and a frequency $f_0 \in (f_1, f_2)$, the analytic signal $\mathring{x}(t)$ can be written, according to the result of Example 2.15, in the form

$$\mathring{x}(t) = \tilde{x}(t)e^{j2\pi f_0 t} \tag{2.172}$$

where $\tilde{x}(t)$ is a (generally complex) signal whose spectrum is zero for $f > f_2 - f_0$ and $f < f_1 - f_0$ (see Fig. 2.16 (b)).

The signal $\tilde{x}(t)$ is called the *complex envelope* associated with the real signal $x(t)$. From (2.172) we have the following representation for a narrowband $x(t)$:

$$\begin{aligned} x(t) &= \Re[\mathring{x}(t)] \\ &= x_c(t)\cos 2\pi f_0 t - x_s(t)\sin 2\pi f_0 t \end{aligned} \tag{2.173}$$

where

$$\begin{aligned} x_c(t) &\triangleq \Re[\tilde{x}(t)] = \Re[\mathring{x}(t)e^{-j2\pi f_0 t}] \\ &= x(t)\cos 2\pi f_0 t + \hat{x}(t)\sin 2\pi f_0 t \end{aligned} \tag{2.174}$$

and

$$\begin{aligned} x_s(t) &\triangleq \Im[\tilde{x}(t)] = \Im[\mathring{x}(t)e^{-j2\pi f_0 t}] \\ &= -x(t)\sin 2\pi f_0 t + \hat{x}(t)\cos 2\pi f_0 t \end{aligned} \tag{2.175}$$

are baseband signals. Equation (2.173) and direct computation prove that $x_c(t)$ and $x_s(t)$ can be obtained from $x(t)$ by using the circuitry shown in Fig. 2.17. There the filters are ideal low-pass.

From (2.172) it is also possible to derive a vector representation of the narrowband signal $x(t)$. To do this we define, at any time instant $t$, a two-dimensional vector whose components are the in-phase and quadrature components of $\tilde{x}(t)$, that is, $x_c(t)$ and $x_s(t)$ (see Fig. 2.18). The magnitude of this vector is

$$A_x(t) \triangleq |\tilde{x}(t)| = \sqrt{x_c^2(t) + x_s^2(t)} \tag{2.176}$$

(see Fig. 2.19), and its phase is

$$\varphi_x(t) \triangleq \arg[\tilde{x}(t)] = \tan^{-1}\frac{x_s(t)}{x_c(t)} \tag{2.177}$$

The time functions $A_x(t)$ and $\varphi_x(t) + 2\pi f_0 t$ are called, respectively, the *instantaneous envelope* and the *instantaneous phase* of $x(t)$. The *instantaneous frequency* of $x(t)$ is defined as $1/2\pi$ times the derivative of the instantaneous phase; that is,

$$f_x(t) \triangleq f_0 + \frac{1}{2\pi}\frac{x_s'(t)x_c(t) - x_s(t)x_c'(t)}{x_c^2(t) + x_s^2(t)} \tag{2.178}$$

where the primes denote time derivatives. From (2.173) to (2.177) the following representation of the narrowband signal $x(t)$ can also be derived:

$$x(t) = A_x(t)\cos[2\pi f_0 t + \varphi_x(t)] \tag{2.179}$$

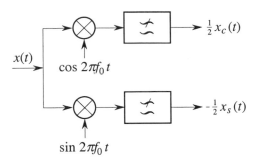

Figure 2.17: *Obtaining the real and imaginary parts of the complex envelope of the narrowband signal $x(t)$.*

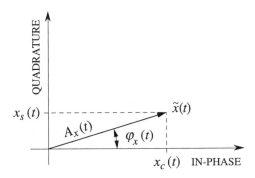

Figure 2.18: *Vector representation of the narrowband signal $x(t)$.*

**Narrowband random processes**

Consider now a real narrowband, WS stationary random process $\nu(t)$, and the complex process

$$\mathring{\nu}(t) \overset{\triangle}{=} \nu(t) + j\hat{\nu}(t). \tag{2.180}$$

The possible representations of $\nu(t)$ are

$$\nu(t) = \Re[\tilde{\nu}(t)e^{j2\pi f_0 t}] \tag{2.181}$$

$$\nu(t) = \nu_c(t)\cos 2\pi f_0 t - \nu_s(t)\sin 2\pi f_0 t \tag{2.182}$$

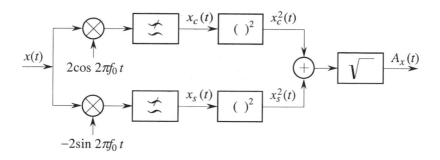

Figure 2.19: *Obtaining the instantaneous envelope of the narrowband signal $x(t)$.*

and

$$\nu(t) = A_\nu(t) \cos[2\pi f_0 t + \varphi_\nu(t)] \tag{2.183}$$

where

$$\begin{aligned}
\tilde{\nu}(t) &= \mathring{\nu}(t)e^{-j2\pi f_0 t} \\
&= \nu_c(t) + j\nu_s(t) \tag{2.184}
\end{aligned}$$

is the complex envelope of $\nu(t)$.

The power spectrum of $\mathring{\nu}(t)$ can be easily evaluated by observing that $\mathring{\nu}(t)$ can be thought of as the output of a linear, time-invariant system with transfer function $2u(f)$ whose input is $\nu(t)$. Thus, its power spectrum equals the power spectrum of $\nu(t)$ times the squared magnitude of the transfer function:

$$\mathcal{G}_{\mathring{\nu}}(f) = 4\mathcal{G}_\nu(f)u^2(f) = 4\mathcal{G}_\nu(f)u(f) \tag{2.185}$$

Equation (2.185) shows that the spectral density of $\mathring{\nu}(t)$ is equal to four times the one-sided spectral density of $\nu(t)$. Consider then the complex envelope $\tilde{\nu}(t)$. From (2.184), its autocorrelation is

$$\begin{aligned}
R_{\tilde{\nu}}(\tau) &= \mathrm{E}[\mathring{\nu}(t+\tau)\mathring{\nu}^*(t)]e^{-j2\pi f_0 \tau} \\
&= R_{\mathring{\nu}}(\tau)e^{-j2\pi f_0 \tau} \tag{2.186}
\end{aligned}$$

and hence

$$\mathcal{G}_{\tilde{\nu}}(f) = \mathcal{G}_{\mathring{\nu}}(f + f_0) \tag{2.187}$$

which shows that the power spectral density of the complex envelope $\tilde{\nu}(t)$ is the version of $\mathcal{G}_{\mathring{\nu}}(f)$ translated around the origin (see Fig. 2.20). Consider finally

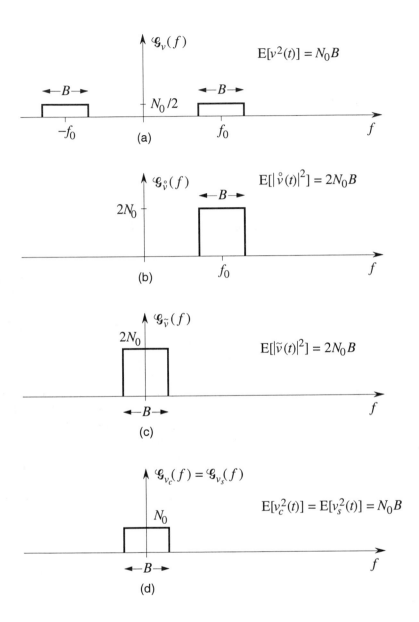

Figure 2.20: *Representations of a narrowband white noise process* $\nu(t)$: *(a) Power spectrum of* $\nu(t)$; *(b) power spectrum of the analytic signal* $\overset{\circ}{\nu}(t)$; *(c) power spectrum of the complex envelope* $\tilde{\nu}(t)$; *(d) power spectra of the real and imaginary parts of* $\tilde{\nu}(t)$.

$\nu_c(t)$ and $\nu_s(t)$, the real and imaginary parts of the complex envelope. It can be shown (see Problem 2.20) that the following equalities hold:

$$R_{\nu_c}(\tau) = R_{\nu_s}(\tau) \qquad (2.188)$$

and

$$E[\nu_c(t+\tau)\nu_s(t)] = -E[\nu_s(t+\tau)\nu_c(t)] \qquad (2.189)$$

Thus,

$$
\begin{aligned}
R_{\tilde{\nu}}(\tau) &= E\{[\nu_c(t+\tau) + j\nu_s(t+\tau)][\nu_c(t) - j\nu_s(t)]\} \\
&= R_{\nu_c}(\tau) + R_{\nu_s}(\tau) + j\{E[\nu_s(t+\tau)\nu_c(t)] - E[\nu_c(t+\tau)\nu_s(t)]\} \\
&= 2[R_{\nu_c}(\tau) + jR_{\nu_s\nu_c}(\tau)]
\end{aligned}
\qquad (2.190)
$$

where

$$R_{\nu_s\nu_c}(\tau) \overset{\triangle}{=} E[\nu_s(t+\tau)\nu_c(t)] \qquad (2.191)$$

From (2.188) to (2.191) we can draw the following conclusions:

(a) As $R_{\tilde{\nu}}(0) = E|\tilde{\nu}(t)|^2$ is a real quantity, Eqs. (2.190) and (2.191) show that

$$E[\nu_s(t)\nu_c(t)] = 0 \qquad (2.192)$$

That is, for any given $t$, $\nu_s(t)$ and $\nu_c(t)$ are uncorrelated RVs. As a special case, if $\nu(t)$ is a Gaussian process, $\nu_s(t)$ and $\nu_c(t)$ are independent RVs for any given $t$.

(b) From (2.185) and (2.186) it follows that

$$E|\tilde{\nu}(t)|^2 = E|\mathring{\nu}(t)|^2 = 2E[\nu^2(t)] \qquad (2.193)$$

Similarly, from (2.188), (2.190), and result (a) we have

$$E|\tilde{\nu}(t)|^2 = 2E[\nu_c(t)]^2 = 2E[\nu_s^2(t)] \qquad (2.194)$$

Thus,

$$E[\nu_c(t)]^2 = E[\nu_s^2(t)] = E[\nu^2(t)] \qquad (2.195)$$

That is, the average power of $\nu_c(t)$ and $\nu_s(t)$ equals that of the original process $\nu(t)$.

(c) If the power spectrum of the process $\nu(t)$ is symmetric around the frequency $f_0$, from (2.187) it follows that the power spectrum of $\tilde{\nu}(t)$ is an even function. This implies that $R_{\tilde{\nu}}(\tau)$ is real for all $\tau$, so (2.190) and (2.191) yield

$$E[\nu_s(t+\tau)\nu_c(t)] = 0 \qquad \text{for all } \tau \qquad (2.196)$$

This means that the processes $\nu_c(t)$ and $\nu_s(t)$ are uncorrelated [or independent when $\nu(t)$ is Gaussian]. Thus, in this situation,

$$\mathcal{G}_{\tilde{\nu}}(f) = 2\mathcal{G}_{\nu_c}(f) = 2\mathcal{G}_{\nu_s}(f) \qquad (2.197)$$

**Example 2.16**   Let $x(t)$ be a bandpass real signal, and let $\mathcal{G}_{\tilde{x}}(f)$ be the power density spectrum of its complex envelope. From (2.185) and (2.187) we have

$$
\begin{aligned}
\mathcal{G}_{\tilde{x}}(f) &= \mathcal{G}_{\hat{x}}(f + f_0) \\
&= 4\mathcal{G}_x(f + f_0)\, u(f + f_0)
\end{aligned}
$$

Recalling the fact that $\mathcal{G}_x(f)$ must be an even function of $f$, the last equality yields

$$
\mathcal{G}_x(f) = \frac{1}{4}[\mathcal{G}_{\tilde{x}}(-f - f_0) + \mathcal{G}_{\tilde{x}}(f - f_0)]
$$

As an example, consider the signal

$$
x(t) = \Re\left[ \sum_{n=-\infty}^{\infty} \alpha_n s(t - nT) \cdot e^{j2\pi f_0 t} \right]
$$

where $\mathrm{E}[\alpha_n] = 0$ and $\mathrm{E}[\alpha_{n+m}\alpha_n^*] = \sigma_\alpha^2 \delta_{0,m}$.

From (2.122)–(2.124) we obtain the power spectrum of the complex envelope of $x(t)$:

$$
\mathcal{G}_{\tilde{x}}(f) = \frac{\sigma_\alpha^2}{T} |S(f)|^2.
$$

Hence, the power spectrum of the signal is

$$
\mathcal{G}_x(f) = \frac{\sigma_\alpha^2}{4T}\{|S(-f - f_0)|^2 + |S(f - f_0)|^2\}.
$$

$\square$

## Narrowband white noise

As we shall see in later chapters, in problems concerning narrowband signals contaminated by additive noise it is usual to assume, as a model for the noise, a Gaussian process with a power density spectrum that is constant in a finite frequency interval and zero elsewhere. This occurs because a truly white noise would have an infinite power (which is physically meaningless), and because any mixture of signal plus noise is always observed at the output of a bandpass filter that is usually not wider than the band occupied by the signal. Thus, in practice, we can assume that the noise has a finite bandwidth, an assumption entailing no loss of accuracy if the noise has a bandwidth much wider than the filter's.

A *narrowband white noise* is a real, zero-mean, stationary random process whose power density spectrum is constant over a finite frequency interval not including the origin. In Fig. 2.20 we showed the power spectrum of a narrowband white noise with a power spectral density $N_0/2$ in the band $B$ centered at $f_0$.

### 2.4.2. Bandpass systems

The complex envelope representation of narrowband signals can be extended to the consideration of bandpass systems (i.e., systems whose response to any input signal is a narrowband signal). In the following we shall see how to characterize the effects of a bandpass system directly in terms of complex envelopes. In other words, assume that $y(t)$ is the response of a bandpass system to the narrowband signal $x(t)$. We want to characterize a system whose response to $\tilde{x}(t)$, the complex envelope of $x(t)$, is exactly $\tilde{y}(t)$, the complex envelope of $y(t)$.

**Bandpass linear systems**

First, consider a bandpass linear, time-invariant system with impulse response $h(t)$ and transfer function $H(f)$. The analytic signal representation of $h(t)$ is $\mathring{h}(t) = h(t) + j\widehat{h}(t)$, which corresponds to the transfer function $\mathring{H}(f) = 2H(f)u(f)$. If $x(t)$ is the narrowband input signal and $y(t)$ the response, the analytic signal $\mathring{y}(t)$ can be obtained by passing $x(t)$ into the cascade of the linear system under consideration and a filter with a transfer function $2u(f)$ (see Fig. 2.21 (a)). In a cascade of linear transformations, the order of the operations can be reversed without altering the final result, so we can substitute the scheme of Fig. 2.21 (b) for that of Fig. 2.21 (a). Next, observe that $\mathring{x}(t)$ has a Fourier transform equal to zero for $f < 0$. Hence, we can substitute a system with transfer function $H(f)$ for another system having a transfer function $H(f)u(f)$ without altering the output. The latter system (see Fig. 2.21 (c)) has an impulse response $\frac{1}{2}\mathring{h}(t)$, input $\mathring{x}(t)$, and output $\mathring{y}(t)$. These signals are related by the convolution integral

$$\mathring{y}(t) = \frac{1}{2}\int_{-\infty}^{\infty} \mathring{h}(\tau)\mathring{x}(t-\tau)\,d\tau \tag{2.198}$$

This equation becomes particularly useful if both $\mathring{x}(t)$ and $\mathring{h}(t)$ are expressed in terms of their complex envelopes. We get

$$\mathring{y}(t) = \frac{1}{2}e^{j2\pi f_0 t}\int_{-\infty}^{\infty} \tilde{h}(\tau)\tilde{x}(t-\tau)\,d\tau \tag{2.199}$$

which shows that $\mathring{y}(t)$ is a narrowband signal, centered at $f_0$, with complex envelope

$$\tilde{y}(t) = \frac{1}{2}\int_{-\infty}^{\infty} \tilde{h}(\tau)\tilde{x}(t-\tau)\,d\tau \tag{2.200}$$

In conclusion, the complex envelope of the response of a bandpass linear, time-invariant system with impulse response $h(t)$ to a given narrowband signal $x(t)$ can be obtained by passing the complex envelope $\tilde{x}(t)$ through the *low-pass*

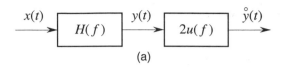

$$x(t) \rightarrow \boxed{H(f)} \xrightarrow{y(t)} \boxed{2u(f)} \xrightarrow{\overset{\circ}{y}(t)}$$
(a)

$$x(t) \rightarrow \boxed{2u(f)} \xrightarrow{\overset{\circ}{x}(t)} \boxed{H(f)} \xrightarrow{\overset{\circ}{y}(t)}$$
(b)

$$x(t) \rightarrow \boxed{2u(f)} \xrightarrow{\overset{\circ}{x}(t)} \boxed{H(f)u(f)} \xrightarrow{\overset{\circ}{y}(t)}$$
(c)

Figure 2.21: *Three equivalent schemes to represent the analytic signal associated with the output of a linear system.*

*equivalent system* whose impulse response is $\frac{1}{2}\tilde{h}(t)$, or, equivalently, whose transfer function is $H(f + f_0)u(f + f_0)$ (see Fig. 2.22). Notice that only if $H(f)$ is symmetric around $f_0$ will the low-pass equivalent system have a *real* impulse response. A nonreal impulse response will induce in the output signal a shift of the phase and a correlation between the in-phase and quadrature components. These effects are usually undesired.

**Example 2.17**  Let
$$x(t) = z(t)e^{j2\pi f_0 t} \tag{2.201}$$
where the Fourier transform of $z(t)$ is zero for $|f| > B$, $B < f_0$. Consider an LRC parallel resonator. Its transfer function is
$$H(f) = \frac{j2\pi f L}{R + j2\pi f L - 4\pi^2 f^2 LC} \tag{2.202}$$
The corresponding impulse response is, for $t > 0$,
$$h(t) = \frac{2\pi f_0}{Q} e^{-\pi f_0 t/2Q} \cos 2\pi f_0 t - \frac{\pi f_0}{Q^2} e^{-\pi f_0 t/2Q} \sin 2\pi f_0 t \tag{2.203}$$

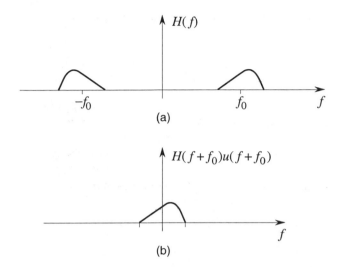

Figure 2.22: *(a) Transfer function of a bandpass linear system; (b) transfer function of a low-pass equivalent linear system.*

where

$$Q \triangleq R\sqrt{\frac{C}{L}}$$

is the "quality factor" of the circuit, and

$$f_0 \triangleq \frac{1}{2\pi\sqrt{LC}}\left(1 - \frac{1}{4Q^2}\right)^{1/2}$$

If $Q \gg 1$, the computation of $\tilde{h}(t)$ becomes very easy. In fact, the second term in the RHS of (2.203) can be disregarded. Additionally, we can safely assume that the exponential factor $\exp\{-\pi f_0 t/2Q\}$ is a bandlimited signal. Thus, from (2.203), we have

$$\overset{\circ}{h}(t) \cong \frac{2\pi f_0}{Q} e^{-\pi f_0 t/2Q}\, e^{j2\pi f_0 t} \qquad t \geq 0, \qquad (2.204)$$

where

$$f_0 = \frac{1}{2\pi\sqrt{LC}}$$

In conclusion,

$$\tilde{h}(t) \cong \frac{2\pi f_0}{Q} e^{-\pi f_0 t/2Q}, \qquad t \geq 0 \qquad (2.205)$$

Notice that the approximations in the computation of $\tilde{h}(t)$ make the Fourier transform of (2.205) symmetric around the origin of the frequency axis.                                    □

**Bandpass memoryless nonlinear systems**

We shall now examine a class of nonlinear systems that are often encountered in radio-frequency transmission. We are especially interested in nonlinear time-invariant systems whose input signal bandwidth is so narrow that the system's behavior is essentially frequency-independent. Moreover, the system is assumed to be bandpass. This in turn means that it can be thought of as being followed by a *zonal filter* whose aim is to stop all the frequency components of the output not close to the center frequency of the input signal. For a simple example of such a system, consider a sinusoidal signal $x(t) = A \cos 2\pi f_0 t$ sent into a time-invariant nonlinear system. Its output includes a sum of several harmonics centered at frequencies $0$, $f_0$, $2f_0$, .... If only the harmonic at $f_0$ is retained at the output, the observed output signal is a sinusoid $y(t) = F(A) \cos[2\pi f_0 t + \varphi(A)]$. If we consider the complex envelopes $\tilde{x}(t) = A$ and $\tilde{y}(t) = F(A) \exp[j\varphi(A)]$, we see that the system operation for sinusoidal inputs can be characterized by the two functions $F(\cdot)$ and $\varphi(\cdot)$. In the following we shall prove that this result holds true even when the input signal is a more general narrowband signal.

Consider a narrowband signal $x(t)$, with a spectrum centered at $f_0$. Its analytic signal representation can be given the form

$$\mathring{x}(t) = A_x(t) \, e^{j[2\pi f_0 t + \varphi_x(t)]} \tag{2.206}$$

where $A_x(t)$ and $\varphi_x(t)$ are baseband signals. Letting

$$\psi_x(t) \stackrel{\triangle}{=} 2\pi f_0 t + \varphi_x(t) \tag{2.207}$$

we rewrite (2.206) as

$$\mathring{x}(t) = A_x(t) \, e^{j\psi_x(t)} \tag{2.208}$$

Consider then the effect of a nonlinear memoryless system whose input-output relationship is assumed to have the form

$$y(t) = S_e[A_x(t) \cos \psi_x(t)] + S_o[A_x(t) \sin \psi_x(t)] \tag{2.209}$$

where $S_e[\cdot]$ is an even function of $\psi_x(t)$, and $S_o[\cdot]$ is an odd function. It is seen that $y(t)$, when expressed as a function of $\psi_x(t)$, is periodic with period $2\pi$. Thus, we can expand $y(t)$ in a Fourier series:

$$y(t) = y[\psi_x(t)] = \sum_{\ell=-\infty}^{\infty} c_\ell[A_x(t)] \, e^{j\ell\psi_x(t)} \tag{2.210}$$

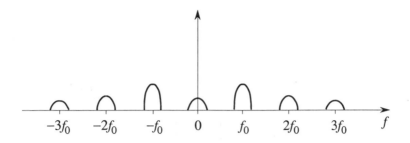

Figure 2.23: *Spectrum of the output of a memoryless nonlinear system whose input is a narrowband signal centered at frequency $f_0$.*

where

$$c_\ell(A) \triangleq \frac{1}{2\pi} \int_0^{2\pi} \{S_e[A\cos\psi] + S_o[A\sin\psi]\} e^{-j\ell\psi} \, d\psi \qquad (2.211)$$

The quantity $c_\ell(A)$ is generally complex. Its real and imaginary part are, respectively,

$$\Re[c_\ell(A)] = \frac{1}{2\pi} \int_0^{2\pi} S_e[A\cos\psi] \cos\ell\psi \, d\psi \qquad (2.212)$$

and

$$\Im[c_\ell(A)] = \frac{1}{2\pi} \int_0^{2\pi} S_o[A\sin\psi] \sin\ell\psi \, d\psi \qquad (2.213)$$

From the definition (2.207) of $\psi_x(t)$, we see how (2.210) expresses the fact that the spectrum of $y(t)$ includes several spectral components, each centered around the frequencies $\pm\ell f_0$, $\ell = 0, 1, \ldots$. Figure 2.23 illustrates qualitatively this situation. Notice that we must assume that the signals $c_\ell[A_x(t)]$ have spectra that do not significantly extend beyond the interval $(-f_0/2, f_0/2)$.

The assumption that the memoryless system is bandpass implies that only one of the spectral components of $y(t)$ can survive at the system output (i.e., that centered at $\pm f_0$). The analytic-signal representation of the output of such a bandpass memoryless system is then

$$\mathring{y}(t) = c[A_x(t)]e^{j\psi_x(t)} \qquad (2.214)$$

where

$$c(A) \triangleq 2c_1(A) \qquad (2.215)$$

As $c(A)$ is generally a complex number, we can put it in the form

$$c(A) = F(A)e^{j\varphi(A)} \qquad (2.216)$$

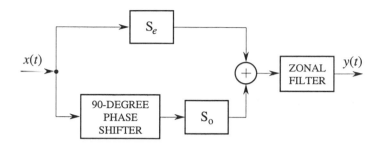

Figure 2.24: *Representation of a bandpass memoryless nonlinear system.*

so that

$$\mathring{y}(t) = F[A_x(t)]e^{j\{\psi_x(t)+\varphi[A_x(t)]\}} \tag{2.217}$$

or, in terms of complex envelopes,

$$\tilde{y}(t) = F[A_x(t)]e^{j\{\varphi_x(t)+\varphi[A_x(t)]\}} \tag{2.218}$$

Comparing the last equation with the complex envelope of the input signal,

$$\tilde{x}(t) = A_x(t)e^{j\varphi_x(t)} \tag{2.219}$$

we can see that the effect of a bandpass memoryless nonlinear system is to alter the amplitude and to shift the phase of the input signal according to a law that depends only on the values of its instantaneous envelope. This shows, in particular, that the system can be characterized by assigning the two functions $F[\cdot]$, $\varphi[\cdot]$, which describe the so-called AM/AM conversion and AM/PM conversion effects of the system (AM denotes amplitude modulation and PM phase modulation). These functions can be determined experimentally by taking as an input signal a single sinusoid with a frequency close to $f_0$ and an envelope $A$, and by measuring, for different values of $A$, the output envelope $F(A)$ and the output phase shift $\varphi(A)$. Notice that, for the validity of this nonlinear system model, the functions $F(A)$ and $\varphi(A)$ should not depend appreciably on the frequency of the test sinusoid as it varies within the range of interest.

Finally, notice that the system we are dealing with can be represented as in Fig. 2.24, where $S_e$ and $S_o$ denote memoryless nonlinear devices. From this scheme it is seen that only if $S_o$ is present can the system show an AM/PM conversion effect.

**Example 2.18 (Polynomial-law devices)**    Consider a nonlinear system whose input-output relationship is

$$y(t) = Cx^\ell(t) \tag{2.220}$$

$\ell$ an integer greater than 1. If $x(t)$ is written in the form

$$
\begin{aligned}
x(t) &= \Re[\tilde{x}(t)e^{j2\pi f_0 t}] \\
&= \frac{1}{2}[\tilde{x}(t)e^{j2\pi f_0 t} + \tilde{x}^*(t)e^{-j2\pi f_0 t}]
\end{aligned} \tag{2.221}
$$

we get

$$x^\ell(t) = \frac{1}{2^\ell} \sum_{k=0}^{\ell} \binom{\ell}{k} [\tilde{x}(t)]^k [\tilde{x}^*(t)]^{\ell-k} e^{j2\pi(2k-\ell)f_0 t} \tag{2.222}$$

When $y(t)$ is filtered through a zonal filter, all its frequency components other than those centered at $\pm f_0$ will be removed. Thus, only the terms with $2k - \ell = \pm 1$ will contribute to the system output. This shows, in particular, that only when $\ell$ is *odd* can the output of the zonal filter be nonzero. For $\ell$ odd, the complex envelope of the system output is then

$$\tilde{y}(t) = \frac{C}{2^{\ell-1}} \binom{\ell}{(\ell+1)/2} \tilde{x}(t) |\tilde{x}(t)|^{\ell-1}. \tag{2.223}$$

More generally, if the system is polynomial, i.e.,

$$y(t) = \sum_{i=1}^{L} a_\ell x^\ell(t), \tag{2.224}$$

we shall get, for $L$ odd,

$$\tilde{y}(t) = \tilde{x}(t) \sum_{m=0}^{(L-1)/2} \frac{a_{2m+1}}{2^{2m}} \binom{2m+1}{m+1} |\tilde{x}(t)|^{2m}. \tag{2.225}$$

Notice that polynomial-law devices with real coefficients never exhibit AM/PM conversion.                                                                                                $\square$

## 2.5.    Discrete representation of continuous signals

In this section we consider the problem of associating a continuous signal with a discrete representation. In other words, we wish to represent a given continuous signal in terms of a (possibly finite) sequence. The representation may be exact or only approximate, in which case it will be chosen on the basis of a compromise between accuracy and simplicity.

As we shall see in later chapters, this representation makes it possible to impart a geometric interpretation to a signal set, and hence to visualize it by extracting from it the features that are relevant when the signals are used for modulation.

## 2.5.1. Orthonormal expansions of finite-energy signals

A fundamental type of discrete representation is based on sets of signals called *orthonormal*. To define these sets, consider first the notion of the scalar product between two finite-energy signals $x(t)$ and $y(t)$: it is denoted by $(x, y)$ and defined as the value of the integral

$$(x, y) \triangleq \int_{-\infty}^{\infty} x(t) y^*(t)\, dt \qquad (2.226)$$

If $X(f)$ and $Y(f)$ denote the Fourier transforms of $x(t)$ and $y(t)$, respectively, and we let

$$(X, Y) = \int_{-\infty}^{\infty} X(f) Y^*(f)\, df \qquad (2.227)$$

*Parseval's equality* relates the scalar products defined in the time and in the frequency domain:

$$(x, y) = (X, Y) \qquad (2.228)$$

If $(x, y) = 0$, or equivalently $(X, Y) = 0$, the signals $x(t)$ and $y(t)$ are called *orthogonal*. From the definitions of scalar product and of orthogonality, it immediately follows that $(x, x) = \mathcal{E}_x$, the energy of $x(t)$, and that the energy of the sum of two orthogonal signals equals the sum of their energies.

Suppose now that we have a sequence $(\psi_i(t))_{i \in I}$ of orthogonal signals; that is,

$$(\psi_i, \psi_j) = \begin{cases} \mathcal{E}_i, & i = j, \\ 0, & i \neq j \end{cases} \qquad (2.229)$$

where I is a finite or countable index set.

If $\mathcal{E}_i = 1$ for all $i \in$ I, the signals of this sequence are called *orthonormal*. Obviously, an orthonormal sequence can be obtained from an orthogonal one by dividing each $\psi_i(t)$ by $\sqrt{\mathcal{E}_i}$. Given an orthonormal sequence, we wish to approximate a given finite-energy signal $x(t)$ with a linear combination of signals belonging to this sequence, that is, with the signal

$$\hat{x}(t) \triangleq \sum_{i \in I} c_i\, \psi_i(t) \qquad (2.230)$$

A suitable criterion for the choice of the constants $c_i$ appearing in (2.230), and hence of the approximation $\hat{x}(t)$, is to minimize the energy of the error signal

$$e(t) \triangleq x(t) - \hat{x}(t) \qquad (2.231)$$

Thus, the task is to minimize

$$\mathcal{E}_e \triangleq \int_{-\infty}^{\infty} |x(t) - \hat{x}(t)|^2\, dt$$

$$\begin{aligned}
&= \mathcal{E}_x + \mathcal{E}_{\hat{x}} - 2\Re(x,\,\hat{x}) \\
&= \mathcal{E}_x + \sum_{i\in I}|c_i|^2 - 2\Re\sum_{i\in I}c_i(x,\,\psi_i)
\end{aligned} \tag{2.232}$$

with respect to $c_i$, $i \in I$. By completing the square, we can also write

$$\mathcal{E}_e = \mathcal{E}_x + \sum_{i\in I}|c_i - (x,\,\psi_i)|^2 - \sum_{i\in I}|(x,\,\psi_i)|^2 \tag{2.233}$$

As the middle term in the RHS of (2.233) is nonnegative, $\mathcal{E}_e$ is minimized if the $c_i$ are chosen such as to render this term equal to zero. This is achieved for

$$c_i = (x,\,\psi_i) = \int_{-\infty}^{\infty} x(t)\psi_i^*(t)\,dt, \qquad i \in I \tag{2.234}$$

The minimum value of $\mathcal{E}_e$ is then given by

$$(\mathcal{E}_e)_{\min} = \mathcal{E}_x - \sum_{i\in I}|c_i|^2 \tag{2.235}$$

When $c_i$, $i \in I$, are computed using (2.234), the signal $\hat{x}(t)$ of (2.230) is called the *projection* of $x(t)$ onto the space spanned by the signals of the sequence $(\psi_i(t))_{i\in I}$, that is, on the set of signals that can be expressed as linear combinations of the $\psi_i(t)$. This denomination stems from the fact that, if (2.234) holds, the error $e(t)$ is orthogonal to every $\psi_i(t)$, $i \in I$, and hence to $\hat{x}(t)$. In fact,

$$\begin{aligned}
(e,\,\psi_i) &= (x - \hat{x},\,\psi_i) \\
&= (x,\,\psi_i) - (\hat{x},\,\psi_i) \\
&= c_i - c_i = 0, \qquad i \in I
\end{aligned}$$

(See Fig. 2.25 for a pictorial interpretation of this property in the case $I = \{1,2\}$.)

An important issue with this theory is the investigation of the conditions under which $(\mathcal{E}_e)_{\min} = 0$. When this happens, the sequence $(\psi_i(t))_{i\in I}$ is said to be *complete* for the signal $x(t)$, and from (2.235) we have the equality

$$\mathcal{E}_x = \sum_{i\in I}|c_i|^2 \tag{2.236}$$

In this case we write

$$x(t) = \sum_{i\in I}c_i\psi_i(t) \tag{2.237}$$

although this equality is not to be interpreted in the sense that its RHS and LHS are equal for every $t$, but rather in the sense that the energy of their difference is

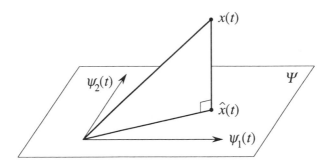

Figure 2.25: $\hat{x}(t)$ *is the projection of* $x(t)$ *onto* $\Psi$, *the signal space spanned by* $\psi_1(t)$ *and* $\psi_2(t)$.

zero. This fact is often expressed by saying that the RHS and LHS of (2.237) are equal *almost everywhere.*

In conclusion, once an orthonormal signal set has been chosen, a signal $x(t)$ can be represented by the sequence $(c_i)_{i \in I}$ defined by (2.234). This representation is exact (in the sense just specified) if the orthonormal set is complete with respect to $x(t)$.

**Example 2.19 (The complex Fourier series)**    The orthonormal sequence

$$\left( \frac{1}{\sqrt{T}} e^{jk2\pi t/T} \right)_{k=-\infty}^{\infty}$$

is complete for every complex signal $x(t)$ defined in the interval $(-T/2, T/2)$ and having bounded variation with finitely many discontinuity points. The expansion

$$x(t) = \frac{1}{\sqrt{T}} \sum_{k=-\infty}^{\infty} c_k e^{jk2\pi t/T} \qquad t \in (-T/2, T/2), \tag{2.238}$$

with

$$c_k = \frac{1}{\sqrt{T}} \int_{-T/2}^{T/2} x(t) e^{-jk2\pi t/T} \, dt \tag{2.239}$$

is the familiar complex Fourier-series representation.                                    $\square$

## Gram-Schmidt procedure

Because of the importance of orthonormal signal sequences, algorithms for constructing these sequences are of interest. One such algorithm, which is computationally convenient because of its iterative nature, is called the *Gram-Schmidt orthogonalization procedure*. Let a sequence $(\phi_i(t))_{i=1}^{N}$ of finite-energy signals be given. We assume these signals to be linearly independent, i.e., to be such that any linear combination $\sum_{i=1}^{N} c_i \phi_i(t)$ is zero almost everywhere only if all the $c_i$ are zero. An orthonormal sequence $(\psi_i(t))_{i=1}^{N}$ is generated by using the following algorithm (see Problem 2.21).

We first define the auxiliary signal $\psi_1'(t)$ equal to $\phi_1(t)$:

$$\psi_1'(t) = \phi_1(t)$$

then we normalize it to obtain the first orthonormal signal:

$$\psi_1(t) = \frac{\psi_1'(t)}{\sqrt{(\psi_1', \psi_1')}}$$

By subtracting from $\phi_2(t)$ its projection onto $\psi_1(t)$ we obtain a signal orthogonal to $\psi_1(t)$, as shown in Fig. 2.26:

$$\psi_2'(t) = \phi_2(t) - (\phi_2, \psi_1)\,\psi_1(t)$$

that we normalize to obtain the second orthonormal signal:

$$\psi_2(t) = \frac{\psi_2'(t)}{\sqrt{(\psi_2', \psi_2')}}$$

By proceeding this way, we obtain the entire set of orthonormal signals. The general step of the algorithm is then:

$$\psi_i'(t) \;=\; \phi_i(t) - \sum_{\ell=1}^{i-1}(\phi_\ell, \psi_\ell)\,\psi_\ell(t) \qquad (2.240)$$

$$\psi_i(t) \;=\; \frac{\psi_i'(t)}{\sqrt{(\psi_i', \psi_i')}} \qquad (2.241)$$

for $i = 1, 2, \ldots, N$ (when $i = 1$ the sum in the first equality is empty).

## Geometric representation of a set of signals

The theory of orthonormal expansions of finite-energy signals shows that a signal $x(t)$ can be represented by the (generally complex) sequence $(c_i)_{i \in I}$ of scalar

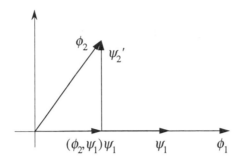

Figure 2.26: *Generating a signal orthogonal to $\psi_1(t)$.*

products (2.234), once an orthonormal sequence that is complete for $x(t)$ has been provided. Now, if we consider a given sequence $(\psi_i(t))_{i=1}^{N}$ of $N$ orthonormal *real* signals, it will be complete for any real $x(t)$ that can be written as a linear combination of the $\psi_i(t)$, that is, in the form

$$x(t) = \sum_{i=1}^{N} x_i\,\psi_i(t) \tag{2.242}$$

Thus, every such signal can be represented by the real $N$-vector $\mathbf{x} \overset{\triangle}{=} (x_1, \ldots, x_N)$, or, equivalently, by a point in the $N$-dimensional Euclidean space (i.e., the space of all real ordered $N$-tuples) whose coordinate axes correspond to the signals $\psi_i(t)$, $i = 1, \ldots, N$.

Consider now a set $\{x_i(t)\}_{i=1}^{M}$ of real signals. Can we find an orthonormal sequence that is complete for these $M$ signals? If so, we can represent $x_1(t)$, $x_2(t), \ldots, x_M(t)$ as $M$ vectors or as $M$ points in a Euclidean space of suitable dimensionality. If the signals in the set $\{x_i(t)\}_{i=1}^{M}$ are linearly independent, it suffices to apply to it the Gram-Schmidt procedure to find such an orthonormal sequence. In fact, (2.240) shows that each of the $\psi_i(t)$ is expressed as a linear combination of signals in $\{x_i(t)\}_{i=1}^{M}$; hence, each of the $x_i(t)$ can be expressed as a linear combination of the $\psi_i(t)$. Suppose, instead, that only $N$ signals in $\{x_i(t)\}_{i=1}^{M}$ are linearly independent, and hence $M - N$ of them can be expressed as linear combinations of the remaining signals. In this case, the Gram-Schmidt procedure can still be used, but it will produce *only $N < M$ nonzero* orthonormal signals. Every $x_i(t)$ is then represented by the $N$-vector

$$\mathbf{x}_i \overset{\triangle}{=} (x_{i1}, x_{i2}, \ldots, x_{iN}) \tag{2.243}$$

where

$$x_{ij} \triangleq (x_i, \psi_j), \quad i = 1, \ldots, M, \quad j = 1, \ldots, N \qquad (2.244)$$

or, equivalently, as a point in the $N$-dimensional Euclidean space whose coordinate axes correspond to the nonzero orthonormal signals found through the Gram-Schmidt procedure. In this situation, we say that the signal set $\{x_i(t)\}_{i=1}^{M}$ *has dimensionality* $N$.

**Example 2.20**   Consider the four signals

$$x_i(t) = \cos\left[\pi t + (i-1)\pi/2\right] \quad t \in (0,2), \quad i = 1,2,3,4 \qquad (2.245)$$

Using the Gram-Schmidt procedure, we get

$$\psi_1(t) = \cos \pi t$$

$$\psi_2(t) = -\sin \pi t$$

and

$$\psi_3(t) = \psi_4(t) = 0$$

which shows that the signal set (2.245) has dimensionality 2 and is represented by the four vectors:

$$\mathbf{x}_1 = (1,\ 0) \quad \mathbf{x}_2 = (0,\ 1), \quad \mathbf{x}_3 = (-1,\ 0) \quad \mathbf{x}_4 = (0,\ -1)$$

The reader should observe that the $M$-signal set

$$x_i(t) = \cos[\pi t + 2(i-1)\pi/M], \quad t \in (0,2), \quad i = 1,2,\ldots,M \qquad (2.246)$$

has dimensionality 2, and can also be represented using the same orthonormal basis.   □

## Computing signal distances and scalar products

Based on the procedure just developed, a real-signal set $\{x_i(t)\}_{i=1}^{M}$ defined for $0 \leq t < T$ can be represented by a set of vectors $\mathbf{x}_i = (x_{i1}, \ldots, x_{iN})$ in the $N$-dimensional Euclidean space. By using (2.242) and orthonormality of the signals $\psi_i(t)$, it can be easily proved that the following holds for any $i = 1, \ldots, M$:

$$\int_0^T x_i^2(t)\, dt = |\mathbf{x}_i|^2 = \sum_{j=1}^{N} x_{ij}^2 \qquad (2.247)$$

which shows that the energy of a signal equals the squared length of the vector representing it. This equivalence between signal energy and distance of the vector from the origin is a very useful relation.

Moreover, we have

$$\int_0^T x_i(t) x_k(t) \, dt = (\mathbf{x}_i, \mathbf{x}_k) = \sum_{j=1}^N x_{ij} x_{kj} \qquad (2.248)$$

As two signals are orthogonal if their scalar product vanishes, we visualize orthogonal signals by two vectors perpendicular to each other.

Finally,

$$\int_0^T [x_i(t) - x_k(t)]^2 \, dt = |\mathbf{x}_i - \mathbf{x}_k|^2 = |\mathbf{x}_i|^2 + |\mathbf{x}_k|^2 - 2(\mathbf{x}_i, \mathbf{x}_k) \qquad (2.249)$$

The latter quantity is the (Euclidean) distance between signals $x_i(t)$, $x_k(t)$, and is equal to the squared distance between the two vectors $\mathbf{x}_i$, $\mathbf{x}_k$.

### Sampling expansion of bandlimited signals

Consider now the set of signals $x(t)$ strictly bandlimited in the frequency interval $(-B, B)$, that is, such that their Fourier transform $X(f)$ is identically zero for $|f| \geq B$. An orthonormal basis for any such $x(t)$ can be found as follows. Expand $X(f)$ in a Fourier series according to (2.238) and (2.239). Then take the inverse Fourier transform to get an expansion for $x(t)$. This procedure yields

$$X(f) = \frac{1}{\sqrt{2B}} \sum_{i=-\infty}^{\infty} c_i \, e^{ji\pi f/B}, \qquad f \in (-B, B) \qquad (2.250)$$

$$c_i = \frac{1}{\sqrt{2B}} \int_{-B}^{B} X(f) e^{-ji\pi f/B} \, df \qquad (2.251)$$

and finally

$$\begin{aligned} x(t) &= \frac{1}{\sqrt{2B}} \sum_{i=-\infty}^{\infty} c_i \int_{-B}^{B} e^{-ji\pi f/B} e^{j2\pi ft} \, df \\ &= \sqrt{2B} \sum_{i=-\infty}^{\infty} c_i \, \frac{\sin 2\pi B[t - i/(2B)]}{2\pi B[t - i/(2B)]} \end{aligned} \qquad (2.252)$$

which is an expansion valid for every $x(t)$ with bandwidth $B$. Observing further that the integral in the RHS of (2.251) is proportional to the inverse Fourier transform of $X(f)$ computed for $t = i/(2B)$, $c_i$ can be put in the form

$$c_i = \frac{1}{\sqrt{2B}} x\left(\frac{i}{2B}\right) \qquad (2.253)$$

This shows that the coefficients of the series expansion (2.252) are the samples of the signal $x(t)$ taken at the time instants $i/(2B)$, $-\infty < i < \infty$. Explicitly, from (2.250) and (2.253) we obtain

$$X(f) = \frac{1}{2B} \sum_{i=-\infty}^{\infty} x\left(\frac{i}{2B}\right) e^{-j2\pi fi/(2B)} \tag{2.254}$$

and hence, by taking the inverse Fourier transform,

$$x(t) = \sum_{i=-\infty}^{\infty} x\left(\frac{i}{2B}\right) \frac{\sin 2\pi B[t - i/(2B)]}{2\pi B[t - i/(2B)]} \tag{2.255}$$

Equation (2.255) shows that every finite-energy signal with bandwidth $B$ can be fully recovered from the knowledge of its samples taken at the rate of $2B$ samples per second. More generally, as any signal bandlimited in $(-B, B)$ is also bandlimited in $(-B', B')$, where $B' > B$, we can say that any finite-energy bandlimited signal can be represented by using the sequence of its samples, provided that they are taken at a rate *not less than* $2B$. This minimum sampling rate of $2B$ is usually called the *Nyquist sampling rate* for $x(t)$. If $x(t)$ is a narrowband signal, it should be observed that it is convenient to apply the sampling expansion (2.255) to its complex envelope instead of the signal itself. This results in a much lower Nyquist frequency and, hence, in a more economical representation.

Observe now that (2.255) can also be written in the form

$$x(t) = \left\{ \sum_{i=-\infty}^{\infty} x\left(\frac{i}{2B}\right) \delta[t - i/(2B)] \right\} * \frac{\sin 2\pi Bt}{2\pi Bt} \tag{2.256}$$

Now, $\sin(2\pi Bt)/(2\pi Bt)$ can be interpreted as the impulse response of a linear, time-invariant system with frequency response

$$H(f) = \begin{cases} 1/(2B), & |f| < B \\ 0, & \text{elsewhere} \end{cases} \tag{2.257}$$

that is, an ideal low-pass filter with cutoff frequency $B$. Thus, (2.256) suggests how to implement a system that recovers $x(t)$ from its samples. The sequence of samples is used to modulate linearly a train of impulses, which is then passed through an ideal low-pass filter (see Fig. 2.27).

A frequency-domain interpretation of the reconstruction of a sampled signal can also be provided. Let the signal $x(t)$ be sampled every $T_s$ seconds, and observe that we can write

$$\sum_{i=-\infty}^{\infty} x(iT_s)\delta(t - iT_s) = x(t) \cdot \sum_{i=-\infty}^{\infty} \delta(t - iT_s) \tag{2.258}$$

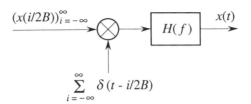

Figure 2.27: *Recovering a bandlimited signal $x(t)$ from its samples. $H(f)$ is an ideal low-pass filter with cutoff frequency $B$.*

The spectrum of this signal is obtained by taking the convolution of $X(f)$ with the Fourier transform of a train of impulses with period $T_s$. This is given by

$$\frac{1}{T_s} \sum_{i=-\infty}^{\infty} \delta(f - i/T_s)$$

[use (2.109)]. Thus, the spectrum of (2.258) is

$$X_s(f) \triangleq \frac{1}{T_s} \sum_{i=-\infty}^{\infty} X\left(f - \frac{i}{T_s}\right) \tag{2.259}$$

which is periodic with period $1/T_s$ (see Fig. 2.28).

The original signal can be recovered from $X_s(f)$ by using the ideal low-pass filter whose transfer function $H(f)$ is shown in Fig. 2.28, provided that the translated copies of $X(f)$ forming $X_s(f)$ do not overlap. This condition holds if and only if $B < \dfrac{1}{T_s} - B$, that is,

$$f_s > 2B \tag{2.260}$$

where $f_s \triangleq 1/T_s$ is the sampling rate.

If (2.260) does not hold (i.e., the signal is sampled at a rate lower than Nyquist's), $x(t)$ cannot be recovered exactly from its samples. The signal obtained at the output of the ideal low-pass filter has the Fourier transform

$$H(f)X_s(f) = \begin{cases} \displaystyle\sum_{i=-\infty}^{\infty} X(f - i/T_s), & |f| < 1/2T_s \\ 0, & \text{elsewhere} \end{cases} \tag{2.261}$$

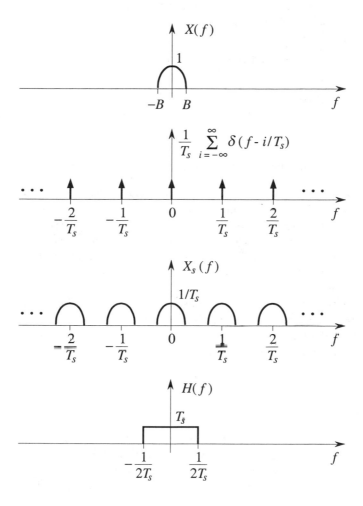

Figure 2.28: *Sampling and reconstructing a bandlimited signal: frequency domain representation.*

It is important to observe that if this situation occurs (i.e., the signal is "undersampled"), even the *phase of the sampling process* affects the shape of the reconstructed signal. Specifically, if the sampled signal is

$$x(t) \sum_{i=-\infty}^{\infty} \delta(t - iT_s + \Theta)$$

where $\Theta$ is a constant smaller than $T_s$, at the output of the low-pass filter we get a signal whose Fourier transform is

$$\sum_{i=-\infty}^{\infty} X\left(f - \frac{i}{T_s}\right) e^{ji2\pi\Theta/T_s}, \qquad |f| < \frac{1}{2T_s} \qquad (2.262)$$

If the bandwidth of $x(t)$ does not exceed $1/(2T_s)$, then (2.262) gives the spectrum of $x(t)$ (as it should). Otherwise, the shape of the signal recovered will also depend on the value of $\Theta$.

### $2BT$-theorem and the uncertainty principle

The sampling expansion (2.255), which is valid for any $x(t)$ bandlimited in the interval $(-B, B)$, when applied to a signal vanishing outside the time interval $(0, T)$ has nonzero terms occurring only for $0 \leq i/(2B) \leq T$ (i.e., for $i = 0, 1, 2, \ldots, 2BT$). Thus, any bandlimited and time limited $x(t)$ is completely specified by $2BT + 1 \sim 2BT$ constants. For real signals, this fact can be summarized by saying that "the space of real signals of duration $T$ and bandwidth $B$ has dimension $2BT$."

However, this argument is fallacious, because *no bandlimited signal* (besides the trivial null signal) *can have a finite duration*. The proof of this property is based on the fact that a signal $x(t)$ whose amplitude spectrum vanishes for $|f| > B$ can be written as

$$x(t) = \int_{-B}^{B} X(f) e^{j2\pi ft} \, df \qquad (2.263)$$

Now, if we allow $t$ in (2.263) to be a complex variable, this extended $x(t)$ is an entire function of $t$. In other words, $x(t)$ has no singularities in the finite $t$ plane, and its Taylor series expansion about every point has an infinite radius of convergence. Thus, any $x(t)$ vanishing on any interval of the time axis would have all its derivatives zero at some interior point of the interval. Hence, its Taylor series expansion would require it to be identically zero.

This impossibility for a signal to be simultaneously bandlimited and time limited is a special case of the *uncertainty principle* for a signal and its Fourier transform. One way to describe this principle is the following (stated without proof). Define two quantities $\alpha \in [0, 1]$ and $\beta \in [0, 1]$ that measure the fraction of the signal energy concentrated in the time interval $(-T/2, T/2)$ and in the frequency interval $(-B, B)$, respectively:

$$\alpha \triangleq \frac{1}{\mathcal{E}_x} \int_{-T/2}^{T/2} |x(t)|^2 \, dt \qquad (2.264)$$

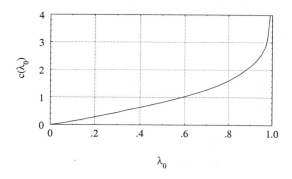

Figure 2.29: *The function $c(\lambda_0)$ of the uncertainty principle.*

and

$$\beta \triangleq \frac{1}{\mathcal{E}_x} \int_{-B}^{B} |X(f)|^2 \, df \qquad (2.265)$$

where $\mathcal{E}_x$ denotes the energy of the signal $x(t)$. The uncertainty principle states that

$$\pi BT \geq c(\lambda_0) \qquad (2.266)$$

where

$$\lambda_0 = \cos^2(\theta_1 + \theta_2), \quad \cos^2 \theta_1 = \alpha, \quad \cos^2 \theta_2 = \beta \qquad (2.267)$$

and the function $c(\,\cdot\,)$ is shown in Fig. 2.29. Notice that $c(\lambda_0) \to \infty$ as $\lambda_0 \to 1$.

With a signal both time limited and bandlimited, we should have $\alpha = \beta = 1$ for a finite product $BT$. But this would be in conflict with (2.266), because in this case $\lambda_0 = 1$, and hence $c(\lambda_0) = \infty$.

**Example 2.21**   For an example of the application of the uncertainty principle, determine the minimum value of the product $\pi BT$ for $\alpha = \beta = 0.95$. From (2.267) we get $\theta_1 = \theta_2 = 0.2255$ and $\lambda_0 = \cos^2 0.4510 = 0.81$. The curve in Fig. 2.29 yields $c(\lambda_0) \cong 1.6$.   □

Let us now return to our $2BT$ theorem. Although it is not strictly true in the form stated at the beginning of this section, it can be reformulated in a more rigorous manner. To this end, we must recognize the inherent physical limitations of measuring equipment, and the consequent inability of measuring an energy smaller than the energy resolution of this equipment. Thus, denote by $\epsilon$ the smallest amount of energy that we could measure. We say that a real signal $x(t)$ is time limited to $(-T/2, T/2)$ *at level* $\epsilon$ if

$$\int_{|t|>T/2} x^2(t)\,dt < \epsilon \tag{2.268}$$

and is bandlimited with bandwidth $B$ *at level* $\epsilon$ if

$$\int_{|f|>B} |X(f)|^2\,df < \epsilon \tag{2.269}$$

Conditions (2.268) and (2.269) indicate that the energy lying outside the time interval $(-T/2, T/2)$ and the frequency range $(-B, B)$ is less than we can measure. Furthermore, a set S of real signals is said to have dimension $N$ *at level* $\epsilon$ if there is a set of $N$ signals $\{\psi_i(t)\}_{i=1}^N$ such that, for each $x(t) \in$ S, there exist $a_1, a_2, \ldots, a_N$ such that

$$\int_{-T/2}^{T/2} \left[ x(t) - \sum_{i=1}^N a_i\psi_i(t) \right]^2 dt < \epsilon \tag{2.270}$$

and there is no set of $N-1$ functions that will approximate every $x(t) \in$ S in this manner. In words, every signal in S can be so well approximated in $(-T/2, T/2)$ by a linear combination of $\psi_1(t), \ldots, \psi_N(t)$ that we could not measure the energy of the difference between the signal and its approximation. With these definitions, we have the following theorem, due to Slepian (1976):

**Theorem 2.1**   Let $S_\epsilon$ be the set of real signals time limited to $(-T/2, T/2)$ at level $\epsilon$ and bandlimited to $(-B, B)$ at level $\epsilon$. Let $N = N(B, T, \epsilon, \epsilon')$ be the approximate dimension of $S_\epsilon$ at level $\epsilon'$. Then, for every $\epsilon' > \epsilon$,

$$\lim_{T\to\infty} \frac{1}{T} N(B, T, \epsilon, \epsilon') = 2B, \qquad \lim_{B\to\infty} \frac{1}{2B} N(B, T, \epsilon, \epsilon') = T \tag{2.271}$$

This "$2BT$-theorem" renders precise the concept that for large $BT$ the space of signals of approximate duration $T$ and approximate bandwidth $B$ has approximate dimension $2BT$. The proof of the theorem will not be reported here: the interested reader is referred to Slepian (1976).

### 2.5.2.  Orthonormal expansions of random signals

We shall now briefly consider the problem of associating a discrete representation with a *random* signal $\xi(t)$. Quite generally, we look for a series expansion of the form

$$\xi(t) = \sum_{i=-\infty}^{\infty} \gamma_i \, \psi_i(t) \tag{2.272}$$

where $\psi_i(t)$, $-\infty < i < \infty$, are deterministic random functions, $\gamma_i$ are random variables, and the equality is to be interpreted in the sense that

$$\lim_{k\to\infty} \mathrm{E} \left| \xi(t) - \sum_{i=-k}^{k} \gamma_i \psi_i(t) \right|^2 = 0 \tag{2.273}$$

Various constraints may be imposed on $\xi(t)$, $\psi_i(t)$, and the random sequence $(\gamma_i)$, thus obtaining different families of expansions. In the *Karhunen-Loève expansion*, $\xi(t)$ is a WS stationary process defined in the finite interval $(0, T)$, $\{\psi_i(t)\}_{i=0}^{\infty}$ is a set of finite-energy orthogonal signals, and the coefficients $\gamma_i$ are uncorrelated random variables.

If the process $\xi(t)$ is bandlimited, in the sense that its power spectrum $\mathcal{G}_\xi(f)$ vanishes outside the interval $(-B, B)$, we have the sampling expansion

$$\xi(t) = \sum_{i=-\infty}^{\infty} \xi\left(\frac{i}{2B}\right) \frac{\sin 2\pi B(t - i/(2B))}{2\pi B(t - i/(2B))} \tag{2.274}$$

The coefficients $\xi(i/(2B))$ are uncorrelated if and only if $\mathcal{G}_\xi(f)$ is constant over $(-B, B)$.

More general classes of series representations of WS stationary random processes were derived by Masry, Liu, and Steiglitz (1968) and Campbell (1969). Similar results were obtained by Cambanis and Liu (1970) for harmonizable processes, and for an even more general class of processes (the "weakly continuous" processes) by Cambanis and Masry (1971).

## 2.6.    Elements of detection theory

In this section we examine the problem of recognizing a signal chosen at random (with known probabilities) from a finite known set $\{s_i(t)\}_{i=1}^{M}$ once it has been perturbed by a random disturbance in the form of a noise process $\nu(t)$ independent of the signal and added to it. More specifically, the problem is to decide which one, among the signals $s_1(t), s_2(t), \ldots, s_M(t)$, has given rise to the observed signal $y(t)$, when it is known that $y(t)$ has the form

$$y(t) = s_j(t) + \nu(t) \tag{2.275}$$

for some $j$, $1 \leq j \leq M$. Signals and noise may be either real or complex (i.e., complex envelopes of narrowband time functions). It will be assumed here that

$\nu(t)$ is a white Gaussian process, with power spectral density $N_0/2$ (real signals) or $2N_0$ (complex envelopes). The signals dealt with have a finite energy and a finite duration. Also, their starting and ending times are known to the observer. We assume that $s_i(t)$, $1 \leq i \leq M$, are defined in the interval $0 \leq t < T$ and that $y(t)$ is observed in the same interval.

   This problem, called a *detection* problem, is central in digital transmission theory. It will be provided further motivation in Chapter 4, which includes a number of applications.

### 2.6.1. Optimum detector: One real signal in noise

We shall consider first, for simplicity's sake, the case in which there are only two signals, one of which is zero. Thus, the task is to decide between the two hypotheses

$$
\begin{aligned}
H_0 : & \qquad y(t) = \nu(t) \\
H_1 : & \qquad y(t) = s(t) + \nu(t)
\end{aligned}
\tag{2.276}
$$

where $s(t)$ is a finite-energy real signal. The decision is based on the observation of $y(t)$ for $0 \leq t < T$, and we want it to be made in such a way that the probability of a wrong decision is minimized. In words, we say that $H_1$ (respectively, $H_0$) is true when the observed signal contains (respectively, does not contain) $s(t)$.

   A basic step in our derivation of the optimum detector is the discrete representation of the signals involved, which allows us to avoid further consideration of time functions. To do this, we expand $y(t)$ in an orthonormal series and represent it using the sequence of its coefficients. As a basis for this expansion, we choose any complete sequence of real signals $(\psi_i(t))_{i=1}^{\infty}$, orthonormal in the interval $(0, T)$ and such that $\psi_1(t) = s(t)/\sqrt{\mathcal{E}_s}$ (see Problem 2.22). Hence, $s(t)$ will be represented by the sequence $(\sqrt{\mathcal{E}_s}, 0, 0, \ldots)$ and $\nu(t)$ by the sequence $(\nu_1, \nu_2, \nu_3, \ldots)$, where

$$
\nu_i \triangleq \int_0^T \nu(t)\psi_i(t)\,dt, \qquad i = 1, 2, \ldots
\tag{2.277}
$$

By direct calculation it can be shown that $\mathrm{E}\{\nu_i\} = 0$, $i = 1, 2, \ldots$, and

$$
\begin{aligned}
\mathrm{E}\{\nu_i \nu_j\} &= \int_0^T \int_0^T \mathrm{E}\{\nu(t)\nu(\tau)\}\psi_i(t)\psi_j(\tau)\,dt\,d\tau \\
&= \frac{N_0}{2} \int_0^T \int_0^T \delta(t - \tau)\psi_i(t)\psi_j(\tau)\,dt\,d\tau \\
&= \frac{N_0}{2} \int_0^T \psi_i(t)\psi_j(\tau)\,dt = \frac{N_0}{2}\delta_{ij} \qquad i, j = 1, 2, \ldots
\end{aligned}
\tag{2.278}
$$

Since $\nu_i$, $i = 1, 2, \ldots$, are Gaussian RVs, (2.278) shows that they are independent. In terms of these discrete representations, we can formulate our decision problem as follows. Decide between the hypotheses

$$
\begin{aligned}
H_0 &: \qquad (Y_i)_{i=1}^{\infty} = (\nu_1, \nu_2, \ldots) \\
H_1 &: \qquad (Y_i)_{i=1}^{\infty} = (\nu_1 + \sqrt{\mathcal{E}_s}, \nu_2, \ldots)
\end{aligned}
\tag{2.279}
$$

on the basis of the observation of the quantities

$$
Y_i \triangleq \int_0^T y(t) \psi_i(t)\, dt, \qquad i = 1, 2, \ldots
\tag{2.280}
$$

Consider now a crucial point. Under both hypotheses $H_0$ and $H_1$, the observed quantities $Y_2$, $Y_3$, ..., are equal to $\nu_2$, $\nu_3$, ..., respectively, and these are independent of each other and $Y_1$. Thus, the observation of $Y_2$, $Y_3$, ..., does not add any information to the decision process, and hence it can be based solely on the observation of

$$
Y_1 \triangleq \frac{1}{\sqrt{\mathcal{E}_s}} \int_0^T y(t) s(t)\, dt
\tag{2.281}
$$

(Notice that the assumption of a Gaussian noise is crucial here. Without it, $Y_2$, $Y_3$, ... would only be *uncorrelated* with $Y_1$, rather than independent of it.)

In conclusion, the problem is reduced to the decision between the two hypotheses

$$
\begin{aligned}
H_0 &: \qquad Y_1 = \nu_1 \\
H_1 &: \qquad Y_1 = \nu_1 + \sqrt{\mathcal{E}_s}
\end{aligned}
\tag{2.282}
$$

upon observation of $Y_1$ as defined in (2.281). The quantity $Y_1$ is called the *sufficient statistics* for deciding between $H_0$ and $H_1$, because it extracts from the observed signal $y(t)$ all that is required to perform the decision. All other information about $y(t)$ is *irrelevant* to the decision process.

Since the decision is based only on the observation of the scalar quantity $Y_1$, the optimum detector will first compute the scalar product (2.281) of the observed signal $y(t)$ and $s(t)$. Then it will choose either $H_0$ or $H_1$ according to the value taken by $Y_1$. If we denote by $S_1$ and $S_0 = R - S_1$ two subsets of the real line R, the decision rule is

$$
\begin{aligned}
\text{choose } H_0 &\qquad \text{if } y_1 \in S_0 \\
\text{choose } H_1 &\qquad \text{if } y_1 \in S_1
\end{aligned}
\tag{2.283}
$$

where $y_1$ is the observed value of the random variable $Y_1$. Hence, the optimum decision rule can be specified by choosing $S_0$ and $S_1$ in such a way that the average error probability is minimized. The error probability is given by

$$P(e) =$$

$$= \int_{S_0} p_1 f_{Y_1|H_1}(y \mid H_1) \, dy + \int_{S_1} p_0 f_{Y_1|H_0}(y \mid H_0) \, dy$$

$$= \int_R p_1 f_{Y_1|H_1}(y \mid H_1) \, dy - \int_{S_1} p_0 f_{Y_1|H_1}(y \mid H_1) \, dy + \int_{S_1} p_0 f_{Y_1|H_0}(y \mid H_0) \, dy$$

$$= p_1 - \int_{S_1} [p_1 f_{Y_1|H_1}(y \mid H_1) - p_0 f_{Y_1|H_0}(y \mid H_0)] \, dy \qquad (2.284)$$

where $p_0 \triangleq P\{H_0\}$, $p_1 \triangleq \{H_1\}$ are the a priori probabilities that $H_0$ is true [i.e., the observed signal does not contain $s(t)$] and $H_1$ is true [i.e., the observed signal contains $s(t)$], respectively. To minimize $P(e)$, we should maximize the contribution to the integral of the term in brackets in the last expression of (2.284). This can be done by including in $S_1$ all the values $y$ taken on by $Y_1$ such that $p_1 f_{Y_1|H_1}(y \mid H_1) > p_0 f_{Y_1|H_0}(y \mid H_0)$ and in $S_0$ the remaining values. Values of $Y_1$ such that the integrand is zero do not affect the value of $P(e)$, and hence may be included in either $S_0$ or $S_1$ arbitrarily. If we define the *likelihood ratio* between hypotheses $H_0$ and $H_1$ as

$$\Lambda(y) \triangleq \frac{f_{Y_1|H_1}(y \mid H_1)}{f_{Y_1|H_0}(y \mid H_0)} \qquad (2.285)$$

the decision rule becomes

$$\text{choose } H_0 \quad \text{if } \Lambda(y_1) \leq \frac{p_0}{p_1}$$

$$\text{choose } H_1 \quad \text{if } \Lambda(y_1) > \frac{p_0}{p_1} \qquad (2.286)$$

In conclusion, the optimum detector consists of a device that computes the likelihood ratio $\Lambda(y_1)$ and compares its value with the threshold $p_0/p_1$. Explicitly, we have

$$\Lambda(y) = \frac{e^{-(y-\sqrt{\mathcal{E}_s})^2/N_0}}{e^{-y^2/N_0}}$$

$$= \exp\left\{\frac{2}{N_0} y \sqrt{\mathcal{E}_s} - \frac{1}{N_0} \mathcal{E}_s\right\} \qquad (2.287)$$

so that, using (2.281),

$$\Lambda(Y_1) = \exp\left\{\frac{2}{N_0} \int_0^T y(t)s(t) \, dt - \frac{1}{N_0} \int_0^T s^2(t) \, dt\right\} \qquad (2.288)$$

Because of the likelihood ratio's structure, it is customary to define the *log-likelihood ratio* as the logarithm of $\Lambda(\cdot)$:

$$\lambda(y) \triangleq \ln \Lambda(y), \qquad (2.289)$$

so (2.288) becomes

$$\lambda(Y_1) = \frac{2}{N_0} \int_0^T y(t)s(t)\, dt - \frac{1}{N_0} \int_0^T s^2(t)\, dt \qquad (2.290)$$

and the decision rule becomes

$$\begin{aligned} \text{choose } H_0 \quad & \text{if } \lambda(y_1) \le \ln \frac{p_0}{p_1} \\ \text{choose } H_1 \quad & \text{if } \lambda(y_1) > \ln \frac{p_0}{p_1} \end{aligned} \qquad (2.291)$$

An important special case occurs when $p_0 = p_1$ (i.e., the two hypotheses are equally likely). In such a case, the decision is made by comparing $\lambda(y_1)$ against a zero threshold. Moreover, from (2.290) it is seen that the value of the constant $N_0$ is not relevant to the decision. Hence, when $p_0 = p_1$ the decision procedure does not depend on the spectral density of the noise. This simplification and the fact that the a priori probabilities $p_0$ and $p_1$ might be unknown justify the frequent use of the simplified decision rule (called the *maximum likelihood*, or ML, rule):

$$\begin{aligned} \text{choose } H_0 \quad & \text{if } \lambda(y_1) \le 0 \\ \text{choose } H_1 \quad & \text{if } \lambda(y_1) > 0 \end{aligned} \qquad (2.292)$$

although it gives minimum error probability only when $p_0 = p_1$. The rule (2.291) is referred to as the *maximum a posteriori probability*, or MAP, rule. The structure of the ML detector is shown in Fig. 2.30.

**Example 2.22 (The integrate and-dump receiver)**   A simple special case of the general ML detector previously considered arises when the signal $s(t)$ has a constant amplitude $A$ in the interval $0 \le t < T$. The task is then to decide between the two hypotheses

$$\begin{aligned} H_0: \quad & y(t) = \nu(t) \\ H_1: \quad & y(t) = A + \nu(t) \end{aligned} \qquad (2.293)$$

upon observation of $y(t)$ for $0 \le t < T$. In this case, $\mathcal{E}_s = A^2 T$, and from (2.281) we have

$$Y_1 = \frac{1}{\sqrt{T}} \int_0^T y(t)\, dt \qquad (2.294)$$

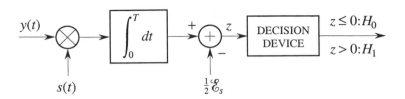

Figure 2.30: *ML detection of a real signal $s(t)$ in white Gaussian noise.*

Equation (2.294) shows that the sufficient statistics for the detection are computed by averaging out the noise from the observed signal. This is obtained by integrating $y(t)$ over the observation interval.

Consider the performance of this detector when $p_0 = p_1$. The RV

$$\nu_1 \triangleq \frac{1}{\sqrt{T}} \int_0^T \nu(t)\, dt \tag{2.295}$$

is Gaussian, with mean zero and variance $N_0/2$. Thus, the error probability under $H_0$ (i.e., the probability of choosing $H_1$ when $H_0$ is true) is

$$
\begin{aligned}
P(e \mid H_0) &= P\{\lambda(Y_1) > 0 \mid H_0\} \\
&= P\left\{\nu_1 > A\sqrt{T}/2\right\} \\
&= \frac{1}{2}\mathrm{erfc}\left(\frac{1}{2}\sqrt{\frac{A^2 T}{N_0}}\right)
\end{aligned}
\tag{2.296}
$$

where $\mathrm{erfc}\,(\cdot)$ is the complementary error function (see Appendix A). Similarly, the error probability under $H_1$ is

$$
\begin{aligned}
P(e \mid H_1) &= P\{\lambda(y_1) \leq 0 \mid H_1\} \\
&= P\left\{\nu_1 < -A\sqrt{T}/2\right\} \\
&= \frac{1}{2}\,\mathrm{erfc}\left(\frac{1}{2}\sqrt{\frac{A^2 T}{N_0}}\right)
\end{aligned}
\tag{2.297}
$$

so that

$$P(e) = P(e \mid H_0)p_0 + P(e \mid H_1)p_1 = \frac{1}{2}\mathrm{erfc}\left(\frac{1}{2}\sqrt{\frac{A^2 T}{N_0}}\right) \tag{2.298}$$

If we define the *signal-to-noise ratio*

$$\eta \triangleq \frac{A^2 T}{N_0} \tag{2.299}$$

it is seen that, as $P(e)$ is a monotone decreasing function of $\eta$, the error probability will decrease by increasing the level $A$, or by increasing the duration $T$ of the observation interval, or by decreasing the noise spectral density.                                      □

## Matched filter

Consider again (2.281). This equation shows that the sufficient statistics can be obtained, apart from a constant factor, as the output at time $t = T$ of a linear, time-invariant filter whose impulse response is

$$h(t) \stackrel{\triangle}{=} s(T - t) \tag{2.300}$$

In fact, with this definition we have

$$y(t) * h(t)|_{t=T} = \int_0^T y(\tau)h(T - \tau)\,d\tau = \int_0^T y(\tau)s(\tau)\,d\tau \tag{2.301}$$

A filter whose impulse response is (2.300), or, equivalently, whose transfer function is

$$H(f) \stackrel{\triangle}{=} S^*(f)e^{-j2\pi fT} \tag{2.302}$$

where $S(f) \stackrel{\triangle}{=} \mathcal{F}[s(t)]$, is called the filter *matched to the signal* $s(t)$. Thus, we can say that a matched filter whose output is sampled at $t = T$ extracts from the observed signal $y(t)$ the sufficient statistics for our decision problem.

An important property of the matched filter is that it maximizes the signal-to-noise ratio at its output, in the following sense. When the filter input is the sum of the signal $s(t)$ plus white noise $\nu(t)$, at time $t = T$ its output will be made up of two terms. The first is the signal part $\int_{-\infty}^{\infty} H(f)S(f)e^{j2\pi fT}\,df$, where $H(f)$ is the transfer function of the filter. The second is the noise part, a Gaussian RV with mean zero and variance $(N_0/2)\int_{-\infty}^{\infty}|H(f)|^2\,df$. If we define the signal-to-noise ratio at the filter output

$$\zeta^2 \stackrel{\triangle}{=} \frac{\left[\int_{-\infty}^{\infty} H(f)S(f)e^{j2\pi fT}\,df\right]^2}{(N_0/2)\int_{-\infty}^{\infty}|H(f)|^2\,df} \tag{2.303}$$

(i.e., the ratio between the instantaneous power of the signal part and the variance of the noise part), we can show that $\zeta^2$ is maximized if $H(f)$ has the form (2.302); that is, if the filter is matched to the signal $s(t)$. The proof is based on Schwarz's inequality, which states that if $A(\cdot)$ and $B(\cdot)$ are two complex functions, then

$$|\textstyle\int AB^*|^2 \leq \int |A|^2 \int |B|^2 \tag{2.304}$$

with equality if and only if $A = \alpha B^*$, where $\alpha$ is any complex constant. Using (2.304) in (2.303), we get

$$\zeta^2 \leq \frac{\int_{-\infty}^{\infty} |H(f)|^2 \, df \int_{-\infty}^{\infty} |S(f)|^2 \, df}{\frac{N_0}{2} \int_{-\infty}^{\infty} |H(f)|^2 \, df} = \frac{2\mathcal{E}_s}{N_0} \qquad (2.305)$$

Thus, the maximum value of the signal-to-noise ratio $\zeta^2$ is obtained for

$$H(f) = \alpha S^*(f) e^{-j2\pi f T} \qquad (2.306)$$

Since $\alpha$ can be any constant, we can set $\alpha = 1$ without loss of optimality, so that the filter sought is indeed the matched filter as defined by (2.302). Notice that this filter may be physically unrealizable, in which case it is necessary to approximate it. Also, its response to the input $s(t)$ is, at time $t = T$:

$$\int_{-\infty}^{\infty} H(f) S(f) e^{j2\pi f T} \, df = \int_{-\infty}^{\infty} |S(f)|^2 \, df = \mathcal{E}_s \qquad (2.307)$$

that is, the energy of $s(t)$.

### 2.6.2.  Optimum detector: $M$ real signals in noise

We now want to solve the most general problem stated at the beginning of this section, that is, to decide among the $M$ hypotheses

$$H_j : \quad y(t) = s_j(t) + \nu(t), \qquad j = 1, 2, \ldots, M \qquad (2.308)$$

upon observation of $y(t)$ in the time interval $(0, T)$. The $M$ real signals $s_j(t)$, $j = 1, \ldots, M$, are known and have a finite duration and a finite energy. Using the Gram-Schmidt procedure, we can determine an orthonormal signal set $\{\psi_i(t)\}_{i=1}^{N}$, $N \leq M$, such that each $s_j(t)$, $j = 1, 2, \ldots, M$, can be expressed as a linear combination of these signals. Also, consider a complete orthonormal signal sequence such that its first $N$ signals are $\psi_1(t), \ldots, \psi_N(t)$ (see Problem 2.22). Denote with $(\psi_i(t))_{i=1}^{\infty}$, this sequence, and define

$$s_{ji} \triangleq \int_{0}^{T} s_j(t) \psi_i(t) \, dt, \quad j = 1, \ldots, M, \quad i = 1, 2, \ldots \qquad (2.309)$$

and $\nu_i$, $Y_i$ as in (2.277) and (2.280), respectively. The decision problem can be formulated in a discrete form as follows. Choose among the $M$ hypotheses

$$H_j : (Y_i)_{i=1}^{\infty} = (s_{j1} + \nu_1, s_{j2} + \nu_2, \ldots, s_{jN} + \nu_N, \nu_{N+1}, \nu_{N+2}, \ldots) \qquad (2.310)$$

$j = 1, 2, \ldots, M$, on the basis of the observation of the values taken by the RVs $Y_1, Y_2, \ldots$. As the noise components $\nu_{N+1}, \nu_{N+2}, \ldots$, are independent of $\nu_1, \ldots, \nu_N$, and of the hypothesis, observation of $Y_{N+1}, Y_{N+2}, \ldots$, does not add any information to the decision process. Thus, it can be based solely on the observation of $Y_1, Y_2, \ldots, Y_N$. By defining the row $N$-vectors $\mathbf{Y} \triangleq [Y_1, Y_2, \ldots, Y_N]$, $\nu \triangleq [\nu_1, \nu_2, \ldots, \nu_N]$, and $\mathbf{s}_j \triangleq [s_{j1}, s_{j2}, \ldots, s_{jN}]$, $j = 1, \ldots, M$, (2.310) can be reduced to the vector form

$$H_j : \quad \mathbf{Y} = \mathbf{s}_j + \nu, \qquad j = 1, 2, \ldots, M \qquad (2.311)$$

Thus, the optimum detector sought for will operate as follows;

$$\text{choose } H_j \text{ if } \mathbf{y} \in S_j \qquad (2.312)$$

where $\mathbf{y}$ denotes the observed value of the random vector $\mathbf{Y}$, and $S_1, S_2, \ldots, S_M$ is a partition of the $N$-dimensional vector space such that the rule (2.312) gives a minimum of the average error probability

$$P(e) = 1 - \sum_{j=1}^{M} p_j \int_{S_j} f_{\mathbf{Y}|H_j}(\mathbf{z} \mid H_j) \, d\mathbf{z}, \qquad (2.313)$$

where $p_j \triangleq P\{H_j\}$, $j = 1, 2, \ldots, M$. It is seen from (2.313) that $P(e)$ is minimized if every $S_j$ is chosen in such a way that

$$\mathbf{z} \in S_j \quad \text{if and only if} \quad p_j f_{\mathbf{Y}|H_j}(\mathbf{z} \mid H_j) = \max_i p_i f_{\mathbf{Y}|H_i}(\mathbf{z} \mid H_i) \qquad (2.314)$$

By combining (2.312) and (2.314), we obtain the MAP decision rule. In this situation the $M$-dimensional regions $S_j$ are called the MAP *decision regions*. In the special case where the hypotheses $H_j$ are equally likely, that is, $p_j = 1/M$, $j = 1, 2, \ldots, M$, (2.314) becomes

$$\mathbf{z} \in S_j \quad \text{if and only if} \quad f_{\mathbf{Y}|H_j}(\mathbf{z} \mid H_j) = \max_i f_{\mathbf{Y}|H_i}(\mathbf{z} \mid H_i) \qquad (2.315)$$

which corresponds to the *maximum-likelihood* (ML) decision rule (accordingly, the $S_i$ are called the ML decision regions). Although it minimizes the average error probability only for equally likely $H_j$, (2.315) is the most used detection rule, so in the following we shall mostly confine our attention to ML detection.

By defining the auxiliary hypothesis

$$H_0 : \quad \mathbf{Y} = \nu \qquad (2.316)$$

(2.315) can also be written in the form

$$\mathbf{z} \in S_j \quad \text{if and only if} \quad \Lambda_j(\mathbf{z}) = \max_i \Lambda_i(\mathbf{z}) \qquad (2.317)$$

where we define the *likelihood ratios*

$$\Lambda_j(\mathbf{y}) \triangleq \frac{f_{\mathbf{Y}|H_j}(\mathbf{y} \mid H_j)}{f_{\mathbf{Y}|H_0}(\mathbf{z} \mid H_0)} \qquad (2.318)$$

Thus, the ML decision rule is

$$\text{choose } H_j \quad \text{if } \Lambda_j(\mathbf{y}) = \max_i \Lambda_i(\mathbf{y}) \qquad (2.319)$$

where, as usual, $\mathbf{y}$ denotes the observed value of $\mathbf{Y}$. That is, the ML detector operates by computing the $M$ likelihood ratios $\Lambda_1(\mathbf{y})$, $\Lambda_2(\mathbf{y})$, $\ldots$, $\Lambda_M(\mathbf{y})$, and then choosing the hypothesis that corresponds to the largest among them. Let us now compute explicitly the likelihood ratios (2.318). By observing that, under hypothesis $H_j$, $j = 0, 1, \ldots, M$, $\mathbf{Y}$ is a Gaussian random vector with mean $\mathbf{s}_j$ (or zero for $j = 0$), independent components, and variance $N_0/2$ for each component, we have, for $j = 1, \ldots, M$,

$$\Lambda_j(\mathbf{y}) = \frac{\exp[-(1/N_0) |\mathbf{y} - \mathbf{s}_j|^2]}{\exp[-(1/N_0) |\mathbf{y}|^2]} = \exp\left\{ \frac{2}{N_0} \mathbf{y}\mathbf{s}'_j - \frac{1}{N_0} |\mathbf{s}_j|^2 \right\} \qquad (2.320)$$

where as usual $|\mathbf{x}|^2 = \mathbf{x}\mathbf{x}' = \sum_{i=1}^N x_i^2$ denotes the squared modulus of the row vector $\mathbf{x}$. Consideration of the log-likelihood ratios

$$\lambda_j(\mathbf{y}) \triangleq \ln \Lambda_j(\mathbf{y}) \qquad (2.321)$$

allows us to rewrite (2.319) in the following simple form:

$$\text{choose } H_j \quad \text{if } \mathbf{y}\mathbf{s}'_j - \frac{1}{2}|\mathbf{s}_j|^2 = \max_i \left\{ \mathbf{y}\mathbf{s}'_i - \frac{1}{2}|\mathbf{s}_i|^2 \right\} \qquad (2.322)$$

A different expression for the log-likelihood ratio can be derived as follows. Because

$$y(t) = \sum_{i=1}^{\infty} Y_i \psi_i(t) \qquad (2.323)$$

and

$$s_j(t) = \sum_{\ell=1}^{N} s_{j\ell} \psi_\ell(t) \qquad (2.324)$$

we have

$$\begin{aligned} \int_0^T y(t) s_j(t)\, dt &= \sum_{i=1}^{\infty} \sum_{\ell=1}^{N} Y_i s_{j\ell} \int_0^T \psi_i(t)\psi_\ell(t)\, dt \\ &= \sum_{i=1}^{N} Y_i s_{ji} \\ &\triangleq \mathbf{Y}\mathbf{s}'_j \end{aligned} \qquad (2.325)$$

and

$$\int_0^T s_j^2(t) = \sum_{\ell=1}^N \sum_{k=1}^N s_{j\ell} s_{jk} \int_0^T \psi_\ell(t) \psi_k(t) \, dt$$

$$= \sum_{\ell=1}^N s_{j\ell}^2$$

$$\stackrel{\triangle}{=} |s_j|^2 \tag{2.326}$$

so that

$$\lambda_j(y) = \frac{2}{N_0} \int_0^T y(t) s_j(t) \, dt - \frac{1}{N_0} \int_0^T s_j^2(t) \, dt, \quad j = 1, \ldots, M \tag{2.327}$$

In Chapter 4 the structures of the optimum detectors based on (2.320) and (2.327) will be reexamined and discussed.

### 2.6.3.   Detection problem for complex signals

We shall now focus our attention on the problem of detecting complex signals in Gaussian noise. This situation occurs when we are dealing with narrowband signals that we want to describe using complex envelopes. Let us first consider the detection of a single complex signal in noise, that is, the decision among the hypotheses

$$H_0 : \quad y(t) = \frac{1}{\sqrt{2}} \nu(t)$$

$$H_1 : \quad y(t) = \frac{1}{\sqrt{2}} s(t) + \frac{1}{\sqrt{2}} \nu(t) \tag{2.328}$$

where $t \in (0, T)$, and $y(t)$, $\nu(t)$, and $s(t)$ are complex envelopes of narrowband signals (for notational simplicity, we omit the tilde). In particular, we have

$$s(t) = s_c(t) + j s_s(t) \tag{2.329}$$

$$\nu(t) = \nu_c(t) + j \nu_s(t) \tag{2.330}$$

$$y(t) = y_c(t) + j y_s(t) \tag{2.331}$$

where $\nu(t)$ is a complex Gaussian noise process with power spectral density $2N_0$, and $\nu_c(t)$, $\nu_s(t)$ are independent, white Gaussian baseband processes with power spectral density $N_0$ (see Fig. 2.20). Hence, $(1/\sqrt{2})\nu_c(t)$ and $(1/\sqrt{2})\nu_s(t)$ have spectral density $N_0/2$, and the energy of $(1/\sqrt{2})s(t)$ is the same as the real signal with which it is associated. Choose now a real orthonormal sequence

$(\psi_{c1}(t),\ \psi_{s1}(t),\ \psi_{c2}(t),\ \psi_{s2}(t),\ldots)$, complete for any real signal in the time interval $(0,\ T)$, with $\psi_{c1}(t) = s_c(t)$ and $\psi_{s1}(t) = s_s(t)$. By formulating our detection problem in a discrete form, we have

$$H_0: \quad (Y_{c1},\ Y_{s1},\ Y_{c2},\ Y_{s2},\ldots) = (\nu_{c1},\ \nu_{s1},\ \nu_{c2},\ \nu_{s2},\ldots) \tag{2.332}$$
$$H_1: \quad (Y_{c1},\ Y_{s1},\ Y_{c2},\ Y_{s2},\ldots) = (s_{c1}+\nu_{c1},\ s_{s1}+\nu_{s1},\ s_{c2}+\nu_{s2},\ s_{s2}+\nu_{s2},\ldots)$$

where

$$s_{ci} \stackrel{\triangle}{=} \frac{1}{\sqrt{2}}\int_0^T s_c(t)\psi_{ci}(t)\,dt \tag{2.333}$$

$$s_{si} \stackrel{\triangle}{=} \frac{1}{\sqrt{2}}\int_0^T s_s(t)\psi_{si}(t)\,dt \tag{2.334}$$

$$\nu_{ci} \stackrel{\triangle}{=} \frac{1}{\sqrt{2}}\int_0^T \nu_c(t)\psi_{ci}(t)\,dt \tag{2.335}$$

$$\nu_{si} \stackrel{\triangle}{=} \frac{1}{\sqrt{2}}\int_0^T \nu_s(t)\psi_{si}(t)\,dt \tag{2.336}$$

$$Y_{ci} \stackrel{\triangle}{=} \frac{1}{\sqrt{2}}\int_0^T y_c(t)\psi_{ci}(t)\,dt \tag{2.337}$$

$$Y_{si} \stackrel{\triangle}{=} \frac{1}{\sqrt{2}}\int_0^T y_s(t)\psi_{si}(t)\,dt \tag{2.338}$$

Discarding the data irrelevant to the decision process, (2.332) can be put in the equivalent form

$$H_0: \quad Y_1 = \nu_1,$$
$$H_1: \quad Y_1 = s_1 + \nu_1, \tag{2.339}$$

where $Y_1 \stackrel{\triangle}{=} Y_{c1} + jY_{s1}$, $\nu_1 \stackrel{\triangle}{=} \nu_{c1} + j\nu_{s1}$, and $s_1 \stackrel{\triangle}{=} s_{c1} + js_{s1}$. In this situation, the decision regions $S_0$ and $S_1$ are two dimensional, and the likelihood ratio

$$\Lambda(y_c,\ y_s) \stackrel{\triangle}{=} \frac{f_{Y_{c1},Y_{s1}|H_1}(y_c,\ y_s\ |\ H_1)}{f_{Y_{c1},Y_{s1}|H_0}(y_c,\ y_s\ |\ H_0)} \tag{2.340}$$

is equal to

$$\begin{aligned}
\Lambda(y_c,\ y_s) &= \exp\left\{\frac{2}{N_0}(y_c s_{c1} + y_s s_{s1}) - \frac{1}{N_0}(s_{c1}^2 + s_{s1}^2)\right\} \\
&= \exp\left\{\frac{2}{N_0}\Re[y^* s_1] - \frac{1}{N_0}|s_1|^2\right\} \tag{2.341}
\end{aligned}$$

where $y \overset{\triangle}{=} y_c + jy_s$. Through computations similar to those that led to (2.327), it can be shown that the likelihood ratio between $H_0$ and $H_1$ can also be written in the form

$$\Lambda(Y_1) = \exp\left\{ \frac{2}{N_0}\Re \int_0^T y^*(t)s(t)\,dt - \frac{1}{N_0} \int_0^T |s(t)|^2\,dt \right\} \qquad (2.342)$$

This result can be extended to the problem of detecting one out of $M$ complex signals in noise. In this case the likelihood ratio among the hypotheses

$$H_j: \quad y(t) = \frac{1}{\sqrt{2}}s_j(t) + \frac{1}{\sqrt{2}}\nu(t), \qquad j = 1, 2, \ldots, M$$

and

$$H_0: \quad y(t) = \frac{1}{\sqrt{2}}\nu(t)$$

(where all the signals are complex) is given by

$$\Lambda_j(y) = \exp\left\{ \frac{2}{N_0}\Re \int_0^T y^*(t)s_j(t)\,dt - \frac{1}{N_0} \int_0^T |s_j(t)|^2\,dt \right\} \qquad (2.343)$$

The proof of (2.343) is left to the reader.

### 2.6.4. Summarizing the detection procedure

From the above discussion we have learned that the detection procedure consists of two basic steps:

1. Computation of the sufficient statistics, which consists of distilling from the observation what is sufficient to make the decision in an optimal way.

2. Use of the sufficient statistics for the detection.

For example, in the case of one real signal in white Gaussian noise the sufficient statistics is $Y_1$ in (2.281), a scalar quantity extracted from the signal $y(t)$ observed. The decision rule is based on $y_1$, the observed value of $Y_1$, and is given for example by (2.291).

This distinction between the two steps of the detection procedure may be especially relevant when a suboptimum detection rule is sought: for example, for simplicity's sake one of the two steps may not be optimum.

## 2.7.　Bibliographical notes

Several excellent books cover the area of signal and system theory. In most of them the reader can find further details regarding the topics covered in this chapter. Continuous-time and discrete-time deterministic signals and systems are treated extensively by Oppenheim, Willsky, and Young (1983). Discrete-time signals and systems are studied, among others, in Schwartz and Shaw (1975), Oppenheim and Schafer (1989), Proakis and Manolakis (1992). Papoulis (1977) covers both continuous- and discrete-time systems and deterministic and random signals. Gallager (1995) covers discrete random processes.

Volterra series were first studied by the Italian mathematician Vito Volterra around 1880 as a generalization of the Taylor series of a function. His work in this area is summarized in Volterra (1959). The application of Volterra series to the analysis of nonlinear systems with memory was suggested by Norbert Wiener. Extensive treatments of Volterra series as applied to the description of nonlinear systems can be found in Schetzen (1980) and Rugh (1981). Basic work in this area is represented by the paper of Flake (1963), whereas a relatively recent good review is found in Schetzen (1981). Applications are covered by Weiner and Spina (1980), and, among others, in the papers by Bedrosian and Rice (1971) and Benedetto, Biglieri, and Daffara (1976 and 1979).

Probability theory and random processes, at the level needed for this book, are covered by Parzen (1962), Papoulis (1965), and Davenport (1970). A comprehensive treatment of cyclostationary processes can be found in the dissertation by Hurd (1969) and in the papers by Gardner and Franks (1975) and Gardner (1978). Complex random processes are covered extensively by Miller (1974). Further details on Markov chains can be found in the classic book by Feller (1968) or in Kemeny and Snell (1960). The two volumes by Gantmacher (1959) on matrix theory include a treatment of Markov chains based on their matrix description. The reader is warned, however, that the nomenclature in Markov chain theory varies in the literature.

Fourier series and Fourier transforms are covered by Bracewell (1978), Dym and McKean (1972), and Papoulis (1962). Arsac (1966) emphasizes generalized functions. The approach to the computation of the power density spectrum of a random process $\xi(t)$ based on the function $\Gamma_\xi(f_1, f_2)$ is described in some detail in Blanc-Lapierre and Fortet (1968) and in Papoulis (1965). Spectral analysis of digital signals based on a Markov chain model was first discussed by Huggins (1957) and Zadeh (1957). Since then, several authors have expanded on the basic results. For a comprehensive and detailed discussion of this topic, see Cariolaro, Pierobon, and Tronca (1983) and Galko and Pasupathy (1981), where the whole treatment is given a firm mathematical basis.

For a more detailed treatment of narrowband signals and bandpass systems than was possible here, the reader is referred to Schwartz, Bennett, and Stein (1966, pp. 29–45) and Franks (1969, pp. 79–97, 195–200), or to the papers of Arens (1957), Dugundji (1958), and Bedrosian (1962). Different possible definitions for the envelope of a narrowband signal are discussed and compared in Rice (1982). Bandpass nonlinear systems are introduced in Blachman (1971) (see also Blachman, 1982).

Orthonormal expansions of finite-energy signals and the Gram-Schmidt orthogonalization procedure are dealt with by Franks (1969), including an introduction to the Karhunen-Loève expansion and to the sampling theorem for random processes.

A profound treatment of detection theory can be found in the classics by Van Trees (1968) and Helstrom (1968). For the computation of the likelihood ratio in signal-detection problems, see also Turin (1969) and Kailath (1971). In the latter paper the case of nonwhite Gaussian noise is treated using the techniques of "reproducing-kernel Hilbert spaces."

## 2.8. Problems

**2.1** A given (discrete or continuous) system may or may not be linear, time-invariant, memoryless, or causal. Determine which of these properties hold and which do not for each of the following systems, described by their input-output relationships. In particular, when a system is not memoryless determine the length if its memory.

(a) $y_n = 2x_n + 1$.

(b) $y_n = nx_n$

(c) $y_n = 1 + \sum_{i=0}^{L} a_i x_{n-i}$

(d) $y_n = x_{\lfloor n/2 \rfloor}$    ($\lfloor z \rfloor \overset{\triangle}{=}$ integer part of $z$)

(e) $y_n = x_n[1 - \delta_n]$,    $\delta_n \overset{\triangle}{=} 1$ for $n = 0$, and 0 elsewhere

(f) $y_n = x_n^*$

(g) $y(t) = 1 + \int_{-\infty}^{t} h(t - \tau)x(\tau)\,d\tau$

(h) $y(t) = \dfrac{dx(t)}{dt}$

(i) $y(t) = x(t - T) - x(t + T)$

(j) $y(t) = \int_{t}^{t+T} x\tau\,d\tau$

(k) $y(t) = x(t)e^{j2\pi f_0 t}$

(l) $y(t) = \int_{-\infty}^{\infty} x(\tau)e^{-j2\pi t\tau}\,d\tau$

**2.2** Find the Fourier transform of the sequence $(x_n)$, for

(a) $y_n = \delta_n,$      $\delta_n \overset{\triangle}{=} 1$ for $n = 0$, and 0 elsewhere

(b) $x_n = \begin{cases} a^n, & n \geq 0 \\ 0, & n < 0 \end{cases}$    $(a < 1)$

(c) $x_n = a^{|n|},$    $(a < 1)$

(d) $x_n = \begin{cases} 1, & 0 \leq n \leq N - 1 \\ 0, & \text{elsewhere} \end{cases}$

(e) $x_n = (-1)^n$

**2.3** Given the discrete linear time-invariant system whose input-output relationship is described by the difference equation

$$y_n = \frac{5}{6} y_{n-1} + \frac{1}{6} y_{n-2} + x_n$$

compute its transfer function $H(f)$ and its impulse response $(h_n)$. Determine the response of the system to the input

$$x_n = \begin{cases} \left(\frac{1}{2}\right)^n, & n \geq 0 \\ 0, & n < 0 \end{cases}$$

**2.4** *Parseval equality:* Prove that

$$\sum_{n=-\infty}^{\infty} |x_n|^2 = \int_{-1/2}^{1/2} |X(f)|^2 \, df$$

where $X(f)$ denotes the Fourier transform of the sequence $(x_n)$.

**2.5** Prove that for a continuous or discrete time-invariant Volterra system there is no loss of generality if it is assumed that the kernels describing the system are symmetric (i.e., any permutation of their arguments leaves the kernels unchanged).

**2.6** Prove that a continuous time-invariant Volterra system is causal if and only if (2.21) holds for all $k$. Provide the corresponding condition for the kernels of a *discrete* time-invariant Volterra system.

**2.7** Find the Volterra kernels for the continuous nonlinear system obtained by cascading a memoryless nonlinearity and a linear time-invariant system with impulse response $h(t)$ (Fig. 2.31). It is assumed that

$$g(x) = \sum_{i=0}^{\infty} a_i x^i$$

Figure 2.31: *See Problem 2.7.*

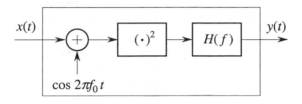

Figure 2.32: *See Problem 2.8.*

**2.8** Find the input-output relationship of the system of Fig. 2.32, where $x(t)$ is a low-pass signal whose spectrum is zero for $|f| > f_1$, and $H(f)$ denotes an ideal bandpass filter centered at $f_0$, $f_0 \gg f_1$.

**2.9** Consider a finite-energy signal $s(t)$ defined for $|t| \leq T/2$, and its Fourier transform $S(f)$. Denoting by $s^{(k)}(t)$ the $k$th derivative of $s(t)$, $[s^{(0)}(t) \stackrel{\triangle}{=} s(t)]$, show that if

$$s^{(k)}\left(\frac{T}{2}\right) = s^{(k)}\left(-\frac{T}{2}\right) = 0 \qquad 0 \leq k \leq K$$

then, as $f \to \infty$

$$|S(f)|^2 = O(f^{-2K-4})$$

Consider then the signal

$$s(t) = \cos\left(\frac{\pi}{T}t + \alpha \sin\frac{4\pi}{T}t\right)$$

and find $\alpha$ so as to get $|S(f)|^2 = O(f^{-6})$.

**2.10** *Discrete matched filter:* Let the input of a discrete linear time-invariant system with transfer function $H(f)$ be the real sequence $(x_n) = (s_n + w_n)$, where $(s_n)$ is a deterministic sequence with Fourier transform $S(f)$, and $(w_n)$ is a sequence of independent, identically distributed random variables. If $(y_n)$ denotes the system response to $(s_n)$ and $(\nu_n)$ the system response to $(w_n)$, find $H(f)$ such as the ratio $y_0^2/\mathrm{E}[\nu_n^2]$ is a maximum.

**2.11** For a homogeneous Markov chain $(\xi_n)$ with transition probability matrix $\mathbf{P}$, given an integer $N$ and an $N$-tuple of integers $k_1 < k_2 < \cdots < k_N$, express the probability $P\{\xi_{k_1} = i_1, \xi_{k_2} = i_2, \ldots, \xi_{k_N} = i_N\}$ in terms of the entries of $\mathbf{P}$ and of the initial state distribution vector $\mathbf{w}^{(0)}$.

**2.12** Prove that, if the transition matrix $\mathbf{P}$ of a fully regular homogeneous Markov chain is *doubly stochastic*, i.e., the sum of its entries in each row and column equals 1, then its stationary distribution vector $\mathbf{w}$ has equal components.

**2.13** Let $\xi(t)$ be a WS stationary random process and $R_\xi(\tau)$ its autocorrelation function. Prove that, given $n \geq 1$, for any $n$-tuple of time instants $\tau_1, \tau_2, \ldots, \tau_n$ and for any $n$-tuple of complex numbers $a_1, a_2, \ldots, a_n$ the following inequality holds:

$$\sum_{i=1}^{n} \sum_{j=1}^{n} a_i^* a_j R_\xi(\tau_i - \tau_j) \geq 0$$

**2.14** Consider the linearly modulated digital signal

$$\xi(t) = \sum_{n=-\infty}^{\infty} \alpha_n s(t - nT)$$

where $(\alpha_n)$ is a sequence of independent, identically distributed random variables taking on values $\pm 1$ with equal probabilities. Compute $\mathrm{E}[\xi^2(t)]$ and show that it is a periodic signal with period $T$.

**2.15** Let $\xi(t)$ be a WS cyclostationary process with period $T$. Consider a random linear system whose output is $\eta(t) = \xi(t - \Theta)$, $\Theta$ a random variable independent of $\xi(t)$ and uniformly distributed in the interval $(0, T)$. Prove that $\eta(t)$ is WS stationary. *Hint:* Consider $\Gamma_\eta(f_1, f_2)$.

**2.16** Evaluate the power density spectrum of the digital signal

$$\xi(t) = \sum_{n=-\infty}^{\infty} s(t - nT; \alpha_n)$$

where $(\alpha_n)$ is a sequence of independent, identically distributed random variables taking on values 1, 2, 3, 4 with equal probabilities,

$$s(t; \alpha_n) \triangleq \beta_n' r(t) + j\beta_n'' r\left(t - \frac{T}{2}\right)$$

with $r(t)$ defined for $t \in (-T/2, T/2)$, and $\beta_n'$, $\beta_n''$ are obtained from $\alpha_n$ according to the following table:

| $\alpha_n$ | $\beta_n'$ | $\beta_n''$ |
|:---:|:---:|:---:|
| 1 | +1 | +1 |
| 2 | −1 | +1 |
| 3 | +1 | −1 |
| 4 | −1 | −1 |

Specialize the result to the cases (which will be treated in detail in later chapters)

$$r(t) = 1 \qquad \text{(offset PSK)}$$

$$r(t) = \cos \frac{\pi}{T} t \qquad \text{(MSK)}$$

$$r(t) = \cos \left( \frac{\pi}{T} t - \frac{1}{4} \sin \frac{4\pi}{T} t \right) \qquad \text{(SFSK)}$$

and plot the resulting power spectra for $|fT| \leq 10$.

**2.17** *FSK digital signals:* Find the power spectral density of the signal

$$\xi(t) = \exp \left\{ j 2\pi f_d \int_0^t q(\tau) \, d\tau \right\}$$

where

$$q(t) = \sum_{n=0}^{\infty} \alpha_n s(t - nT),$$

$(\alpha_n)$ is a sequence of independent, identically distributed random variables taking on values $\pm 1$ with equal probabilities, and

$$s(t) = \begin{cases} 1, & 0 \leq t < T \\ 0, & \text{elsewhere} \end{cases}$$

(This refers to CPFSK modulated signals. They will be treated in Chapter 6.)

**2.18** Evaluate the power spectral density of the digital signal

$$\xi(t) = \exp \left\{ j \sum_{n=-\infty}^{\infty} \alpha_n g(t - nT) \right\}$$

where $g(t)$ is a signal with duration $\tau \leq T$, and $(\alpha_n)$ is a sequence of independent, identically distributed random variables taking on the $M$ values $(\pi/M)(2i - 1)$, $i = 1, \ldots, M$, with equal probabilities.

**2.19** Prove the following properties of the Hilbert transform $\hat{x}(t)$ of the signal $x(t)$:

(a) If $x(t)$ is an even function of $t$, then $\hat{x}(t)$ is an odd function.

(b) If $x(t)$ is an odd function of $t$, then $\hat{x}(t)$ is an even function.

(c) The Hilbert transform of $\hat{x}(t)$ is equal to $-x(t)$.

(d) The energy of $\hat{x}(t)$ is equal to the energy of $x(t)$.

(e) The energy of $x(t) + j\hat{x}(t)$ is equal to twice the energy of $x(t)$.

(f) $R_{\hat{x}\hat{x}}(\tau) = R_{xx}(\tau)$.

**(g)** $E[\hat{x}(t + \tau)x(t)]$ is equal to $\hat{R}_{xx}(\tau)$, the Hilbert transform of the autocorrelation function of $x(t)$, and $E[x(t + \tau)\hat{x}(t)]$ is equal to $-\hat{R}_{xx}(\tau)$.

**2.20** Prove the equalities (2.188) and (2.189). *Hint:* Compute $R_{\nu\nu}(\tau)$ by using (2.182).

**2.21** Show that the Gram-Schmidt procedure (2.240)–(2.241) generates an orthonormal set of signals.

**2.22** Consider a given orthonormal signal set $\{\psi_i(t)\}_{i=1}^N$. Prove that it is possible to find a *complete* orthonormal signal sequence such that its first elements are $\psi_1(t)$, $\ldots, \psi_N(t)$.

**2.24** Let $x_c(t)$ be a continuous-time signal and $(x_n)$ the sequence of its samples taken every $T$: that is, $x_n \triangleq x_c(nT)$. If $X(f)$ denotes the Fourier transform of the sequence $(x_n)$, and $X_s(f)$ the Fourier transform of the continuous-time signal

$$x_s(t) \triangleq x_c(t) \sum_{n=-\infty}^{\infty} \delta(t - nT)$$

prove that

$$X_s(f) = X(fT).$$

**2.25** Generalize (2.259) to the case in which the sampling waveform, instead of being a train of ideal impulses, is the periodic signal $\sum_{n=-\infty}^{\infty} p(t - nT)$, where $p(t)$ is a rectangular pulse with duration $\tau < T$. Can the original signal $x(t)$ still be recovered exactly from the product signal $x(t) \sum_{n=-\infty}^{\infty} p(t - nT)$?

**2.26** *Matched filter for nonwhite noise:* Consider a continuous linear time-invariant system whose input is the sum of a deterministic signal $s(t)$ and a WS stationary noise $\nu(t)$ whose power spectral density $\mathcal{G}_\nu(f)$ is nonzero for all $f$. Find the transfer function of the system that maximizes the ratio between the instantaneous power of the signal part and the variance of the noise part at its output.

# Basic results from information theory

In this chapter we deal with information sources and communication channels. The main part of the treatment is devoted to the discrete case. Only at the end of the chapter do we present a brief description of continuous sources and channels, aimed at obtaining the capacity of the bandlimited Gaussian channel.

The first part of the chapter defines a discrete stationary source and shows how the quantity of information that is emitted from the source can be measured. In general, the source output (the *message*) consists of a sequence of symbols chosen from a finite set, the *alphabet* of the source. A probability distribution is associated with the source alphabet, and a probabilistic mechanism governs the emission of successive symbols in the message. Generally, different messages convey different quantities of information; thus an average information quantity, or *entropy*, must be defined for the source. The unit of measure for the information is taken to be the *bit*, that is, the information provided by the emission of one among two equally likely symbols. The entropy of the source represents the minimum average number of binary symbols (*digits*) that is necessary to represent each symbol in the message. The source output can thus be replaced by a string of binary symbols conveying the same quantity of information and having an average number of digits per symbol of the original source as close as desired to the source entropy. The block in the system that implements this function is called the *source encoder*.

The communication channel is the physical medium used to connect the source of information with its user. In the second part of the chapter we define discrete memoryless channels and study their properties. Discrete memoryless channels are specified by a probability law linking symbols of the channel input alphabet to symbols of the channel output alphabet. A basic point is the knowledge of the maximum average information flow that can reliably pass through the channel. This leads to the definition of the *channel capacity* and to the problem of computing it. Both topics are addressed in this chapter.

The final part of the chapter is devoted to a presentation of the channel coding theorem and its converse. They provide a link between the concepts of entropy of a source and capacity of a channel and assess precisely what reliable transmission means and how it can be achieved.

The main goal of this chapter is to provide a general frame to the subsequent material, which deals with specific aspects of data transmission systems. It also assesses the theoretical limits in performance that can be obtained over a binary channel and an additive Gaussian channel.

## 3.1.   Introduction

The goal of every communication system is the reproduction of a message emitted from a source into a place where the user of the information is located. The distance between the source and the user may be either considerable, as in the case of intercontinental transmission, or very small, as in the storage and retrieval of data using the disk unit of a computer (in this case, the distance between transmitter and receiver may be considered in *time*). However, irrespective of distance, there exists between the source and the user a communicating channel affected by various disturbances, like *noise* and distortions.

The presence of the disturbed channel makes the exact reproduction of the message emitted from the source at the user's premises an impossible achievement. Nevertheless, the designer of a communication system will always be asked to provide the user with an "as close as possible" replica of the original message. A closer insight into the characteristics of the user better specifies, case by case, the meaning of "as close as possible," that is, the specification of a *user-oriented* criterion of acceptability. For example, in the case of speech communication in the area of service communications, one is normally satisfied when the listener can understand the semantic content of what the speaker is saying. Quite often, however, in the domain of public telephone services, the listener wishes to recognize the identity and mood of the speaker through the pitch and inflection of his or her voice, and this gives rise to a more stringent criterion of acceptability. Hence, as illustrated in these examples, different user requirements may lead to different criteria of acceptability and, consequently, to different bandwidth requirements for speech transmission.

As explained, the problem of noise in the communication channel creates the need for user-sensitive specifications of criteria of acceptability in the design of

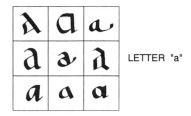

Figure 3.1: *Class of equivalence relative to the letter "a" in handwritten texts.*

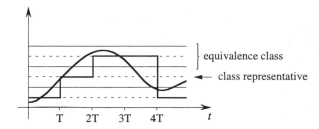

Figure 3.2: *Quantization process in PCM.*

communication systems. This can be accomplished in the following fashion. The possible source outputs in a given time interval are partitioned into *equivalence classes* on the basis of a certain criterion of acceptability. This permits one to regard such source outputs as a set of equivalence classes where the source outputs residing in the same equivalence class are indistinguishable with respect to the acceptability criterion. Thus, the communication system, in this regard, can be reduced to the transmission of the specific class to which the source output belongs in each successive time interval.

In Fig. 3.1, one possible class of equivalence is depicted for the transmission of written texts, where the criterion of acceptability is merely the semantic intelligibility of the message. The class represents different ways of writing the letter "a".

Another well-known example is the quantization process performed in connection with *pulse-coded modulation* (PCM). In Fig. 3.2, the process is schematically outlined.

The source waveform is first sampled every $T$ seconds. Each sample is then quantized, (that is, the closest value in a finite pre-selected set is substituted for it), and kept constant for $T$ seconds.

From here on we shall assume that the criterion of acceptability had been established, yielding a finite number of equivalence classes, say $M$, in a specified time interval. The transmission of information then consists in the communication of a sequence of integer numbers chosen in the set $\{1, 2, \ldots, M\}$ from the source to the user. The user, upon receiving the indication of the equivalence class, generates the representative (which he knows) of the class so as to restore an information close to the original.

## 3.2.  Discrete stationary sources

Consider a finite alphabet X formed by the $M$ symbols $\{x_i\}_{i=1}^M$ and define a message as a sequence of symbols such as $(x_n)_{n=0}^\infty$. A discrete stochastic source is a device emitting messages through the selection of symbols from the alphabet X according to a probability distribution $\{p_i\}_{i=1}^M$, where $p_i \triangleq P(x_i)$. From a probabilistic point of view, one can regard the whole set of messages as a discrete random process, that is, a sequence $(\xi_n)_{n=0}^\infty$ of random variables (RVs), each taking values in the set X with the probability distribution $\{p_i\}$.

  We shall assume that the source is stationary; that is,

$$P\{\xi_{i_1} = x_1, \ldots, \xi_{i_k} = x_k\} = P\{\xi_{i_1+h} = x_1, \ldots, \xi_{i_k+h} = x_k\} \qquad (3.1)$$

for all nonnegative integers $i_1, \ldots, i_k, h$ and all $x_1, \ldots, x_k \in$ X. In this case, the message sequence forms a discrete-time stationary random process with the properties described in Chapter 2.

### 3.2.1.  A measure of information: entropy of the source alphabet

The quantity of information carried by one particular symbol of the source alphabet is strictly related to its uncertainty. Increased uncertainty should correspond to more information. As an example, the letter size in a newspaper headline is larger when the news is unexpected like "Life found on Mars!" than in the case of "A new government in Italy." It is then fairly natural that the information content of the $i$th symbol, denoted by $I(x_i)$, be a decreasing function of its probability

$$I(x_j) > I(x_i), \quad \text{if } p_j < p_i \qquad (3.2)$$

and that the information content associated with the emission of two independent symbols be the sum of the two individual informations:

$$\text{If } P(x_i, x_j) = P(x_i)P(x_j) \text{ then } I(x_i, x_j) = I(x_i) + I(x_j) \qquad (3.3)$$

A definition of the information content satisfying both (3.2) and (3.3) is

$$I(x_i) \triangleq \log_a \left( \frac{1}{p_i} \right) \qquad (3.4)$$

In (3.4) the base of the logarithm (indicated with $a$) is unspecified. Its choice determines the unit of measure assigned to the information content. If the natural (base $e$) logarithm is used, then the unit is called *nat*. When the base is 2, the unit is widely known as *bit* (a contraction of the words "binary digit"). The use of bit is based on the fact that the correct identification of one out of two equally likely symbols conveys an amount of information equal to $I(x_1) = I(x_2) = \log_2 2 = 1$ bit. Unless otherwise specified, we shall use the base 2 in this chapter and write *log* to mean $log_2$.

The definition (3.4) allows one to associate with each symbol of the source alphabet its information content. A characterization of the whole alphabet can be obtained by defining the *average information content* of X

$$H(X) \triangleq \sum_{i=1}^{M} p_i I(x_i) = \sum_{i=1}^{M} p_i \log \left( \frac{1}{p_i} \right) \tag{3.5}$$

which is called the *entropy* of the source alphabet and is measured in bit/symbol.

**Example 3.1**  The source alphabet consists of four possible symbols with probabilities $p_1 = \frac{1}{2}, p_2 = \frac{1}{4}, p_3 = p_4 = \frac{1}{8}$. To compute the entropy of the source alphabet, we apply definition (3.5)

$$H(X) = \frac{1}{2} \log 2 + \frac{1}{4} \log 4 + 2\frac{1}{8} \log 8 = 1.75 \text{ bit/symbol}$$

If the source alphabet consists of $M$ equally likely symbols, we have

$$H(X) = \sum_{i=1}^{M} \frac{1}{M} \log M = \log M \text{ bit/symbol}$$

When the source alphabet consists of two symbols with probabilities $p$ and $q = 1 - p$ the alphabet entropy is

$$H(X) = p \log \frac{1}{p} + (1 - p) \log \left( \frac{1}{1 - p} \right) \triangleq H(p) \tag{3.6}$$

In Fig. 3.3 the function $H(p)$ is plotted.

It can be seen that the maximum occurs for $p = 0.5$, that is, when the two symbols are equally likely. □

The last result of Example 3.1, that is, the maximization of the source entropy for equally likely symbols, is fairly general, as will be stated in the following theorem.

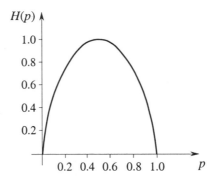

Figure 3.3: *Plot of the entropy function $H(p)$ of a binary source with $P(x_1) = p$ and $P(x_2) = 1 - p$.*

**Theorem 3.1**

The entropy H(X) of a source alphabet with $M$ symbols satisfies the inequality

$$H(X) \leq \log M \tag{3.7}$$

with equality when the symbols are equally likely. $\triangledown$

**Proof of Theorem 3.1**

To prove the theorem, consider the difference

$$H(X) - \log M = \sum_{i=1}^{M} p_i \log \left( \frac{1}{p_i} \right) - \sum_{i=1}^{M} p_i \log M = \sum_{i=1}^{M} p_i \log \left( \frac{1}{p_i M} \right) \tag{3.8}$$

Making use of the inequality

$$\ln y \leq y - 1 \tag{3.9}$$

in the RHS of (3.8), we obtain

$$H(X) - \log M \leq \log e \sum_{i=1}^{M} \left( \frac{1}{M} - p_i \right) = 0$$

**QED**

### 3.2.2. Coding of the source alphabet

For a given source, we are now able to compute the information content of each symbol in the source alphabet and the entropy of the alphabet itself. Suppose now that we want to transmit each symbol using a binary channel, that is, a channel able to communicate only binary symbols. Before being delivered to the channel, each symbol must be represented by a finite string of digits, called the *code word*. Leaving aside the problem of possible channel errors, efficient communication would involve transmitting a symbol in the shortest possible time, which, in turn, means representing it with a code word as short as possible. As usual, we are interested in average quantities, so our goal will be that of minimizing the average length of a code word

$$\bar{n} \triangleq \mathrm{E}\{n\} = \sum_{i=1}^{M} p_i n_i \tag{3.10}$$

where $n_i$ is the length (number of digits) of the code word representing the symbol $x_i$, and $n$ is the random variable representing its length (that is, assuming the value $n_i$ with probability $p_i$, $i = 1, 2, ..., M$).

The minimization of (3.10) must be accomplished according to an important constraint on the assignment of code words to the alphabet symbols. To understand the necessity of this constraint, consider the following code:

| Symbol | Code word |
|--------|-----------|
| $x_1$ | 0 |
| $x_2$ | 01 |
| $x_3$ | 10 |
| $x_4$ | 100 |

In it, the binary sequence 010010 could correspond to any one of the five messages $x_1 x_3 x_2 x_1$, $x_1 x_3 x_1 x_3$, $x_1 x_4 x_3$, $x_2 x_1 x_1 x_3$, or $x_2 x_1 x_2 x_1$. The code is ambiguous, or *not uniquely decipherable*. It then seems natural to require that the code be uniquely decipherable, which means that every finite sequence of binary digits corresponds to, at most, one message. A condition that ensures unique decipherability is to require that no code word be a prefix of a longer code word. Codes satisfying this constraint are called *prefix codes*. The codes described in the sequel are of this kind.

A very useful graphical representation of a code satisfying the prefix constraint is that which associates to each code word a terminal node in a binary tree, like the one of Fig. 3.4.

Starting from the root of the tree, the two branches leading to the first-order nodes correspond to the choice between 0 and 1 as the first digit in the code

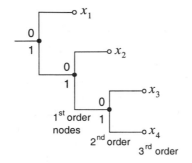

| Symbols | Code words |
|---------|------------|
| $x_1$   | 0          |
| $x_2$   | 10         |
| $x_3$   | 110        |
| $x_4$   | 111        |

Figure 3.4: *Binary tree associated with a binary source code.*

words. The two branches stemming from each of the first-order nodes correspond to the second digit of the code words, and so on. Since code words are assigned only to terminal nodes, no code word can be a prefix of another code word. A tree is said to be of order $n$ if it contains nodes up to the $n$-th order. A necessary and sufficient condition for a given code to satisfy the prefix constraint is given in the following theorem.

**Theorem 3.2**

**Kraft inequality**. A necessary and sufficient condition for the existence of a binary prefix code with word lengths $n_1, n_2, \ldots, n_M$ is the following:

$$\sum_{i=1}^{M} 2^{-n_i} \leq 1 \tag{3.11}$$

$\nabla$

**Proof of Theorem 3.2**

We prove first that (3.11) is a necessary condition. Since the code satisfies the prefix constraint, it is embedded in a tree of order

$$n = \max(n_1, n_2, \ldots, n_M)$$

The presence in the tree of a terminal node of order $n_i$ eliminates $2^{n-n_i}$ of the possible nodes of order $n$. Thus, for the code to be embedded in the tree, the sum of all nodes of order $n$ eliminated by terminal nodes associated with code words must be less than or equal to the number of nodes of order $n$ in the tree; that is,

$$\sum_{i=1}^{M} 2^{n-n_i} \leq 2^n \tag{3.12}$$

Dividing both sides of the last inequality by $2^n$ yields (3.11).

To prove that (3.11) is a sufficient condition for the existence of a prefix code, let us assume that the $n_i$'s are arranged in nondecreasing order, $n_1 \leq n_2 \leq \ldots \leq n_M$. Choose as the first terminal node in the code tree any node of order $n_1$ in a tree of order $n_M$ containing all branches. All nodes in the tree of each order greater than or equal to $n_1$ are still available for use as terminal nodes in the code tree, except for the fraction $2^{-n_1}$ that stems from the chosen node. Next, choose any available node of order $n_2$ as the next terminal node in the code tree. All nodes in the tree of each order greater than or equal to $n_2$ are still available except for the fraction $2^{-n_1} + 2^{-n_2}$ that stem from either of the two chosen nodes. Continuing in this way, after the assignment of the $j$-th terminal node in the code tree, the fraction of nodes eliminated by previous choices is $\sum_{i=1}^{j} 2^{-n_i}$. From (3.11), this fraction is always strictly less than 1 for $j < M$, and thus there is always a node available to be assigned to the next code word. **QED**

Since we are using a binary code, the maximum information content of each digit in the code word is 1 bit. So the average information content in each code word is, at most, equal to $\bar{n}$. On the other hand, to uniquely specify a symbol of the source alphabet, we need an average amount of information equal to $H(X)$ bits. Hence we can intuitively conclude that

$$\bar{n} \geq H(X) \tag{3.13}$$

Comparing the definitions (3.5) and (3.10) of $H(X)$ and $\bar{n}$, it can be seen that the condition (3.13) can be satisfied with the equal sign if and only if (the "if" part is straightforward, for the "only if" proof see Fano (1961)):

$$p_i = 2^{-n_i} \quad i = 1, 2, \ldots, M \tag{3.14}$$

In this case, (3.11) also becomes an equality.

**Example 3.2**   The following is an example of a code satisfying (3.13) with the equal sign and obeying the prefix constraint.

| Symbol | $p_i$ | Code word |
|--------|-------|-----------|
| $x_1$ | $\frac{1}{2}$ | 1 |
| $x_2$ | $\frac{1}{4}$ | 00 |
| $x_3$ | $\frac{1}{8}$ | 010 |
| $x_4$ | $\frac{1}{16}$ | 0110 |
| $x_5$ | $\frac{1}{16}$ | 0111 |

Computing the value of $\bar{n}$ defined in (3.10), one obtains

$$\bar{n} = \mathrm{H}(X) = \frac{15}{8}$$

$\square$

In general, condition (3.14) with $n_i$ integers is not satisfied. So we cannot hope to attain the lower bound for $\bar{n}$ as in the previous example. However, a code satisfying the prefix constraint can be found whose $\bar{n}$ obeys the following theorem.

## Theorem 3.3

A binary code satisfying the prefix constraint can be found for any source alphabet of entropy $\mathrm{H}(X)$ whose average code word length $\bar{n}$ satisfies the inequality

$$\mathrm{H}(X) \leq \bar{n} < \mathrm{H}(X) + 1$$

$\triangledown$

## Proof of Theorem 3.3

An intuitive proof of the lower bound has already been given when introducing (3.13). Let us now choose for the code word representing the symbol $x_i$ a number of bits $n_i$ corresponding to the smallest integer greater than or equal to $I(x_i)$. So we have

$$I(x_i) \leq n_i < I(x_i) + 1 \tag{3.15}$$

Multiplying (3.15) by $p_i$ and summing over $i$, we obtain

$$\mathrm{H}(X) \leq \bar{n} < \mathrm{H}(X) + 1$$

To complete the proof of the theorem, we have to show that the code satisfies the prefix constraint, that is, the lengths $n_i$'s of the code words obey the Kraft inequality (3.11). Recalling the definition (3.4) of $I(x_i)$, the left-hand inequality of (3.15) leads to $p_i \geq 2^{-n_i}$; so, summing over $i$, we obtain

$$\sum_{i=1}^{M} 2^{-n_i} \leq \sum_{i=1}^{M} p_i = 1 \tag{3.16}$$

**QED**

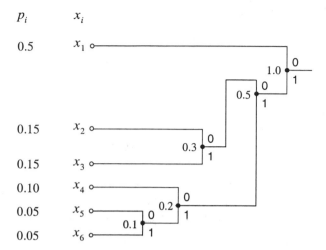

Figure 3.5: *Tree generated by the Huffman encoding procedure for a source with six symbols.*

The last step in our description of the source alphabet coding is the *construction* of a code uniquely decipherable that minimizes the average code word length. We shall present a method for the construction of such optimal codes due to Huffman. The proof of optimality will be omitted; the interested reader can find it in any book specifically devoted to information theory, as for example McEliece (1977). The Huffman procedure will be described step by step. The reader is referred to Fig.3.5, in which the steps can be spotted in the tree originated by the encoding procedure.

**Step 1** Have the $M$ symbols ordered according to nonincreasing values of their probabilities.

**Step 2** Group the last two symbols $x_{M-1}$ and $x_M$ into an equivalent "symbol," with probability $p_{M-1} + p_M$.

**Step 3** Repeat steps 1 and 2 until only one "symbol" is left.

**Step 4** Looking at the tree originated by the preceding steps (see Fig. 3.5), associate the binary symbols 0 and 1 to each pair of branches departing from intermediate nodes. The code word of each symbol can be read as the binary sequence encountered when starting from the root of the tree and reaching the terminal node associated with the symbol at hand.

For the example of Fig. 3.5, the code words obtained using the Huffman procedure are

| Symbol | Code word |
|:------:|:---------:|
| $x_1$  |     0     |
| $x_2$  |    100    |
| $x_3$  |    101    |
| $x_4$  |    110    |
| $x_5$  |   1110    |
| $x_6$  |   1111    |

The average length $\bar{n}$ and the entropy $H(X)$ are 2.1 digits/symbol and 2.086 bits/symbol, respectively; they satisfy Theorem 3.3, and no other code can do better.

**Example 3.3**   For the code shown in Fig. 3.5, use the tree to decode the received sequence 110010110010. Starting from the root of the tree, we follow the branches at each intermediate node according to the binary digits in the received sequence, until a terminal node (and hence a symbol) is reached. Then we restart the procedure. The decoded sequence is $x_4 x_1 x_3 x_2 x_4$.                                                  □

The reader is invited to repeat the decoding procedure of Example 3.3, assuming that an error had been introduced by the channel in the first position. This permits us to verify the catastrophic effect of error propagation in these variable-length codes. On the other hand, the goal of the source coding is the reduction of redundancy of the source alphabet, and not the protection against channel errors. This is the scope of channel encoding, as we will see in Chapters 10-12.

So far, we have seen how a code word can be efficiently assigned to each symbol $x_i$ of the source alphabet X. The main result is represented by Theorem 3.3. In fact, the lower bound of the Theorem can be approached as closely as desired if we are allowed to encode *blocks of symbols* instead of single symbols. Suppose we take a sequence of independent observations of X and assign a code word to the resulting group of symbols. In other words, we construct a code for a new alphabet $Y \equiv X^{\nu}$ containing $M^{\nu}$ symbols, denoted $y_i$. The probability of $y_i$ is then given by the product of the probabilities corresponding to the $\nu$ symbols of X that specify $y_i$. By Theorem 3.3, we can construct a code for Y whose average code word length $\bar{n}_{\nu}$ satisfies

$$H(Y) \leq \bar{n}_{\nu} < H(Y) + 1 \tag{3.17}$$

Every symbol in Y is made by $\nu$ independent symbols of the original alphabet X; so the entropy of Y is $H(Y) = \nu H(X)$ (see Problem 3.2). Thus, from (3.17) we get

$$H(X) \leq \frac{\bar{n}_\nu}{\nu} < H(X) + \frac{1}{\nu} \tag{3.18}$$

But $\bar{n}_\nu/\nu$ is the average number of digits/symbol of X; so, from (3.18), it follows that it can be made arbitrarily close to $H(X)$ by choosing $\nu$ sufficiently large.

The *efficiency* $\epsilon$ of a code is defined as

$$\epsilon \triangleq \frac{\nu H(X)}{\bar{n}_\nu} \tag{3.19}$$

and its *redundancy* is $(1-\epsilon)$.

**Example 3.4**   Given the source alphabet $X = \{x_1, x_2, x_3\}$, with $p_1 = 0.5$, $p_2 = 0.3$, and $p_3 = 0.2$, we want to construct the new alphabet $Y = X^2 = \{y_1, y_2, \ldots, y_9\}$, obtained by grouping the symbols $x_i$ two by two.

| Symbol | Code word |
|---|---|
| $y_1 = x_1 x_1$ | $P(y_1) = P(x_1 x_1) = P(x_1)P(x_1) = 0.25$ |
| $y_2 = x_1 x_2$ | $P(y_2) = 0.15$ |
| $y_3 = x_2 x_1$ | $P(y_3) = 0.15$ |
| $y_4 = x_1 x_3$ | $P(y_4) = 0.10$ |
| $y_5 = x_3 x_1$ | $P(y_5) = 0.10$ |
| $y_6 = x_2 x_2$ | $P(y_6) = 0.09$ |
| $y_7 = x_3 x_2$ | $P(y_7) = 0.06$ |
| $y_8 = x_2 x_3$ | $P(y_8) = 0.06$ |
| $y_9 = x_3 x_3$ | $P(y_9) = 0.04$ |

The reader is invited to construct the Huffman codes for block lengths $\nu = 1$ and $\nu = 2$ and compare the average numbers of digits/symbol obtained in both cases, using the preceding definition of code efficiency.   $\square$

### 3.2.3.   Entropy of stationary sources

Although our definition of a discrete stationary source is fairly general, we have so far considered in detail only the information content and the encoding of the source alphabet. Even when describing the achievement of the block encoding of the source, we made the assumption of independence between the symbols forming each block. Of course, when the messages emitted by the source are actually a sequence of independent random variables, then the results obtained for the source alphabet also hold true for the source message. In practice, however, this is rarely the case. Thus we need to extend our definition of the information content of the source alphabet to the information content of the source, which will involve consideration of the statistical dependence between symbols in a message. Let us consider a message emitted by the source, like $(x_n)_{n=0}^{\infty}$, and try to compute the average information needed to specify each symbol $x_n$ in the message. The information content of the first symbol $x_0$ is, of course, the entropy $H(X_0)$[1]

$$H(X_0) = \sum_{i=1}^{M} p_i \log \left( \frac{1}{p_i} \right)$$

The information content of the second symbol $x_1$, having specified $x_0$, is the *conditional entropy* $H(X_1 \mid X_0)$ based on the conditional information $I(x \mid y) \triangleq \log \left( 1/P(x \mid y) \right)$

$$H(X_1 \mid X_0) \triangleq \sum_{X_0} \sum_{X_1} P(x_0, x_1) I(x_1 \mid x_0)$$

$$= \sum_{X_0} \sum_{X_1} P(x_0, x_1) \log \left( \frac{1}{P(x_1 \mid x_0)} \right) \qquad (3.20)$$

In general, the information content of the $i$th symbol, given the previous $h$ symbols in the message, is obtained as

$$H(X_i \mid X_{i-1}, \ldots, X_{i-h}) \qquad\qquad\qquad\qquad\qquad\qquad (3.21)$$

$$= \sum_{X_{i-h}} \cdots \sum_{X_i} P(x_{i-h}, \ldots, x_i) \cdot \log \left( \frac{1}{P(x_i \mid x_{i-1}, \ldots, x_{i-h})} \right), \quad 1 \le h \le i$$

It thus seems fairly intuitive to define the information content of the source, or its entropy $H_{\infty}(X)$, as the information content of any symbol produced by the source, given that we have observed all previous symbols. Given a stationary information source $(\xi_n)_{n=0}^{\infty}$, its entropy $H_{\infty}(X)$ is then defined as

$$H_{\infty}(X) \triangleq \lim_{n \to \infty} H(X_n \mid X_{n-1}, \ldots, X_0)$$

To gain a deeper insight into the meaning and properties of $H_{\infty}(X)$, we shall prove the following theorem.

---

[1] We are using the notation $X_i$ to denote the alphabet pertaining to the $i$-th symbol in the message. Usually all $X_i$'s refer to the same set $X$; nevertheless, it is notationally convenient to keep them distinct.

**Theorem 3.4**

The conditional entropy $H(X_1 \mid X_0)$ satisfies the inequality

$$H(X_1 \mid X_0) \leq H(X_1) \tag{3.22}$$

▽

**Proof of Theorem 3.4**

To prove the theorem, consider the difference

$$H(X_1 \mid X_0) - H(X_1) = \sum_{X_0} \sum_{X_1} P(x_0, x_1) \log \left( \frac{P(x_1)}{P(x_1 \mid x_0)} \right)$$

and use in the RHS the inequality (3.9) so as to get

$$H(X_1 \mid X_0) - H(X_1) \leq \log e \sum_{X_0} \sum_{X_1} P(x_0, x_1) \left[ \frac{P(x_1)}{P(x_1 \mid x_0)} - 1 \right] = 0$$

**QED**

The relationship (3.22) becomes an equality when $\xi_1$ and $\xi_0$ are independent random variables. In this case, in fact, $P(x_1 \mid x_0) = P(x_1)$. A shrewd extension of Theorem 3.4 and the exploitation of the stationarity of the sequence $(\xi_n)_{n=0}^{\infty}$ allow one to write

$$\begin{aligned} H(X_n \mid X_{n-1}, \ldots, X_0) \; &\leq \; H(X_n \mid X_{n-1}, \ldots, X_1) \\ &= \; H(X_{n-1} \mid X_{n-2}, \ldots, X_0) \end{aligned} \tag{3.23}$$

So the sequence $H(X_n \mid X_{n-1}, \ldots, X_0)$, $n = 1, 2, \ldots$, is nonincreasing, and since the terms of the sequence are nonnegative, the limit $H_{\infty}(X)$ exists. Moreover, it satisfies the following inequality:

$$0 \leq H_{\infty}(X) \leq H(X) \tag{3.24}$$

where the RHS inequality becomes an equality when the symbols in the sequence are independent.

The entropy of an information source is difficult to compute in most cases. We will describe how this is achieved for a particular class of sources, the *stationary Markov sources*. A stationary Markov source is an information source whose output can be modeled as a finite-state, fully regular Markov chain (see Section 2.2.1). The properties of a stationary Markov source can be described as follows:

(i) At the beginning of each symbol interval, the source is in one of $q$ possible states $\{S_j\}_{j=1}^q$. During each symbol interval, the source changes state, say from $S_j$ to $S_k$, according to a *transition probability* $p_{jk}$ whose value is independent of the particular symbol interval.

(ii) The change of the state is accompanied by the emission of a symbol $x_i$, chosen from the source alphabet X, which depends only on the present state $S_j$ and the next state $S_k$.

(iii) The state $S_j$ and the emitted symbol $x_i$ uniquely determine the next state $S_k$.

In other words, the current symbol emitted by the source depends on the past symbols only through the state of the source. The stationary Markov model for information sources is a useful approximation in many physical situations. The interested reader is referred to Ash (1967, Chapter 6) for a detailed exposition of the subject. Here, we will illustrate the concept with an example.

**Example 3.5**   Let a stationary information source $(\xi_n)_{n=0}^\infty$ be characterized by the property $P(x_n \mid x_{n-1}, \ldots, x_0) = P(x_n \mid x_{n-1})$; that is, each symbol in the sequence depends only on the previous one. We assume that the alphabet X is formed by three symbols, say the letters $A, B$, and $C$. The probabilities $P(x_n \mid x_{n-1})$ are given as follows:

| $x_n/x_{n-1}$ | $A$ | $B$ | $C$ |
|---|---|---|---|
| $A$ | 0.2 | 0.4 | 0.4 |
| $B$ | 0.3 | 0.5 | 0.2 |
| $C$ | 0.6 | 0.1 | 0.3 |

This source can be represented by using the directed graph of Fig. 3.6, where each state represents the last emitted symbol and the transitions are identified by their probabilities and the presently emitted symbols. It can be verified that this source satisfies properties (i), (ii), and (iii), and thus it is a stationary Markov source.                    □

Let us compute now the source entropy $H_\infty(X)$. Defining the *entropy of the state* $S_j$ as

$$H(S_j) \triangleq \sum_{k=1}^{M_j} p_{jk} \log \left( \frac{1}{p_{jk}} \right)$$

where $M_j$ represents the number of symbols available at the state $S_j$, and denoting by $\{w_i\}_{i=1}^q$ the components of the stationary distribution vector (2.49), the following basic theorem can be proved (see, e.g., Ash, 1967, Chapter 6).

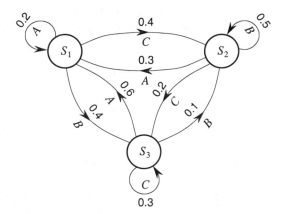

Figure 3.6: *Graph representation of a stationary Markov source with alphabet* $X = \{A, B, C\}$.

## Theorem 3.5

The entropy $H_\infty(X)$ of a stationary Markov source is given by

$$H_\infty(X) = \sum_{j=1}^{q} w_j H(S_j) \qquad (3.25)$$

$\triangledown$

**Example 3.6**    With reference to the Markov source of Example 3.5, computing the stationary distribution vector and applying (3.25) we obtain the entropy $H_\infty(X) = 1.441$.
$\square$

### Encoding stationary Markov sources

The inequalities (3.23) and (3.24) show that the average information content of a source emitting nonindependent symbols decreases as the message length increases. If we define the entropy of a block of successive symbols of a source message as

$$H(X^\nu) \triangleq H(X_0, X_1, \ldots, X_{\nu-1}) \qquad (3.26)$$

$$= \sum_{X_0} \sum_{X_1} \cdots \sum_{X_{\nu-1}} P(x_0, x_1, \ldots, x_{\nu-1}) \log\left(\frac{1}{P(x_0, x_1, \ldots, x_{\nu-1})}\right)$$

and apply the way of reasoning that led to (3.18), we find that a code can be devised for blocks of $\nu$ consecutive source symbols whose average number of bits per symbol satisfies the inequality

$$\frac{H(X^{\nu})}{\nu} \leq \frac{\bar{n}^{\nu}}{\nu} < \frac{H(X^{\nu})}{\nu} + \frac{1}{\nu}$$

Moreover, using (3.23) and some easy algebra, it can be proved that the sequence $H(X^{\nu})/\nu$, $\nu = 1, 2, \ldots$, is nonincreasing, and its limit is $H_{\infty}(X)$; that is,

$$\lim_{\nu \to \infty} \frac{H(X^{\nu})}{\nu} = H_{\infty}(X)$$

Thus we can see that increasing $\nu$ (the block length) makes the code more efficient at each step, and, as $\nu$ goes to infinity, the average length $\bar{n}_{\nu}/\nu$ approaches the source entropy $H_{\infty}(X)$ as close as desired; that is,

$$H_{\infty}(X) \leq \frac{\bar{n}^{\nu}}{\nu} < H_{\infty}(X) + O(\nu^{-1}), \quad \nu \to \infty$$

The price that must be paid for this increased efficiency lies in the complexity of the encoder, whose input alphabet size increases exponentially with $\nu$, and in the decoding delay. In fact, before obtaining the first symbol in every block, one must wait for the decoding of the entire block of $\nu$ symbols at the output of the source decoder.

Turning our attention to the particular case of Markov sources, we can apply the Huffman procedure to encode the symbols of the alphabet *for each state* $S_j$. This may require using a different set of code words for each state of the source. The performance of such a coding procedure is easily obtained. Using Theorem 3.3 and denoting by $\bar{n}(S_j)$ the average number of digits/symbol of the alphabet used in the state $S_j$, we obtain

$$H(S_j) \leq \bar{n}(S_j) < H(S_j) + 1$$

Thus, the average length of a code word is

$$\bar{n} = \sum_{j=1}^{q} w_j \bar{n}(S_j)$$

and satisfies the inequality

$$H_{\infty}(X) \leq \bar{n} < H_{\infty}(X) + 1$$

where (3.25) has been taken into account.

**Example 3.7**    Let us use the state-dependent Huffman procedure to encode the Markov source of Example 3.5 and compute its efficiency. Using the tree-encoding procedure for the three symbols that can be emitted from the source in any state, we obtain the following code

|   | $S_1$ | $S_2$ | $S_3$ |
|---|---|---|---|
| $A$ | 11 | 10 | 0 |
| $B$ | 10 | 0 | 11 |
| $C$ | 0 | 11 | 10 |

The average number of digits/symbol is

$$\bar{n} = \sum_{j=1}^{3} w_j \bar{n}(S_j) = 1.5054$$

Using the result of Example 3.6 we can compute the efficiency $\epsilon_\infty$ of the code, which is defined as

$$\epsilon_\infty \triangleq \frac{H_\infty(X)}{\bar{n}}$$

and is equal to 0.957 in this case.                                                        □

The source encoders we have described so far require the knowledge of the source statistics, something that is often completely, or at least partially, unavailable in practice. As a consequence, *universal* coding schemes have been deeply studied, which encode efficiently a broad class of sources in an adaptive fashion. The best known, and widely applied, scheme is the Lempel-Ziv algorithm. This important algorithm is not described here for space reasons; the interested reader is referred to the original paper (Ziv and Lempel, 1977), or, for a general and comprehensive treatment of source encoding algorithms, to the book by Bell *et al.* (1990).

### Information rate of a stationary source

In our definition of a discrete stationary source at the beginning of Section 3.2, *time* was not taken into account. To overcome this, we need to place the events forming a source message in correspondence with a sequence of points on the time axis. In particular, let us assume that the source emits the symbols forming a message at equally spaced time instants, and that the time period between two consecutive emissions is $T_s$. Thus we can define the *average information rate* of the source, $R_s$, as

$$R_s \triangleq \frac{H_\infty(X)}{T_s} \ \text{bit/s} \tag{3.27}$$

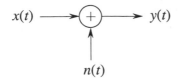

$$x(t) \longrightarrow \boxed{+} \longrightarrow y(t)$$

$$n(t)$$

Figure 3.7: *Model of the additive noise channel.*

As we shall see later, the appearance of time in our paradigm is strictly related to the bandwidth of the channel that will be used to convey the information.

## 3.3. Communication channels

The communication channel is the physical medium used to connect the source of information (in general, the transmitter) and its user (the receiver). As we saw in Chapter 1, according to the block diagram of Fig. 1.4, different kinds of channels can be specified, depending on the sections of the system we are observing. Between the output of the modulator and the input of the demodulator, for example, we have a *continuous channel*, which can be modeled in its simplest form by the additive channel shown in Fig. 3.7.

In it, $x(t)$ is the information signal emitted by the modulator, $n(t)$ represents the noise added to the signal on the channel, and $y(t)$ is the received signal. The channel is completely characterized by the probability distribution of the noise. If we now observe the block diagram of Fig. 1.4 between the channel encoder output and the decoder input, we have a *discrete channel*, which accepts symbols $x_i$ belonging to the input alphabet X of the channel encoder and returns symbols $y_j$ belonging to its own output alphabet Y. When X and Y contain the same symbols, $y_j$ is an estimate of the $j$th transmitted symbol $x_j$.

In the following, we shall see how to characterize a communication channel and how to compute the rate at which the information can be reliably transmitted through it.

### 3.3.1. Discrete memoryless channel

A discrete channel is characterized by an input alphabet $X = \{x_i\}_{i=1}^{N_X}$, an output alphabet $Y = \{y_j\}_{j=1}^{N_Y}$, and by a set of conditional probabilities

$$p_{ij}, \qquad i = 1, 2, \ldots, N_X, \qquad j = 1, 2, \ldots, N_Y$$

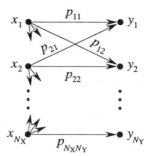

Figure 3.8: *Model of a discrete memoryless channel.*

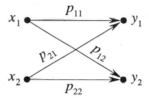

Figure 3.9: *The binary channel model.*

where $p_{ij} \triangleq P(y_j \mid x_i)$ represents the probability of receiving the symbol $y_j$ given that the symbol $x_i$ has been transmitted. We assume that the channel is *memoryless*, that is

$$P(y_1, \dots, y_n \mid x_1, \dots, x_n) = \prod_{i=1}^{n} P(y_j \mid x_i)$$

where $x_1, \dots, x_n$ and $y_1, \dots, y_n$ represent $n$ consecutive transmitted and received symbols, respectively. A graphical model of the discrete memoryless channel is shown in Fig. 3.8.

Each arrow represents a transition from one of the symbols of the input alphabet to one of the symbols of the output alphabet, that is, the transmission of a symbol belonging to X and the reception of a symbol belonging to Y. Each transition is labeled with its conditional probability. The sum of all the transition probabilities labeling the arrows stemming from the same input symbol is equal to 1.

**Example 3.8 The binary channel** It is a special case of the discrete channel when $N_X = N_Y = 2$, as depicted in Fig. 3.9. The average error probability $P(e)$ is defined as

$$P(e) \overset{\triangle}{=} P(x_1, y_2) + P(x_2, y_1) \qquad (3.28)$$

and can be computed as

$$P(e) = P(x_1)P(y_2 \mid x_1) + P(x_2)P(y_1 \mid x_2) = P(x_1)p_{12} + P(x_2)p_{21} \qquad (3.29)$$

If the two transition probabilities $p_{12}$ and $p_{21}$ are equal, say, to $p$, the channel is called *binary symmetric channel* (BSC) and (3.29) becomes

$$P(e) = p[P(x_1) + P(x_2)] = p$$

□

It is customary to arrange the conditional probabilities $\{p_{ij}\}$ into the channel matrix

$$\mathbf{P} \overset{\triangle}{=} \begin{bmatrix} p_{11} & p_{12} & \cdots & p_{1N_Y} \\ p_{21} & p_{22} & \cdots & p_{2N_Y} \\ \cdot & \cdots & \cdots & \cdot \\ \cdot & \cdots & \cdots & \cdot \\ & \cdots & \cdots & \\ p_{N_X1} & p_{N_X2} & \cdots & p_{N_X N_Y} \end{bmatrix} \qquad (3.30)$$

In (3.30) the numbers $p_{ij}$ represent probabilities, so they satisfy the inequality $0 \leq p_{ij} \leq 1$, and, obviously, the relationship

$$\sum_{j=1}^{N_Y} p_{ij} = 1, \qquad i = 1, 2, \ldots, N_X$$

That is, the sum of the elements in each row of $\mathbf{P}$ is 1. The average error probability is defined by extension of (3.28) for $N_Y = N_X = N$, as

$$\begin{aligned} P(e) \overset{\triangle}{=} \sum_{\substack{i=1 \\ }}^{N} \sum_{\substack{j=1 \\ j \neq i}}^{N} P(x_i, y_j) &= \sum_{i=1}^{N} P(x_i) \sum_{\substack{j=1 \\ j \neq i}}^{N} P(y_j \mid x_i) \\ &= \sum_{i=1}^{N} P(x_i) \sum_{\substack{j=1 \\ j \neq i}}^{N} p_{ij} \\ &= \sum_{i=1}^{N} P(x_i)(1 - p_{ii}) \qquad (3.31) \end{aligned}$$

whereas the probability of correctly receiving the symbol transmitted over the channel is given by

$$P(c) \overset{\triangle}{=} 1 - P(e) = \sum_{i=1}^{N} P(x_i)P(y_i \mid x_i) = \sum_{i=1}^{N} P(x_i)p_{ii} \qquad (3.32)$$

Particular forms of $\mathbf{P}$ lead to cases of interest.

**Example 3.9 The noiseless channel**   For this channel, we have $N_X = N_Y = N$, and the conditional probabilities $p_{ij}$ satisfy the relationship

$$p_{ij} = \begin{cases} 1, & i = j \\ 0, & i \neq j \end{cases} \tag{3.33}$$

In words, the symbols of the input alphabet are in a one-to-one correspondence with the symbols of the output alphabet. It can be easily verified that in this case $P(e) = 0$, as it is intuitive.    □

**Example 3.10 The useless channel**   For this channel, we have $N_X = N_Y = N$, and the output symbols are independent from the input ones, or

$$P(y_j \mid x_i) = P(y_j), \qquad \forall j, i \tag{3.34}$$

Regarding matrix $\mathbf{P}$, it is evident that (3.34) is verified if and only if $\mathbf{P}$ has identical rows.    □

A noiseless channel and a useless channel represent extremes of possible channel behavior. The output symbol of a noiseless channel uniquely specifies the input symbol, whereas a useless channel completely "scrambles" all input symbols, so that the received symbol gives no useful information to decide upon the transmitted one.

**Example 3.11 The symmetric channel**   For this channel, each row of the matrix $\mathbf{P}$ contains the same set of numbers $\{p_j\}_{j=1}^{N_Y}$, and each column contains the same set of numbers $\{q_i\}_{i=1}^{N_X}$. The following matrices provide examples of symmetric channels

$$\mathbf{P} = \begin{bmatrix} 1/2 & 1/3 & 1/6 \\ 1/6 & 1/2 & 1/3 \\ 1/3 & 1/6 & 1/2 \end{bmatrix}, \quad \mathbf{P} = \begin{bmatrix} 1/3 & 1/3 & 1/6 & 1/6 \\ 1/6 & 1/6 & 1/3 & 1/3 \end{bmatrix}$$

□

According to the input and output channel alphabets X and Y and to their probabilistic dependence specified by the channel matrix $\mathbf{P}$, we can define five entropies.

**(i)** The *input* entropy H(X),

$$H(X) \triangleq \sum_{i=1}^{N_X} P(x_i) \log \left( \frac{1}{P(x_i)} \right) \quad \text{bit/symbol} \tag{3.35}$$

which measures the average information content of the input alphabet.

**(ii)** The *output* entropy H(Y),

$$H(Y) \triangleq \sum_{j=1}^{N_Y} P(y_j) \log \left( \frac{1}{P(y_j)} \right) \quad \text{bit/symbol} \qquad (3.36)$$

which measures the average information content of the output alphabet.

**(iii)** The *joint* entropy H(X,Y),

$$H(X, Y) \triangleq \sum_{i=1}^{N_X} \sum_{j=1}^{N_Y} P(x_i, y_j) \log \left( \frac{1}{P(x_i, y_j)} \right) \quad \text{bit/(symbol pair)}$$
$$(3.37)$$

which measures the average information content of a pair of input and output symbols, or the average uncertainty of the communication system formed by the input alphabet, the channel, and the output alphabet as a whole.

**(iv)** The *conditional* entropy H(Y | X),

$$H(Y \mid X) \triangleq \sum_{i=1}^{N_X} \sum_{j=1}^{N_Y} P(x_i, y_j) \log \left( \frac{1}{P(y_j \mid x_i)} \right) \quad \text{bit/symbol} \qquad (3.38)$$

which measures the average information quantity needed to specify the output symbol $y$ when the input symbol $x$ is known.

**(v)** The *conditional* entropy H(X | Y),

$$H(X \mid Y) \triangleq \sum_{i=1}^{N_X} \sum_{j=1}^{N_Y} P(x_i, y_j) \log \left( \frac{1}{P(x_i \mid y_j)} \right) \quad \text{bit/symbol} \qquad (3.39)$$

which measures the average information quantity needed to specify the input symbol $x$ when the output (or received) symbol $y$ is known. This conditional entropy represents the average amount of information that has been lost on the channel, and it is called *equivocation*. The term equivocation seems appropriate if one realizes that for a noiseless channel $H(X \mid Y) = 0$ (the received symbol uniquely determines the transmitted one), whereas for a useless channel we find that $H(X \mid Y) = H(X)$. In this case the uncertainty about the transmitted symbol remains unaffected by the reception of an output symbol (all the information has been lost on the channel).

Using these definitions and (3.22), it can be verified that the following relationships between the entropies just defined hold true:

$$H(X, Y) = H(Y, X) = H(X) + H(Y \mid X) = H(Y) + H(X \mid Y) \qquad (3.40)$$

$$H(X \mid Y) \le H(X) \tag{3.41}$$

$$H(Y \mid X) \le H(Y) \tag{3.42}$$

The reader is invited to verify some of the following results by applying the definitions and properties (3.35) to (3.42) (see Problem 3.17).

**Example 3.9 (continued)**

$$H(X \mid Y) = 0 \tag{3.43}$$

$$H(Y \mid X) = 0 \tag{3.44}$$

and

$$H(X, Y) = H(X)H(Y) \tag{3.45}$$

□

**Example 3.10 (continued)**

$$H(X \mid Y) = H(X) \tag{3.46}$$

$$H(Y \mid X) = H(Y) \tag{3.47}$$

and

$$H(X, Y) = H(X) + H(Y) \tag{3.48}$$

Equation (3.46) says that all transmitted information is lost on the channel. □

**Example 3.11 (continued)** An important property of the symmetric channel is that $H(Y \mid X)$ is independent of the input probabilities $P(x_i)$ and depends only on the channel matrix **P**. To show this, let us write

$$H(Y \mid X) = \sum_{i=1}^{N_X} P(x_i)H(Y \mid x_i) \tag{3.49}$$

where

$$H(Y \mid x_i) \triangleq \sum_{j=1}^{N_Y} p_{ij} \log \left( \frac{1}{p_{ij}} \right) \tag{3.50}$$

According to the definition of symmetry, all the rows of **P** are permutations of the same set of numbers $\{p_j\}_{j=1}^{N_Y}$. Thus

$$H(Y \mid x_i) = \sum_{j=1}^{N_Y} p_j \log \left( \frac{1}{p_j} \right) \tag{3.51}$$

and inserting (3.51) into (3.49) gives

$$\text{H}(Y \mid X) = \sum_{i=1}^{N_X} P(x_i) \sum_{j=1}^{N_Y} p_j \log\left(\frac{1}{p_j}\right) = \sum_{j=1}^{N_Y} p_j \log\left(\frac{1}{p_j}\right) \tag{3.52}$$

which does not depend on the input probabilities $P(x_i)$, $i = 1, \ldots, N_X$.                    □

### 3.3.2.    Capacity of the discrete memoryless channel

We have seen that a part of the information H(X) that must be transmitted over the channel is lost because of the noise present in the channel itself. This part is measured by the channel equivocation H(X | Y). Thus, it seems natural to define the *average information flow* (also known as *mutual information* between X and Y) I(X; Y) through the channel as

$$\text{I}(X; Y) \stackrel{\triangle}{=} \text{H}(X) - \text{H}(X \mid Y) \text{ bit/symbol} \tag{3.53}$$

Using (3.40), the following alternative forms can be derived:

$$\text{I}(X; Y) = \text{H}(Y) - \text{H}(Y \mid X) = \text{H}(X) + \text{H}(Y) - \text{H}(X, Y) \tag{3.54}$$

Comparing (3.53) and the first equality of (3.54), it is apparent that $I(X; Y) = I(Y; X)$.

**Example 3.12**    Let us compute $I(X; Y)$ for the BSC with error probability $p = 0.1$ and equally likely input symbols. Because $P(x_1) = P(x_2) = 0.5$, the output symbols $y_1$ and $y_2$ are also equally likely. Thus we have

$$\text{H}(X) = \text{H}(Y) = 1 \text{ bit/symbol}$$

To compute $I(X; Y)$ using (3.54), we need the joint entropy $H(X, Y)$ given by (3.37). The joint probabilities $P(x_i, y_j)$ are easily computed as

$$P(x_1, y_1) = P(x_1)P(y_1 \mid x_1) = 0.5 \cdot 0.9 = 0.45$$
$$P(x_1, y_2) = 0.05$$
$$P(x_2, y_1) = 0.05$$
$$P(x_2, y_2) = 0.45$$

Thus we have
$$\text{H}(X, Y) = 1.469$$

and in conclusion

$$\text{I}(X; Y) = 1 + 1 - 1.469 = 0.531 \text{ bit/symbol}$$

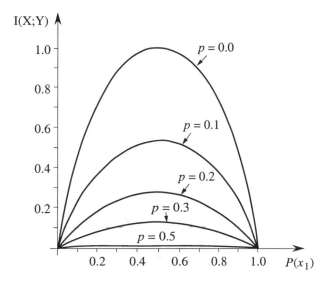

Figure 3.10: *Plot of the average information flow through a BSC as a function of the input probability* $P(x_1)$. *The error probability p of the channel is the parameter.*

The result shows that almost one half of the information is lost on the channel. How does this result compare with the intuitive remark that, on the average, only ten percent of the received bits are in error if $p = 0.1$ ?                    □

Let us consider again the BSC and see how $I(X; Y)$ depends on the probability distribution of the input symbols. Using the form

$$I(X; Y) = H(Y) - H(Y \mid X)$$

and computing $H(Y \mid X)$ using (3.49), we get

$$I(X; Y) = H(Y) - H(p) \qquad (3.55)$$

where $H(p)$ was defined in (3.6). In Fig. 3.10 the plot of $I(X; Y)$ versus $P(x_1)$ for different values of $p$ is shown. It can be observed that the maximum value of $I(X; Y)$, no matter what the value of $p$ is, is obtained for $P(x_1) = 0.5$, that is, when the input symbols are equally likely. Then, fixing $P(x_1) = 0.5$, we obtain a value for $I(X; Y)$ that depends only on the channel and represents the *maximum information flow* through a BSC. It is given by

$$\max_{P(x)} I(X; Y) = 1 - H(p) = 1 + p \log p + (1 - p) \log(1 - p) \qquad (3.56)$$

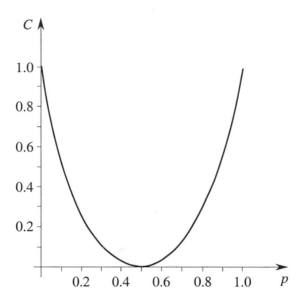

Figure 3.11: *Capacity of the BSC as a function of the error probability p.*

where $P(x)$ is the set of all possible probability distributions of the input symbols. The maximum value of $I(X; Y)$ is called *channel capacity* $C$ and is plotted in Fig. 3.11. The capacity is maximum when $p$ is equal to 0 or equal to 1, since both these situations lead to a noiseless channel. For $p = 0.5$, the capacity is zero, since the output symbols turn out to be independent from the input symbols, and no information can flow through the channel. Note that, due to the symmetry of the channel, $H(p) = H(1 - p)$.

Based on the result obtained for the BSC, we can now define similarly the capacity $C$ of a discrete memoryless channel as the maximum information $I(X; Y)$ that can be transmitted through the channel. Recalling the first equality of (3.54), we obtain

$$C \triangleq \max_{P(x)} I(X; Y) = \max_{P(x)} \sum_{i=1}^{N_X} \sum_{j=1}^{N_Y} P(x_i)p_{ij} \log \left( \frac{p_{ij}}{\sum_{k=1}^{N_X} P(x_k)p_{kj}} \right) \qquad (3.57)$$

The meaning of the channel capacity and its significance are not completely apparent so far. It will be proved later in this chapter that reliable transmission through the channel is not possible when the average number of bits per channel symbol is greater than the channel capacity. The analytical computation of the channel capacity is difficult in most cases. However, numerical algorithms are available, such as those due to Arimoto and Blahut (see Viterbi and Omura,

1979, Appendix 3C).  It becomes simpler in the particular cases of Examples 3.9 to 3.11.

**Example 3.9 (continued)**   Using the result (3.43), we can write

$$C = \max_{P(x)} \left[ H(X) - H(X \mid Y) \right] = \max_{P(x)} H(X) = \log N_X$$

Hence no information is lost on the channel.  As a matter of fact, the information flow through the channel equals the average quantity of information H(X) that is needed to specify an input symbol. □

**Example 3.10 (continued)**   Using (3.46), we get

$$C = H(X) - H(X \mid Y) = 0$$

Hence no information can flow through the channel. □

**Example 3.11 (continued)**   We have proved that the conditional entropy $H(Y \mid X)$ does not depend on the input probability distribution. Thus, the problem of maximizing $I(X; Y) = H(Y) - H(Y \mid X)$ reduces to the problem of maximizing the output entropy $H(Y)$. We know that $H(Y) \leq \log N_Y$, where the equal sign refers to the case of equally likely outputs, that is,

$$P(y_j) = \frac{1}{N_Y}, \qquad j = 1, \dots, N_Y$$

We prove that the output symbols are equally likely when the inputs are equally likely. In fact, if

$$P(x_i) = \frac{1}{N_X}, \qquad i = 1, \dots, N_X$$

we have

$$P(y_j) = \sum_{i=1}^{N_X} P(x_i, y_j) = \sum_{i=1}^{N_X} P(x_i)p_{ij} = \frac{1}{N_X} \sum_{i=1}^{N_X} p_{ij}$$

But the term $\sum_{i=1}^{N_X} p_{ij}$ is the sum of the entries of the $j$th column of the channel matrix **P**, and, by definition of symmetric channels, it does not depend on $j$. Thus all symbols $y \in Y$ have the same probability, and the capacity of a symmetric channel is given by

$$C = \log N_Y + \sum_{j=1}^{N_Y} p_j \log p_j \quad \text{bit/symbol} \tag{3.58}$$

□

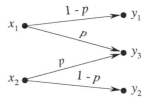

Figure 3.12: *The binary erasure channel model.*

**Example 3.13**   The capacity of the symmetric channel whose matrix is

$$P = \begin{bmatrix} 1/3 & 1/3 & 1/6 & 1/6 \\ 1/6 & 1/6 & 1/3 & 1/3 \end{bmatrix}$$

can be computed using (3.58) with $N_Y = 4$, and yields

$$C = 2 + 2(\frac{1}{3}\log\frac{1}{3} + \frac{1}{6}\log\frac{1}{6}) \simeq 0.082 \text{ bit/symbol}$$

□

**Example 3.14**   Consider a channel with $N_X = N_Y = N$ and probabilities $p_{ij} \in P$ given by

$$p_{ij} = \begin{cases} 1 - p, & i = j \\ p/(N-1), & i \neq j \end{cases}$$

The rows and columns of $P$ are in this case permutations of the $N$ numbers

$$\left(1 - p, \frac{p}{N-1}, \dots, \frac{p}{N-1}\right)$$

Thus the channel is symmetric, and its capacity $C$ is given by

$$C = \log N + (1 - p)\log(1 - p) + p\log\left(\frac{p}{N-1}\right) \tag{3.59}$$

The capacity of the BSC is obtained as a particular case of (3.59) with $N = 2$.   □

**Example 3.15   The binary erasure channel (BEC)** Consider the channel of Fig. 3.12. The outputs $y_1$ and $y_2$ correspond to the input symbols $x_1$ and $x_2$, whereas $y_3$ refers to an ambiguous output for which no decision about the transmitted symbol will be taken.

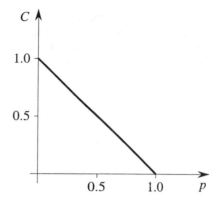

Figure 3.13: *The capacity of the binary erasure channel.*

This model represents a practical digital transmission system (see Problem 3.21). Let us compute its capacity. The channel matrix is the following:

$$\mathbf{P} = \begin{bmatrix} 1-p & 0 & p \\ 0 & 1-p & p \end{bmatrix}$$

It does not satisfy the symmetry conditions, so (3.58) cannot be applied. Starting from the first equality of (3.54), it is straightforward to show that $I(X; Y)$ is given by (3.55) also in this case. Therefore, the capacity $C$ is obtained for the input distribution that maximizes H(Y). Denoting $P(x_1)$ by $\alpha$ and computing H(Y), we get

$$
\begin{aligned}
H(Y) = \;&-p\log p - (1-p)\log(1-p) \\
&-\alpha(l-p)\log\alpha - (1-\alpha)(1-p)\log(1-\alpha)
\end{aligned}
\tag{3.60}
$$

which is seen to have a maximum for $\alpha = 0.5$. So the capacity is obtained for equally likely input symbols, and it is given by

$$C = 1 - p \tag{3.61}$$

Comparing the plot of $C$ shown in Fig. 3.13 with the capacity of the BSC of Fig. 3.11, we can see that erasing the received symbol when the information is not reliable can improve the information flow through the channel. This is true even in a more realistic situation when the probabilities $P(y_2 \mid x_1)$ and $P(y_1 \mid x_2)$ are different from zero (see Problem 3.21). $\qquad\qquad\square$

We have seen that the channel capacity of the BEC is achieved with equally likely input symbols, although the channel is not symmetric according to our definition. However, by inspection of Fig. 3.12, we can see that the structure of the channel exhibits a clear symmetry with respect to the inputs. To cope with situations like this, it is possible to generalize the definition of symmetric channels. The reader is referred to Gallager (1968, Chapter 4) for useful theorems on the computation of channel capacity in some particular cases of interest.

### 3.3.3. Equivocation and error probability

In Section 3.3.2 we have defined the average error probability $P(e)$ of a discrete channel and its equivocation $H(X \mid Y)$. Both can be used as measures of the channel quality, and certainly they are not independent quantities. In the following, we shall derive a relationship between them. Let us refer to a channel matrix $\mathbf{P}$, with $N_X = N_Y = N$. Recalling the definition of error probability already given in (3.31)

$$P(e) = \sum_{i=1}^{N} \sum_{\substack{j=1 \\ j \neq i}}^{N} P(x_i, y_j) \tag{3.62}$$

Let us define now the entropy $H(e)$ as

$$H(e) \triangleq -P(e) \log P(e) - [1 - P(e)] \log[1 - P(e)] \tag{3.63}$$

that is, consider $H(e)$ as the entropy of a binary alphabet with symbol probabilities $P(e)$ and $1 - P(e)$, which corresponds to the amount of information needed to specify if an error has occurred during the transmission on a channel with error probability $P(e)$. We can prove the following theorem.

### Theorem 3.6

**Fano's inequality**. Given a discrete memoryless channel whose input and output alphabets X and Y have the same number $N$ of symbols, and with error probability $P(e)$, the following inequality holds:

$$H(X \mid Y) \leq H(e) + P(e) \log(N - 1) \tag{3.64}$$

$\triangledown$

### Proof of Theorem 3.6

To prove the theorem, we use the definition (3.39) of the equivocation $H(X \mid Y)$ to write

$$H(X \mid Y) = \sum_{i=1}^{N} \sum_{\substack{j=1 \\ j \neq i}}^{N} P(x_i, y_j) \log \left( \frac{1}{P(x_i \mid y_j)} \right) + \sum_{i=1}^{N} P(x_i, y_i) \log \left( \frac{1}{P(x_i \mid y_i)} \right)$$

$$\tag{3.65}$$

and the definition (3.31) of $P(e)$ to get

$$H(X \mid Y) - P(e) \log(N - 1) - H(e) \tag{3.66}$$
$$= \sum_{\substack{i=1}}^{N} \sum_{\substack{j=1 \\ j \neq i}}^{N} P(x_i, y_j) \log \left( \frac{P(e)}{(N-1)P(x_i \mid y_j)} \right) + \sum_{i=1}^{N} P(x_i, y_i) \log \left( \frac{1 - P(e)}{P(x_i \mid y_i)} \right)$$

Applying now the inequality (3.9) to the RHS of (3.66), we obtain

$$H(X \mid Y) \quad - \quad P(e) \log(N - 1) - H(e)$$
$$\leq \quad \log e \left\{ \sum_{\substack{i=1}}^{N} \sum_{\substack{j=1 \\ j \neq i}}^{N} P(x_i, y_j) \left[ \frac{P(e)}{(N-1)P(x_i \mid y_j)} - 1 \right] \right.$$
$$\left. + \sum_{i=1}^{N} P(x_i, y_i) \log \left[ \frac{1 - P(e)}{P(x_i \mid y_i)} - 1 \right] \right\} \tag{3.67}$$
$$= \quad \log e \left\{ \frac{P(e)}{N-1} \sum_{\substack{i=1}}^{N} \sum_{\substack{j=1 \\ j \neq i}}^{N} P(y_j) - \sum_{\substack{i=1}}^{N} \sum_{\substack{j=1 \\ j \neq i}}^{N} P(x_i, y_j) \right.$$
$$\left. + [1 - P(e)] \sum_{i=1}^{N} P(y_i) - \sum_{i=1}^{N} P(x_i, y_i) \right\}$$
$$= \quad \log e \{ P(e) - P(e) + [1 - P(e)] - [1 - P(e)] \} = 0$$

**QED**

The inequality (3.64) can also be given an intuitive interpretation. Detecting whether or not an error has occurred, upon receiving a symbol $y \in Y$, removes an uncertainty equal to $H(e)$. If no error occurred, the remaining uncertainty about the transmitted symbol is zero. If an error occurred, an event that has probability $P(e)$, we still have to decide which of the remaining $N - 1$ symbols has been transmitted to make a correct decision. The uncertainty about this choice cannot exceed $\log(N - 1)$.

In Fig. 3.14 the function $H(e) + P(e) \log(N - 1)$ is plotted versus $P(e)$. Since $H(X \mid Y) = H(X) - I(X; Y)$, the theorem provides a lower bound to the error probability in terms of the excess of entropy of the input alphabet X with respect to the information flow through the channel. Considering now that $I(X; Y) \leq C$, (3.64) can be written as

$$H(X) - C \leq H(e) + P(e) \log(N - 1) \tag{3.68}$$

The curve $C + H(e) + P(e) \log(N-1) = 0$ is reported in Fig. 3.15. It can be seen by inspection that the region of the allowed pairs $(P(e), H(X))$ contains points

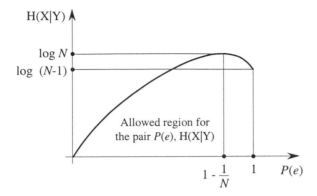

Figure 3.14: *Plot of the function H(e) + P(e) log (N-1) versus P(e).*

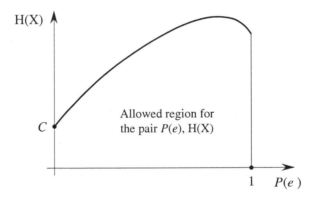

Figure 3.15: *Plot of the function C + H(e) + P(e) log (N-1) versus P(e).*

with abscissa $P(e) = 0$ only if $H(X) \leq C$. In other words, if *the entropy of the input alphabet exceeds the channel capacity, it is impossible to transmit the information through the channel with arbitrarily small error probability.* This result is a simplified version of the converse to the fundamental theorem of information theory, that will be discussed later on. If we identify the input alphabet of the channel with the output alphabet of the source encoder, the previously described situation refers to a communication system in which the symbols at the output of the source encoder are sent directly through the channel: no channel encoding is performed. We shall now include the channel encoder into our system and extend the previous results.

Let us consider the system shown in Fig. 3.16 in which the block labeled

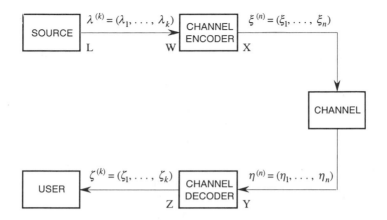

Figure 3.16: *Model of a discrete communication system.*

"source" represents the cascade of the source itself and the source encoder. Suppose for simplicity that the output of the source is a sequence of binary symbols emitted every $T_s$ seconds. The channel encoder is a binary *block encoder* (to be described in detail in Chapter 10). It transforms blocks of $k$ consecutive source digits (*words*) into blocks of $n$ digits belonging to the input channel alphabet X. An encoding rate $R_c$ can be defined as

$$R_c \triangleq \frac{k}{n} \tag{3.69}$$

Since $n$ symbols must be transmitted over the channel every $k \cdot T_s$ seconds, the channel must be used every $T_c = R_c T_s$ seconds. Denoting with W the set of $2^k$ messages at the input of the channel encoder and with Z the set of $2^k$ messages at the output of the channel decoder, we can apply the Fano inequality to these two sets, obtaining

$$\mathrm{H}(\mathrm{W} \mid \mathrm{Z}) \leq H_w(e) + P_w(e) \log(2^k - 1) \tag{3.70}$$

where $H_w(e)$ is the entropy of a binary alphabet with symbol probabilities $P_w(e)$ and $1 - P_w(e)$, and where the subscript $w$ in $P_w(e)$ denotes "word" and $P_w(e)$ represents the average probability of decoding a word erroneously, that is, of incorrectly recognizing the transmitted code word. Moreover, since $\mathrm{H}(\mathrm{W} \mid \mathrm{Z}) = \mathrm{H}(\mathrm{W}) - \mathrm{I}(\mathrm{W}; \mathrm{Z})$, and taking into account that the following inequality holds true (*data-processing theorem*; see Viterbi and Omura, 1979, Chapter 1 for a proof; roughly speaking, it states that it is impossible to increase the information content of a message by processing it in some way),

$$\mathrm{I}(\mathrm{W}; \mathrm{Z}) \leq \mathrm{I}(\mathrm{X}; \mathrm{Y})$$

we get

$$H(W \mid Z) \geq H(W) - I(X; Y) \tag{3.71}$$

Since the transmission of each block of $k$ bits involves using $n$ times the channel, we can also write

$$I(X; Y) \leq nC \tag{3.72}$$

so that, inserting (3.72) into (3.71) and the result into (3.70), we obtain

$$H(W) - nC \leq H(e) + P_w(e) \log(2^k - 1) \tag{3.73}$$

The inequality (3.73) is the *converse to the coding theorem*. Since the alphabet W is obtained by grouping $k$ consecutive symbols at the output of the source, the entropy H(W) is given by

$$H(W) = kH_\infty(L) \tag{3.74}$$

where $H_\infty(L)$ is the entropy of the source. Thus, inequality (3.73) states that the probability of erroneously decoding a sequence of $k$ source symbols cannot be made arbitrarily small when the encoding rate $R_c$ is greater than the ratio $C/H_\infty(L)$. A lower bound to the error probability can be derived from (3.73) and (3.74) as follows:

$$P_w(e) \geq \frac{kH_\infty(L) - nC - H(e)}{\log(2^k - 1)} > \frac{kH_\infty(L) - nC - 1}{k} = H_\infty(L) - \frac{C}{R_c} - \frac{1}{k}$$

Now, letting $k$ and $n$ go to infinity and keeping constant their ratio $R_c$ yields

$$P_w(e) > H_\infty(L) - \frac{C}{R_c} \tag{3.75}$$

Previous considerations refer to the word error probability. We want to extend them to the bit error probability, that is, the probability that a source binary symbol will be delivered erroneously to the user. With reference to the notations of Fig. 3.16, the bit error probability is defined as $P_b(e) \triangleq P[\zeta_i \neq \lambda_i]$, *i.e.*, the probability that a single source digit is in error after channel decoding. Assuming for simplicity that the binary source symbols are independent, identically distributed, equally likely RVs, we can apply Theorem 3.6 to obtain

$$I(\lambda_i; \zeta_i) = H(\lambda_i) - H(\lambda_i \mid \zeta_i) = 1 - H(\lambda_i \mid \zeta_i) \geq 1 - H_b(e) \tag{3.76}$$

where $H_b(e)$ is the entropy of a binary alphabet with symbol probabilities $P_b(e)$ and $1 - P_b(e)$. Moreover, using (3.76) it can be proved that

$$I(W \mid Z) \geq \sum_{i=1}^{k} I(\lambda_i; \zeta_i) \geq k[1 - H_b(e)] \tag{3.77}$$

Finally, observing that, owing to the data processing theorem, the following chain of inequalities holds :

$$I(W;Z) \leq I(X;Y) \leq nC \tag{3.78}$$

Combining it with (3.77), we obtain

$$R_c = \frac{k}{n} \leq \frac{C}{1 - H_b(e)} \tag{3.79}$$

The bound (3.79) is a decreasing function of the bit error probability through the denominator of the RHS. This is not surprising, as it merely means that the more reliably we want to communicate, the slower we must communicate.

From (3.79) we can also derive a lower bound to the bit error probability, analogous to (3.75), in the form

$$P_b(e) \geq H_b^{-1}(1 - \frac{C}{R_c}) \tag{3.80}$$

which states once again that we cannot communicate reliably at rates above the channel capacity.

### Channel coding theorem

We have seen in the previous subsection that there exists a lower bound to the error probability, different from zero, when the encoding rate $R_c$ is greater than the channel capacity $C$. This is the "negative" result known as the converse to the coding theorem. When the encoding rate $R_c$ is smaller than $C$ the system behavior is dictated by the *channel coding theorem*, which will be stated here without proof. It was proved in 1948 by C. E. Shannon, and the interested reader is referred to his original paper (Shannon, 1948) or to one of the many books available, for example, Gallager (1968, Chapter 5).

### Theorem 3.7

Given a binary information source, with entropy $H_\infty(L)$ bits/symbol and a discrete, memoryless channel with capacity $C$, there exists a code of rate $R_c = k/n$ for which the word error probability is bounded by

$$P_w(e) < 2^{-nE(R)}, \qquad R = R_c H_\infty(L) \tag{3.81}$$

where $E(R)$ is a convex $\cup$, decreasing, nonnegative function of $R$ for $0 \leq R \leq C$.
$\triangledown$

A typical behavior of the function $E(R)$ is shown in Fig. 3.17. Based on (3.81), we can undertake three different actions to improve the performance of a digital communication system.

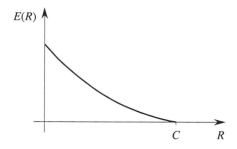

Figure 3.17: *Typical behavior of the function $E(R)$.*

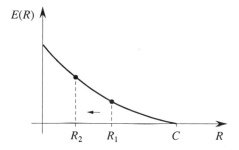

Figure 3.18: *Increasing the value of the function $E(R)$ by decreasing $R_c = k/n$.*

**(i)** Decrease $R$ by decreasing $R_c = k/n$. This means increasing the *redundancy* of the code and, for a given source emission rate, using the channel more often. In other words, we need a channel with a larger bandwidth (see Chapter 5 for a thorough discussion of this point). What happens is shown in Fig. 3.18. We move from $R_1$ to $R_2$, so that $E(R)$ increases and the RHS in (3.81) decreases.

**(ii)** Increase the channel capacity $C$ by increasing the signal-to-noise ratio over the channel. This situation is depicted in Fig. 3.19. The operating point moves from the previous function $E_1(R)$ to the new function $E_2(R)$, thus decreasing the RHS of (3.81).

**(iii)** Increase $n$, while keeping the ratio $R_c = k/n$ constant. This third approach does not require any intervention on the bandwidth and/or signal-to-noise ratio of the channel. It allows one to improve the performance of the communication system by simply increasing the length of the code words, and thus at the expense of a greater complexity of the encoder-decoder pair and of a longer delay in reconstructing the decoded sequence.

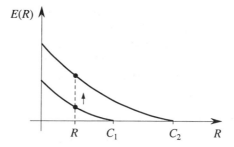

Figure 3.19: *Increasing the value of the function $E(R)$ by increasing the capacity $C$ of the channel.*

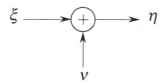

Figure 3.20: *Block diagram of a time-discrete additive Gaussian channel.*

While (i) and (ii) were well-known remedies to counteract the disturbances in a communication system, the use of the third way is one of the major achievements of Shannon's theory.

### 3.3.4.   Additive Gaussian channel

In this section we shall consider a communication channel that is *amplitude-continuous* and *time-continuous*. Starting from the discrete channels considered so far, this situation will be approached in two steps. First, we shall examine the time-discrete, amplitude-continuous Gaussian channel shown in Fig. 3.20. Every $T_s$ seconds the source transmits a symbol chosen from a possibly uncountable alphabet. The channel disturbance has the form of an unwanted noise added to the signal to be transmitted. The assumption that the noise is Gaussian, which is highly desirable from a mathematical point of view, also turns out to be reasonable in a wide variety of physical settings. After the analysis of this simplified case has been completed, we shall extend the results to the time-continuous channel, in which the transmission of information will be allowed to be continuous in time. The derivation of the main result will be done in a heuristic manner, avoiding all mathematical subtleties. The unsatisfied reader is invited to quench his/her thirst for rigor in Ash (1967, Chapter 3). For the channel of Fig. 3.20, we have $\eta = \xi + \nu$, where $\xi$ and $\eta$ are RVs representing the input and output symbols and taking values $x \in X$, $y \in Y$ with the probability density functions (pdf's) $f_\xi(x)$ and $f_\eta(y)$, respectively, and $\nu$ is a zero-mean Gaussian RV representing the noise. Denoting the noise variance with $\sigma_\nu^2$, we have $\nu \sim \mathcal{N}(0, \sigma_\nu^2)$, and the output $\eta$, given $\xi = x$, is $\mathcal{N}(x, \sigma_\nu^2)$. Our goal is the evaluation of the channel capacity; thus we need to extend our definition of the information measure to the continuous case.

**Measure of information in the continuous case**

Let $\xi$ be a continuous RV taking values in X with pdf $f_\xi(x)$. We define its entropy H(X) as

$$H(X) \triangleq - \int_{-\infty}^{\infty} f_\xi(x) \log f_\xi(x) \, dx \qquad (3.82)$$

Although (3.82) seems a straightforward extension of the definition (3.5) given in the discrete case, some differences arise. The main one consists in the fact that H(X) defined in (3.82) may be arbitrarily large, positive, or negative (see Problem 3.30). In the same way as for (3.82), we can define for two random variables $\xi$ and $\eta$ having a joint probability density function $f_{\xi\eta}(x, y)$, the joint entropy H(X,Y), and the conditional entropies H(X | Y) and H(Y | X) as

$$H(X, Y) \triangleq - \int_{-\infty}^{\infty} \int_{-\infty}^{\infty} f_{\xi\eta}(x, y) \log f_{\xi\eta}(x, y) \, dx \, dy \qquad (3.83)$$

$$H(X \mid Y) \triangleq - \int_{-\infty}^{\infty} \int_{-\infty}^{\infty} f_{\xi\eta}(x, y) \log f_{\xi|\eta}(x \mid y) \, dx \, dy \qquad (3.84)$$

$$H(Y \mid X) \triangleq - \int_{-\infty}^{\infty} \int_{-\infty}^{\infty} f_{\xi\eta}(x, y) \log f_{\eta|\xi}(y \mid x) \, dx \, dy \qquad (3.85)$$

Assuming now that both H(X) and H(Y) are finite, the following relationships hold true, as in the discrete case:

$$H(X, Y) \le H(X) + H(Y)$$

$$H(X, Y) = H(X) + H(Y \mid X) = H(Y) + H(X \mid Y)$$

$$H(Y \mid X) \le H(Y), \qquad H(X \mid Y) \le H(X)$$

In all the preceding relationships, inequalities become equalities if $\xi$ and $\eta$ are statistically independent.

In the discrete case, we have proved (Theorem 3.1) that the entropy is maximized by equally likely symbols. In the continuous case, the following theorem holds.

**Theorem 3.8**

Let $\xi$ be a continuous RV with pdf $f_\xi(x)$. If $\xi$ has finite variance $\sigma_\xi^2$, then H(X) exists and satisfies the inequality

$$H(X) \leq \frac{1}{2}\log(2\pi e \sigma_\xi^2) \tag{3.86}$$

with equality if and only if $\xi \sim \mathcal{N}(\mu, \sigma_\nu^2)$ . $\triangledown$

For the proof, see Problem 3.31.

**Capacity of the discrete-time Gaussian channel**

Suppose that the continuous RV's $\xi$ and $\eta$ represent the input and output symbols for the channel of Fig. 3.20. As we did for the discrete channel, we define the average information flow through the channel as

$$I(X; Y) \overset{\triangle}{=} H(X) - H(X \mid Y) = I(Y; X) = H(Y) - H(Y \mid X) \tag{3.87}$$

and the channel capacity $C$ as

$$C \overset{\triangle}{=} \max_{f_\xi(x)} I(X; Y) \tag{3.88}$$

We know that, given $\xi = x$, $\eta \sim \mathcal{N}(x, \sigma_\nu^2)$. Thus

$$H(Y \mid X) = 2\log(2\pi e \sigma_\xi^2)$$

and

$$C = \max_{f_\xi(x)} H(Y) = \log(2\pi e \sigma_\xi^2) \tag{3.89}$$

By Theorem 3.8, H(Y) is maximum when $\eta$ is Gaussian, and this in turn happens if and only if $\xi$ is Gaussian. Therefore, the capacity $C$ is attained for a Gaussian input $\xi$, say $\xi \sim \mathcal{N}(0, \sigma_\xi^2)$, and its value is given by

$$C = \frac{1}{2}\log\left[2\pi e(\sigma_\xi^2 + \sigma_\nu^2)\right] - \frac{1}{2}\log\left(2\pi e \sigma_\nu^2\right) = \frac{1}{2}\log\left(1 + \frac{\sigma_\xi^2}{\sigma_\nu^2}\right) \tag{3.90}$$

**Capacity of the bandlimited Gaussian channel**

We have treated up to this point only time-discrete channels. On the other hand, many channels of practical interest are time-continuous, in the sense that their inputs and outputs are time-continuous functions. To extend the result (3.90) to this new situation, let us suppose that the channel input signals are strictly band-limited to the frequency interval $(-B, B)$ Hz. Then, by the sampling theorem (see Section 2.5.1), we can represent each signal using (at least) $2B$ samples per second. Each sample has a variance $\sigma_\xi^2$ equal to the signal power $\mathcal{P}$. Moreover, the noise is assumed to be a white Gaussian random process, with two-sided power spectral density $N_0/2$, sampled every $1/2B$ seconds. Hence its power is $\sigma_\nu^2 = (N_0/2) \cdot (2B) = N_0 B$. As a consequence, the result (3.90) becomes

$$C = \frac{1}{2} \log \left( 1 + \frac{\mathcal{P}}{N_0 B} \right) \quad \text{bit/(channel use)} \tag{3.91}$$

Recalling the definition (3.27) of the source rate $R_s$ in bits/s, we can define the *energy per information bit* of a transmitter with average power $\mathcal{P}$

$$\mathcal{E}_b \triangleq \frac{\mathcal{P}}{R_s} \quad \text{J/bit} \tag{3.92}$$

Substituting (3.92) into (3.91) yields

$$C = \frac{1}{2} \log \left( 1 + \frac{\mathcal{E}_b R_s}{N_0 B} \right) \quad \text{bit/(channel use)} \tag{3.93}$$

If the system includes a channel encoder with rate $R_c$, we shall present to the channel a bit flow with rate $R_s/R_c$. We will see in Chapter 6 that the minimum value of $B$ required to transmit reliably this rate of information (the so called *Nyquist bandwidth*) is $B = \frac{R_s}{2R_c}$, so that (3.93) becomes

$$C = \frac{1}{2} \log \left( 1 + 2R_c \mathcal{E}_b/N_0 \right) \quad \text{bit/(channel use)} \tag{3.94}$$

Finally, by means of (3.79), we obtain the inequality

$$R_c \leq \frac{\log \left( 1 + 2R_c \mathcal{E}_b/N_0 \right)}{2[1 - H_b(e)]} \tag{3.95}$$

Equation (3.95), with equality substituted for the inequality, yields the relationship between the bit error probability and the signal-to-noise ratio per information bit, for a given code rate. It is plotted in Figure 3.21 for various values of $R_c$. For a given code rate, only the region above the respective curve is admissible for the pair $(P_b(e), \mathcal{E}_b/N_0)$.

Letting $P_b(e) = 0$, i.e., $H_b(e) = 0$ in (3.91), and solving with respect to $\mathcal{E}_b/N_0$, yields

$$\frac{\mathcal{E}_b}{N_0} \geq \frac{2^{2R_c} - 1}{2R_c} \tag{3.96}$$

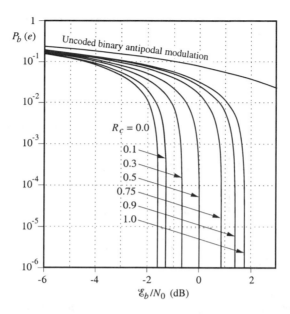

Figure 3.21: *Plot of the points $(\mathcal{E}_b/N_0, P_b(e))$ satisfying (3.95) with equality for various values of the code rate $R_c$. The curve pertaining to the uncoded binary antipodal modulation (see Chapter 4) is also shown.*

Equation (3.96), with equality substituted for the inequality, yields the relationship between the signal-to-noise ratio per bit which is required to obtain a reliable transmission as a function of the code rate. It is plotted as curve (A) in Figure 3.22, where we also plotted two other curves. The first of these, curve (B), stems from the capacity of the Gaussian channel constrained to a binary input. Such a capacity, which is denoted as $C_{2,c}$ to make explicit the continuous unconstrained output, has been derived in Chapter 4 of the book by McEliece (1977). It is given by

$$C_{2,c} = \frac{2R_c\mathcal{E}_b}{N_0} - \frac{1}{\sqrt{2\pi}} \int_{-\infty}^{\infty} e^{-y^2/2} \log \cosh \left( \frac{2R_c\mathcal{E}_b}{N_0} + y\sqrt{\frac{2R_c\mathcal{E}_b}{N_0}} \right) dy \quad (3.97)$$

For small signal-to-noise ratios, the capacity $C_{2,c}$ approaches the capacity $C$ of the unconstrained Gaussian channel, which proves that binary-input quantization does not hurt for low signal-to-noise ratios. The third curve, curve (C), refers to the capacity of the binary-input, binary-output channel obtained from a Gaussian channel through a double binary quantization of both input and output. Its capacity, denoted with obvious notation as $C_{2,2}$, coincides with the capacity of

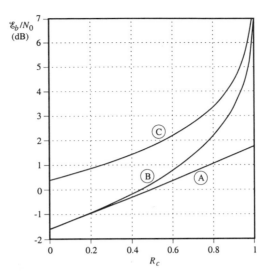

Figure 3.22: *Plot of the points $(R_c, \mathcal{E}_b/N_0)$ satisfying (3.96) (curve (A), unconstrained Gaussian channel), and derived from (3.97) (curve (B), binary-input Gaussian channel), and from (3.56) (curve (C), binary-input, binary-output Gaussian channel) versus the code rate $R_c$.*

the binary symmetric channel given in (3.56)

$$C_{2,2} = 1 + p \log p + (1 - p) \log(1 - p) \tag{3.98}$$

with the following expression for the transition probability $p$, obtained assuming the most efficient modulation scheme (binary antipodal, see Chapter 4) over the additive Gaussian noise channel and a code with rate $R_c$ (see Chapter 10):

$$p = \frac{1}{2} \operatorname{erfc}\left( \sqrt{\frac{R_c \mathcal{E}_b}{N_0}} \right) \tag{3.99}$$

Comparing curve (A) with curve (C), we notice a degradation of about 2 dB for low signal-to-noise ratios; this is the price to be paid for using binary output quantization.

To pass from the capacity expressed in bit per channel use to its expression in bit per second, we simply need the consideration that we need $2B$ samples per second to represent a signal with bandwidth $B$ Hz, so that we are using $2B$ times per second a discrete-time Gaussian channel with capacity $C$ given by (3.91). Thus, finally, we obtain the capacity $C_s$ in bit/s of a bandlimited white Gaussian channel as

$$C_s = B \log \left( 1 + \frac{\mathcal{P}}{N_0 B} \right) \quad \text{bit/s} \tag{3.100}$$

The result (3.100) is of fundamental importance, since it gives the upper limit that can be reached when information is to be reliably transmitted over Gaussian channels. The designer of a digital communication system tries to choose the system parameters in such a way as to approach the capacity $C_s$ as closely as possible with a preassigned error probability. In Chapter 5 the modulation schemes most widely used in digital transmission systems will be compared, among themselves and with respect to the limit given by $C_s$ in (3.100).

**Example 3.16** Most of today's telephone connections use a transmission medium that is almost entirely digital, with a fairly short analog local loop connecting the subscriber at one end to this medium via a digital central office. Such high quality telephone lines have a signal-to-noise ratio of about 37 dB, *i.e.*, $\mathcal{P}/N_0 B = 10^{3.7}$, and a bandwidth of about 3500 Hz.

Computing the capacity $C_s$ of such a channel, considered in a first approximation as an additive Gaussian channel, we obtain, through (3.100)

$$C_s = 3500 \log(1 + 10^{3.7}) = 43,020 \quad \text{bit/s}$$

Let us compare this theoretical capacity with the data rate achievable by today's off-the-shelf modulator-demodulators (modems). The last standard approved by the International Telecommunications Union is contained in the Recommendation V.34. It is a modem using multilevel amplitude and phase modulation combined with channel coding and shaping to provide roughly a 5 dB gain. The highest data rate provided by V.34 modems is 33,600 bit/s and represents a dramatic improvement over the previous V.32 bis standard (1990) of 14,400 bit/s. The gap between the achieved data rate of 33,600 bit/s and the channel capacity (43,020 bit/s) is due to several reasons, such as the nonGaussianness of the channel, other disturbances, implementation losses, etc.

The highly sophisticated V.34 modem incorporates most of the state-of-the-art theoretical achievements of the last years in the field, like *trellis-coded modulation* (see Chapter 12), *signal shaping* through shell mapping (see Chapter 5), *adaptive equalization* and *precoding* (see Chapter 8).

Presently, the V.34 modems are being superseded in the market by PCM (pulse-coded modulation) modems that advertise speeds around 50 kbit/s. These modems are not based on the classic additive Gaussian channel model, but rather exploit the fact that in many applications a digital connection to the network can be made (for further information, see Humblet and Troulis, 1996). □

From (3.93), the capacity in bits per second can also be expressed as a function of the source rate $R_s$ and of the energy per information bit $\mathcal{E}_b$ as

$$C_s = B \log \left( 1 + \frac{\mathcal{E}_b R_s}{N_0 B} \right) \tag{3.101}$$

We can state and prove the following theorem.

**Theorem 3.9**

To transmit information on a additive white Gaussian noise channel with two-sided noise power spectral density $N_0/2$ W/Hz, any digital communication system requires an energy per bit satisfying the following inequality:

$$\mathcal{E}_b \geq \frac{N_0}{\log e} = 0.693 N_0 \tag{3.102}$$

$\bigtriangledown$

**Proof of Theorem 3.9**

The theorem concerns unlimited-bandwidth channels. Thus, taking the limit of the RHS of (3.101) for $B \to \infty$, the limit yields (see Problem 3.32)

$$C_s = \frac{\mathcal{E}_b R_s}{N_0} \log e$$

Considering now that $R_s \leq C_s$, we obtain (3.102). **QED**

Inequality (3.102) expressed in decibels provides an absolute lower limit to the signal-to-noise ratio of every communication system to operate reliably (that is, with $P_b(e) \to 0$)

$$\frac{\mathcal{E}_b}{N_0} \geq -1.59 \text{ dB} \tag{3.103}$$

In Chapters 4 and 5, we shall see that the most efficient binary modulation scheme (2-PAM or 2-PSK) requires a much greater signal-to-noise ratio to operate at low error probabilities, like 10.5 dB at $P(e) = 10^{-6}$. The range between -1.59 and 10.5 dB is the vast region where coding can be used to improve the power efficiency of digital communication systems (see also Chapter 10, 11, and 12).

Theorem 3.9 assumed no bandwidth limitation. When the channel is band-limited, instead, we can define the spectral efficiency $r$ of the modulation scheme

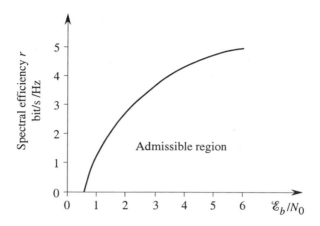

Figure 3.23: *Plot of the points $(\mathcal{E}_b/N_0, r)$ satisfying Theorem 3.10 with equality.*

employed (see Chapter 5, which also contains a detailed discussion on how to define the *bandwidth*)

$$r \triangleq \frac{R_s}{B} \text{ bit/sec/Hz} \qquad (3.104)$$

and prove the following theorem providing a lower bound to the signal-to-noise ratio as a function of $r$.

**Theorem 3.10**

To transmit information reliably on an additive white Gaussian noise channel with spectral efficiency $r$ any digital communication system requires a signal-to-noise ratio satisfying the following inequality

$$\frac{\mathcal{E}_b}{N_0} \geq \frac{2^r - 1}{r} \qquad (3.105)$$

$\triangledown$

**Proof of Theorem 3.10**

Starting from (3.101), include $R_s \leq C_s$, and solve for $\mathcal{E}_b/N_0$. **QED**

In Fig. 3.23 we report the function $r$ that satisfies (3.105) with equality versus $\mathcal{E}_b/N_0$. Any communication system can be described by a point lying below the curve, and for any point below the curve a communication system can be designed whose bit error probability is as small as desired. The challenge of system designers is precisely to approach as close as possible this curve.

**Example 3.17**   Consider a source with rate $R_s$ whose output enters a block channel encoder with rate $R_c = 0.5$. To compute the lower bound (3.105) we need the spectral efficiency $r$. The data rate at the output of the channel encoder (and at the input of the channel) is equal to $R_s/R_c$ digits/s, $R_c$ being the encoding rate. Using the *Nyquist* bandwidth (see Chapter 7) $B = R_s/(2R_c)$, we get

$$r = \frac{R_s}{B} = \frac{R_s}{\frac{R_s}{2R_c}} = 2R_c = 1 \ .$$

Substituting $r = 1$ in (3.105) yields

$$\frac{\mathcal{E}_b}{N_0} \geq 0 \ \text{dB} \ .$$

□

## 3.4.   Bibliographical notes

This chapter has no presumptions of originality, as its material simply summarizes results discussed in greater depth in the many excellent textbooks available on the subject of information theory. All of them stem from the pioneering work of C. E. Shannon that was published in his fundamental paper of 1948 (Shannon, 1948); see also the collection of all papers by Shannon edited by Sloane and Wyner (1993).

As students first, as researchers and teachers later, the authors have been especially familiar with the classical books by Fano (1961), Ash (1967), Gallager (1968), and Cover and Thomas (1991). We are indebted to these books for the development of topics in this chapter. In the following, we give some suggestions to the reader wishing to go deeper into the subject.

Berger (1971) wrote an advanced book dealing wholly with the source coding theorem, its generalizations, and its practical applications. Chapter 3 of McEliece (1977) is devoted to a modern and original presentation of discrete memoryless sources and their rate-distortion functions. Source coding and recent advances in rate-distortion theory are also treated extensively in Viterbi and Omura (1979, Chapters 7 and 8). A comprehensive and highly informative book on source encoding is the one by Bell *et al.* (1990). In this chapter, we have described the Huffman source coding algorithm. Although it has found ubiquitous applications, the Huffman coding procedure has some drawbacks: the source statistics must be known, and, because of the code word's variable length, there is a mismatch between source and channel rates that requires buffering at the transmitter. Moreover, the algorithm is not designed to take advantage of long sequences of the same characters. An alternative source coding technique remedies some of the deficiencies of the Huffman coding algorithm. It is the Lempel-Ziv algorithm, described in Ziv and Lempel (1977). The Lempel-Ziv algorithm is intrinsically adaptive to the source and can efficiently encode frequently-occurring groups of source symbols.

For the subject of discrete channels with memory (not covered in this chapter), see Ash (1967, Chapter 7), Gallager (1968, Chapter 4), and Viterbi and Omura (1979, Chapter 2). The continuous-time Gaussian channel, briefly mentioned here, is treated in detail in Fano (1961, Chapter 5), Gallager (1968, Chapter 8), and Ash (1967, Chapter 8).

## 3.5.  Problems

*Problems marked with an asterisk should be solved with the aid of a computer.*

**3.1** For the third source alphabet of Example 3.1 show by direct differentiation that the entropy has a maximum for $p = 0.5$.

**3.2** For the source of Problem 3.1, consider sequences of two outputs as a single output of an extended source with alphabet

$$\mathrm{X}^2 = \{(x_1, x_1), (x_1, x_2), (x_2, x_1), (x_2, x_2)\}$$

Under the hypothesis that consecutive outputs from the source are statistically independent, show directly that $\mathrm{H}(\mathrm{X}^2) = 2 \cdot \mathrm{H}(\mathrm{X})$. Generalize the result to the case of an extended source $X^\nu$.

**3.3** A source emits a sequence of independent symbols from an alphabet X consisting of five symbols $x_1, \ldots, x_5$, with probabilities $\frac{1}{4}, \frac{1}{8}, \frac{1}{8}, \frac{3}{16}, \frac{5}{16}$, respectively. Find the entropy of the source alphabet.

**3.4** A black and white TV picture consists of 525 lines of picture information. Assume that each line consists of 525 picture elements and that each element can have 256 different brightness levels. Pictures are repeated at the rate of 30 per second. What is the average rate of information conveyed by a TV set assuming independence?

**3.5** Consider two discrete sources with alphabets $X_1$ and $X_2$, having $M_1$ and $M_2$ symbols, respectively, and probability distributions $\{p_i\}_{i=1}^{M_1}$ and $\{q_i\}_{i=1}^{M_2}$. From these sources a new source is formed, with $M_1 + M_2$ symbols: the first $M_1$ symbols have probability distribution $\{\lambda p_i\}_{i=1}^{M_1}$, while the last $M_2$ symbols have probability distribution $\{(1-\lambda)q_i\}_{i=1}^{M_2}$, $0 \leq \lambda \leq 1$. Find the entropy of the new source and the value of $\lambda$ that maximizes it.

**3.6** Given a source of alphabet X with $M$ symbols and probability distribution $\{p_i\}_{i=1}^{M}$, group the last $m$ symbols to form a new source X' with $M' = M - m + 1$ symbols and probability distribution $\{q_i\}_{i=1}^{M'}$ such that

$$q_i = p_i, \quad i = 1, 2, \ldots, M' - 1$$

$$q_{M'} = p_{M'} + p_{M'+1} + \ldots + p_M$$

Show that the entropy of the new source satisfies the inequality $H(X') \leq H(X)$.

**3.7** A binary source with alphabet X and symbols $\{x_1, x_2\}$ has probabilities $P\{x_1\} = 0.1$ and $P\{x_2\} = 0.9$. Construct the Huffman codes corresponding to the sources $X^\nu$, $(\nu = 1, 2, 3, 4)$, obtained by grouping the outputs of X in words of length $\nu$. For every value of $\nu$, find the efficiency of the encoding scheme.

**3.8** Given the source with alphabet X having symbols $\{x_i\}_{i=1}^{8}$, and probability distribution

$$\left\{ \tfrac{1}{2}, \tfrac{1}{4}, \tfrac{1}{8}, \tfrac{1}{16}, \tfrac{1}{64}, \tfrac{1}{64}, \tfrac{1}{64}, \tfrac{1}{64} \right\}$$

find three binary codes satisfying the prefix constraint such that:

(a) The average number of digits/symbol $\bar{n}$ is minimized;

(b) The maximum number of digits in every code word is minimized;

(c) The average number of digits/symbol $\bar{n}$ is minimized subject to the constraint that the maximum length of the code words is 4.

**3.9** Consider a source alphabet with $N$ symbols, and two probability distributions $\{p_1, p_2, p_3, \ldots, p_N\}$ and $\{p_1', p_2', p_3, \ldots, p_N\}$ where

$$\left. \begin{array}{l} p_1' = p_1 - \Delta p \\ p_2' = p_2 + \Delta p \end{array} \right\}, \quad \Delta p > 0, \quad p_1 > p_2$$

Show that the entropy of the alphabet is greater for the second probability distribution provided that $p_1' > p_2'$. *Hint:* Applying the inequality $\ln x \leq x - 1$, with $x = q_i/p_i$, show that the following inequality holds true:

$$- \sum_i p_i \log p_i \leq - \sum_i p_i \log q_i$$

where $\{p_i\}_i^N$ and $\{q_i\}_i^N$ are probability distributions.

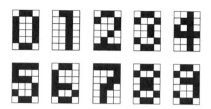

Figure 3.24: *See Problem 3.10*

**3.10** In Fig. 3.24, the representations of symbols $0, 1, \ldots, 9$ are shown. A two-dimensional field with $4 \times 6$ positions (white or black) is used. Find the redundancy of this code under the hypothesis that the symbols are equally likely.

**3.11** A binary source with alphabet X having symbols $\{x_1, x_2\}$ has a probability distribution $\{p_1 = 0.005, p_2 = 0.995\}$. The outputs from the source are grouped in blocks of 100 each, and a code word is associated only with those blocks containing no more than three symbols $x_1$. Assuming that the symbols from the source are statistically independent

    (a) Find the minimum code word length for a fixed-length code;

    (b) Find the probability that a block is not encoded.

**3.12** A source has an alphabet of four symbols. The probabilities of the symbols and two possible sets of binary code words for the source are as follows:

| Symbol | $P(x_i)$ | Code I | Code II |
|--------|----------|--------|---------|
| $x_1$  | 0.4      | 1      | 1       |
| $x_2$  | 0.3      | 01     | 10      |
| $x_3$  | 0.2      | 001    | 100     |
| $x_4$  | 0.1      | 000    | 1000    |

For each code, answer the following questions:

    (a) Does the code satisfy the prefix condition?

    (b) Is the code uniquely decipherable? .

    (c) What is the mutual information provided about the source symbol by the specification of the first digit of the code word?

**3.13** For the source of Example 3.5, compute the entropy and check the result of Example 3.6.

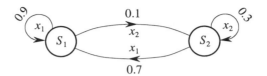

Figure 3.25: *See Problem 3.14*

**3.14** It is desired to re-encode more efficiently the output of the Markov source illustrated in Figure 3.25.

(a) Find the entropy $H_\infty(X)$ of the source.

(b) Construct the Markov diagram for pairs of symbols.

(c) Construct the optimum binary code for all pairs of symbols and evaluate the resulting code efficiency.

**3.15** Consider a stationary source with a ternary alphabet $X = \{x_1, x_2, x_3\}$, for which the probability of each symbol depends only on the preceding symbol. The probabilities of the possible ordered symbol pairs are given in the following table:

| $x_{n-1}/x_n$ | $x_1$ | $x_2$ | $x_3$ |
|---|---|---|---|
| $x_1$ | 0.20 | 0.05 | 0.15 |
| $x_2$ | 0.15 | 0.05 | 0.10 |
| $x_3$ | 0.05 | 0.20 | 0.05 |

Determine the optimum binary code words and the resulting code efficiency for the following encoding schemes:

- The sequence is divided into successive pairs of symbols and each pair is represented by a code word.

- A Markov model for the source is devised, and a state-dependent code is used to encode each symbol.

**3.16** The state diagram of a Markov source is given in Fig. 3.26, and the symbol probabilities for each state are as follows:

|  | $S_1$ | $S_2$ | $S_3$ | $S_4$ |
|---|---|---|---|---|
| $x_1$ | 0.7 | 0.3 | 0.5 | 0.3 |
| $x_2$ | 0.125 | 0.5 | 0.1 | 0.5 |
| $x_3$ | 0.075 | 0 | 0.1 | 0 |
| $x_4$ | 0.1 | 0.2 | 0.3 | 0.2 |

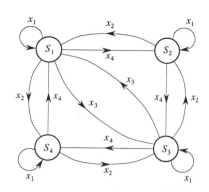

Figure 3.26: *See Problem 3.16*

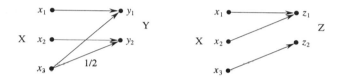

Figure 3.27: *See Problem 3.18*

(a) Evaluate the source entropy $H_\infty(X)$.

(b) Construct an optimum set of binary code words for each state and evaluate the resulting code efficiency.

**3.17** Prove the results (3.43) to (3.48).

**3.18** The symbols from a source with alphabet $X = \{x_1, x_2, x_3\}$ and probability distribution $\{\frac{1}{4}, \frac{1}{4}, \frac{1}{2}\}$ are sent independently through the channels shown in Fig. 3.27.

Evaluate $H(X), H(Y), H(Z), H(X \mid Y)$, and $H(X \mid Z)$.

**3.19** A channel with input alphabet X, output alphabet Y, and channel matrix $\mathbf{P}_1$ is cascaded with a channel with input alphabet Y, output alphabet Z, and channel matrix $\mathbf{P}_2$ (Fig. 3.28).

Under the hypothesis of independent transmissions over the two channels, find the channel matrix $\mathbf{P}$ of the equivalent channel with input X and output Z.

**3.20** Prove that the cascade of $n$ BSC's is still a BSC. Under the hypothesis that the $n$ channels are equal, evaluate the channel matrix and the error probability of the equivalent channel and let $n \to \infty$.

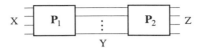

Figure 3.28: *See Problem 3.19*

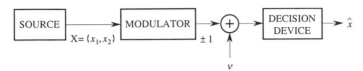

Figure 3.29: *See Problem 3.21*

**3.21** Consider the transmission system shown in Fig. 3.29. A modulator associates with the two symbols $x_1, x_2$ emitted by the two source voltages of $+1$ and $-1$ V, respectively. A Gaussian noise represented by the RV $\nu$ is added to the modulator output, with $\nu \sim \mathcal{N}(0, 1)$. The decision device can operate in two ways:

(a) It compares the received voltage with the threshold zero and decides that $x_1$ has been transmitted when the threshold is exceeded or that $x_2$ has been transmitted when the threshold is not exceeded.

(b) It compares the received voltage with two thresholds, $+\delta$ and $-\delta$. When $+\delta$ is exceeded, it decides for $x_1$; when $-\delta$ is not exceeded, it decides for $x_2$; if the voltage lies between $-\delta$ and $+\delta$, it does not decide and erases the symbol.

Compute and plot the capacity of the two discrete channels resulting from the application of decision schemes (a) and (b). Plot the second one as a function of $\delta$.

**3.22** For the two situations depicted in Fig. 3.30, show that $I(X; Y) = I(X; Z)$. Verify also that in the first case $H(Y) > H(Z)$, whereas in the second case $H(Y) < H(Z)$.

**3.23** Compute the capacity $C$ of the channel shown in Fig. 3.31.

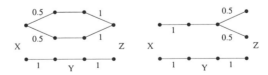

Figure 3.30: *See Problem 3.22*

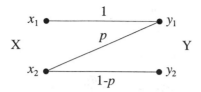

Figure 3.31: *See Problem 3.23*

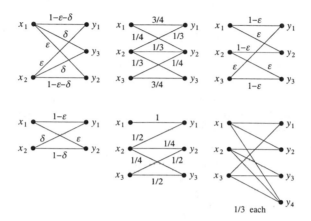

Figure 3.32: *See Problem 3.27*

Cascading $n$ such channels, compute the capacity $C_n$ of the equivalent channel and let $n \to \infty$.

**3.24** Consider the cascade of two BSC channels. Prove that the capacity of the equivalent channel cannot exceed the capacity of each single channel.

**3.25** Consider the BEC channel with channel matrix given in (3.60). Compute and plot $H(X \mid Y)$ as a function of $P(x_1)$.

**3.26** (*) Write a computer program implementing the Arimoto-Blahut algorithm (see Viterbi and Omura, 1979, Appendix 3C) to find the capacity of a discrete channel.

**3.27** Find the capacity and an optimizing input probability assignment for each of the discrete channels shown in Fig. 3.32 (Gallager, 1968).

**3.28** (*) Compute the capacity of the channels of Problem 3.27 using the program developed in Problem 3.26 and compare the results with those obtained analytically in Problem 3.27.

**3.29** Let $\xi$ be a continuous RV uniformly distributed in the interval $X = (0, b)$. Compute its entropy $H(X)$.

**3.30** Show that the entropy $H(X)$ of the RV $\xi$ having probability density function:

$$f_\xi(x) = \begin{cases} 1/[x(\log x)^2], & x \geq e \\ 0, & x < e \end{cases}$$

is infinite.

**3.31** Prove Theorem 3.8. *Hint*: Prove first that the following inequality holds:

$$-\int_{-\infty}^{\infty} f_\xi(x) \log[f_\xi(x)]dx \leq -\int_{-\infty}^{\infty} f_\xi(x) \log[f_\eta(x)] \ dx$$

where $f_\xi(x)$ and $f_\eta(x)$ are arbitrary probability density functions. Then apply it by considering an arbitrary probability density function $f_\xi(x)$ with finite variance $\sigma_\xi^2$, and a Gaussian density function

$$f_\eta(x) = \frac{1}{\sqrt{2\pi}\sigma_\xi} \exp\left[-\frac{(x-\mu)^2}{2\sigma_\xi^2}\right]$$

**3.32** Compute the upper limit of the capacity (3.100) of a bandlimited Gaussian channel as $B \to \infty$.

**3.33** Prove the inequality (3.105) of Theorem 3.10.

# 4

# Waveform transmission over the Gaussian channel

In this chapter we introduce digital modulation as a way of delivering to the user digital information generated by the source. A physical communication channel is available, which may consist of a pair of wires, a coaxial cable, an optical fiber, a radio link, or a combination of these. Therefore, it is necessary to convert the sequence of source symbols into waveforms that match the physical properties of the transmission medium. What is called *digital modulation* (or *digital signaling*) is indeed the mapping of digital sequences into a set of waveforms.

The *digital modulator* is the functional device that achieves such mapping. In its simplest possible form the mapping is one-to-one between binary digits and a set of two waveforms. This type of transmission is called *binary modulation*, or *binary signaling*. More generally, the modulator may map into waveforms blocks of $h$ binary digits at a time, and hence need a set of $M = 2^h$ different waveforms. This type of transmission is called *M-ary* (or *multilevel*) modulation.

All physical channels corrupt the information-bearing waveforms with different impairments such as distortions, interferences, and various types of noise. At the receiving side, the corrupted waveforms are processed by the *digital demodulator*. Its task is inverse to the modulator, since it estimates which particular waveform was actually transmitted by the modulator, and hence recovers from it an estimate of the source information. In order for the source information to be delivered to the user as reliably as possible, the design of the demodulator must account for the impairments introduced by the channel. This chapter deals with only one of these impairments, namely, Gaussian noise added to the signal.

A central role in this chapter is played by the evaluation of the performance of the modulator-demodulator pair. We are interested in assessing how well a

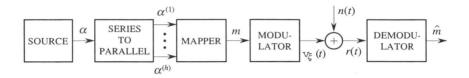

Figure 4.1: *The channel model assumed in this chapter.*

given modulation scheme does its job of carrying information through a channel in a reliable way by making the most efficient use of the basic resources available, namely, power, bandwidth, and complexity. While these concepts will be made clear later on, we mention from the onset of this chapter that a good digital modulation/demodulation scheme should deliver information to the end user with a low error probability, low bandwidth occupancy for a given transmission rate, and low power expenditure. In addition, the complexity, and hence the cost, of the modulator/demodulator pair (in short, of the *modem*) should be made as low as possible.

## 4.1.  Introduction

The channel model considered in this chapter is reproduced in Fig. 4.1. A source produces a sequence of independent, identically distributed binary symbols with an information rate $R_s = 1/T_s$ bit/s (the subscript $s$ stands for source). These binary digits are grouped in blocks of length $h$, that shall be referred to as *source symbols*. The $M = 2^h$ symbols occur with probabilities $p_i$, $i = 1, \ldots, M$, assigned to the set of symbols $\{m_i\}_{i=1}^M$. For simplicity's sake, we assume that all the symbols are equally likely, so that $p_i = 1/M$ for all $i$. However, the reader should be warned that this assumption may not hold in some instances, as for example during the transmission of a preamble message intended to establish a communication, or the like.

   In the simplest form of a digital modulation scheme, the modulator maps each symbol onto a set of $M$ waveforms, that are transmitted sequentially over the channel. The resulting composite signal generated by the modulator is written $v_\xi(t)$, where the subscript $\xi$ denotes the entire sequence of source symbols. Since the transmission of one symbol requires a time $T = hT_s$, the rate $1/T$ at which the signals are transmitted over the channel is called the *signaling rate*.

This is given by

$$\frac{1}{T} = \frac{1}{hT_s} = \frac{R_s}{\log_2 M}$$

and is measured in ($M$-ary) symbols per second.

A great variety of digital modulation schemes (or mapping rules) is available, and an effort will be made here to present them in a unifying conceptual frame. We start with some simple examples.

**Example 4.1**   The modulator uses $M$ signals $\{s_i(t)\}_{i=1}^M$ with duration $T = hT_s$. Fig. 4.2 shows an example of transmitted waveforms for $M = 2$, $M = 4$, and $M = 8$. For $M = 2$ we have

$$s_1(t) = +A, \qquad s_2(t) = -A, \qquad 0 \le t < T_s$$

For $M = 4$ we have

$$s_1(t) = -3A, \quad s_2(t) = -A, \quad s_3(t) = +A, \quad s_4(t) = +3A, \qquad 0 \le t < 2T_s$$

and for $M = 8$ we have

$$s_i(t) = (2i - 9)A, \quad i = 1, \ldots, 8, \qquad 0 \le t < 3T_s$$

We observe that in general each signal has duration $hT_s$, i.e., for a given source rate its time span is proportional to $h$. ☐

A basic point may be raised from the sheer consideration of this simple example. That is, how should we pick one among these various modulation schemes? Or, does any of them perform better than the others? As we shall see, we cannot say in general that there is an optimum choice of $M$: rather, this choice is the result of a tradeoff between complexity and power and bandwidth efficiency.

In this example it turns out that, when the source rate $1/T_s$ is kept constant, the waveforms with $M = 8$ require less bandwidth than those with $M = 2$, since the pulse duration in the former case is longer. In general, increasing $M$ reduces the bandwidth occupancy, and hence increases the bandwidth efficiency of the modulation scheme. On the other hand, if the average signal power spent for transmission is kept constant as $M$ varies, we see that in the presence of noise it will be a harder task for a demodulator to distinguish between signals when $M$ is large, because their amplitude levels are closer. Making signal levels as separated in the case $M = 8$ as they are for $M = 2$, and hence keeping the error probability at about the same level, would require increasing the average signal power. Thus, increasing $M$ decreases the power efficiency of this modulation scheme.

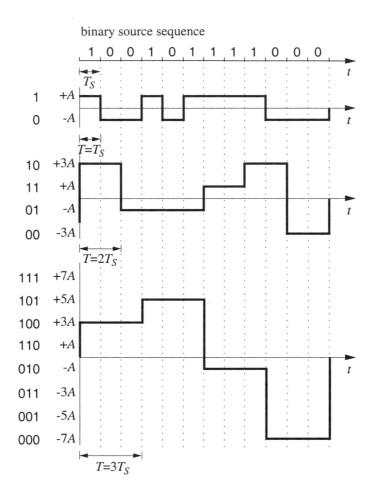

Figure 4.2: *Three examples of memoryless modulation schemes. From top to bottom: Binary modulation, quaternary modulation, and octonary modulation.*

The tradeoff involved in the selection of $M$ should be evident now. If this modulation scheme is to be used for a transmission system where bandwidth is at a premium (called a "bandwidth-limited" system) we should use a larger value of $M$. If power is at a premium (a "power-limited" system) we should use a lower value of $M$. Other considerations—among them, implementation complexity—occur as well, but the above describes the basic tradeoff the system designer is faced with.

The simple example above describes one among the various modulation schemes available to the designer, one which is suitable for a baseband system. For radio systems, a favorite scheme consists of associating messages to $M$ possible phases of a sinusoid, called the *carrier*. This modulation scheme, called $M$-ary *phase-shift keying* (PSK), will be described at length in the following. Here we limit ourselves to observe that it transmits over the channel a constant-envelope signal, and hence is a good choice for applications in which power amplifiers are operated at or near saturation for best power efficiency, and hence are nonlinear because of AM/AM and AM/PM conversions (see Section 2.4.2). This nonlinearity would distort any signal with a time-varying envelope.

We hasten to observe here that the modulation model considered so far, i.e., one in which there is a memoryless, one-to-one correspondence among source symbols and modulator signals, although it is the most important, is by no mean the only one. To motivate what we call "modulations with memory," we describe here a simple special case of *continuous-phase modulation* (CPM), that will be discussed in more depth in Chapter 6.

### 4.1.1.  A simple modulation scheme with memory

Consider 2-PSK, i.e., a binary modulation scheme in which binary source symbols "0" and "1" are associated with two phases, 0 and $\pi$, of a sinusoid. The signal sent through the channel is shown in Fig. 4.3. Its power density spectrum can be easily calculated by using the techniques described in Chapter 2. Here it suffices to say that the resulting spectral occupancy may be just too wide for certain applications (e.g., some mobile-radio systems). Now, Fourier theory suggests to us a reason for this: it is known that the presence of discontinuities in a signal widens its spectrum, and the PSK signal exhibits jumps in its phase at each occurrence of a pair 01 or 10 in the source sequence.

Based on this observation, we expect that a narrower spectrum will be obtained if these discontinuities are smoothed in some way. This is obtained for example as shown in Fig. 4.4. In this modulation scheme the transmission of a source "0" makes the phase of a sinusoid increase *linearly* by $\pi/2$, while the transmission of a "1" makes it decrease by the same amount. The resulting phase trajectory is now continuous, and hence we expect it to yield a narrower spectrum than 2-PSK. Notice that we can do even more than linear phase transitions: smoother phase trajectories can be obtained by shaping them in other forms, as we shall see in our general presentation of CPM (Chapter 6).

At this point we can add the observation that this modulation scheme is not memoryless anymore: in fact, rather than associating two different waveforms to the binary source symbols, at time $kT$ (say) we transmit a waveform whose

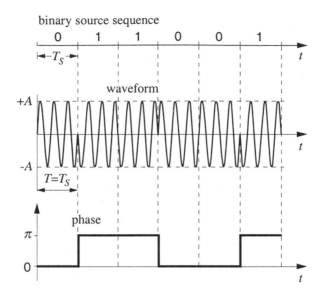

Figure 4.3: *Evolution of the transmitted signal and of the phase in binary PSK.*

shape depends not only on the source symbol emitted at $kT$, but also on the phase reached by the signal at the same instant. We say that the modulator *has memory*. As we shall see, the presence of this memory has a strong impact on the structure of the optimum demodulator.

### 4.1.2.  Coherent vs. incoherent demodulation

We are now ready to start our discussion of the demodulators' structures and their performance, but first we need to introduce a further classification. As mentioned before, throughout this chapter we shall be considering that the signal received at the output of the channel is corrupted by AWGN $n(t)$ with power spectral density $N_0/2$. Two different cases will be considered here, giving rise to different families of demodulators. In the first, we assume that the receiver has complete knowledge of the set of possible transmitted signals. We call this receiver a *coherent receiver*. We can write the received signal in the form

$$r(t) = v_\xi(t) + n(t) \tag{4.1}$$

In the second case we consider a typical situation arising in bandpass communication systems. In these we may not be able to assume that the signals used by

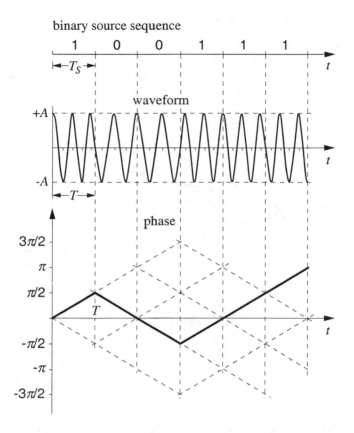

Figure 4.4: *A simple modulation scheme with memory. The transmitted signal is phase-continuous.*

the modulator are fully known, because the exact phase of the carrier sinusoid used by the modulator is unknown. We write the received signal in the form

$$r(t) = v_\xi(t; \theta) + n(t) \tag{4.2}$$

where $\theta$ is a random variable modeling the uncertainty on the phase angle of the transmitted signal. A demodulator that operates without any knowledge of $\theta$ will be called *incoherent*.

### 4.1.3.  Symbol error probability

The purpose of the demodulator is to process the received signal $r(t)$ to produce an estimate $\widehat{\boldsymbol{\xi}}$ of the transmitted sequence $\boldsymbol{\xi}$, and consequently an estimate $\widehat{m}_k$ of each transmitted symbol. The performance of the modulator/demodulator pair will be evaluated through the *symbol error probability*

$$P(e) \triangleq P\{\widehat{\xi}_k \neq \xi_k\} \tag{4.3}$$

We are interested in the demodulator that achieves the minimum value of $P(e)$. We call it *optimum* in this sense. For simplicity, we shall assume throughout the chapter that the transmitted messages are equally likely, so that, as shown in Section 2.6, minimum error probability is achieved with a maximum likelihood (ML) receiver. We shall deal with three different cases:

1. Memoryless modulators and coherent receivers. Each generated waveform has a duration strictly limited to the time interval $T$, and the modulator is memoryless.

2. Memoryless modulators and incoherent receivers. Each generated waveform still has a duration limited to the interval $T$ and the modulator is again memoryless, but the receiver has an uncertainty due to a random phase angle as in (4.2).

3. Modulators with memory. These will be considered in Chapter 6.

## 4.2.  Memoryless modulation and coherent demodulation

The memoryless nature of the modulation process implies that the waveforms available at the modulator are strictly limited to the time interval $T$. While we shall restrict ourselves to this situation throughout this chapter, we hasten to say that the theory developed here is also valid in some instances when those waveforms have a longer duration. In fact, this theory also applies, *mutatis mutandis*, whenever the quantities obtained at the channel output by processing the received signal in a single symbol interval are sufficient statistics for the decision on that symbol. This situation occurs, for instance, with the infinite-duration raised-cosine pulses of Chapter 7.

The demodulator outputs one realization of the random process

$$v_\xi(t) = \sum_{k=0}^{K-1} s(t - kT; \xi_k), \qquad 0 \leq t < KT \tag{4.4}$$

depending on the $K$-symbol sequence $\boldsymbol{\xi}$. The log-likelihood ratio for $\boldsymbol{\xi}$, based on the observation of the noisy signal $r(t) = v_\xi(t) + n(t)$, is given by (see Section 2.6.1)

$$\lambda_\xi = \frac{2}{N_0} \int_0^{KT} r(t) v_\xi(t)\, dt - \frac{1}{N_0} \int_0^{KT} v_\xi^2(t)\, dt \qquad (4.5)$$

The optimum demodulator chooses the sequence $\widehat{\boldsymbol{\xi}}$ that maximizes $\lambda_\xi$ in (4.5); that is,

$$\widehat{\boldsymbol{\xi}}: \quad \lambda_{\widehat{\xi}} = \max_\xi \lambda_\xi \qquad (4.6)$$

In the following, with a slight abuse of notation we shall denote by $\lambda_\xi$ the quantity (4.5) multiplied by the inessential constant $N_0/2$. Thus, insertion of (4.4) into (4.5) shows that the ML sequence must maximize the quantity

$$\lambda_\xi \triangleq \int_0^{KT} r(t) \sum_{k=0}^{K-1} s(t - kT;\ \xi_k)\, dt - \frac{1}{2} \int_0^{KT} \left[\sum_{k=0}^{K-1} s(t - kT;\ \xi_k)\right]^2 dt \quad (4.7)$$

By recalling that $s(t;\ \xi_k)$ has duration $T$, (4.7) can be rewritten in the form

$$\lambda_\xi = \sum_{k=0}^{K-1} \lambda_{\xi_k} \qquad (4.8)$$

where

$$\lambda_{\xi_k} \triangleq \int_{kT}^{(k+1)T} r(t) s(t - kT;\ \xi_k)\, dt - \frac{1}{2} \int_{kT}^{(k+1)T} s^2(t;\ \xi_k)\, dt \qquad (4.9)$$

form a sequence of independent random variables under our assumptions that $\xi_k$ are independent and the noise is white.

From (4.8) we can conclude that the ML sequence $\widehat{\boldsymbol{\xi}}$ is obtained as an ML symbol-by-symbol decision, i.e., in each time interval $T$ the quantities $\lambda_{\xi_k}$ in (4.9) are maximized separately. In fact, under our assumptions the maximum value of the sum (4.8) corresponds to the sum of the maximum values of its components (see Appendix F). Considering this fact, without any loss of generality from now on we shall examine modulator and demodulator by restricting ourselves to the time interval $(0, T)$ corresponding to $k = 0$.

The quantities in (4.9) are a set of sufficient statistics of the received signal $r(t)$. This, as discussed in Section 2.6, means that all we need to know about the received signal $r(t)$ to allow an ML decision is contained in these quantities. Since the RV $\xi_0$ can take on $M$ different values, each signal $s(t; \xi_0)$ comes from a set $\{s_i(t)\}_{i=1}^M$ of different waveforms of duration $T$. Therefore, the RV $\lambda_{\xi_0}$ becomes, for $i = 1, 2, \ldots, M$,

$$\lambda_i = \int_0^T r(t) s_i(t)\, dt + c_i \qquad (4.10)$$

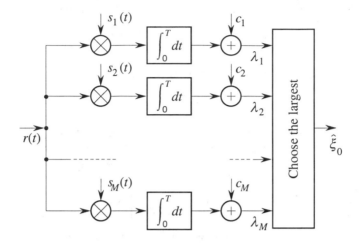

Figure 4.5: *Correlator implementation of the optimum coherent receiver for memoryless modulation and transmission over the AWGN channel.*

where

$$c_i \triangleq -\frac{1}{2}\int_0^T s_i^2(t)\, dt = -\frac{1}{2}\mathcal{E}_i \qquad (4.11)$$

and $\mathcal{E}_i$ denotes the energy of the $i$th signal.

The block diagram of the ML demodulator is shown in Fig. 4.5, in the form usually referred to as a *correlation demodulator*. For simplicity it refers to the demodulation of the first symbol $\xi_0$.

An equivalent method to get the quantities (4.10) is to replace the bank of correlators with a bank of $M$ filters, each matched to one of the signals $\{s_i(t)\}_{i=1}^M$. The filter matched to $s_i(t)$ has an impulse response $h_i(t) = s_i(T - t)$, so that the output of this filter at $t = T$, when the input is $r(t)$, gives exactly the integral in (4.10). The block diagram of this *matched-filter demodulator* is shown in Fig. 4.6. This version of the optimum demodulator shows that a bank of $M$ matched filters supplies the sufficient statistics for our decision problem.

A simpler version of the optimum demodulator can be obtained by representing the signals $\{s_i(t)\}_{i=1}^M$ in the orthonormal basis $\{\psi_j(t)\}_{j=1}^N$, $N \leq M$, by using the Gram-Schmidt procedure (see Section 2.5). We get

$$s_i(t) = \sum_{j=1}^N s_{ij}\psi_j(t), \quad i = 1, 2, \ldots, M, \quad 0 \leq t < T \qquad (4.12)$$

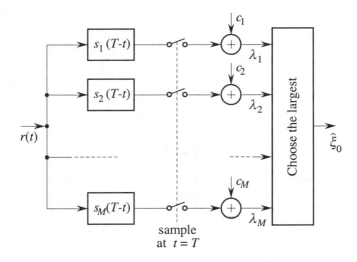

Figure 4.6: *Matched-filter implementation of the optimum coherent receiver for memoryless modulation and transmission over the AWGN channel.*

By inserting (4.12) in (4.10), after some algebra we get

$$\lambda_i = \sum_{j=1}^{N} s_{ij} r_j - \frac{1}{2} \sum_{j=1}^{N} s_{ij}^2, \quad i = 1, 2, \ldots, M \qquad (4.13)$$

where

$$r_j \triangleq \int_0^T r(t) \psi_j(t) \, dt \qquad (4.14)$$

Thus, to construct the sufficient statistics needed by the demodulator we may use the $N$ quantities $r_j$, $j = 1, \ldots, N$, the projections of $r(t)$ onto the orthonormal basis $\{\psi_j(t)\}_{j=1}^{N}$ spanning the $N$-dimensional signal space. The components of the received signal that are orthogonal to this space are irrelevant to the decision process. The block diagram of the demodulator based on (4.13) is shown in Fig. 4.7. There is no difference in performance between this demodulator and that of Fig. 4.5; however, it contains only $N$ rather than $M$ correlators, which entails a reduction in complexity, which is considerable when $M$ is much larger than $N$.

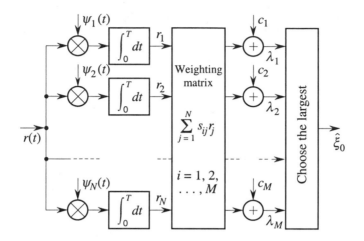

Figure 4.7: *Correlator implementation of the optimum coherent receiver for memory-less modulation and transmission over the AWGN channel. The modulator signals are represented as linear combinations of $N$ basis signals $\{\psi_j(t)\}_{j=1}^N$.*

**Example 4.2**  Let us reconsider in this example the binary modulation scheme of Example 4.1. By defining the unit-energy function

$$\psi(t) \triangleq \begin{cases} 1/\sqrt{T}, & 0 \le t < T, \\ 0, & \text{elsewhere} \end{cases}$$

we can write the two elements of the binary signal set in the form (4.12), that is, for $0 \le t < T$,

$$\begin{aligned} s_1(t) &= A\sqrt{T}\,\psi(t) \\ s_2(t) &= -A\sqrt{T}\,\psi(t) \end{aligned} \tag{4.15}$$

The computation of the quantities in (4.13) yields

$$\lambda_1 = A\sqrt{T}r - \frac{1}{2}A^2T \qquad \lambda_2 = -A\sqrt{T}r - \frac{1}{2}A^2T \tag{4.16}$$

where

$$r = \frac{1}{\sqrt{T}} \int_0^T r(t)\,dt \tag{4.17}$$

The sufficient statistics is represented now by the RV $r$, the component of $r(t)$ along $\psi(t)$. Inspection of (4.16) leads us to conclude that the decision is based on the sign of

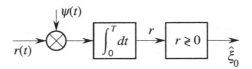

Figure 4.8: *Optimum coherent receiver for a binary modulation scheme.*

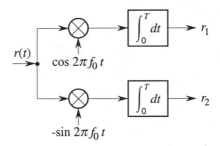

Figure 4.9: *Demodulating quaternary PSK. Only two correlators are needed to determine the quantities $r_1$ and $r_2$ upon which the optimum decision can be based.*

the value taken on by $r$: for positive $r$ we have $\lambda_1 > \lambda_2$, whereas for negative $r$ we have $\lambda_2 > \lambda_1$. The block diagram of the optimum demodulator is shown in Fig. 4.8. □

**Example 4.3 (Quaternary PSK)** As we have seen in Example 2.20 the quaternary PSK signal set has 2 dimensions. If the signals are, for $0 \leq t < T$,

$$s_i(t) = A\cos[2\pi f_0 t + (i-1)\pi/2], \qquad i = 1, 2, 3, 4$$

and the carrier frequency $f_0$ is much larger than $1/T$, the inverse of the signal duration, then the orthonormal basis functions are $\psi_1(t) = \sqrt{2/T}\cos 2\pi f_0 t$ and $\psi_2(t) = -\sqrt{2/T}\sin 2\pi f_0 t$. The energy of the four signals is the same, so that the demodulator decision can be based on the two quantities $r_1$ and $r_2$, obtained as shown in Fig. 4.9. It can also be seen that only the signs of $r_1$ and $r_2$ are relevant to the decision process, so that the multiplicative factor $\sqrt{2/T}$ is actually immaterial. The reader may want to observe that the demodulator structure of Fig. 4.9 works with any value of $M$, since every $M$-ary PSK signal set has two dimensions. □

### 4.2.1. Geometric interpretation of the optimum demodulator

By defining the three vectors

$$\mathbf{r} \overset{\triangle}{=} (r_1, r_2, \ldots, r_N)$$

$$\mathbf{s}_j \overset{\triangle}{=} (s_{j1}, s_{j2}, \ldots, s_{jN})$$

$$\mathbf{n} \overset{\triangle}{=} (n_1, n_2, \ldots, n_N)$$

where

$$n_j \overset{\triangle}{=} \int_0^T n(t)\psi_j(t)\, dt \qquad (4.18)$$

we represent the signals $r(t)$, $s(t)$ and $n(t)$ as points in an $N$-dimensional Euclidean space. The coordinates of these points are the projections of the corresponding signals onto the basis of the space, and, as we know, these are all we need to make a decision in the optimal way. We may write

$$\mathbf{r} = \mathbf{s}_j + \mathbf{n} \qquad j = 1, 2, \ldots, M \qquad (4.19)$$

to express the fact that in the additive white-Gaussian-noise channel the signal vector is perturbed by a Gaussian-noise vector, with independent components, to generate the observation $\mathbf{r}$.

Now, recall that the sufficient statistics to be used by the demodulator are

$$\lambda_i = \int_0^T r(t)s_i(t)\, dt - \frac{1}{2}\int_0^T s_i^2(t)\, dt$$

If we complete the square by subtracting the term

$$\frac{1}{2}\int_0^T r^2(t)\, dt$$

(which is independent of $i$ and hence does not alter the decision) we obtain the new sufficient statistics

$$l_i' = -\int_0^T [r(t) - s_i(t)]^2\, dt$$

The significance of the last equality lies in the fact that now, by using (2.249), we can interpret the maximization of $l_i'$ over $i$ as the search for the value of $i$ that minimizes $\int_0^T [r(t) - s_i(t)]^2\, dt$. Now, since the components of $r(t)$ lying outside of the signal space spanned by the $s_i(t)$ are irrelevant to the decision

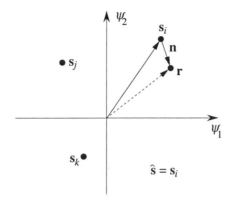

Figure 4.10: *Geometry of the minimum-distance decision rule, corresponding to optimum demodulation.*

(see Section 2.6.1) this minimization is tantamount to the minimization of the squared Euclidean distance $|\mathbf{r} - \mathbf{s}_i|^2$ between $\mathbf{r}$ and one of the vectors $\mathbf{s}_i$, where $\mathbf{r}$ represents the projection of $r(t)$ in that signal space.

In geometrical terms, *the optimum (ML) demodulator looks for the transmitted signal vector which lies closer to the received signal vector:* it is a minimum-distance demodulator. We may interpret this by saying that in a sense the optimum demodulator *trusts the channel*: that is, it assumes that the transmitted signal is the one most similar (in a Euclidean-distance sense) to the received waveform.

This geometrical view is illustrated in Fig. 4.10 for a two-dimensional signal space. The receiver decides in favor of $\mathbf{s}_i$ because this is the signal closest to the received vector $\mathbf{r}$.

**Decision regions**

Let us push our geometrical interpretation a little further with the aid of Fig. 4.11. Each point of the $N$-dimensional Euclidean space $\mathbf{R}^N$ is a possible received vector $\mathbf{r}$, and the demodulator can be thought of as a (many-to-one) mapping of the received vectors into the signal vectors. Specifically, denote by $R_i, i = 1, \ldots, M$, the regions of $\mathbf{R}^N$ such that if $\mathbf{r}$ lands in $\mathbf{R}^N$ then the optimum demodulator's choice is $\mathbf{s}_i$. These regions form a partition of $\mathbf{R}^N$, and are called *ML decision*

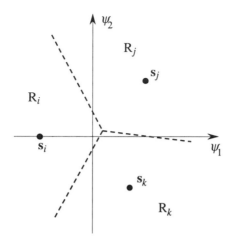

Figure 4.11: *Maximum-likelihood decision regions in a two-dimensional signal set with three points* $s_i$, $s_j$, *and* $s_k$.

*regions* or *Voronoi regions*. Formally,

$$R_i \triangleq \{ \mathbf{r} : |\mathbf{r} - s_i| = \min_j |\mathbf{r} - s_j| \} \tag{4.20}$$

The ML decision rule can therefore be put in the form

$$\text{choose} \quad \widehat{s} = s_i \quad \text{whenever} \quad \mathbf{r} \in R_i \tag{4.21}$$

which is interpreted by saying that the demodulator partitions $\mathbf{R}^N$ into $M$ decision regions $R_i$, the sets of points closer to $s_i$ than to any other signal-vector point. These regions are bounded by hyperplanes that are the loci of the points equidistant from two neighbor signals.

We note in passing that in the event $|\mathbf{r} - s_i| = |\mathbf{r} - s_j|$ (i.e., when the received-signal point lies on a boundary hyperplane) the ambiguity between $s_i$ and $s_j$ may be resolved without loss of optimality by tossing a coin. In fact, this event occurs with probability zero, and the error probability is not affected by the decision made when it occurs.

## Summary of optimum demodulation

Before proceeding with the evaluation of the error probability for optimum demodulators, we summarize here our assumptions and the different aspects under which these demodulators may be seen. We have assumed that the signal received in the interval $t \in (0, T)$ is

$$r(t) = s_i(t) + n(t)$$

where $i$ is an integer in the set $\{1, \ldots, M\}$, that the source chooses at random, with probability $1/M$ for all $i$ and independently of the choices made in other time intervals. Also, $n(t)$ is Gaussian noise with flat power spectral density $N_0/2$.

The optimum demodulator operates equivalently in one of the following forms:

1. It looks for the maximum over $j$ among the $M$ quantities

$$\int_0^T r(t) s_j(t)\, dt - \frac{1}{2}\mathcal{E}_j$$

   where $\mathcal{E}_j$ denotes the energy of the signal $s_j(t)$. A demodulator operating this way must calculate every $T$ seconds these $M$ integrals (by using correlators, or matched filters) before searching for the largest.

2. It looks for the maximum over $j$ among the $M$ quantities

$$\sum_{k=1}^N s_{jk} r_k - \frac{1}{2}\mathcal{E}_j$$

   which are constructed by computing once for all the quantities

$$s_{jk} = \int_0^T s_j(t)\psi_k(t)\, dt$$

   and every $T$ seconds the $N$ quantities

$$r_k = \int_0^T r(t)\psi_k(t)\, dt$$

   The latter computation requires $N$ matched filters or correlators, which makes this second demodulator more attractive when $N < M$.

3. By defining (and computing) the $N$-vectors $\mathbf{r} = (r_1, \ldots, r_N)$ and $\mathbf{s}_j = (s_{j1}, \ldots, s_{jN})$, it looks for the vector $\mathbf{s}_j$ that minimizes the squared Euclidean distance $|\mathbf{r} - \mathbf{s}_j|^2$.

4. By defining the $M$ decision regions

$$R_i \stackrel{\triangle}{=} \{\mathbf{r} : |\mathbf{r} - \mathbf{s}_i| = \min_j |\mathbf{r} - \mathbf{s}_j|\}$$

   it searches for $R_j$ that contains the received vector $\mathbf{r}$.

### 4.2.2.  Error probability evaluation

Under the usual assumption of equally likely symbols, the symbol error probability (4.3) can be written as

$$P(e) = 1 - P(c) = 1 - \frac{1}{M} \sum_{j=1}^{M} P(c \mid \mathbf{s}_j) \tag{4.22}$$

where $P(c \mid \mathbf{s}_j)$ is the probability of a correct decision when the transmitted signal is $\mathbf{s}_j$. Thus, the computation of $P(e)$ requires the computation of the set of probabilities $\{P(c \mid \mathbf{s}_j)\}_{j=1}^{M}$. Similarly, we can write, with obvious meaning of the symbols,

$$P(e) = \frac{1}{M} \sum_{j=1}^{M} P(e \mid \mathbf{s}_j) \tag{4.23}$$

Now, a correct decision on $\mathbf{s}_j$ occurs whenever the noise vector $\mathbf{n}$ does not move $\mathbf{s}_j$ out of its decision region $\mathrm{R}_j$: thus,

$$P(c \mid \mathbf{s}_j) = P\{\mathbf{r} \in \mathrm{R}_j \mid \mathbf{s}_j\} \tag{4.24}$$

Further, observe that, given $\mathbf{s}_j$, $\mathbf{r}$ is a conditionally Gaussian random vector with independent components, variance $N_0/2$ along each component, and mean value $\mathbf{s}_j$: in fact, $\mathbf{r}$ is generated by adding to $\mathbf{s}_j$ a Gaussian zero-mean noise vector $\mathbf{n}$ with independent components. Consequently, from (4.24) we obtain $P(c \mid \mathbf{s}_j)$ by integrating over $\mathrm{R}_j$ the probability density function of $\mathbf{r}$ given $\mathbf{s}_j$. We have

$$P(c \mid \mathbf{s}_j) = \int_{\mathrm{R}_j} \frac{1}{(\pi N_0)^{N/2}} e^{-|\mathbf{r} - \mathbf{s}_j|^2/N_0} \, d\mathbf{r} \tag{4.25}$$

The last equality shows that the error probability is expressed in the form of an integral extended to a region centered at $\mathbf{s}_j$. The integrand function has a spherical symmetry around $\mathbf{s}_j$, that is, it takes constant values over spherical surfaces centered at $\mathbf{s}_j$. This observation allows us to conclude that the error probability $P(c \mid \mathbf{s}_j)$ depends only on the shape and the size of the decision region $\mathrm{R}_j$, and not on its location in space. Every transformation that modifies the signal constellation by leaving its decision regions invariant in shape and size does not change its error probability. Thus, rotations, translations, or reflections of the signal constellation do not change $P(e)$ (at least for an *optimum* demodulator).

For example, the three constellations of Fig. 4.12 have the same error probability. However, we hasten to observe that this does not imply that they be all equivalent from the point of view of communication efficiency. In fact, we are

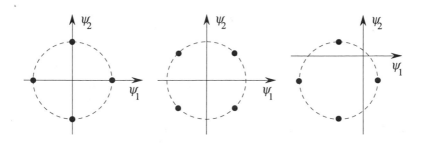

Figure 4.12: *Three quaternary signal constellations that have the same error probability for transmission over the AWGN channel.*

interested in achieving a given $P(e)$, i.e., a preassigned reliability, with the lowest expenditure of energy. The average energy required for transmission of a constellation with equally-likely signal vectors $\{s_i\}_{i=1}^M$ is

$$\mathcal{E} = \frac{1}{M} \sum_{i=1}^M \mathcal{E}_i = \frac{1}{M} \sum_{i=1}^M |s_i|^2 \tag{4.26}$$

We recognize that (4.26) is precisely the definition of the moment of inertia around the origin for a set of $M$ equal point masses located at the signal points. Thus, $\mathcal{E}$ is minimized if their center of gravity is at the origin. This condition can be stated mathematically as

$$\sum_{i=1}^M s_i = 0 \tag{4.27}$$

so that a signal set satisfying (4.27) requires the minimum average energy. (Since the third constellation of Fig. 4.12 uses, on the average, more energy than the other two, it should be regarded as less efficient.)

**Uniform signal sets**

We conclude these considerations about error probability by defining a *geometrically uniform signal set* as one whose decision regions are all congruent, in the sense that all of them can be obtained from a single one by translations and rotations. For this signal set all the signals generated by the modulator are on an equal footing, in the sense that they have the same error probability: for all pairs $i, j$ we have $P(e \mid s_i) = P(e \mid s_j)$, and consequently

$$P(e) = \frac{1}{M} \sum_{i=1}^M P(e \mid s_i) = P(e \mid s_i) \tag{4.28}$$

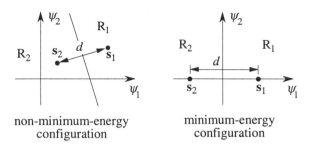

non-minimum-energy
configuration

minimum-energy
configuration

Figure 4.13: *Geometric representation of a binary signaling scheme. General configuration and minimum-energy (antipodal) configuration. The two configurations have the same error probability, but different energies.*

for any $i$. We say that this signal constellation has the *uniform error property*.

### 4.2.3.  Exact calculation of error probability

In spite of the fact that (4.25) has a compact expression, its calculation is, in but a few instances, a formidable task that cannot be carried out to obtain a closed-form solution. In this section we deal with these simpler (yet very important) cases.

#### Binary signals

Whenever the modulator has only two signals, i.e., when $M = 2$, the error probability can be computed in closed form. A general configuration of two signal vectors is shown in Fig. 4.13. The two decision regions are the half-planes separated by the axis of the segment joining the two signal points. Since they are congruent, every binary signal set is uniform.

To compute its error probability, it is convenient to modify the signal-space basis so that $\psi_1(t)$ is parallel to the line joining $s_1$ with $s_2$, and the midpoint of the two signal vectors is at the origin. This can be accomplished by rotating and translating the signal set, which, as we know, does not change $P(e)$. We have

$$
\begin{aligned}
P(e) &= P(e \mid s_1) \\
     &= P\{\mathbf{r} \in R_2 \mid s_1\} \\
     &= P\{r_1 < 0 \mid s_1\} \\
     &= P\{d/2 + n_1 < 0\}
\end{aligned}
$$

$$= P\{n_1 < -d/2\}$$

$$= \frac{1}{2}\text{erfc}\left(\frac{d}{2\sqrt{N_0}}\right) \tag{4.29}$$

where the last equality stems from a result of Appendix A, and $d$ denotes the Euclidean distance between $s_1$ and $s_2$:

$$d^2 = |s_1 - s_2|^2 = \int_0^T |s_1(t) - s_2(t)|^2 \, dt \tag{4.30}$$

The last equality in (4.29) shows the important fact that in the coherent demodulation of two equally likely signals transmitted on the AWGN channel the error probability depends only on the Euclidean distance between the two signals, and not on their other features.

Let us compute this distance by relating it to the signal set. From (4.30) we have

$$d^2 = \mathcal{E}_1 + \mathcal{E}_2 - 2\int_0^T s_1(t)s_2(t)\, dt \tag{4.31}$$

By defining the *correlation coefficient* of the two signals as their normalized scalar product

$$\rho \triangleq \frac{1}{\sqrt{\mathcal{E}_1\mathcal{E}_2}} \int_0^T s_1(t)s_2(t)\, dt \tag{4.32}$$

we can also write

$$d^2 = \mathcal{E}_1 + \mathcal{E}_2 - 2\rho\sqrt{\mathcal{E}_1\mathcal{E}_2} \tag{4.33}$$

By combining (4.33) and (4.29) we can compute $P(e)$ in a closed form that depends on the energy of the signals, their correlation coefficient, and the power spectral density of the noise. Here we specialize this general result to the case of equal-energy signals, i.e., $\mathcal{E}_1 = \mathcal{E}_2 = \mathcal{E}$. We obtain

$$P(e) = \frac{1}{2}\text{erfc}\left(\sqrt{\frac{\mathcal{E}(1-\rho)}{2N_0}}\right) \tag{4.34}$$

**Antipodal signals.**    Since, by Schwarz's inequality

$$\left|\int_0^T s_1(t)s_2(t)\, dt\right| \le \sqrt{\mathcal{E}_1\mathcal{E}_2}$$

the correlation coefficient takes values $|\rho| \le 1$. The maximum value of error probability is achieved by $\rho = 1$, which corresponds to $s_1 = s_2$, a situation hardly attractive for signal transmission as it yields $P(e) = 1/2$. On the contrary, $\rho = -1$, corresponding to

$$s_1(t) = -s_2(t) \tag{4.35}$$

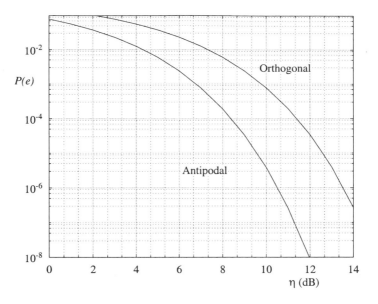

Figure 4.14: *Error probability as a function of $\eta \triangleq \mathcal{E}/N_0$ for binary antipodal and orthogonal signals.*

(and hence to $s_1 = -s_2$) provides the minimum error probability (equivalently, the minimum energy expenditure for a given $P(e)$). Signals such that (4.35) holds are called *antipodal*. For these

$$P(e) = \frac{1}{2}\text{erfc}\left(\sqrt{\frac{\mathcal{E}}{N_0}}\right) \qquad (4.36)$$

The corresponding curve is shown in Fig. 4.14.

**Binary orthogonal signals.**   Two orthogonal signals are shown in Fig. 4.15. In this case $\rho = 0$, and therefore the error probability is given by

$$P(e) = \frac{1}{2}\text{erfc}\left(\sqrt{\frac{\mathcal{E}}{2N_0}}\right) \qquad (4.37)$$

Its curve is plotted in Fig. 4.14. There is a 3-dB penalty in the signal energy to be paid with respect to the antipodal case. In fact, by comparing (4.36) with (4.37), we see that to achieve the same error probability with orthogonal signals as for antipodal signals, the energy of the latter must be doubled because of the factor 2 showing up in the argument of the error function.

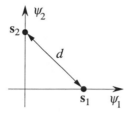

Figure 4.15: *Geometric representation of binary orthogonal signals.*

**A bandpass binary constellation.** Consider a binary constellation generated by shifting up and down by an amount $f_d$ the frequency of a carrier $\cos(2\pi f_0 t)$:

$$s_1(t) = A\cos\{2\pi(f_0 - f_d)t\}, \quad s_2(t) = A\cos\{2\pi(f_0 + f_d)t\}, \qquad 0 \le t < T$$

We assume that $(f_0 \pm f_d)T \gg 1$, so that we have approximately

$$\mathcal{E} = \int_0^T s_1^2(t)\,dt = \int_0^T s_2^2(t)\,dt = \frac{A^2 T}{2}$$

The correlation coefficient is, from its definition (4.32):

$$\rho = \frac{\sin 4\pi f_d T}{4\pi f_d T}$$

The behavior of $\rho$ as a function of $2\pi f_d T$ is shown as in Fig. 4.16. Eq. (4.34) yields the minimum error probability when $\rho$ achieves its minimum value, that is, when $\rho \simeq -0.22$. In this situation,

$$P(e) \simeq \frac{1}{2}\operatorname{erfc}\left(\sqrt{0.61\frac{\mathcal{E}}{N_0}}\right)$$

**Rectangular signal sets**

The integral (4.25) yields a closed-form $P(e)$ also for two-dimensional signal constellations whose decision regions are bounded by orthogonal straight lines parallel to the coordinate axes. Let us start with a simple example of this calculation.

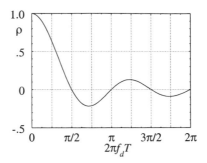

Figure 4.16: *Correlation coefficient of the two signals* $s_1(t) = A\cos[2\pi(f_0 - f_d)t]$ *and* $s_2(t) = A\cos[2\pi(f_0 + f_d)t]$.

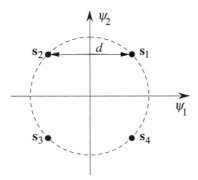

Figure 4.17: *Geometric representation of 4-PSK signals.*

**Error probability of 4-PSK.**    This signal constellation is shown in Fig. 4.17. This includes four points located symmetrically on a circumference. Its decision regions are the four quadrants of the plane, so that this constellation is uniform. Consequently, by using the independence of the two noise components $n_1$ and $n_2$, we have

$$
\begin{aligned}
P(e) &= P(e \mid \mathbf{s}_1)\\
&= 1 - P(c \mid \mathbf{s}_1)\\
&= 1 - P\{n_1 > -d/2,\, n_2 > -d/2\}\\
&= 1 - P\{n_1 > -d/2\}\, P\{n_2 > -d/2\}
\end{aligned}
$$

$$= 1 - q^2$$

where (see Appendix A)

$$q = P\{n_1 > -d/2\} = \frac{1}{2}\text{erfc}\left(-\frac{d}{2\sqrt{N_0}}\right) = 1 - p$$

with

$$p = \frac{1}{2}\text{erfc}\left(\frac{d}{2\sqrt{N_0}}\right) \tag{4.38}$$

and $d$ the Euclidean distance between two neighboring signals. In conclusion,

$$P(e) = 1 - (1 - p)^2 = 2p - p^2 \tag{4.39}$$

where $p$ is given by (4.38).

A more complex situation occurs when the decision regions of a two-dimensional constellation are still bounded by straight lines parallel to the coordinate axes, but they are not congruent. The calculations that follow illustrate this case, corresponding to a signal set which is not geometrically uniform.

**Error probability of 16-QAM.**    This signal constellation is shown in Fig. 4.18. It consists of 16 points located in the plane to form a square grid. It has three different types of decision regions, namely, that pertaining to the four corner signals $s_1$, $s_4$, $s_{13}$, and $s_{16}$, that pertaining to the eight signals $s_2$, $s_3$, $s_5$, $s_8$, $s_9$, $s_{12}$, $s_{14}$, and $s_{15}$, and that pertaining to the four internal signals $s_6$, $s_7$, $s_{10}$, and $s_{11}$. By defining

$$q_1 \triangleq P(c \mid s_1)$$

$$q_2 \triangleq P(c \mid s_2)$$

and

$$q_6 \triangleq P(c \mid s_6)$$

it should be an easy matter to derive from (4.22) that in this case

$$P(c) = \frac{1}{16}(4q_1 + 8q_2 + 4q_6) \tag{4.40}$$

Specifically, we have

$$q_1 = P\{n_1 < d/2\}\, P\{n_2 > -d/2\} = (1 - p)^2$$

where $p$ is again as in (4.38). Moreover,

$$q_2 = P\{-d/2 < n_1 < d/2\}P\{n_2 > -d/2\}$$

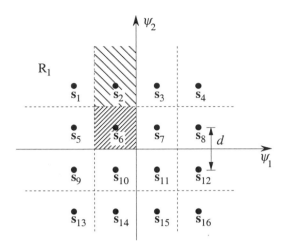

Figure 4.18: *A two-dimensional signal set with 16 points and decision regions bounded by straight lines parallel to the coordinate axes: 16-QAM.*

with

$$P\{-d/2 < n_1 < d/2\} = P\{n_1 > -d/2\} - P\{n_1 > d/2\} = (1-p)-p = 1-2p$$

and

$$P\{n_2 > -d/2\} = 1 - p$$

Thus,

$$q_2 = (1 - 2p)(1 - p)$$

Finally,

$$q_6 = P\{-d/2 < n_1 < d/2\}P\{-d/2 < n_2 < d/2\} = (1 - 2p)^2$$

In conclusion, from the latter calculations and (4.40) we obtain

$$P(c) = \frac{1}{4}[(1 - p)^2 + 2(1 - 2p)(1 - p) + (1 - 2p)^2] = 1 - 3p + \frac{9}{4}p^2 \quad (4.41)$$

and hence

$$P(e) = 3p - \frac{9}{4}p^2 \quad (4.42)$$

We observe here, also for later use, that for $p \ll 1$, i.e., $p^2 \ll p$ (a situation that should always occur for reliable transmission) we have

$$P(e) \approx 3p \quad (4.43)$$

It is interesting to interpret (4.43). The term $p$, defined in (4.38), represents the probability that in binary modulation a signal will be mistaken for another signal lying at distance $d$ from it. Further, observe that in this constellation the four corner signals similar to $s_1$ have 2 nearest neighbors (i.e., signals at distance $d$ away), the eight signals similar to $s_2$ have 3 nearest neighbors, and the four signals similar to $s_6$ have 4 nearest neighbors. The average number of nearest neighbors in this constellation is then

$$\bar{\nu} = \frac{1}{16}(4 \times 2 + 8 \times 3 + 4 \times 4) = 3$$

Thus, it is tempting to interpret (4.43) by saying that the error probability is approximately equal, for low enough noise, to the product of the binary error probability $p$ computed for the minimum distance of the constellation, multiplied by a factor equal to the average number of signals at the minimum distance. This satisfies the intuition that, when the noise is low, an error will occur by mistaking the transmitted signal for one of its nearest neighbors. The larger the number of these nearest neighbors, the larger the error probability.

If we return for a moment to the error probability for 4-PSK we can see that this interpretation makes sense also in that case. In fact, from (4.39) we have

$$P(e) \approx 2p$$

and every 4-PSK signal has exactly two nearest neighbors at distance $d$. Later on we shall prove that this approximation is valid in general.

**Error probability of orthogonal signal sets.** Another important signal configuration allows one to obtain an expression for $P(e)$ which is nearly closed-form. This is the set of $M$ orthogonal signals with equal energies, that is,

$$(\mathbf{s}_i, \mathbf{s}_j) = \begin{cases} 0, & i \neq j, \\ \mathcal{E}, & i = j \end{cases}$$

This signal set has dimensionality $N = M$. The two-dimensional case was shown in Fig. 4.15, whereas the case $M = 3$ is shown in Fig. 4.19. The decision regions may be hard to visualize for $M > 2$, but the decision rule can be described in a simple way. Assume that $s_1$ is transmitted, and consider the two-dimensional space spanned by $s_1$ and $s_k$, for any $k = 2, \ldots, M$ (Fig. 4.20). It is seen that the received signal point $\mathbf{r}$ belongs to the decision region $R_1$ when the received signal component $r_1$ is greater than $r_k$, $k = 2, \ldots, M$. Thus, in the $M$-dimensional space the decision region $R_1$ is bounded by the hyperplanes $r_1 = r_2, r_1 = r_3, \ldots, r_1 = r_M$. The same argument holds for the other decision

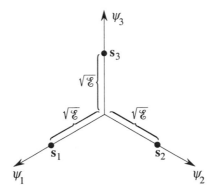

Figure 4.19: *Geometric representation of three orthogonal signals.*

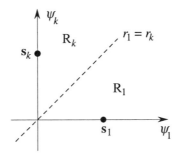

Figure 4.20: *Decision regions for a pair of orthogonal signals.*

regions, which are congruent and thus make this signal constellation geometrically uniform.

Consequently, we have

$$P(c) = P(c \mid \mathbf{s}_1) = P\{r_1 > r_2, r_1 > r_3, \ldots, r_1 > r_M \mid \mathbf{s}_1\} \qquad (4.44)$$

To compute this probability we observe that, when $\mathbf{s}_1$ is transmitted, the random variables $r_1, \ldots, r_M$ are independent Gaussian with equal variance $N_0/2$ and mean values

$$\mathrm{E}[r_i \mid \mathbf{s}_1] = \begin{cases} \sqrt{\mathcal{E}}, & i = 1, \\ 0, & i \neq 1 \end{cases}$$

The main difficulty in this computation arises from the fact that the events $r_1 > r_i$ are *not* independent. However, they are *conditionally* independent for given $r_1$, so that we can write

$$
\begin{aligned}
P(c) &= \mathrm{E}_{r_1}[P\{r_1 > r_2, \, r_1 > r_3, \ldots, r_1 > r_M \mid s_1, \, r_1\}] \\
&= \mathrm{E}_{r_1}\left\{\left[1 - \frac{1}{2}\mathrm{erfc}\left(\frac{r_1}{\sqrt{N_0}}\right)\right]^{M-1}\right\}.
\end{aligned}
$$

Our final step is taken by observing that the conditional pdf of $r_1$ is given by $(\pi N_0)^{-1/2}\exp[-(\alpha - \sqrt{\mathcal{E}})^2/N_0]$. We obtain the equation

$$
P(e) = 1 - \frac{1}{\sqrt{\pi}}\int_{-\infty}^{\infty} e^{-(x-\sqrt{\mathcal{E}/N_0})^2}\left[1 - \frac{1}{2}\mathrm{erfc}(x)\right]^{M-1}dx \qquad (4.45)
$$

The integral in (4.45) cannot be further simplified, but it can be easily computed numerically.

A brief discussion of this result allows a first glance at the problem of comparing modulation schemes with different values of $M$ (Chapter 5 contains a thorough discussion of this point). By looking at the integral in the right-hand side of (4.45), one may observe that, since the quantity in square brackets is smaller than 1, a small value of $M$ gives a smaller error probability for the same value of $\mathcal{E}/N_0$. On the other hand, notice that the greater is $M$, the higher is the information content of each signal, which in fact conveys $h = \log_2 M$ bits. The transmission of a single binary digit requires an energy $\mathcal{E}/\log_2 M$. Given this, a reasonable question is: for a given value of the noise power spectral density $N_0/2$, what happens to the error probability $P(e)$ when $M$ is increased but the energy expenditure per transmitted bit, $\mathcal{E}_b = \mathcal{E}/\log_2 M$, is kept constant? The answer is obtained by plotting $P(e)$ vs. the latter quantity. From Fig. 4.21 it can be seen that, at least for low error probabilities, increasing the size $M$ of the signal set requires less energy per bit to obtain the same error probability.

## 4.3. Approximations and bounds to P(e)

In most practical cases the probability of error cannot be computed in closed form. An example was already encountered in the case of orthogonal signals. Another example stems from $M$-PSK, with $M = 2^h$ and $h > 2$. When an exact, closed-form expression is not available, we resort to approximations (we are especially interested in approximations that are good for low error-probability values) or to bounds. Approximations and bounds are useful only if they require simple computations. Moreover, we require the bounds to be "tight," that is, that the gap between upper and lower bound be small enough as to give a reasonable approximation to the unknown true value.

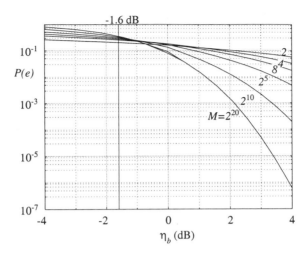

Figure 4.21: *Error probability for coherent detection of M orthogonal signals. Here* $\eta_b \overset{\triangle}{=} \mathcal{E}_b/N_0$.

### 4.3.1. An *ad hoc* technique: Bounding *P(e)* for M-PSK

Consider $M$-PSK, whose $M$ signal points are evenly distributed along a circumference with radius $\sqrt{\mathcal{E}}$. Fig. 4.22 shows 8-PSK, along with the decision region of signal $s_1$. Since the constellation is uniform irrespectively of $M$, we write

$$P(e) = P\{\mathbf{r} \notin R_1 \mid s_1\} = P\{\mathbf{r} \in S_1 \mid s_1\}$$

where $S_1$ is the complement of $R_1$. Now, observe that $S_1$ is the union of the two half-planes $S_1'$ and $S_1''$ (Fig. 4.23). We can write the pair of inequalities

$$P\{\mathbf{r} \in S_1' \mid s_1\} \leq P(e) \leq P\{\mathbf{r} \in S_1' \mid s_1\} + P\{\mathbf{r} \in S_1'' \mid s_1\} \qquad (4.46)$$

The inequality on the right stems from the fact that the probability of a union of events cannot exceed the sum of the probabilities of the single events. That on the left stems from the fact that $S_1'$ is a subset of $S_1$ (obviously, we could use $S_1''$ instead, but the end result would not change).

To compute the probabilities in (4.46) we choose to represent the noise vector with a coordinate system $(\psi_1', \psi_2')$ obtained by rotating the original axes by $\pi/M$. This operation does not change the noise statistics—which has a spherical symmetry— but simplifies our calculations. Fig. 4.24 shows that the probability

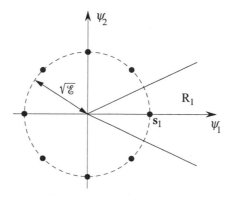

Figure 4.22: *Geometric representation of M-PSK for M = 8. The decision region of* s₁ *is also shown.*

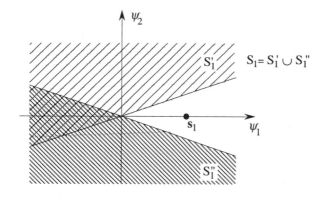

Figure 4.23: S₁, *the complement of the decision region* R₁, *can be expressed as the union of the two half-planes* S₁' *and* S₁''.

that the received vector be in $S_1'$ does not depend on the component $n_1$, and is equal to the probability that $n_2$ take on a value exceeding $\sqrt{\mathcal{E}} \sin \pi/M$. Thus,

$$P\{\mathbf{r} \in S_1' \mid s_1\} = \frac{1}{2}\text{erfc}\left(\sqrt{\frac{\mathcal{E}}{N_0}}\,\sin\frac{\pi}{M}\right) \tag{4.47}$$

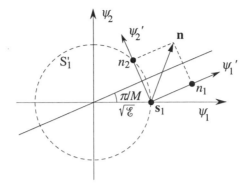

Figure 4.24: *Rotating the coordinate axes: the noise component along $\psi_1'$ becomes irrelevant.*

Because of the symmetry of the problem, the probability $P\{\mathbf{r} \in S_1'' \mid s_1\}$ takes on the same value as above, and consequently from (4.46) and (4.47) we have the following upper and lower bounds to $P(e)$:

$$\frac{1}{2}\text{erfc}\left(\sqrt{\frac{\mathcal{E}}{N_0}}\sin\frac{\pi}{M}\right) \leq P(e) \leq \text{erfc}\left(\sqrt{\frac{\mathcal{E}}{N_0}}\sin\frac{\pi}{M}\right) \qquad (4.48)$$

We observe that upper and lower bounds in (4.48) differ only by factor of 2, which is usually adequate for applications. Moreover, notice that the upper bound has the form $\bar{\nu} \cdot 0.5\,\text{erfc}(d/2\sqrt{N_0})$, where $d = 2\sqrt{\mathcal{E}}\sin\pi/M$ is the minimum distance between signal points in $M$-PSK and $\bar{\nu} = 2$ is the number of nearest neighbors. Based on our discussion in previous section, which will be made more precise soon, we may expect that the upper bound to $P(e)$ in (4.48) be a good approximation to the true value. In fact, the upper bound turns out to be closer to $P(e)$ than the lower bound.

### 4.3.2.   The union bound

We start by defining a quantity which will prove central in all the discussions that follow, the *pairwise error probability* $P\{s_i \to s_j\}$. This is the probability that, when $s_i$ is transmitted, $s_j$ will be closer than $s_i$ to the received vector $\mathbf{r}$, i.e., $s_j$ will be preferred to $s_i$ by the demodulator. The reason for its name is that if the transmission system uses only two signals, viz., $s_i$ and $s_j$, then $P(e \mid s_i) = P\{s_i \to s_j\}$. This can be easily computed from (4.29):

$$P\{s_i \to s_j\} = \frac{1}{2}\text{erfc}\left(\frac{d_{ij}}{2\sqrt{N_0}}\right) \qquad (4.49)$$

where

$$d_{ij} = |\mathbf{s}_i - \mathbf{s}_j|$$

is the Euclidean distance between $\mathbf{s}_i$ and $\mathbf{s}_j$.

For a general $M$-ary modulation scheme, if $\mathbf{s}_i$ is transmitted an error occurs if one or more of the signals other than $\mathbf{s}_i$ are preferred to it by the demodulator. Since the probability of a union of events cannot exceed the sum of the individual probabilities, we have the *union bound*

$$P(e \mid \mathbf{s}_i) \leq \sum_{\mathbf{s}_j \neq \mathbf{s}_i} P\{\mathbf{s}_i \to \mathbf{s}_j\} \tag{4.50}$$

By combining the results (4.49) and (4.50) we obtain the union bound in the explicit form:

$$P(e \mid \mathbf{s}_i) \leq \sum_{j \neq i} \frac{1}{2} \operatorname{erfc}\left(\frac{d_{ij}}{2\sqrt{N_0}}\right) \tag{4.51}$$

By further averaging (4.51) over the signal set, we get

$$P(e) \leq \frac{1}{M} \sum_{i=1}^{M} \sum_{j \neq i} \frac{1}{2} \operatorname{erfc}\left(\frac{d_{ij}}{2\sqrt{N_0}}\right) \tag{4.52}$$

We notice that for the computation of (4.52) it suffices to know all the distances $d_{ij}$ among signals in the constellation.

An important observation is that the union bound becomes tighter and tighter as $N_0$ decreases, i.e., when $P(e)$ decreases, so that for low enough error probabilities it provides a good approximation to their exact values. On the other hand, for large $P(e)$ values its approximation may be very loose. (Actually, nothing prevents the value of a union bound to exceed 1 for high noise values.) The latter fact may not be overly bad in system design, because, after all, if we use an upper bound in lieu of the true $P(e)$ we keep ourselves on the safe side.

### 4.3.3.    The union-Bhattacharyya bound

A simpler form of the union bound (4.52) can be obtained by using a bound to the pairwise error probability in lieu of its exact value. Since (Appendix A)

$$\frac{1}{2} \operatorname{erfc}\left(\frac{d_{ij}}{2\sqrt{N_0}}\right) \leq \exp\left\{-\frac{d_{ij}^2}{4N_0}\right\}$$

from (4.52) we obtain the *union-Bhattacharyya bound*

$$P(e) \leq \frac{1}{M} \sum_{i=1}^{M} \sum_{j \neq i} \exp\left\{-\frac{d_{ij}^2}{4N_0}\right\} \tag{4.53}$$

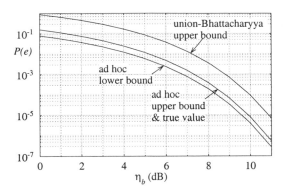

Figure 4.25: *Error probability of 4-PSK. Comparison of true value,* ad hoc *bounds, and union-Bhattacharyya upper bound. Here* $\eta_b \triangleq \mathcal{E}_b/N_0$.

Use of (4.53) may be convenient for signal sets whose *distance enumerator* can be computed in closed form. This is defined as the polynomial $T(Z)$ in the indeterminate $Z$ such that the presence of a term $\alpha Z^\delta$ indicates that among the squared distances $d_{ij}^2$, $i, j = 1, \ldots, M$, $j \neq i$, there are $\alpha$ of them taking the value $\delta$. Then the union-Bhattacharyya bound takes the especially simple form

$$P(e) \leq \frac{1}{M} T(Z) \Big|_{Z = e^{-1/4N_0}} \qquad (4.54)$$

**Example 4.4**   As an example, consider 4-PSK as in Fig. 4.17. Every signal has two neighbors at squared distance $d^2$, and one at squared distance $2d^2$. Thus, the distance enumerator is

$$T(Z) = 4(2Z^{d^2} + Z^{2d^2})$$

and the union-Bhattacharyya bound gives

$$P(e) \leq 2e^{-d^2/4N_0} + e^{-d^2/2N_0}$$

For illustration's sake, this is plotted in Fig. 4.25 as a function of $\mathcal{E}_b/N_0$, where $\mathcal{E}_b$ is the energy per bit, equal to $\mathcal{E}/2$.                                    □

The usefulness of the union-Bhattacharyya bound will appear in full in our discussion of error probabilities of convolutional codes and trellis-coded modulation (Chapters 11–12).

### 4.3.4. A looser upper bound

From the union bound we may obtain a simpler but looser bound that requires only the knowledge of a single parameter of the signal constellation. This derives from the observation that the function erfc($\cdot$) decreases monotonically as its argument increases, so that if we define the *minimum (Euclidean) distance* of the constellation as the smallest distance between any two signals:

$$d_{\min} = \min_{i \neq j} d_{ij} \tag{4.55}$$

we have

$$\text{erfc}\left(\frac{d_{ij}}{2\sqrt{N_0}}\right) \leq \text{erfc}\left(\frac{d_{\min}}{2\sqrt{N_0}}\right)$$

and consequently from (4.52) we obtain

$$
\begin{aligned}
P(e) &\leq \frac{1}{M} \sum_{i=1}^{M} \sum_{j \neq i} \frac{1}{2} \text{erfc}\left(\frac{d_{\min}}{2\sqrt{N_0}}\right) \\
&= \frac{M-1}{2} \text{erfc}\left(\frac{d_{\min}}{2\sqrt{N_0}}\right).
\end{aligned}
\tag{4.56}
$$

where the last equality derives from the observation that the two summations involve $M(M-1)$ terms. This upper bound is obviously looser than the union bound from which it derives. Its simplicity derives from its depending on a single parameter of the constellation, $d_{\min}$.

### 4.3.5. A lower bound

As we have seen in our derivation of the union bound, when $s_i$ is transmitted an error occurs if one or more of the following events occur: "The demodulator prefers to $s_i$ the signal $s_j$, $j \neq i$." These events have probabilities $P(s_i \rightarrow s_j)$.

Now, if an event is the union of sub-events, its probability is lower-bounded by each one of the probabilities of these sub-events, so that, for all $j \neq i$,

$$P(e \mid s_i) \geq P(s_i \rightarrow s_j)$$

Consequently, we have

$$
\begin{aligned}
P(e \mid s_i) &\geq \max_{j \neq i} P(s_i \rightarrow s_j) \\
&= \max_{j \neq i} \frac{1}{2} \text{erfc}\left(\frac{d_{ij}}{2\sqrt{N_0}}\right).
\end{aligned}
\tag{4.57}
$$

Since erfc$(\,\cdot\,)$ is monotone decreasing, its maximum value is achieved when its argument is minimum, so that we may write

$$P(e \mid \mathbf{s}_i) \geq \begin{cases} \dfrac{1}{2}\mathrm{erfc}\left(\dfrac{d_{\min}}{2\sqrt{N_0}}\right), & \text{if } \mathbf{s}_i \text{ has at least one signal at } d_{\min}, \\ 0, & \text{otherwise.} \end{cases} \tag{4.58}$$

By averaging (4.58) over the signal set, we obtain

$$\begin{aligned} P(e) &= \frac{1}{M}\sum_{i=1}^{M} P(e \mid \mathbf{s}_i) \\ &\geq \frac{\nu_{\min}}{M}\frac{1}{2}\mathrm{erfc}\left(\frac{d_{\min}}{2\sqrt{N_0}}\right) \end{aligned} \tag{4.59}$$

where $\nu_{\min}$ denotes the number of signals that have at least one neighbor at distance $d_{\min}$. Notice that $\nu_{\min}/M$ is the fraction of such signals.

**Example 4.5**   Consider $M$-ary orthogonal signals. Here *all* signals have a neighbor at distance $d_{\min} = \sqrt{2\mathcal{E}}$, so that $\nu_{\min} = M$, and we have the lower bound

$$P(e) \geq \frac{1}{2}\mathrm{erfc}\left(\sqrt{\frac{\mathcal{E}}{2N_0}}\right)$$

By comparing this lower bound with the union upper bound, obtained from (4.52) by observing that $d_{ij} = d_{\min}$ for all $i$ and $j \neq i$:

$$P(e) \leq \frac{M-1}{2}\,\mathrm{erfc}\left(\sqrt{\frac{\mathcal{E}}{2N_0}}\right)$$

we see that the difference between the two bounds increases with $M$.           □

### 4.3.6.   Significance of $d_{\min}$

From the definition of $\nu_{\min}$ we immediately have

$$\nu_{\min} \geq 2$$

so that (4.59) yields

$$P(e) \geq \frac{1}{M}\,\mathrm{erfc}\left(\frac{d_{\min}}{2\sqrt{N_0}}\right) \tag{4.60}$$

By combining this simplest form of lower bound with the correspondingly simple upper bound (4.56), that is,

$$\frac{1}{M}\operatorname{erfc}\left(\frac{d_{\min}}{2\sqrt{N_0}}\right) \le P(e) \le \frac{M-1}{2}\operatorname{erfc}\left(\frac{d_{\min}}{2\sqrt{N_0}}\right) \tag{4.61}$$

we may appreciate the significance of the parameter $d_{\min}$. In fact, knowledge of it allows us to obtain both upper and lower bounds to the error probability of any $M$-ary signal set, these bounds differing only for a multiplicative constant. For this reason, we may say that $d_{\min}$ is the single most important parameter that determines the quality of a signal constellation, especially for low noise. In these conditions, constellations with the same average energy may be compared on the basis of their minimum distances, because it is expected that the one with the largest value of $d_{\min}$ has the lowest error probability. Thus, maximization of $d_{\min}$ has become a popular design criterion for signal constellations.

We hasten to add a word of caution here: especially when the error probability is not very small, maximizing $d_{\min}$ may not be tantamount to minimizing $P(e)$. The next subsection reveals that the number of nearest neighbors also plays a relevant role in determining $P(e)$ (see also Problem 4.17 at the end of this Chapter).

### 4.3.7.   An approximation to error probability

A useful approximation to $P(e \mid s_i)$, especially valid for intermediate $P(e)$ values, will now be derived. This approximation was anticipated at the end of our derivation of the exact error probability of 16-QAM.

Recall the union bound in the form (4.51):

$$P(e \mid s_i) \le \sum_{j \ne i} \frac{1}{2}\operatorname{erfc}\left(\frac{d_{ij}}{2\sqrt{N_0}}\right) \tag{4.62}$$

As $N_0 \to 0$, or, equivalently, as $d_{ij}/\sqrt{N_0}$ grows to infinity, the function $\operatorname{erfc}(\cdot)$ becomes very steep (in a logarithmic scale), so that in the summation at the right-hand side of (4.62) the only significant terms are those whose argument includes $d_{\min}$. If $\nu_i$ denotes the number of signals at distance $d_{\min}$ from $s_i$, we have

$$P(e \mid s_i) \overset{\sim}{\le} \frac{\nu_i}{2}\operatorname{erfc}\left(\frac{d_{\min}}{2\sqrt{N_0}}\right) \tag{4.63}$$

where the quirky notation $\overset{\sim}{\le}$ means an approximate upper bound, one that becomes closer and closer to a true upper bound as $N_0$ (and hence $P(e)$) approaches zero. By averaging over the transmitted signals we obtain

$$P(e) = \frac{1}{M} \sum_{i=1}^{M} P(e \mid \mathbf{s}_i)$$

$$\underset{\sim}{\leqq} \frac{\bar{\nu}}{2} \text{erfc}\left(\frac{d_{\min}}{2\sqrt{N_0}}\right) \tag{4.64}$$

where

$$\bar{\nu} \triangleq \frac{1}{M} \sum_{i=1}^{M} \nu_i \tag{4.65}$$

is the average number of nearest neighbors of the signals in the constellation.

From (4.64) we can see that, besides $d_{\min}$, at least another parameter should be accounted for to evaluate the performance of a signal constellation at intermediate values of $P(e)$. This is the average number of nearest neighbors in the constellation. As a rule of thumb, we may say that, for $P(e)$ values around $10^{-6}$, doubling the value of $\bar{\nu}$ is equivalent to losing about .2 dB in the ratio $d_{\min}^2/N_0$. (See Problem 4.18 at the end of this Chapter.) The latter quantity, as we shall see in the next chapter, is related to the signal-to-noise ratio.

## 4.4.   Incoherent demodulation of bandpass signals

The case that will be addressed in this section arises in practical situations where bandpass signals are transmitted, and the demodulator does not have a precise knowledge of the phase of the oscillator that generates the carrier signal at the transmitter side. Consequently, there is an uncertainty at the receiver side, modeled by a random phase $\theta$ of which only the pdf is assumed to be known.

The modulator is assumed to be memoryless, with signals limited to a time interval of duration $T$. As in our analysis of coherent demodulation, without loss of generality we restrict our attention to the signal transmitted in the interval $(0, T)$, that we model by using analytic-signal notations as developed in Section 2.4:

$$s(t; \xi_0) = \Re\left[\tilde{s}(t; \xi_0)e^{j2\pi f_0 t}\right] \tag{4.66}$$

where the complex-envelope signals $\tilde{s}(t; \xi_0)$ are chosen from the set $\{\tilde{s}_i(t)\}_{i=1}^{M}$.

To account for the fact that the receiver knows the carrier frequency $f_0$, but not its phase, we write the signal received in $(0, T)$ as

$$r(t) = s(t; \xi_0, \theta) + n(t) \tag{4.67}$$

where we have defined

$$s(t; \xi_0, \theta) = \Re\left[\tilde{s}(t; \xi_0)e^{j(2\pi f_0 t + \theta)}\right] \tag{4.68}$$

The coherent-demodulator solution discussed so far in this chapter implies estimating $\theta$, then using its value for demodulation. Here we examine a different solution, which consists of designing a demodulator that operates without the assumption that $\theta$ is known. To achieve this goal, we first evaluate the likelihood ratios of the transmitted signals conditioned on the value of $\theta$, and then average over it to obtain likelihood ratios independent of the carrier-phase value.

The conditional likelihood ratio can be written as

$$\Lambda_i(\theta) = \exp\left\{\frac{2}{N_0}\int_0^T r(t)s_i(t;\,\theta)\,dt - \frac{1}{N_0}\int_0^T s_i^2(t;\,\theta)\,dt\right\}, \quad i = 1, 2, \ldots, M \tag{4.69}$$

where

$$s_i(t;\theta) \triangleq \Re\{\tilde{s}_i(t)e^{j\theta}e^{j2\pi f_0 t}\} \tag{4.70}$$

By taking the expectation of $\Lambda_i(\theta)$ with respect to $\theta$, i.e., by multiplying it by the pdf $f_\theta(\,\cdot\,)$ and integrating from $-\pi$ to $\pi$, we obtain the unconditional likelihood ratio:

$$\Lambda_i = \exp\left(-\frac{\mathcal{E}_i}{N_0}\right)\int_{-\pi}^{\pi}\exp\left\{\frac{2}{N_0}\int_0^T r(t)s_i(t;\,z)\,dt\right\}f_\theta(z)\,dz \tag{4.71}$$

where $\mathcal{E}_i$ denotes the energy of the signal $s_i(t;\,\theta)$, which does not depend on the phase $\theta$.

The latter expression may be simplified by defining the complex quantities

$$L_i \triangleq \frac{1}{\sqrt{\mathcal{E}_i}}\int_0^T r(t)\tilde{s}_i^*(t)e^{-j2\pi f_0 t}\,dt \tag{4.72}$$

which allow us to write

$$\Lambda_i = \exp\left(-\frac{\mathcal{E}_i}{N_0}\right)\int_{-\pi}^{\pi}\exp\left\{\Re\left[\frac{2\sqrt{\mathcal{E}_i}}{N_0}L_i^* e^{jz}\right]\right\}f_\theta(z)\,dz, \quad i = 1, 2, \ldots, M \tag{4.73}$$

These are the quantities upon which the optimum incoherent demodulator operates to make its decisions.

When there is no a priori information about the distribution of $\theta$, this lack of knowledge is reflected by the choice of the uniform pdf, that is,

$$f_\theta(z) = \frac{1}{2\pi}, \quad -\pi \le z < \pi$$

By using this pdf in (4.73), and recalling the definition of the modified Bessel function of the first kind, $I_0(\,\cdot\,)$ (see Appendix A), we get

$$\Lambda_i = \exp\left(-\frac{\mathcal{E}_i}{N_0}\right)I_0\left(\frac{2\sqrt{\mathcal{E}_i}}{N_0}|L_i|\right), \quad i = 1, 2, \ldots, M \tag{4.74}$$

and the log-likelihood ratio for each decision becomes

$$\lambda_i = \ln \Lambda_i = \ln I_0 \left( \frac{2\sqrt{\mathcal{E}_i}}{N_0} |L_i| \right) + c_i, \qquad i = 1, 2, \ldots, M \qquad (4.75)$$

where $c_i \overset{\triangle}{=} -\mathcal{E}_i / N_0$.

Thus, the key quantities that the demodulator has to compute are the magnitudes of $L_i$. Let us examine in some detail how this can be done. From definition (4.72), we have

$$L_i e^{-j\theta} = \frac{1}{\sqrt{\mathcal{E}_i}} \int_0^T r(t) \left[ \tilde{s}_i^*(t) e^{-j\theta} e^{-j2\pi f_0 t} \right] dt$$

Due to definition (4.70), the term in square brackets in the integrand is the conjugate of the analytic signal associated with $s_i(t; \theta)$. Thus, by defining $\hat{s}_i(t; \theta)$ as the Hilbert transform of the latter, we have

$$L_i e^{-j\theta} = \frac{1}{\sqrt{\mathcal{E}_i}} \int_0^T r(t) \left[ s_i(t; \theta) - j\hat{s}_i(t; \theta) \right] dt$$

It follows that $|L_i|^2$ may be constructed by summing the squares of the real part

$$\frac{1}{\sqrt{\mathcal{E}_i}} \int_0^T r(t) s_i(t; \theta) \, dt$$

and of the imaginary part

$$\frac{1}{\sqrt{\mathcal{E}_i}} \int_0^T r(t) \hat{s}_i(t; \theta) \, dt$$

The signals $s_i(t; \theta)$, affected by the uncertainty $\theta$, are available to the demodulator. Their Hilbert transforms can be generated by using a device that shifts these by $\pi/2$.

### 4.4.1.  Equal-energy signals

An important special case occurs when equal-energy signals are transmitted. In this case the constants $c_i$ are all equal and can be omitted from (4.75). Then, since $I_0(\cdot)$ is monotone increasing for nonnegative arguments, the ML decisions can be based on the simpler quantities

$$\ell_i \overset{\triangle}{=} |L_i|^2, \qquad i = 1, 2, \ldots, M \qquad (4.76)$$

Notice also the important fact that if the energies are not equal, the value of $N_0$ should also be known for optimum demodulation.

The block diagram of the optimum incoherent demodulator for equal-energy signaling is shown in Fig. 4.26. This is called a *correlation demodulator*.

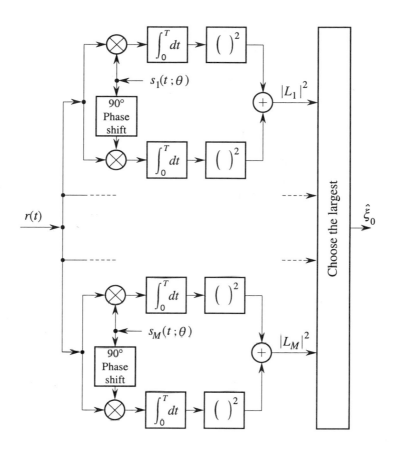

Figure 4.26: *Block diagram of the correlator receiver for incoherent demodulation of bandpass signals with equal energies.*

### 4.4.2.    On-off signaling

Consider binary modulation with the signals

$$s_1(t) = 0,$$
$$s_2(t) = \sqrt{\frac{2\mathcal{E}}{T}} \cos 2\pi f_0 t, \quad 0 \le t < T. \tag{4.77}$$

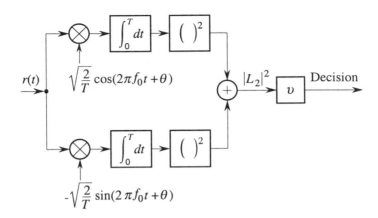

Figure 4.27: *Optimum receiver for the case of incoherent demodulation of binary on-off signals.*

Here $s_1(t)$ has energy zero, while $s_2(t)$ has energy $\mathcal{E}$. From (4.75) we get

$$\lambda_1 = \ln I_0(0) + 0 = \ln 1 = 0$$

and

$$\lambda_2 = \ln I_0 \left( \frac{2\sqrt{\mathcal{E}}}{N_0} |L_2| \right) - \frac{\mathcal{E}}{N_0}$$

Comparing $\lambda_1$ to $\lambda_2$ to choose the largest is tantamount to comparing $\lambda_2$ with 0, or, equivalently, $|L_2|^2$ against a suitable threshold value $\nu$, according to the block diagram of Fig. 4.27.

We now determine $\nu$. $|L_2|^2$ takes value $\nu$ when the two signals $s_1(t)$ and $s_2(t)$ are equally likely, i.e, when

$$\ln I_0 \left( \frac{2\sqrt{\mathcal{E}}}{N_0} |L_2| \right) - \frac{\mathcal{E}}{N_0} = 0$$

Thus, $\nu$ is the solution of the equation

$$I_0 \left( \frac{2\sqrt{\mathcal{E}}}{N_0} \sqrt{\nu} \right) = \exp(\mathcal{E}/N_0)$$

which also shows that the optimum threshold, say $\nu_{\mathrm{opt}}$, depends on the values of $\mathcal{E}$ and of $N_0$ (the estimate of the latter may not be an easy task). For small enough

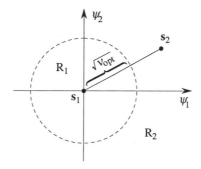

Figure 4.28: *Optimum decision regions for binary on-off signaling and incoherent demodulation.*

$N_0$ (which corresponds, as we shall see, to small enough error probability) we may use the approximation $I_0(x) \approx e^x$, which gives

$$\sqrt{\nu_{\text{opt}}} \approx \frac{\sqrt{\mathcal{E}}}{2} \tag{4.78}$$

**Error probability**

Since the demodulator operation is based on the envelope of the received signal, it should be immediately realized that the boundary between the two decision regions is a circle of radius $\sqrt{\nu_{\text{opt}}}$ (see Fig.4.28). The two coordinates of the received vector are given by

$$r_1 = \sqrt{\frac{2}{T}} \int_0^T r(t) \cos 2\pi f_0 t \, dt$$

and

$$r_2 = -\sqrt{\frac{2}{T}} \int_0^T r(t) \sin 2\pi f_0 t \, dt$$

as it can be seen by observing that $L_2 = r_1 + jr_2$.

Given a transmitted signal $s_i(t)$ and a given value of $\theta$, these two random variables are conditionally Gaussian and independent with variance $N_0/2$. When $s_1(t)$ is transmitted $E[r_1 \mid s_1, \theta] = E[r_2 \mid s_1, \theta] = 0$, while when $s_2(t)$ is transmitted we have

$$E[r_1 \mid s_2, \theta] = \sqrt{\mathcal{E}} \cos \theta$$
$$E[r_2 \mid s_2, \theta] = \sqrt{\mathcal{E}} \sin \theta.$$

The error probability $P(e \mid s_1)$ is the probability that a point with coordinates $(r_1, r_2)$ will be out of the circle with radius $\sqrt{\nu_{\text{opt}}}$, that is,

$$P(e \mid s_1) = \frac{1}{\pi N_0} \iint_{R_2} \exp\left(-\frac{\alpha_1^2 + \alpha_2^2}{N_0}\right) d\alpha_1 \, d\alpha_2 \qquad (4.79)$$

Moving to polar coordinates $\alpha_1 = \rho \cos \phi$ and $\alpha_2 = \rho \sin \phi$, we have

$$P(e \mid s_1) = \frac{1}{\pi N_0} \int_0^{2\pi} d\phi \int_{\sqrt{\nu_{\text{opt}}}}^{\infty} e^{-\rho^2/N_0} \rho \, d\rho = e^{-\nu_{\text{opt}}/N_0} \qquad (4.80)$$

Similarly,

$$P(e \mid s_2) = \frac{1}{\pi N_0} \iint_{R_1} \exp\left\{\frac{(\alpha_1 - \sqrt{\mathcal{E}} \cos \theta)^2 + (\alpha_2 - \sqrt{\mathcal{E}} \sin \theta)^2}{N_0}\right\} d\alpha_1 \, d\alpha_2 \qquad (4.81)$$

Moving again to polar coordinates, after integration with respect to the angular coordinate $\phi$ we get

$$P(e \mid s_2) = \frac{2}{N_0} \int_0^{\sqrt{\nu_{\text{opt}}}} \rho \exp\left\{-\frac{\rho^2 + \mathcal{E}}{N_0}\right\} I_0\left(\frac{2\rho\sqrt{\mathcal{E}}}{N_0}\right) d\rho \qquad (4.82)$$

an integral that can be expressed in closed form in terms of the Marcum's $Q$-function (see Appendix A). Specifically, we obtain

$$P(e \mid s_2) = 1 - Q\left(\sqrt{\frac{2\mathcal{E}}{N_0}}, \sqrt{\frac{2\nu_{\text{opt}}}{N_0}}\right) \qquad (4.83)$$

Combination of (4.80) with (4.83), yields the final expression for the error probability:

$$P(e) = \frac{1}{2}\left\{\exp\left(-\frac{\nu_{\text{opt}}}{N_0}\right) + 1 - Q\left(\sqrt{\frac{2\mathcal{E}}{N_0}}, \sqrt{\frac{2\nu_{\text{opt}}}{N_0}}\right)\right\} \qquad (4.84)$$

Notice that the above expression for error probability also holds for non-optimum threshold values, i.e., with $\nu_{\text{opt}}$ changed into $\nu$.

### 4.4.3.  Equal-energy binary signals

Assume two signals with equal energy $\mathcal{E}$ and correlation coefficient

$$\rho = \frac{1}{2\mathcal{E}} \int_0^T \tilde{s}_1(t) \tilde{s}_2^*(t) \, dt \qquad (4.85)$$

(Notice the difference between this definition and (4.32). The factor 2 in the denominator of (4.85) accounts for the fact that the energy of the complex envelope is twice the energy of the real signal, so that we obtain $\rho = 1$ when $\tilde{s}_1(t) = \tilde{s}_2(t)$).

When $s_1(t)$ is transmitted, the received signal $r(t)$ has complex envelope

$$\tilde{r}(t) \triangleq \tilde{s}_1(t)e^{j\theta} + \tilde{n}(t) \tag{4.86}$$

so that

$$r(t) = \Re\{\tilde{r}(t)e^{j2\pi f_0 t}\} = \frac{1}{2}\left[\tilde{r}(t)e^{j2\pi f_0 t} + \tilde{r}^*(t)e^{-j2\pi f_0 t}\right] \tag{4.87}$$

By assuming $f_0 T \gg 1$, the terms at $2f_0$ can be dropped from (4.72), so that we obtain, for $i = 1, 2$,

$$L_i = \frac{1}{2\sqrt{\mathcal{E}}}\int_0^T \tilde{r}(t)\tilde{s}_i^*(t)\, dt \tag{4.88}$$

Notice that, since the sufficient statistics (4.76) is based on the squared magnitude of $|L_i|$, there is no loss of optimality if we use instead of $L_i$ the rotated quantity $L_i e^{-j\theta}$. Using (4.86) and (4.85) in (4.88), we get

$$
\begin{aligned}
L_1 e^{-j\theta} &= \sqrt{\mathcal{E}} + n_1, \\
L_2 e^{-j\theta} &= \rho\sqrt{\mathcal{E}} + n_2
\end{aligned}
\tag{4.89}
$$

where, given $\theta$, $n_1$ and $n_2$ are conditionally Gaussian complex random variables, defined by

$$n_i \triangleq \frac{1}{2\sqrt{\mathcal{E}}}e^{-j\theta}\int_0^T \tilde{n}(t)\tilde{s}_i^*(t)\, dt$$

and hence such that

$$
\begin{aligned}
\mathrm{E}\{n_i \mid \theta\} &= 0, & &\tag{4.90}\\
\mathrm{E}\{|n_i|^2 \mid \theta\} &= N_0, & i = 1, 2, &\tag{4.91}\\
\mathrm{E}\{n_1^* n_2 \mid \theta\} &= N_0\rho &&\tag{4.92}
\end{aligned}
$$

Notice from last equation that $n_1$ and $n_2$ need not be independent, due to the fact that they are obtained by projecting white Gaussian noise on axes that may not be orthogonal.

Thus,

$$P(e \mid s_1) = P\{|L_2|^2 > |L_1|^2 \mid s_1\} = P\{|L_2 e^{-j\theta}| > |L_1 e^{-j\theta}| \mid s_1\} \tag{4.93}$$

Calculation of (4.93) is complicated by the fact that the two random variables involved are generally not independent, as observed before. However, by using

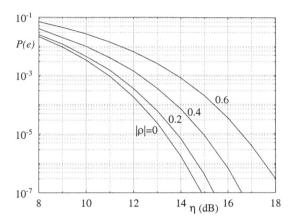

Figure 4.29: *Error probability for incoherent detection of binary signals. Here $\eta = \mathcal{E}/N_0$.*

again the Marcum $Q$-function (see Problem 4.15) we can obtain a closed-form expression for $P(e \mid s_1)$. Moreover, it is an easy matter to check that $P(e \mid s_1) = P(e \mid s_2)$, so that

$$P(e) = Q\left(\sqrt{a\frac{\mathcal{E}}{N_0}}, \sqrt{b\frac{\mathcal{E}}{N_0}}\right) - \frac{1}{2}\exp\left(-\frac{\mathcal{E}}{N_0}\frac{a+b}{2}\right)I_0\left(\frac{\mathcal{E}}{N_0}\sqrt{ab}\right) \quad (4.94)$$

where

$$a \triangleq \frac{1}{2}(1 - \sqrt{1 - |\rho|^2}),$$
$$b \triangleq \frac{1}{2}(1 + \sqrt{1 - |\rho|^2})$$

This error probability is plotted versus $\mathcal{E}/N_0$ in Fig. 4.29. The minimum value of $P(e)$ is achieved when $|\rho| = 0$, i.e., for orthogonal signals. In this case, we have, by using a result from Appendix A,

$$P(e) = Q\left(0, \sqrt{\frac{\mathcal{E}}{N_0}}\right) - \frac{1}{2}e^{-\mathcal{E}/2N_0} = \frac{1}{2}e^{-\mathcal{E}/2N_0} \quad (4.95)$$

### 4.4.4. Equal-energy $M$-ary orthogonal signals

The results of the last subsection can be easily generalized to a constellation of $M$ orthogonal signals. These may be conveniently generated by picking $M$ sinusoidal signals with duration $T$, with frequencies located symmetrically around a carrier frequency $f_0$, and such that the correlation between any pair of signals is zero.

By duplicating the calculations that lead to (4.89), we have, under the assumption that $s_i(t)$ was transmitted, and for $j = 1, \ldots, M$,

$$L_j e^{-j\theta} = \begin{cases} \sqrt{\mathcal{E}} + n_i, & j = i, \\ n_j, & j \neq i. \end{cases} \tag{4.96}$$

Now the noise components $n_i$ are independent. If we define the normalized envelope $R_j$ as

$$R_j \triangleq \sqrt{\frac{2}{N_0}} |L_j| \tag{4.97}$$

then this has a conditional pdf which is a Rice pdf when $i = j$, and Rayleigh pdf when $i \neq j$, viz., for $\alpha \geq 0$,

$$f_{R_j | s_i}(\alpha \mid s_i) = \begin{cases} \alpha \exp\left\{ -\frac{1}{2}\left( \alpha^2 + \frac{2\mathcal{E}}{N_0} \right) \right\} I_0\left( \alpha\sqrt{\frac{2\mathcal{E}}{N_0}} \right), & j = i, \\ \alpha \exp\left( -\frac{\alpha^2}{2} \right), & j \neq i \end{cases} \tag{4.98}$$

We now have

$$P(c \mid s_i) = P\{R_i = \max_j R_j \mid s_i\} \tag{4.99}$$

As a consequence of the independence of the noise components in (4.96), the envelopes $R_j$ are independent as well. Thus, from (4.98) we have, by duplicating the arguments that led us to (4.45),

$$P(c \mid s_i) = \int_0^\infty \alpha \exp\left[ -\frac{1}{2}\left( \alpha^2 + \frac{2\mathcal{E}}{N_0} \right) \right]$$
$$I_0\left( \alpha\sqrt{\frac{2\mathcal{E}}{N_0}} \right) \left[ 1 - \exp\left( -\frac{\alpha^2}{2} \right) \right]^{M-1} d\alpha \tag{4.100}$$

We observe that the RHS of (4.100) is independent of the transmitted vector $s_i$. Therefore, it provides the unconditional probability $P(c)$ of a correct decision. Moreover, this can be brought to a closed form by using the binomial expansion for the bracketed term raised to power $(M - 1)$ and then integrating termwise. The final expression for the error probability is found to be

$$P(e) = \frac{1}{M} \exp\left( -\frac{\mathcal{E}}{2N_0} \right) \sum_{i=2}^{M} \binom{M}{i} (-1)^i \exp\left[ \frac{(2-i)\mathcal{E}}{2iN_0} \right] \tag{4.101}$$

For $M = 2$ this result agrees with (4.95), as it should. The curves of (4.101) are shown in Fig. 4.30 as a function of $\mathcal{E}/N_0$.

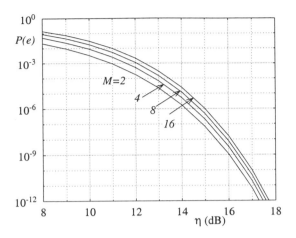

Figure 4.30: *Error probability for incoherent detection of M orthogonal signals. Here* $\eta = \mathcal{E}/N_0$.

## 4.5. Bibliographical notes

Much of the material in this chapter is classical in detection and modulation theory and, as such, it can be found in most of the textbooks available on this subject.

The authors are indebted to the excellent books by Wozencraft and Jacobs (1965) and Van Trees (1968). The first, in particular, emphasizes the geometric viewpoint. For the theory of signal spaces, the book by Franks (1969) is a recommended reading. The paper by Arthurs and Dym (1962) contains a clear presentation, in a geometric context, of the problems of coherent and incoherent demodulation of memoryless signals.

A more detailed analysis of the phase coherence of the receiver and of its effects on the demodulated signals can be found in the books by Viterbi (1966) and Simon, Hinedi, and Lindsey (1995) and in Viterbi (1965). The problem of incoherent demodulation of two equal-energy signals appears in Helstrom (1958); here we have followed closely the general derivation presented in the book by Schwartz, Bennett, and Stein (1966).

## 4.6. Problems

*Problems marked with an asterisk should be solved with the aid of a computer.*

**4.1.** Given an orthonormal basis $\{\psi_j(t)\}_{j=1}^N$, show that the RVs $n_i$ defined in (4.18) are independent, zero-mean Gaussian with covariance $N_0/2$, where $N_0/2$ is the power spectral density of the white Gaussian random process $n(t)$.

**4.2.** Consider the three binary modulation schemes for the AWGN channel whose signals, defined in $(0, T)$, are the following:

$$s_1(t) = 0, \qquad s_2(t) = \sqrt{2\mathcal{E}/T}\sin 2\pi f_2 t$$

$$s_1(t) = \sqrt{2\mathcal{E}/T}\sin 2\pi f_1 t, \qquad s_2(t) = \sqrt{2\mathcal{E}/T}\sin 2\pi f_2 t$$

and

$$s_1(t) = \sqrt{2\mathcal{E}/T}\sin 2\pi f_1 t, \qquad s_2(t) = -\sqrt{2\mathcal{E}/T}\sin 2\pi f_1 t$$

Assume $f_1 - f_2 = n/T$ and $f_1 = m/T$, $n$ and $m$ two nonzero integers.

(a) Represent geometrically each scheme.

(b) Compute their error probabilities.

(c) Comment on the relative efficiency of each scheme with regard to the utilization of the average transmitted energy.

**4.3.** Consider a binary antipodal modulation scheme whose signals have distance $d$ and are not equally likely. Define $p_1 = P(s_1)$ and $p_2 = P(s_2)$, the a priori probabilities of the two signals, and show that the optimum demodulator of Fig. 4.8 must set its threshold to the value

$$\frac{N_0}{2d}\ln\frac{p_2}{p_1}$$

In words, the boundary of the two decision regions is not at the origin, but is shifted closer to the signal with the lower probability. Find the resulting expression of error probability.

**4.4.** Assume a binary antipodal modulation scheme.

(a) Evaluate the error probability of the demodulator of Fig. 4.8 when the decision threshold has an offset $A$ with respect to the optimum (zero) value.

*(b) For some values of $A/\sqrt{\mathcal{E}}$ plot error probability curves, and compare with Fig. 4.15.

**4.5.** In this problem we analyze the degradation in performance of a binary demodulator due to the use of a filter different from the optimum matched filter. The system is shown in Fig. 4.31. Assume

$$s_1(t) = 0,$$

$$s_2(t) = \sqrt{\frac{\mathcal{E}}{T}}, \quad 0 \le t < T$$

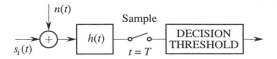

Figure 4.31: *A suboptimum receiver.*

and $n(t)$ a Gaussian noise process with power spectral density $N_0/2$. The optimum demodulator would require a matched filter, i.e, one with impulse response

$$h_{\text{opt}}(t) = \sqrt{\frac{1}{T}}, \quad 0 \le t < T$$

Assume instead an approximation of a simple *RC* filter with impulse response $h(t) = e^{-At} u_T(t)$.

(a) Compute the error probability of this nonoptimum demodulator.

(b) Find the value of $A$ that minimizes the error probability found in (a).

(c) Evaluate the increase of transmitted energy required to get the same error probability as with the optimum demodulator.

**4.6.** Two antipodal signals are transmitted over the AWGN channel. The optimum receiver achieves an error probability of 0.1 when $\sqrt{N_0} = 1$.

(a) Compute the capacity of the binary symmetric channel generated by this transmission scheme (see Section 3.3.2).

*(b) Modify the receiver by introducing two thresholds at $\pm A$, so that an erasure is declared when the received signal component has an absolute value less than $A$. Derive the discrete equivalent binary erasure channel and compute its capacity as a function of $A$.

Compare the two cases, and comment on them.

**4.7.** Assume a binary modulation scheme with signals, defined over $(0, T)$,

$$s_1(t) = 0,$$
$$s_2(t) = \sqrt{\frac{2\mathcal{E}}{T}} \sin 2\pi f_2 t, \quad f_2 = m/T, \quad m \text{ an integer}$$

and consider coherent and incoherent demodulation.

(a) Obtain for each scheme its equivalent binary discrete channel, and determine its parameters as functions of $\mathcal{E}/N_0$.

∗(b) For both schemes, plot the channel capacity versus $\mathcal{E}/N_0$.

**4.8.** The two equally likely signals, defined in $(0, T)$,

$$s_1(t) = \sqrt{\frac{2\mathcal{E}}{T}} \cos 2\pi f_1 t,$$

$$s_2(t) = \sqrt{\frac{2\mathcal{E}}{T}} \cos 2\pi (f_1 + \Delta f) t$$

are transmitted over an AWGN channel with noise power spectral density $N_0/2$. Assume $T = 2$ ms and $f_1 = 1$ MHz. Consider the two cases $\Delta f = 500$ Hz and $\Delta f = 1$ kHz.

(a) Compute the error probability for coherent demodulation.

(b) Compute the error probability for incoherent demodulation.

**4.9.** (Wozencraft and Jacobs, 1965) The eight equally likely signals shown in Fig. 4.32 are transmitted over an AWGN channel with noise power spectral density $N_0/2$.

(a) Compute the error probability achieved by the optimum coherent demodulator.

(b) Interpret the result by using a binary symmetric channel model.

**4.10.** (Lindsey and Simon, 1973) Consider the following set of $M$ equally likely signals, with $f_0 T \gg 1$:

$$s_i(t) = \sqrt{\frac{2}{T}} a_i \cos 2\pi f_0 t - \sqrt{\frac{2}{T}} b_i \sin 2\pi f_0 t, \quad 0 \le t < T, \quad i = 1, 2, \ldots, M$$

The average energy of this signal set is

$$\mathcal{E} = \frac{1}{2M} \sum_{i=1}^{M} (a_i^2 + b_i^2)$$

(a) Show that the union bound (4.52) can be put in the form

$$P(e) \le \frac{1}{M} \sum_{i=1}^{M} \sum_{j \ne i} \text{erfc} \left\{ \sqrt{\frac{\mathcal{E}}{N_0}} \left( \frac{d_{jk}}{2\sqrt{(1/M) \sum_{i=1}^{M} |\mathbf{s}_i|^2}} \right) \right\}$$

∗(b) Consider the four signal sets of Fig. 4.33 and compare their error probabilities by using the expression found in part (a).

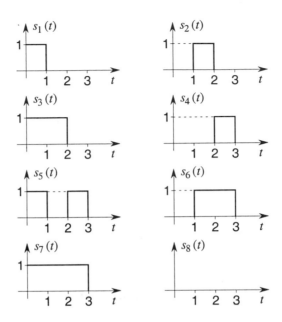

Figure 4.32: *Signal set of an octonary modulation scheme.*

**4.11.** Define a configuration of *biorthogonal signals* as a set of $M = 2N$ signals in an $N$-dimensional signal space obtained by augmenting an original orthogonal signal set with the opposite $-\mathbf{s}_i$ of each signal $\mathbf{s}_i$. The case of $N = 2$ is shown in Fig. 4.34. Paralleling the calculations that lead to (4.45), show that for biorthogonal signals

$$P(c) = \frac{1}{\sqrt{\pi N_0}} \int_0^\infty e^{-(\alpha - \sqrt{\mathcal{E}})^2 / N_0} \left[ \text{erf}\left( \frac{\alpha}{\sqrt{N_0}} \right) \right]^{(M/2)-1} d\alpha$$

**4.12.** Given a set of $M$ equally likely orthogonal signals of energy $\mathcal{E}$, show that a signal set with the same error probability but minimum average energy can be obtained by translating its origin by

$$\mathbf{a} = \frac{1}{M} \sum_{i=1}^{M} \mathbf{s}_i$$

The resulting set of signals $\{\mathbf{s}_i - \mathbf{a}\}_{i=1}^{M}$ is called a *simplex*. Show that the simplex set has an average energy $\mathcal{E}(1 - 1/M)$.

**4.13.** (Wozencraft and Jacobs, 1965) Assume that a set of $M$ equal-energy signals sat-

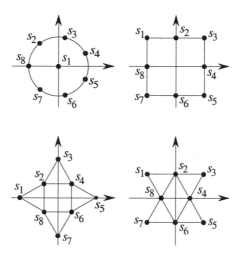

Figure 4.33: *Four octonary signal sets.*

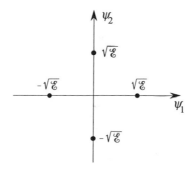

Figure 4.34: *A biorthogonal signal set with $M = 4$, $N = 2$.*

isfies the condition

$$(\mathbf{s}_i, \mathbf{s}_j) = \begin{cases} \mathcal{E}, & i = j \\ \rho\mathcal{E}, & i \neq j. \end{cases}$$

These signals are said to be *equally correlated*.

(a) Prove that

$$-\frac{1}{M-1} \leq \rho \leq 1$$

*Hint:* Consider $\left| \sum_{i=1}^{M} \mathbf{s}_i \right|^2$.

(b) Verify that the minimum value of $\rho$ is achieved by a simplex set (see Problem 4.12).

(c) Prove that, for any $\rho$, the signal set has the same error probability as the simplex signal set with energy

$$\mathcal{E}_s = \mathcal{E}\left(1 - \frac{1}{M}\right)(1 - \rho)$$

*Hint:* Consider the set $\{\mathbf{s}_i - \mathbf{a}\}_{i=1}^{M}$, with $\mathbf{a} = (1/M)\sum_{i=1}^{M}\mathbf{s}_i$.

(d) Verify, as a consequence of part (c), that this signal set has the same error probability as an orthogonal signal set with energy

$$\mathcal{E}_0 = \mathcal{E}(1 - \rho)$$

**4.14.** Prove (4.90)–(4.92).

**4.15.** This problem consists of a step-by-step derivation of (4.94). Define the two RVs

$$z_1 = \sqrt{\mathcal{E}} + n_1, \qquad z_2 = \sqrt{\mathcal{E}} + n_2$$

and apply the following linear transformation:

$$t_1 = \frac{1}{2}\{z_1(1 + K) + z_2(1 - K)e^{-j\phi}\}$$

$$t_2 = \frac{1}{2}\{z_1(1 - K) + z_2(1 + K)e^{-j\phi}\}$$

where

$$K = \sqrt{\frac{1 + |\rho|}{1 - |\rho|}}, \quad \cos\phi = \frac{\Re(\rho)}{|\rho|}, \quad \sin\phi = \frac{\Im(\rho)}{|\rho|}$$

(a) Show that the two RVs $t_1$ and $t_2$ are Gaussian and independent.

(b) Show that $R_1 = |t_1|$ and $R_2 = |t_2|$ are independently distributed Rician random variables with pdf given by

$$f_{R_i}(r) = \frac{r}{\sigma_i^2}\exp\left\{-\frac{r^2 + a_i^2}{2\sigma_i^2}\right\}I_0\left(\frac{a_i r}{\sigma_i^2}\right)$$

with $0 \le r < \infty$, $i = 1, 2$, and

$$a_i \triangleq |\mathrm{E}[t_i]|, \qquad \sigma_i^2 \triangleq \frac{1}{2}\mathrm{E}[|t_i - \mathrm{E}[t_i]|^2]$$

(c) Show that (4.93) can be rewritten as

$$P(e) = P\{|z_2|^2 > |z_1|^2\} = P\{|t_2|^2 > |t_1|^2\} = P\{R_2 > R_1\}$$

*Hint:* Write $|t_i|^2 = t_i^* t_i$.

(d) Use the results of Appendix A to get

$$P(e) = Q(\sqrt{a}, \sqrt{b}) - \frac{\nu^2}{1 + \nu^2} \exp\left(-\frac{a+b}{2}\right) I_0(\sqrt{ab})$$

where

$$a \triangleq \frac{a_2^2}{\sigma_1^2 + \sigma_2^2}, \qquad b \triangleq \frac{a_1^2}{\sigma_1^2 + \sigma_2^2}, \qquad \nu^2 = \frac{\sigma_1^2}{\sigma_2^2}$$

(e) Finally, use the definitions of part (b) to show that in our case

$$a = \frac{\mathcal{E}}{2N_0}(1 - \sqrt{1 - |\rho^2|}), \qquad b = \frac{\mathcal{E}}{2N_0}(1 + \sqrt{1 - |\rho^2|}), \qquad \nu^2 = 1$$

**4.16.** Assume an $M$-ary modulation scheme with signals given by

$$s_i(t) = A_i \sqrt{\frac{2}{T}} \cos 2\pi f_0 t, \quad 0 \le t < T, \quad i = 1, 2, \ldots, M$$

and $A_i = (i-1)d$. Consider an incoherent envelope demodulator that uses the following nonoptimum thresholds

$$b_1 = 0,$$
$$b_i = \sqrt{\frac{2}{N_0}}\left(i - \frac{3}{2}\right)d, \quad i = 1, 2, \ldots, M,$$
$$b_{M+1} = \infty$$

(a) By extending to this case the analysis that led to the calculation of error probability for on-off signals, show that for $i = 1, 2, \ldots, M$,

$$P(c \mid s_i) = Q\left[(i-1)d\sqrt{\frac{2}{N_0}}, b_i\right] - Q\left[(i-1)d\sqrt{\frac{2}{N_0}}, b_{i+1}\right]$$

(b) Show that, with $\mathcal{E}$ the average energy of the signal set, we have

$$d^2 = \frac{6\mathcal{E}}{(M-1)(2M-1)}$$

*Hint:* Use the equality

$$\sum_{i=0}^{n-1} i^2 = \frac{1}{6} n(n-1)(2n-1)$$

*(c) Plot the error probability

$$P(e) = 1 - \frac{1}{M} \sum_{i=1}^{M} P(c \mid s_i)$$

versus the ratio $\mathcal{E}/N_0$.

**4.17.** This problem shows that the maximization of the minimum Euclidean distance does not necessarily lead to a signal constellation with minimum error probability over the AWGN channel.

Consider the unit-energy one-dimensional quaternary constellation with signal vectors $s_1 = a$, $s_2 = -a$, $s_3 = b$, and $s_4 = -b$, where $a^2 + b^2 = 1/2$ and $b > a$.

(a) Compute the exact value of the error probability for coherent demodulation.

*(b) Determine the optimum values of $a$ and $b$ as a function of $N_0$, the power spectral density of the noise. In particular, verify that as $N_0 \to 0$ the minimum distance $d_{\min}$ of the constellation is maximized, while for large $N_0$ the best constellation has $a \to 0$.

**4.18.** By using (4.64) and an exponential approximation to the complementary error function, prove that for $P(e)$ values around $10^{-6}$ doubling the value of $\bar{\nu}$ is equivalent to losing about 0.2 dB in the ratio $d_{\min}^2/N_0$.

**4.19.** Consider the binary transmission system based on the signal pair

$$s_1(t) = \begin{cases} A\cos[2\pi(f_0 - f_d)t] & 0 \le t < \tau, \\ A\cos[2\pi f_0 t - 2\pi f_d \tau] & \tau \le t < T \end{cases}$$

$$s_2(t) = \begin{cases} A\cos[2\pi(f_0 + f_d)t] & 0 \le t < \tau, \\ A\cos[2\pi f_0 t + 2\pi f_d \tau] & \tau \le t < T \end{cases}$$

where $\tau < T$. For a given value of $f_d$, find the value of the correlation coefficient $\rho$ that minimizes the error probability, and the corresponding value of $P(e)$.

**4.20.** (*) Consider incoherent detection of $M$-ary orthogonal signals. Compare numerically the resulting error probability with that of coherent detection, and observe the performance degradation due to the lack of knowledge of the carrier phase. By focusing on the binary case, observe how this degradation becomes monotonically smaller as the ratio $\mathcal{E}/N_0$ increases.

**5**

# Digital modulation schemes

This chapter is devoted to the study of a number of important classes of digital modulation schemes. The concepts and the tools developed in Chapter 4 will be extensively used to analyze their performance. Transmission over the additive Gaussian noise channel is assumed throughout this chapter. The effect of other impairments other than Gaussian noise, viz., intersymbol interference and fading, will be examined in later chapters, while modulations aimed specifically at the wireless channel will be dealt with in Chapter 6.

Here we aim at assessing how each modulation scheme uses the resources available, that is, power, bandwidth, and complexity, to achieve a preassigned performance quality as expressed in terms of error probability. Several constraints and theoretical limitations generate conflicts among the designer's desired goals. Therefore, the whole conceptual framework of this Chapter is finalized at clarifying the tradeoffs that are fundamental to the choice of a modulation scheme.

## 5.1. Bandwidth, power, error probability

As in Chapter 4, we assume transmission over the additive white Gaussian noise (AWGN) channel with a two-sided noise power spectral density $N_0/2$. We denote by $\mathcal{P}$ the average power of the digital signal at the receiver front-end. As we assume for simplicity that the transmission channel introduces no attenuation, then $\mathcal{P}$ is also the average power of the signal observed at the transmitter output.

In this section we define the parameters that will be useful to assess the performance of a digital modulation scheme, that is, bandwidth (and bandwidth efficiency), signal-to-noise ratio (and power efficiency), and error probability.

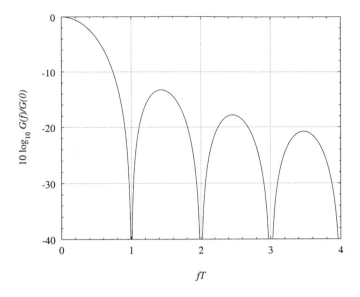

Figure 5.1: *Power density spectrum of a PSK signal.*

### 5.1.1. Bandwidth

Consider first, for motivation's sake, an $M$-PSK signal whose elementary wave-forms have duration $T$ and amplitude $A$. The power density spectrum of such a signal is, as computed in Section 2.3,

$$\mathcal{G}_v(f) = \frac{1}{4}[G(-f - f_0) + G(f - f_0)] \tag{5.1}$$

where

$$G(f) = A^2 T \left(\frac{\sin \pi f T}{\pi f T}\right)^2 \tag{5.2}$$

The latter function, plotted in Fig. 5.1 on a logarithmic scale, is seen to consist of a main lobe surrounded by smaller sidelobes. Spectra of signals obtained at the output of different modulators have a similar appearance. If the spectrum had a finite support, i.e., it were nonzero only on a finite frequency interval, then it would be an easy matter to define the spectrum occupancy as the width of its support. However, for digital modulations employing finite-duration elementary waveforms (as is the case of the schemes described in Chapter 4), the power densities extend in frequency from $-\infty$ to $\infty$, thus making it necessary to stipulate a conventional definition of bandwidth.

The bandwidth of a real signal accounts only for the positive frequencies of its spectrum. Then we may use a number of different definitions:

(a) *Null-to-null bandwidth.* This measures the width of the main spectral lobe. It is simple to evaluate whenever the first two nulls around the carrier frequency enclose the main lobe, which in turn contains most of the signal power.

(b) *Fractional power-containment bandwidth.* This is the frequency interval that contains $(1 - \epsilon)$ of the signal power in positive frequencies (which is 50% of the total signal power). This definition is useful for wireless systems that share a common frequency band: for example, if a signal has 99.9% of its power in the bandwidth $B$ allocated to it, then 0.1% of its power falls out of $B$, thus interfering with adjacent channels.

(c) *Bounded power-spectral-density bandwidth.* The criterion that specifies this bandwidth states that everywhere outside $B$ the power spectral density does not exceed a certain threshold (for example, 50 dB below its maximum value).

(d) *Equivalent noise bandwidth.* Originally defined for linear, time-invariant systems (see (2.89)), this measures the dispersion of the power spectral density around the carrier frequency.

The definitions above depend on the modulation scheme and on the specific signals used to implement it. Since in the following we shall be interested in a comparison among modulation schemes that leaves out of consideration the actual signals and focuses instead on the geometric features of the signal constellations, it is convenient to use an "abstract" definition of bandwidth. Let us recall from the $2BT$-theorem of Chapter 2 that the dimensionality of a set of signals with duration $T$ and bandwidth $W$ is approximately $N = 2WT$. This motivates our definition of the "Shannon bandwidth" of a signal set with $N$ dimensions as

$$W = \frac{N}{2T} \tag{5.3}$$

This bandwidth can of course be expressed in Hz, but it may be more appropriate in several instances to express it in *dimensions per second*. The Shannon bandwidth is the minimum amount of bandwidth that the signal *needs*, in contrast to the definitions above. Any of them, which can be called *Fourier bandwidths* of the modulated signal, expresses the amount of bandwidth that the signal actually *uses*. In most cases, Shannon bandwidth and Fourier bandwidth differ little: however, there are examples of modulated signals ("spread-spectrum" signals) whose Fourier bandwidth is much larger than their Shannon bandwidth.

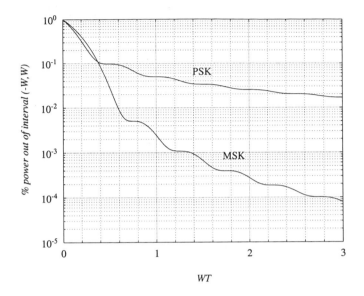

Figure 5.2: *Power-containment bandwidth of PSK and MSK.*

**Example 5.1 (PSK)**   An $M$-PSK signal has 2 dimensions, so that $W = 1/T$. Its null-to-null bandwidth is $2/T$, its equivalent noise bandwidth is $1/T$, while its power-containment bandwidth is plotted in Fig. 5.2. In the same figure the power-containment bandwidth of another modulation scheme, MSK, to be described in Chapter 6, is also shown for future reference.                                                    □

Note that in general, for any sensible definition of the bandwidth $W$, we have $W = \alpha/T$, which reflects the fundamental fact from Fourier theory that the time duration of a signal is inversely proportional to its bandwidth occupancy. The actual value of $\alpha$ depends on the definition of bandwidth and on the signals used by the modulator.

### 5.1.2.  Signal-to-noise ratio

In our discussion of Section 4.1 we have seen that the information rate of the source, $R_s$, is related to the number of waveforms used by the memoryless modulator, $M$, and to the duration of these waveforms, $T$, by the equality

$$R_s = \frac{\log_2 M}{T} \qquad (5.4)$$

This is the rate in bit/s that can be accepted by the modulator. The average power expended by the modulator is

$$\mathcal{P} = \frac{\mathcal{E}}{T}$$

where $\mathcal{E}$ is the average energy of the modulator signals. Each signal carries $\log_2 M$ information bits. Thus, defining $\mathcal{E}_b$ as the average energy expended by the modulator to transmit one bit, so that $\mathcal{E} = \mathcal{E}_b \log_2 M$, we have

$$\mathcal{P} = \mathcal{E}_b \frac{\log_2 M}{T} = \mathcal{E}_b R_s \qquad (5.5)$$

We define the *signal-to-noise ratio* as the ratio between the average signal power and the average noise power. The latter equals $(N_0/2) \cdot 2W = N_0 W$, where now $W$ is the equivalent noise bandwidth of the receiving filter, i.e., of the filter at the receiving front-end whose task, in our channel model, is to limit the noise power while leaving the signal undistorted. We have

$$\frac{\mathcal{P}}{N_0 W} = \frac{\mathcal{E}_b}{N_0} \frac{R_s}{W} \qquad (5.6)$$

**Bandwidth efficiency and asymptotic power efficiency**

Expression (5.6) shows that the signal-to-noise ratio is the product of two quantities, viz., $\mathcal{E}_b/N_0$, the energy per bit divided by twice the power spectral density, and $R_s/W$, the *bandwidth* (or *spectral*) *efficiency* of a modulation scheme. In fact, the latter, measured in bit/s/Hz, tells us how many bits per second are transmitted in a given bandwidth $W$. For example, if a system transmits data at a rate of 9,600 bit/s in a 4,800 Hz-wide system, then its spectral efficiency is $R_s = 2$ bit/s/Hz. The higher the bandwidth efficiency, the more efficient the use of the available bandwidth made by the modulation scheme.

We also observe that if $W$ denotes the Shannon bandwidth then $R_s/W$ may also be measured in bit/dimension. We have

$$\frac{R_s}{W} = 2 \frac{\log_2 M}{N} \qquad (5.7)$$

Since two-dimensional modulation schemes are especially important in applications, often the spectral efficiency is measured in bits per *dimension pair*.

We now define the *asymptotic power efficiency* $\gamma$ of a modulation scheme as follows. From (4.65) we know that for high signal-to-noise ratios the error probability is approximated by a complementary error function whose argument is $d_{min}/2\sqrt{N_0}$. Define $\gamma$ as the quantity satisfying

$$\sqrt{\gamma \frac{\mathcal{E}_b}{N_0}} = \frac{d_{min}}{2\sqrt{N_0}}$$

that is,

$$\gamma = \frac{d_{min}^2}{4\mathcal{E}_b} \qquad (5.8)$$

In words, $\gamma$ expresses how efficiently a modulation scheme makes use of the available signal energy to generate a given minimum distance. Thus we may say that, at least for high signal-to-noise ratios, a modulation scheme is better than another (having a comparable average number of nearest neighbors $\bar{\nu}$) if its asymptotic power efficiency is greater.

For example, the antipodal binary modulation of Chapter 4 has $\sqrt{\mathcal{E}} = \sqrt{\mathcal{E}_b}$ and $d_{min} = 2\sqrt{\mathcal{E}}$, so that $\gamma = 1$. This may serve as a baseline figure.

### 5.1.3.  Error probability

Most of the calculations in Chapter 4 were based on symbol error probability. To allow comparisons among modulation schemes with different values of $M$, and hence whose signals carry different numbers of bits, a better performance measure is the *bit error probability* $P_b(e)$, often also referred to as *bit-error rate* (BER). This is the probability that a bit emitted by the source will be received erroneously by the user.

In general, it can be said that the calculation of $P(e)$ is a far simpler task than the calculation of $P_b(e)$. Moreover, the latter depends also on the mapping of the source bits onto the signals in the modulator's constellation. A simple bound on $P_b(e)$ can be derived by observing that, since each signal carries $\log_2 M$ bits, one symbol error produces at least one bit error and at most $\log_2 M$ bit errors. Therefore,

$$\frac{P(e)}{\log_2 M} \leq P_b(e) \leq P(e) \qquad (5.9)$$

Since (5.9) is valid in general, we should try to keep $P_b(e)$ as close as possible to its lower bound. One way of achieving this goal is to choose the mapping in such a way that, whenever a symbol error occurs, the signal erroneously chosen by the demodulator differs from the transmitted one by the least number of bits. Since for high signal-to-noise ratios we may expect that errors occur by mistaking a signal for one of its nearest neighbors, then a reasonable pick is a mapping such that neighboring signal points correspond to binary sequences that differ in only one digit. When this is achieved we say that the signals are *Gray-mapped*, and we approximate $P_b(e)$ by its lower bound in (5.9). In the following we shall provide examples of Gray-mapped signal constellations, but we hasten to observe here that exact Gray-mapping is not possible for every conceivable constellation.

### 5.1.4. Trade-offs in the selection of a modulation scheme

In summary, the evaluation of a modulation scheme may be based on the following three parameters: the bit error probability $P_b(e)$, the signal-to-noise ratio $\mathcal{E}_b/N_0$ necessary to achieve $P_b(e)$, and the bandwidth efficiency $R_s/W$. The first tells us about the reliability of the transmission, the second measures the efficiency in power expenditure, and the third measures how efficiently the modulation scheme makes use of the bandwidth. For low error probabilities, we may simply consider the asymptotic power efficiency $\gamma$ and the bandwidth efficiency.

The ideal system achieves a small $P_b(e)$ with a low $\mathcal{E}_b/N_0$ and a high $R_s/W$: now, Shannon's theory as discussed in Chapter 3 places bounds on the values of these parameters that can be achieved by any modulation scheme. In addition, complexity considerations force us to move further apart from the theoretical limits. Consequently, complexity should also be introduced among the parameters that force the trade-off in the selection of a modulation scheme.

## 5.2.   Pulse-amplitude modulation (PAM)

This is a linear modulation scheme, also referred to as *amplitude-shift keying* (ASK). A sequence $\boldsymbol{\xi}$ of $K$ source symbols is carried by the signal

$$v_\xi(t) = \sum_{k=0}^{K-1} \xi_k\, s(t - kT), \qquad 0 \le t < KT \tag{5.10}$$

where the RVs $\xi_k$ take on values in the set of equally-spaced amplitudes $\{a_i\}_{i=1}^M$ given by

$$a_i = (2i - 1 - M)\frac{d}{2}, \qquad i = 1, 2, \ldots, M \tag{5.11}$$

Consequently, the waveforms used by the modulator are a set of scalar multiples of a single waveform: $\{s_i(t)\}_{i=1}^M = \{a_i\}_{i=1}^M s(t)$. If $s(t)$ is a unit-energy pulse, it plays the role of a basis signal in an orthonormal expansion, which shows that this signal set is one-dimensional. The geometrical representation of PAM signal sets for $d = 2$, $M = 4$ and $M = 8$, is shown in Fig. 5.3, where the signals are Gray-mapped.

The simplest form of optimum demodulator has only one correlator, or matched filter (with impulse response $s(T - t)$). Its output is sampled, then compared to a

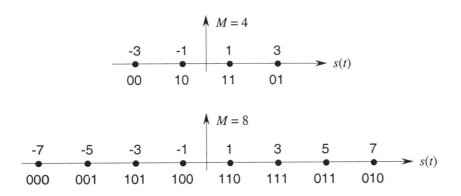

Figure 5.3: *Geometrical representation of Gray-mapped PAM signal sets.*

set of $M-1$ thresholds, located at the midpoints of adjacent signal points. The result of these comparisons provides the minimum-distance (and hence maximum-likelihood) decision.

### 5.2.1.  Error probability

The symbol error probability of PAM with coherent demodulation can be evaluated as shown in Section 4.2.3. Explicitly, we have the probability of a correct decision

$$P(c) = \frac{1}{M}\left[2q_1 + (M-2)q_2\right]$$

where $q_1$ is the correct-decision probability for the two outer points of the constellation, and $q_2$ is the same probability for the $(M-2)$ inner points. By defining

$$p = \frac{1}{2}\,\text{erfc}\left(\frac{d}{2\sqrt{N_0}}\right)$$

we have $q_1 = 1 - p$ and $q_2 = 1 - 2p$, so that

$$P(c) = 1 - 2p\frac{M-1}{M}$$

and finally

$$P(e) = \frac{M-1}{M}\,\text{erfc}\left(\frac{d}{2\sqrt{N_0}}\right)$$

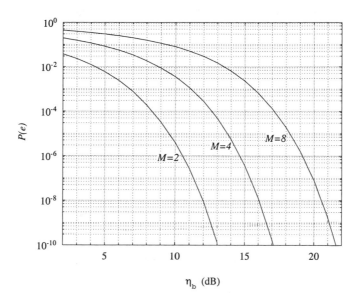

Figure 5.4: *Symbol error probabilities of M-ary PAM. Here* $\eta_b = \mathcal{E}_b/N_0$.

To express $P(e)$ as a function of $\mathcal{E}_b/N_0$, observe that the average signal energy is, from (5.11),

$$\mathcal{E} = \frac{1}{M}\sum_{i=1}^{M} a_i^2 = \frac{d^2}{4M}\sum_{i=1}^{M}(2i - 1 - M)^2 = \frac{M^2 - 1}{12}d^2 \qquad (5.12)$$

Thus

$$P(e) = \frac{M - 1}{M}\,\text{erfc}\left(\sqrt{\frac{3\log_2 M}{M^2 - 1}\frac{\mathcal{E}_b}{N_0}}\right) \qquad (5.13)$$

The error probabilities for several values of $M$ are plotted in Fig. 5.4. The asymptotic power efficiency is the factor multiplying $\mathcal{E}_b/N_0$ in the argument of erfc($\cdot$). For this scheme we have

$$\gamma_{\text{PAM}} = \frac{3\log_2 M}{M^2 - 1} \qquad (5.14)$$

which can be seen to decrease when $M$ increases.

In PAM the average energy of the transmitted signal differs from the *peak energy* $\mathcal{E}_p$, which is the energy of the maximum-amplitude signal. When design constraints are put on the peak transmitted power, we may want to express $P(e)$ in terms of $\mathcal{E}_p = (M-1)^2 d^2/4$. From (5.12) we obtain

$$\frac{\mathcal{E}_p}{\mathcal{E}} = 3\,\frac{M-1}{M+1} \tag{5.15}$$

For example, for $M = 4$ we find that $\mathcal{E}_p$ is 2.55 dB larger than $\mathcal{E}$.

### 5.2.2.  Power spectrum and bandwidth efficiency

The power spectral density of the PAM signal is obtained from (2.125) as

$$\mathcal{G}_v(f) = \frac{\mathcal{E}}{T}|S(f)|^2$$

where $S(f)$ is the Fourier transform of $s(t)$. Notice that here and in the following, when dealing with power spectral densities, we extend the summation in (5.10) from $-\infty$ to $\infty$, so as to avoid edge effects and render the signal wide-sense cyclostationary.

The Shannon bandwidth of this modulation scheme is $W = 1/2T$, so that its bandwidth efficiency is

$$\left(\frac{R_s}{W}\right)_{\mathrm{PAM}} = 2\,\log_2 M \tag{5.16}$$

This increases with $M$.

In conclusion, for PAM, increasing $M$ improves bandwidth efficiency but decreases power efficiency.

## 5.3.  Phase-shift keying (PSK)

This is a linear modulation scheme in which the source symbols shift the phase of a carrier signal. A sequence of $K$ symbols is represented by the signal

$$v_\xi(t) = \Re\left\{\sum_{k=0}^{K-1} \xi_k s(t-kT)e^{j2\pi f_0 t}\right\}, \qquad 0 \le t < KT \tag{5.17}$$

where $\xi_k = e^{j\phi_k}$, and each discrete phase $\phi_k$ takes values in the set

$$\left\{\frac{2\pi}{M}(i-1) + \Phi\right\}_{i=1}^{M} \tag{5.18}$$

with $\Phi$ an arbitrary constant phase. In the following the modulator waveform $s(t)$ is assumed to be $u_T(t)$, a rectangular pulse of amplitude $A$ and duration

$T$, so that the envelope of a PSK signal is constant (but other waveforms are possible). We can write explicitly

$$v_\xi(t) = A \sum_{k=0}^{K-1} u_T(t - kT) \cos(2\pi f_0 t + \phi_k) \qquad (5.19)$$

$$= I(t) \cos 2\pi f_0 t - Q(t) \sin 2\pi f_0 t \qquad (5.20)$$

where we have defined the *in-phase* and *quadrature* components of the PSK signal:

$$I(t) \triangleq A \sum_{k=0}^{K-1} \cos \phi_k \, u_T(t - kT)$$

$$Q(t) \triangleq A \sum_{k=0}^{K-1} \sin \phi_k \, u_T(t - kT)$$

The PSK signal set is represented geometrically in Fig. 5.5 for $M = 2$, $M = 4$, and $M = 8$. In all cases the signals are Gray-mapped. We have seen in Example 2.20 that the PSK signal set is two-dimensional. The modulators of 2-PSK and 4-PSK are shown in Figs. 5.6 and 5.7.

### 5.3.1. Error probability

Consider coherent demodulation of PSK. For illustration purposes, the structure of the demodulator of 4-PSK is shown in Fig. 5.8.

### Binary PSK

The exact error probability of binary PSK is determined by observing that 2-PSK is an antipodal modulation scheme. Hence, by using the result (4.37), and observing that for a binary scheme $\mathcal{E}_b = \mathcal{E}$, we obtain

$$P(e) = \frac{1}{2} \text{erfc} \left( \sqrt{\frac{\mathcal{E}_b}{N_0}} \right) \qquad (5.21)$$

### Quaternary PSK

The error probability of quaternary PSK (4-PSK, or QPSK) was determined explicitly in Subsection 4.2.3. By observing that now $\mathcal{E}_b = \mathcal{E}/2$ and $d_{\min} = \sqrt{2\mathcal{E}} = 2\mathcal{E}_b$, from (4.39)–(4.40) we have

$$P(e) = \text{erfc} \left( \sqrt{\frac{\mathcal{E}_b}{N_0}} \right) - \frac{1}{4} \left[ \text{erfc} \left( \sqrt{\frac{\mathcal{E}_b}{N_0}} \right) \right]^2 \qquad (5.22)$$

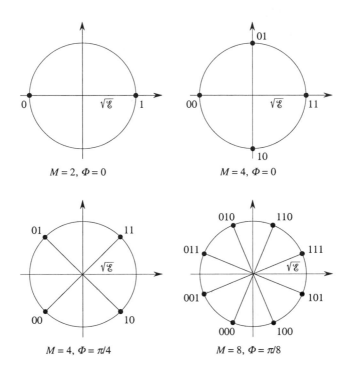

Figure 5.5: *Geometrical representation of Gray-mapped PSK signal sets.*

### $M$-ary PSK

For general $M$-ary PSK we may use the upper and lower bound to $P(e)$ derived in Subsection 4.3.1. Here $\mathcal{E}_b = \mathcal{E} / \log_2 M$, and from (4.49) we have

$$\frac{1}{2} \operatorname{erfc}\left(\sqrt{\frac{\mathcal{E}_b}{N_0} \log_2 M} \sin \frac{\pi}{M}\right) \leq P(e) \leq \operatorname{erfc}\left(\sqrt{\frac{\mathcal{E}_b}{N_0} \log_2 M} \sin \frac{\pi}{M}\right) \quad (5.23)$$

The error probabilities for several values of $M$ are plotted in Fig. 5.9.

The asymptotic power efficiency of PSK is given by

$$\gamma_{\text{PSK}} = \sin^2 \frac{\pi}{M} \cdot \log_2 M \quad (5.24)$$

which can be seen to decrease as $M$ increases, $M > 2$. (Notice how for both $M = 2$ and $M = 4$ we have $\gamma = 1$).

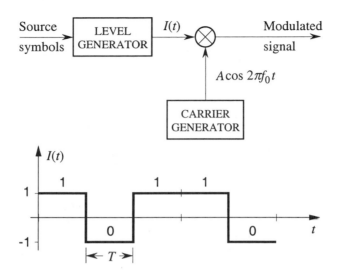

Figure 5.6: *Binary PSK modulator.*

### 5.3.2.   Power spectrum and bandwidth efficiency

The power spectral density of the PSK signal is expressed by (5.1)–(5.2). Since $N = 2$, the Shannon bandwidth of this modulation scheme is $W = 1/T$, so that its bandwidth efficiency is

$$\left(\frac{R_s}{W}\right)_{\text{PSK}} = \log_2 M \tag{5.25}$$

This increases with $M$.

In conclusion, for PSK (as for PAM) increasing $M$ improves bandwidth efficiency but decreases power efficiency.

## 5.4.   Quadrature amplitude modulation (QAM)

This is a linear modulation scheme such that the source symbols determine the amplitude as well as the phase of a carrier signal. Contrary to PSK, the signal envelope is not constant. A sequence of $K$ symbols is represented by the signal

$$v_\xi(t) = \Re\left\{\sum_{k=0}^{K-1} \xi_k s(t - kT) e^{j2\pi f_0 t}\right\}, \qquad 0 \le t < KT \tag{5.26}$$

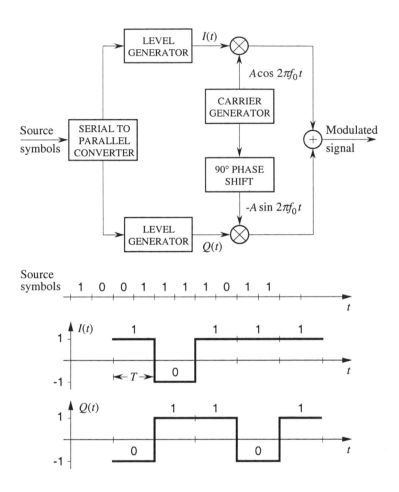

Figure 5.7: *Quaternary PSK modulator.*

where the discrete RV $\xi_k$ is defined as

$$\xi_k \triangleq \xi_k' + j\xi_k'' = A_k e^{j\phi_k}$$

and $s(t)$ is a baseband complex signal with duration $T$. When the latter is a rectangular pulse of unit amplitude, i.e., $s(t) = u_T(t)$, we can rewrite (5.26) as

$$v_\xi(t) = \sum_{k=0}^{K-1} \{\xi_k' \cos 2\pi f_0 t - \xi_k'' \sin 2\pi f_0 t\} u_T(t - kT) \qquad (5.27)$$

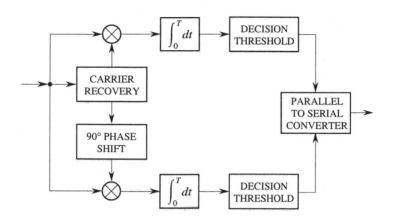

Figure 5.8: *Quaternary PSK demodulator.*

which expresses the transmitted signal in the form of a pair of orthogonal carriers modulated by a set of discrete amplitudes. This family of signal constellations is two-dimensional, and its modulator and demodulator have the same structure of those of PSK.

Several QAM families may be selected. Fig. 5.10 shows two constellations with $M = 8$ and $M = 16$. These are obtained by choosing a set of discrete amplitude levels (in number of 2 and 4, respectively), and four equally-spaced phase values in each. Another choice, way more popular, consists of picking an infinite grid of regularly spaced points, with coordinates $(n_1 + 1/2, \, n_2 + 1/2)$, $n_1$ and $n_2$ two relative integers (that is, $n_1 \in \mathbf{Z}$ and $n_2 \in \mathbf{Z}$), and carving out of it a finite constellation with $M$ points (for example, under the constraint that the average energy be minimized).

This infinite grid can be thought of as generated by translating the so-called *square lattice* $\mathbf{Z}^2$ with points $(n_1, \, n_2)$, $n_1$ and $n_2$ any two relative integers. This lattice, as well as its translated version, has minimum distance 1. Three square constellations obtained from the translated lattice $\mathbf{Z}^2 + (\frac{1}{2}, \frac{1}{2})$ with $M = 4$, $M = 16$, and $M = 64$ are shown in Fig. 5.11. When $M$ is not a power of 4, the corresponding constellation is not square, and can be given the shape of a cross to reduce its minimum average energy. Two examples of these "cross constellations" are shown in Fig. 5.12 for $M = 32$ and $M = 128$.

Another constellation can be obtained by picking signal points from the lattice $D_2$. This is derived from $\mathbf{Z}^2$ by removing one out of two points in a checker-

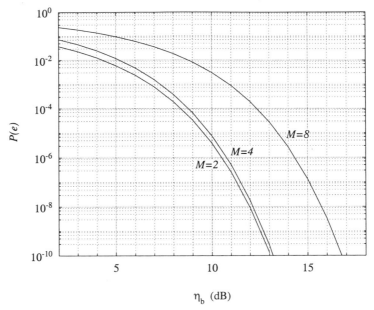

Figure 5.9: *Symbol error probabilities of PSK. Here* $\eta_b = \mathcal{E}_b/N_0$.

board fashion, so that the minimum distance among the remaining points is maximized. In formulas, $D_2$ is the set of the integer pairs $(n_1, n_2)$ with even sum $n_1 + n_2$. The lattice $D_2$, whose minimum distance is $\sqrt{2}$, and an 8-point constellation carved from $D_2 + (\frac{1}{2}, \frac{1}{2})$ are shown in Figs. 5.13 and 5.14, respectively.

### 5.4.1.   Error probability

#### Square constellations carved from $\mathbf{Z}^2 + (\frac{1}{2}, \frac{1}{2})$

The symbol error probability of $M$-points square constellations carved from $\mathbf{Z}^2 + (\frac{1}{2}, \frac{1}{2})$ can be easily derived from the observation that they consist of the cross-product of two independent PAM constellations with $\sqrt{M}$ signals each and an average energy one half that of the QAM constellation (so that $\mathcal{E}_b$ is the same for both constellations). To see this, note that a square constellation can be demodulated independently on the two coordinate axes, corresponding to the in-phase and quadrature components.

Thus, the probability of correct detection in this $M$-signal QAM equals the square of the probability of a correct detection for a PAM constellation with $\sqrt{M}$

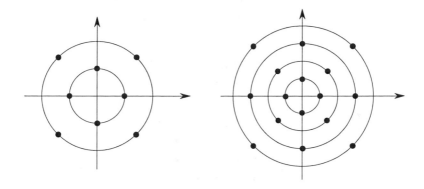

Figure 5.10: *Two QAM constellations with M = 8 and M = 16.*

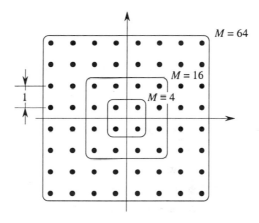

Figure 5.11: *Three square QAM constellations carved from the translated lattice* $\mathbf{Z} + \left(\frac{1}{2}, \frac{1}{2}\right).$

signals obtained by projecting the former on one coordinate axis. If $p$ denotes the symbol error probability in each PAM constellation, we have

$$P(e) = 1 - (1 - p)^2 = 2p - p^2 \tag{5.28}$$

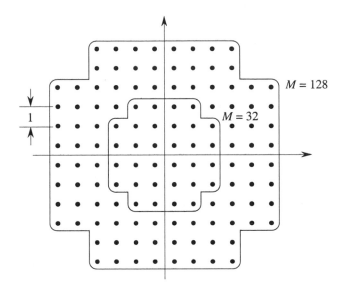

Figure 5.12: *Two cross constellations carved from the translated lattice* $\mathbf{Z} + (\frac{1}{2}, \frac{1}{2})$.

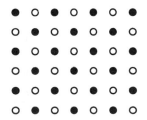

Figure 5.13: *The lattice* $D_2$, *obtained from the square lattice by removing every second point in a checkerboard fashion.*

where, from (5.13) with $M$ changed into $\sqrt{M}$,

$$p = \left(1 - \frac{1}{\sqrt{M}}\right) \text{erfc} \left(\sqrt{\frac{3 \log_2 M}{2(M-1)} \frac{\mathcal{E}_b}{N_0}}\right) \qquad (5.29)$$

A simple upper bound to $P(e)$ (which is also an approximation useful for

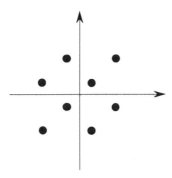

Figure 5.14: *An 8-point signal constellation carved from* $D_2 + (\frac{1}{2}, \frac{1}{2})$.

large $M$ and large $\mathcal{E}_b/N_0$) can be obtained by writing $P(e) < 2p$ and observing that $(1 - 1/\sqrt{M}) < 1$ in (5.28)–(5.29). Equivalently, we may observe again that a square constellation can be thought of as the product of two PAM with $\sqrt{M}$ signals and half the energy. By using (5.12) we obtain

$$\mathcal{E} = \frac{M - 1}{6} \, d_{\min}^2$$

Moreover, the average number of nearest neighbors is approximately 4 (it is lower for the outer points of the constellation, and exactly 4 for all inner points). Thus, from (4.64) we have

$$P(e) \lesssim 2 \operatorname{erfc} \left( \sqrt{\frac{3 \log_2 M}{2(M - 1)} \frac{\mathcal{E}_b}{N_0}} \right) \qquad (5.30)$$

**Cross constellations carved from $\mathbf{Z}^2 + (\frac{1}{2}, \frac{1}{2})$**

To construct a cross constellation with $M = 2^{5+2\mu}$ signals we may use the 32-square template of Fig. 5.15. We scale it, by partitioning each square into four squares, $\mu$ times. The cross constellation is then the set of $2^{5+2\mu}$ points from $\mathbf{Z}^2 + (\frac{1}{2}, \frac{1}{2})$ located in the middle of the resulting squares.

For the error probability of cross constellations no exact result is available. However, for large enough $M$ the approximation (5.30) still holds. In fact, the average number of nearest neighbors is still about 4, while the average energy is slightly lower than for square constellations. To justify the latter statement, we

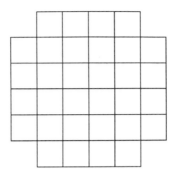

Figure 5.15: *Template for a 32-point QAM cross constellation.*

observe that for large $M$ the average energy of the constellation, that is,

$$\mathcal{E} = \frac{1}{M} \sum_{i=1}^{M} |\mathbf{s}_i|^2$$

can be thought of as the discrete approximation to the integral

$$\mathcal{I} = \frac{1}{M} \int |\mathbf{x}|^2 \, d\mathbf{x} \tag{5.31}$$

where the integration is performed over the domain that encloses the constellation. For example, if it is performed over the square with side $\sqrt{M} d_{\min}$ (and hence area $M d_{\min}^2$) that encloses the square constellation with $M$ points, it yields the approximation $\mathcal{E} \approx M d_{\min}^2/6$. For the cross constellation described above, we obtain $\mathcal{E} \approx 31/32 \cdot M d_{\min}^2/6$, a good approximation even for moderate values of $M$. For example, $M = 128$ yields the approximate value $20.67 \, d_{\min}^2$, while the true value obtained from direct computation is $\mathcal{E} = 20.5 \, d_{\min}^2$. We conclude that the average energy of a cross constellation is lower than that of a square constellation by a factor $31/32$, i.e., $0.14$ dB.

The error probabilities for square constellations with several values of $M$ are plotted in Fig. 5.16.

### 5.4.2.  Asymptotic power efficiency

From (5.30) the asymptotic power efficiency of QAM is given by

$$\gamma_{\text{QAM}} = \frac{3}{2} \frac{\log_2 M}{M - 1} \tag{5.32}$$

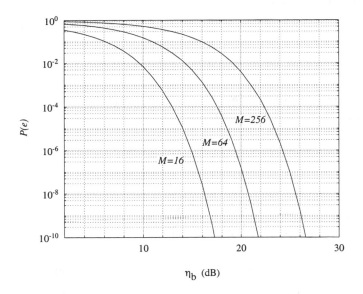

Figure 5.16: *Symbol error probabilities of square constellations carved from* $\mathbf{Z}^2 +$ $(\frac{1}{2}, \frac{1}{2})$. *Here* $\eta_b = \mathcal{E}_b/N_0$.

which can be seen to decrease when $M$ increases.

It is instructive to compare the asymptotic power efficiencies of PSK and QAM. We expect the latter to be larger, since QAM does not suffer from the constraint of having all its signal points on a circumference for constant envelope. By taking the ratio between (5.32) and (5.24) we obtain

$$\frac{\gamma_{QAM}}{\gamma_{PSK}} = \frac{3}{2} \frac{1}{(M-1)\sin^2 \pi/M}$$

For large $M$, this ratio approaches

$$\frac{3}{2} \frac{M^2}{(M-1)\pi^2} \approx 0.152 \, M$$

and hence increases linearly with $M$. For example, for $M = 64$ the ratio between the two efficiencies is close to 10 dB, and for $M = 256$ is close to 15.9 dB.

### 5.4.3.  Power spectrum and bandwidth efficiency

The power spectral density of the QAM signal is given by (5.1), where now

$$\mathcal{G}(f) = \frac{\mathrm{E}[|\xi_k|^2]}{T}\,|\tilde{S}(f)|^2 \tag{5.33}$$

and $\tilde{S}(f)$ is the Fourier transform of the complex signal $\tilde{s}(t)$.

Since $N = 2$, the Shannon bandwidth of this modulation scheme is $W = 1/T$, so that its bandwidth efficiency is the same as for PSK:

$$\left(\frac{R_s}{W}\right)_{\mathrm{QAM}} = \log_2 M \tag{5.34}$$

As for PAM and PSK, this increases with $M$.

In conclusion, for QAM, increasing $M$ improves bandwidth efficiency but decreases power efficiency.

### 5.4.4.  QAM and the capacity of the two-dimensional channel

We have seen that, in the absence of a constant-energy constraint, QAM is more efficient than PSK. In this section we shall examine how this modulation scheme compares with the limiting performance predicted by information theory.

Recall from Chapter 3 the expression of the capacity of the additive white Gaussian noise (AWGN) channel, expressed in bit/s/Hz:

$$\frac{C}{W} = \log_2\left(1 + \frac{R_s}{W}\frac{\mathcal{E}_b}{N_0}\right) = \log_2(1 + \mathrm{SNR}) \tag{5.35}$$

By recalling (5.34) we may write

$$\mathrm{SNR} = \log_2 M \frac{\mathcal{E}_b}{N_0}$$

Let us define the normalized SNR as

$$\mathrm{SNR}_0 = \mathrm{SNR}\,2^{-r_2} \tag{5.36}$$

where $r_2$ denotes the transmission rate in bits per dimension pair. Now, consider QAM with $M = 2^{r_2}$ equally likely signals. Its symbol error probability over the AWGN channel is given by (5.30), which can be rewritten, for large $M$, in the form

$$P(e) \approx 2\,\mathrm{erfc}\left(\sqrt{\frac{3}{2M}\mathrm{SNR}}\right) = 2\,\mathrm{erfc}\left(\sqrt{\frac{3}{2}\mathrm{SNR}_0}\right) \tag{5.37}$$

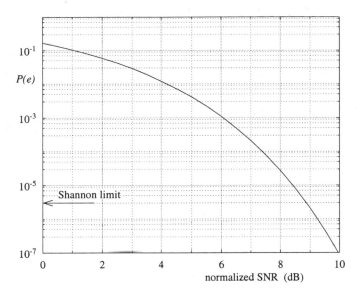

Figure 5.17: *Symbol error probability vs. normalized SNR for QAM. The Shannon limit is also shown.*

This expression of error probability can be compared with the Shannon capacity bound derived from (5.35). This states that for large SNR there exists a coding/ modulation scheme which achieves, over the AWGN channel, arbitrarily low error probabilities provided that

$$\frac{R_s}{W} < \frac{C}{W} = \log_2(1 + \text{SNR}) \approx \log_2 \text{SNR} \qquad (5.38)$$

i.e.,

$$\log_2(\text{SNR}/M) > 1 \qquad (5.39)$$

This implies that arbitrarily low error probabilities can be achieved for any normalized $\text{SNR}_0 > 1$ (i.e., 0 dB). Fig. 5.17 compares the symbol error probability achieved by QAM with the Shannon bound. From this figure we may observe that for an error rate of the order of $10^{-3}$ there is a SNR gap of about 6 dB, which becomes 7.5 dB at $10^{-4}$ and about 9 dB at $10^{-9}$. As we shall see later, most of this gap can be filled by a combination of coded modulation and shaping.

## 5.5.   Orthogonal frequency-shift keying (FSK)

This is a nonlinear modulation scheme such that the source symbols determine the frequency of a constant-envelope carrier. Specifically, it is assumed that the modulator consists of a set of $M$ separate oscillators tuned to the desired frequencies.[1]

A sequence of $K$ symbols is represented by the signal

$$v_\xi(t) = \Re\left\{ A \sum_{k=0}^{K-1} u_T(t - kT)e^{j2\pi f_d \xi_k(t-kT)}e^{j2\pi f_o t} \right\}, \qquad 0 \le t < KT \quad (5.40)$$

where the discrete RV $\xi_k$ takes values in the set $\{2i - 1 - M\}_{i=1}^{M}$, and hence $2f_d$ is the separation between adjacent frequencies.

The transmitter uses the signals

$$\begin{aligned} s_i(t) &= A\cos 2\pi f_i t, \qquad 0 \le t < T \\ f_i &= f_0 + (2i - 1 - M)f_d, \qquad i = 1, 2, \ldots, M \end{aligned} \quad (5.41)$$

They have common energy $\mathcal{E} = A^2 T/2$, and a constant envelope.

By choosing appropriately the frequency separation, the signals can be made orthogonal. Specifically, we have

$$\begin{aligned} \int_0^T s_i(t)s_j(t)\, dt &= A^2 \int_0^T \cos 2\pi f_i t \cdot \cos 2\pi f_j t\, dt \\ &= \frac{A^2}{2} \int_0^T \cos 2\pi(f_i + f_j)t\, dt + \frac{A^2}{2} \int_0^T \cos 2\pi(f_i - f_j)t\, dt \\ &= \frac{A^2 T}{2} \frac{\sin 2\pi(f_i + f_j)T}{2\pi(f_i + f_j)T} + \frac{A^2 T}{2} \frac{\sin 4\pi(i - j)f_d T}{4\pi(i - j)f_d T} \end{aligned}$$

We assume that the product $f_0 T$ of carrier frequency and symbol interval is so large that the first term in the last expression can be disregarded. Thus, the scalar product of two distinct waveforms is zero whenever $4\pi f_d T$ is a nonzero multiple of $\pi$. The minimum frequency separation yielding orthogonal signals is

$$2f_d = \frac{1}{2T} \quad (5.42)$$

---

[1] Another practical possibility is the use of a single oscillator whose frequency is modulated by source bits. The resulting FSK signal is phase-continuous, and the absence of abrupt phase transitions yields a narrower power density spectrum. See Chapter 6 for further details.

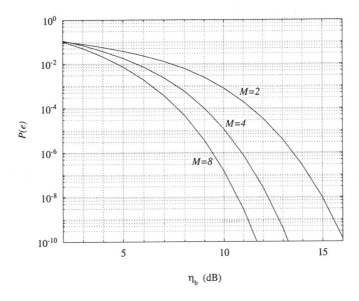

Figure 5.18: *Symbol error probability of orthogonal FSK. Here* $\eta_b = \mathcal{E}_b/N_0$.

### 5.5.1.  Error probability

By bounding the symbol error probability with the union bound of Example 4.5, we obtain

$$P(e) \leq \frac{M-1}{2} \operatorname{erfc}\left(\sqrt{\frac{\log_2 M}{2} \frac{\mathcal{E}_b}{N_0}}\right) \tag{5.43}$$

(See Fig. 5.18).

For this modulation scheme the bit error probability can be easily related to $P(e)$. Choose a position in the $(\log_2 M)$-tuple of bits associated with any signal. Then $M/2$ signals have a 0 there, and $M/2$ have a 1. We have an error in that position if the demodulator selects one out of the $M/2$ signals with the wrong bit there. Now, when the demodulator makes an error, all the $M-1$ incorrect signals have the same probability of being selected, because all are at the same distance $d_{\min}$ from the transmitted signal and the noise is spherically symmetric. In conclusion,

$$P_b(e) = \frac{M/2}{M-1} P(e)$$

### 5.5.2.  Asymptotic power efficiency

From (5.43) the asymptotic power efficiency of orthogonal FSK is given by

$$\gamma_{\text{FSK}} = \frac{1}{2} \log_2 M \qquad\qquad (5.44)$$

and increases with $M$.

### 5.5.3.  Power spectrum and bandwidth efficiency

By using eqs. (2.164) and (2.165) under the assumption that all signals are equally likely to be transmitted, we obtain the power density spectrum of the FSK signal as

$$\mathcal{G}_\xi^{(c)}(f) = \frac{1}{MT} \left\{ \sum_{i=1}^{M} |S_i(f)|^2 - \frac{1}{M} \left| \sum_{i=1}^{M} S_i(f) \right|^2 \right\}$$

and

$$\mathcal{G}_\xi^{(d)}(f) = \frac{1}{(MT)^2} \left| \sum_{i=1}^{M} S_i(f) \right|^2 \sum_{m=-\infty}^{\infty} \delta\left( f - \frac{m}{T} \right)$$

where as usual $S_i(f)$ denotes the Fourier transform of signal $s_i(t)$, $i = 1, \ldots, M$.

**Example 5.2**  Consider binary FSK signaling, with

$$s_1(t) = A \cos 2\pi (f_0 - f_d)t \qquad \text{and} \qquad s_2(t) = A \cos 2\pi (f_0 + f_d)t$$

The corresponding complex envelopes are

$$\tilde{s}_1(t) = A e^{-j2\pi f_d t} \qquad \text{and} \qquad \tilde{s}_2(t) = A e^{j2\pi f_d t}$$

and their Fourier transforms are

$$\tilde{S}_1(f) = g(f + f_d) \qquad \text{and} \qquad \tilde{S}_2(f) = g(f - f_d)$$

where $g(f)$ is the transform of the rectangular pulse $A u_T(t)$:

$$g(f) = AT \frac{\sin \pi f T}{\pi f T} e^{-j\pi f T}$$

The power density spectrum of the complex envelope of the modulated signal is given by

$$\mathcal{G}^{(c)}(f) = \frac{1}{4T} \left[ |g(f + f_d)|^2 + |g(f - f_d)|^2 - 2\Re \left\{ g(f + f_d)g^*(f - f_d) \right\} \right]$$

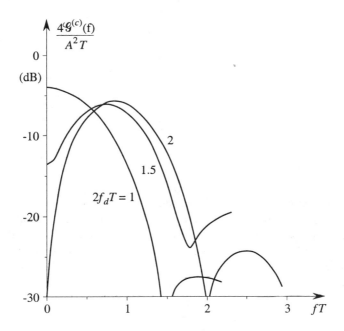

Figure 5.19: *Power density spectrum of binary FSK (continuous part).*

and

$$\mathcal{G}^{(d)}(f) =$$

$$\frac{1}{4T^2}\left[|g(f+f_d)|^2 + |g(f-f_d)|^2 + 2\Re\{g(f+f_d)g^*(f-f_d)\}\right]\sum_{m=-\infty}^{\infty}\delta\left(f-\frac{m}{T}\right)$$

Finally, the power spectrum of the real signal can be obtained from

$$\mathcal{G}_v(f) = \frac{1}{4}\left\{\mathcal{G}^{(c)}(f-f_0) + \mathcal{G}^{(c)}(-f-f_0)\mathcal{G}^{(d)}(f-f_0) + \mathcal{G}^{(d)}(-f-f_0)\right\}$$

Curves showing the continuous part of the spectrum of FSK for some values of $2f_dT$ can be found in Fig. 5.19. Table 5.1 lists the amplitudes of the line spectrum for the same values of $2f_dT$. □

In interpreting the results described above for the power density spectrum of FSK, one should keep in mind that they were obtained under the fundamental assumption, made in this chapter, that the modulation scheme is *memoryless*. This

| $fT$ | $2f_dT = 1.0$ | $2f_dT = 1.5$ | $2f_dT = 2.0$ |
|------|---------------|---------------|---------------|
| 0    |               | $-13.4$       |               |
| 1    | $-7.4$        | $-6.8$        | $-6.0$        |
| 2    | $-15.4$       | $-20.1$       |               |
| 3    | $-19.2$       | $-24.6$       |               |
| 4    | $-21.8$       | $-27.5$       |               |
| 5    | $-23.8$       | $-29.6$       |               |
| 6    | $-25.4$       | $-31.3$       |               |
| 7    | $-26.8$       | $-32.7$       |               |

Table 5.1: *Discrete power density spectrum for the signals of Example 5.2: Coefficients of the line components in dB.*

implies, as briefly mentioned at the beginning of this section, that the pulses at different frequencies be generated independently: in particular, no phase dependence among them is assumed. This model fits a situation in which the modulated signal is obtained by switching at a rate $1/T$ among $M$ different oscillators that generate signals with the same initial phase in each symbol interval. This may occur when the signal waveforms are generated digitally.

**Bandwidth efficiency.** Here $N = M$, so that the Shannon bandwidth of this modulation scheme is $W = M/2T$, and its bandwidth efficiency is

$$\left(\frac{R_s}{W}\right)_{\text{FSK}} = 2\,\frac{\log_2 M}{M} \tag{5.45}$$

We note that, unlike with PAM, PSK, and QAM, by increasing $M$ the bandwidth efficiency of orthogonal FSK decreases. On the other hand, its power efficiency increases.

## 5.6. Multidimensional signal constellations: Lattices

In this section we focus our attention on a special case of multidimensional signals, those generated by *lattices*. Lattices are infinite constellations with a high degree of symmetry, and lattice signaling, which has recently been receiving considerable attention for theoretical analyses as well as for applications, is deemed to provide excellent trade-offs between performance and implementation complexity for digital communication systems. We give here an overview of some of the aspects of lattice theory that are most relevant to applications.

The idea is to consider an $N$-dimensional lattice $\Lambda$, and to carve a finite set of signals $S$ out of it by retaining only the elements of $\Lambda$ that lie in a finite region

$\mathcal{R}$. One then derives the properties of $\mathcal{S}$ from the properties of $\Lambda$ and those of $\mathcal{R}$.

We start by listing some parameters useful for assessing the quality of a multidimensional constellation.

**Bit rate and minimum distance.**   Let $\mathcal{S}$ denote an $N$-dimensional signal constellation, $|\mathcal{S}|$ the number of its points, and $q(\mathbf{x})$ the probability of transmitting the signal $\mathbf{x} \in \mathcal{S}$. Its *bit rate* is the number of bits per dimension carried by each signal, that is, $\log_2 |\mathcal{S}|/N$. Its *normalized bit rate* is the number of bits carried *per dimension pair*:

$$\beta = \frac{2\log_2 |\mathcal{S}|}{N} \tag{5.46}$$

The *normalized minimum distance* of the constellation $\mathcal{S}$ is the ratio between $d_{\min}^2$ and the average signal energy $\mathcal{E}$, where

$$\mathcal{E} = \frac{1}{|\mathcal{S}|} \sum_{\mathbf{x} \in \mathcal{S}} q(\mathbf{x})|\mathbf{x}|^2$$

**Figure of merit.**   The *constellation figure of merit* of $\mathcal{S}$ is the ratio between $d_{\min}^2$ and the average energy of $\mathcal{S}$ *per dimension pair*:

$$\mathrm{CFM}(\mathcal{S}) = \frac{d_{\min}^2}{2\mathcal{E}/N}$$

**The constituent constellation.**   Assume the $N$-dimensional constellation to be generated by transmitting $N/\nu$ consecutive $\nu$-dimensional elementary signals (in typical wireline modems $\nu = 2$, with QAM as elementary signals). If the projections of $\mathcal{S}$ onto coordinate pairs $(1,2), (3,4), \ldots, (N-1, N)$ are identical, we call this common constellation the *constituent 2-D constellation* of $\mathcal{S}$. This is denoted by $\mathcal{S}_2$, and the induced probability distribution by $q_2(\cdot)$. We desire that the size $|\mathcal{S}_2|$ be as small as possible. With this definition of $\mathcal{S}_2$ we have

$$|\mathcal{S}| \leq |\mathcal{S}_2|^{N/2} \tag{5.47}$$

which shows that $|\mathcal{S}_2|$ is lower bounded by

$$|\mathcal{S}_2| \geq |\mathcal{S}|^{2/N} = 2^{2\log_2 |\mathcal{S}|/N} = 2^\beta \tag{5.48}$$

Thus, we may define the *constellation expansion ratio* of $\mathcal{S}$ as

$$\mathrm{CER}(\mathcal{S}) = \frac{|\mathcal{S}_2|}{2^\beta} = \frac{|\mathcal{S}_2|}{|\mathcal{S}|^{2/N}} \geq 1 \tag{5.49}$$

In designing a multidimensional constellation, one should keep its expansion ratio as close as possible to the lower bound of 1.

**Peak-to-average energy ratio.** The peak-to-average energy ratio (PAR) is a measure of the dynamic range of the signals transmitted by a two-dimensional modem, and it measures the sensitivity of a signal constellation to nonlinearities and other signal-dependent perturbations. Peak power is measured in the constituent (two-dimensional) constellation. The PAR is given by

$$\text{PAR}\,(\mathcal{C}) = \frac{\epsilon_{\max}^2}{\mathcal{E}_2},$$

where $\epsilon_{\max}^2$ is the peak energy of the signals in $\mathcal{S}_2$, and $\mathcal{E}_2$ is their average energy:

$$\epsilon_{\max}^2 = \max_{\mathbf{x} \in \mathcal{S}_2} |\mathbf{x}|^2$$

$$\mathcal{E}_2 = \frac{1}{|\mathcal{S}_2|} \sum_{\mathbf{x} \in \mathcal{S}_2} q_2(\mathbf{x})|\mathbf{x}|^2.$$

**Example 5.3** The baseline one-dimensional PAM constellation (with $M \geq 2$ signals) has a bit rate per dimension pair

$$\beta = 2 \log_2 M \geq 2$$

and a figure of merit, from (5.12):

$$\text{CFM}_0 = \frac{d_{\min}^2}{2\mathcal{E}} \approx \frac{6}{M^2}$$

$\square$

**Example 5.4** Consider the infinite set $\Lambda$ of 4-dimensional signals with semi-integer coordinates. We denote this set $\mathbf{Z}^4 + (\frac{1}{2}, \frac{1}{2}, \frac{1}{2}, \frac{1}{2})$. Within this set we carve a finite constellation $S$ obtained as follows. We choose a representative signal in $\Lambda$, and form a class of signals by applying to it all permutations and all changes of the signs of their coordinates. All the vectors in one class have the same energy as their representative. The union of a finite number of classes gives a constellation. We consider here the constellation with $|\mathcal{S}| = 512$ obtained as the union of the 7 classes shown in Table 5.2. The projections of $S$ on two dimensions are identical. The constituent constellation $\mathcal{S}_2$ is shown in Fig. 5.20. It should also be observed (the calculations are left as an exercise) that if the signals in $S$ are equally likely, this is not the case for the signals in $\mathcal{S}_2$. $\square$

| Representative s | $\|\mathbf{s}\|^2$ | number of signals in the class |
|---|---|---|
| $(\frac{1}{2},\frac{1}{2},\frac{1}{2},\frac{1}{2})$ | 1 | 16 |
| $(\frac{3}{2},\frac{1}{2},\frac{1}{2},\frac{1}{2})$ | 3 | 64 |
| $(\frac{3}{2},\frac{3}{2},\frac{1}{2},\frac{1}{2})$ | 5 | 96 |
| $(\frac{5}{2},\frac{1}{2},\frac{1}{2},\frac{1}{2})$ | 7 | 64 |
| $(\frac{3}{2},\frac{3}{2},\frac{3}{2},\frac{1}{2})$ | 7 | 64 |
| $(\frac{5}{2},\frac{3}{2},\frac{1}{2},\frac{1}{2})$ | 9 | 192 |
| $(\frac{3}{2},\frac{3}{2},\frac{3}{2},\frac{3}{2})$ | 9 | 16 |

Table 5.2: *Construction of a 4-D constellation with 512 signals carved from* $\mathbf{Z}^4 +$ $(\frac{1}{2},\frac{1}{2},\frac{1}{2},\frac{1}{2})$.

### 5.6.1.   Lattice constellations

In general, a lattice $\Lambda$ in the Euclidean $N$-space $\mathbf{R}^N$ is defined as an infinite set of $N$-vectors closed under ordinary addition and multiplication by integers. This "group" property makes a lattice look the same no matter from which one of its points it is observed. The simplest lattice is the only one-dimensional lattice $\mathbf{Z}$, the set of relative integers.

A *basis* for $\Lambda$ is a set of $m$ vectors $\mathbf{a}_1, \cdots, \mathbf{a}_m$ in $\mathbf{R}^N$ such that

$$\Lambda = \sum_{i=1}^{m} x_i \mathbf{a}_i, \qquad x_i \in \mathbf{Z}$$

In words, each lattice point can be expressed as a linear combination, with integer coefficients, of $m$ basis vectors. Under these conditions $\Lambda$ is said to be $m$-dimensional (usually we have $m = N$).

Two lattices are *equivalent* if one of them can be obtained from the other by a rotation, reflection, or scaling. If $\Lambda$ is equivalent to $\Lambda'$, we write

$$\Lambda \cong \Lambda'$$

If $d_{\min}$ is the minimum distance between any two points in the lattice, the *kissing number* $\tau$ is the number of adjacent lattice points located at $d_{\min}$, i.e.,

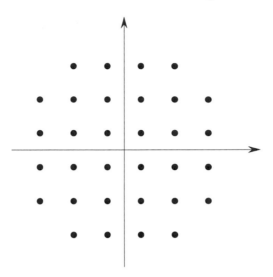

Figure 5.20: *2-dimensional constituent constellation.*

the number of nearest neighbors of any lattice point. This name comes from the fact that if each lattice point is surrounded by a sphere with radius $d_{\min}/2$, these spheres will touch, or "kiss," $\tau$ similar spheres.

**The coding gain of a lattice**

Each lattice point has a polyhedron surrounding it that contains the points of the $N$-dimensional space closer to it than to any other lattice point. This is called the Voronoi region of the lattice. The regularity of the lattice entails that all Voronoi regions are congruent. The *coding gain* $\gamma_c(\Lambda)$ of the lattice $\Lambda$ is defined as

$$\gamma_c(\Lambda) = \frac{d_{\min}^2}{V(\Lambda)^{2/N}} \tag{5.50}$$

where $V(\Lambda)$ is the *fundamental lattice volume*. This is defined as the volume of the Voronoi region of any of the lattice points, or, equivalently, the reciprocal of the number of lattice points per unit volume (for example, $V(\mathbf{Z}^N) = 1$). The main properties of $\gamma_c(\Lambda)$ are listed in Forney (1988, pp. 1128–1129).

**Transformation of lattices**

Given a lattice $\Lambda$ with vectors $\mathbf{x}$, new lattices can be generated by the following operations.

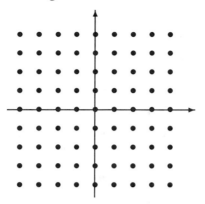

Figure 5.21: *The lattice* $\mathbf{Z}^2$.

- *Scaling:* If $r$ is any real number, then $r\Lambda$ is the lattice with vectors $r\mathbf{x}$.

- *Orthogonal transformation:* If $T$ is a scaled orthogonal transformation of $\mathbf{R}^n$, then $T\Lambda$ is the lattice with vectors $T\mathbf{x}$.

- *Direct product:* The $n$-fold direct product of $\Lambda$ with itself, i.e., the set of all $nN$-tuples $(\mathbf{x}_1, \mathbf{x}_2, \cdots, \mathbf{x}_n)$, where each $\mathbf{x}_i$ is in $\Lambda$, is a lattice denoted by $\Lambda^n$.

### 5.6.2.    Examples of lattices

#### The lattice $\mathbf{Z}^N$

The set $\mathbf{Z}^N$ of all $N$-tuples with integer coordinates is called the *cubic lattice*, or *integer lattice*. Its Voronoi region is a hypercube with unit edge length. Its minimum distance is $d_{\min} = 1$, and its kissing number is $\tau = 2N$. For example, $\mathbf{Z}^2$ is shown in Fig. 5.21.

#### The $N$-dimensional lattice $A_N$

$A_N$ is the set of all vectors with $(N + 1)$ integer coordinates whose sum is zero. This lattice may be viewed as the intersection of $\mathbf{Z}^{N+1}$ and a hyperplane cutting the origin. Its minimum distance is $d_{\min} = \sqrt{2}$, and its kissing number is $\tau = N(N+1)$. Fig. 5.22 shows $A_2$, called the "hexagonal" lattice because its Voronoi regions are hexagons.

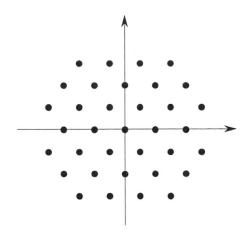

Figure 5.22: *The 2-dimensional hexagonal lattice $A_2$.*

### The $N$-dimensional lattice $D_N$

$D_N$ is the set of all $N$-dimensional points whose integer coordinates have an even sum. It may be viewed as a punctured version of $\mathbf{Z}^N$, in which the points are colored alternately black and white with a checkerboard coloring, and the white points (those with odd sums) are removed (see Fig. 5.13). We have $d_{\min} = \sqrt{2}$, and $\tau = 2N(N-1)$.

$D_4$ represents the densest lattice packing in $\mathbf{R}^4$. This means that if unit-radius, 4-dimensional spheres with centers in the lattice points are used to pack $\mathbf{R}^4$, then $D_4$ is the lattice with the largest number of spheres per unit volume.

### The Gosset lattice $E_8$

$E_8$ consists of the points

$$\{(x_1, \cdots, x_8) : \quad \forall x_i \in \mathbf{Z} \text{ or } \forall x_i \in \mathbf{Z} + \frac{1}{2}, \quad \sum_{i=1}^{8} x_i \equiv 0 \mod 2\}$$

In words, $E_8$ consists of the 8-vectors whose components are all integers, or all halves of odd integers, and whose sum is even.

This lattice has $d_{\min} = \sqrt{2}$ and $\tau = 240$. The 240 nearest neighbors of the origin (the point $0^8$) are the 112 points obtained by permuting the components of $(\pm 1)^2 0^6$, and the 128 points $(\pm \frac{1}{2})^8$, where the number of minus signs is even.

If we build a sphere with radius $\sqrt{2}/2$ centered at any point of $E_8$, we obtain an arrangement of spheres in the 8-dimensional space such that 240 spheres touch any given sphere. It can be shown that it is impossible to do better in $\mathbf{R}^8$, and that the only 8-D lattice with $\tau = 240$ is $E_8$.

### Other lattices

The description and the properties of other important lattices, like the 16-dimensional Barnes-Wall lattice $\Lambda_{16}$ and the 24-dimensional Leech lattice $\Lambda_{24}$, are outside the scope of this book and can be found, for example, in Chapter 4 of Conway and Sloane (1988).

## 5.7.   Carving a signal constellation out of a lattice

We now study how a finite signal constellation $\mathcal{S}$ can be obtained from an infinite lattice $\Lambda$, and some of the properties of the resulting constellation. We shall denote with $\mathcal{S}(\Lambda, \mathcal{R})$ a constellation obtained from $\Lambda$ (or from its translate $\Lambda + \mathbf{a}$) by retaining only the points that fall in the region $\mathcal{R}$ with volume $V(\mathcal{R})$. The resulting constellation has

$$|\mathcal{S}| \approx \frac{V(\mathcal{R})}{V(\Lambda)}$$

points, provided that $V(\mathcal{R}) \gg V(\Lambda)$, i.e., that $|\mathcal{S}|$ is large enough.

In order to express the figure of merit of the constellation $\mathcal{S}(\Lambda, \mathcal{R})$, we need to introduce the definition of the *shape gain* $\gamma_s(\mathcal{R})$ of the region $\mathcal{R}$. This is defined as the reduction in average energy (per dimension pair) required by a constellation bounded by $\mathcal{R}$ compared to that which would be required by a constellation bounded by an $N$-dimensional cube of the same volume $V(\mathcal{R})$. In formulas, the shape gain is the ratio between the *normalized second moment* of any $N$-dimensional cube (which is equal to $1/12$) and the normalized second moment of $\mathcal{R}$:

$$\gamma_s(\mathcal{R}) = \frac{1/12}{m_2(\mathcal{R})} \tag{5.51}$$

where

$$m_2(\mathcal{R}) = \frac{\int_{\mathcal{R}} |\mathbf{r}|^2 \, d\mathbf{r}}{N V(\mathcal{R})^{1+2/N}} \tag{5.52}$$

The main properties of $\gamma_s(\mathcal{R})$ are listed in Forney and Wei (1989).

Here we can quote without proof the following important result: The figure of merit of the constellation $\mathcal{S}(\Lambda, \mathcal{R})$ having normalized bit rate $\beta$ is given by

$$\mathrm{CFM}(\mathcal{S}) \approx \mathrm{CFM}_0 \cdot \gamma_c(\Lambda) \cdot \gamma_s(\mathcal{R}) \tag{5.53}$$

| Name | $\Lambda$ | $N$ | Kissing number | $\gamma_c(\Lambda)$ (dB) |
|------|-----------|-----|----------------|--------------------------|
| Integer lattice | $\mathbf{Z}$ | 1 | 2 | 0.00 |
| Cubic lattice | $\mathbf{Z}^N$ | $N$ | $2N$ | 0.00 |
| Hexagonal lattice | $A_2$ | 2 | 6 | 0.62 |
| Schläfli | $D_4$ | 4 | 24 | 1.51 |
| Gosset | $E_8$ | 8 | 240 | 3.01 |
| Barnes-Wall | $\Lambda_{16}$ | 16 | 4,320 | 4.52 |
| Leech | $\Lambda_{24}$ | 24 | 196,560 | 6.02 |

Table 5.3: *Parameters of important lattices.*

where $CFM_0$ is the figure of merit of the one-dimensional PAM constellation (chosen as the baseline), $\gamma_c(\Lambda)$ is the coding gain of the lattice $\Lambda$ (see (5.50)), and $\gamma_s(\mathcal{R})$ is the shape gain of the region $\mathcal{R}$. The approximation holds for large constellations.

This result shows that, at least for large constellations, the gain from shaping by the region $\mathcal{R}$ is almost completely decoupled from the coding gain due to $\Lambda$ — or, more generally, the gain due to the use of a code. Thus, for a good design it makes sense to optimize separately $\gamma_c(\Lambda)$ (i.e., the choice of the lattice) and $\gamma_s(\mathcal{R})$ (i.e., the choice of the region).

The values of $\gamma_c$ for some important lattices are summarized in Table 5.3.

### 5.7.1. Spherical constellations

The maximum shape gain achieved by an $N$-dimensional region $\mathcal{R}$ is that of a sphere $\Sigma$. If $R$ is its radius and $N = 2n$, it has

$$V(\Sigma) = \frac{\pi^n R^{2n}}{n!}$$

and

$$\int_\Sigma |\mathbf{r}|^2 \, d\mathbf{r} = \frac{n}{n+1} R^2 V(\Sigma)$$

so that

$$\gamma_s(\Sigma) = \frac{\pi(n+1)}{6(n!)^{1/n}}$$

As $N \to \infty$, $\gamma_s(\Sigma)$ approaches $\pi e/6$, or 1.53 dB. The last figure is thus the maximum achievable shaping gain. A problem with spherical constellations is that the complexity of the encoding procedure (mapping input symbols to signals) may be too high.

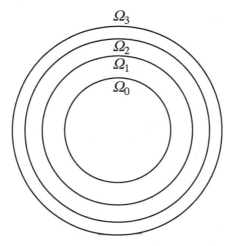

Figure 5.23: *Partitioning a 2-D constellation into equal-size rings.*

The main goal of $N$ dimensional constellation design is to obtain a shape gain as close to that of the $N$ sphere as possible, while maintaining reasonable implementation complexity and other desirable constellation features.

### 5.7.2. Shell mapping

In *shell mapping* (Laroia, Farvardin, and Tretter, 1994), the component two-dimensional signal constellation $S_2$ is partitioned into $M$ rings $\Omega_0, \ldots, \Omega_{M-1}$, each containing the same number of points (Fig. 5.23). Then some of the bits are used to select rings, while other select points in those rings. An important feature of shell mapping is that it integrates shaping and mapping.

The basic idea is as follows. Each of the $M$ rings is assigned a "cost" $c_i$, $i = 0, \cdots, M - 1$, which approximates the average energy of the points in the ring ($c_i = i$ provides a good approximation). Let $S_2^N$ denote the $2N$-dimensional constellation consisting of all possible combinations of signal points in the component 2-D constellation $S_2$. To send $b$ bits in $N$ symbols, we use the $2^b$ lowest-cost points in $S_2^N$, where the cost of a point is the sum of the costs in the 2-D rings. This is done with a sequence of table look-ups and arithmetic operations.

The shell-mapping method naturally chooses a $2^b$-point signal constellation $S$ that approximates a $2N$-dimensional sphere. For example, when $b/N = 8$ and $|S| = 256$, with shell mapping we obtain a shape gain of 0.2 dB. For $|S| = 320$ we obtain a shape gain of 0.8 dB. By using larger constellations, shaping gains approaching 1.0 dB can be obtained.

### 5.7.3.  Error probability

Assume that a finite constellation carved from a lattice is used for digital transmission over the AWGN channel. From (4.65) we have

$$P(e) \overset{\sim}{\leq} \frac{\tau}{2} \text{erfc} \left( \frac{d_{\min}}{2N_0} \right) \tag{5.54}$$

On the other hand, since all lattice points have at least one neighbor at distance $d_{\min}$, (4.60) yields

$$P(e) \geq \frac{1}{2} \text{erfc} \left( \frac{d_{\min}}{2N_0} \right) \tag{5.55}$$

By comparing the last two inequalities, we can see that the lower bound and the approximate upper bound to error probability for lattice signaling differ by a factor $\tau$, which may be rather large for high-dimensional lattices (see Table 5.3 above). When a finite constellation is used, (5.54) can still be considered a good approximation to $P(e)$ if the constellation size and the signal-to-noise ratio are both large.

## 5.8.   No perfect carrier-phase recovery

So far in this chapter we have assumed that coherent demodulation is performed, i.e., that the carrier phase is perfectly known at the receiver. When this assumption is not valid because no carrier phase estimation is performed, an alternative detection method, described in Chapter 4, is incoherent demodulation. In this section we shall consider a situation in which the receiver achieves an imperfect knowledge of the carrier phase. To understand how this can occur, consider for illustration sake the transmission of a 4-PSK signal with phases $(0, \pm\pi/2, \pi)$. Observation of the received signal under an imperfect synchronization between transmitter and receiver carrier-frequency generators will show that its phases belong to the set $\{\theta, \pm\pi/2 + \theta, \pi + \theta\}$. Based upon this observation, it may seem at first that one can easily align the received phases with those transmitted: it suffices to shift the former by $-\theta$. However, things are not so simple: in fact, any shift $\theta + k\pi/2$, $k$ any integer, produces the same received signal constellation, so that sheer observation of the latter is not sufficient to estimate the rotation induced by the carrier-phase misalignment between transmitter and receiver. We may say that this misalignment has the form $\theta + k\pi/2$, where $\theta$ can be estimated (techniques for doing so will be discussed in Chapter 9), while the remaining term, the *phase ambiguity*, remains to be corrected. In general, with $M$-PSK the phase ambiguity introduced by this process, which we denote by $\varphi$, is equal to a multiple of $2\pi/M$.

Two techniques can be used to resolve this phase ambiguity:

1. A *preamble*, i.e., a fixed symbol sequence known by the receiver, is sent at the beginning of the transmission (and whenever it may be assumed that the receiver carrier has lost its coherence). Observation of this sequence is used to remove the ambiguity: in fact, any nonzero value of $\varphi$ will cause the received symbols to differ from the fixed preamble.

2. *Differential!encoding* is used, i.e., the information to be transmitted is associated with phase differences rather than with absolute phases.

Here we describe the latter technique, which generates the so-called differentially-encoded PSK. The model of a channel introducing the phase ambiguity $\varphi$ can be easily constructed by assuming that, when the phase sequence $(\theta_n)_{n=0}^{\infty}$ is transmitted, the corresponding received phases are $(\theta_n + \varphi)_{n=0}^{\infty}$. (We neglect the noise here.) We start our discussion with a simple example.

**Example 5.5**   Consider transmission of binary PSK. Assume the transmitted phases to be

$$(0, 0, \pi, 0, \pi, \pi, 0, 0, \pi, 0, \cdots)$$

If the channel is affected by the phase ambiguity $\varphi = \pi$, we receive

$$(\pi, \pi, 0, \pi, 0, 0, \pi, \pi, 0, \pi, \cdots)$$

and hence all of the received bits differ from those transmitted.                    □

The example above shows that, while all the bits are received erroneously (which incidentally would be detected by observing that the preamble received differs from the preamble transmitted), the phase transitions between adjacent bits are left invariant by the presence of an ambiguity. Thus, if the information bits are associated with these differences rather than with absolute phases, the ambiguity has no effect on the information received.

**Example 5.5 (Continued)**   Before modulation, transform the source phases $(\theta_n)_{n=0}^{\infty}$ into the *differentially encoded* phase sequence $(\theta_n^*)_{n=0}^{\infty}$ according to the rule

$$\theta_n^* = \theta_n + \theta_{n-1}^* \qquad \mathrm{mod}\ \pi \tag{5.56}$$

where it is assumed that $\theta_{-1}^* = 0$. Thus, if the uncoded phase sequence is

$$(0, \pi, \pi, \pi, 0, \pi, 0, \pi, \pi, \cdots)$$

the differentially encoded sequence is

$$(0, 0, \pi, 0, \pi, \pi, 0, 0, \pi, 0, \cdots)$$

Assume again that the channel is affected by the phase ambiguity $\varphi = \pi$. We receive

$$(\pi, \pi, 0, \pi, 0, 0, \pi, \pi, 0, \pi, \cdots)$$

Now, we *differentially decode* the received phase sequence by inverting (5.56):

$$\hat{\theta}_n = \hat{\theta}_n^* + \hat{\theta}_{n-1}^* \qquad \mathrm{mod}\ \pi \tag{5.57}$$

where a hat denotes phase estimates. We obtain the phase sequence

$$(0, \pi, \pi, \pi, 0, \pi, 0, \pi, \pi, \cdots)$$

as it should be.

It can be seen that (5.56) and (5.57) correspond to modulo-2 operations on the source bits corresponding to the BPSK phases.                                                          □


To explain in general terms the differential coding/encoding procedure, we use here $z$-transform notations, which corresponds to describing a semi-infinite sequence $(x_n)_{n=0}^\infty$ through the power series $X(z) \overset{\triangle}{=} \sum_{n=0}^\infty x_n z^{-n}$. With this notation, we write the transmitted phase sequence as

$$\Theta(z) = \theta_0 + \theta_1 z^{-1} + \theta_2 z^{-2} + \cdots$$

and the received sequence as

$$
\begin{aligned}
\Theta(z) + \Phi(z) &= (\theta_0 + \varphi) + (\theta_1 + \varphi)z^{-1} + (\theta_2 + \varphi)z^{-2} + \cdots \\
&= \Theta(z) + \varphi(1 + z^{-1} + z^{-2} + \cdots) \\
&= \Theta(z) + \frac{\varphi}{1 - z^{-1}}
\end{aligned}
\tag{5.58}
$$

(we are still neglecting the effect of the additive Gaussian noise).

To get rid of the ambiguity term $\varphi/(1 - z^{-1})$ we may multiply the received signal (5.58) by $(1 - z^{-1})$. This is accomplished by the circuit shown in Fig. 5.24, called a *differential decoder*. In the time domain, this circuit subtracts from the phase received at any instant the phase that was received in the preceding symbol period: since both phases are affected by the same ambiguity $\varphi$, this is removed by the difference (except for the phase at time 0). The received sequence is now $(1 - z^{-1})\Theta(z) + \varphi$, which shows that the ambiguity is now removed (except at the initial time $n = 0$, as reflected by the term $\varphi$ multiplying $z^0$). Now, the information term $\Theta(z)$ is multiplied by $(1 - z^{-1})$: to recover it exactly we

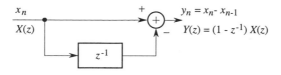

Figure 5.24: *Differential decoder.*

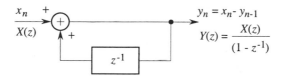

Figure 5.25: *Differential encoder.*

must divide $\Theta(z)$, before transmission, by $(1 - z^{-1})$, as shown in Fig. 5.24. This operation is called *differential encoding*.   The overall channel, including differential encoder and decoder, is shown in Fig. 5.26.

In conclusion, with differential encoding the receiver may consist of a co-herent demodulator (with imperfect phase recovery) followed by a differential decoder, as described hereafter. Another possibility is to incorporate the differ-ential decoder into the demodulator, i.e., to design a demodulator which directly outputs a phase difference. The latter avoids any estimate of the carrier phase. In the following we analyze the error performance of both receivers applied to PSK.

### 5.8.1.   Coherent demodulation of differentially-encoded PSK (DCPSK)

What is new here with respect to nondifferential PSK is that each decision made on a transmitted symbol requires a pair of $M$-ary phase decisions. Let us denote the phases received in two adjacent intervals $\beta_{k-1}$ and $\beta_k$. Introduce the phase $\psi_k$ representing the received signal's phase displacement from the transmitted

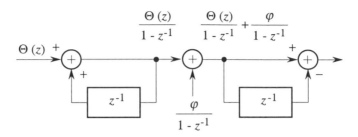

Figure 5.26: *Differential encoding and decoding for a channel affected by a phase ambiguity $\varphi$.*

one, including the effect of the phase ambiguity $\varphi$:

$$\psi_k \overset{\triangle}{=} \beta_k - (\theta_k + \varphi)$$

To evaluate the symbol error probability, we consider all possible ways of making a correct decision on the $k$th transmitted symbol. A correct decision is made if and only if the difference between $\psi_k$ and $\psi_{k-1}$ is close enough to zero. Specifically, a correct decision is made if and only if one of the $M$ following exhaustive and mutually exclusive events occurs:

$$c_i \overset{\triangle}{=} \left\{ \frac{2i-1}{M}\pi \leq \psi_{k-1} < \frac{2i+1}{M}\pi, \ \frac{2i-1}{M}\pi \leq \psi_k < \frac{2i+1}{M}\pi \right\}$$

for $i = 0, 1, \ldots, (M-1)$.

To evaluate the probability of these events, notice that the RVs $\psi_k$ and $\psi_{k-1}$ are statistically independent (they only depend on independent noise samples). If their pdf is denoted by $f_\psi(\cdot)$, and

$$p_i \overset{\triangle}{=} \int_{(2i-1)\pi/M}^{(2i+1)\pi/M} f_\psi(x)\,dx \tag{5.59}$$

we have

$$P(c_i) = p_i^2$$

and finally

$$P(e) = 1 - \sum_{i=0}^{M-1} p_i^2 \tag{5.60}$$

A pictorial interpretation of the quantities involved in (5.60) is shown in Fig. 5.27. The pdf $f_\psi(\cdot)$ is obtained in Problem 5.6.

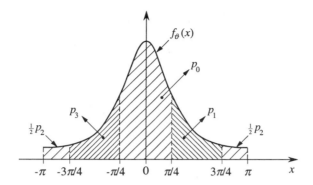

Figure 5.27: *Pictorial interpretation of the quantities involved in the derivation of error probability for differentially-encoded quaternary PSK.*

It is interesting to relate the symbol error probability (5.60), corresponding to differentially-encoded, coherently-demodulated PSK (DECPSK), to that of coherently-demodulated PSK (CPSK). Due to our definition of probabilities $p_i$, we may write the latter in the form

$$P(e)|_{\text{CPSK}} = 1 - p_0 \tag{5.61}$$

Introducing (5.61) into (5.60), we obtain

$$P(e)|_{\text{DECPSK}} = 2\ P(e)|_{\text{CPSK}} \left\{ 1 - \frac{1}{2}\ P(e)|_{\text{CPSK}} - \frac{1}{2\ P(e)|_{\text{CPSK}}} \sum_{i=1}^{M-1} p_i^2 \right\} \tag{5.62}$$

For high signal-to-noise ratios, $p_0$ is the dominant term in the right-hand side of (5.60), and (5.62) becomes

$$P(e)|_{\text{DECPSK}} \approx 2\ P(e)|_{\text{CPSK}} \tag{5.63}$$

Thus, for low error probabilities differential encoding doubles the symbol error probability. This is the price to pay for removing the phase ambiguity. It should be intuitive that with DECPSK symbol errors tend to occur in pairs: in fact, when a demodulated absolute phase is mistaken, it causes an error in two adjacent intervals.

**Example 5.6** Specialization of (5.62) to binary and quaternary PSK yields formally simple results. With $M = 2$, the only term in the summation of (5.62) is $p_1$, which from (5.59) becomes

$$p_1 = \int_{\pi/2}^{3\pi/2} f_\psi(x)\,dx = 2\int_{\pi/2}^{\pi} f_\psi(x)\,dx = P(e)|_{\text{CPSK}}$$

and therefore

$$
\begin{aligned}
P(e)|_{\text{DECPSK}} &= 2\,P(e)|_{\text{CPSK}}\left\{1 - P(e)|_{\text{CPSK}}\right\} \\
&= \left(\text{erfc}\sqrt{\frac{\mathcal{E}_b}{N_0}}\right)\left(1 - \frac{1}{2}\text{erfc}\sqrt{\frac{\mathcal{E}_b}{N_0}}\right)
\end{aligned}
$$

For quaternary signals, by observing that the in-phase and quadrature signals are independent, from (4.38)–(4.39) we obtain

$$
\begin{aligned}
P(e)|_{\text{DECPSK}} &= 2\text{erfc}\sqrt{\frac{\mathcal{E}_b}{N_0}} - 2\left(\text{erfc}\sqrt{\frac{\mathcal{E}_b}{N_0}}\right)^2 \\
&\quad + \left(\text{erfc}\sqrt{\frac{\mathcal{E}_b}{N_0}}\right)^3 - \frac{1}{4}\left(\text{erfc}\sqrt{\frac{\mathcal{E}_b}{N_0}}\right)^4
\end{aligned}
$$

$\square$

### 5.8.2.   Differentially-coherent demodulation of differentially encoded PSK

Coherent demodulation of PSK requires the local generation of a reference carrier. This may be undesirable either because of the additional complexity required or because some applications do not afford sufficient time for carrier acquisition. An approach that avoids the need for a reference carrier consists in accomplishing the demodulation by looking at the phases of the received signal in two adjacent symbol intervals and estimating their difference. If the information phases have been differentially encoded at the transmitter, then the observed phase difference at the receiver allows the recovery of the information and the removal of the phase ambiguity. The signal are still in the form (5.20), but now the information is encoded into phase differences, taking values in the set

$$\left\{\frac{2\pi}{M}(i-1) + \Phi\right\}_{i=1}^{M} \tag{5.64}$$

with $\Phi$ equal either to 0 or to $\pi/M$.

The demodulator's block diagram is shown in Fig. 5.28. It can be proved to be optimum, in the ML sense, for the estimation of the phase differences of the received signal (see Problem 5.7). The phase ambiguity of the received signal is

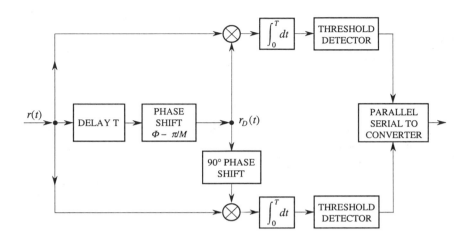

Figure 5.28: *Block diagram of the differentially-coherent demodulator of PSK.*

removed under the assumption that it remains constant in two adjacent symbol intervals.

To evaluate the error probability of this receiver we resort to a bounding technique. Let the received signal in the $k$th time interval be written as

$$r(t) = \Re \left\{ \tilde{r}(t) e^{j2\pi f_0 t} \right\} \qquad (5.65)$$

where

$$\tilde{r}(t) = A e^{j\phi_k} + \tilde{n}(t), \qquad kT \le t < (k+1)T$$

and $\tilde{n}(t)$ is the complex envelope of a narrowband Gaussian noise as described in Section 2.4.

The delayed and shifted replica of $r(t)$ can be be written as

$$r_D(t) \stackrel{\triangle}{=} \Re \left\{ \tilde{r}(t-T) e^{j(\Phi - \pi/M)} e^{j2\pi f_0 t} \right\} = \Re \left\{ \tilde{r}_D(t) e^{j2\pi f_0 t} \right\} \qquad (5.66)$$

where

$$\tilde{r}_D(t) \stackrel{\triangle}{=} A e^{j(\phi_{k-1} + \Phi - \pi/M)} + \tilde{n}(t-T) e^{j(\Phi - \pi/M)}$$

and in (5.66) we have assumed that $f_0 T$ is an integer. If this were not the case, the phase shifter of Fig. 5.28 should be adjusted to compensate for this phase shift. The receiver bases its decision on the difference between the phases of the signals (5.65) and (5.66), that is, on

$$\Delta\beta_k \triangleq \arg[\tilde{r}(t)\tilde{r}_D^*(t)], \qquad kT \le t < (k+1)T \qquad (5.67)$$

When we send a phase difference belonging to the set (5.64), the values taken on by $\Delta\beta_k$, in the absence of noise, are in the set

$$\left\{ \frac{2\pi}{M}\left(i - \frac{1}{2}\right)\right\}_{i=1}^{M} \qquad (5.68)$$

A correct decision is made if the point representing the signal $\tilde{r}(t)\tilde{r}_D^*(t)$ lies inside a sector of width $2\pi/M$ centered around the correct value of $\Delta\beta_k$.

We can observe that the problem exhibits the same kind of symmetry as for coherent detection of PSK: therefore, we may compute the error probability by limiting our consideration to $i = 1$ in (5.68). By proceeding as in Section 4.3.1, we obtain the upper bound

$$P(e) < P\left\{\gamma_2 < \Delta\beta_k < \gamma_2 + \pi\right\} + P\left\{\gamma_1 - \pi < \Delta\beta_k < \gamma_1\right\} \qquad (5.69)$$

where we have defined the two phase thresholds

$$\gamma_1 = 0, \qquad \gamma_2 = \frac{2\pi}{M} \qquad (5.70)$$

For computational purposes that we shall clarify soon, inequality (5.69) can be given the following form:

$$\begin{aligned} P(e) \;\; < \;\; & 1 - P\left\{-\frac{3\pi}{2} < \Delta\beta_k - \gamma_2 - \frac{\pi}{2} < -\frac{\pi}{2}\right\} \\ & + P\left\{-\frac{3\pi}{2} < \Delta\beta_k - \gamma_1 - \frac{\pi}{2} < -\frac{\pi}{2}\right\} \end{aligned} \qquad (5.71)$$

Let us now define, at the $k$th decision instant $t = t_k$, $kT \le t_k < (k+1)T$, the following RVs:

$$\begin{aligned} z_1 &\triangleq \tilde{r}(t_k) = Ae^{j\phi_k} + \tilde{n}(t_k) \\ z_2 &\triangleq j\tilde{r}_D(t_k)e^{j\gamma_2} = je^{j(\Phi+\pi/M)}\left\{Ae^{j\phi_{k-1}} + \tilde{n}(t_k - T)\right\} \\ z_3 &\triangleq j\tilde{r}_D(t_k)e^{j\gamma_1} = je^{j(\Phi-\pi/M)}\left\{Ae^{j\phi_{k-1}} + \tilde{n}(t_k - T)\right\} \end{aligned} \qquad (5.72)$$

It is immediately verified that

$$\Delta\beta_k - \frac{2\pi}{M} - \frac{\pi}{2} = \arg[z_1 z_2^*], \qquad \Delta\beta_k - \frac{\pi}{2} = \arg[z_1 z_3^*] \qquad (5.73)$$

Using (5.73), we can finally express (5.71) in the form

$$P(e) < 1 - P\{\Re(z_1 z_2^*) < 0\} + P\{\Re(z_1 z_3^*) < 0\} \qquad (5.74)$$

By using the identity

$$\Re(z_i z_j^*) = \left|\frac{z_i + z_j}{2}\right|^2 - \left|\frac{z_i - z_j}{2}\right|^2 \triangleq |\xi_{ij}|^2 - |\eta_{ij}|^2 \qquad (5.75)$$

in (5.74), we obtain

$$P(e) < 1 - P\{|\xi_{12}| < |\eta_{12}|\} + P\{|\xi_{13}| < |\eta_{13}|\} \qquad (5.76)$$

The four random variables involved in the right-hand side of (5.76) are independent and have a Rice distribution. With the details of the computation deferred to Problem 5.8, we obtain

$$P(e) < 1 + Q\left(\sqrt{\frac{\mathcal{E}_b}{N_0}}b_M, \sqrt{\frac{\mathcal{E}_b}{N_0}}a_M\right) - Q\left(\sqrt{\frac{\mathcal{E}_b}{N_0}}a_M, \sqrt{\frac{\mathcal{E}_b}{N_0}}b_M\right) \qquad (5.77)$$

where

$$a_M \triangleq \log_2 M\left(1 + \sin\frac{\pi}{M}\right), \qquad b_M \triangleq \log_2 M\left(1 - \sin\frac{\pi}{M}\right)$$

and $Q(\cdot, \cdot)$ is the Marcum $Q$ function (see Appendix A). For high values of $\mathcal{E}_b/N_0$, the bound (5.77) is very tight. When $M$ also is large, by using the asymptotic expansion, valid for $b \gg 1$ and $b \gg b - a$:

$$Q(a, b) \sim \frac{1}{2}\operatorname{erfc}\left(\frac{b - a}{\sqrt{2}}\right)$$

we obtain

$$P(e) \sim \operatorname{erfc}\left(\frac{\log_2 M}{2}\frac{\mathcal{E}_b}{N_0}\sin\frac{\pi}{M}\right)$$

Specialization of the above results to the binary case leads to an especially simple result. In this case the exact value of $P(e)$ can be written as

$$P(e) = P\{\pi < \Delta\beta_k < 2\pi\} \qquad (5.78)$$

and therefore, since (Appendix A)

$$Q(x, 0) = 1, \qquad Q(0, x) = e^{x^2/2}$$

we have, from the same calculations that led to (5.77):

$$P(e) = \frac{1}{2}e^{-\mathcal{E}_b/N_0} \qquad (5.79)$$

Error probability curves are plotted in Fig. 5.29.

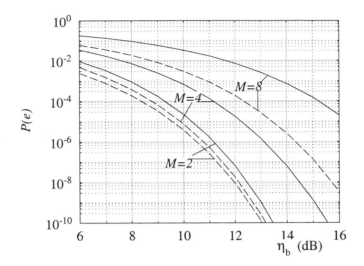

Figure 5.29: *Symbol error probability of differential demodulation of differentially-encoded M-ary PSK (solid lines). Error probabilities of CPSK are also shown for comparison (dashed lines). Here $\eta_b = \mathcal{E}_b/N_0$.*

### 5.8.3.  Incoherent demodulation of orthogonal FSK

Incoherent demodulators results in an even simpler implementation than for differentially-coherent detection. For error probability results related to orthogonal FSK we refer the reader to our discussion in Section 4.4.

## 5.9.  Digital modulation trade-offs

As mentioned at the onset of this chapter, the choice of a digital modulation scheme aims at the best trade-off among error probability, bandwidth efficiency, power efficiency, and complexity. This section summarizes the results of the chapter. For our purposes, it is interesting to compare the performance of actual modulations with the ultimate performance achievable over the AWGN channel. The latter is obtained from the channel capacity formula (5.35), which yields, when $R_s = C$:

$$\frac{\mathcal{E}_b}{N_0} = \frac{2^{C/W} - 1}{C/W} \tag{5.80}$$

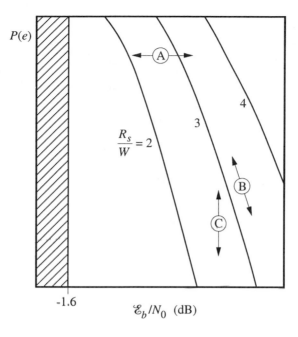

Figure 5.30: *Error-plane performance plot for PAM, PSK, QAM.*

This expression shows that $\mathcal{E}_b/N_0$ must increase exponentially with $C/W$. On the other hand, if $C/W \to 0$, then

$$\frac{\mathcal{E}_b}{N_0} = \lim_{C/W \to 0} \frac{2^{C/W} - 1}{C/W} = \ln 2 \qquad (5.81)$$

which is $-1.6$ dB. This result indicates that reliable transmission, with as low an error probability as desired, is possible over the AWGN channel only if $\mathcal{E}_b/N_0 > -1.6$ dB.

Fig. 5.30 shows qualitatively the trade-offs among error probability, power efficiency, and bandwidth efficiency. Reference is made here to modulations like PAM, PSK, and QAM, such that when $M$ is increased their power efficiency decreases, while their bandwidth efficiency increases. Moving along line A (that is, changing the value of $M$ while $P(e)$ remains constant) can be viewed as trading power efficiency for bandwidth efficiency. Similarly, line B shows how $P(e)$ can be traded off for $\mathcal{E}_b/N_0$, with fixed bandwidth. Movement along line C illustrates trading bandwidth efficiency for error probability, while $\mathcal{E}_b/N_0$ is fixed. Notice that movement along B corresponds to changing the signal energy, while movement along A or C requires changing the modulation size (which typically implies a different equipment complexity).

The *bandwidth efficiency-power efficiency* chart is a useful tool for the comparison of different modulation schemes. By selecting a value of bit error probability, a modulation/demodulation scheme can be represented in this chart as a point, whose abscissa is the value of $\mathcal{E}_b/N_0$ necessary to achieve such $P_b(e)$ and whose ordinate is its bandwidth efficiency $R_s/W$. Fig. 5.31 shows such chart for $P_b(e) = 10^{-5}$. The Shannon capacity curve shows the bound to reliable transmission of any conceivable modulation scheme. It is customary to divide the achievable region in this chart in a bandwidth-limited region ($R_s/W > 1$) and a power-limited region ($R_s/W < 1$). In the former, the system bandwidth is at a premium, and should be traded for power (i.e., $\mathcal{E}_b/N_0$). PSK and QAM are effective modulation schemes in this region, as their bandwidth efficiency increases by increasing the size of their constellation. On the other hand, FSK schemes operate in the power-limited region: they make inefficient use of bandwidth, but trade bandwidth for a reduction of $\mathcal{E}_b/N_0$ necessary to achieve a given $P_b(e)$.

The choice of a modulation scheme that achieves a target $P_b(e)$ will be guided by this chart.

## 5.10. Bibliographical notes

The discussion on bandwidth in Section 5.1.1 is taken from Amoroso (1980). The definition of Shannon bandwidth is due to Massey (1995). Further details on modulation performance, receiver implementation, etc., can be found, among others, in the book by Simon, Hinedi, and Lindsey (1995). The paper by Forney *et al.* (1984) contains many details and advances on QAM, as well an extensive bibliography on the subject. The material of Section 5.4.4 is taken from Forney and Ungerboeck (1998).

The literature on lattices is extremely abundant, but most of it is written by mathematicians for mathematicians. A thorough treatment of this topic can be found in the encyclopedic book by Conway and Sloane (1988) from which most of the material in this chapter was taken; but see also Forney (1988).

The presentation of the final section on digital modulation trade-offs was inspired by the tutorial paper by Sklar (1983).

## 5.11. Problems

*Problems marked with an asterisk should be solved with the aid of a computer.*

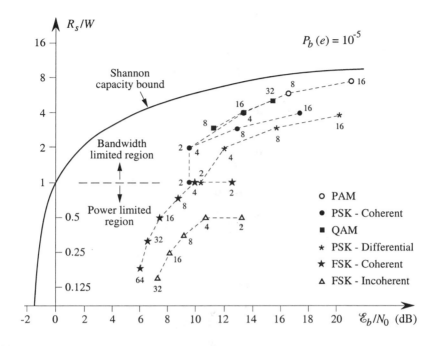

Figure 5.31: *Bandwidth efficiency–power efficiency chart of modulation schemes. Here W is the Shannon bandwidth, and $P_b(e) = 10^{-5}$.*

**5.1.** A coherent $M$-ary PAM transmission scheme is used with the constraint that the peak power of each signal be not greater than 1 mW. The noise power spectral density $N_0/2$ is 0.25 $\mu$W/Hz. A bit error probability $P_b(e) < 10^{-6}$ is required.

    **(a)** Compute the maximum possible transmission speed in bit/s for $M = 2$, $M = 4$, and $M = 8$.

    **(b)** Which one of the three schemes requires the minimum value of $\mathcal{E}_b$ (energy per bit)?

**5.2.** Consider the 64-point signal constellation of Fig. 5.32. This is obtained from a square constellation by moving to the axes the four points with highest energy. The resulting constellation is "more circular" and hence more power-efficient than the mother constellation. Evaluate the amount of improvement in power efficiency with respect to the square constellation.

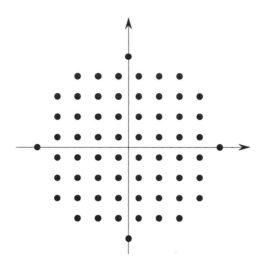

Figure 5.32: *A 64-point constellation.*

**5.3.**    **(a)** Prove that the error probability for $M$-ary PSK can be written in the form

$$P(e) = \frac{1}{\pi} \int_0^{\pi(1-1/M)} \exp\left[-\frac{\sin^2 \pi/M}{\sin^2 \theta} \frac{\mathcal{E}}{N_0}\right] d\theta \qquad (5.82)$$

   **(b)** By comparing the above equation with (5.21), derive an integral expression for the complementary error function.

**5.4.** Assume that we want to transmit $(h + 1/2)$ bits per symbol. To this purpose we use a signal constellation with $2^h$ inner points taken from the square lattice and $2^{h-1}$ outer points drawn from the same lattice with the goal of maximum symmetry and minimum average power. Two examples are shown in Fig. 5.33 for $h = 4$ (24 points) and $h = 5$ (48 points). The transmission goes as follows.

   1. Group the source bits into blocks of $2h + 1$ bits to be sent with two wave-forms ($h + 1/2$ bits per symbol).

   2. The first bit determines whether or not any outer point is to be used.

   3. If not, the remaining $2h$ bits are used to select two inner points for trans-mission in two successive periods.

   4. If yes, one additional bit selects which of the two signals should be an outer point and the remaining $h - 1$ and $h$ bits select, respectively, the outer and the inner point to be sent.

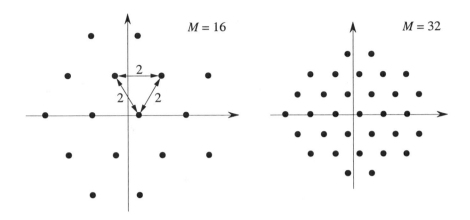

Figure 5.33: *Two constellations for the transmission of* $(h + 1/2)$ *bits per symbol.*

On the average, one outer point is transmitted every other four transmissions.

(a) Compute the average energy of the two signal sets of Fig. 5.33 and verify that the transmission of the additional $1/2$ bit per symbol requires about 1.5 dB more energy.

(b) Generalize this scheme to transmit $h + 2^{-t}$ bit/symbol, with $t$ an integer greater than 1.

**5.5.** Consider an 8-PSK constellation. By assuming a large signal-to-noise ratio, show that some mappings of three bits to PSK signals yield a different $P_b(e)$ for different bits (this effect is known as "unequal error protection"). Find the mappings that yield equal error probabilities for the three bits, and the mappings that yield the largest ratio between the maximum and the minimum bit error probability.

**5.6.** Consider transmission of a PSK signal with energy $\mathcal{E}$ over the AWGN channel with power spectral density $N_0/2$.

(a) Prove that the probability density function of the difference $\psi$ between the transmitted and the received phase is given by

$$f_\psi(x) =$$
$$\frac{1}{2\pi} e^{-\mathcal{E}/N_0} \left\{ 1 + \sqrt{\frac{\pi \mathcal{E}}{N_0}} \cos x \, e^{(\mathcal{E}/N_0)\cos^2 x} \left[ 1 + \mathrm{erf}\left( \sqrt{\frac{\mathcal{E}}{N_0}} \cos x \right) \right] \right\}$$

for $-\pi \leq x < \pi$.

(b) By using the approximation, valid for $x \gg 1$,

$$\text{erf}(x) \approx 1 - \frac{1}{\sqrt{\pi x}}$$

derive from the above an approximate expression for the error probability of coherently-demodulated $M$-ary PSK.

5.7. This problem is proposed in order to show that the demodulator of Fig. 5.28 is a maximum-likelihood detector of phase differences.

(a) First, consider (2.315) and notice that the ML receiver looks for the maximum conditional pdf $f_{\mathbf{r}|\mathbf{s}_j}(\mathbf{r} \mid \mathbf{s}_j)$, where $\mathbf{r}$ is the vector of the received signal that contains the sufficient statistics.

(b) The pair of received signals used in the detection of the transmitted signal $\mathbf{s}_j$ is the following:

$$r_1(t) = \sqrt{\frac{2\mathcal{E}}{T}} \cos(2\pi f_0 t + \varphi) + n(t)$$

$$r_2(t) = \sqrt{\frac{2\mathcal{E}}{T}} \cos\left[2\pi f_0 t + \varphi + \frac{2\pi}{M}(j-1) + \Phi\right] + n(t-T)$$

where $n(t)$ is a white Gaussian noise. Notice that the hypothesis on signal $\mathbf{s}_j$ means that the phase difference is

$$\Delta\phi_j = \frac{2\pi}{M}(j-1) + \Phi$$

(c) Show that the received signal points coordinates in the plane with axes

$$\psi_1(t) = \sqrt{\frac{2}{T}} \cos 2\pi f_0 t, \qquad \psi_2(t) = -\sqrt{\frac{2}{T}} \sin 2\pi f_0 t$$

have the form

$$
\begin{aligned}
r_{11} &= \sqrt{\mathcal{E}} \cos \varphi + n_{11} \\
r_{12} &= \sqrt{\mathcal{E}} \sin \varphi + n_{12} \\
r_{21} &= \sqrt{\mathcal{E}} \cos\left[\varphi + \frac{2\pi}{M}(j-1) + \Phi\right] + n_{21} \\
r_{22} &= \sqrt{\mathcal{E}} \sin\left[\varphi + \frac{2\pi}{M}(j-1) + \Phi\right] + n_{22}
\end{aligned}
$$

(d) Define the vector $\mathbf{r} \triangleq [r_{11}, r_{12}, r_{21}, r_{22}]$ of the received signals and show that the conditional pdf is given by

$$f_{\mathbf{r}|\mathbf{s}_j,\varphi}(\mathbf{r} \mid \mathbf{s}_j, \varphi) = A \exp(C \cos \varphi + D \sin \varphi)$$

where

$$A \triangleq \left(\frac{1}{\pi N_0}\right)^2 \exp\left\{-\frac{1}{N_0}(|\mathbf{r}|^2 + 2\mathcal{E})\right\}$$

$$C \triangleq \frac{2\sqrt{\mathcal{E}}}{N_0}(r_{11} + r_{21}\cos\Delta\phi_j + r_{22}\sin\Delta\phi_j)$$

$$D \triangleq \frac{2\sqrt{\mathcal{E}}}{N_0}(r_{12} - r_{21}\sin\Delta\phi_j + r_{22}\cos\Delta\phi_j)$$

(e) Assume that the phase ambiguity $\varphi$ has a uniform distribution and average it out. You will get

$$f_{\mathbf{r}|s_j}(\mathbf{r} \mid s_j) = AI_0(\sqrt{C^2 + D^2})$$

where $I_0(\cdot)$ is the modified Bessel function of order 0. Choosing the maximum pdf is equivalent to choosing the maximum of $(C^2 + D^2)$, since $A$ does not depend on $\Delta\phi_j$ and $I_0(\cdot)$ is a monotone increasing function of its argument when the latter is nonnegative.

(f) Switch to polar coordinates, and show that

$$C^2 + D^2 = \frac{8\mathcal{E}\rho}{N_0}[\rho + \cos(\Delta\beta - \Delta\phi_j)]$$

where

$$\rho = |\mathbf{r}_1| = |\mathbf{r}_2|$$

$$\Delta\beta = \arg(\mathbf{r}_2) - \arg(\mathbf{r}_1)$$

and

$$\mathbf{r}_i = [r_{i1}, r_{i2}]$$

(g) Decide on the optimality of the receiver.

**5.8.** In this problem we outline the computations that yield the result (5.77). Let us start from (5.74).

(a) Compute first $P\{\Re(z_1 z_2^*) < 0\}$. Define the two Gaussian random variables

$$\xi_{12} = \frac{z_1 + z_2}{2} \qquad \eta_{12} = \frac{z_1 - z_2}{2}$$

and show that they are independent under the assumption

$$E\{\tilde{n}^*(t_k)\tilde{n}(t_k - T)\} = 0$$

**(b)** The random variables $R_1 \overset{\triangle}{=} |\xi_{12}|$ and $R_2 \overset{\triangle}{=} |\eta_{12}|$ are independent Rician with pdf given by

$$f_{R_i}(x) = \frac{x}{\sigma_i^2} \exp - \frac{a_i^2 + x^2}{2\sigma_i^2} I_0 \left( \frac{a_i x}{\sigma_i^2} \right)$$

with $0 \le x < \infty$, $i = 1, 2$, and

$$a_1^2 = |E\{\xi_{12}\}|^2 = \frac{A^2}{2} \left( 1 - \sin \frac{\pi}{M} \right)$$

$$a_2^2 = |E\{\eta_{12}\}|^2 = \frac{A^2}{2} \left( 1 + \sin \frac{\pi}{M} \right)$$

$$\sigma_1^2 = \frac{1}{2} E\{|\xi_{12} - E\{\xi_{12}\}|^2\} = \frac{N_0}{2T}$$

$$\sigma_2^2 = \frac{1}{2} E\{|\eta_{12} - E\{\eta_{12}\}|^2\} = \frac{N_0}{2T} = \sigma_1^2$$

**(c)** Use formulas of Appendix A to get

$$P\{\Re(z_1 z_2^*) < 0\} = \frac{1}{2}[1 - Q(\sqrt{b}, \sqrt{a}) + Q(\sqrt{a}, \sqrt{b})]$$

where

$$a \overset{\triangle}{=} \frac{a_2^2}{\sigma_1^2 + \sigma_2^2}, \qquad b \overset{\triangle}{=} \frac{a_1^2}{\sigma_1^2 + \sigma_2^2}$$

**(d)** Following the same procedure, show that

$$P\{\Re(z_1 z_3^*) < 0\} = \frac{1}{2}[1 - Q(\sqrt{a}, \sqrt{b}) + Q(\sqrt{b}, \sqrt{a})]$$

**(e)** Conclude and obtain (5.77).

**5.9.** Consider octonary PSK with coherent demodulation and Gray mapping as in Fig. 5.5.

**(a)** Draw a block diagram of the optimum receiver with a logic device that makes decisions based only on the sign of the received signal components.

**(b)** (*) Compute the capacity of the equivalent discrete transmission channel by using for $P(e)$ the upper bound (5.23). Assume that $\mathcal{E}/N_0$ is so high that only errors between signals that are adjacent in the signal space may be taken into account. Compare this capacity with that of binary and quaternary PSK as a function of $\mathcal{E}_b/N_0$.

**(c)** Compute the exact (closed-form) expression of the bit error probability $P_b(e)$.

**5.10.** With this problem we want to show that PAM achieves capacity as $R_s/W \to \infty$. Use first the expression (3.100) for channel capacity, and show that if the transmitted power is increased by a factor $4^n$ (i.e., $\mathcal{P}' = 4^n \mathcal{P}$) then

$$\left(\frac{C_s}{W}\right)' \cong \frac{C_s}{W} + 2n$$

That is, the bandwidth efficiency increases by $2n$ bit/s/Hz. Take now the error probability (5.13) and show that the same increase of power corresponds to

$$\left(\frac{R_s}{W}\right)' \cong \frac{R_s}{W} + 2n$$

Reach a conclusion as $n \to \infty$.

**5.11.** With this problem we want to show that orthogonal FSK achieves capacity as $R_s/W \to 0$. Start from (4,45), and write the error probability in the form

$$P(e) = \frac{1}{\sqrt{2\pi}} \int_{-\infty}^{\infty} \left\{ 1 - \left[ 1 - \frac{1}{2}\text{erfc}\left(\frac{x}{\sqrt{2}}\right) \right]^{M-1} \right\} e^{-(x-\sqrt{2\eta})^2/2} \, dx$$

where $\eta \triangleq \mathcal{E}/N_0$. Use the following two bounds:

$$\left\{ 1 - \left[ 1 - \frac{1}{2}\text{erfc}\left(\frac{x}{\sqrt{2}}\right) \right]^{M-1} \right\} \leq \frac{M-1}{2}\text{erfc}\left(\frac{x}{\sqrt{2}}\right) < M e^{-x^2/2} \quad x \text{ large}$$

$$\left\{ 1 - \left[ 1 - \frac{1}{2}\text{erfc}\left(\frac{x}{\sqrt{2}}\right) \right]^{M-1} \right\} \leq 1 \quad x \text{ small}$$

Therefore,

$$P(e) < \frac{1}{2} \int_{-\infty}^{x_0} e^{-(x-\sqrt{2\eta})^2/2} \, dx + \frac{M}{\sqrt{2\pi}} \int_{x_0}^{\infty} e^{-x^2/2} e^{-(x-\sqrt{2\eta})^2/2} \, dx$$

Optimize $x_0$, and show that $x_0 = \sqrt{2\log_2 M \ln 2}$. Using simple exponential bounds for the two integrals, show that

$$P(e) < \begin{cases} 2\exp\{-\log_2 M(\eta_b - 2\ln 2)/2\}, & \eta/4 \geq \ln M \\ 2\exp\{-\log_2 M(\sqrt{\eta_b} - \sqrt{\ln 2})/2\}, & \eta/4 < \ln M \leq \eta \end{cases}$$

where $\eta_b = \eta/\log_2 M$. Notice that $P(e) \to 0$ as $M \to \infty$ provided that $\eta_b > \ln 2$ ("Shannon bound").

# Modulations for the wireless channel

This chapter describes a number of digital modulation schemes that may be thought of as being derived from PSK, in the sense that they retain the single most attractive feature of PSK—its constant envelope—but reduce certain undesired effects. One of these is the bandwidth occupancy of PSK, which may be excessive for applications, like wireless systems, that call for a highly efficient use of bandwidth. As we shall see, constraining phase-shift keying to preserve phase continuity may have a beneficial effect on spectrum occupancy: the resulting modulation scheme, called CPM (for *continuous-phase modulation*) will be described later in this chapter. Another undesired effect follows from filtering a PSK signal. Consider for simplicity the QPSK modulator described in Chapter 5. Here the carrier phase changes only once every $2T_s$ seconds, $1/T_s$ being the source binary rate. When only one of the two quadrature components, either in-phase ($I$) or quadrature ($Q$), changes its sign, a phase shift of $\pm 90°$ occurs. A change in both components generates a phase shift of $180°$. These phase jumps, which are ideally instantaneous, are shown in the phasor diagram of Fig. 6.1(a). Usually, the transmitted QPSK signal is bandlimited by a bandpass filter so as to reduce the out-of-band spectral sidelobes and prevent interference with adjacent channels; moreover, any practical modulator will exhibit reactive components which generate a filtering effect. A consequence of this filtering is that the bandlimited QPSK signal no longer exhibits a constant envelope. In fact, the occasional $180°$ phase shifts occur now in a nonzero time and cause the envelope to approach zero, as shown qualitatively in Fig. 6.2. This effect is highly undesirable when the signal undergoes nonlinear power amplification (see Chapter 14 for further details). Actually, a nonlinear amplifier operated at saturation tends

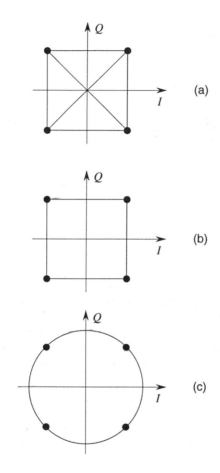

Figure 6.1: *Phasor diagrams of (a) QPSK signals; (b) OQPSK signals; (c) MSK signals.*

to restore the constant envelope of the signal, but at the same time it enhances the out-of-band spectral sidelobes. Thus, the filtering action at the transmitter is destroyed.

We proceed now with the description of a family of modulation schemes, derived from quaternary PSK and intended to limit this deleterious effect.

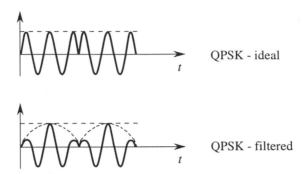

Figure 6.2: *Qualitative description of filtering effects over a QPSK signal. The dashed line shows the envelope.*

## 6.1. Variations on the QPSK theme

### 6.1.1. Offset QPSK

A reduction of the envelope fluctuations of QPSK signals is made possible by the simple device of delaying the $Q$-channel digits by $T_s$ seconds relative to the $I$-channel, as shown in Fig. 6.3. The resulting modulation scheme is called *offset QPSK* (OQPSK), or sometimes *staggered QPSK*, because the two quadrature components are offset in time by a bit period $T_s$. This solution eliminates the possibility of 180° phase changes. In fact, phase changes of only ±90° can occur every $T_s$. This feature is shown pictorially in the phasor diagram of Fig. 6.1(b). As a result, the ratio of the maximum to the minimum value of the envelope of *filtered* OQPSK signals is $\sqrt{2}$ with this simplified model, while for standard QPSK it is infinity. Therefore, it may be expected that the undesired envelope variations of QPSK due to filtering are greatly reduced, as is the dynamic range required from the power amplifier.

The complex envelope of the transmitted signal can be written as follows. Define $\Xi_k = (\xi_{2k}, \xi_{2k+1})$; then

$$v_\xi(t) = A \sum_k s(t - 2kT_s; \Xi_k) \tag{6.1}$$

where

$$s(t; \Xi_k) = \xi_{2k} f(t) + j\xi_{2k+1} f(t - T_s) \tag{6.2}$$

$$f(t) = u_{2T_s}(t) = \begin{cases} 1, & -T_s \leq t < T_s \\ 0, & \text{elsewhere} \end{cases} \tag{6.3}$$

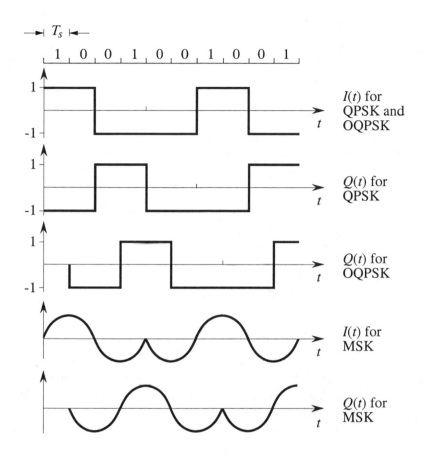

Figure 6.3: *In-phase and quadrature components of QPSK, OQPSK, and MSK signals.*

and the $\xi_k$'s are independent random variables taking values $\pm 1$ with equal probabilities. Eqs. (6.1)–(6.3) imply that the data sequence $(\xi_k)$ is split into even symbols $\xi_{2k}$ and odd symbols $\xi_{2k+1}$. These determine the sign of the shaping waveforms, which are translates of a unit square pulse with duration $2T_s$. The coherent receiver for OQPSK is identical to that of QPSK, with the only change being the delay of the $I$-stream by $T_s$ so that the two pulses carrying the even and odd symbols are realigned in time. Consequently, the error performance of this modulation is identical to that of QPSK.

**Power spectral density**

The power spectral density of the transmitted signal is not affected by the delay incurred by one of the quadrature components, and hence is the same as for QPSK. To prove this, observe that transforming (6.2) we obtain

$$S(f; \Xi_k) = (\xi_{2k} + j\xi_{2k+1}e^{-j2\pi fT_s})F(f) \tag{6.4}$$

We are in the conditions of validity of (2.164)–(2.165), where $S_i(f)$, $i = 1, 2, 3, 4$, denote the four realizations of $S(f; \Xi_k)$. Recall that the binary source symbols are independent and equally likely. Then, by observing that the four different realizations of $S(f; \Xi_k)$ sum to zero, we have from (6.4)

$$\mathcal{G}_\xi(f) = \mathcal{G}_\xi^{(c)}(f) = \frac{A^2}{8T_s} \sum_{i=1}^{4} |S_i(f)|^2 = \frac{A^2}{T_s}|F(f)|^2$$

and finally

$$\mathcal{G}_\xi(f) = 4A^2T_s \left(\frac{\sin 2\pi fT_s}{2\pi fT_s}\right)^2 \tag{6.5}$$

### 6.1.2.  Minimum-shift keying (MSK)

This modulation scheme will be discussed in greater length later on, and the reason for its name will be explained. For the time being, we view it as a modification of OQPSK, obtained by shaping the transmitted pulse. The modulated signal retains the form (6.1)–(6.2), but now

$$f(t) = \cos\left(\frac{\pi t}{2T_s}\right), \qquad -T_s \le t \le T_s \tag{6.6}$$

so that

$$f(t - T_s) = \sin\left(\frac{\pi t}{2T_s}\right), \qquad 0 \le t \le 2T_s \tag{6.7}$$

Therefore, MSK is a form of offset QPSK with a half-sinusoid amplitude shaping pulse. The shaping waveforms cause the phase transitions shown in Fig. 6.1(c).

**Power spectral density**

By duplicating the calculations done for OPSK, where now

$$F(f) = T_s \frac{4}{\pi} \frac{\cos 2\pi fT_s}{1 - 16f^2T_s^2}$$

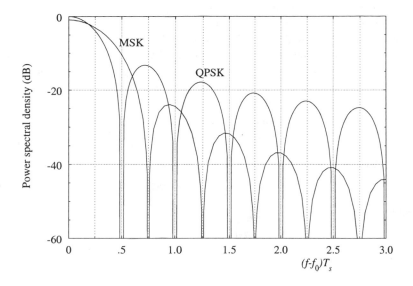

Figure 6.4: *Power density spectra of QPSK (and OPSK), and MSK.*

we obtain the power density spectrum

$$\mathcal{G}_\xi(f) = T_s \frac{16A^2}{\pi^2} \left( \frac{\cos 2\pi f T_s}{1 - 16 f^2 T_s^2} \right)^2 \tag{6.8}$$

The power spectrum of the real MSK signal is shown in Fig. 6.4 along with the spectrum of OPSK (and hence of QPSK). It can be noticed how the main lobe of MSK is wider than that of OPSK, and its sidelobes decrease more steadily than those of OPSK. In fact, from the analytical expression of the spectra, it is seen that as the frequency increases the power density spectrum of OPSK decreases as $1/f^2$, while that of MSK decreases as $1/f^4$. The power-containment bandwidths of MSK and PSK are compared in Fig. 5.2.

**MSK as a digital frequency modulation**

We shall now show that MSK can be viewed as a special form of FSK. To do this, let us focus for a moment our attention on the real part of the complex envelope of the transmitted signal, corresponding to the even-index symbol sequence $(\xi_{2k})$. In our formulation of MSK, transmission of the all-1 sequence corresponds to an in-phase part of the complex envelope consisting of a train of positive arcs of a sinusoid. Now, consider a slightly different version of MSK, in which the values taken on by $\xi_{2k}$ are alternately changed in sign. The transmitted signal that corresponds to the all-1 sequence is now the "freewheeling" sinusoid $\cos \pi t/2T_s$. A similar consideration holds for the imaginary part. Now, the signal resulting from this modified version of MSK retains the same features of the original version: the only difference between the two is a change in sign of every other source symbols, which does not alter the statistical properties of the source sequence, and hence of the modulated signal.

We can write the real signal transmitted in this modified form of MSK as

$$v_\xi(t) = I(t) \cos 2\pi f_0 t - Q(t) \sin 2\pi f_0 t \tag{6.9}$$

where now

$$I(t) = A \cos \left( \frac{\pi t}{2T_s} \right) \sum_k \xi_{2k}\, u_{2T_s}(t - 2kT_s) \tag{6.10}$$

and

$$Q(t) = A \sin \left( \frac{\pi t}{2T_s} \right) \sum_k \xi_{2k+1}\, u_{2T_s}[t - (2k-1)T_s] \tag{6.11}$$

By defining

$$\zeta_I(t) \stackrel{\triangle}{=} \sum_k \xi_{2k}\, u_{2T_s}(t - 2kT_s) \tag{6.12}$$

and

$$\zeta_Q(t) \stackrel{\triangle}{=} \sum_k \xi_{2k+1}\, u_{2T_s}[t - (2k-1)T_s] \tag{6.13}$$

we can write, after some simple trigonometry,

$$v_\xi(t) = \frac{A}{2}\zeta_I(t)[\cos 2\pi f_2 t + \cos 2\pi f_1 t] - \frac{A}{2}\zeta_Q(t)[\cos 2\pi f_2 t - \cos 2\pi f_1 t] \tag{6.14}$$

where

$$f_1 \stackrel{\triangle}{=} f_0 + \frac{1}{4T_s}, \qquad f_2 \stackrel{\triangle}{=} f_0 - \frac{1}{4T_s} \tag{6.15}$$

Now, observe that both $\zeta_I(t)$ and $\zeta_Q(t)$ keep a constant value (depending on the source symbol) over a $2T_s$-interval, and that over a $T_s$-interval this pair can take on four possible values, corresponding to the four waveforms for $v_\xi(t)$ shown in Table 6.1.

This shows that MSK can be interpreted as a form of frequency-shift keying with frequencies $f_1$ and $f_2$; notice further that two signals with the frequencies $f_1$ and $f_2$ given in (6.15) are orthogonal, and their frequency spacing is *the smallest* for orthogonality, as shown in (5.42) (see also Fig. 4.16). This is the reason why this modulation scheme is called *minimum*-shift keying.

Perusal of Table 6.1 also shows that in an interval with duration $T_s$ there are four possible signals: each of them has two other signals orthogonal to it, and one antipodal. This is a biorthogonal signaling scheme, not unlike QPSK.

| $\zeta_I(t)$ | $\zeta_Q(t)$ | $v_\xi(t)$ |
|:---:|:---:|:---:|
| $+1$ | $+1$ | $A\cos 2\pi f_1 t$ |
| $+1$ | $-1$ | $A\cos 2\pi f_2 t$ |
| $-1$ | $+1$ | $-A\cos 2\pi f_2 t$ |
| $-1$ | $-1$ | $-A\cos 2\pi f_1 t$ |

Table 6.1: *Shapes of $v_\xi(t)$ during an interval of duration $T_s$ as a function of the pair of source symbols emitted during the same interval.*

### 6.1.3.   Pseudo-octonary QPSK ($\pi/4$-QPSK)

This is a modulation scheme derived from QPSK, offering a tradeoff between the latter and OQPSK in terms of density of phase transitions.

While QPSK uses four carrier phases and has a maximum phase transition of $180°$, and OQPSK has a maximum phase transition of $90°$, pseudo-octonary QPSK, usually referred to as $\pi/4$-QPSK, uses 8 phases to carry 2 information bits per modulated symbol, and has a maximum phase transition between two adjacent symbols of $135°$. This modulation is easily amenable to differentially-coherent demodulation.

The idea here is to use two different QPSK signal constellations shifted by $\pi/4$, and to move from one to the other in every symbol interval. This guarantees a phase transition of at least $\pi/4$ in each interval, which eases symbol synchronization (see Chapter 9).

The complex envelope of the transmitted signal is

$$v_\xi(t) = A \sum_k s(t - 2kT_s; \xi_k)$$

with

$$s(t; \xi_k) = A e^{j(\varphi_k + k\pi/4)} f(t) \tag{6.16}$$

where $\varphi_k \in \{0, \pm\pi/2, \pi\}$, and $f(t) = u_{2T_s}(t)$. The signal may be differentially encoded, allowing it to be differentially demodulated. In this case it is called $\pi/4$-DQPSK. Let $\Delta_k = \theta_k - \theta_{k-1}$ denote the difference between the phases transmitted in two adjacent intervals: this difference takes on the four values $\{\pm\pi/4, \pm3\pi/4\}$. The signal-space diagrams for OQPSK and $\pi/4$-QPSK are shown in Fig. 6.5, with the continuous lines indicating the possible transitions among phases. Notice that the phase transitions never pass through the origin.

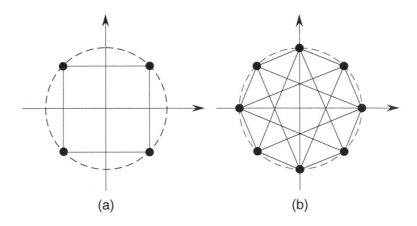

$$(a) \qquad\qquad\qquad (b)$$

Figure 6.5: *Signal-space diagrams of OQPSK (a) and $\pi/4$-QPSK (b).*

**Power density spectrum**

By using (6.16) it is easy to show that the power density spectrum is the same as for QPSK. In fact, by duplicating the spectrum calculation for OQPSK, we have

$$
\begin{aligned}
\mathrm{E}[S(f;\xi_{k+\ell})S^*(f;\xi_k)] &= \mathrm{E}\left[e^{j(\varphi_{k+\ell}+(k+\ell)\pi/4)}e^{-j(\varphi_k+k\pi/4)}\right]|F(f)|^2 \\
&= \mathrm{E}\left[e^{j(\phi_{k+\ell}-\phi_k)}\right]e^{j\ell\pi/4}|F(f)|^2 \\
&= \begin{cases} |F(f)|^2, & \ell=0 \\ 0, & \ell\neq 0 \end{cases}
\end{aligned}
$$

**Demodulation**

Coherent demodulation can be achieved by feeding a standard QPSK demodulator with a received signal sequence shifted by $\pi/4$ every $2T_s$. A differential demodulation scheme is shown in Fig. 6.6. The received signal, after being bandpass-filtered, is sent to a coherent demodulator which separates the in-phase and quadrature components and samples them synchronously to derive the two sequences $(w_k)$ and $(z_k)$. These samples are then processed to obtain the two new sequences

$$x_k = w_k w_{k-1} + z_k z_{k-1}$$

and

$$y_k = z_k w_{k-1} - w_k z_{k-1}$$

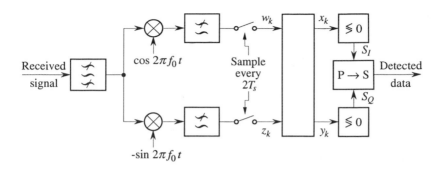

Figure 6.6: *Differential demodulation of $\pi/4$-DQPSK.*

to be used for decisions. With $\theta_0$ denoting the unknown initial carrier phase, which (as usual with differential detection) is assumed to remain practically constant during a pair of symbol intervals, we may write, under the simplifying assumption that there is no noise, $w_k = \cos(\theta_k - \theta_0)$ and $z_k = \sin(\theta_k - \theta_0)$, so that

$$
\begin{aligned}
x_k &= \cos(\theta_k - \theta_0)\cos(\theta_{k-1} - \theta_0) + \sin(\theta_k - \theta_0)\sin(\theta_{k-1} - \theta_0) \\
&= \cos(\theta_k - \theta_{k-1}) \\
&= \cos \Delta_k
\end{aligned}
$$

and, similarly,

$$
\begin{aligned}
y_k &= \sin(\theta_k - \theta_0)\cos(\theta_{k-1} - \theta_0) - \cos(\theta_k - \theta_0)\sin(\theta_{k-1} - \theta_0) \\
&= \sin(\theta_k - \theta_{k-1}) \\
&= \sin \Delta_k
\end{aligned}
$$

The decision device outputs $S_I = 1$ if $x_k > 0$ (and $S_I = 0$ otherwise), $S_Q = 1$ if $y_k > 0$ (and $S_Q = 0$ otherwise). A parallel-to-serial converter (denoted $P \to S$ in Fig. 6.6) outputs the stream of detected binary data.

## 6.2. Continuous-phase modulation

We now describe an exceedingly general family of modulations, which retain the basic feature of PSK (and of FSK) of having a constant envelope, while decreasing the spectrum occupancy of the latter by smoothing the phase transitions of the transmitted signal.

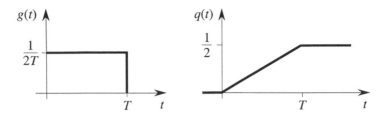

Figure 6.7: *Time pulse $g(t)$ with duration $T$ and area 1/2, and its integral $q(t)$.*

The general expression for a constant-envelope, phase-modulated signal is

$$v_\xi(t) = \sqrt{\frac{2\mathcal{E}_s}{T}} \cos[2\pi f_0 t + \theta(t; \boldsymbol{\xi}) + \theta_0] \qquad (6.17)$$

where $\mathcal{E}_s$ is the energy per symbol of the signal, $\boldsymbol{\xi}$ denotes the source-symbol sequence, $T$ is the symbol interval, and $\theta_0$ is the initial phase of the carrier. If we examine the evolution of the information-bearing phase $\theta(t; \boldsymbol{\xi})$ for PSK, we note that this is a piecewise-constant function with jumps taking place at every phase transition. A major cause for the wide spectral occupancy of PSK can be seen in the discontinuities in its phase function: in fact, smoother signals have a more compact spectrum. To reduce spectral occupancy, one may think of smoothing out the phase discontinuities, which is precisely the idea underlying continuous-phase modulation (CPM). Here the phase $\theta(t; \boldsymbol{\xi})$ is generated as the *integral* of another time function: by choosing the latter regular enough (i.e., without delta functions), a continuous phase is easily obtained.

For example, let us start with an $M$-ary PAM signal

$$x_\xi(t) = \sum_{n=0}^{\infty} \xi_n g(t - nT) \qquad (6.18)$$

where $\xi_n = \pm 1, \pm 3, \ldots, \pm(M-1)$ and $g(t)$ is a rectangular signal with duration $T$; its area is chosen to be 1/2 for later convenience (see Fig. 6.7). Next, generate a signal whose instantaneous frequency (apart from a factor $2\pi$, this is the derivative of the signal phase) is

$$f_0 + 2 f_d T \, x_\xi(t)$$

where $f_d$ is the peak frequency deviation when the signals are binary ($\xi_n = \pm 1$). This is equivalent to generating the modulated signal

$$v_\xi(t) = \sqrt{\frac{2\mathcal{E}_s}{T}} \cos\left[2\pi f_0 t + 4\pi f_d T \int_0^t x_\xi(\tau) \, d\tau + \theta_0\right] \qquad (6.19)$$

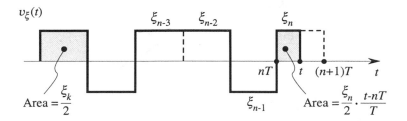

Figure 6.8: *Evolution of the phase of a CPM signal in one symbol interval.*

By comparing (6.17) with (6.19), we have, apart from the initial carrier phase:

$$\theta(t; \boldsymbol{\xi}) = 4\pi f_d T \int_0^t x_\xi(\tau) \, d\tau \qquad (6.20)$$

Let us determine the evolution of the phase $\theta(t; \boldsymbol{\xi})$ within the symbol interval $[nT, (n+1)T]$. From Fig. 6.8 we have, by computing the integral in the latter equation,

$$\theta(t; \boldsymbol{\xi}) = 4\pi f_d T \cdot \frac{1}{2} \sum_{k=0}^{n-1} \xi_k + 4\pi f_d T \cdot \frac{\xi_n}{2} \frac{t - nT}{T} \qquad (6.21)$$

that is,

$$\theta(t; \boldsymbol{\xi}) = \theta_n + 2\pi h \xi_n q(t - nT) \qquad (6.22)$$

where we have made the positions

$$\theta_n = \pi h \sum_{k=0}^{n-1} \xi_k \qquad (6.23)$$

with $h = 2f_d T$ (the "modulation index") and $q(t)$ the integral of $g(t)$ (see Fig. 6.7).

The signal we have obtained by combining (6.17) with (6.22) is a special case of CPM called continuous-phase FSK (CPFSK). The reason for its name is that the instantaneous frequency of $v_\xi(t)$ varies every $T$ according to the source symbol, while its phase is continuous. The most general version of CPM will be described later: it allows $g(t)$ to take a nonrectangular shape and a duration greater than $T$. However, before doing this we take a closer look at this simpler version.

As a further concession to simplicity, let us assume a binary source, that is, $\xi_k = \pm 1$. The *phase tree* of binary CPFSK is shown in Fig. 6.9, under the

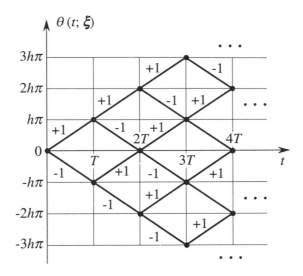

Figure 6.9: *Evolution of the phase of binary CPFSK.*

assumption that at $t = 0$ the phase is 0 (that is, $\theta_0 = 0$). This is the ensemble of all the phase trajectories, and completely describes the signal except for the carrier frequency $f_0$ and its initial phase $\theta_0$. It is seen from the figure that the phase is continuous. At the beginning of each symbol interval it is allowed to increase (if $\xi_k = +1$) or to decrease (if $\xi_k = -1$) by the quantity $h\pi$. If for example $h = 1/2$, at the end of each symbol interval the possible values taken on by the signal phase are (after reduction modulo $2\pi$) 0, $\pm\pi/2$, and $\pi = -\pi$.

The phases that differ by an integer multiple of $2\pi$ are physically indistinguishable. Thus, when after reduction mod $2\pi$ the values taken on by $\theta(t; \boldsymbol{\xi})$ at the end of each interval are of finite number, the phase tree can be made to collapse into a *phase trellis*, as shown in Fig. 6.10 for $h = 1/2$. This figure should be interpreted as wrapped on a cylinder, because the ordinates $-\pi$ and $\pi$ are actually one and the same, as are the pairs of points labeled A and B.

### 6.2.1. Time-varying vs. time-invariant trellises

From Fig. 6.10 we observe that at the end of each symbol interval only two phase values are alternately allowed: either $\pm\pi/2$ or $0, \pi$. This phase trellis is time-varying, in the sense that the phase trajectories in the even-numbered symbol intervals are not time translations of those in odd-numbered symbol intervals.

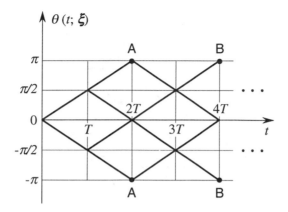

Figure 6.10: *Trellis of the phase of binary CPFSK with* $h = 1/2$.

Now, this can be a nuisance when the trellis is used for demodulation. As we shall soon see, the Viterbi algorithm can be used for demodulating CPM, and the complexity of the algorithm increases if the trellis is not time-invariant.

One way of making the trellis time-invariant is the following. If we measure the phase relative to the lowest phase trajectory in Fig. 6.10, this new "tilted" phase, defined by

$$\psi(t; \boldsymbol{\xi}) = \theta(t; \boldsymbol{\xi}) + \frac{\pi t}{2T}$$

has the phase tree and the phase trellis shown in Fig. 6.11 and 6.12, respectively. The trellis of Fig. 6.12 is now time invariant, i.e., the phase trajectories in any two symbol intervals (after a $T$-second transient due to our constraint of having zero phase at the origin) are time translates of one another.

A different method to obtain a time-invariant trellis is the following. Observe from (6.22) that $\theta(t; \boldsymbol{\xi})$ depends separately on $\theta_n$ and on $\xi_n$: the phase starts at value $\theta_n$ at time $nT$, and the value of $\xi_n$ forces its transition to the value $\theta_{n+1} = \theta_n + \pi h \xi_n$ at time $(n+1)T$. Thus, we may list as states all the values that $\theta_n$ can possibly take, irrespective of the value of $n$; they are joined by branches labeled by the values of $\xi_n$. It can be seen that this "natural" construction leads to a higher number of states than with the tilted-phase trellis: for example, binary CPFSK with $h = 1/2$ has a tilted-phase trellis with two states (Fig. 6.12), while it has 4 states (corresponding to the phases 0, $\pm\pi/2$, $\pi$) with the latter construction.

What can be done with the tilted-phase trellis can also be done with the natu-

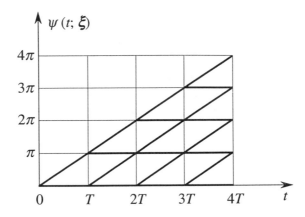

Figure 6.11: *Tilted-phase tree of binary CPFSK with* $h = 1/2$.

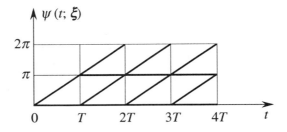

Figure 6.12: *Tilted-phase trellis of binary CPFSK with* $h = 1/2$.

ral phase trellis, because the two are equivalent. However, the use of the former can often simplify analyses and realizations of CPM (Rimoldi, 1988). In the following, the tilted-phase trellis will be used to evaluate the power-density spectrum of a special case of CPM, and to derive a number of equivalent realizations of MSK receivers.

## 6.2.2.  General CPM

We are now ready to generalize the simple case examined so far, and derive a more general version of CPM. Since our goal is to smooth out the phase trajectories in order to achieve a more compact power spectrum for the modulated signal, we may consider phase transitions that are even smoother than those provided by CPFSK. Specifically, while the transmitted signal retains the form (6.17)–(6.18), now the "frequency pulse" $g(t)$, a time function with area $1/2$, is not necessarily rectangular, and its duration is generally $LT$, $L$ an integer $\geq 1$. If $L = 1$, we talk of *full response* CPM, otherwise of *partial response* CPM. The integral of $g(t)$ is the "phase pulse" $q(t)$

$$q(t) = \int_0^t g(\tau)\,d\tau$$

which takes value zero for $t \leq 0$ and $1/2$ for $t \geq LT$.

By generalizing previous calculations, from (6.20) we have, for the evolution of $\theta(t; \boldsymbol{\xi})$ in the interval $nT \leq t \leq (n+1)T$:

$$
\begin{aligned}
\theta(t; \boldsymbol{\xi}) &= 2\pi h \sum_{k=0}^{n} \xi_k q(t - kT) \\
&= \pi h \sum_{k=0}^{n-L} \xi_k + 2\pi h \sum_{k=n-L+1}^{n} \xi_k q(t - kT) \\
&\triangleq \theta_n + \phi_n(t)
\end{aligned}
\tag{6.24}
$$

In the second line of last equation, the first term represents the contribution to the phase at $t$ of the "exhausted" phase pulses, while the second term describes the behavior of the pulses that are still evolving toward their final value $1/2$. Let us further analyze $\phi_n(t)$. We have

$$
\begin{aligned}
\phi_n(t) &\triangleq 2\pi h \sum_{k=n-L+1}^{n} \xi_k q(t - kT) \\
&= 2\pi h \sum_{k=n-L+1}^{n-1} \xi_k q(t - kT) + 2\pi h \xi_n q(t - nT)
\end{aligned}
\tag{6.25}
$$

We observe that the first term in the last line is a function of the past source symbols $(\xi_{n-1}, \xi_{n-2}, \ldots, \xi_{n-L+1})$, while the remaining term is a function of $\xi_n$ alone, the present symbol. Using the expression just derived, we can again represent the evolution of the phase $\theta(t, \boldsymbol{\xi})$ by using a trellis. However, in the present case $\theta_n$ is not sufficient to describe this evolution, due to the contribution of "non-exhausted" pulses. A trellis whose states are in one-to-one correspondence with the values taken on by the phase at the end of each symbol interval is also not sufficient. Here we need, according to (6.24), a *state-trellis*, which describes the transitions among the states

$$\cdot\ \sigma_n = (\theta_n, \xi_{n-1}, \xi_{n-2}, \ldots, \xi_{n-L+1}) \tag{6.26}$$

The value of $\xi_n$ forces a transition between a pair of states $\sigma_n \to \sigma_{n+1}$.

How many states are there in the state-trellis? The $L - 1$ variables $\xi_{n-1}, \ldots, \xi_{n-L+1}$, that we assume to be $M$-ary and statistically independent, can take on a total of $M^{L-1}$ values. This figure is to be multiplied by the number of values that $\theta_n$ can possibly take to yield the number of states.

Assume that $h$ is a rational number:

$$h = \frac{m}{p}$$

with $m$ and $p$ relatively prime integers. We see from (6.24) that $\theta_n$ is a sum of integers multiplied by $\pi h$. Such a sum can take on all the positive and negative integer values. Consider first the case of $m$ even. The sequence of possible values of $\theta_n$ is:

$$0, \quad \pi\frac{m}{p}, \quad 2\pi\frac{m}{p}, \quad 3\pi\frac{m}{p}, \cdots$$

After $p$ different terms, (the last among them being $(p-1)\pi m/p$), we obtain $p\pi m/p = m\pi$, which is congruent to 0 under our assumption of even $m$. Similarly, for odd $m$, we have the sequence

$$0, \quad \pi\frac{m}{p}, \quad \cdots \quad (2p-1)\pi\frac{m}{p}$$

that includes $2p$ terms, which then keep on reproducing themselves mod $2\pi$. In conclusion, the number of states is given by

$$S = \begin{cases} pM^{L-1} & m \text{ even} \\ 2pM^{L-1} & m \text{ odd} \end{cases} \tag{6.27}$$

**Example 6.1**   The state-trellis of a binary full-response CPM scheme (that is, $M = 2$, $L = 1$) with $h = 1/2$ (that is, $m = 1$ and $p = 2$) has $S = 4 \times 2^0 = 4$ states. If, with $M = 2$, $L$ is increased to 2 and the modulation index is changed to $h = 3/4$, the number of states is increased to $S = 16$. Specifically, $\sigma_n = (\theta_n, \xi_{n-1})$, with $\theta_n$ taking on the 8 values $0, \pm\pi/4, \pm\pi/2, \pm 3\pi/4$, and $\pi$, while $\xi_{n-1} = \pm 1$.   □

The final step in studying the state-trellis of CPM is taken by examining the structure of its transitions from one state to the next. By recalling from (6.24) that

$$\theta_n = \pi h \sum_{k=0}^{n-L} \xi_k$$

the new state becomes

$$\begin{aligned} \sigma_{n+1} &= (\theta_{n+1}, \xi_n, \xi_{n-1}, \cdots \xi_{n-L+2}) \\ &= (\theta_n + \pi h \xi_{n-L+1}, \xi_n, \xi_{n-1}, \cdots \xi_{n-L+2}) \end{aligned}$$

**Example 6.1 (Continued)**    Consider again the case of Example 6.1 above, with $h = 3/4$, $M = 2$, and $L = 2$. Assume that $\theta_n = 3\pi/4$ and $\xi_{n-1} = -1$, that is, $\sigma_n = (3\pi/4, -1)$. We have

$$\sigma_{n+1} = (3\pi/4 + \pi h\xi_{n-1}, \xi_n) = (0, \xi_n).$$

This means that the new state can be either $(0, +1)$ or $(0, -1)$, according to the value of the symbol output by the source at time $n$. Fig. 6.13 shows the complete trellis, constructed by repeating, *mutatis mutandis*, the calculations above.                    □


**Tilted-phase trellis for general CPM**

Although only the "natural" trellis was considered in previous calculations, a tilted-phase trellis for general CPM can also be constructed. The tilted phase is defined as

$$\psi(t; \xi) \triangleq \theta(t; \xi) + \pi h(M - 1)t/T \qquad (6.28)$$

Consideration of the *modified data sequence*

$$\chi_k = [\xi_k + (M - 1)]/2 \qquad (6.29)$$

taking values in the set $\{0, 1, \cdots, M - 1\}$, allows one to obtain a time-invariant trellis, as before (see Problem 6.9 for details).

### 6.2.3.    Power spectrum of full-response CPM

In this section we deal with the power spectral density of CPM. Since the general theory is rather complicated, here for simplicity's sake we limit ourselves to $M$-ary full-response ($L = 1$) CPM whose modulation index $h$ is a rational number of the form $h = J/M$. Let us write the complex envelope of the signal $v_\xi(t)$ in (6.17):

$$\tilde{v}_\xi(t) = \sqrt{\frac{2\mathcal{E}_s}{T}} \exp j[\theta(t; \xi)] \qquad (6.30)$$

By introducing the tilted phase (6.28) and the modified data sequence (6.29), we can also write

$$\tilde{v}_\chi(t) = \sqrt{\frac{2\mathcal{E}_s}{T}} \exp j[2\pi f_h t + \psi(t; \chi)] \qquad (6.31)$$

where

$$f_h \triangleq -h(M - 1)/2T$$

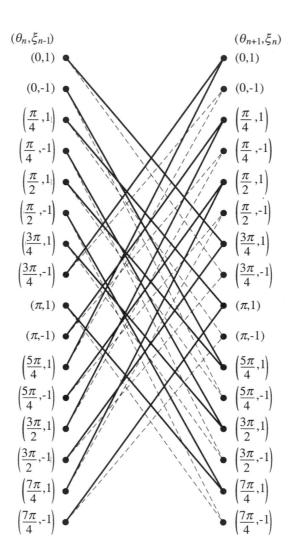

Figure 6.13: *State-trellis of a CPM scheme with* $M = 2$, $L = 2$, *and* $h = 3/4$.

We use the technique described in Section 2.3, which requires only the straightforward computation of a few relevant quantities and makes the computation rather easy.

Observe that for $L = 1$ the tilted phase (6.28) may be written in the form

$$\psi(t; \psi_n, \chi_n) = 2\pi h \psi_n + 4\pi h \chi_n q(t) + w(t) \qquad (6.32)$$

where

$$w(t) \triangleq \pi h(M - 1)t/T - 2\pi h(M - 1)q(t) \qquad (6.33)$$

is a symbol-independent term. The transitions among the states $\psi_n$ are driven by the symbols $\chi_n$, and we can write

$$\psi_{n+1} = \psi_n + \chi_n \qquad (6.34)$$

where the sum has to be reduced modulo the number $S$ of states, which we label from 0 to $S - 1$. Under our assumption of independent and equally likely source symbols, the state sequence of CPM described by its tilted-phase trellis forms a fully regular Markov chain with $S$ states.

Consider the calculation of the quantities defined in Section 2.3. From (6.34) we may see that the matrices $\mathbf{E}_k$ have the form

$$\mathbf{E}_k = \mathbf{E}_1^k, \qquad k = 0, 1, \ldots, S - 1$$

where $\mathbf{E}_1$ is the matrix whose effect on a vector with $S$ components is to cyclically shift its components by one step to the left. Thus

$$\mathbf{E}_S = \mathbf{E}_0 = \mathbf{I}$$

where $\mathbf{I}$ denotes the $S \times S$ identity matrix, and among the $M$ matrices $\mathbf{E}_k$ there are $M/S$ matrices equal to $\mathbf{E}_1$, $M/S$ matrices equal to $\mathbf{E}_2$, and so on. Thus, the transition probability matrix of the Markov chain is

$$\mathbf{P} \triangleq \frac{1}{M} \sum_{k=0}^{M-1} \mathbf{E}_k = \frac{1}{M} \cdot \frac{M}{S} \sum_{k=0}^{S-1} \mathbf{E}_k = \frac{1}{S} \mathbf{J}$$

where $\mathbf{J}$ denotes the $S \times S$ matrix all of whose elements are 1. Since $\mathbf{PP} = \mathbf{P}$, we have

$$\mathbf{P}^\infty = \mathbf{P} = \begin{bmatrix} \mathbf{w} \\ \mathbf{w} \\ \vdots \\ \mathbf{w} \end{bmatrix}$$

where

$$\mathbf{w} = \frac{1}{S}\begin{bmatrix} 1 & 1 & \cdots & 1 \end{bmatrix}$$

Similarly, we get

$$\mathbf{D} = \frac{1}{S}\mathbf{I}$$

We also obtain

$$[s_k]_i \triangleq \mathcal{F}\{\exp[\psi(t; \psi_n, \chi_n)]\} = \epsilon_h^i \alpha_k(f)$$

for $i = 0, \cdots, S - 1$ and $k = 0, \cdots, M - 1$, where

$$\epsilon_h \triangleq e^{j2\pi h}$$

(notice that $\epsilon_h^S = 1$) and

$$\alpha_k(f) \triangleq \mathcal{F}\left\{ Ae^{j[2\pi f_h t + 4\pi hkq(t) + w(t)]} \right\} \tag{6.35}$$

We can write

$$s_k(f) = \alpha_k(f)\begin{bmatrix} 1 & \epsilon_h & \epsilon_h^2 & \cdots & \epsilon_h^{S-1} \end{bmatrix}$$

In conclusion, we obtain

$$\mu(f) = 0$$

$$c_0(f) = \frac{1}{M}\sum_{k=0}^{M-1} |\alpha_k(f)|^2$$

$$\mathbf{c}_1(f) = \frac{1}{SM}\begin{bmatrix} 1 & \epsilon_k & \cdots & \epsilon_h^{S-1} \end{bmatrix}\sum_{k=0}^{M-1} \alpha_k(f)\epsilon_h^{S-k}$$

$$\mathbf{c}_2(f) = \frac{1}{M}\begin{bmatrix} 1 & \epsilon_k & \cdots & \epsilon_h^{S-1} \end{bmatrix}\sum_{k=0}^{M-1} \alpha_k(f)$$

and

$$\mathbf{A}(f) = \left(\mathbf{I} - \frac{1}{S}\mathbf{J}\right)e^{-j2\pi fT}$$

Hence, there is no discrete spectrum, and

$$\begin{aligned}
\mathcal{G}_\xi(f) &= \frac{1}{T}\left[\frac{1}{M}\sum_{k=0}^{M-1} |\alpha_k(f)|^2\right] \\
&+ \frac{2}{T}\Re\left\{ e^{-j2\pi(f-f_h)T}\frac{1}{M}\sum_{k=0}^{M-1} \epsilon_h^k \alpha_k^*(f)\frac{1}{M}\sum_{\ell=0}^{M-1} \alpha_\ell(f)\right\} \tag{6.36}
\end{aligned}$$

If desired, (6.36) can be given a more compact form by defining the vector

$$\boldsymbol{\alpha}(f) = [\alpha_0(f), \cdots, \alpha_{M-1}(f)]$$

and the matrix $\mathbf{H}(f)$ whose entry $i, j$ is $\delta_{ij} + (2/M)e^{-j2\pi(f-f_h)T}\epsilon_h^i$, for $i, j = 0, \ldots, M-1$, and with $\delta_{ij}$ denoting the Kronecker symbol. These definitions yield

$$\mathcal{G}_\xi(f) = \frac{1}{TM}\Re[\boldsymbol{\alpha}(f)\mathbf{H}\boldsymbol{\alpha}^\dagger(f)] \tag{6.37}$$

Notice that the calculation of (6.37) requires only the computation of the Fourier transforms $\alpha_k(f)$ defined in (6.35).

**Example 6.2 (CPFSK)**   Eq. (6.36) can be specialized to CPFSK by letting $q(t) = t/2T$, and consequently $w(t) = 0$ from (6.33). Direct calculation shows that

$$\alpha_k(f) = a_0(fT - kh)e^{-j\pi((f-f_h)T-kh)}$$

where

$$a_0(fT) = AT\frac{\sin\pi(f-f_h)T}{\pi(f-f_h)T}$$

Thus,

$$\begin{aligned}
\mathcal{G}(f) = {} & \frac{1}{T}\left[\frac{1}{M}\sum_{k=0}^{M-1}a_0^2(fT-kh)\right] \\
& + \frac{2}{T}\Re\left\{e^{-j2\pi(f-f_h)T}\frac{1}{M}\sum_{k=0}^{M-1}\epsilon_h^k a_0(fT-kh)\right. \\
& \times\left.\frac{1}{M}\sum_{\ell=0}^{M-1}\epsilon_h^{\ell-k}a_0(fT-k\ell)\right\}
\end{aligned} \tag{6.38}$$

$\square$

The general case of partial-response CPM will not be dealt with in detail (see the bibliographical notes at the end of this chapter for references to the computation of general spectra). Here it suffices to observe that the power spectrum of CPM generally depends on the values of $h$, $L$, and $M$, and on the shape of the frequency pulse $g(t)$. Small values of $h$ generate a small bandwidth occupancy, as do smooth pulses. For example, the raised-cosine pulse

$$g(t) = \left\{\begin{array}{ll} \dfrac{1}{2LT}\left(1 - \cos\dfrac{2\pi t}{LT}\right), & 0 \le t < LT \\ 0, & \text{otherwise} \end{array}\right.$$

| Frequency pulse | $L$ | $h$ | 99% bandwidth | 99.9% bandwidth |
|---|---|---|---|---|
| Rectangular | 1 | 0.5 | 1.2 | 2.8 |
| Rectangular | 2 | 0.5 | 0.9 | 1.9 |
| Raised-cosine | 2 | 0.5 | 1.1 | 1.6 |
| Rectangular | 2 | 0.7 | 1.2 | 2.1 |
| Raised-cosine | 2 | 0.7 | 1.1 | 1.9 |

Table 6.2: *Power-containment bandwidth as a fraction of* $1/T_s$ *for some CPM schemes.*

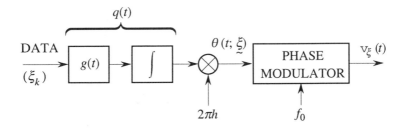

Figure 6.14: *Structure of a modulator for continuous-phase signals.*

$L$ an integer $\geq 1$, results into a CPM scheme whose bandwidth occupancy is smaller than that of CPFSK. As $L$ increases, the pulse $g(t)$ becomes smoother, and the bandwidth occupancy is reduced, at the price of an increase of the number of trellis states, and hence of the demodulator complexity if the Viterbi algorithm is used (see below). Some values of power-containment bandwidth are shown in Table 6.2.

### 6.2.4.  Modulators for CPM

A general modulator for CPM is shown in Fig. 6.14. The data, after modulating the phase pulse $q(t)$ and being multiplied by $2\pi h$, are sent into a phase modulator (a device which outputs a signal whose phase is equal to its input). The output of the latter is the CPM signal. The observation that taking the derivative of a signal phase, and dividing it by $2\pi$, yields the instantaneous frequency of the signal, leads to the alternative structure shown in Fig. 6.15, which contains a frequency modulator (i.e., a voltage-controlled oscillator—see Chapter 9).

Other forms of modulators are possible—some of them will be described at the end of this chapter for the special case of MSK.

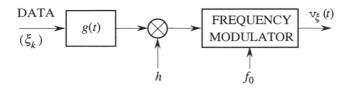

Figure 6.15: *Alternative structure of a modulator for continuous-phase signals.*

### 6.2.5.  Demodulating CPM

From the theory developed in Section 4.2, we understand that optimum demodulation of CPM in additive white Gaussian noise consists of selecting the data sequence $\xi$ that, upon observation of the received signal $r(t)$, minimizes the integral

$$\int_{\mathcal{I}} \left\{ r(t) - \sqrt{\frac{2\mathcal{E}_s}{T}} \cos[2\pi f_0 t + \theta(t;\,\xi)] \right\}^2 dt \qquad (6.39)$$

where $\mathcal{I}$ is the observation interval. We can represent this graphically by looking at the trellis that represents, for a given CPM scheme, the set of allowable phase trajectories. There is a one-to-one correspondence between an allowable phase trajectory and a data sequence $\xi$: thus, the problem of demodulating CPM can be viewed as the problem of choosing, among all the phase trajectories, the one *closest* to the phase trajectory of $r(t)$, in the sense that (6.39) is minimized. Fig. 6.16 illustrates this qualitatively.

The evolution of the modulated-signal phase can be described through its state-trellis, and the demodulation performed through the Viterbi algorithm. We start by expanding the square in (6.39): by doing this we are left with three terms, only one of which depends on $\xi$ under the usual assumption of a large enough carrier frequency. Thus, minimizing (6.39) is equivalent to maximizing the scalar product

$$\int_{\mathcal{I}} r(t) \cos[2\pi f_0 t + \theta(t;\,\xi)]\, dt \qquad (6.40)$$

Let us first split this term into a sum of contributions, each coming from one symbol interval $[nT, (n+1)T]$:

$$\int_{nT}^{(n+1)T} r(t) \cos[2\pi f_0 t + \theta(t;\,\xi)]\, dt$$

After observing $r(t)$, we may label each branch of the CPM state-trellis with the value of the above integral. We are left with a trellis all of whose branches

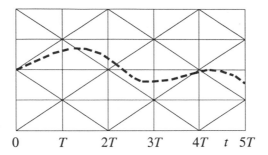

Figure 6.16: *Continuous lines: Possible phase trajectories of a CPM signal. Dashed line: Phase trajectory of the received signal* $r(t)$.

carry a label, and the demodulation problem consists of choosing the trajectory in the CPM trellis whose sum of labels is a maximum. This problem can be solved by using the Viterbi algorithm as described in Appendix F. An important consequence of this is the fact that the demodulator complexity is proportional to the number of trellis states.

Later in this chapter we shall examine the demodulator operation in more depth in one special case.

## Truncating the Viterbi algorithm

Consider the transmitted signal corresponding to a sequence $\boldsymbol{\xi} = (\xi_0, \ldots, \xi_{K-1})$ of information symbols taking values in the set $\{\pm 1, \pm 3, \ldots, \pm(M-1)\}$. The Viterbi algorithm outputs the maximum-likelihood decision on the whole transmitted sequence after observing the whole received signal. Since $K$, and hence the observation length, may be so large as to make impractical both the path storage and the decision delay implied by ideal implementation of the Viterbi algorithm, the truncated version of the algorithm (see Appendix F) is used in practice. That is, practical Viterbi processors *force* a decision at time $t = NT$, $N < K$. This is done by making a decision on the first source symbol $\xi_0$ based on the observation of the signal received up to $t = NT$. After the decision is made for $\xi_0$, the process is repeated for $\xi_1$ at time $t = (N+1)T$, and so forth.

**Error probability of CPM**

As we learned in Section 4.3, the pairwise error probability of the two signals associated with the symbol sequences $\xi$ and $\hat{\xi}$ depends on the minimum Euclidean distance between the two signals. That is, recalling (6.17) we have

$$
\begin{aligned}
d^2(v_\xi(t), v_{\hat{\xi}}(t)) &\triangleq \frac{2\mathcal{E}_s}{T} \int_0^{KT} [v_\xi(t) - v_{\hat{\xi}}(t)]^2 \, dt \\
&= \int_0^{KT} v_\xi^2(t) \, dt + \int_0^{KT} v_{\hat{\xi}}^2(t) \, dt \\
&\quad - \frac{4\mathcal{E}_s}{T} \int_0^{KT} \cos[2\pi f_0 t + \theta(t; \xi)] \cos[2\pi f_0 t + \theta(t; \hat{\xi})] \, dt \\
&= 2K\mathcal{E}_s - \frac{2\mathcal{E}_s}{T} \int_0^{KT} \cos[\theta(t; \xi) - \theta(t; \hat{\xi})] \, dt \\
&= \frac{2\mathcal{E}_s}{T} \int_0^{KT} \left\{ 1 - \cos[\theta(t; \xi) - \theta(t; \hat{\xi})] \right\} \, dt
\end{aligned}
\tag{6.41}
$$

Since from (6.22) and (6.23) $\theta(t; \xi)$ is related linearly to the source symbols $\xi$, we may also write

$$
d^2(v_\xi(t), v_{\hat{\xi}}(t)) = \frac{2\mathcal{E}_s}{T} \int_0^{KT} \left\{ 1 - \cos[\theta(t; \xi - \hat{\xi})] \right\} \, dt
\tag{6.42}
$$

Thus, by defining the minimum Euclidean distance as the limit

$$
d_{\min}^2 \triangleq \lim_{K \to \infty} \min_{\xi \neq \hat{\xi}} d^2[v_\xi(t), v_{\hat{\xi}}(t)]
\tag{6.43}
$$

the error probability of CPM can be expressed, following (4.64), as

$$
P(e) \approx \frac{\bar{\nu}}{2} \mathrm{erfc} \left( \sqrt{\frac{\mathcal{E}_b}{N_0} \delta_{\min}^2} \right)
\tag{6.44}
$$

where we have defined the normalized squared minimum distance

$$
\delta_{\min}^2 \triangleq \frac{1}{4\mathcal{E}_b} d_{\min}^2 = \frac{1}{4\mathcal{E}_s} d_{\min}^2 \log_2 M
$$

and $\bar{\nu}$ is the average number of signals at distance $d_{\min}^2$ from any signal. The approximation (6.44) is increasingly better as $\mathcal{E}_b/N_0$ grows.

**Computing $d_{\min}$**

Let us examine the problem of calculating $d_{\min}$, the quantity that dominates the error performance of CPM for high signal-to-noise ratios. Consider a pair of phase trajectories $\theta(t; \boldsymbol{\xi})$ and $\theta(t; \widehat{\boldsymbol{\xi}})$ that differ in the first symbol (that is, $\xi_0 \neq \widehat{\xi}_0$). After some time intervals in which symbols may or may not differ, eventually the two trajectories merge and coincide. We define this situation as a *merge*. In general, it can be seen that for frequency pulses $g(t)$ of duration $LT$ a merge cannot occur before $(L + 1)$ symbol intervals from the start of the phase trajectory. Now, $d_{\min}$ cannot exceed the Euclidean distance $d_B$ between two signals whose phase trajectories merge: actually, other merges are possible after $(L + 2)$ or more symbol intervals from the start, and nothing prevents the corresponding squared distance to be smaller than that corresponding to the first merge. Only for full-response CPM is the value of $d_{\min}$ found for the first merge, so that $d_{\min} = d_B$. (see Aulin and Sundberg, 1981)

**Example 6.3.**    Assume $L = 1$ and, for the moment, $M = 2$. The two symbol sequences

$$\begin{aligned} \boldsymbol{\xi} &= +1, -1, \xi_2, \xi_3, \ldots \\ \widehat{\boldsymbol{\xi}} &= -1, +1, \xi_2, \xi_3, \ldots \end{aligned}$$

correspond to two phase trajectories that merge after 2 symbol intervals. Assume rectangular pulses, and a modulation index $h$. From (6.42) we have

$$\begin{aligned} d_B^2 &= \frac{2\mathcal{E}_s}{T} \int_0^{2T} \left\{ 1 - \cos[\theta(t; \boldsymbol{\xi} - \widehat{\boldsymbol{\xi}})] \right\} dt \\ &= 4\mathcal{E}_s \left( 1 - \frac{\sin 2\pi h}{2\pi h} \right) \end{aligned}$$

It can be verified that this bound is maximized when $h = 0.715$, which gives $d_B^2 = 4.87\mathcal{E}_s$. For $h = 1/2$ (which corresponds to MSK, as we shall see soon) we have $d_B^2 = 4\mathcal{E}_s$, and hence the normalized value $\delta_B^2 \overset{\triangle}{=} d_B^2 / 4\mathcal{E}_b = 1$. With $L = 1$, it can be shown that $d_{\min}^2 = d_B^2$: this proves that the asymptotic power efficiency of MSK is the same as that of traditional binary and quaternary PSK.

For $M > 2$, $d_B^2$ can be obtained by considering the two symbol sequences

$$\begin{aligned} \boldsymbol{\xi} &= +\ell, -\ell, \xi_2, \xi_3, \ldots \\ \widehat{\boldsymbol{\xi}} &= -\ell, +\ell, \xi_2, \xi_3, \ldots \end{aligned}$$

which yields the value

$$d_B^2 = 4\mathcal{E}_s \min_{1 \leq k \leq M-1} \left\{ 1 - \frac{\sin 2k\pi h}{2k\pi h} \right\}$$

For example, with $h = 1/2$ we obtain $\delta_B^2 = \log_2 M$.                    □

We hasten to recall here that $d_B^2$ is only an upper bound to $d_{\min}^2$, and that in several cases it is not actually achieved. In general, $d_{\min}^2$ depends on $M$, $h$, the frequency-pulse shape, and $L$ in a very complex way. The text of Anderson *et al.* (1986) provides many examples of actual values of $d_{\min}^2$. Sundberg (1986) also provides charts showing the tradeoff between $d_{\min}^2$ and bandwidth occupancy of several CPM schemes. A numerical algorithm for computing the minimum distance is described in Saxena (1983): this is an application of the general algorithm described in Chapter 14.

## 6.3. MSK and its multiple avatars

In this Section we examine in more depth a special case of CPM, namely MSK, and we show how it can be equivalently represented as shaped offset PSK or as CPFSK with $h = 1/2$. Demodulation of MSK can be obtained with several structures, which provide further insight into this modulation scheme.

### 6.3.1. MSK as CPFSK

We show here that MSK can be viewed as a special case of CPFSK, and hence of CPM. Specifically, assume a binary CPM with $L = 1$, $g(t)$ a rectangular pulse, and $h = 1/2$. By combining (6.17), (6.22), and (6.23), and by observing that $q(t) = t/2T$ for $0 \leq t \leq T$, we obtain

$$v_\xi(t) = \sqrt{\frac{2\mathcal{E}_s}{T}} \cos\left[2\pi f_0 t + \frac{\pi}{2}\sum_{k=0}^{n-1}\xi_k + \frac{\pi}{2}\xi_n\frac{t-nT}{T}\right] \qquad (6.45)$$

or, equivalently,

$$v_\xi(t) = \sqrt{\frac{2\mathcal{E}_s}{T}} \cos\left[\left(2\pi f_0 + \frac{\pi}{2T}\xi_n\right)t + \frac{\pi}{2}\sum_{k=0}^{n-1}\xi_k - \frac{\pi n}{2}\xi_n\right] \qquad (6.46)$$

The latter form shows once again that $v_\xi(t)$ is a frequency-modulated signal, obtained by shifting the carrier frequency up or down by an amount $1/4T$. After some algebra, (6.46) can be transformed into

$$v_\xi(t) = \sqrt{\frac{2\mathcal{E}_s}{T}}[I_n f(t - 2nT)\cos 2\pi f_0 t - Q_n f(t - T - 2nT)\sin 2\pi f_0 t] \qquad (6.47)$$

for $nT \leq t \leq (n+1)T$, where

$$I_n = -Q_{n-1}\xi_{2n-1} \qquad \text{and} \qquad Q_n = I_n\xi_{2n}$$

and
$$f(t) = \cos\left(\frac{\pi t}{2T}\right), \qquad -T \le t \le T$$

so that
$$f(t - T) = \sin\left(\frac{\pi t}{2T}\right), \qquad 0 \le t \le 2T$$

Since under our assumptions $(\xi_n)$ is a sequence of independent random variables taking on values $\pm 1$, then the two sequences $(I_n)$ and $(Q_n)$ take on the same values, are mutually independent, and have independent components. We conclude that the signal (6.47) is mathematically equivalent to an offset QPSK with sinusoidally-shaped pulses, i.e., to MSK as in (6.1), (6.2), and (6.6).

Obviously, from the above result we may obtain the power density spectrum of MSK as a special case of (6.38), but this would come at the price of a considerable amount of algebra, which is saved if the spectrum is calculated as we did to derive (6.8).

## A modulator for MSK

Use once again the tilted-phase representation (6.28) and the modified data sequence (6.29), which in our case yield

$$v_\xi(t) = \sqrt{\frac{2\mathcal{E}_s}{T}} \cos[2\pi f_0 t + \theta(t; \boldsymbol{\xi})], \qquad t \ge 0 \qquad (6.48)$$

where

$$\theta(t; \boldsymbol{\xi}) = \theta(t; \psi_n, \chi_n) = \pi\psi_n + \pi\frac{t - nT}{T}\chi_n, \qquad nT \le t < (n+1)T \quad (6.49)$$

is the information-carrying phase in the $n$th symbol interval, $\chi_n \in \{0, 1\}$ is the modified data sequence, and

$$\psi_n = \sum_{i=0}^{n-1} \chi_i \qquad (6.50)$$

with the sum reduced mod 2, represents the tilted-phase trellis state during the $n$th interval. By convention, $\psi_0 = 0$, and in general $\psi_n \in \{0, 1\}$. Eq. (6.49) shows that in any symbol interval the information-carrying phase $\theta(t; \boldsymbol{\xi})$ either remains constant (when $\chi_n = 0$) or increases linearly by $\pi$. We may interpret $\pi\psi_n$ as the initial phase in the $n$th interval, and $\pi\chi_n$ as the phase increment.

Eq. (6.49) makes explicit the fact that the signal transmitted during the $n$th interval is completely specified by the source symbol $\chi_n$ and by the state $\psi_n$. This suggests the implementation of the MSK modulator shown in Fig. 6.17. In this figure, the modulator is memoryless because at any given time it outputs a

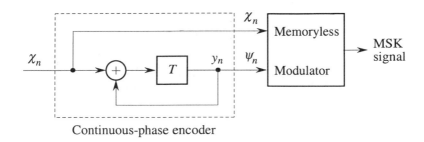

Figure 6.17: *Implementation of MSK modulator: MSK as CPFSK.*

| $\psi_n$ | $\psi_{n+1}$ | output signal |
|:---:|:---:|:---|
| 0 | 0 | $s_0(t) = \sqrt{\frac{2\mathcal{E}}{T}} \cos 2\pi f_0 t$ |
| 0 | 1 | $s_1(t) = \sqrt{\frac{2\mathcal{E}}{T}} \cos 2\pi f_1 t$ |
| 1 | 1 | $s_2(t) = -s_0(t)$ |
| 1 | 0 | $s_3(t) = -s_1(t)$ |

Table 6.3: *Input-output relationship for the memoryless mapper in DMSK. Here $f_1 \triangleq f_0 + 1/2T$.*

signal depending only on its inputs at the same time; it can be viewed as a table look-up device.

The discrete system whose input is the source symbol $\chi_n$ and whose output is the pair $(\chi_n, \psi_n)$ has transfer function $[1, z^{-1}/(1 - z^{-1})]$. (This is expressed in terms of $z$-transform—the reader more familiar with $D$-transforms should simply change $z^{-1}$ into $D$.)

**Differential MSK**

If we precode MSK by passing the source symbols into the discrete system with transfer function $1 - z^{-1}$ (the precoder), the continuous-phase encoder of Fig. 6.17 is changed into one whose transfer function is $[1 - z^{-1}, z^{-1}]$, or, equivalently, $[1 + z^{-1}, z^{-1}]$ (because modulo-2 addition and subtraction are the same operation). The resulting modulator is shown in Fig. 6.18, while the relation between source inputs and mapper outputs is shown in Table 6.3. What we obtain is called differential MSK (DMSK). In a sense, DMSK is more natural than MSK: in fact, up to a time shift the information sequence equals the

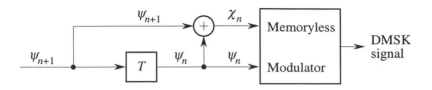

Figure 6.18: *Implementation of DMSK modulator.*

state sequence. For this reason in the balance of this section we shall focus our attention on DMSK. However, any transmitter and receiver for DMSK can be transformed into a transmitter and receiver for MSK by doing a simple invertible operation on the information sequence: specifically, by pre-multiplying the input by $1/(1 - z^{-1})$ and post-multiplying the output by $1 - z^{-1}$, respectively. Observe finally that by looking at the transmitted signal one cannot distinguish between MSK and DMSK, as the waveforms generated are the same in both modulations.

**State trellis diagram**

Here we use the tilted-phase trellis diagram, which allows us to represent MSK with only two states. This trellis for both MSK and DMSK is shown in Fig. 6.19. It has branches labeled by the transmitted signals, and describes the operation of the transmitter. The trellis of Fig. 6.20 shows the branch metrics, and hence is used by the receiver for maximum-likelihood detection. The correspondence between state pairs and transmitted symbols is summarized in Table 6.3.

The branch metric $\lambda_n(s_i)$, $i = 0, 1, 2, 3$, is the correlation between the received signal $r(t)$ and $s_i(t - nT)$, namely,

$$\lambda_n(s_i) = \int_{nT}^{(n+1)T} r(t) s_i(t - nT) \, dt, \qquad i = 0, 1, 2, 3$$

**6.3.2.  Massey's implementation**

Here we show how MSK can be seen as a modulation scheme whose optimum demodulator needs to process the received signal over only two symbol intervals. Specifically, we derive sufficient statistics for the estimate of $\psi_n$ that is obtained from the received signal in three adjacent symbol intervals, i.e., in $(n - 1)T \leq t < (n + 1)T$. This derivation is based on the fact that $s_2(t) = -s_0(t)$ and $s_3(t) = -s_1(t)$, which imply

$$\lambda_n(s_2) = -\lambda_n(s_0) \quad \text{and} \quad \lambda_n(s_3) = -\lambda_n(s_1) \tag{6.51}$$

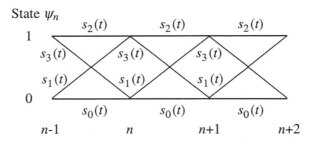

Figure 6.19: *Trellis diagram of MSK and DSMK. Branches are labeled by transmitted signals.*

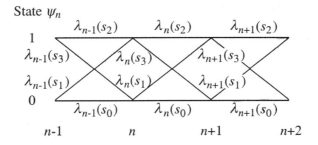

Figure 6.20: *Trellis diagram of MSK and DSMK. Branches are labeled by branch metrics.*

Now, assume that a genie has informed the receiver that $\psi_{n-1} = 0$ and $\psi_{n+1} = 0$. In these conditions, the maximum-likelihood receiver does its job by choosing between the two signal pairs corresponding to the pairs of branches joining those two states, viz., $(s_0, s_0)$ and $(s_1, s_3)$ (see Fig. 6.20). The decision rule between the two alternatives is

$$\hat{\psi}_n = 0 \quad \text{if and only if} \quad \lambda_{n-1}(s_0) + \lambda_n(s_0) \geq \lambda_{n-1}(s_1) + \lambda_n(s_3) \quad (6.52)$$

Thus, this genie-aided receiver needs only to process the received signal over two symbol intervals before making a decision on the most likely trellis state. We now show that the genie information is irrelevant, and hence (6.52) is always a maximum-likelihood rule for the estimate of $\psi_n$. Suppose for example that the

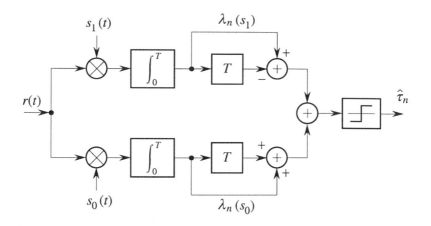

Figure 6.21: *Massey's optimal receiver for DMSK.*

genie information to the receiver was that $\psi_{n-1} = 0$ and $\psi_{n+1} = 1$. The choice here is between the signal pairs $(s_0, s_1)$ and $(s_1, s_2)$, which corresponds to the decision rule

$$\widehat{\psi}_n = 0 \qquad \text{if and only if} \qquad \lambda_{n-1}(s_0) + \lambda_n(s_1) \geq \lambda_{n-1}(s_1) + \lambda_n(s_2) \quad (6.53)$$

which, because of (6.51), is equivalent to (6.52). A similar conclusion holds for the case $\psi_{n-1} = 1$ and $\psi_{n+1} = 0$, as well as for the case $\psi_{n-1} = 1$ and $\psi_{n+1} = 1$.

It is convenient from now on to replace $\psi_n \in \{0, 1\}$ with $\tau_n \in \{\pm 1\}$, defined by $\tau_n \stackrel{\triangle}{=} 1 - 2\psi_n$. The rule (6.52) then becomes

$$\widehat{\tau}_n = 1 \qquad \text{if and only if} \qquad \lambda_{n-1}(s_0) + \lambda_n(s_0) - \lambda_{n-1}(s_1) + \lambda_n(s_1) \geq 0 \quad (6.54)$$

This rule is implemented by the receiver of Fig. 6.21. This in turn suggests the implementation of the DMSK transmitter in the form shown in Fig. 6.22.

### 6.3.3. Rimoldi's implementation

The task of the correlators in Massey's receiver is to compute $\lambda_n(s_0)$ and $\lambda_n(s_1)$. Another possibility, first suggested by Rimoldi (1994), is to have correlators computing the two quantities $\lambda_n(s_0) \pm \lambda_n(s_1)$, as in Fig. (6.23). The corresponding transmitter structure is shown in Fig. 6.24. An interesting interpretation of this transmitter structure is that the information symbol $\tau_n$ is transmitted twice: first it amplitude modulates $[s_0(t) - s_1(t)]$, then it amplitude modulates the signal, orthogonal to the former one, $[s_0(t) + s_1(t)]$.

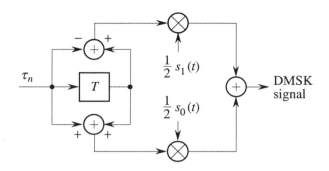

Figure 6.22: *Massey's transmitter for DMSK.*

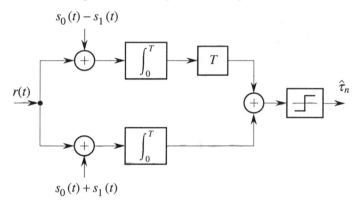

Figure 6.23: *Rimoldi's receiver for DMSK.*

### 6.3.4.    De Buda's implementation

A third way of implementing (6.54) is to compute the left-hand side of that inequality by means of a single integral as follows:

$$\lambda_{n-1}(s_0) + \lambda_n(s_0) - \lambda_{n-1}(s_1) + \lambda_n(s_1) = \int_{(n-1)T}^{(n+1)T} r(t)s[t - (n-1)T]\, dt \quad (6.55)$$

where

$$s(t) \overset{\triangle}{=} s_0(t) - s_1(t) + s_0(t - T) + s_1(t - T) \quad (6.56)$$

A difficulty here is that the integration in (6.55) must be carried out on two adjacent symbol intervals, which cannot be done with a single integrator (it can

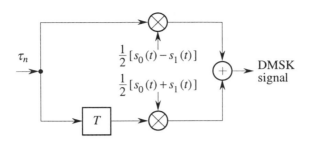

Figure 6.24: *Rimoldi's transmitter for DMSK.*

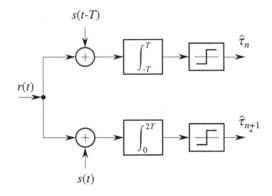

Figure 6.25: *De Buda's receiver for DMSK.*

only provide the result for odd—or even—symbol intervals). This difficulty can be overcome by using a second integral to obtain (6.55) for even—or odd— symbol intervals. The resulting receiver is shown in Fig. 6.25.

### 6.3.5.  Amoroso and Kivett's implementation

In previous implementation, a second integrator was called for because of the impossibility of computing the left-hand side of (6.54) for every $n$ by means of a single integrator. This difficulty can be circumvented by using a matched filter

Figure 6.26: *Amoroso and Kivett's receiver for DMSK.*

with impulse response

$$h(t) \triangleq \begin{cases} s(2T - t), & t \in (0, 2T) \\ 0, & \text{otherwise} \end{cases}$$ (6.57)

The resulting receiver is shown in Fig. 6.26.

## 6.4. GMSK

Gaussian MSK (GMSK) is a CPM scheme with $L > 1$, whose design is aimed at obtaining an especially compact spectrum, and hence a modulation scheme applicable to wireless systems. It has $h = 1/2$ and $M = 2$, like MSK, but the frequency pulse $g(t)$ is selected here by passing the rectangular pulse of MSK, which has duration $T$, through a filter whose impulse response is Gaussian, viz.,

$$h(t) = \frac{\sqrt{\pi}}{T} \frac{\beta}{\alpha} \exp\left\{ -\pi^2 \frac{\beta^2}{\alpha^2} \frac{t^2}{T^2} \right\}$$ (6.58)

where

$$\alpha^2 \triangleq \frac{\ln 2}{2}$$

or, equivalently, whose transfer function is

$$H(f) = \exp\left\{ -\frac{\alpha^2}{\beta^2} f^2 T^2 \right\}$$

(in practice, $g(t)$ will be truncated in a window with suitable finite duration). The actual filtered pulse $g(t)$ is obtained by integrating $h(t)$ in the interval $(t - T/2, t + T/2)$, which yields

$$g(t) = \frac{1}{2} \left[ \text{erf} \left( \frac{\pi}{T} \frac{\beta}{\alpha} (t + T/2) \right) - \text{erf} \left( \frac{\pi}{T} \frac{\beta}{\alpha} (t - T/2) \right) \right]$$

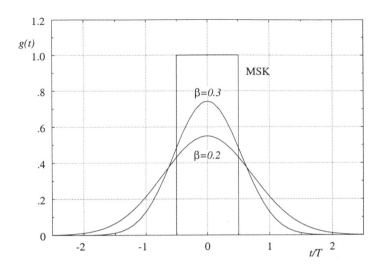

Figure 6.27: *Shape of a Gaussian-filtered rectangular frequency pulse, for three values of the bandwidth-controlling parameter $\beta$: MSK corresponds to $\beta \to \infty$.*

It is easily seen that $\int_{-\infty}^{\infty} h(t)\, dt = 1$, and that the variance of $h(t)$, interpreted as a probability density function, is proportional to $T^2/\beta^2$. Thus, by decreasing the value of the parameter $\beta$ the variance is increased, i.e., the smoothness of the filtered frequency pulse increases, and the bandwidth occupancy decreases. Fig. 6.27 shows the filtered frequency pulse with $\beta = \infty$, which corresponds to MSK, with $\beta = 0.3$ (which was selected for the second-generation European cellular radio standard GSM), and with $\beta = 0.2$ (which was selected for the wireless standard called DECT). The actual value of $\beta$ is selected as a result of a compromise between spectral occupancy, which calls for low values, and complexity of the Viterbi demodulator, which calls for high values. In fact, a smoother pulse obtained by decreasing $\beta$ results in a longer duration, and hence in an increase of the number of states of the demodulator.

The calculation of the power spectral density of GMSK, as in general that of CPM with $L > 1$, is complicated, and will not be discussed here (see, e.g., Garrison, 1975). Table 6.4 shows the power-containment bandwidth of GMSK for some values of $\beta$. It can be observed that a smaller value of $\beta$ results into a modulated signal spectrally more compact.

| $WT_s$ | 90% | 99% | 99.9% | 99.99% |
|---|---|---|---|---|
| GMSK ($\beta = 0.2$) | 0.52 | 0.79 | 0.99 | 1.22 |
| GMSK ($\beta = 0.25$) | 0.57 | 0.86 | 1.09 | 1.37 |
| GMSK ($\beta = 0.5$) | 0.69 | 1.04 | 1.33 | 2.08 |
| MSK | 0.78 | 1.20 | 2.76 | 5.60 |

Table 6.4: *Fractional power-containment bandwidth as a fraction of* $1/T_s$ *for GMSK and MSK.*

## 6.5. Bibliographical notes

Gronemeyer and McBride (1976) compare the performance of MSK and OQPSK. Most of the results on CPM can be found in the book by Anderson, Aulin, and Sundberg (1986). The representation of CPM using the tilted phase is due to Rimoldi (1988), although the choice of phase reference which makes the trellis time-invariant was introduced by Amoroso and Kivett (1977) and by Morales-Moreno and Pasupathy (1984). The derivation of the power density spectrum of full-response CPM with modulation index $h = J/M$ is taken from Biglieri and Visintin (1990). For a more general calculation, see pp. 209 ff. of Proakis (1995) or Anderson, Aulin, and Sundberg (1986). Multi-$h$ CPM is a type of CPM where the modulation index $h$ in each symbol interval is cyclically picked from a set $\{h_1, h_2, \ldots, h_H\}$ of rational numbers. Multi-$h$ CPM may exhibit a more compact spectrum than single-$h$ CPM. Tamed-frequency modulation (TFM) is another special type of partial-response CPM, introduced by De Jager and Dekker (1978).

Although in our presentation we have described only maximum-likelihood detection of CPM, other simpler demodulators have been proposed. Symbol-by-symbol detectors were described by de Buda (1972) for coherent demodulation, and by Osborne and Luntz (1974) and Schonhoff (1976) for noncoherent demodulation. The presentation of the equivalence between MSK and CPFSK in Section 6.3.1 is drawn from Stüber (1996).

MSK was invented by Doelz and Heald (1961). Later, De Buda (1972) and Amoroso and Kivett (1977) introduced "fast FSK" and "serial MSK," respectively. While the original invention introduced MSK as OQPSK with shaping done by a "full-wave rectified sine wave," Pasupathy (1979) defined another version of MSK with shaping by "unrectified sine wave." (The latter version is sometimes—and curiously—referred to as MSK-Type I, with the former one being called MSK-Type II.) Pasupathy's MSK actually turns out to be the OQPSK version of Amoroso and Kivett's serial MSK, a fact pointed out and proved by Peebles (1987). A coded-modulation view of MSK is described in Leib and Pasupathy (1993), where Ungerboeck's set-partitioning concept (see Chapter 12) is applied to provide some novel insight about MSK. The same paper summarizes how different versions of MSK have different focus on continuity of phase (such as squared Euclidean distance being proportional to Hamming distance, etc.). Our presentation of the various forms of MSK transmitters and receivers closely follows Rimoldi (1994), which in turn draws from de Buda (1972), Amoroso and Kivett (1977), and Massey (1980).

GMSK was introduced in Murota and Hirade, 1981. In that paper, power density spectrum, eye pattern, and error probability plots are obtained experimentally or by computer simulation.

## 6.6. Problems

**6.1** Compute the power spectral density of SFSK, obtained by letting

$$f'(t) = f''(t) = \cos\left(\frac{\pi}{T_s} - \frac{1}{4}\sin\frac{4\pi}{T_s}t\right)$$

**6.2** As $f \to \infty$, the asymptotic behavior of the power spectral density of offset PSK, MSK, and SFSK (Problem 6.1) is $O(f^{-2})$, $O(f^{-4})$, and $O(f^{-6})$, respectively. How can this asymptotic behavior be inferred from the expression of $f(t)$ in (6.2)? Derive a general form of a pulse $f(t)$ which gives a power spectrum decreasing asymptotically as $f^{-K}$.

**6.3** Using (6.38), prove that for $M$ large enough the power density spectrum of $M$-ary CPFSK with modulation index $h = J/M$ depends on $J$ but not on $M$.

**6.4** Derive a demodulator for $\pi/4$-DQPSK, equivalent in performance to that shown in Fig. 6.6, which does not include local oscillators but requires a delay line.

**6.5** Derive the squared Euclidean distance $d_B^2$ for partial-response CPM with rectangular pulses and $L = 2$. Compare the values obtained by considering the merges at $t = 3T$ and those at $t = 4T$.

**6.6** Derive an explicit expression for the signal $s(t)$ in (6.56), and use it to prove that de Buda's receiver implementation is equivalent to the implementation of a receiver for MSK as shaped offset PSK. Derive also a transmitter based on de Buda's implementation.

**6.7** Derive a transmitter for MSK based on Amoroso and Kivett's serial implementation.

**6.8** In Amoroso and Kivett's implementation of the MSK receiver (Fig. 6.26) the signal $s(t)$ has duration $2T$, and therefore pulse translates overlap. Show that this has no effect on the matched filter.

**6.9** Prove that, with the definitions (6.28) and (6.29), after an initial transient the tilted-phase trellis is time-invariant.

# Intersymbol interference channels

Chapter 3 has set the theoretical limits of digital modulation schemes transmitted over the additive white Gaussian noise (AWGN) channel. In Chapters 4 and 5 we have derived the optimal receiver structures for the most common modulation schemes, evaluated their performance over the AWGN channel, and shown how they compare to the theoretical limits.

Although in many practical situations the AWGN channel is not a realistic model, in most cases the performance of the different modulation schemes on such a channel can be considered as an upper bound of the actual performance. Moreover, in a Gaussian noise environment, the symbol error probability depends only on one parameter, the signal-to-noise ratio. Thus, meaningful comparisons among different modulation schemes can be obtained with only a moderate computational effort. Finally, it is hoped (and often true) that the hierarchy among different systems obtained on an AWGN channel is maintained over real channels, although the absolute performance may change.

In this chapter we shall consider a more realistic model of the system that includes additional impairments degrading the overall performance. Emphasis will be placed on the intersymbol interference (ISI) caused by linear distortion introduced by the finite bandwidth and the nonideal characteristics of the devices used in the system, such as filters and amplifiers. In addition to ISI, other factors affecting the system performance will be given some consideration, such as cochannel and interchannel interferences, which arise in systems sharing a common medium (e.g., in frequency-division multiplexing, FDM).

We shall focus on memoryless coherent modulation schemes whose representative signal points lie in a one- or two-dimensional space. This choice permits a unified treatment of different modulation schemes exploiting the concept of analytic signal introduced in Section 2.4, and encompasses a wide range of practical cases. The first part of the chapter is focused on the performance analysis of digital transmission schemes in the presence of ISI. The first section of the chapter presents a unifying analysis of coherent digital systems. Besides its effects on performance, ISI also complicates their computation. The second section of the chapter deals with this subject. It presents some methods that permit a reasonably fast and accurate evaluation of the error probability; the analytical details can be found in Appendix E.

The second part of the chapter examines some design problems related to the transmission of linearly modulated signals over a time-dispersive channel, that is, a channel perturbed by ISI. The first problem we take into consideration is the design of a system in which the receiver is constrained to the form of a linear filter followed by a sampler and a detector that makes decisions on a sample-by-sample basis. Two design criteria will be considered under this constraint. The first is the elimination of ISI from the sequence of samples to be processed by the detector, and the second is the minimization of the joint effects of ISI and noise on the same sequence.

If the receiver structure is not constrained, an optimum receiver can be designed performing maximum-likelihood (ML) estimation of the information sequence. This is the subject of the third part of this chapter.

## 7.1.  Analysis of coherent digital systems

In this section we analyze the coherent modulation schemes whose signal points lie in one or two-dimensional spaces, like PAM, $M$-PSK, and QAM. Their performance on the AWGN channel, as well as modulator's and demodulator's block diagrams, were described in Chapter 5.

A block diagram of the transmission system we consider here is shown in Fig. 7.1. The bit stream at the output of the information source is first sent to a serial-to-parallel converter that groups the binary digits in blocks of length $h$. Then the signal enters the modulator, which performs a memoryless mapping between the $M = 2^h$ input sequences and its alphabet of $M$ waveforms. A waveform is emitted by the modulator every $T$. As we know from Chapter 2 and Chapter 4, each waveform can be represented in this case as a point in a one or two-dimensional Euclidean space, characterized by two real coordinates or, equivalently, by a complex number. The modulated signal is transmitted over the channel, in which Gaussian noise is added. The bandpass filter in Fig.7.1 represents, without loss of generality, the cascade of the transmitter filter, the channel filter, and the receiving filter.[1] The received signal is fed to the carrier

---

[1]Obviously, the fact of including the receiving filter into the bandpass filter of Fig. 7.1 modifies the spectral density of the additive Gaussian noise, which is not white anymore, as it has

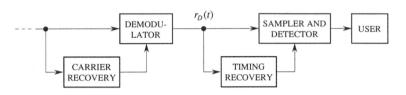

Figure 7.1: *Block diagram of a transmission system with coherent receiver.*

recovery device, which supplies the reference carrier to the coherent demodulator. The main features of the carrier recovery devices will be studied later in Chapter 9. Here, we assume that the recovered carrier phase is affected by a *jitter* $\theta(t)$, whose variations with time are so slow that it can be considered as a random variable (RV) with known probability density function (pdf). Successively, the demodulated signal is sampled at the symbol rate $1/T$, in correspondence of the sampling instant provided by the timing recovery device (this topic too will be treated in Chapter 9). Finally, the detector makes a decision on which signal was transmitted based on samples thus obtained and according to decision regions tailored to the particular modulation scheme.

A general representation of the modulated signal $v(t)$ is as follows:

$$v(t) = v_P(t) \cos 2\pi f_0 t - v_Q(t) \sin 2\pi f_0 t \tag{7.1}$$

where $f_0$ is the carrier frequency and[2]

$$v_P(t) = \sum_n a_{Pn} s_P(t - nT) - \sum_n a_{Qn} s_Q(t - nT) \tag{7.2}$$

$$v_Q(t) = \sum_n a_{Qn} s_P(t - nT) + \sum_n a_{Pn} s_Q(t - nT) \tag{7.3}$$

In (7.2) and (7.3), $a_{Pn}$ and $a_{Qn}$ are the coordinates of the signal point in the $n$-th signaling interval $[nT, (n + 1)T]$ and can take $M$ values in the sets $A_P$ and $A_Q$. The waveforms $s_P(t)$ and $s_Q(t)$ are suitable baseband shaping functions. The representation (7.1) includes the case of a baseband signal, for which $f_0 = 0$ and $a_{Qn} = 0$.

---

been filtered by the receiving filter.

[2]From here on the symbol $\sum_n$ will denote summation over all integers $n$ from $-\infty$ to $+\infty$.

| Modulation scheme | $A_P$ | $A_Q$ | $s_P(t)$ | $s_Q(t)$ |
|---|---|---|---|---|
| PAM-DSB | $\{(2k - M - 1)d/2\}_{k=1}^{M}$ | $0$ | $s(t)$ | $0$ |
| PAM-SSB | $\{(2k - M - 1)d/2\}_{k=1}^{M}$ | $0$ | $s(t)$ | $\hat{s}(t)$ |
| CPSK | $A \cos \phi_k$ | $A \sin \phi_k$ | $s(t)$ | $0$ |
| QAM | $A_k \cos \phi_k$ | $A_k \sin \phi_k$ | $s(t)$ | $0$ |

Table 7.1: *Coordinates of signal points and shaping functions for coherent modulation schemes.*

**Example 7.1** In the case of $M$-ary PAM modulation (see Section 5.2), we have $A_P = \{(2k - M - 1)d/2\}_{k=1}^{M}$, $s_Q(t) = 0$, and, for instance, $s_P(t) = u_T(t)$, where $u_T(t)$ is a rectangular waveform of unit amplitude in $(0, T)$ and zero elsewhere. □

The sets $A_P, A_Q$ and $s_P(t), s_Q(t)$ for the different modulation schemes are reported in Table 7.1., where $A_k, \phi_k$ represent the amplitude and phase of the two-dimensional signal points. In the case of a single-sideband pulse amplitude modulation (PAM-SSB), i.e., of a bandpass PAM signal in which only half of the bandwidth around the carrier is transmitted, $s_Q(t)$ is obtained as the Hilbert transform of $s_P(t)$. According to the theory of the complex envelope representation of bandpass signals, developed in Section 2.4, we can represent $v(t)$ in (7.1) by its complex envelope $\tilde{v}(t)$

$$\tilde{v}(t) = v_P(t) + jv_Q(t) \tag{7.4}$$

Moreover, the bandpass filtering operated by the channel on $v(t)$ can be represented by the filtering operated by the low-pass equivalent channel on $\tilde{v}(t)$. Thus, defining

$$\tilde{g}(t) \overset{\triangle}{=} g_P(t) + jg_Q(t) \tag{7.5}$$

so that $g(t) = \Re\left[\tilde{g}(t)e^{j2\pi f_0 t}\right]$ is the impulse response of the bandpass filter in Fig. 7.1, we can write the complex envelope of the received signal $r(t)$ as

$$\tilde{r}(t) = \frac{1}{2}\tilde{v}(t) * \tilde{g}(t) + \tilde{n}(t) \tag{7.6}$$

where $\tilde{n}(t) = n_P(t) + jn_Q(t)$ is the complex envelope of the bandpass Gaussian noise process, and $n_P(t), n_Q(t)$ are baseband Gaussian processes whose samples are Gaussian RVs with zero mean and variance $\sigma_n^2$. The variance $\sigma_n^2$ is obtained as $N_0 B_{eq}$, $N_0/2$ being the two-sided power spectral density of the white noise and

$B_{eq}$ the equivalent noise bandwidth of the receiving filter. Using now (7.2),(7.3), and (7.4), we can write

$$\tilde{v}(t) = \sum_n a_n \tilde{s}(t - nT) \tag{7.7}$$

having defined

$$a_n \triangleq a_{Pn} + j a_{Qn} \tag{7.8}$$

$$\tilde{s}(t) \triangleq s_P(t) + j s_Q(t) \tag{7.9}$$

Thus, finally, the received signal $\tilde{r}(t)$ can be given the following expression:

$$\tilde{r}(t) = \sum_n a_n \tilde{h}(t - nT) + \tilde{n}(t) \tag{7.10}$$

where

$$\tilde{h}(t) \triangleq h_P(t) + j h_Q(t) = \frac{1}{2}\tilde{s}(t) * \tilde{g}(t) \tag{7.11}$$

From here on, $(a_n)$ is assumed to be a sequence of independent identically distributed RVs. Recalling (7.5) and (7.8), the convolution in (7.11) gives rise to

$$h_P(t) = \frac{1}{2}[s_P(t) * g_P(t) - s_Q(t) * s_Q(t)] \tag{7.12}$$

$$h_Q(t) = \frac{1}{2}[s_P(t) * g_Q(t) + s_Q(t) * g_P(t)] \tag{7.13}$$

**Example 7.2**   Consider a PAM transmission and a bandpass filter whose transfer function $G(f)$ satisfies the following symmetry conditions for every $f$:

$$G_R^+(f_0 + f) = G_R^+(f_0 - f)$$
$$G_I^+(f_0 + f) = -G_I^+(f_0 - f)$$

where $G_R$ and $G_I$ are the real and imaginary parts of the transfer function

$$G^+(f) \triangleq \begin{cases} G(f), & f \geq 0 \\ 0, & f < 0 \end{cases}$$

To compute $\tilde{h}(t)$ according to (7.11), we need $\tilde{g}(t)$, that is, the inverse Fourier transform of $G^+(f + f_0)$. But $G^+(f + f_0)$ exhibits the symmetries of $G^+(f)$ around the origin $f = 0$, and this makes $\tilde{g}(t) = g_P(t)$ real. Thus we have

$$\tilde{h}(t) = \frac{1}{2}s(t) * g_P(t), \quad \text{(real), for PAM-DSB}$$

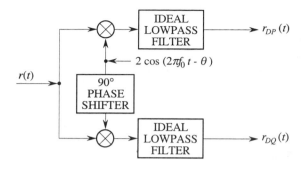

Figure 7.2: *Demodulation of a two-dimensional modulation scheme.*

and

$$\tilde{h}(t) = \frac{1}{2}[s(t) + j\hat{s}(t)] * g_P(t), \quad \text{for PAM-SSB}$$

□

The actual signal that enters the demodulator of Fig. 7.1 can be obtained simply as $r(t) = \Re\{\tilde{r}(t)e^{j2\pi f_o t}\}$. The task performed by the demodulator is represented by the block diagram of Fig. 7.2.

In fact, this is the general form of the demodulator, which simplifies and reduces to the upper branch when a one-dimensional modulation scheme like PAM is used.[3] It can be proved (see Problem 7.1) that the two outputs $r_{DP}(t)$ and $r_{DQ}(t)$ from the branches of the demodulator of Fig. 7.2 are the same as the outputs of the system shown in Fig. 7.3. At its input, the complex envelope of $r(t)$ is presented.

The presence of nonideal low-pass filters in the demodulator could also be easily accounted for by including their transfer functions in the overall low-pass equivalent filter represented by $\tilde{h}(t)$ in (7.11). Then we can immediately write the expressions of the demodulated signals. For the sake of clarity, let us consider separately the one and two-dimensional cases.

**PAM modulation**

The demodulated signal is given by

$$
\begin{aligned}
r_D(t) &\triangleq r_{DP}(t) = \mathcal{R}\{\tilde{r}(t)e^{j\theta}\} \\
&= \cos\theta \sum_n a_n h_P(t - nT) - \sin\theta \sum_n a_n h_Q(t - nT) + \nu_P(t)
\end{aligned}
\tag{7.14}
$$

---

[3] Also for PAM, however, two branches of the demodulator are needed when a single-sideband (PAM-SSB) modulation is used.

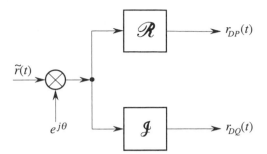

Figure 7.3: *Equivalent demodulator, in the complex envelope representation, of a two-dimensional modulation scheme.*

In (7.14), for given $\theta$, the baseband random process $\nu_P(t) = n_P(t)\cos\theta - n_Q(t)\sin\theta$ is Gaussian, with zero mean and variance $\sigma_n^2$, like $n_P(t)$ and $n_Q(t)$ (see Problem 7.2). Equation (7.14) makes evident the different sensitivities of DSB and SSB to the phase jitter. In fact, suppose that the channel transfer function $G(f)$ satisfies the symmetry conditions of Example 7.2. Then, $h_Q(t)$ is equal to zero for the PAM-DSB modulation. In this case, the presence of the phase jitter reduces to an attenuation of the received signal by $\cos\theta$. However, for SSB systems, $h_Q(t)$ is not zero. Thus, the second summation in the RHS of (7.14) contributes to the performance degradation.

**Two-dimensional modulations**

The two demodulated signals are given by

$$r_{DP}(t) \triangleq \Re\{\tilde{r}(t)e^{j\theta}\} = \sum_n \Re\{a_n\tilde{h}(t-nT)e^{j\theta}\} + \nu_P(t) \qquad (7.15)$$

$$r_{DQ}(t) \triangleq \Im\{\tilde{r}(t)e^{j\theta}\} = \sum_n \Im\{a_n\tilde{h}(t-nT)e^{j\theta}\} + \nu_Q(t) \qquad (7.16)$$

where $\nu_P(t)$ is the same as before, and $\nu_Q(t) = n_P(t)\sin\theta + n_Q(t)\cos\theta$ is a conditionally Gaussian baseband process with zero mean and variance $\sigma_n^2$. Moreover, samples of $\nu_P(t)$ and $\nu_Q(t)$, taken at the same time instant, are conditionally independent RVs (see Problem 7.2).

**Example 7.3**  Consider a QAM system without phase jitter ($\theta = 0$) and a band-pass transfer function $G(f)$ exhibiting the symmetries of Example 7.2. Using (7.12) and (7.13), we have

$$h_P(t) = \frac{1}{2}s(t) * g_P(t)$$

$$h_Q(t) = 0$$

and, from (7.15) and (7.16),

$$r_{DP}(t) = \sum_n a_{Pn} h_P(t - nT) + n_P(t)$$

$$r_{DQ}(t) = \sum_n a_{Qn} h_P(t - nT) + n_Q(t)$$

$\square$

In the detector, decisions on the transmitted $a_n$ are taken by comparing sampled values of $r_{DP}(t)$ and $r_{DQ}(t)$ (or only $r_{DP}(t)$ in the PAM case) with suitable thresholds. In other words, the receiver is the same as the one described in Chapter 4 for the Gaussian channel. The sampling times form a sequence $(t_0 + iT)_{i=-\infty}^{\infty}$, where $0 \le t_0 \le T$ is the optimum (in some sense) timing instant depending on the impulse response $\tilde{h}(t)$. Assuming that the sequence $(a_n)$ is stationary, the processes $r_{DP}(t)$ and $r_{DQ}(t)$ are cyclostationary random processes with period $T$ (see Section 2.2.2). Thus, the performance of the system does not depend on the particular signaling interval. We shall consider the sampling instant $t_0$.

The following shorthand notation will be used in this chapter for all the time functions:

$$y_n \overset{\triangle}{=} y(t_0 - nT) , \quad \text{for all integers } n \qquad (7.17)$$

The sampled demodulated signals are then given by the following expressions:

*PAM modulation*

$$r_{D0} = a_0(h_{P0} \cos \theta - h_{Q0} \sin \theta) + \sum_{n \ne 0} a_n(h_{Pn} \cos \theta - h_{Qn} \sin \theta) + \nu_{P0} \quad (7.18)$$

*Two-dimensional modulations*

$$r_{DP0} = r_{DP}(a_0) + \sum_{n \ne 0} r_{DP}(a_n) + \nu_{P0} \qquad (7.19)$$

$$r_{DQ0} = r_{DQ}(a_0) + \sum_{n \ne 0} r_{DQ}(a_n) + \nu_{P0} \qquad (7.20)$$

where

$$r_{DP}(a_n) = (a_{Pn} h_{Pn} - a_{Qn} h_{Qn}) \cos \theta \qquad (7.21)$$
$$\quad - (a_{Pn} h_{Qn} + a_{Qn} h_{Pn}) \sin \theta$$
$$r_{DQ}(a_n) = (a_{Pn} h_{Qn} + a_{Qn} h_{Pn}) \cos \theta \qquad (7.22)$$
$$\quad + (a_{Pn} h_{Pn} - a_{Qn} h_{Qn}) \sin \theta$$

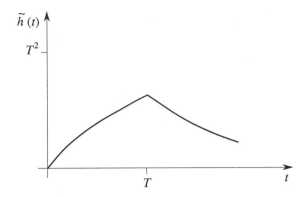

Figure 7.4: *Example of low-pass equivalent impulse response.*

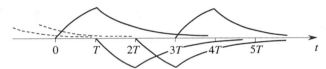

Figure 7.5: *Successive component waveforms of the received signal.*

In (7.18), (7.19), and (7.20), the term with $n = 0$ has been given special consideration as it contains the required information about the symbol $a_0$ on which we are deciding. The summations in the RHS represent the contribution (unwanted!) to the sample taken at $t = t_0$ of the past and future symbols in the sequence $(a_n)$. These terms are called *intersymbol interference* (ISI), and may represent a major cause of impairment to system performance. Looking at (7.18)–(7.22), an important fact can be observed. Even in the absence of phase jitter, we have an interaction between the in-phase and quadrature channels whenever $h_Q(t)$ is not zero at the sampling instants. This happens when the transfer function of the channel $G(f)$ does not satisfy the symmetry conditions of Example 7.2.

**Example 7.4**    Consider a binary PAM system, with $d = 2$ and $s_P(t) = u_T(t)$, transmitted over a channel with $\tilde{g}(t) = 2Te^{-t/T}$ , $t \geq 0$. Using (7.11), we have (see Fig. 7.4)

$$\tilde{h}(t) = \begin{cases} T^2(1 - e^{-t/T}), \ 0 \leq t \leq T \\ T^2(e - 1)e^{-t/T}, \ t \geq T \end{cases}$$

Using now (7.14), and assuming that the transmitted sequence $(a_n)$ is $+1, -1, -1, +1$, and $\theta = 0$, we obtain, by summing the various contributions of Fig. 7.5, and in the absence of noise, the received signal $r_D(t)$ shown in Fig. 7.6. □

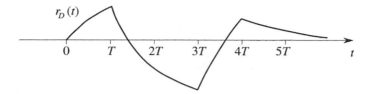

Figure 7.6: *Received signal for the impulse response of Fig. 7.4 corresponding to the binary data sequence +1,−1,−1,+1, in the absence of noise.*

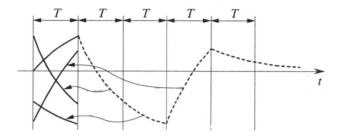

Figure 7.7: *Construction of the eye diagram for the signal of Fig. 7.4.*

An effective way of displaying the qualitative effects of ISI is the construction of the *eye pattern*, or *eye diagram*. It consists of slicing the demodulated signal (in the absence of noise) in segments of $T$ seconds duration and superimposing the various slices in the interval $(0, T)$ as in Fig. 7.7, which refers to Example 7.4. The eye diagram is obtained by observing the data signal through an oscilloscope, whose time axis is synchronized at the symbol rate. For a binary PAM modulation, the typical aspect of the eye pattern is as in Fig. 7.8, where the sampling instant is shown to correspond with the maximum eye opening, yielding the greatest protection against the noise. In Fig. 7.8, the *amplitude peak distortion* is also indicated. It is defined as the maximum value assumed by the ISI over all the possible transmitted sequences $(a_n)$. Using (7.14), with $\theta = 0$, we can write it as

$$D_P \triangleq \max_{(a_n)} \sum_{n \neq 0} a_n h_{Pn} = \sum_{n \neq 0} |h_{Pn}| \qquad (7.23)$$

The concept of eye diagram and peak distortion can be generalized to the multilevel PAM and two-dimensional modulation systems. The general form of the overall low-pass equivalent impulse response $\tilde{h}(t)$ is shown in Table 7.2, together with the expressions of $r_{DP0}$ and $r_{DQ0}$.

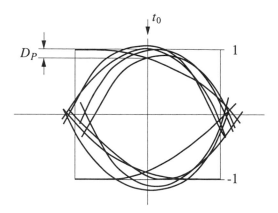

Figure 7.8: *Example of eye diagram for binary PAM modulation.*

| Modulation scheme | $h_P$ | $h_Q$ | $r_{DP0}$ | $r_{DQ0}$ |
|---|---|---|---|---|
| PAM - DSB | $s * g_P$ | $s * g_Q$ | $\sum_n a_n(h_{Pn}\cos\theta - h_{Qn}\sin\theta) + v_{P0}$ | $0$ |
| PAM - SSB | $s * g_P - \hat{s} * g_Q$ | $s * g_Q + \hat{s} * g_P$ | $\sum_n a_n(h_{Pn}\cos\theta - h_{Qn}\sin\theta) + v_{P0}$ | $0$ |
| CPSK, AM - PM | $s * g_P$ | $s * g_Q$ | $\sum_n (a_{Pn}h_{Pn} - a_{Qn}h_{Qn})\cos\theta$ $-\sum_n (a_{Pn}h_{Qn} + a_{Qn}h_{Pn})\sin\theta + v_{P0}$ | $\sum_n (a_{Pn}h_{Qn} + a_{Qn}h_{Pn})\cos\theta$ $+\sum_n (a_{Pn}h_{Pn} - a_{Qn}h_{Qn})\sin\theta + v_{Q0}$ |

Table 7.2: *Low-pass equivalent impulse responses and in-phase and quadrature samples of the received signal for coherent modulation schemes. The acronyms DSB and SSB mean double sideband and single sideband, respectively.*

We have proved that the system shown in Fig. 7.9 permits us to obtain the real and imaginary parts of the demodulated signal $r_D(t)$ of Fig. 7.1. Note that in Fig. 7.9 the modulating and demodulating carriers have disappeared. Besides its great simplicity and conciseness, this result proves to be very useful in the computer simulation of bandpass digital transmission systems. In fact, using the model of Fig. 7.9, the frequency at which signals must be sampled before being processed by the computer is related to the bandwidth of the *modulating signal*, and not to the *carrier frequency*, which is usually much larger.

To conclude this part, let us summarize step by step how the signal analysis we have just described can be done. This analysis is the preliminary step in the computation of the error probability, as we shall see in the next sections.

**Step 1** Given the modulation scheme and the shaping filter $s(t)$, use Table 7.1.

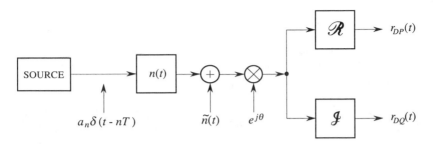

Figure 7.9: *Equivalent block diagram of a linear digital transmission system using complex envelope representation.*

to obtain the transmitted signal $v(t)$ of (7.2) to (7.4).

**Step 2**  Cascade the bandpass transmitter, channel, and receiver filters to obtain the transfer function $G(f)$.

**Step 3**  Compute the low-pass equivalent impulse response $\tilde{g}(t)$ by taking the inverse Fourier transform of $G^+(f + f_0)$, and use it in the convolutions (7.12) and (7.13) to obtain the real and imaginary parts of $\tilde{h}(t)$.

**Step 4**  Find the expressions of $r_{DP0}$ and $r_{DQ0}$ in Table 7.1. as a function of $a_{Pn}$, $a_{Qn}$, (Table 7.1.), and $h_P$, $h_Q$ computed in Step 3.

The computational tools normally used in a digital computer to evaluate the convolutions in Step 3 are the fast Fourier transform in the frequency domain and the state variable technique or the bilinear $z$-transform in the time domain. The interested reader is invited to consult the Bibliographical Notes at the end of this chapter for relevant references.

## 7.2.   Evaluation of the error probability

The received signal, after demodulation and sampling, enters the decision device, which locates it in one of the decision regions and chooses the corresponding point in the signal space as the transmitted one. In practice, the decision regions are coincident with the optimum ones under the criterion of minimum Euclidean distance (i.e., optimum for the AWGN channel) described in Section 4.2. Thus, computing the error probability for a given transmitted signal point entails evaluating the probability that the point $(r_{DP0}, r_{DQ0})$ lies in a suitable two-dimensional region, depending on the particular modulation scheme adopted. The computation is usually performed in two steps:

**Step 1** Compute the probability that a point lies in a two-dimensional (or one-dimensional in the PAM case) region, taking advantage of the fact that the RVs $r_{DP0}$ and $r_{DQ0}$ are conditionally independent Gaussian RVs. This probability is precisely the error probability conditioned on ISI and phase jitter $\theta$.

**Step 2** Compute the expectation of the result obtained in Step 1 with respect to (a) the two RVs (one for PAM) representing the ISI affecting the in-phase and quadrature components of the received signal and (b) the random phase $\theta$.

As we shall see, Step 1 can be achieved analytically in an exact or approximate manner for almost all coherent modulation schemes. Indeed, there is no difference from the AWGN channel case discussed in detail in Chapters 4 and 5. What really complicates the computation is Step 2. Although in most cases all the values assumed by the ISI RVs could be exhaustively enumerated and the conditional error probabilities computed, such a procedure, in fact, may often take an extremely long time, and hence be impractical. For the sake of simplicity, we shall first verify this conclusion and show how to circumvent it with reference to the PAM transmission system.

### 7.2.1. PAM modulation

The sampled received signal's expression was given in (7.18), where the symbols forming the sequence $(a_n)$ can take the values shown in Table 7.1. with equal probabilities $1/M$. With the following shorthand notations:

$$h_n(\theta) \triangleq h_{Pn} \cos \theta - h_{Qn} \sin \theta \tag{7.24}$$

$$X_n(\theta) \triangleq a_n h_n(\theta) \tag{7.25}$$

$$X(\theta) \triangleq \sum_{n \neq 0} X_n(\theta) \tag{7.26}$$

the received signal $r_{D0}$ becomes

$$r_{D0} = a_0 h_0(\theta) + X(\theta) + \nu_{P0}$$

Assuming $M$ even and $h_0(\theta) > 0$, the decisions at the receiver are made by comparing $r_{D0}$ with the following thresholds:

$$-\left(\frac{M}{2} - 1\right)dh_0(\theta), \dots, -dh_0(\theta), \dots, \left(\frac{M}{2} - 1\right)dh_0(\theta)$$

Thus, following the analysis of Section 5.2 and Steps 1 and 2, the error probability is easily expressed as

$$P(e) = \mathrm{E}_\theta \mathrm{E}_X \{ P(e \mid \theta, X) \} = \frac{M-1}{M} \mathrm{E}_\theta \mathrm{E}_X \left\{ \mathrm{erfc} \left[ \frac{(d/2)h_0(\theta) - X(\theta)}{\sqrt{2}\sigma_n} \right] \right\}$$
$$(7.27)$$

where $\mathrm{E}_X$ and $\mathrm{E}_\theta$ denotes average over the RVs $X$, representing the ISI, and $\theta$, respectively. We shall discuss later how to perform the average with respect to $\theta$. The problem, then, is the computation of the conditional expectation with respect to the RV $X(\theta)$ for a given $\theta$, that is, of the integral

$$I \triangleq \int_{\mathcal{X}} \mathrm{erfc} \left[ \frac{(d/2)h_0 - X}{\sqrt{2}\sigma_n} \right] f_X(x) dx \qquad (7.28)$$

For simplicity, in (7.28) we have dropped the coefficient $(M-1)/M$ and the dependence on $\theta$. In the integral (7.28), $\mathcal{X}$ and $f_X(x)$ represent, respectively, the range and the pdf of the RV $X$.

### Some facts about the RV $X$

Looking at (7.25) and (7.26), the RV X is seen to be the sum of a number, say $N$, of RVs $X_n$. The number $N$ depends on the duration of the impulse response $h(t)$ through its samples $h_n$. In principle, $N$ may be infinite. However, in practice, only a finite number of samples significantly contribute to the performance degradation. A thorough discussion on the convergence of $X$ to a random variable, and on the existence of a pdf for it, can be found in Campbell and Wittke (1997) and the references therein.

The structure of $X$ (see (7.25) and (7.26)) is such that one is tempted to invoke the central limit theorem and assume that it converges to a Gaussian RV as $N \to \infty$. Unfortunately, the central limit theorem cannot be applied as, in practice, the range of $X$ is almost always a bounded interval, and its variance is limited (see Loève, 1963, p. 277). In fact, the largest value taken by $X$ cannot exceed

$$x_{\sup} \triangleq (M-1)\frac{d}{2} \sum_{n \neq 0} |h_n| \qquad (7.29)$$

The value $x_{\sup}$ is assumed by $X$ with our assumption that $(a_n)$ is a sequence of independent RVs. When $N$ is infinite, $x_{\sup}$ is still bounded if the asymptotic decay of the impulse response $h(t)$ is faster than $1/t$. In the practice this is always the case. What happens when we try to apply the central limit theorem to this case is shown in the following example.

**Example 7.5**   Consider a PAM transmission system for which $x_{\sup} < h_0 d/2$. This means that the eye pattern of the system is open, i.e., that the peak distortion is less than half the separation between two adjacent signal levels, and, consequently, that we can transmit with zero error probability in the absence of noise. Applying the central limit theorem (which leads to the *Gaussian assumption* for $X$), the RV $X$ is treated like a Gaussian RV, with zero mean and variance $\sigma_X^2 = \mathrm{E}\{X^2\}$, independent of the noise. Thus, the sum $X + \nu_{P0} \sim \mathcal{N}(0, \sigma_X^2 + \sigma_n^2)$, and the integral in (7.28) becomes

$$I_G = \mathrm{erfc}\left(\frac{dh_0}{2\sqrt{2}\sqrt{\sigma_X^2 + \sigma_n^2}}\right)$$

Now, increasing the signal-to-noise ratio in the channel by letting $\sigma_n \to 0$, we get

$$\lim_{\sigma_n \to 0} I_G = \mathrm{erfc}\left(\frac{dh_0}{2\sqrt{2}\sigma_X}\right)$$

which leads to an asymptotic error probability value different from zero (*error floor*). This clearly contrasts with the hypothesis $x_{\sup} < h_0 d/2$. However, when ISI is small, this asymptotic value may be so low that in the region of interest the curve for the Gaussian assumption gives a reasonable approximation of the error probability.   □

### Exact value of the integral $I$

Henceforth, we shall suppose that $N$ is finite. Although this is not always true, in practice it is possible to find a finite $N$ large enough to make immaterial the error due to the truncation of $h(t)$. In Prabhu (1971), the problem of bounding the error due to the impulse response truncation was examined.

The RV $X$ is then a discrete RV, assuming values $\{x_i\}_{i=1}^{L}$ with probabilities $\{p_i\}_{i=1}^{L}$, and its pdf $f_X(x)$ can be written as

$$f_X(x) = \sum_{i=1}^{L} p_i \delta(x - x_i) \tag{7.30}$$

Inserting (7.30) into (7.28), we immediately get

$$I = \sum_{i=1}^{L} p_i \mathrm{erfc}\left(\frac{(d/2)h_0 - x_i}{\sqrt{2}\sigma_n}\right) \tag{7.31}$$

and the problem is solved. The ease in obtaining the *true value* of $I$ should nevertheless make the reader suspicious. In fact, what often renders (7.31) very complex to compute is the number $L$, which can be extremely large. Suppose, for example, that we have an octonary PAM with a channel memory $N = 20$. Then $L$ is given by

$$L = M^N = 8^{20} \simeq 1.15 \cdot 10^{18}$$

If we could use an extremely fast computer able to compute a million complementary error functions in 1 second, it would take only slightly less than 42 thousand years to compute the exact value of $I$. That alone seems a good motivation for the large amount of research done in this area in the seventies and later.

Many methods have been proposed in the literature to obtain approximations of $I$ in (7.28), with different trade-offs between accuracy and computer time. Here, we propose the simplest upper bound, known as the *worst-case bound*, and the *Gauss quadrature rules* (GQR) method, described in Appendix E, since it has emerged as one of the most efficient in approximating integrals like $I$ in (7.28).

**Worst-case bound**

The worst-case bound is an upper bound to $I$ in (7.28) computed through the substitution of the RV $X$ with the constant value $x_{\text{sup}}$ defined in (7.29). Thus, we have

$$I \leq \text{erfc}\left[\frac{(d/2)h_0 - x_{\text{sup}}}{\sqrt{2}\sigma_n}\right] \qquad (7.32)$$

Since erfc $(\cdot)$ is a monotonically decreasing function, the RHS of (7.32) is clearly an upper bound to the RHS of (7.28). The term $(d/2)h_0 - x_{\text{sup}}$ is precisely the semi-opening of the eye diagram at the sampling instant. The worst-case bound is very easily computed. The approximation involved is reasonable when one interfering sample is dominant with respect to the others. Otherwise, the bound becomes too loose. A better upper bound based on the Chernoff bound is described in Saltzberg (1968) (see also Problem 7.3) and will be used later in the examples.

**The Gauss quadrature rules technique**

The method of GQR is described in detail in Appendix E. Its use has now become classical, owing to its being one of the best compromises between accuracy and computer time. Essentially, it allows one to compute an approximation of $I$ in (7.28) in the form

$$I \simeq \sum_{j=1}^{J} w_j \text{erfc}\left[\frac{(d/2)h_0 - x_j}{\sqrt{2}\sigma_n}\right] \qquad (7.33)$$

The $\{x_j\}_{j=1}^J$ and $\{w_j\}_{j=1}^J$ are called, respectively, the *abscissas* and the *weights* of the quadrature rule. They can be obtained through a numerical algorithm based on the knowledge of the first $2J$ moments of RV $X$. Comparing (7.33) with the exact value (7.31) of $I$, one immediately realizes the similarity. The great difference lies in the value of $J$ in (7.33), which is usually much less than the value of $L$ in (7.31). The tightness of the approximation depends on $J$ (i.e., on the number $2J$ of known moments). Computational experience shows that a value of $J$ between 5 and 10 leads to very good approximations of the true value of $I$. The same method of GQR can be used to evaluate the average with respect to the RV $\theta$ in (7.27), once the moments of $\theta$ are known. An efficient algorithm to evaluate the moments $\mu_k$ of the RV $X$

$$\mu_k \triangleq E\{X^k\}, \quad k = 1, 2, \ldots, 2J \tag{7.34}$$

without resorting to the pdf of $X$ is explained in Appendix E.

**Example 7.6**   In this example the methods described to compute the error probability in the presence of ISI will be applied, for the sake of comparison, to the case of binary PAM transmission, with $\theta = 0$ and

$$h_P(t) = \frac{\sin(\pi t/T)}{\pi t/T} \tag{7.35}$$

The impulse response of (7.35) is that of an ideal low-pass filter with cutoff frequency $1/(2T)$. The transfer function of the filter satisfies the Nyquist criterion (see Section 7.3), and, thus, it does not give rise to ISI when properly sampled at the time instants $t = 0, \pm T, \pm 2T, \ldots$. We will suppose that the timing recovery circuit is not ideal, so the sampling instants will be $t_n = t_0 + nT$, $n = -\infty, \ldots, \infty$, with $t_0 \neq 0$, and we define the normalized sampling time deviation $\Delta \triangleq t_0/T$.

The methods discussed for computing the error probability are the worst-case bound (curve labeled (1) in Fig. 7.10), the Chernoff bound (curves labeled (2)), the series expansion described in Appendix E (curve labeled (3)), and the GQR (curve labeled (4)). In Figure 7.10 the error probability is plotted as a function of $\Delta$ for a signal-to-noise ratio at the nominal sampling instant ($t_0 = 0$) SNR $\triangleq 1/(2\sigma_n^2)$ of 15 dB. The impulse response has been truncated to $N = 50$. The curve (3), relative to the series expansion method, stops at $\Delta = 0.15$, since the summation of the series exhibits numerical instability for larger values of $\Delta$. This is visualized in Figure 7.11, where the exact error probability, computed through (7.31) for $N = 10$, and the error probability estimated either with the series expansion or with the GQR method are reported for $\Delta = 0.2$ as a function of $J$, the number of terms used in the series or the GQR. The curve giving the results of the series expansion method ends with $J = 8$, since the successive nine-term approximation yields a negative value for $P(e)$. The processing time required for

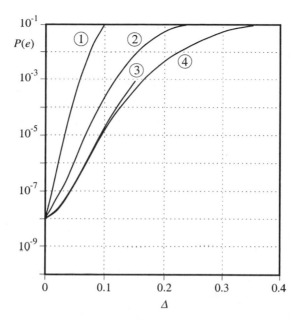

Figure 7.10: *Error probability for binary PAM as a function of the normalized sampling time deviation $\Delta$. The impulse response (with 50 interfering samples) is that of an ideal low-pass filter with cutoff frequency $1/(2T)$. The labels of the curves are as follows: (1) worst-case bound, (2) Chernoff bound, (3) series expansion method, (4) GQR method, SNR = 15 dB.*

the computation on a desk-top computer is less than a few seconds for all the methods described. It is practically constant with $N$ for the worst-case bound, whereas with the other methods it grows linearly with $N$.                                      □

### 7.2.2.    Two-dimensional modulation schemes

Expressions of the sampled in-phase and quadrature received signals were given in (7.19) and (7.20). The error probability will involve in general, as a final step, the average with respect to the RV $\theta$, as for PAM. For simplicity, let us assume $\theta = 0$. With the following shorthand notations:

$$X_P \triangleq \sum_{n \neq 0} r_{DP}(a_n) = \sum_{n \neq 0} (a_{Pn} h_{Pn} - a_{Qn} h_{Qn}) \qquad (7.36)$$

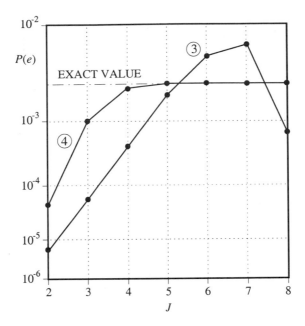

Figure 7.11: *Error probability of the system of Example 7.6 for* $\Delta = 0.2$ *as a function of the number of terms used in the series expansion or in the GQR. Labels of curves as in Figure 7.10. The exact value is also given.*

$$X_Q \triangleq \sum_{n \neq 0} r_{DQ}(a_n) = \sum_{n \neq 0} (a_{Pn} h_{Qn} + a_{Qn} h_{Pn}) \qquad (7.37)$$

the in-phase and quadrature received signals become

$$r_{DP0} = r_{DP}(a_0) + X_P + \nu_{P0} \qquad (7.38)$$
$$r_{DQ0} = r_{DQ}(a_0) + X_Q + \nu_{Q0} \qquad (7.39)$$

The decisions at the receiver are made through a rule that partitions the two-dimensional space of the received signal points into $M$ regions $R_k$. The error probability can be immediately derived from (4.24) in the form

$$P(e) = 1 - \frac{1}{M} \sum_{a_k \in \mathcal{A}} P\{\mathbf{r}_{D0} \in R_k \mid a_0 = \alpha_k\} \qquad (7.40)$$

where $\mathbf{r}_{D0}$ is the received vector with components $r_{DP0}, r_{DQ0}$, and $\mathcal{A} = \{\alpha_k\}_{k=1}^{M}$ is the set of values assumed by $a_0$. The probabilities in the RHS of (7.40) can be computed in two steps

$$P\{\mathbf{r}_{D0} \in \mathrm{R}_k \mid a_0 = \alpha_k\} = \mathrm{E}_{X_P, X_Q} P\{\mathbf{r}_{D0} \in \mathrm{R}_k \mid a_0 = \alpha_k, X_P, X_Q\} \quad (7.41)$$

The first step consists in evaluating the conditional probability in the right-hand side of (7.41). The received vector $\mathbf{r}_{D0}$, conditioned on $\alpha_k, X_P, X_Q$, is a Gaussian vector with independent components $r_{DP0}$ and $r_{DQ0}$. Thus, the evaluation of (7.41) involves integration of a bivariate Gaussian RV with independent components within the region $\mathrm{R}_k$. This problem has been discussed in Chapter 5 for the most important two-dimensional coherent modulation schemes. If we define

$$D_k(\alpha_k, X_P, X_Q) \triangleq P\{\mathbf{r}_{D0} \in \mathrm{R}_k \mid a_0 = \alpha_k, X_P, X_Q\}$$

the second step to get the probability in the LHS of (7.41) becomes the evaluation of the integral

$$I_k(\alpha_k) \triangleq \int\!\!\int_{\mathcal{X}} D_k(\alpha_k, X_P, X_Q) f_{X_P X_Q}(x_P, x_Q) dx_P dx_Q \quad (7.42)$$

where $\mathcal{X}$ and $f_{X_P X_Q}(x_P, X_Q)$ represent the joint range and pdf of $X_P$ and $X_Q$, respectively.

In Appendix E the method of cubature rules is outlined to approximate integrals like (7.42) on the basis of the knowledge of a certain number of joint moments of the RVs $X_P$ and $X_Q$. These moments can be computed using an extension of the recursive algorithm already explained for the one-dimensional case (see Problem 7.6). In some cases, owing to the symmetry of the modulation scheme, the two-dimensional problem can be reduced to the product of two one-dimensional problems, or even to a single one-dimensional problem. An example is provided by the case of $M$-ary phase modulation $M$-PSK.

**$M$-PSK modulation**

The complete symmetry of the signal set allows us to simplify the error probability (7.40) as

$$P(e) = 1 - P\{\mathbf{r}_{D0} \in \mathrm{R}_1 \mid a_0 = A\} \quad (7.43)$$
$$= 1 - P\left\{-\frac{\pi}{M} < \phi_{D0} \leq \frac{\pi}{M}\right\}$$

In (7.43) we have assumed that the phase zero has been transmitted (see Table 7.1.), and have defined the phase of the received vector $r_{D0}$ as

$$\phi_{D0} \triangleq \tan^{-1} \frac{r_{DQ0}}{r_{DP0}} \quad (7.44)$$

A straightforward extension of the bounding technique that led to (5.23) for the AWGN channel results in the following bounds for the error probability:

$$\max(I_1, I_2) \leq P(e) \leq I_1 + I_2 \qquad (7.45)$$

where

$$I_1 = \frac{1}{2} \int_{\mathcal{L}} \operatorname{erfc}\left(\frac{\lambda_0^+ + \lambda}{\sqrt{2}\sigma_n}\right) f_\Lambda(\lambda) d\lambda \qquad (7.46)$$

$$I_2 = \frac{1}{2} \int_{\mathcal{L}} \operatorname{erfc}\left(\frac{\lambda_0^- + \lambda}{\sqrt{2}\sigma_n}\right) f_\Lambda(\lambda) d\lambda \qquad (7.47)$$

$\Lambda$ is the random variable accounting for ISI

$$\Lambda \triangleq A \sum_{n \neq 0} \left[ h_{Pn} \sin\left(\frac{\pi}{M} + \phi_n\right) + h_{Qn} \cos\left(\frac{\pi}{M} + \phi_n\right) \right] \qquad (7.48)$$

and

$$\lambda_0^\pm \triangleq A \left( h_{P0} \sin\frac{\pi}{M} \pm h_{Q0} \cos\frac{\pi}{M} \right) \qquad (7.49)$$

Looking at (7.48) and (7.49), we can see that the evaluation of the bounds to the error probability for $M$-PSK modulation has been reduced to the computation of two one-dimensional integrals like (7.28). Thus, all the methods introduced in the PAM case directly apply; in particular, we can apply the GQR method.

**Example 7.7**   Consider a binary PSK modulation scheme that uses a channel modeled as a third-order Butterworth filter (see Example 2.3) with 3-dB bandwidth $B_0$. In Figure 7.12 the error probability computed using the GQR technique is plotted as a function of the number of points $J$ of the quadrature formula.

The dashed line is the exact value of $P(e)$ obtained by means of (7.31). The number of interfering samples has been chosen equal to 20. It can be seen that even with a small value of $J$ the GQR offers a high accuracy. The difference in the computer times needed to obtain the two curves (the exact and the GQR ones) of Figure (7.12) is enormous, and such as to prevent the use of the direct enumeration when the number of phases increases. In Figure 7.13 the signal-to-noise ratio $\mathcal{E}_b/N_0$ necessary to obtain an error probability of $10^{-6}$ for quaternary PSK is plotted as a function of the normalized bandwidth $2B_0T$. The two curves refer to the Chernoff bound and to the GQR methods. The asymptotic value represents the case of no ISI. It can be seen that the Chernoff bound is rather loose, and leads to an asymptotic difference of about 1 dB in signal-to-noise ratio.

□

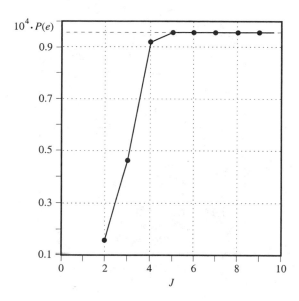

Figure 7.12: *Error probability as a function of the number of points of the quadrature rule for a binary PSK system with third-order Butterworth filter. The dashed line represents the exact value.*

**Example 7.8**    In this example we want to show how the computational techniques that have been presented in this chapter can be extended to the analysis and design of a digital transmission system operating in a Frequency-Division-Multiplexing (FDM) multichannel environment. The system model employing $M$-PSK modulation is presented in Figure 7.14. The attention is focused on one particular channel (the *useful* channel), disturbed by two adjacent channels, working at the same signaling rate, giving rise to *interchannel* interference, and by one channel at the same frequency. This schematic model suits wireless communication systems employing FDMA and frequency reuse (for example, the widely used GSM standard), or any fixed point-to-point system employing FDM and making use of two orthogonal polarizations to increase the bandwidth efficiency (like in some radio-relay links). The transmitter filters are assumed to have the same transfer function, except for the frequency location. In other words, let

$$G_i^+(f) \triangleq G_0^+(f + if_d), \quad i = -1, 0, 1 \tag{7.50}$$

be the transfer function of the $i$-th channel transmitter filter for positive frequencies, where $f_d$ is the frequency spacing between two adjacent channels. For simplicity, we shall assume that $G_i(f)$ satisfies the symmetry conditions of Example 7.2 with respect

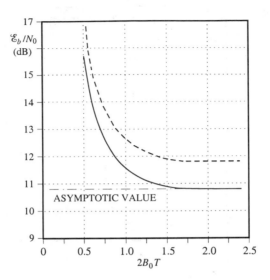

Figure 7.13: *Signal-to-noise ratio $\mathcal{E}_b/N_0$ necessary to obtain a symbol error probability $P(e) = 10^{-6}$ for a quaternary PSK system with third-order Butterworth filter as a function of the normalized 3-dB bandwidth $2B_0T$. The dashed line refers to the Chernoff bound and the continuous one to the Gauss quadrature rules.*

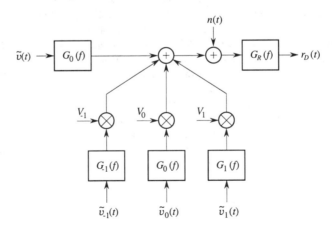

Figure 7.14: *Block diagram modeling a 4-channel FDM system. The figure shows the useful channel, two adjacent and one co-frequency interfering channels.*

to its center frequency $f_0 + i f_d$, $f_0$ being the carrier frequency of the useful signal. If

$$\tilde{v}(t) = \sum_n \exp(j\phi_n) s(t - nT) \tag{7.51}$$

is the complex envelope of the useful signal at the modulator output, the $i$-th interfering signal can be written as

$$\tilde{v}_i(t) = V_i \sum_{m_i} \exp[j\phi_{m_i} + j(2\pi f_d t + \theta_i)] s(t - \tau_i - m_i T) \tag{7.52}$$

where the meaning of the symbols is as follows:

- $V_i$ is the magnitude of the signal in the $i$th channel.

- $\tau_i$ accounts for the possible misalignment of signaling intervals in different channels. It may be modeled as a uniformly distributed RV in the interval $(0, T)$.

- $\theta_i$ is a RV uniformly distributed in the interval $(0, 2\pi)$ and accounts for the lack of coherence among the different carriers.

- $(\phi_{m_i})$ is the sequence of information phases pertaining to the $i$-th channel.

The bounding technique described for the case of $M$-PSK with ISI can be applied here for estimating the error probability. Moreover, the GQR method can also handle this situation, provided that the error probability conditioned on given values of $\tau_i$ and $\theta_i$ is first computed and the averages over $\tau_i$ and $\theta_i$ are performed later using standard quadrature rules. From the system engineer's viewpoint, the main design parameters are the frequency spacing $f_d$ between adjacent channels, the amount of co-channel interference that the system can tolerate, the choice of the transmitter and receiver filters (types and bandwidths), and the signal-to-noise ratio required to get a desired value of the error probability. The choice of these parameters is usually accomplished through a cut-and-try approach, which requires repeated analyses of the system and, hence, the availability of a tool to quickly evaluate system performance.

As usual in PSK, we shall assume that the shaping function $s(t)$ is rectangular. Both the transmitter and receiver filters are assumed to be Butterworth. Consider now the following parameters defining the system:

- $n_T, n_R$: the order of transmitter and receiver filters, respectively.

- $(B_{eq}T)_T, (B_{eq}T)_R$: equivalent noise bandwidths of the transmitter and receiver filters normalized to the symbol period $T$.

- $D = f_d T$: frequency spacing between two adjacent channels normalized to the symbol period $T$.

- $\mathcal{E}_b/N_0$: signal-to-noise ratio per transmitted bit of information, $N_0/2$ being the two-sided power spectral density of the noise.

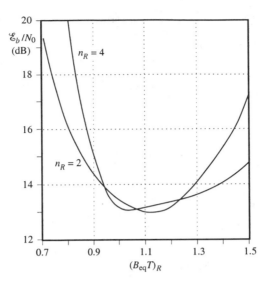

Figure 7.15: *Quaternary PSK system: signal-to-noise ratio $\mathcal{E}_b/N_0$ necessary to obtain a $P(e) = 10^{-6}$ as a function of the normalized equivalent noise bandwidth of the receiving filter. The parameters are as follows: $n_T = 6$, $(B_{eq}T)_T = 2.5$, and $D = 1.6$.*

The results that follow have been obtained by choosing as the sampling instant $t_0$ the time value corresponding to the maximum of the impulse response of the overall system without interchannel and cochannel interferences.

**Interchannel interference**

Two symmetrically located interfering channels are present at the same power level as the one interfered with. The modulation is assumed to be quaternary PSK. The first parameter considered for optimization is the normalized bandwidth of the receiver filter. In Figure 7.15 the signal-to-noise ratio necessary to obtain an error probability equal to $10^{-6}$ is plotted as a function of the normalized receiver filter bandwidth. The symbol intervals in the three channels are first assumed to be time-aligned (i.e., $\tau_i = 0$ for both interfering channels). It can be seen that a value of the normalized bandwidth around 1.1 is optimum. In the remaining curves of this example, the normalized receiver bandwidth will be assumed equal to 1.1. Let us now consider the choice of the channel spacing. In Figure 7.16 the signal-to-noise ratio necessary to obtain an error probability of $10^{-6}$ is plotted as a function of the normalized channel spacing $D$. The three curves refer to different values of the transmitter filter bandwidths (the value $\infty$ means absence of the transmitter filter). It is seen that the presence of a transmitter filter with bandwidth equal to 2.4 significantly improves the performance of the system.

This result is confirmed by Figure 7.17, where the only difference is represented by the fact that there is a random misalignment among the modulating bit streams. Thus

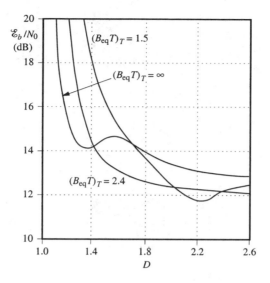

Figure 7.16: *Quaternary PSK system: signal-to-noise ratio $\mathcal{E}_b/N_0$ necessary to obtain a $P(e) = 10^{-6}$ as a function of the normalized frequency displacement of two symmetrically located interfering channels modulated by time-aligned bit streams. The parameters are as follows: $n_T = 6, n_R = 2, (B_{eq}T)_R = 1.1$.*

the final error probability is evaluated through an average over the RV $\tau_i$.

**Cochannel interference**

Finally, in Figure 7.18, the presence of one interfering channel at the same frequency as the useful one is considered. The modulating bit stream on the interfering channel is supposed to have a random misalignment. The curves plot the signal-to-noise ratio necessary to obtain an error rate of $10^{-6}$ as a function of the attenuation of the interfering channel. It is seen that the attenuation has to be of the order of 14, 16, or 20 dB for the cases of binary, quaternary, and octonary PSK, respectively, to ensure a negligible performance degradation as compared with the case of no interference. □

# 7.3. Eliminating intersymbol interference: the Nyquist criterion

In this section we will derive the conditions under which intersymbol interference (ISI) can be eliminated in a linearly modulated (one- or two-dimensional)

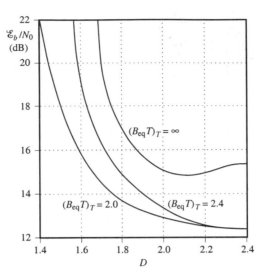

Figure 7.17: *Quaternary PSK system: same situation as in Figure 7.16 except for the random misalignment between the modulating bit streams of the interfering and interfered channels.*

system. Consider the transmission model of Fig. 7.19.

The source and the modulator are modeled assuming that the data to be transmitted form a stationary random sequence $(a_l)$ of independent, identically distributed (iid) real or complex random variables (RVs) with zero mean and variance

$$\sigma_a^2 \triangleq \mathrm{E}|a_\ell|^2 \tag{7.53}$$

The data sequence $(a_\ell)$ is sent to a linear modulator. For mathematical convenience, as it has been done in the previous part of the chapter, this is modeled as the cascade of a modulator having the ideal impulse $\delta(t)$ as its basic waveform, and of a shaping filter with an impulse response $s(t)$ and a frequency response $S(f)$. The number of symbols to be transmitted per second (i.e., the signaling rate) is denoted by $1/T$. Thus, the modulated signal is $\sum_{\ell=-\infty}^{\infty} a_\ell\delta(t - \ell T)$, and the signal sent to the channel is $\sum_{\ell=-\infty}^{\infty} a_\ell s(t - \ell T)$.

The channel section is represented by a time-invariant linear system having known transfer function $C(f)$ and impulse response $c(t)$ and a generator of additive noise. The noise process $w(t)$ is assumed to be Gaussian, independent of the data sequence, to have zero mean, finite power, and a known power density spectrum $G_w(f)$. Thus, the signal observed at the output of the channel section

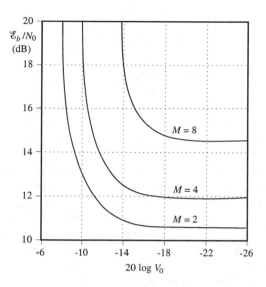

Figure 7.18: *Cochannel interference effects in binary, quaternary and octonary PSK systems. The signal-to-noise ratio $\mathcal{E}_b/N_0$ necessary to obtain a $P(e) = 10^{-6}$ is plotted as a function of the attenuation of the interfering channel. The modulating bit streams are assumed to be randomly misaligned. The parameters are as follows: $n_R = 2, (B_{eq}T)_T = \infty, (B_{eq}T)_R = 1.1$.*

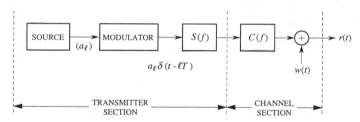

Figure 7.19: *Model for the transmission of linearly modulated data over a time-dispersive channel.*

can be written as

$$r(t) = \sum_{\ell=-\infty}^{\infty} a_\ell p(t - \ell T) + w(t) \tag{7.54}$$

where $p(t)$ is the response of the noiseless part of the channel to the waveform $s(t)$ or, equivalently, the convolution

$$p(t) = s(t) * c(t) \tag{7.55}$$

Figure 7.20: *Sampling receiver for the transmission system of Fig. 7.19.*

Our aim is the design of a receiver (see Fig. 7.20) having the form of a linear filter (hereafter referred to as the receiving filter) followed by a sampler. After linear filtering, the received signal is sampled every $T$ and the resulting sequence $(x_\ell)$ is sent to a detector. The detector makes decisions, on a sample-by-sample basis, according to the minimum-distance rule described in Section 4.2. The criterion considered in the design of the receiver shown in Fig. 7.20 concerns the elimination of ISI from the sampled sequence $(x_\ell)$. Such a criterion, known as the *Nyquist criterion*, will define the constraints on the overall system transfer function $S(f)C(f)U(f)$. As should be obvious, the elimination of ISI only concerns the cascade $S(f)C(f)U(f)$, leaving open the choice of how to partition the overall transfer function between transmitter and receiver (i.e., how to choose $S(f)$ and $U(f)$ once the product $S(f)C(f)U(f)$ has been specified). One can then give the burden of eliminating ISI to the receiving filter $U(f)$, or choose both $S(f)$ and $U(f)$ so as to meet the specified needs for their product: in this case, $S(f)$ and $U(f)$ can be chosen so as to minimize the effects of additive noise at the detector input, and hence to minimize the probability of error for the transmission system under the constraint of no ISI.

With reference to the transmission system shown in Fig. 7.19 and the sampling receiver of Fig. 7.20, denote by $q(t)$ the convolution

$$q(t) = p(t) * u(t) = s(t) * c(t) * u(t) \tag{7.56}$$

and by $n(t)$ the convolution

$$n(t) = w(t) * u(t) \tag{7.57}$$

where $u(t)$ is the impulse response of the receiving filter. Thus, at the sampler input we have

$$x(t) = \sum_{k=-\infty}^{\infty} a_k q(t - kT) + n(t) \tag{7.58}$$

and hence, at its output

$$x_\ell = \sum_{k=-\infty}^{\infty} a_k q_{\ell-k} + n_k \tag{7.59}$$

where the signal and noise samples are defined by

$$x_\ell \stackrel{\triangle}{=} x(t_0 + \ell T), \quad q_\ell \stackrel{\triangle}{=} q(t_0 + \ell T), \quad n_\ell \stackrel{\triangle}{=} n(t_0 + \ell T) \tag{7.60}$$

and $t_0 + \ell T$, $-\infty < \ell < \infty$, are the sampling instants. In what follows, for the sake of clarity, we will assume $t_0 = 0$. For error-free transmission, allowing for a delay of $D$ symbol intervals between transmission and reception of a given symbol, we must satisfy the condition that $x_\ell$ is equal to $a_{\ell-D}$. However, from (7.59) we obtain

$$x_\ell = q_D a_{\ell-D} + \sum_{k \neq \ell - D} a_k q_{\ell-k} + n_\ell \tag{7.61}$$

The factor $q_D$ of (7.61) is a complex number representing a constant change of scale, and possibly a phase shift if the channel is bandpass (see Section 7.1): under the hypothesis of a known channel, it can be easily compensated for. Thus, we assume $q_D = 1$. The second term of (7.61) represents the contribution of ISI. As noted previously in this chapter, it depends on the entire transmitted sequence $(a_k)$, as weighted by the samples $q_{\ell-k}$ of the impulse response of the overall channel. This is the effect of the tails and precursors of the waveforms overlapping the one carrying the information symbol $a_{\ell-D}$. The third term in (7.61) represents the effect of the additive noise. The sample sequence $(x_\ell)$ must be processed to get an estimate $(\hat{a}_\ell)$ of the transmitted symbols sequence. Of course, a reasonable way to do this is to perform symbol-by-symbol decisions (i.e., to use only $x_\ell$ to obtain an estimate of $a_{\ell-D}$, $-\infty < \ell < \infty$). This procedure is the simplest, but suboptimum as the samples $x_\ell$ given by (7.61) are correlated due to the effect of ISI. Hence, for an optimum decision the whole sequence $(x_\ell)$ should be processed. In the framework proposed, what seems at first a reasonable approach to the problem of optimizing the transmission system is trying to eliminate the ISI term in (7.61). If this is achieved, the problem is reduced to a situation in which only additive Gaussian noise is present. Hence, a symbol-by-symbol decision rule based on the minimum distance is optimum under the constraint of no ISI. We shall examine this solution.

To avoid the appearance of the ISI term in (7.61) the overall channel impulse response sample sequence $(q_l)$ should satisfy the condition

$$q_\ell = \begin{cases} 0, & \ell \neq D \\ 1, & \ell = D \end{cases} \tag{7.62}$$

This condition can also be expressed by observing that, with $\Delta_T(t)$ denoting a periodic train of delta functions spaced $T$ apart, that is,

$$\Delta_T(t) \triangleq \sum_{k=-\infty}^{\infty} \delta(t - kT) \tag{7.63}$$

Equation (7.62) is equivalent to

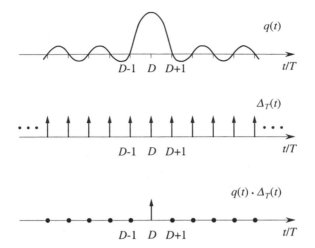

Figure 7.21: *Eliminating ISI from the samples of the channel output.*

$$q(t) \cdot \Delta_T(t) = \delta(t - DT) \tag{7.64}$$

(see Fig. 7.21). Taking the Fourier transform of both sides of (7.64), with the definition

$$Q(f) \triangleq \mathcal{F}[q(t)] = S(f)C(f)U(f) \tag{7.65}$$

we get

$$\frac{1}{T}Q(f) * \Delta_{\frac{1}{T}}(f) = \exp(-j2\pi f DT) \tag{7.66}$$

The effect of convolving $Q(f)$ with the train $\Delta_{1/T}(f)$ of spectral lines spaced $1/T$ Hz apart is to obtain a train of replicas of $Q(f)$ spaced $1/T$ Hz apart (Fig. 7.22). By denoting this convolution by $Q_{\text{eq}}(f)$:

$$Q_{\text{eq}}(f) \triangleq \sum_{k=-\infty}^{\infty} Q(f + \frac{k}{T}) \tag{7.67}$$

Eq. (7.66) requires that $Q_{\text{eq}}(f)$ have a constant magnitude and a linear phase.[4] It is easily seen that, for any $Q(f)$, $Q_{\text{eq}}(f)$ is a periodic function of $f$ with period $1/T$. Thus, without loss of generality, we can confine our consideration of this

---

[4]It may be worthwhile to notice that the condition of constant magnitude and linear phase ensures no distortion also in the *analog* domain, where, on the other hand, the condition concerns the *true* transfer function $Q(f)$, instead of its aliased version $Q_{\text{eq}}(f)$.

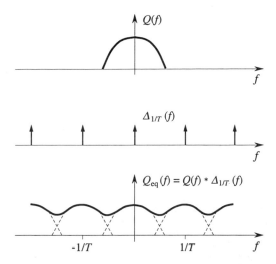

Figure 7.22: *Convolution of $Q(f)$ with a train of spectral lines in the frequency domain.*

function to the fundamental interval $[-1/(2T), 1/(2T)]$, and express condition (7.66) in the form

$$Q_{eq}(f) = T \cdot \exp(-j2\pi f DT), \quad |f| \leq \frac{1}{2T} \tag{7.68}$$

Condition (7.68) for the removal of ISI is called the (first) *Nyquist criterion* and the interval $[-1/(2T), 1/(2T)]$ the *Nyquist interval*. This criterion says that, if the frequency response $Q(f)$ of the overall channel is cut in slices of width $1/T$ and these are piled up in the Nyquist interval with the proper phases (see Fig. 7.23), ISI is eliminated from the sample sequence $(x_\ell)$ when the resulting *equivalent spectrum* $Q_{eq}(f)$ has a constant magnitude and a linear phase. Looking at the achievable data rate, if the modulator uses $M$ amplitude levels and a baseband (or single-sideband) transmission, we can transmit up to $\log_2 M$ bits in a $1/(2T)$ bandwidth without ISI.

### 7.3.1. The raised-cosine spectrum

If $Q(f)$ is nonzero outside the Nyquist interval, many classes of responses satisfy (7.68). Thus, the Nyquist criterion does not uniquely specify the shape of the frequency response $Q(f)$. On the contrary, if $Q(f)$ is limited to an interval smaller than Nyquist's, it is impossible for (7.68) to hold. Thus, ISI cannot be removed from the received signal. If $Q(f)$ is exactly bandlimited in the Nyquist

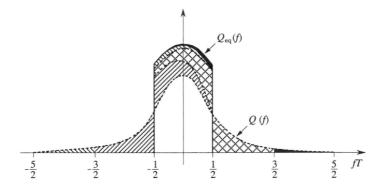

Figure 7.23: *Construction of $Q_{eq}(f)$ in the Nyquist interval $[-1/(2T), 1/(2T)]$. Here $Q_{eq}(f)$ is assumed to be real.*

interval, (7.68) requires that

$$Q(f) = \begin{cases} Q_{eq}(f) = T \cdot \exp(-j2\pi f DT), & |f| \leq \frac{1}{2T} \\ 0, & \text{elsewhere} \end{cases} \qquad (7.69)$$

That is, the only transfer function $Q(f)$ satisfying the Nyquist criterion is the "brickwall" frequency response of the ideal low-pass filter with delay $DT$.

With $Q(f)$ as in (7.69), the overall channel impulse response $q(t)$ becomes

$$q(t) = \frac{\sin \pi(t/T - D)}{\pi(t/T - D)} \qquad (7.70)$$

a noncausal function (for any finite $D$) that decays for large $t$ as $1/t$. The transfer function (7.69) poses two serious problem. First, it is not physically realizable because of its sudden instantaneous jump to 0 at $f = 1/(2T)$ (as the Latin saying goes, *natura non facit saltus*). The second drawback comes from the fact that every real-world system will exhibit errors in the timing synchronization causing erroneous sampling times. Even a minimum sampling error would cause the eye pattern to close simply because the series $\sum_{k=-\infty}^{\infty} q(\tau + kT)$ is not absolutely summable for $\tau \neq 0$ when $q(t)$ is as in (7.70) (see Section 7.2.1). For this reason, it becomes mandatory to trade a wider bandwidth for a reduced sensitivity to inaccuracies in sampling times (and possibly for an easier implementation). Since it is recognized that the problem with the impulse response (7.70) is due to its slow rate of decay, and since the rate of decay of a pulse is intimately related to the discontinuities of its Fourier transform, it is reasonable to investigate classes of responses that satisfy the Nyquist criterion with a minimum of discontinuities, considering also the discontinuities in the derivatives. This can be

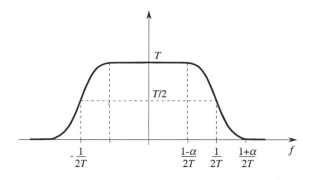

Figure 7.24: *Example of a real $Q(f)$ satisfying the Nyquist criterion.*

obtained, for example, as shown in Fig. 7.24. Let $\alpha$, $0 \leq \alpha \leq 1$, be the allowed relative amount of bandwidth in excess of Nyquist's; that is, let $Q(f)$ be strictly bandlimited to the interval $|f| \leq (1 + \alpha)/(2T)$. Letting $D = 0$ for simplicity, choose

**(a)** $Q(f) = T$,   for $|f| \leq (1 - \alpha)/(2T)$;

**(b)** $Q(f)$ real, decaying from $T$ to zero for $(1-\alpha)/(2T) \leq |f| \leq (1+\alpha)/(2T)$, and exhibiting symmetry with respect to the points of abscissa $\pm 1/(2T)$ and ordinate $T/2$. This *roll-off spectrum* must be chosen in such a way that it presents a minimum of discontinuities at $|f| = (1 + \alpha)/(2T)$, the band edges.

The choice of a sinusoidal form for the roll-off spectrum leads to the *raised cosine* transfer function defined as follows:

$$Q(f) = \begin{cases} T, & |f| \leq \dfrac{1 - \alpha}{2T} \\ \dfrac{T}{2}\left\{1 - \cos\left[\dfrac{\pi T}{\alpha}\left(f - \dfrac{1+\alpha}{2T}\right)\right]\right\}, & \dfrac{1 - \alpha}{2T} \leq |f| \leq \dfrac{1+\alpha}{2T} \\ 0, & |f| \geq \dfrac{1+\alpha}{2T} \end{cases}$$

(7.71)

The impulse response corresponding to a raised cosine spectrum is

$$q(t) = \frac{\sin(\pi t/T)}{\pi t/T} \cdot \frac{\cos(\alpha \pi t/T)}{1 - (2\pi t/T)^2}$$

(7.72)

and decays asymptotically as $1/t^3$ for $t \to \infty$.

Fig. 7.25 shows the raised cosine spectra and the corresponding impulse responses for $\alpha = 0.25$, 0.5 and 1.0. In Fig. 7.26 we show the inner envelopes of

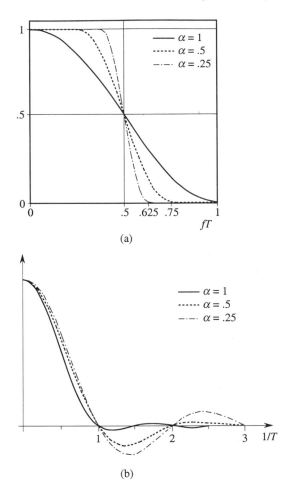

Figure 7.25: *(a) Raised cosine spectra; (b) impulse response of raised cosine filters.*

the corresponding eye patterns for binary transmission with symbols $\pm1$. It is seen from Fig. 7.26 that the immunity to erroneous sampling instants increases with $\alpha$. In particular, with a 100% roll-off, open-eye transmission is possible even with a sampling time error approaching $0.5\ T$ in absolute value. With smaller values of $\alpha$, the margin against erroneous sampling decreases, and is zero when $\alpha = 0$ (corresponding to the brickwall frequency response).

Notice also that $q(t)$ in (7.72) is not causal and hence not physically realizable. However, approximate realizations can be obtained by considering a delay

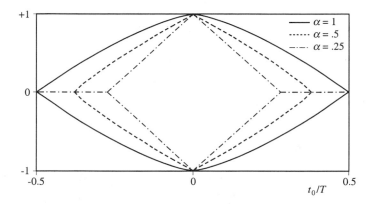

Figure 7.26: *Inner envelopes of eye patterns resulting from antipodal binary transmission over a channel with a raised-cosine transfer function.*

$D$ so large that a causal approximation to $q(t - D)$ gives a performance satisfactorily close to that predicted by the theory. Raised cosine spectra are often considered for practical modem applications.

### 7.3.2.    Optimum design of the shaping and receiving filters

Assume now that $Q(f)$ has been chosen so as to satisfy the Nyquist criterion, so that freedom from ISI is assured by taking the shaping filter and the receiving filter such that

$$S(f)C(f)U(f) = Q(f) \qquad (7.73)$$

Thus, for a given $C(f)$ the actual design of $S(f)$ and $U(f)$ still leaves a degree of freedom, as only their product is specified. This freedom can be taken advantage of by imposing one further condition, that is, the minimization of the effect of the noise at the sampler input or, equivalently, the minimization of the error probability (in fact, in the absence of ISI, errors are caused only by the additive noise).

The average noise power at the receiving filter output is, from (7.57)

$$\sigma_n^2 = \int_{-\infty}^{\infty} G_w(f)|U(f)|^2 df \qquad (7.74)$$

Minimization of $\sigma_n^2$ without constraints would lead to the trivial solution $|U(f)| = 0$, which is not compatible with (7.73). To avoid this situation, we constrain the signal power at the channel input to a finite value, which poses a constraint on $S(f)$. This, in turn, prevents (through (7.73)) $U(f)$ from assuming too small values. Since the overall channel frequency response is the fixed function $Q(f)$, the signal power spectral density at the shaping filter output is, from (2.128),

$$\frac{\sigma_a^2}{T}|S(f)|^2 = \frac{\sigma_a^2}{T}\frac{|Q(f)|^2}{|C(f)U(f)|^2} \tag{7.75}$$

and the corresponding signal power is

$$\mathcal{P} = \frac{\sigma_a^2}{T}\int_{-\infty}^{\infty}\frac{|Q(f)|^2}{|C(f)U(f)|^2}df \tag{7.76}$$

Minimization of $\sigma_n^2$ under the constraint (7.76) can be performed using the Lagrange-multiplier and variational techniques (see Appendix C). Omitting an unessential factor, the minimizing $U(f)$ is given by the equation

$$|U(f)| = \frac{|Q(f)|^{1/2}}{G_w^{1/4}(f)|C(f)|^{1/2}} \tag{7.77}$$

and the corresponding shaping filter is obtained through

$$S(f) = \frac{Q(f)}{C(f)U(f)} \tag{7.78}$$

In (7.77) and (7.78) it is assumed that $Q(f)$ is zero at those frequencies for which the denominators are zero. Notice that the phase characteristics of $U(f)$ are not specified, and are therefore arbitrary (of course, $S(f)$ in (7.78) is such that $Q(f)$ has a linear phase, as required by the Nyquist criterion). In the special case of white noise and $C(f) =$ constant, it is seen from (7.77) and (7.78) that $U(f)$ and $S(f)$ can be identical apart from an irrelevant scale factor, so only one design has to be implemented for both filters.

## 7.4. Mean-square error optimization

In the last section we saw how a system free of ISI can be designed. After choosing the overall channel transfer function, the optimum design of shaping and receiving filters was achieved by minimizing the noise power at the sampler's input. Although this procedure sounds reasonable, it does not guarantee minimization of the error probability. In fact, it might happen that, by trading a small ISI for a lower additive noise power, a better error performance is obtained. On the other hand, system optimization under the criterion of a minimum error probability is a rather complex task. This suggests that we look for a criterion leading to a more manageable problem.

Thus, in this section we shall consider the mean-square error (MSE) criterion for system optimization; this choice allows ISI and noise to be taken jointly into account, and in most practical situations leads to values of error probability very close to their minimum.

Consider again the system model shown in Figs. 7.19 and 7.20. Instead of constraining the noiseless samples to be equal to the transmitted symbols, we can take into account the presence of additive noise and try to minimize the mean-squared difference between the sequence of transmitted symbols $(a_\ell)$ and the sampler outputs $(x_\ell)$. By allowing for a channel delay of $D$ symbol intervals, we shall determine the shaping filter $S(f)$ and the receiving filter $U(f)$ so that the mean-square value of

$$\epsilon_\ell \overset{\triangle}{=} x_\ell - a_{\ell-D} \qquad (7.79)$$

is minimized. This will result in a system that, although not specifically designed for optimum error performance, should provide a satisfactory performance even in terms of error probability.

We begin by deriving an expression for the MSE at the detector input, defined as

$$\mathcal{E} \overset{\triangle}{=} \mathrm{E}|\epsilon_\ell|^2 = \mathrm{E}|x_\ell - a_{\ell-D}|^2 \qquad (7.80)$$

From (7.61), $\epsilon_\ell$ can be given the form

$$\epsilon_\ell = a_{\ell-D}(q_D - 1) + \sum_{k \neq \ell-D} a_k q_{\ell-k} + n_\ell \qquad (7.81)$$

so that, due to the independence of the terms summed up in the RHS, we obtain

$$\begin{aligned}
\mathcal{E} &= \sigma_a^2 |q_D - 1|^2 + \sigma_a^2 \sum_{k \neq D} |q_k|^2 + \sigma_n^2 \\
&= \sigma_a^2 [1 - 2\Re(q_D)] + \sigma_a^2 \sum_{k=-\infty}^{\infty} |q_k|^2 + \sigma_n^2
\end{aligned} \qquad (7.82)$$

Now we want to express $\mathcal{E}$ by using frequency-domain quantities. By assuming as usual that $t_0$ is equal to zero, we get

$$q_k = \int_{-\infty}^{\infty} Q(f) e^{j2\pi fkT} df \qquad (7.83)$$

and consequently, by direct calculation,

$$\sum_{k=-\infty}^{\infty} |q_k|^2 = \frac{1}{T} \sum_{k=-\infty}^{\infty} \int_{-\infty}^{\infty} Q^*(f) Q\left(f + \frac{k}{T}\right) df \qquad (7.84)$$

Thus, (7.82) can be rewritten, using also (7.74), in the form

$$\mathcal{E} = \sigma_a^2 \left[ 1 - 2\Re \int_{-\infty}^{\infty} Q(f) e^{j2\pi f DT} df \right. \tag{7.85}$$

$$\left. + \frac{1}{T} \int_{-\infty}^{\infty} Q^*(f) \sum_{k=-\infty}^{\infty} Q(f + k/T) df \right] + \int_{-\infty}^{\infty} G_w(f) |U(f)|^2 df$$

We observe that the MSE is the sum of two terms. The first (enclosed in square brackets) represents the overall channel ISI power, while the second represents the contribution of the additive noise power. These terms are not independent, as any change in $U(f)$ would also affect $Q(f)$. Qualitatively, it can be said that, if the bandwidth of the receiving filter $U(f)$ is reduced in order to decrease the value of the noise term, this will result in a corresponding increase of the overall channel ISI.

**Example 7.9**   Consider a baseband transmission system with white Gaussian noise having power spectral density $N_0/2$, data with $\sigma_a^2 = 1$, $s(t) = u_T(t)$, a channel modeled through a fourth-order low-pass Butterworth filter with 3-dB frequency $B_C$, and a second-order low-pass Butterworth receiving filter with 3-dB frequency $B_U$. In Fig. 7.27 the dashed lines represent the contribution of the noise (which augments with increasing $B_U$ and $N_0$) and the continuous lines the contribution of the overall ISI MSE (which augments with decreasing $B_U$ and $B_C$). The total MSE $\mathcal{E}$ is obtained by summing up the two contributions, which results in a minimum for an optimum value of $B_U$.   $\square$

### 7.4.1.   Optimizing the receiving filter

We shall now consider the selection of a transfer function $U(f)$ that gives a minimum for $\mathcal{E}$ when $S(f)$, as well as $C(f)$, are given. By using (7.85) and applying standard variational techniques (see Appendix C), it can be proved that a necessary and sufficient condition for $U(f)$ to minimize $\mathcal{E}$ is that

$$\frac{1}{T} S^*(f) C^*(f) \sum_{k=-\infty}^{\infty} S\left(f + \frac{k}{T}\right) C\left(f + \frac{k}{T}\right) U\left(f + \frac{k}{T}\right) +$$

$$+ \frac{1}{\sigma_a^2} G_w(f) U(f) = S^*(f) C^*(f) = S^*(f) C^*(f) e^{-j2\pi f DT} \tag{7.86}$$

be satisfied. In spite of its formidable appearance, (7.86) is amenable to a closed-form solution, which in turn admits an interesting interpretation. To see this, let us first show that the optimum receiving filter, say $U_{\mathrm{opt}}(f)$, has the following

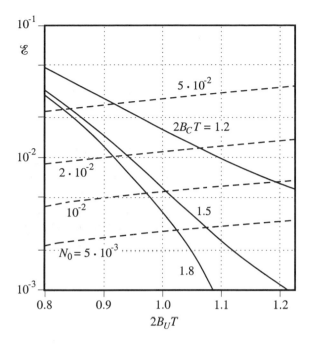

Figure 7.27: *Contributions to mean-square error of intersymbol interference (continuous line) and additive noise (dashed line) in the situation of Example 7.9.*

expression

$$U_{\text{opt}}(f) = \frac{P^*(f)}{G_w(f)}\Gamma(f) \tag{7.87}$$

where $\Gamma(f)$ is a periodic function with period $1/T$, and

$$P(f) \overset{\triangle}{=} S(f)C(f) \tag{7.88}$$

is the transfer function of the cascade of the shaping filter and the channel (we assume here for simplicity that $G_w(f)$ is nonzero everywhere). By substituting (7.87) into (7.86) and observing that $\Gamma(f + k/T) = \Gamma(f)$ for all $k$ due to its periodicity, we get

$$\frac{1}{T}\sum_{-\infty}^{\infty}\frac{|P(f+k/T)|^2}{G_w(f+k/T)} + \frac{1}{\sigma_a^2}P^*(f)\Gamma(f) = P^*(f)e^{-j2\pi fDT} \tag{7.89}$$

For all the frequencies at which $P(f)$ vanishes, (7.86) shows that $U_{\text{opt}}(f)$ must also be zero, so that (7.87) is true. For $P(f) \neq 0$, (7.89) gives

$$\Gamma(f) = \frac{\sigma_a^2 e^{-j2\pi f DT}}{1 + \sigma_a^2 L(f)} \tag{7.90}$$

where

$$L(f) = \frac{1}{T} \sum_{k=-\infty}^{\infty} \frac{|P(f + k/T)|^2}{G_w(f + k/T)} \tag{7.91}$$

is periodic with period $1/T$, as required. This shows that the solution to (7.86) has the form (7.87). Insight into the behavior of the optimum receiving filter can be gained by considering the special case of a channel bandlimited to the Nyquist interval $[-1/(2T), 1/(2T)]$. In this case, (7.91) specializes to

$$L(f) = \frac{1}{T} \frac{|P(f)|^2}{G_w(f)} \tag{7.92}$$

and, from (7.87) and (7.90) we get

$$U_{\text{opt}}(f) = \frac{\sigma_a^2 P^*(f)}{G_w(f) + (\sigma_a^2/T)|P(f)|^2} e^{-j2\pi f DT} \tag{7.93}$$

Equation (7.93) shows that, in the absence of noise, the optimum receiving filter is simply the inverse of $P(f)$. This is an obvious result, since in this situation ISI is the only contribution to the MSE, and, in turn, can be reduced to zero by forcing the overall channel to a flat frequency response in the Nyquist band. However, when $G_w(f) \neq 0$, elimination of ISI does not provide the best solution. On the contrary, for spectral regions where the denominator of the RHS of (7.93) is dominated by $G_w(f)$, $U_{\text{opt}}(f)$ (apart from a scale factor and a delay term) approaches the matched filter characteristics $P^*(f)/G_w(f)$.

   More generally, for a channel not constrained to have a zero transfer function outside the Nyquist interval, (7.87) can be interpreted by observing that $P^*(f)/G_w(f)$ is the transfer function of a filter matched to the impulse response $p(t)$ of the cascade of the shaping filter and the channel. Also, $\Gamma(f)$, being a periodical transfer function with period $1/T$, can be thought of as the transfer function of a transversal filter whose taps are spaced $T$ seconds apart. Thus, we can affirm that the optimum receiving filter is the cascade of a matched filter and a transversal filter. The former reduces the noise effects and provides the principal correction factor when the signal-to-noise ratio is small. The latter reduces ISI and in the situation of high signal-to-noise ratio attempts to suppress it.

### 7.4.2. Performance of the optimum receiving filter

Let us now evaluate the MSE of a system in which $S(f)$ and $C(f)$ are given and $U(f)$ has been optimized. Substituting (7.87) for $U(f)$ in (7.85) and using (7.92) and (7.93), we get, after algebraic manipulations,

$$\mathcal{E} = \sigma_a^2 \left[ 1 - \int_{-\infty}^{\infty} \frac{|P(f)|^2}{G_w(f)} \frac{\sigma_a^2}{1 + \sigma_a^2 L(f)} df \right] \tag{7.94}$$

For a more compact form of the error expression, the integral appearing in (7.94) is rewritten as follows:

$$
\begin{aligned}
&\int_{-\infty}^{\infty} \frac{|P(f)|^2}{G_w(f)} \frac{\sigma_a^2}{1 + \sigma_a^2 L(f)} df \\
&= \sum_{k=-\infty}^{\infty} \int_{(2k-1)/(2T)}^{(2k+1)/(2T)} \frac{|P(f)|^2}{G_w(f)} \frac{\sigma_a^2}{1 + \sigma_a^2 L(f)} df \\
&= \int_{-1/(2T)}^{1/(2T)} \left( \sum_{k=-\infty}^{\infty} \frac{|P(f+k/T)|^2}{G_w(f+k/T)} \right) \frac{\sigma_a^2}{1 + \sigma_a^2 L(f)} df
\end{aligned}
\tag{7.95}
$$

Also, using (7.91), we can express (7.95) in the form

$$\mathcal{E} = \sigma_a^2 \left[ 1 - T \int_{-1/(2T)}^{1/(2T)} \frac{\sigma_a^2 L(f)}{1 + \sigma_a^2 L(f)} df \right] \tag{7.96}$$

and, finally

$$\mathcal{E} = T \int_{-1/(2T)}^{1/(2T)} \frac{\sigma_a^2}{1 + \sigma_a^2 L(f)} df \tag{7.97}$$

which, in conjunction with (7.91), is the expression of the MSE achievable by optimizing the receiving filter $U(f)$ for a given channel and a given shaping filter.

**Example 7.10** Let us consider again Example 7.9, in which the goal is to optimize the receiving filter. We assume here $B_C T = 0.6$. The MSE for such a system is depicted in Fig. 7.28. The dotted line refers to a second-order Butterworth receiving filter whose bandwidth has been chosen so as to minimize $\mathcal{E}$, while the dashed line refers to the optimum receiving filter given by (7.87). It can be observed that the effectiveness of the optimization increases as the noise power spectral density $N_0$ decreases (i.e., the system performance is limited by ISI rather than by additive noise). □

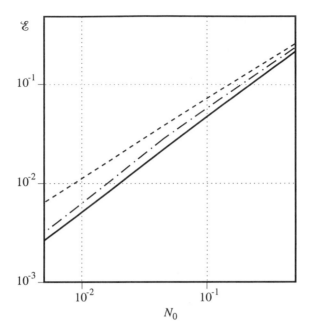

Figure 7.28: *Performance of the transmission system of Examples 7.9 to 7.11: MSE $\mathcal{E}$ versus the noise power spectral density $N_0$. Dotted line: the receiving filter is second-order Butterworth with 3-dB bandwidth chosen to minimize $\mathcal{E}$. Dashed line: the receiving filter has the optimum transfer function given by Eq. (7.87). Continuous line: both shaping and receiving filters are optimum in the MSE sense.*

### 7.4.3. Optimizing the shaping filter

The final step toward system optimization can now be taken by looking for the optimum shaping filter $S(f)$. To do this, $\mathcal{E}$ in (7.85) should be minimized with respect to $S(f)$ subject to the power constraint at the shaping filter output

$$\mathcal{P} = \frac{\sigma_a^2}{T} \int_{-\infty}^{\infty} |S(f)|^2 df \qquad (7.98)$$

which is the same as in (7.76).

The resulting equation, expressing a necessary and sufficient condition for $S(f)$ to minimize $\mathcal{E}$, does not seem amenable to a simple closed-form solution. Hence, we shall omit the details of the derivation and restrict ourselves to a general description of the solution and an example of its application.

The optimum shaping filter transfer function can be obtained as follows:

**Step 1** For every $f$ in the Nyquist interval $[-1/(2T), 1/(2T)]$, determine the integer $k_f$ such that $|C(f + k/T)|^2/G_w(f + k/T)$ takes on its maximum value with respect to $k$. We shall define F as the set of frequencies that can be written in the form $f + k_f/T$, $f \in [-1/(2T), 1/(2T)]$.

**Step 2** Choose $\lambda > 0$ and define the subset $F_\lambda$ of F such that, for $f \in F_\lambda$

$$\frac{|C(f)|^2}{G_w(f)} > \frac{1}{\lambda} \tag{7.99}$$

**Step 3** Take

$$|S_{\text{opt}}|^2 = \begin{cases} -\dfrac{TG_w(f)}{\sigma_a^2|C(f)|^2} + \dfrac{T}{\sigma_a^2|C(f)|}\sqrt{\lambda G_w(f)}, & f \in F_\lambda \\ 0, & \text{elsewhere} \end{cases} \tag{7.100}$$

Then compute $U(f)$ according to (7.87) and (7.88) and choose the phases of $S_{\text{opt}}(f), U(f)$ so that $Q(f) = S_{\text{opt}}(f)C(f)U(f)$ is real and positive. Inspection of (7.85) demonstrates that the MSE depends on the phase of $S(f)U(f)$, but is independent of the way it is distributed between $S(f)$ and $U(f)$.

**Step 4** Evaluate the resulting average channel input power by substituting for $|S(f)|^2$ in (7.98) the expression obtained from (7.100). The value computed will generally be different from the constraint value $\mathcal{P}$, so steps (2) to (4) should be repeated for different values of $\lambda$ until the average channel input power is equal to $\mathcal{P}$.

From this procedure, it is seen that the optimum shaping filter, and hence the whole channel transfer function $Q(f)$, is generally bandlimited to the frequency set F, which has measure $1/T$ (this set is usually referred to as a generalized Nyquist set). The pulses transmitted through the channel have their energy confined in this set, whose frequencies are chosen, according to Step 1, in order to afford the largest possible contribution to $L(f)$ in (7.91), and hence by rendering $L(f)$ as large as possible to minimize the RHS of (7.97). This simply shows that the pulse energy must be allocated at those frequencies where the channel performs better in the sense that $|C(f)|$ is large and/or $G_w(f)$ is small. In Step 2, the set F is further reduced in order to preserve only those frequencies at which the ratio $|C(f)|^2/G_w(f)$ lies above a certain level depending on the value of $\lambda$. Actually, $\lambda$, a Lagrange multiplier in the constrained optimization problem, turns out to be proportional to the signal-to-noise ratio, defined as the ratio of the average transmitted power to the average noise power at the output of the receiving filter.

**Example 7.11**   We consider again the situations described in Examples 7.9 and 7.10, and try to optimize the shaping filter. The noise is white, and $|C(f)|^2$ is assumed to be a monotonically decreasing function of $|f|$ (in fact, the function has the form $|C(f)|^2 = [1 + (f/B_C)^8]^{-1}$). Thus, it follows that $k_f$, as already defined in Step 1, is always zero, and hence F$= [-1/(2T), 1/(2T)]$. Furthermore, $F_\lambda = [-1/(2T'), 1/(2T')]$, where for high $\lambda$ values (i.e., high signal-to-noise ratios) $T' = T$, while for low $\lambda$ values (i.e., in the situation that (7.99) does not hold for all $f \in$ F), $T' > T$. Figure 7.28 shows the MSE obtained after optimizing both the shaping and the receiving filter in the situation dealt with here (continuous line).                                             □

### 7.4.4.   Information-theoretic optimization

In Section 3 we have derived the capacity of the additive Gaussian channel, under the hypothesis of band-limited white Gaussian noise. Here, we will show how to design the transfer function of the transmitting filter $S(f)$ in Fig. 7.19 in order to maximize the capacity of the channel.

Consider the system represented in Fig. 7.19, where $w(t)$ is additive Gaussian noise with power spectral density $G_w(f)$. We want to find the transfer function $S(f)$ of the shaping (transmitting) filter that maximizes the average mutual information (see Section 3.3) between channel input and output, subject to the power constraint (7.98) at the shaping filter output, here rewritten in the form

$$\int_{-\infty}^{\infty} |S(f)|^2 df \leq \frac{\mathcal{P}T}{\sigma_a^2} \tag{7.101}$$

Define the "channel signal-to-noise ratio function" (already used in the previous section)

$$\eta(f) \triangleq \frac{|C(f)|^2}{G_w(f)} \tag{7.102}$$

and consider, as already discussed in the previous section, that the preferred transmission bandwidth is the one in which $\eta(f)$ is large. This leads to the following formal result for $|S(f)|^2$, known as the *water-pouring, or water-filling* solution (Gallager, 1968, Chapter 8):

$$\frac{\sigma_a^2}{T}|S(f)|_{\text{opt}}^2 = \begin{cases} K - \dfrac{1}{\eta(f)}, & f \in B \\ 0, & f \notin B \end{cases} \tag{7.103}$$

where $B$ is the *capacity-achieving bandwidth*, i.e., the following range of frequencies $f$:

$$B \triangleq \{f : \eta(f) \geq 1/K\} \tag{7.104}$$

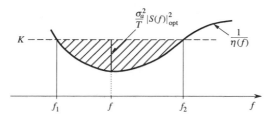

Figure 7.29: *Water-pouring model for optimal energy distribution.*

and $K$ is a constant chosen so as to satisfy the constraint (7.101), i.e., the solution to

$$\int_{f \in B} \{K - 1/[\eta(f)]\} \, df = \mathcal{P} \tag{7.105}$$

In practice, $B$ is often a continuous frequency interval $(f_1, f_2)$ (see Fig. 7.29), and the optimum shaping filter is the one that "pours" more signal power where the noise-to-signal ratio is lower, and avoids sending signal power when the noise-to-signal ratio exceeds a certain threshold ($K$) that depends on the available signal power. Capacity of the channel is achieved when the transmitted signal has a Gaussian statistic with power spectral density given by (7.103), and is equal to[5]

$$C = \frac{1}{2} \int_B \log_2 \left[ \frac{K|C(f)|^2}{G_w(f)} \right] df = \frac{1}{2} \int_B \log_2 \left[ 1 + \frac{\sigma_a^2}{T} |S(f)|^2_{\text{opt}} \eta(f) \right] df \ \text{bit/s} \tag{7.106}$$

Comparing (7.105), (7.103) and (7.106) with Fig. 7.29, we see that the power $\mathcal{P}$ is given by the area of the shaded region, and that the appropriate power spectral density is given by the height of the shaded region at any given $f$. This is the reason for the water-filling interpretation, since we can think of $1/\eta(f)$ as being the bottom of a container of unit depth, and of pouring in an amount of water $\mathcal{P}$. Assuming the region to be connected, we see that the water (power) will distribute itself in such a way as to achieve capacity.

The water-pouring argument leads to a practical application for the system called orthogonal frequency-division multiplexing (OFDM), where the channel bandwidth $B$ is divided into, say, $N$, subbands of width $\Delta f$ around frequency $f_i$, $i = 1, \ldots, N$, in such a way that the channel signal-to-noise ratio function $\eta(f)$ and, consequently, $|S(f)|^2$, are nearly constant in each subband (see Bingham, 1990). In these conditions, each subband can be considered as an ideal

---

[5]The reader is invited to compare (7.106) with the capacity (3.91) of the additive white Gaussian noise channel.

band-limited AWGN channel with bandwidth $\Delta f$, in which an optimal power

$$\mathcal{P}_i = [(|S(f_i)|^2)(\Delta f)\sigma_a^2]/T$$

is transmitted, with capacity

$$C_i = \log_2[1 + \mathcal{P}_i\eta(f_i)] \quad \text{bit/s} \tag{7.107}$$

The total expended power is then approximately $\mathcal{P}$, and the total capacity $\simeq C$. To approach the capacity, a code with rate matched to the capacity $C_i$ should be used in the $i$th subband. We have thus a situation in which the transmitter has a knowledge of the signal-to-noise ratio across the channel bandwidth (knowledge that can be acquired, for example, through some probe signals, like a comb of frequencies $f_i$), and adjusts the transmitted power (and possibly the size of the signal constellation) in each subband based on the water-pouring solution (7.103).

## 7.5.  Maximum-likelihood sequence receiver

In this section an entirely different approach will be adopted in the design of the optimum receiver for the system of Fig. 7.19.[6] In particular, we shall apply the theory of ML reception outlined in Chapters 2 and 4 to a channel with ISI and demonstrate that this approach provides a conceptually simple (although not always practical) solution to the optimization problem. Our assumptions are that the noise $w(t)$ is white and that the filters $S(f)$, $C(f)$ have a finite-length impulse response. A consequence of the latter assumption is that, before the addition of the noise, the waveforms at the channel output, as considered in any finite time interval, can only take a finite number of shapes (this number can be very large, but conceptually this is not a hindrance). Thus, the ML reception of a finite-length message is equivalent to the detection of one out of a finite set of waveforms in AWGN, so the theory developed in Sections 2.6 and 4.2 is valid. In particular, the optimum receiver consists of a bank of matched filters, one for each waveform. Their outputs are sampled at the end of the transmission, and the largest sample is used to select the most likely symbol sequence.

In practice, however, this solution would be unacceptable due to its excessive complexity. In fact, for a message of $K$ $M$-ary symbols, $M^K$ matched filters might be necessary, with about $M^K$ comparisons to be performed to select their largest sampled output. Thus, to provide a practical solution to the problem of ML reception, we must overcome several difficulties that appear intrinsic to it.

---

[6]We continue here to consider the case of linear modulations and systems; this assumption, however, is not strictly necessary, and will be removed in Chapter 14.

The first is the complexity induced by the large number of matched filters needed. The second is that induced by the large number of comparisons necessary to make a decision. The third is the size of the memory required to store all the possible transmitted sequences and the delay involved in the detection process. As we shall see, satisfactory solutions can be found for these difficulties. In fact, only one matched filter is sufficient, due to the channel linearity. Furthermore, we can specify an algorithm whereby the number of computations necessary for the selection of the most likely symbol sequence and the memory size grow only linearly with respect to the message length $K$. Also, a suboptimum version of this algorithm can be adopted that allows decisions to be made about the first transmitted symbols with a fixed delay, without waiting for the whole sequence to be received.

### 7.5.1. Maximum-likelihood sequence detection using the Viterbi algorithm

The key to the ML receiver design is the expression of the log-likelihood ratio for the detection of the finite sequence of symbols a $\overset{\triangle}{=} (a_0, a_1, \ldots, a_{K-1})$ based on the observation of the waveform

$$r(t) \overset{\triangle}{=} \sum_{\ell=0}^{K-1} a_\ell p(t - \ell T) + w(t), \quad t \in I \tag{7.108}$$

where I is a time interval long enough to ensure that $p(t), p(t-T), \ldots, p[t-(K-1)T]$ are identically zero outside it. Definition (7.108) is derived from (7.54) by considering a finite symbol sequence instead of an infinite one. Moreover, for simplicity, we deal with real signals only. The extension to the complex case is straightforward and requires only some minor changes of notation (see Section 2.6). In (7.108), $w(t)$ denotes white Gaussian noise with power spectral density $N_0/2$. We also assume that the sequence length $K$ is large enough to disregard certain end effects. This concept will be made more precise when the need arises. The log-likelihood ratio for a is then

$$\lambda_{\mathbf{a}} = \frac{2}{N_0} \int_I v_{\mathbf{a}}(t) r(t) dt - \frac{1}{N_0} \int_I v_{\mathbf{a}}^2 dt \tag{7.109}$$

where $v_{\mathbf{a}}(t)$ is the noiseless waveform corresponding to the symbol sequence a:

$$v_{\mathbf{a}} \overset{\triangle}{=} \sum_{\ell=0}^{K-1} a_\ell p(t - \ell T) \tag{7.110}$$

Using (7.110), we can rewrite (7.109) in the form

$$\lambda_{\mathbf{a}} = \frac{2}{N_0} \cdot \sum_{\ell=0}^{K-1} a_\ell \int_I p(t - \ell T) r(t) dt$$

$$- \frac{1}{N_0} \sum_{\ell=0}^{K-1} \sum_{m=0}^{K-1} a_\ell a_m \int_I p(t - \ell T) p(t - mT) dt \qquad (7.111)$$

For notational simplicity, it is convenient to define the following quantities:

$$Z_\ell \triangleq \int_I p(t - \ell T) r(t) dt \qquad (7.112)$$

and

$$s_{\ell-m} \triangleq \int_I p(t - \ell T) p(t - mT) dt \qquad (7.113)$$

With the assumption of a finite duration for the waveform $p(t)$, say

$$p(t) = 0, \quad t < 0, \ t > (L+1)T \qquad (7.114)$$

(the value $L$, $L \ll K$, will be referred to hereafter as the *memory* of the channel), the sequence $Z_\ell$, $\ell = 0, 1, \ldots, K - 1$, can be obtained by sampling at times $(L + \ell + 1)T$ the output of a filter matched to the waveform $p(t)$ when its input is the received signal $r(t)$. Notice also that (7.114) implies that I= $[0, (K + L)T]$. Strictly speaking, the RHS of (7.113) depends on $\ell$ and $m$ separately. However, we assume that the choice of $K$ makes the interval I long enough for it to depend on $\ell - m$ only. Moreover, due to the assumption of a finite-memory channel, $s_{\ell-m}$ can be nonzero only for a finite set of values of $\ell - m$. In fact, we have

$$s_k = 0, \quad |k| \geq L + 1 \qquad (7.115)$$

Finally, observe that, under the hypothesis of a known function $p(t)$, the values of $s_k$ are also known. Use now (7.112) and (7.113) in (7.111). Upon multiplication by the constant factor $N_0$, it is seen that the ML sequence â is the one that minimizes the quantity[7]

$$\lambda_{\mathbf{a}} \triangleq -2 \sum_{\ell=0}^{K-1} a_\ell Z_\ell + \sum_{\ell=0}^{K-1} \sum_{m=0}^{K-1} a_\ell a_m s_{\ell-m} \qquad (7.116)$$

Now we observe that one of the results anticipated at the beginning of this section can be proved. In fact, all we need in order to compute $\lambda_{\mathbf{a}}$ for every vector a is the sample sequence $(Z_\ell)_{\ell=0}^{K-1}$ obtained at the output of a single matched filter. Precisely, this set of samples provides a sufficient statistics for $r(t)$. This means that all we need to know about the received signal is contained in these samples.

The ML decision requires $\lambda_{\mathbf{a}}$ to be minimized over the whole set of possible sequences a. Thus, the matched filter must be followed by a processor, the *ML sequence detector*, determining as the most likely transmitted data sequence, say

---

[7]With a slight abuse of notation, we keep using $\lambda$ for the normalized log-likelihood ratio.

â, the one minimizing $\lambda_a$. The direct computation of $\lambda_a$ for all possible a to find the minimum is impractical due to the sheer number of computations involved. However, a sequential algorithm is available that performs such a selection in a computationally efficient manner. This is the celebrated *Viterbi algorithm* described in Appendix F, and already used in Chapter 6. It performs the minimization of a function of several variables and is applicable to minimization problems that can be formulated as the search for the minimum-length path in a finite trellis. The significance of the Viterbi algorithm is that the number of computations required for the ML detection of a sequence of length $K$ grows only linearly with $K$.

We shall now show how the Viterbi algorithm can be applied to our problem. (From now on we shall assume that the reader is familiar with Appendix F.) Essentially, our task is to show that $\lambda_a$ can be reduced to a sum of terms, each one corresponding to the label of a branch in a suitable trellis diagram.

To do this, the first step is to rewrite $\lambda_a$, as defined in (7.116), in the form

$$
\lambda_a = \left\{ -2 \left( \sum_{\ell=0}^{K-2} a_\ell Z_\ell \right) + \sum_{\ell=0}^{K-2} \sum_{m=0}^{K-2} a_\ell a_m s_{\ell-m} \right\} \tag{7.117}
$$
$$
+ \left\{ -2(a_{K-1} Z_{K-1}) + 2 \left( a_{K-1} \sum_{m=K-L-1}^{K-2} a_m s_{K-1-m} \right) + a_{K-1}^2 s_0 \right\}
$$

where (7.115) and the property $s_{-\ell} = s_\ell$ have been used. In (7.117) we have decomposed $\lambda_a$ into the sum of two bracketed terms. The first is similar to the RHS of (7.116) (the only change is the upper summation limit), and the second is a function only of the $L+1$ symbols $a_{K-L-1}, a_{K-L}, \ldots, a_{K-1}$, not of the entire vector a. Our decomposition of $\lambda_a$ into a sum of functions suitable for the application of the Viterbi algorithm will be based on repeated application of such decompositions. At this point it is convenient to define the variables

$$
\sigma_\ell \triangleq (a_{\ell-1}, a_{\ell-2}, \ldots, a_{\ell-L}), \quad \ell = L, \ldots, K \tag{7.118}
$$

and the quantities

$$
U_{k+1}(\sigma_L, \ldots, \sigma_{k+1}) \tag{7.119}
$$
$$
\triangleq -2 \left( \sum_{\ell=0}^{k} a_\ell Z_\ell \right) + \sum_{\ell=0}^{k} \sum_{m=0}^{k} a_\ell a_m s_{\ell-m}, \quad k = L-1, \ldots, K-1
$$

$$
V_{k+1}(\sigma_k, \sigma_{k+1}) \tag{7.120}
$$
$$
\triangleq -2(a_k Z_k) + 2 \left( a_k \sum_{m=k-L}^{k-1} a_m s_{k-m} \right) + a_k^2 s_0, \quad k = L, \ldots, K-1
$$

Now, observing that

$$\lambda_{\mathbf{a}} = U_K(\sigma_L, \ldots, \sigma_K) \tag{7.121}$$

we can rewrite (7.117) in the form

$$U_K(\sigma_L, \ldots, \sigma_K) = U_{K-1}(\sigma_L, \ldots, \sigma_{K-1}) + V_K(\sigma_{K-1}, \sigma_K)$$

and generalize the latter to show that

$$U_{k+1}(\sigma_L, \ldots, \sigma_{k+1}) = U_k(\sigma_L, \ldots, \sigma_k) \tag{7.122}$$
$$+ V_{k+1}(\sigma_k, \sigma_{k+1}), \quad k = L, \ldots, K-1$$

Repeated application of (7.122) yields

$$\begin{aligned}\lambda_{\mathbf{a}} &= U_L(\sigma_L) + V_{L+1}(\sigma_L, \sigma_{L+1}) \\ &+ V_{L+2}(\sigma_{L+1}, \sigma_{L+2}) + \cdots + V_K(\sigma_{K-1}, \sigma_K)\end{aligned} \tag{7.123}$$

which is the required decomposition.

Our next step will be to exhibit a trellis such that we can associate with its branches the values taken on by the functions $V_{k+1}(\sigma_k, \sigma_{k+1})$, $k = L, \ldots, K-1$. This task is simplified by a proper interpretation of the meaning of the variables $\sigma_k$ defined in (7.118).

Recall that we have assumed the channel to have a finite memory $L$. This assumption is expressed mathematically by (7.114), and can be interpreted in the following manner. At any given time $t$, the received signal $r(t)$ defined in (7.108) depends on a set of $L + 1$ consecutive symbols, say $a_\ell, a_{\ell-1}, \ldots, a_{\ell-L}$. The last $L$ of these symbols has been defined to form $\sigma_\ell$.

This is then called the *state of the channel* at time $t$. The transmission of the symbol $a_\ell$ when the channel state is $\sigma_\ell$ will then bring the channel to the succeeding state $\sigma_{\ell+1} = (a_\ell, a_{\ell+1}, \ldots, a_{\ell-L+1})$, and so forth, for symbols $a_{\ell+1}, a_{\ell+2}, \ldots$. Thus, we have set a one-to-one correspondence between the sequence of transmitted symbols $a_0, a_1, \ldots, a_{K-1}$ and the sequence of states $\sigma_L, \ldots, \sigma_K$. Therefore, the problem of selecting the most likely symbol sequence is equivalent to that of selecting the most likely sequence of states. This can also be seen directly from (7.123).

We are now able to define the trellis structure needed for the application of the Viterbi algorithm. For each value of the index $\ell$, $\ell = L, L + 1, \ldots, K$, associate a set of $M^L$ nodes where each corresponds to a value of $\sigma_\ell$. Each node has $M$ branches stemming from it, one for each value taken by $a_\ell$. Also, the branches represent the transition from the state $\sigma_\ell$ to the next state $\sigma_{\ell+1}$ as shown in Fig. 7.30. An example will help clarify these procedures.

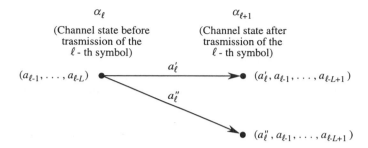

Figure 7.30: *Transition from one state to the next and construction of the trellis diagram for application of the Viterbi algorithm; $a'_\ell$ and $a''_\ell$ are two possible values taken on by the $\ell$th data symbol.*

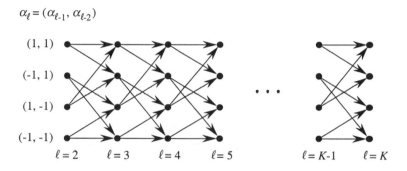

Figure 7.31: *Trellis diagram for the situation of Example 7.12.*

**Example 7.12**  Assume a binary baseband modulation with symbols $\pm 1$ and a channel with a finite memory $L = 2$. The trellis for this situation consists of the states $\sigma_\ell = (a_{\ell-1}, a_{\ell-2})$. Each state can assume four values for each $\ell$ (see Fig. 7.31), and the branches joining adjacent states represent the structure of the state vectors. For instance, the two branches stemming from the state $(a_\ell, a_{\ell-1}) = (1, -1)$ connect it to the allowable successor states $(a_{\ell+1}, a_\ell) = (-1, 1)$ or $(1, 1)$, corresponding to the symbols $a_{\ell+1} = -1$ and $1$, respectively. Conversely, given the state $(a_{\ell+1}, a_\ell) = (1, -1)$ there are two allowable predecessor states $(a_\ell, a_{\ell-1}) = (1, 1)$ and $(1, -1)$, corresponding to the symbols $a_{\ell-1} = 1$ and $-1$, respectively.                                      □

The ML detection problem has now been reduced to the selection of a path through the trellis just described once the branches joining states $\sigma_\ell$ and $\sigma_{\ell+1}$ have been assigned the values taken by the function $V_{\ell+1}(\sigma_\ell, \sigma_{\ell+1})$, usually referred to as the metric. The minimum-metric path corresponds to the most likely sequence of states, and hence to the most likely sequence of symbols. The Viterbi algorithm is then applicable as follows:

**Step 1** Observe the values of $Z_0$ at time $(L+1)T$, $Z_1$ at time $(L+2)T, \ldots$, and $Z_{L-1}$ at time $2LT$. Let $l = L$ and use (7.119) to compute $U_L(\sigma_L)$ for each value of $\sigma_L$. Store the values of $U_L(\sigma_L)$.

**Step 2** Let $\ell \to \ell + 1$. Observe the value of $Z_\ell$ at time $(L + \ell + 1)T$, and use (7.120) to compute $V_{\ell+1}(\sigma_\ell, \sigma_{\ell+1})$ for each pair of states $\sigma_\ell, \sigma_{\ell+1}$ such that the transition from $\sigma_\ell$ to $\sigma_{\ell+1}$ is allowed by the trellis structure.

**Step 3** For each state $\sigma_{\ell+1}$, compute

$$u_{\ell+1}(\sigma_{\ell+1}) \triangleq \min_{\sigma_\ell}[u_\ell(\sigma_\ell) + V_{\ell+l}(\sigma_\ell, \sigma_{\ell+1})] \qquad (7.124)$$

where the minimum is taken over the values of $\sigma_\ell$ compatible with $\sigma_{\ell+1}$, and $u_L(\cdot) \triangleq U_L(\cdot)$. The quantity $u_{\ell+l}(\sigma_{\ell+1})$ is the minimum length of the paths leading to $\sigma_{\ell+1}$; store this quantity and this path with minimum length for each value of $\sigma_{\ell+1}$. If $\ell = K$, go to Step 5.

**Step 4** Go to Step 2.

**Step 5** Compute $\min_{\sigma_K} u_K(\sigma_K)$; this is the minimum length of the paths through the trellis. The minimum-length path corresponds to the most likely sequence of states.

**Example 7.13** Consider the situation of Example 7.12 and assume $s_0 = 1$, $s_1 = s_{-1} = 0.4$, $s_2 = s_{-2} = -0.2$, $s_k = 0$, $|k| > 2$. Assume also that $K = 8$, and $Z_0 = 1.0$, $Z_1 = -1.2$, $Z_2 = 0.5$, $Z_3 = -1.5$, $Z_4 = -0.2$, $Z_5 = 1.0$, $Z_6 = 0.8$, and $Z_7 = 0.9$. Upon reception of the matched filter outputs $Z_0$ and $Z_1$, $U_2(\sigma_2)$ can be computed for the four values of $\sigma_2$; we get

$$U_2(-1, -1) = 2.4, \ U_2(1, -1) = 5.6, \ U_2(-1, 1) = -3.2, \ U_2(1, 1) = 3.2$$

Then, after receiving each value of $Z_\ell$, $\ell = 2, \ldots, 7$, $V_{\ell+1}(\sigma_\ell, \sigma_{\ell+1})$ can be computed. The corresponding values are shown in Fig. 7.32 together with those of $u_\ell(\sigma_\ell)$, $\ell = 2$.

The minimum-length paths stored in Step 3 of the algorithm are shown by the solid lines. Application of the Viterbi algorithm shows that the ML path joins the states $(-1, -1)$, $(1, -1)$, $(-1, 1)$, $(1, -1)$, $(-1, 1)$, $(1, -1)$, and $(-1, 1)$, corresponding to the data sequence $-1, -1, 1, -1, 1, -1, 1, -1$.                                                                  □

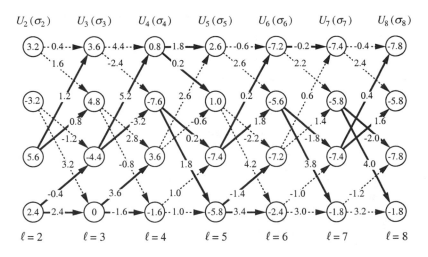

Figure 7.32: *Branch and path metrics for the trellis diagram of Example 7.13.*

### 7.5.2.    Error probability for the maximum-likelihood sequence receiver

In the following, the performance of the ML sequence receiver will be evaluated by computing upper and lower bounds to the probability of a symbol error

$$P(e) \overset{\triangle}{=} P(\hat{a}_\ell \neq a_\ell) \tag{7.125}$$

where $a_\ell$ denotes the $\ell$th transmitted symbol and $\hat{a}_\ell$ its estimate. Strictly speaking, this probability is a function of the index $\ell$ as our model is not stationary due to the consideration of a finite symbol sequence $a_0, a_1, \ldots, a_{K-1}$. However, under the usual assumption that $K$ is large enough, we shall disregard this difficulty and assume that the RHS in (7.125) does not depend on $\ell$. Since the ML sequence detection can be viewed as the choice of a path in the state trellis, for errors to occur it is necessary that the ML path diverge for a certain index, say $\ell_1$, from the path representing the transmitted symbol sequence, and remerge later, say for index $\ell_1 + H$. When this happens, we say that an error event of length $H - 1$ has taken place (Fig. 7.33). The concept of error event specifies mathematically the fact that, when a sequence is estimated, symbol errors do not occur independently, but in finite clumps (*bursts*). If we define

$$e_\ell \overset{\triangle}{=} \hat{a}_\ell - a_\ell, \quad \ell = 0, 1, \ldots, K - 1 \tag{7.126}$$

and recall definition (7.118), it is seen that an error event starting at index $\ell_1$ and extending up to index $\ell_1 + H$, say

$$\{\hat{\sigma}_\ell = \sigma_\ell, \ell = \ell_1, \ell_1 + H; \ \hat{\sigma}_\ell \neq \sigma_\ell, \ell_1 < \ell < \ell_1 + H\}$$

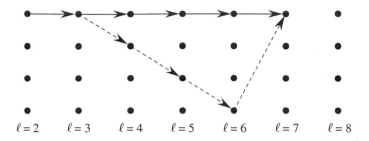

$\ell=2 \qquad \ell=3 \qquad \ell=4 \qquad \ell=5 \qquad \ell=6 \qquad \ell=7 \qquad \ell=8$

Figure 7.33: *An error event of length 3. The continuous line is the path corresponding to the transmitted sequence of states. The dashed line corresponds to the estimated sequence of states.*

corresponds to a sequence of symbol errors

$$\mathbf{e} \overset{\triangle}{=} (e_{\ell_1}, e_{\ell_1+1,}, \ldots, e_{\ell_1+H-L}), \quad e_{\ell_1} \neq 0, \; e_{\ell_1+H-L} \neq 0$$

(Incidentally, this shows that $H \geq L$, i.e., the error events are always at least as long as the channel memory.) If we let U be the set of all nonzero error events, $w(\mathbf{e})$ the number of decision errors entailed by the error event e (i.e., the number of nonzero entries in e), and $P\{\mathbf{e}\}$ the probability of e to occur, we have

$$P(e) = \sum_{\mathbf{e} \in U} w(\mathbf{e}) P\{\mathbf{e}\} \tag{7.127}$$

Since the exact computation of error probability using (7.127) does not seem feasible, we shall resort to evaluation of upper and lower bounds to $P(e)$.

### An upper bound to *P(e)*

Computation of an upper bound to $P(e)$ will be based on the approximate evaluation of $P\{\mathbf{e}\}$ in (7.127). Let A(e) be the event that the transmitted sequence a of data symbols (with the same length of e) is compatible with the occurrence of e. Then, for e to occur, it is necessary that A(e) occur and that a + e have a likelihood greater than any other possible sequence of source symbols, including a. Since this latter event is included in the event $\{\lambda_{a+e} > \lambda_a\}$, the probability of e can be upper bounded as follows:

$$P\mathbf{e} \leq P\{\lambda_{\mathbf{a+e}} > \lambda_{\mathbf{a}} \mid A(\mathbf{e})\} \cdot P\{A(\mathbf{e})\} \tag{7.128}$$

We shall now proceed to evaluate separately the two factors in the RHS of (7.128)...
For a stationary sequence of independent source symbols, we have, for an error

event of length $H$,

$$P\{A(e)\} = \prod_{\ell=0}^{H-L} P\{A(e_l)\} \tag{7.129}$$

**Example 7.14**   For example, if $a_\ell$ takes on values $\pm 1$ with equal probabilities, only $a_\ell = +1$ is compatible with $e_\ell = 0$, only $a_\ell = -1$ is compatible with $e_\ell = +2$, and both values $a_\ell = \pm 1$ are compatible with $e_\ell = 0$. Thus, $P\{A(2)\} = P\{A(-2)\} = \frac{1}{2}$ and $P\{A(0)\} = 1$, which yields

$$P\{A(e)\} = \prod_{\ell=0}^{H-L} \left(1 - \frac{|e_\ell|}{4}\right) \tag{7.130}$$

More generally, it can be easily proved that, if $a_\ell$ can take on the $M$ values $-M + 1, -M+3, \ldots, M-1$ with equal probabilities, we have

$$P\{A(e)\} = \prod_{\ell=0}^{H-L} \left(1 - \frac{|e_\ell|}{2M}\right) \tag{7.131}$$

□

Consider then the event $\{\lambda_{a+e} > \lambda_a \mid A(e)\}$.  By recalling (7.111) and using (7.112) and (7.113), the inequality $\{\lambda_{a+e} > \lambda_a\}$ can be rewritten, after some algebraic manipulations, in the form

$$d^2(e) \leq 2 \left(\sum_{\ell=0}^{H-L} e_\ell \nu_\ell\right) \tag{7.132}$$

where

$$d^2(e) \triangleq \sum_{\ell=0}^{H-L} \sum_{m=0}^{H-L} e_\ell e_m s_{\ell-m} \tag{7.133}$$

is called the *distance of the error event*, and

$$\nu_\ell \triangleq \int_I p(t - \ell T) w(t)dt, \quad \ell = 0, 1, \ldots, H-L \tag{7.134}$$

are Gaussian random variables with zero mean and covariance

$$E\{\nu_\ell \nu_k\} = \frac{N_0}{2} s_{\ell-k} \tag{7.135}$$

The RHS of (7.132) turns out to be a Gaussian RV with zero mean and variance $(N_0/2)d^2(e)$.  Therefore,

$$P\{\lambda_{a+e} > \lambda_a \mid A(e)\} = \frac{1}{2}\mathrm{erfc}\left(\frac{d(e)}{2\sqrt{N_0}}\right) \tag{7.136}$$

Using finally (7.127), (7.128) and (7.136), we get

$$P(e) \le \sum_{e \in U} w(e) \frac{1}{2} \mathrm{erfc} \left( \frac{d(e)}{2\sqrt{N_0}} \right) P\{A(e)\} \qquad (7.137)$$

where $P\{A(e)\}$ can be computed through (7.129). Equation (7.137), in spite of the considerable effort spent to derive it, is still not in a usable form, mainly because of the difficulty involved in enumerating the elements of U. Thus, we shall resort to an approximation of the RHS of (7.137), valid for small values of $N_0$ and based on the steep decrease of the function erfc($\cdot$). It suffices to observe that the terms in the summation (7.137) will be dominated, as $N_0 \to 0$, by the terms involving the smallest value of $d(e)$, which we shall denote by $d_{\min}$

$$d_{\min} \overset{\triangle}{=} \min_{e \in U} d(e) \qquad (7.138)$$

Hence we get the approximate upper bound

$$P(e) \overset{\sim}{\le} \frac{1}{2} \psi'(d_{\min}) \mathrm{erfc} \left( \frac{d_{\min}}{2\sqrt{N_0}} \right) \qquad (7.139)$$

where

$$\psi'(d_{\min}) \overset{\triangle}{=} \sum_{e \in U(d_{\min})} w(e) P\{A(e)\} \qquad (7.140)$$

and $U(d_{\min})$ is the subset of U including the error events with distance $d_{\min}$. Notice, in particular, that $\psi'(d_{\min})$ is a constant independent of $N_0$.

**Example 7.15**   Consider a binary transmission system with equally likely symbols $\pm 1$, and assume that $U(d_{\min})$ includes the two error sequences $(+2, -2)$ and $(-2, +2)$. The set $A(+2, -2)$ includes only the data sequence $(-1, +1)$, and $A(-2, +2)$ includes only $(+1, -1)$. Thus,

$$\psi'(d_{\min}) = 2 \cdot \frac{1}{4} + 2 \cdot \frac{1}{4} = 1$$

In the same conditions, if $U(d_{\min}) = \{(-2), (+2)\}$, we have $A(-2) = \{(+1)\}$ and $A(+2) = (-1)$, so

$$\psi'(d_{\min}) = \frac{1}{2} + \frac{1}{2} = 1$$

$\square$

**A lower bound to $P(e)$**

We shall now proceed to evaluate a lower bound to the symbol error probability. To do this, we consider the ideal situation in which the detection process is aided by a genie supplying to the receiver some side information on the transmitted symbols. If the receiver makes its decisions by exploiting optimally the genie information, it is clear that it cannot be outperformed by any receiver working without the genie's aid. Thus, if $P_G(e)$ denotes the symbol error probability achieved by the genie-aided receiver, we have, for every real-life receiver (and hence for the ML sequence detector),

$$P(e) \geq P_G(e) \qquad (7.141)$$

Assume that the genie operates as follows: when the sequence $\mathbf{a} = (a_0, \ldots, a_{K-1})$ is transmitted, he chooses at random another sequence $\mathbf{a}' = \mathbf{a} + \mathbf{e}$, which has an error on the $\ell$th symbol $a_\ell$ (and possibly others). Then he tells the receiver that either $\mathbf{a}$ or $\mathbf{a}'$ was transmitted. In this situation, the task of the receiver is to choose one out of two known signals perturbed by white Gaussian noise. This can be achieved optimally with a probability of error

$$\frac{1}{2}\text{erfc}\left(\frac{d(\mathbf{e})}{2\sqrt{N_0}}\right)$$

Thus, the probability that the genie-aided receiver makes an incorrect decision on $a_\ell$ is

$$P_G(e) = \sum_{\mathbf{e} \in U} \frac{1}{2}\text{erfc}\left(\frac{d(\mathbf{e})}{2\sqrt{N_0}}\right) P\{A(\mathbf{e})\} \qquad (7.142)$$

where $A(\mathbf{e})$ can now be interpreted as the event that the data sequence chosen by the genie is compatible with $\mathbf{e}$. Equivalently, $P\{A(\mathbf{e})\}$ is the ratio between the number of sequences $\mathbf{a}'$ such that $\mathbf{a}' = \mathbf{a} + \mathbf{e}$ for some $\mathbf{a}$, and the total number of data sequences with length $K$.

By combining (7.141) and (7.142), and discarding from the summation in the RHS of (7.142) all those sequences $\mathbf{e}$ for which $d(\mathbf{e}) > d_{\min}$, we have

$$P(e) \geq \frac{1}{2}\psi''(d_{\min})\text{erfc}\left(\frac{d_{\min}}{2\sqrt{N_0}}\right) \qquad (7.143)$$

where

$$\psi''(d_{\min}) \triangleq \sum_{\mathbf{e} \in U} (d_{\min}) P\{A(\mathbf{e})\} \qquad (7.144)$$

is the probability that a data sequence chosen at random has a sequence $\mathbf{e}$ compatible with it and such that $d(\mathbf{e}) = d_{\min}$. The result (7.143) generalizes the lower bound (4.63) obtained for symbol-by-symbol detection.

**Example 7.16**    Under the same conditions as for Example 7.15, if $U(d_{\min}) = \{(-2,$ $+2), (+2, -2)\}$, we get $\psi''(d_{\min}) = \frac{1}{4} + \frac{1}{4} = \frac{1}{2}$, and if $U(d_{\min}) = \{(-2), (+2)\}$, we get $\psi''(d_{\min}) = \frac{1}{2} + \frac{1}{2} = 1$.                                                                    □

### 7.5.3.   Significance of $d_{\min}$ and its computation

The results obtained so far in this section, and in particular the upper and lower bounds (7.139) and (7.143) on error probability, show that the key parameter for the performance evaluation of the ML sequence detector is the minimum distance $d_{\min}$ defined by (7.138) and (7.133) or, equivalently, by

$$d_{\min}^2 \overset{\triangle}{=} \min_{e \neq 0} \sum_\ell \sum_m e_\ell e_m s_{\ell-m} \tag{7.145}$$

or

$$d_{\min} \overset{\triangle}{=} \min_{a' \neq a} \int_I [v_a(t) - v_{a'}(t)]^2 dt \tag{7.146}$$

where (7.110) has been used. In words, $d_{\min}^2$ can be viewed from (7.145) as arising from minimization over error patterns, or from (7.146) as the square of the smallest possible distance between distinct signals at the output of the deterministic part of the channel. It is easily seen that, in the special case of transmission of independent symbols over the AWGN channel without intersymbol interference, (7.146) reduces to the minimum distance between signal points, as defined in Section 4.3. Also, it is interesting to observe that inequality (7.143) provides a bound to the symbol error probability of *any* real-life receiver that can be conceived to detect a sequence of data transmitted on a channel with intersymbol interference. Thus, computation of $d_{\min}$ provides an important parameter for judging the quality of the channel itself.

The direct computation of $d_{\min}$ involves a minimization problem that may be hard to solve, as the number of relevant error patterns e or, equivalently, of symbol sequence pairs $(a, a')$ that have to be tested can be very large.

Tree-search algorithms for the determination of $d_{\min}$ have been proposed (see Fredricsson, 1974; Messerschmitt, 1973; Anderson and Foschini, 1975). For channels with a short memory, a convenient procedure to find $d_{\min}$ has been proposed by Anderson and Foschini (1975). This procedure, stemming from an approach combining functional analysis and computer search, is based on the selection, from the full set of error sequences e, of a small subset of crucial sequences such that at least one element of the subset attains the minimum distance. As a result, $d_{\min}$ can be obtained from (7.145) by evaluating $\sum_\ell \sum_m e_\ell e_m s_{\ell-m}$ for every element of this subset and choosing the smallest value found.

An efficient algorithm for the computation of $d_{\min}$ based on dynamic programming will be described in Chapter 12 in the context of trellis-coded modulation.

**Example 7.17** Consider an $M$-ary baseband transmission with symbols $-M+1, -M+3, \ldots, M-1$ and a channel bandlimited to $[-1/(2T), 1(2T)]$ with $p_0 = 1$, $p_1 = -1$. Thus, we have $s_0 = 2T$, $s_1 = s_{-1} = -T$, and $s_\ell = 0$, $|\ell| > 1$. In this case (see Problem 7.21), $d_{\min}^2 = 8T$, which is achieved by the error events $(\pm 2, \pm 2, \ldots, \pm 2)$, with length $m = 1, 2, \ldots$. Hence, using (7.131) and (7.140), we obtain

$$\psi'(d_{\min}) = 2 \sum_{m=1}^{\infty} m \cdot \left(1 - \frac{1}{M}\right) = 2M(M-1)$$

Similarly, (7.144) yields

$$\psi''(d_{\min}) = 2 \sum_{m=1}^{\infty} \left(1 - \frac{1}{M}\right) = 2(M-1)$$

The symbol error probability is then bounded as follows:

$$(M-1)\mathrm{erfc}\left(\sqrt{\frac{2T}{N_0}}\right) \leq P(e) \leq M(M-1)\mathrm{erfc}\left(\sqrt{\frac{2T}{N_0}}\right)$$

For a channel without ISI, $M$-ary transmission using pulses with energy $2T$ would result in an error probability of

$$P(e) = \frac{M-1}{M}\mathrm{erfc}\left(\sqrt{\frac{2T}{N_0}}\right)$$

$\square$

## 7.5.4. Implementation of maximum-likelihood sequence detectors

Even with present-day technology, the implementation of an ML sequence detector can be difficult in high-speed data transmission due to the processing requirements of the Viterbi algorithm. In fact, the number of values taken by the state variables $\sigma_\ell$, and hence the number of quantities to be stored and processed per received symbol, grows as $M^L$, and the demand on the processor speed increases with the symbol rate for a given value of $M^L$. For binary symbols and a very short channel memory (say, $L = 1$ to 3), there may not be any problem with this complexity for low-speed data transmission (see for example the case of the wireless communication standard GSM). However, for many real-life channels,

$M^L$ can be so large as to make implementation of a Viterbi receiver unfeasible even at low data rates.

Also, a truly optimum receiver delays its decision on the symbol sequence until it has been received in its entirety. In certain cases, a decision can be made *before* the entire sequence $(Z_\ell)_{\ell=0}^{K-1}$ has been observed and processed; this occurs when during the computations it is seen that all the $M^L$ trellis paths that have been stored leading to the nodes corresponding to state $\sigma_{\ell_2}$ (say) pass through a single node corresponding to state $\sigma_{\ell_1}$, $\ell_1 < \ell_2$. In this situation, it is said that a *merge* has occurred for $\ell = \ell_1$, and a decision can be made on the first states from $\sigma_L$ to $\sigma_{\ell_1}$. For example, in Fig. 7.32 a merge occurs for $\ell = 4$ in the state $(-1, 1)$; this is detected for $\ell = 6$, and a decision can be taken on the states $\sigma_2, \sigma_3, \sigma_4$.

In general, merges occur at random, and in certain unfortunate cases they may never occur during the transmission of a finite sequence. Thus, in practice, it is necessary to force decisions about the first transmitted symbols when the area allocated for the paths' storage is liable to be exceeded. Qureshi (1973) has shown by analysis and computer simulation that in most practical situations the probability of additional errors due to premature decisions becomes irrelevant if the decisions are made after a reasonable delay. In many cases, it will be sufficient to choose a delay just beyond twice the channel memory $L$, provided of course that the decisions are made by selecting the sequence that has the greatest likelihood at the moment of the decisions. Some systems require that the data be organized in bursts: in those cases, each burst is decoded independently, and the decoding delay problem disappears, provided that the bursts are not too long.

To limit the receiver complexity due to the channel memory length, an approach that has often been adopted is to use a linear filter preceding the optimum receiver in order to reduce the channel memory to a small value. With this prefilter taking care of limiting $L$, the Viterbi algorithm can be implemented with a tolerable complexity. However, any linear filtering of the received signal will also affect the noise. Thus, any attempt to compensate for the nulls or near nulls in the channel equivalent transfer function results in prefilter characteristics that, by trying to invert the channel transfer function, will increase the noise power at the receiver input. Thus, linear prefiltering designed to condition optimally the channel impulse response should also take into account the output noise variance. To do this, the desired frequency response of the combination channel-prefilter should be close to the channel's in those frequency intervals where the channel cannot be equalized without excessive noise enhancement.

Several solutions to the problem of optimum prefilter design have been proposed. Qureshi and Newhall (1973) use a mean-square criterion to force the overall response of the channel plus the prefilter to approximate a truncated version of the original channel impulse response. Falconer and Magee (1973) show how the desired response can be chosen to minimize the noise variance. The approach of Messerschmitt (1974) is to minimize the noise variance while keeping the first nonzero sample of the desired response fixed. Beare's (1978) design method results in a transfer function for the cascade of the channel and the prefilter that is as close as possible to that of the original channel under the constraint of the memory length. Notice that the process of truncating the impulse response of the channel will never be perfect, so that the receiver will ignore some of the input ISI. The performance of this "mismatched" receiver has been considered by Divsalar (1978). McLane (1980) has derived an upper bound to the bit error probability due to this residual neglected channel memory. Other approaches to the reduction of complexity of the optimum receiver have been taken. Vermeulen and Hellman (1974) and Foschini (1977) consider the choice of a reduced-state trellis in order to simplify the Viterbi algorithm; Lee and Hill (1977) embed a decision-feedback equalizer (see Chapter 8) into the receiver structure.

## 7.6. Bibliographical notes

The state-variable method to simulate linear filtering using a digital computer can be found in Smith (1977) and in Jeruchim *et al.* (1992). The $z$ transform and the FFT methods to simulate linear filtering in the time and frequency domains, respectively, are described in Chapters 4 and 6 of Rabiner and Gold (1975). The problem of digital transmission systems performance evaluation in the presence of additive Gaussian noise and ISI has received considerable research attention from the late 1960s. The first approach was to find an upper bound to the error probability using the Chernoff inequality (Saltzberg, 1968; Lugannani, 1969). Other authors computed the error probability using a Hermite polynomials series expansion (Ho and Yeh, 1970, 1971) or a Gram-Charlier expansion (Shimbo and Celebiler, 1971). A different bounding technique based on the first moments of the RV representing the ISI has been described by Glave (1972) and refined by Matthews (1973). The Gauss quadrature rule approach to the evaluation of the error probability was first proposed by Benedetto, De Vincentiis, and Luvison (1973). Upper and lower bounds based on moments and related to the Gauss quadrature rules approach have been proposed by Yao and Biglieri (1980) (see also Appendix E). Algorithms for the recursive computation of the moments of the ISI RV are described in (Prabhu, 1971) for the case of independent data sequences and in Cariolaro and Pupolin (1975) for the case of correlated data sequences.

Although tailored for PAM modulation, almost all the aforementioned methods have been applied to the evaluation of the symbol error probability of coherent and noncoherent modulation schemes in the presence of ISI and adjacent channel interferences. A useful reference for these applications can be found in the second part of the IEEE reprints collection edited by Stavroulakis (1980).

The paper by Nyquist (1928), a classic in the field of data transmission, and the subsequent paper by Gibby and Smith (1965), include the formulation of what has been named the Nyquist criterion. The generalization of Nyquist criterion to a situation in which the transmission is assumed to be affected by both ISI and crosstalk interference was considered by Shnidman (1967) and Smith (1968).

The design of signal pulses subject to criteria other than the elimination of ISI has been the subject of several studies. Chalk (1950) finds the time-limited pulse shape that minimizes adjacent-channel interference. Spaulding (1969) considers the design of networks whose response simultaneously minimizes ISI and bandwidth occupancy; his procedure generates better results than the approximation of raised-cosine responses. Mueller (1973) designs a transversal filter whose impulse response is constrained to give zero ISI and has minimum out-of-band energy. Mueller's theory has been generalized by Boutin *et al.* (1982). Franks (1968) selects pulses that minimize the effect of the ISI resulting from a small deviation from the proper timing instants $t_0 + \ell T$, $-\infty < \ell < \infty$.

For a receiver with the structure shown in Fig. 7.20, the most natural approach to the optimization of the filter $U(f)$ is to choose the error probability as a performance criterion. This was done by Aaron and Tufts (1966), whereas Yao (1972) provided a more efficient computational technique. A simpler approach is to constrain the ISI to be zero and then minimize the error probability, as described in Section 7.3. This was considered by Lucky *et al.* (1968). Yet another approach is to maximize the signal-to-noise ratio at the sampling instants (George, 1965).

Joint optimization of shaping and receiving filters under a minimum MSE criterion was considered by Smith (1968) and Berger and Tufts (1967) (our handling of the issue follows closely the latter paper). A different derivation of Berger and Tufts's results was obtained by Hansler (1971). Ericson (1971 and 1973) proved that for every reasonable optimization criterion the optimum shaping filter is bandlimited, and that the optimum receiving filter can be realized as the cascade of a matched filter and a tapped-delay line.

Nonlinear receivers have also been studied. Since maximum a posteriori or ML detection seems at first to lead to a receiver complexity that grows exponentially with the length $K$ of the sequence to be detected, sequential algorithms were investigated in order to reduce this complexity. Chang and Hancock (1966) developed a sequential algorithm for a maximum a posteriori sequence detection whose complexity grows only linearly with $K$. A different algorithm with the same complexity has been proposed in Bahl *et al.* (1974) (see Appendix F). Abend and Fritchman (1970) obtained a similar algorithm for symbol-by-symbol detection. The idea of using the Viterbi algorithm for ML detection of data sequences for baseband transmission channels was developed, independently and almost simultaneously, by Forney (1972), Kobayashi (1971), and Omura (1971). The case of complex symbols (i.e., carrier-modulated signals) was considered by Ungerboeck (1974), whereas Foschini (1975) provided a mathematically rigorous derivation of error probability bounds. Our treatment follows those of Forney (1972), Ungerboeck (1974), Foschini (1975), and Hayes (1975).

## 7.7. Problems

*Problems marked with an asterisk should be solved with the aid of a computer.*

**7.1** Show that the outputs of the block diagram of Fig. 7.2 are the same as those of the block diagram of Fig. 7.3.

**7.2** Given two independent Gaussian random processes $n_P(t)$ and $n_Q(t)$ with zero mean and equal variance $\sigma^2$, find the first-order pdf of the processes $\nu_P(t) = n_P(t)\cos\theta - n_Q(t)\sin\theta$ and $\nu_Q(t) = n_P(t)\sin\theta + n_Q(t)\cos\theta$, where $\theta$ is a constant. Prove that samples of $\nu_P(t)$ and $\nu_Q(t)$ taken at the same time instant are statistically independent.

**7.3** (*) Write a computer program implementing the recursive algorithm described in Section 7.2.1 to evaluate the ISI moments in the case of multilevel PAM transmission.

**7.4** (*) Write a computer program implementing the algorithm described in Golub and Welsch (1969) (see also Appendix E) to construct a Gauss quadrature rule starting from the first $2J$ moments of a RV.

**7.5** (*) Use the programs available from Problems 7.3 and 7.4 to evaluate the error probability for an octonary PAM system with ISI due to a raised cosine impulse response $h(t)$ with a roll-off factor $\alpha = 0.5$ (for the impulse response of raised cosine type, see (7.71), in the presence of a normalized sampling time deviation of 0.05, 0.1, 0.15, and 0.2. Assume an SNR that yields an error probability of $10^{-6}$ at the nominal sampling instants and use a 15-sample approximation for the impulse response.

**7.6** Find a recursive algorithm for the computation of the joint moments of the RVs $X_P$ and $X_Q$ defined in (7.36) and (7.37) assuming that the $a_n$'s are iid discrete RVs with known moments. *Hint:* Generalize the procedure described in Section 7.2.1 for a single RV.

**7.7** (*) For the same case of Problem 7.5, compute the Chernoff bound to the error probability extending the method described in Saltzberg (1968) to the multilevel case.

**7.8** (*) For the same case of Problem 7.5, compute the error probability using the series expansion method described in Ho and Yeh (1970) and in Appendix E.

**7.9** Particularize the result (7.41) to the case of 16-QAM modulation with $\theta = 0$.

**7.10** Extend the program developed in Problem 7.3 to the case of an $M$-PSK modulation scheme.

**7.11** (*) Using the program developed in Problem 7.11 and the results of Problems 7.4 and 7.10, compute the error probability for a quaternary PSK modulation using a second-order Butterworth filter impulse response $h(t)$ (with a normalized 3-dB bandwidth of 1.1) as a function of the phase offset $\theta$. Assume the signal-to-noise ratio that yields an error probability of $10^{-6}$ in ideal conditions (no ISI) and truncate the impulse response to 10 samples.

**7.12** Consider the transmission of binary antipodal PAM signals over a linear channel perturbed by additive Gaussian noise and an intersymbol interference $X = \sum_{i=1}^{N} a_i h_i$. Assume $d = 2$, denote the resulting error probability by $P_N(e)$, and the error probability without intersymbol interference by $P_0(e)$.

  **(a)** Prove that, if the eye pattern is open, i.e., $h_0 > \sum_{i=1}^{N} |h_i|$, then we have $P_0(e) \le P_N(e)$, i.e., intersymbol interference increases the error probability.

  **(b)** Generalize the result of (a) by deriving an inequality involving $P_{N'}(e)$, $N' < N$ (i.e., the error probability obtained by retaining only $N'$ out of $N$ interfering samples).

  **(c)** Show, through an example, that if the eye pattern is *not* open, we may have $P_0(e) > P_N(e)$.

**7.13** Consider a raised-cosine transfer function $Q(f)$ with roll-off $\alpha$ and its powers $Q^\gamma(f)$, $0 < \gamma < 1$.

  **(a)** Compute the equivalent noise bandwidth of $Q^\gamma(f)$ for several values of $\alpha$ and $\gamma$ [see (2.89)].

  **(b)** Compute the error probability in a binary baseband PAM system modeled as in Figs. 7.19 and 7.20 with $S(f) = \beta Q^\gamma(f)$, $U(f) = (1/\beta)Q^{1-\gamma}(f)$, symbols $a_n = \pm 1$, rate 1200 bits/s, and a white Gaussian noise with power spectral density $G_w(f) = 10^{-5}$ W/Hz. The constant $\beta$ is chosen so as to have a unit power at the output of the shaping filter.

**7.14** In the system of Figs. 7.19 and 7.20, for given transfer functions $S(f)$ and $C(f)$, choose the receiving filter $U(f)$ so as to maximize at its output, for a given sampling instant, the ratio between the instantaneous signal power and the average power of ISI plus noise. Show that this filter can be implemented in the form of a matched filter cascaded to a transversal filter.

**7.15** Consider a binary PAM digital transmission system. Data must be transmitted at a rate of 9600 bits/s with a bit error probability lower than $10^{-5}$. The channel transfer function is given by

$$C(f) = \begin{cases} 1, & |f| < 6000 \text{ Hz} \\ 0, & \text{elsewhere} \end{cases} \qquad (7.147)$$

The noise is white Gaussian with a power spectral density $G_w(f) = 10^{-6}$ W/Hz. Choose the shaping filter $S(f)$ and the receiving filter $U(f)$ so as to minimize the average transmitted power while getting rid of the intersymbol interference at the sampling instants. Compute the signal power at the output of the shaping filter.

**7.16** In a binary baseband PAM system modeled as in Figs. 7.19 and 7.20, the cascade of $S(f)$, $C(f)$, and $U(f)$ has a raised-cosine response with roll-off $\alpha$, $0 < \alpha < 1$. The sampling instants are affected by a constant offset of 5 percent with respect to the nominal values, so ISI is present. Assuming that the transmitted symbols are $\pm 1$ and that the noise is white Gaussian with a power spectral density $G_w(f) = 10^{-6}$ W/Hz, compute the bit error probability of the system as a function of $\alpha$ using one of the techniques described in Section 7.2.

**7.17** Consider a bandpass transmission system operating at a signaling rate of $1/T$ on a channel with a flat transfer function $C(f)$. The shaping filter is fourth-order Butterworth with a 3-dB bandwidth $B_S$ and the receiving filter is second-order Butterworth with 3-dB bandwidth $B_U$. Assuming a white Gaussian noise, determine, for $1.2 < B_S T < 2$, the values of $B_U T$ that give the minimum bit error probability for an $M$-ary coherent PSK modulation ($M = 2$, 4, and 8). For every situation, choose the signal-to-noise ratio $\mathcal{E}/N_0$ so that this minimum probability is $10^{-6}$.

**7.18** Consider a binary baseband PAM data-transmission system operating at a rate of 4800 bits/s and modeled as in Figs. 7.19 and 7.20. Assume $s(t) = u_T(t)$,

$$C(f) = \frac{1}{1 + jf/f_c}, \quad f_c = 2400 Hz$$

and a white Gaussian noise with power spectral density $G_w(f) = 10^{-7}$ W/Hz.

(a) Determine the shape of the receiving filter that minimizes the bit error probability while removing ISI at the sampling instants.

**(b)** Determine the shape of the receiving filter that minimizes the mean-square error at the sampler's output.

**(c) (\*)** Compare the error probabilities obtained with the systems designed in parts (a) and (b).

**7.19** Consider a filter with impulse response

$$s(t) = \sum_{i=-N_m}^{N_m} s_i b(t - iT/N)$$

($N$ an integer $> 1$). This can be modeled as a linear transversal filter cascaded to a linear system with impulse response $b(t)$. Define the $(2N_m + 1)$-dimensional vectors

$$\mathbf{s} \overset{\triangle}{=} [s_{-N_m}, \ldots, s_0, \ldots, s_{N_m}]'$$

and

$$\mathbf{z} \overset{\triangle}{=} [z^{-N_m}, \ldots, z^0, \ldots, z^{N_m}]'$$

with $z \overset{\triangle}{=} \exp(j2\pi fT/N)$. Assume that $b(t)$ has energy $\mathcal{E}$ and a duration $\leq T/N$.

**(a)** Show that if $s(t)$ is the impulse response of the shaping filter of a data-transmission system with independent and zero-mean symbols $(a_n)$ and a signaling rate $1/T$, the power density spectrum of its output signal can be written in the form

$$\frac{\sigma_a^2}{T}|B(f)|^2 \mathbf{s}^\dagger \mathbf{z} \mathbf{z}^\dagger \mathbf{s}$$

where $B(f)$ denotes the Fourier transform of $b(t)$.

**(b)** Let $\mathcal{P}_F$ denote the power of the signal at the output of the shaping filter in the frequency interval $(-F, F)$. Show that the shaping filter coefficients vector $\mathbf{s}$ that maximizes the ratio $\mathcal{P}_F/\mathcal{P}_\infty$ (relative power in the frequency interval $(-F, F)$) is the eigenvector of a symmetric matrix $\mathbf{R}$ corresponding to the largest eigenvalue $\lambda_{\max}$. Determine the entries of $\mathbf{R}$, and show that $\lambda_{\max}$ coincides with the maximum value of $\mathcal{P}_F/\mathcal{P}_\infty$.

**(c)** Assume that $s(t)$ satisfies the Nyquist criterion for intersymbol interference-free transmission; that is,

$$s_{\pm i/N} = 0, \quad \text{for } i = \pm 1, \pm 2, \ldots, \pm m$$

How can this constraint be included in part (b)?

**(d) (\*)** Assuming $b(t) = \delta(t)$, derive the shaping filter that gives a Nyquist-type response with maximum relative power in the frequency interval $|f| < 2400$ Hz for $m = 8$ (impulse response limited in duration to $|t| < 8T$) and $N = 4$ (four samples per signaling interval $T$) (Mueller, 1973).

**7.20** Prove that the joint optimization of the shaping filter $S(f)$ and the receiving filter $U(f)$ under the power constraint (7.98) leads to filters that are strictly bandlimited to a generalized Nyquist set of measure $1/T$.

**7.21** Assume that the channel transfer function $P(f)$ is bandlimited in the Nyquist interval $(-1/2, 1/2)$, and denote by $p_m$, $-\infty < m < \infty$, the samples, taken every second, of its response.

(a) Derive an expression of the minimum distance for this channel in terms of the discrete convolution between the sequence $(p_m)$ and the sequence $(e_m)$ of symbol errors.

(b) Using the result obtained in part (a), derive the minimum distance for a channel with memory $L = 1$ when the data symbols take on the values $0, 1, \ldots, M - 1$.

(c) Consider a channel whose impulse response samples are $p_m = 1/\sqrt{n}$, $m = 0, 1, \ldots, n - 1$, and binary symbols $\pm 1$. Derive the minimum distance for this channel, and verify that $d_{\min} \to 0$ as $n \to \infty$.

**7.22** Show that, with the notations of Section 7.5, the inequality

$$\sum_{m \neq 0} |s_m| \leq s_0 \qquad (7.148)$$

is a sufficient condition for the nonexistence of error events whose distance is smaller than that achieved by a single error. *Hint:* If $\delta_0$ denotes the minimum nonzero value of $|e_m|$, show that

$$\left| \sum_{k=0}^{H} e_{m+k} e_k \right| \leq \sum_{k=0}^{H} e_k^2 - \delta_0^2, \quad m \neq 0$$

and use this result to prove that, if (7.148) holds, $d_{\min}^2$ cannot be smaller than $\delta_0^2 s_0$, the minimum distance achieved by a single error.

# Adaptive receivers and channel equalization

The theory developed in Chapter 7, devoted to the design of an optimum receiver in the presence of channel distortion, was based on the assumption of a linear channel and of the exact knowledge of its impulse response (or transfer function). While the first assumption is reasonable in many situations (and we shall see in Chapter 14 what to do when it is not), the latter assumption is often far from realistic. In fact, whereas it is generally true that the designer knows the basic features of the channel, this knowledge may not be accurate enough to allow system optimization. This occurs, for example, when the channel, although time-invariant, is selected at random from an ensemble, a circumstance typical of dial-up telephone lines. Another possibility is that the channel varies randomly with time. This is typical of certain radio channels affected by fading, to be described in some depth in Chapter 13. A consequence of the above is that the receiver designed to cope with the effects of intersymbol interference (ISI) and additive noise should be self-optimizing or *adaptive*. That is, its parameters should be automatically adjusted to an optimum operating point, and should possibly keep track of the changing conditions.

Two philosophies can be the rationale behind the design of an adaptive receiver. The first, described in Fig. 8.1, assumes that the relevant channel parameters are first estimated, then fed to a detector which is optimum (or suboptimum, if the complexity of the latter is too large to be acceptable) for those parameters. This can be, for example, a Viterbi detector, which for ideal operation needs the channel impulse response samples to be known. Another approach is depicted in Fig. 8.2. Here a device, called an *equalizer*, compensates for the unwanted channel features, and presents the detector with a sequence of samples that have

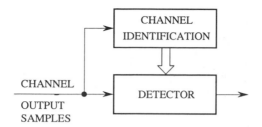

Figure 8.1: *Scheme of an adaptive receiver based on the identification of channel parameters. Here the detector is matched to the channel.*

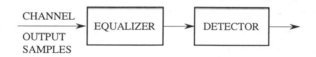

Figure 8.2: *Scheme of an adaptive receiver based on channel equalization. Here the channel (including the equalizer) is matched to the detector.*

been "mopped up" from ISI. The term "equalization" describes a set of operation intended to eliminate ISI. Hence, the cascade of channel and equalizer is seen by the detector as close to an AWGN channel, and consequently the detectors described in Chapter 4 are close to optimum in this situation.

## 8.1. Channel model

Throughout this chapter, unless otherwise specified we shall assume a channel model similar to that of Chapter 7. With reference to Fig. 8.3, the source symbols $(a_\ell)$ form a stationary sequence of identically distributed, uncorrelated, zero-mean complex random variables with $\mathrm{E}[|a_\ell|^2] = 1$ (the assumption of complex symbols is equivalent to assuming one- or two-dimensional modulations). The modulation scheme and the channel are both linear. The source sequence modulates a train of ideal impulses, then passes through a transmission filter. After addition of noise, the signal at the output of the channel is first passed through a receiving filter, then sampled every $T$ to produce the sample sequence $(x_\ell)$, and finally processed before detection.

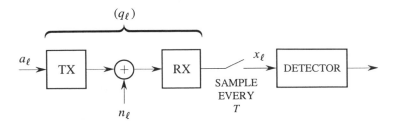

Figure 8.3: *Channel model for channel identification or equalization.*

We denote by $(q_\ell)$ the impulse response of the discrete system (the "discrete overall channel") that responds to the source sequence $(a_\ell)$ with the sequence $(x_\ell)$. For mathematical convenience, it is also assumed that this system is time-invariant, and has a finite memory $L$. Thus, the relationship between the sequences $(a_\ell)$ and $(x_\ell)$ is expressed by

$$x_\ell = \sum_{i=0}^{L} q_i a_{\ell-i} + n_\ell \tag{8.1}$$

where the noise samples of the sequence $(n_\ell)$ at the output of the receiving filter have mean zero and $E[|n_\ell|^2] = \sigma_n^2$. These are independent of the source symbols.

## 8.2. Channel identification

The effect of this linear, finite-memory channel on any input sequence $(a_\ell)$ is described in full by the $L + 1$ complex numbers $q_0, \ldots, q_L$. Hence, with our assumptions, identifying the channel is tantamount to estimating their values.

Eq. (8.1) can be rewritten in a compact vector form by introducing the column vectors

$$\mathbf{q} \triangleq [q_0, q_1, \ldots, q_L]' \tag{8.2}$$

and

$$\mathbf{a}_\ell \triangleq [a_\ell, a_{\ell-1}, \ldots, a_{\ell-L}]' \tag{8.3}$$

(the latter is the vector of source symbols that affect the output at time $\ell$). With these notations, (8.1) takes the form

$$x_\ell = \mathbf{q}'\mathbf{a}_\ell + n_\ell \tag{8.4}$$

so that the problem of identifying the channel can be formulated as follows: derive an estimate of the vector **q** based upon the observation of the received sequence $(x_\ell)$ and the knowledge of the transmitted symbol sequence $(a_\ell)$.

### 8.2.1. Using a channel-sounding sequence

A simple solution to the channel identification problem consists of sending, before transmission, a *channel-sounding* data sequence with length $N$ and noise-like correlation properties, i.e., such that

$$\frac{1}{N} \sum_{i=0}^{N-1} a_{i+k} a_i^* \approx \begin{cases} 1, & k = 0 \\ 0, & k \neq 0 \end{cases} \tag{8.5}$$

for $N$ large enough. Disregard the effect of noise for simplicity. Upon reception of the sample sequence $(x_\ell)$, the receiver computes the correlation (equivalent to a discrete matched filtering)

$$y_\ell = \sum_{i=0}^{N-1} x_{\ell+i} a_i^*$$

Owing to (8.1) and to (8.5), we obtain

$$y_\ell = \sum_{j=0}^{L} q_j \left[ \sum_{i=0}^{N-1} a_{\ell+i-j} a_i^* \right] \approx N q_\ell \tag{8.6}$$

so that, apart from an unessential scale factor, $y_\ell$ approximates the $\ell$th sample of the channel impulse response. This technique is used for example in the GSM standard for digital cellular telephony, where a channel-sounding sequence (the "midamble") in inserted in the center of each data burst. If the channel impulse response remains constant for the burst duration, optimum MLSE can in principle be achieved.

### 8.2.2. Mean-square error channel identification

Suppose that the information data sequence $(a_\ell)$ is known at the receiver. If $\hat{\mathbf{q}}$ denotes an estimate of the vector in (8.2), we can construct a sequence $(\hat{x}_\ell)$ approximating the true received sequence as follows:

$$\hat{x}_\ell = \hat{\mathbf{q}}' \mathbf{a}_\ell \tag{8.7}$$

and measure the accuracy of the channel estimate by computing the mean square value of the error $|x_\ell - \hat{x}_\ell|$. By observing that, due to our assumptions on the source sequence, $E[a_\ell^* a_\ell']$ equals the identity matrix, this is given by

$$
\begin{aligned}
E[|x_\ell - \hat{x}_\ell|^2] &= (\mathbf{q} - \hat{\mathbf{q}})^\dagger E[a_\ell^* a_\ell'](\mathbf{q} - \hat{\mathbf{q}}) + \sigma_n^2 \\
&= |\mathbf{q} - \hat{\mathbf{q}}|^2 + \sigma_n^2
\end{aligned}
\tag{8.8}
$$

Thus, the minimum achievable mean-square error (MSE) is $\sigma_n^2$, which corresponds to the situation $\hat{\mathbf{q}} = \mathbf{q}$, i.e., to perfect identification.

**Algorithms for minimum-MSE channel identification**

Since the condition $\hat{\mathbf{q}} = \mathbf{q}$ occurs when the MSE (8.8) is minimized, a sensible identification algorithm may be based on the search for this minimum. Numerical analysis offers several minimization algorithms: however, not all of them are applicable in a situation where the computations should be done in a very short time, with limited complexity, and in the presence of channel noise and of roundoff errors due to digital implementation of the algorithm. An algorithm which is widely used is the steepest-descent, or *gradient*, algorithm. To understand its behavior, assume that $\mathbf{q}$ is a real vector, and observe that the MSE, a quadratic function of the estimated vector $\hat{\mathbf{q}}$, can be viewed geometrically as a bowl-shaped surface in the $(L+1)$-dimensional space. As the minimum of MSE corresponds to the bowl's bottom, minimizing MSE is equivalent to seeking this bottom. In the gradient algorithm, one starts by choosing arbitrarily an "initial" vector $\hat{\mathbf{q}}^{(0)}$, which corresponds to a point of the surface. The $(L+1)$-component gradient vector of the MSE with respect to $\hat{\mathbf{q}}$ is then computed at this point. As the negative of the gradient is parallel to the direction of steepest descent, a step is performed on the surface in the direction of the negative of the gradient. If the step is short enough, the new point $\hat{\mathbf{q}}^{(1)}$ will be closer to the bottom of the surface, and hence result in a better estimate of the channel. Now, the gradient is computed at $\hat{\mathbf{q}}^{(1)}$ and the procedure repeated. Under certain conditions (to be explored later), this process will eventually converge to the bottom of the bowl (where the gradient is zero) regardless of the choice of the initial point.

Consider now (8.8) with a complex channel response. The gradient of the MSE with respect to $\hat{\mathbf{q}}$ is $-2(\mathbf{q} - \hat{\mathbf{q}})$ (see Section B.5), a *linear* function of the overall channel's estimated impulse response. This result is due to our choice of MSE as the function to be minimized: in principle, any convex function would do the job, but only the choice of a quadratic function yields a linear gradient. The gradient algorithm will then take the form

$$
\hat{\mathbf{q}}^{(n+1)} = \hat{\mathbf{q}}^{(n)} + \alpha(\mathbf{q} - \hat{\mathbf{q}}^{(n)}), \qquad n = 0, 1, \dots
\tag{8.9}
$$

where $\hat{\mathbf{q}}^{(n)}$ denotes the value assumed by the estimated impulse response at the $n$th iteration step, and $\alpha$ is a positive constant small enough to ensure convergence of the iterative procedure. The significance of $\alpha$ will be discussed shortly.

It should be observed that (8.9) is not in an appropriate form for implementing the identification algorithm. In fact, it involves the vector q, which is obviously not available. Using (8.7) and recalling the independence of the data $a_\ell$, we can change (8.9) to

$$\hat{q}^{(n+1)} = \hat{q}^{(n)} + \alpha E[(x_n - \hat{x}_n)a_n^*] \tag{8.10}$$

which is expressed in terms of the observable quantities $x_n$, $\hat{x}_n$, and of the vector $a_n$, assumed to be known. The difficulty now is that exact evaluation of the expectation in the RHS of (8.10) is not practically achievable. In fact, its explicit computation requires knowledge of the channel impulse response, which is not available. The expectation could be approximated in the form of a time average, computed over a sufficiently long time interval: but this would prevent real-time operation. Thus, in most implementations the expectation in (8.10) is simply removed, and only its argument kept.

Before discussing this approximation problem, we analyze the performance of algorithm (8.9) or (8.10). The analysis is relatively simple, and provides more than a bit of insight into the behavior of the implementable algorithm, to be described afterwards.

**Gradient algorithm**

Consider again (8.9), rewritten in the form

$$\hat{q}^{(n+1)} = (1 - \alpha)\hat{q}^{(n)} + \alpha q \tag{8.11}$$

By subtracting q from both sides, and defining the *estimation error* $\epsilon^{(n)}$ at the $n$th iteration as

$$\epsilon^{(n)} = \hat{q}^{(n)} - q \tag{8.12}$$

we get the simple first-order homogeneous recursion describing the evolution of $\epsilon^{(n)}$:

$$\epsilon^{(n+1)} = (1 - \alpha)\epsilon^{(n)}, \qquad n = 0, 1, \ldots \tag{8.13}$$

which has the solution

$$\epsilon^{(n)} = (1 - \alpha)^n \epsilon^{(0)} \tag{8.14}$$

From (8.14) it follows that $|\epsilon^{(n)}| \to 0$ as $n \to \infty$, i.e., the algorithm converges for any $\epsilon^{(0)}$, if $|1 - \alpha|^n \to 0$ as $n \to \infty$. The latter condition simply means that we must have $|1 - \alpha| < 1$, i.e., $0 < \alpha < 2$.

Notice also that the choice $\alpha = 1$ would make the algorithm converge in a single step. However, even if this algorithm were implementable (and we know it is not, because the value of the expectation in (8.10) is not available), in the presence of round-off errors it would be advisable to pick a different value for $\alpha$. In fact, iterations average out the effect of these errors.

**Stochastic-gradient algorithm**

Let us now consider an approximation of (8.9) in a form useful for real-time implementation. The simplest such approximation, and by far the most widely used, is obtained by disregarding the expectation operator in (8.10). We obtain the new algorithm, usually denoted as the *stochastic-gradient algorithm*,

$$\hat{\mathbf{q}}^{(n+1)} = \hat{\mathbf{q}}^{(n)} + \alpha(x_n - \hat{x}_n)\mathbf{a}_n^* \qquad (8.15)$$

The second term in the RHS of (8.15) obviously has the same expected value as the corresponding term in the gradient algorithm. Hence, it can be viewed as an unbiased estimate of that quantity.

Implementation of (8.15) is shown in Fig. 8.4 for real data symbols. Each iteration is performed every discrete time instant, i.e., every $T$. When a new source symbol $a_n$ is fed into the tapped delay line (TDL), the channel output estimate $\hat{x}_n$ is obtained by combining linearly $a_n, \ldots, a_{n-L}$ according to (8.7). The error signal $x_n - \hat{x}_n$ is then formed, and, after multiplication by the scaling factor $\alpha$, this value is also multiplied by the content of the TDL, expressed by the vector $\mathbf{a}_n$. The resulting values are used finally to update the accumulators containing the actual estimates $\hat{q}_0, \ldots, \hat{q}_L$.

Consider now the convergence of this algorithm. Its study is far more difficult that with the "true-gradient" algorithm (8.9). In fact, we have from (8.15)

$$\boldsymbol{\epsilon}^{(n+1)} = \left(\mathbf{I} - \alpha\,\mathbf{a}_n^*\mathbf{a}_n'\right)\boldsymbol{\epsilon}^{(n)} \qquad (8.16)$$

a version of (8.13) with the scalar $(1 - \alpha)$ changed into the matrix $(\mathbf{I} - \alpha\,\mathbf{a}_n^*\mathbf{a}_n')$ ($\mathbf{I}$ denotes the identity matrix). A complication arises here from $\mathbf{a}_n^*\mathbf{a}_n'$ being a matrix with random entries, which in turn are not statistically independent of $\boldsymbol{\epsilon}^{(n)}$. We cannot proceed further without an approximation: a widely employed one, which makes the convergence analysis mathematically tractable, namely, the *independence assumption*. This assumes $(\mathbf{a}_n)$ to be a sequence of iid zero-mean vectors. In spite of its being rather crude, it provides a convergence analysis whose results are in close agreement with those of experiments and simulations, provided that the step size $\alpha$ is sufficiently small (see, e.g., Ungerboeck, 1972; Widrow *et al.*, 1976; Gitlin and Weinstein, 1979; Mazo, 1979; Jones *et al.*, 1982).

Since in (8.16) $\boldsymbol{\epsilon}^{(n)}$ depends only on the sequence $a_0, a_1, \ldots, a_{n-1}$, the independence assumption entails that $\boldsymbol{\epsilon}^{(n)}$ must be independent of $\mathbf{a}_n$. Thus, if we take the expectation of both sides of (8.16), we obtain

$$\mathrm{E}[\boldsymbol{\epsilon}^{(n)}] = (1 - \alpha)\mathrm{E}[\boldsymbol{\epsilon}^{(n)}] \qquad (8.17)$$

No random quantity appears in (8.17), so we can repeat the convergence analysis carried out for the true gradient algorithm. This enables us to conclude

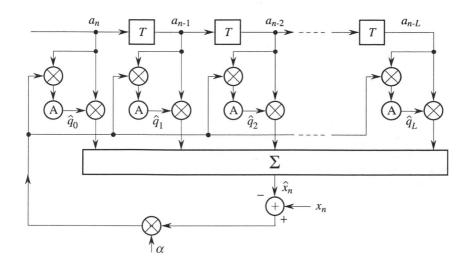

Figure 8.4: *Implementation of the stochastic-gradient algorithm (real source symbols $a_n$ are assumed). The blocks labeled $T$ denote delay elements; the blocks labeled A denote accumulators.*

that the average error vector $\mathrm{E}[\epsilon^{(n)}]$ tends to the null vector, as $n \to \infty$, if $0 < \alpha < 2$. Nonetheless, this result is incomplete, as the behavior of $\mathrm{E}[\epsilon^{(n)}]$ does not provide us with a complete picture of the convergence of the algorithm. In fact, nothing prevents this vector from being very close to the null vector while $\epsilon^{(n)}$ itself exhibits large deviations around its average. A deeper analysis of the behavior of $\epsilon^{(n)}$ is therefore called for. This can be obtained by studying the quadratic error $\mathrm{E}|\epsilon^{(n)}|^2$.

If we define

$$\mathcal{E}^{(n)} \triangleq \mathrm{E}|\epsilon^{(n)}|^2 = \mathrm{E}[\epsilon^{(n)\dagger}\epsilon^{(n)}] \tag{8.18}$$

from (8.16) we obtain the following recursive equation:

$$\mathcal{E}^{(n+1)} = \mathrm{E}[\epsilon^{(n)\dagger}(\mathbf{I} - \alpha\mathbf{a}_n^*\mathbf{a}_n')^2\epsilon^{(n)}] \tag{8.19}$$

By defining

$$k_a \triangleq \mathrm{E}[|a_n|^4] \tag{8.20}$$

we have

$$\mathrm{E}[(\mathbf{a}_n^*\mathbf{a}_n')^2] = (k_a + L)\mathbf{I} \tag{8.21}$$

and consequently from (8.19)

$$\mathcal{E}^{(n+1)} = \mathrm{E}[\epsilon^{(n)\dagger}(1 - 2\alpha + \alpha^2(k_a + L))\epsilon^{(n)}] \tag{8.22}$$

Repeated application of the latter yields

$$\mathcal{E}^{(n)} = [1 - 2\alpha + \alpha^2(k_a + L)]^n \mathcal{E}^{(0)} \tag{8.23}$$

so that the quadratic-error convergence is assured provided that the quantity in square brackets is less than 1.

**Example 8.1**   A simple special case occurs when the random variables $a_n$ take on values $\pm 1$ with equal probabilities, and hence $k_a = 1$. We obtain

$$\mathcal{E}^{(n)} = [1 - 2\alpha + \alpha^2(L + 1)]^n \, \mathcal{E}^{(0)}$$

The quantity in brackets can be minimized with respect to the choice of $\alpha$, yielding

$$\mathcal{E}^{(n)} = \left(\frac{L}{L+1}\right)^n \mathcal{E}^{(0)}$$

Notice how convergence is slowed down as $L$ increases.                              $\square$

## Some further problems

Several other features of automatic channel identification may be worth discussing here. However, since several of them are in common with adaptive equalization, which we shall examine in the balance of this chapter, to avoid unnecessary duplications we postpone their discussion. Here we restrict ourselves to three problems, the first two because of their relevance, and the third because it is typical of adaptive channel identification.

The first issue concerns the behavior of the stochastic-gradient algorithm when the channel is not stationary, i.e., when its impulse response changes with time. This is the problem of *adaptive* identification. If changes occur slowly enough with respect to the signaling rate, we can expect that the algorithm will allow the channel estimate to track continually the channel features. Another problem relates to the assumption that the source data sequence $(a_n)$ is known to the receiver. It can be solved by first sending through the channel a known sequence, which is expected to provide a reasonably good channel estimate. Afterwards, the receiver should provide a sufficiently good performance to assume that most of its estimates of transmitted symbols are correct. In this situation, the assumption that $(a_n)$ is known becomes reasonable.

Finally, consider the effect on automatic identification of inaccurate knowledge of the true channel memory span $L$. Clearly, if the TDL of Fig. 8.4 has a number $\widehat{L}$ of delay elements larger than $L$, then at the end of the identification process $\widehat{L} - L$ tap weights will take on zero values in identification. If instead $\widehat{L} < L$, that is, the number of delay elements in the TDL is smaller than the channel memory, it can be shown (see Problem 8.2) that $\widehat{L} + 1$ of the channel impulse response samples can still be identified correctly under the assumption that the data symbols be uncorrelated and have mean zero.

### 8.2.3. Blind channel identification

The identification techniques described so far are based on the existence of a data sequence known at the receiver: channel identification is feasible since both input and output samples are known. When the channel is varying, to make identification adaptive, this training data sequence has to be sent periodically to update the channel estimates, thus reducing the effective transmission rate because a fraction of transmission time is wasted for a training sequence. Another class of identification techniques, called *blind*, do not require the transmission of a preassigned data sequence. Instead, the statistical properties of the transmitted signal are exploited to carry out the identification without the receiver having access to the symbols being transmitted.

A number of blind techniques are based on higher-order statistics of the channel output (see, for example, Giannakis, Inouye, and Mendel, 1989; Giannakis and Mendel, 1989; and Mendel, 1991). These algorithms, besides being computationally intensive, suffer from the fact that the estimates of higher-order statistics usually converge more slowly than those of second-order statistics, and hence the process may be too slow for applications involving rapid channel variations such as those in mobile radio communications.

### Second-order algorithms

A more recent class of identification algorithms use only second-order statistics, and hence exhibit a faster convergence; if certain mild conditions on the frequency response of the channel are satisfied, only the autocorrelation function of the oversampled channel output needs to be evaluated for channel identification (see Tong *et al.*, 1994; Tong *et al.*, 1995; Tugnait, 1995; Moulines *et al.*, 1995; and Buisán and Biglieri, 1996).

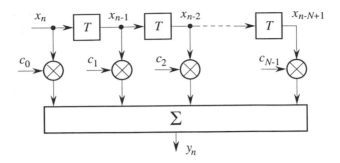

Figure 8.5: *The TDL equalizer.*

## 8.3. Channel equalization

From Chapter 7 we know how the optimum linear receiving filter is made: it consists of a matched filter, a sampler, and an infinite-length TDL filter. Thus, it makes sense to approximate this optimum filter as follows. Rather than a filter matched to $(q_\ell)$, which may be unknown, we use a filter matched to the *transmitted* signal, or a compromise filter matched to a representative of an entire class of received signals. The infinite-length TDL is approximated with a finite one, which is simple to implement, to analyze, and to adjust, mainly because of the linear relation between its tap weights and its output.

In this section we study an equalizer based on an algorithm for automatically adjusting the coefficients of such a TDL. In the following we keep on assuming that the data symbols $a_n$ are uncorrelated and have mean zero. We define $\sigma_a^2 \triangleq \mathrm{E}[|a_n|^2]$.

The TDL equalizer operating on the samples $(x_n)$ of the received signal is shown in Fig. 8.5. This structure, also called a *transversal filter*, has $N - 1$ delay elements, $N$ taps, $N$ weight accumulators, and $N$ multipliers. It stores $N$ samples that are linearly combined to produce the equalizer output

$$y_n = \mathbf{c}'\mathbf{x}_n \qquad (8.24)$$

where $\mathbf{c}$ is the tap-weight vector

$$\mathbf{c} = [c_0,\, c_1,\, \ldots,\, c_{N-1}]' \qquad (8.25)$$

and $\mathbf{x}_n$ denotes the TDL's content at discrete time $n$:

$$\mathbf{x}_n = [x_n,\, x_{n-1},\, \ldots,\, x_{n-N+1}]' \qquad (8.26)$$

Ideally, we would like the sequence $(y_n)$ at the output of the equalizer to reproduce the sequence $(a_n)$ of transmitted data symbols, except perhaps for a finite delay $D$. As we cannot expect to achieve this, even in the absence of noise, with a finite-length TDL, a reasonable goal would be to find c so as to minimize a suitable distortion measure, with the constraints of the equalizer length and the delay $D$. If we choose as a distortion measure the ubiquitous mean-square error between $y_n$ (what we get) and $a_{n-D}$ (what we would like to get), with $D$ a fixed integer, we have to minimize the MSE

$$\mathcal{E}(\mathbf{c}) \triangleq \mathrm{E}[|y_n - a_{n-D}|^2] = \mathbf{c}^\dagger \mathbf{X} \mathbf{c} - 2\Re[\mathbf{c}^\dagger \mathbf{v}] + \sigma_a^2 \qquad (8.27)$$

where $\mathbf{v}$ is a vector expressing the correlations between the source symbols and the channel outputs:

$$\mathbf{v} \triangleq \mathrm{E}[a_{n-D}\mathbf{x}_n^*] \qquad (8.28)$$

and $\mathbf{X}$ is the autocorrelation matrix of the samples stored in the TDL:

$$\mathbf{X} \triangleq \mathrm{E}[\mathbf{x}_n^* \mathbf{x}_n'] \qquad (8.29)$$

The latter matrix is assumed to be positive definite. In fact, for every complex $N$-vector a we have

$$\mathbf{a}^\dagger \mathbf{X} \mathbf{a} = \mathrm{E}[|\mathbf{a}' \mathbf{x}_n|^2] \geq 0 \qquad (8.30)$$

which shows that $\mathbf{X}$ is at least nonnegative definite. Moreover, the RHS of (8.30) can be viewed as the average power at the output of an equalizer with tap weight vector a. This power cannot be zero because of the random noise added to the samples at the channel's output. In the absence of noise and for a nonzero a, (8.30) can be zero only if the samples $x_n$ are linearly dependent. With this exception (that we want to discard), $\mathbf{a}^\dagger \mathbf{X} \mathbf{a} > 0$, i.e., $\mathbf{X}$ is positive definite, and hence invertible.

To define the value of the tap-weight vector c that minimizes (8.27), we must find the value of c that yields a null value of the gradient of $\mathcal{E}(\mathbf{c})$. This can be computed by using the results of Examples B.5 and B.6:

$$\nabla \mathcal{E}(\mathbf{c}) = 2(\mathbf{X}\mathbf{c} - \mathbf{v}) \qquad (8.31)$$

As the gradient has a unique zero for

$$\mathbf{c} = \mathbf{c}_{\mathrm{opt}} \triangleq \mathbf{X}^{-1}\mathbf{v} \qquad (8.32)$$

we obtain the minimum MSE

$$\mathcal{E}_{\mathrm{min}} \triangleq \mathcal{E}(\mathbf{c}_{\mathrm{opt}}) = \sigma_a^2 - \mathbf{v}^\dagger \mathbf{X}^{-1} \mathbf{v} \qquad (8.33)$$

An alternative form for $\mathcal{E}(\mathbf{c})$ is then

$$\mathcal{E}(\mathbf{c}) = \mathcal{E}_{\min} + (\mathbf{c} - \mathbf{c}_{\text{opt}})^\dagger \mathbf{X}(\mathbf{c} - \mathbf{c}_{\text{opt}}) \qquad (8.34)$$

This explicitly shows the quadratic nature of the functional $\mathcal{E}(\mathbf{c})$, and separates the contribution of the minimum achievable MSE, $\mathcal{E}_{\min}$, from the term reflecting the nonoptimum weight setting.

### 8.3.1. Performance of the infinitely long equalizer

We shall now analyze the performance of the optimum equalizer under the assumption that the number of its tap weights tends to infinity. Mathematically, this allows us to approximate the Toeplitz matrix $\mathbf{X}$ with a circulant matrix whose rows are cyclic shifts of one of them (see Section B.3 for the relevant definitions). This approximation entails neglecting that the true $\mathbf{X}$ differs from a circulant matrix in the lower-left and the upper-right corners. Now, eigenvalues and eigenvectors of a circulant matrix have closed-form expressions: thus, we shall be able to derive explicitly $\mathbf{c}_{\text{opt}}$ and $\mathcal{E}_{\min}$.

Consider the diagonal decomposition of $\mathbf{X}$ in the form (see (B.45))

$$\mathbf{X} = \mathbf{U}\boldsymbol{\Lambda}\mathbf{U}^{-1} \qquad (8.35)$$

where $\boldsymbol{\Lambda}$ is the diagonal matrix of the eigenvalues of $\mathbf{X}$, say $\mu_0, \ldots, \mu_{N-1}$, and $\mathbf{U}$ is the $N \times N$ unitary matrix of its eigenvectors (see Section B.4). As $\mathbf{X}$ is circulant, from Example B.3 the entries of $\mathbf{U}$ are

$$u_{ik} = \frac{1}{\sqrt{N}} w^{ik}, \qquad i = 0, 1, \ldots, N-1, \qquad k = 0, 1, \ldots, N-1 \qquad (8.36)$$

where

$$w \triangleq e^{j2\pi/N}. \qquad (8.37)$$

Consider then the eigenvalues of $\mathbf{X}$. From (B.39) and the definition of $\mathbf{X}$, we obtain, as $N \to \infty$,

$$\mu_i = \sum_{\ell=-N/2}^{N/2} \mathrm{E}[x_{n+\ell} x_n^*] \, e^{j2\pi\ell i/N} \qquad (8.38)$$

Define now the signal $x_{\text{eq}}(t)$ as the time function, bandlimited in the frequency interval $(-1/2T, 1/2T)$, whose sequence of samples, taken every $T$, is the equalizer input $(x_n)$. From the sampling expansion (2.254) we obtain the Fourier transform

$$X_{\text{eq}}(f) = T \sum_{n=-\infty}^{\infty} x_n \, e^{-j2\pi fnT}, \qquad 0 \le f < \frac{1}{T}. \qquad (8.39)$$

By applying the techniques introduced in Section 2.3.1, we can derive the power density spectrum of $x_{\mathrm{eq}}(t)$:

$$\mathcal{G}_{\mathrm{eq}}(f) = T \sum_{\ell=-\infty}^{\infty} \mathrm{E}[x_{n+\ell} x_n^*] \, \mathrm{e}^{-j2\pi f \ell T}, \qquad 0 \le f < \frac{1}{T} \qquad (8.40)$$

The comparison of (8.38) with (8.40) shows that, as $N \to \infty$,

$$\mu_i = \frac{1}{T} \mathcal{G}_{\mathrm{eq}}\left(\frac{i}{NT}\right), \qquad i = 0, 1, \ldots, N-1; \qquad (8.41)$$

that is, apart from an unessential constant, the eigenvalues of $\mathbf{X}$ are the values that the power spectrum of the bandlimited signal $x_{\mathrm{eq}}(t)$ takes on at equally spaced frequencies $i/NT$. Let us now express $\mathcal{G}_{\mathrm{eq}}(f)$ in terms of the channel parameters. Using (7.58) and results from Section 2.3.1, we obtain

$$\mu_i = \frac{\sigma_a^2}{T^2} \left| Q_{\mathrm{eq}}\left(\frac{i}{NT}\right) \right|^2 + \frac{1}{T} G_n\left(\frac{i}{NT}\right) \qquad (8.42)$$

where $Q_{\mathrm{eq}}(f)$ is defined as in (7.67), and $G_n(f)$ is the noise power spectral density at the receiver filter output. For simplicity, we have assumed that the noise is bandlimited in $(-1/2T, 1/2T)$: otherwise we should write, in lieu of $G_n(f)$, its "aliased" version $\sum_k G_n(g + k/T)$.

We are now ready to compute the performance of the infinite-length equalizer. Using (8.32) and (8.35), we get for the optimum tap-weight vector

$$\mathbf{c}_{\mathrm{opt}} = \mathbf{U}\mathbf{\Lambda}^{-1}\mathbf{U}^\dagger\mathbf{v} \qquad (8.43)$$

Define

$$C_{\mathrm{opt}}(f) \overset{\triangle}{=} \frac{1}{\sqrt{N}} \sum_{i=0}^{N-1} [\mathbf{c}_{\mathrm{opt}}]_i \mathrm{e}^{-j2\pi f i T} \qquad (8.44)$$

and

$$V(f) \overset{\triangle}{=} \frac{1}{\sqrt{N}} \sum_{i=0}^{N-1} [\mathbf{v}]_i \mathrm{e}^{-j2\pi f i T} \qquad (8.45)$$

Due to (8.36) and (8.37), premultiplication of vector $\mathbf{c}_{\mathrm{opt}}$ by $\sqrt{N}\mathbf{U}^\dagger$ yields a vector whose components are $C_{\mathrm{opt}}(k/NT)$, $k = 0, \ldots, N-1$. Similarly, premultiplication of $\mathbf{v}$ by the same matrix yields a vector whose components are $V(k/NT)$, $k = 0, \ldots, N-1$. Consequently, by observing that $\mathbf{U}^\dagger\mathbf{U} = \mathbf{I}$, from (8.43) we obtain

$$C_{\mathrm{opt}}\left(\frac{k}{NT}\right) = \frac{1}{\mu_k} V\left(\frac{k}{NT}\right), \qquad k = 0, 1, \ldots, N-1 \qquad (8.46)$$

In (8.42) the $\mu_k$ have been expressed in terms of the channel parameters; we derive now an analogous expression for the $V(k/NT)$. If we write $x_\ell$ explicitly as in (8.1), from definition (8.28) we obtain

$$\mathbf{v} = \sigma_a^2 \begin{bmatrix} q_D^* \\ q_{D-1}^* \\ \vdots \\ q_{D-N+1}^* \end{bmatrix} \tag{8.47}$$

where the $q_\ell$ are the samples of the impulse response of the overall channel preceding the sampler. Then

$$
\begin{aligned}
V(f) &= \sigma_a^2 \sum_{i=0}^{N-1} q_{D-i}^* e^{-j2\pi f i T} \\
&= \sigma_a^2 \sum_{i=0}^{N-1} \int_{-\infty}^{\infty} Q^*(f') e^{-j2\pi f' DT} e^{-j2\pi (f-f')iT} \, df'
\end{aligned}
$$

and, as $N \to \infty$, using equality (2.109),

$$
\begin{aligned}
V(f) &= \frac{\sigma_a^2}{T} \sum_{m=-\infty}^{\infty} \int_{-\infty}^{\infty} Q^*(f') e^{-j2\pi f' DT} \delta\left(f - f' - \frac{m}{T}\right) df' \\
&= \frac{\sigma_a^2}{T} \sum_{m=-\infty}^{\infty} Q^*\left(f - \frac{m}{T}\right) e^{-j2\pi f DT} \\
&= \frac{\sigma_a^2}{T} Q_{\text{eq}}^*(f) e^{-j2\pi f DT} \tag{8.48}
\end{aligned}
$$

By combining (8.41), (8.46), and (8.48), we get finally, for $k = 0, 1, \ldots, N-1$,

$$C_{\text{opt}}\left(\frac{k}{NT}\right) = \frac{\sigma_a^2 Q_{\text{eq}}^*(k/NT)}{G_n(k/NT) + (\sigma_a^2/T)|Q_{\text{eq}}(k/NT)|^2} e^{-j2\pi k D/N} \tag{8.49}$$

As $N \to \infty$, we can assume that the transfer function of the optimum infinitely long equalizer is obtained from the former expression by writing $f$ in lieu of $i/NT$:

$$C_{\text{opt}}(f) = \frac{\sigma_a^2 Q_{\text{eq}}^*(f)}{G_n(f) + (\sigma_a^2/T)|Q_{\text{eq}}(f)|^2} e^{-j2\pi f DT}, \qquad 0 \le f < \frac{1}{T} \tag{8.50}$$

By comparing this result with (7.93) it can be seen that, under the minimum-MSE criterion, $C_{\text{opt}}(f)$ is the optimum receiving filter when $Q_{\text{eq}}(f) = Q(f)$, that is, when the channel is bandlimited in the interval $(-1/2T, 1/2T)$. If the latter condition is not fulfilled, the TDL equalizer fails to be the optimum filter.

We shall return to this point later, in our discussion of the fractionally-spaced equalizer.

We are finally ready to evaluate the infinite-length equalizer performance. The quadratic form appearing in (8.33) is computed using (8.35):

$$
\begin{aligned}
\mathbf{v}^\dagger \mathbf{X}^{-1} \mathbf{v} &= \mathbf{v}^\dagger \mathbf{U} \mathbf{\Lambda}^{-1} \mathbf{U}^\dagger \mathbf{v} \\
&= \frac{1}{N} \sum_{i=0}^{N-1} \frac{|V(i/NT)|^2}{\mu_i} \\
&= \frac{1}{N} \sum_{i=0}^{N-1} \frac{\sigma_a^4 |Q_{\mathrm{eq}}(i/NT)|^2}{T G_n(i/NT) + \sigma_a^2 |Q_{\mathrm{eq}}(i/NT)|^2}
\end{aligned} \tag{8.51}
$$

By taking the limit of (8.51), which can be done by using the Toeplitz distribution theorem (B.36), we have, as $N \to \infty$,

$$
\mathbf{v}^\dagger \mathbf{X}^{-1} \mathbf{v} = T \int_{-1/2T}^{1/2T} \frac{\sigma_a^4 |Q_{\mathrm{eq}}(f)|^2}{T G_n(f) + \sigma_a^2 |Q_{\mathrm{eq}}(f)|^2} \, df \tag{8.52}
$$

and finally

$$
\mathcal{E}_{\min} = T \int_{-1/2T}^{1/2T} \frac{\sigma_a^2 G_n(f)}{G_n(f) + (\sigma_a^2/T) |Q_{\mathrm{eq}}(f)|^2} \, df. \tag{8.53}
$$

This is the minimum MSE achievable by using an infinitely long TDL equalizer. Equation (8.53) shows how it is related to the noise power spectral density and to the channel transfer function (recall that the latter includes the receiving filter, which may not be a filter matched to the received signal). For finite-length equalizers, it is difficult to say much about $\mathcal{E}$ without resorting to numerical computations, which, as we know, involve the transmission filter, the channel response, the noise power spectral density, the receiving filter, the delay $D$, and the number $N$ of equalizer coefficients.

### Some considerations

A few interesting considerations can be derived from (8.50) and (8.53), as follows:

(a) In the absence of noise, (8.50) reduces to

$$
C_{\mathrm{opt}}(f) = T Q_{\mathrm{eq}}^{-1}(f) e^{-j2\pi f DT}
$$

which shows that, apart from a delay $DT$, the equalizer has a transfer function proportional to the inverse of $Q_{\mathrm{eq}}(f)$. This shows that the optimum equalizer inverts the channel transfer function in the frequency interval

$(-1/2T, 1/2T)$ (it cannot operate on a wider interval because the samples are taken every $T$: more on this later). When noise dominates, i.e., when $G_n(f) \gg (\sigma_a^2/T)|Q_{eq}(f)|^2$, the equalizer transfer function is proportional to $Q_{eq}^*(f)/G_n(f)$, i.e., the optimum equalizer is a matched filter (see Problem 2.26.)

(b) The maximum value that $\mathcal{E}_{min}$ can attain is $\sigma_a^2$: from (8.53) we see that this circumstance corresponds to a channel with a null transfer function: $|Q_{eq}(f)| = 0$. This trivial result suggests that in the presence of noise the performance of the linear equalizer described here is limited by large depressions in the frequency response of the channel. This fact can also be understood by observing that the equalizer will try to compensate for a deep null by synthesizing a large gain at the corresponding frequencies. But this large gain will enhance the effect of the noise at the same frequency, thus preventing perfect compensation or even leading to serious performance degradation.

(c) Certain channels (e.g., the multipath radio channel of Chapter 13) may exhibit nulls in their frequency response. Other channels, not having deep spectral depressions in their transfer function $Q(f)$, may exhibit them in their "aliased" version $Q_{eq}(f)$ (e.g., the telephone channel). It happens that when the channel bandwidth exceeds the Nyquist interval $(-1/2T, 1/2T)$, the choice of the sampling instant, which obviously does not affect $|Q(f)|$, does indeed affect $|Q_{eq}(f)|$ (this was discussed in Section 2.5.1). Thus, for a channel whose frequency response extends beyond the Nyquist interval, inappropriate choice of the sampling epochs can produce nulls in the equivalent channel response. In this case, the linear TDL equalizer described above may be inadequate to compensate for ISI.

(d) The minimum MSE (8.53) of the infinitely long equalizer does not show dependence on the allowed delay $D$. However, the performance of a finite equalizer does depend on that choice.

### 8.3.2. Gradient algorithm for equalization

We shall now describe a gradient algorithm for the automatic adjustment of the tap-weight values vector $c$ to its optimum value. As the gradient of the MSE was shown in (8.31) to be proportional to $\mathbf{Xc} - \mathbf{v}$, the gradient algorithm for equalization takes the form

$$\mathbf{c}^{(n+1)} = \mathbf{c}^{(n)} - \alpha(\mathbf{Xc}^{(n)} - \mathbf{v}) = \mathbf{c}^{(n)} - \alpha E[(y_n - a_{n-D})\mathbf{x}_n^*] \qquad (8.54)$$

where $c^{(n)}$ denotes the tap-weight vector at the $n$th iteration step, and $\alpha$ is a positive constant small enough to insure convergence. The difficulty in the implementation of (8.54) is that the average cannot be computed in real time (see the discussion following (8.10)). Thus, as for automatic channel identification, we must resort to a stochastic gradient algorithm. However, before doing that it is expedient to analyze the convergence properties of the "true-gradient" algorithm (8.54). By duplicating arguments used in connection with the convergence analysis of the channel identification algorithm (8.10), it can be proved that the tap-weight error at the $n$th iteration,

$$\epsilon^{(n)} \triangleq c^{(n)} - c_{\text{opt}} \tag{8.55}$$

satisfies the recursion

$$\epsilon^{(n+1)} = (\mathbf{I} - \alpha\mathbf{X})^n \, \epsilon^{(n)}, \qquad n = 0, 1, \dots \tag{8.56}$$

so that

$$\epsilon^{(n)} = (\mathbf{I} - \alpha\mathbf{X})^n \, \epsilon^{(0)} \tag{8.57}$$

and convergence of the tap-weight error is assured for any $\epsilon^{(0)}$ provided that

$$0 < \alpha < 2/\mu_{\max} \tag{8.58}$$

where $\mu_{\max}$ denotes the largest eigenvalue of the matrix $\mathbf{X}$. (Notice that the convergence analysis here is far more complex than for channel identification: this is due to the fact that the role of the source symbols is taken here by the channel outputs $x_n$, whose autocorrelation is not the identity matrix anymore).

Define now the excess MSE at the $n$th iteration step, that is, when $c = c^{(n)}$. We have, using (8.34),

$$\Delta^{(n)} \triangleq \mathcal{E}(c^{(n)}) - \mathcal{E}_{\min} = \epsilon^{(n)\dagger}\mathbf{X}\epsilon^{(n)} \tag{8.59}$$

Substitution of (8.57) in (8.59) yields

$$\Delta^{(n)} = \epsilon^{(0)\dagger}(\mathbf{I} - \alpha\mathbf{X})^n\mathbf{X}(\mathbf{I} - \alpha\mathbf{X})^n\epsilon^{(0)} \tag{8.60}$$

or, observing that $\mathbf{X}$ commutes with $\mathbf{I} - \alpha\mathbf{X}$, and hence with any of its powers,

$$\Delta^{(n)} = \epsilon^{(0)\dagger}(\mathbf{I} - \alpha\mathbf{X})^{2n}\mathbf{X}\epsilon^{(0)} \tag{8.61}$$

Our next step in the analysis of the convergence of $\Delta^{(n)}$ is based on the diagonal decomposition of $\mathbf{X}$ as in (8.35):

$$\mathbf{X} = \mathbf{U}\boldsymbol{\Lambda}\mathbf{U}^{-1} \tag{8.62}$$

This yields

$$(\mathbf{I} - \alpha\mathbf{X})^{2n}\mathbf{X} = \mathbf{U}(\mathbf{I} - \alpha\Lambda)^{2n}\Lambda\mathbf{U}^{-1} \tag{8.63}$$

so that we finally get

$$\Delta^{(n)} = \sum_{i=0}^{N-1} \beta_i^{(0)}\mu_i(1 - \alpha\mu_i)^{2n} \tag{8.64}$$

where $\beta_i^{(0)}$, $i = 0, 1, \ldots, N - 1$, is the squared magnitude of the $i$th element of the vector $\mathbf{U}^{\dagger}\boldsymbol{\epsilon}^{(0)}$. From (8.64) it is seen that $\Delta^{(n)}$ can be decomposed as the sum of $N$ exponentials, all of them decaying to zero if (8.58) holds. For fast convergence, we may choose the step size $\alpha$ so as to speed up the convergence of the slowest-decaying exponential term in (8.64). This is done by minimizing the quantity

$$r(\alpha) \triangleq \max_i |1 - \alpha\mu_i| \tag{8.65}$$

This "optimum" $\alpha$ satisfies the condition

$$1 - \alpha\mu_{\min} = -(1 - \alpha\mu_{\max}) \tag{8.66}$$

(see Fig. 8.6), where $\mu_{\min}$ is the smallest eigenvalue of $\mathbf{X}$. Thus

$$\alpha_{\text{opt}} = \frac{2}{\mu_{\max} + \mu_{\min}} \tag{8.67}$$

and

$$r(\alpha_{\text{opt}}) = \frac{\mu_{\max}/\mu_{\min} - 1}{\mu_{\max}/\mu_{\min} + 1} \tag{8.68}$$

Thus, we have proved that the maximum convergence speed is, in a sense, dominated by the eigenvalue spread $\mu_{\max}/\mu_{\min}$. In fact, the smaller this value is, the faster the convergence of the true-gradient algorithm that can be achieved by a suitable choice of the step size $\alpha$. A fast rate of convergence of the algorithm allows the equalizer to converge closely enough to its optimum setting in a short time; moreover, in a nonstationary environment channel variations can be tracked.

### Stochastic-gradient algorithm for equalization

We shall now examine the stochastic-gradient version of algorithm (8.54):

$$\mathbf{c}^{(n+1)} = \mathbf{c}^{(n)} - \alpha(y_n - a_{n-D})\mathbf{x}_n^* \tag{8.69}$$

Fig. 8.7 shows how this algorithm can actually be implemented. For simplicity, real signals are assumed in this figure. However, it should be kept in mind

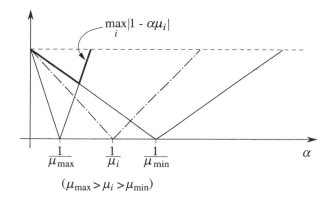

Figure 8.6: *Choice of the "optimum" value of the step size $\alpha$.*

that the quantities involved are generally complex, which implies that four real multiplications are involved in each multiplier. We assume, for the moment, that the source symbols $a_n$ necessary to compute the RHS of (8.69) are known at the receiver, and disregard the dashed box (to be described later). Every $T$ one iteration is performed. When a new sample $x_n$ enters the TDL, a value $y_n$ is computed by combining linearly the $N$ samples contained in the TDL according to (8.24). After subtraction of $a_{n-D}$, the stochastic gradient is formed by multiplying this "error signal" by the samples $x_{n-i}$, $i = 0, 1, \ldots, N - 1$. The values obtained, after rescaling by a factor $-\alpha$, are added to the values of the tap weights stored in their accumulators so as to provide their updated versions.

To analyze the convergence properties of the stochastic-gradient algorithm for equalization we must resort again, *faute de mieux*, to the simplifications allowed by the independence assumption. In our situation this consists in assuming that $(\mathbf{x}_n)$ is a sequence of zero-mean iid vectors. Again, this simplification, which does not make much sense mathematically, offers results validated experimentally.

Consider first the tap-weight vector $\mathbf{c}^{(n)}$. From (8.69), it depends on $\mathbf{x}_0, \mathbf{x}_1, \ldots, \mathbf{x}_{n-1}$. Given the independence assumption, $\mathbf{c}^{(n)}$ is independent of $\mathbf{x}_n$; thus, averaging both sides of (8.69) and using (8.55), we obtain

$$\mathrm{E}[\boldsymbol{\epsilon}^{(n+1)}] = (\mathbf{I} - \alpha \mathbf{X}) \, \mathrm{E}[\boldsymbol{\epsilon}^{(n)}] \qquad (8.70)$$

This recursion shows that the average tap-weight error $\mathrm{E}[\boldsymbol{\epsilon}^{(n)}]$ converges to zero subject to condition (8.58) (this also shows the usefulness of the true-gradient-algorithm analysis). Thus, if (8.58) is satisfied, the stochastic-gradient algorithm

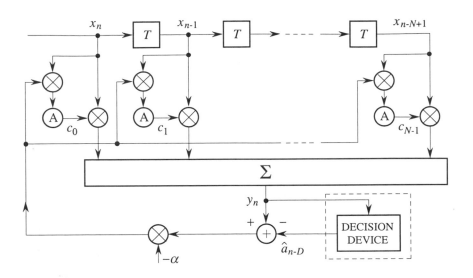

Figure 8.7: *Implementation of the stochastic-gradient algorithm for channel equalization (real signals are assumed).*

is stable on the average (i.e., the *average* tap-weight vector will converge to its optimum value irrespective of the initial tap-weight setting).

Consider next the evolution of the mean-square error. If we define the excess MSE $\delta^{(n)}$ as

$$\delta^{(n)} \triangleq \mathrm{E}\,|y_n - a_{n-D}|^2 - \mathrm{E}\,|y_{\mathrm{opt}}^{(n)} - a_{n-D}|^2 \tag{8.71}$$

where

$$y_{\mathrm{opt}}^{(n)} \triangleq \mathbf{c}'_{\mathrm{opt}}\mathbf{x}_n \tag{8.72}$$

we have, after some manipulations,

$$\delta^{(n)} = \mathrm{E}\,|y_n - y_{\mathrm{opt}}^{(n)}|^2 + 2\Re\mathrm{E}\,[(\mathbf{c}^{(n)} - \mathbf{c}_{\mathrm{opt}})^\dagger \mathbf{x}_n^*(y_{\mathrm{opt}}^{(n)} - a_{n-D})] \tag{8.73}$$

The second term in the RHS of (8.73) is zero. To show this, use (8.69) and the independence assumption to verify that $\mathbf{c}^{(n)}$ depends on $\mathbf{x}_m$ and $a_{m-D}$ only for $m < n$, and hence that

$$\mathrm{E}\,[(\mathbf{c}^{(n)} - \mathbf{c}_{\mathrm{opt}})^\dagger \mathbf{x}_n^*(y_{\mathrm{opt}}^{(n)} - a_{n-D})] = \mathrm{E}\,[(\mathbf{c}^{(n)} - \mathbf{c}_{\mathrm{opt}})^\dagger]\mathrm{E}\,[\mathbf{x}_n^*(y_{\mathrm{opt}}^{(n)} - a_{n-D})] \tag{8.74}$$

Moreover,

$$\mathrm{E}\,[\mathbf{x}_n^*(y_{\mathrm{opt}}^{(n)} - a_{n-D})] = \mathrm{E}\,[\mathbf{x}_n^*\mathbf{x}'_n]\,\mathbf{c}_{\mathrm{opt}} - \mathbf{v} = \mathbf{X}\mathbf{c}_{\mathrm{opt}} - \mathbf{v} = \mathbf{0} \tag{8.75}$$

In conclusion, we have

$$\delta^{(n)} = \mathrm{E}\, |y_n - y_{\mathrm{opt}}^{(n)}|^2 = \mathrm{E}\,[\epsilon^{(n)\dagger}\mathbf{x}_n^*\mathbf{x}_n'\epsilon^{(n)}] = \mathrm{E}\,[\epsilon^{(n)\dagger}\mathbf{X}\epsilon^{(n)}] \qquad (8.76)$$

where the independence assumption has again been invoked in the last equality, and $\epsilon^{(n)}$ is defined as in (8.55). It is worthwhile emphasizing that (8.76) is, in general, invalid if both $(\mathbf{x}_n)$ and $(a_n)$ are not independent sequences.

To proceed further, we first rewrite the stochastic-gradient recursion (8.69) in the form

$$\epsilon^{(n+1)} = \mathbf{A}_n\epsilon^{(n)} + \mathbf{b}_n \qquad (8.77)$$

where

$$\mathbf{A}_n \overset{\triangle}{=} \mathbf{I} - \alpha\mathbf{x}_n^*\mathbf{x}_n' \qquad (8.78)$$

and

$$\mathbf{b}_n \overset{\triangle}{=} -\alpha(y_{\mathrm{opt}} - a_{n-D})\mathbf{x}_n^* \qquad (8.79)$$

Thus, from (8.76) we obtain

$$\delta^{(n-1)} = \mathrm{E}\,\{\epsilon^{(n)\dagger}\mathrm{E}\,[\mathbf{A}_n\mathbf{X}\mathbf{A}_n]\epsilon^{(n)}\} + 2\mathrm{E}\,[\epsilon^{(n)}]\,\mathrm{E}\,[\mathbf{A}_n\mathbf{X}\mathbf{b}_n] + \mathrm{E}\,[\mathbf{b}_n^\dagger\mathbf{X}\mathbf{b}_n] \quad (8.80)$$

To simplify the analysis, we shall limit ourselves to the case in which the components of the vector $\mathbf{x}_n$ are independent, and $y_{\mathrm{opt}} - a_{n-D}$ is also independent of $\mathbf{x}_n$. Thus, recalling that $\mathrm{E}[a_n] = 0$ (which implies $\mathrm{E}[\mathbf{x}_n] = 0$), we have $\mathbf{X} = \sigma_x^2\mathbf{I}$, where $\sigma_x^2 \overset{\triangle}{=} \mathrm{E}\{|x_n|^2\}$, and the middle term in (8.80) vanishes. In fact, we have

$$\mathrm{E}\,[\mathbf{A}_n\mathbf{X}\mathbf{b}_n] = 0 \qquad (8.81)$$

Furthermore, the last term in (8.80) reduces to

$$\mathrm{E}[\mathbf{b}_n^\dagger\mathbf{X}\mathbf{b}_n] = \alpha^2\mathrm{E}\,|y_{\mathrm{opt}}^{(n)} - a_{n-D}|^2\,\mathrm{E}\,[\mathbf{x}_n'\mathbf{X}\mathbf{x}_n^*] = \alpha^2 N\mathcal{E}_{\mathrm{min}}\sigma_x^4 \qquad (8.82)$$

Finally, the matrix in the first term of the RHS of (8.80) reduces to

$$\mathrm{E}\,[\mathbf{A}_n\mathbf{X}\mathbf{A}_n] = [(1 - \alpha\sigma_x^2)^2 + \alpha^2\rho]\,\sigma_x^2\mathbf{I} \qquad (8.83)$$

where

$$\rho\mathbf{I} \overset{\triangle}{=} \mathrm{E}\,[\mathbf{x}_n^*\mathbf{x}_n']^2 - \sigma_x^4\mathbf{I} = [\mathrm{E}\,|x_n|^4 + (N-2)\sigma_x^4]\mathbf{I} \qquad (8.84)$$

From the substitution of (8.81)–(8.84) into (8.80) and the observation that under our assumptions $\mathrm{E}[\epsilon^{(n)\dagger}\epsilon^{(n)}] = \sigma_x^{-2}\delta^{(n)}$, we finally obtain

$$\delta^{(n+1)} = \gamma\delta^n + \alpha^2 N\mathcal{E}_{\mathrm{min}}\sigma_x^4 \qquad (8.85)$$

where

$$\gamma = (1 - \alpha\sigma_x^2)^2 + \alpha^2[\mathrm{E}|x_n|^4 + (N-2)\sigma_x^4] \qquad (8.86)$$

Thus, $\gamma < 1$ turns out to be a necessary and sufficient condition for the convergence of the excess MSE $\delta_n^{(n)}$. We can see, for example, that the convergence is adversely affected by the number $N$ of tap weights in the equalizer, as well as by the fourth absolute moment of the received samples $\mathrm{E}\{|x_n|^4\}$. Also, if $\gamma < 1$ and $n \to \infty$, the excess MSE tends to its residual value

$$\delta^{(\infty)} \triangleq \frac{\alpha^2 N \mathcal{E}_{\min} \sigma_x^4}{1 - \gamma} \qquad (8.87)$$

Notice in particular from (8.87) that the excess mean-square error does not tend to zero, as in the true-gradient algorithm, but to a value approximately proportional to $\alpha^2$, at least for small $\gamma$ values. This shows, for example, that the choice of the step size $\alpha$ in the stochastic-gradient algorithm entails a trade-off between fast convergence and small residual MSE.

We conclude this section with the observation that, by using an iterative algorithm, the equalizer can work adaptively by tracking and compensating for channel changes, provided that they are sufficiently slow with respect to the settling time of the equalizer.

## 8.4.   Fractionally-spaced equalizers

We have assumed so far that the signal $x(t)$ received at the channel's output is first filtered and then sampled every $T$ before being sent to the TDL with adjustable weights and elementary delays $T$. This is an optimum procedure, as we have discussed in Section 7.4, if the equalizer is preceded by a filter matched to the channel-distorted transmitted pulse. In practice, when the channel is unknown the best we can do is to match the receiver filter to the undistorted transmitted pulse, or to a compromise representative of an entire class of distorted signals. This change, however, is far from being innocuous. We know from Section 2.5 that the process of sampling a signal at rate $1/T$ superimposes its spectral components spaced $1/T$ Hz apart (the "aliasing" effect), and hence makes the equalizer performance very sensitive to the choice of the sampling time, because this can cause the appearance of deep nulls in the equivalent channel transfer function.

Another way of seeing this is by observing that the transfer function of the $T$-spaced equalizer is periodic with period $1/T$; thus, spectral components of the incoming signal lying at frequencies spaced $1/T$ apart cannot be processed independently by adjusting the tap weights. Moreover, this periodicity does not allow the noise-frequency components lying outside the interval $(-1/2T, 1/2T)$ to be suppressed. This task is assigned to the receiver filter preceding the equalizer, the one which should, optimally, be a matched filter.

Instead, assume that $x(t)$ is sampled every $T' < T$, and consequently the TDL elementary delay is $T'$. If $\beta$ denotes the excess bandwidth, that is, if the received signal $x(t)$ is confined to the frequency interval $[-(1 + \beta)/2T, (1 + \beta)/2T]$, we can choose $T' \le (1 + \beta)^{-1}T$. With this choice, the equalizer transfer function becomes sufficiently large to accommodate the whole signal spectrum. Hence, $Q_{\mathrm{eq}}(f) = Q(f)$; the sampling instant becomes irrelevant, and the appearance of deep nulls caused by a badly chosen sampling instant is avoided. Finally, from (8.50) we see that the equalizer provides the optimum (MSE) receiving filter, thus avoiding the need for a separate matched filter to suppress the noise (an anti-aliasing filter will be sufficient).

It must be kept in mind that the signal at the output of the equalizer is still sampled at rate $1/T$. But, since its input is sampled at $1/T'$, the equalizer acts on the received signal before aliasing its frequency components. In summary, we can say that a $T$-spaced TDL with symbol-rate sampling cannot perform matched filtering, while a $T'$-spaced TDL can incorporate the functions of a matched filter and of an equalizer. Equalizers based on this principle are called *fractionally-spaced*, and were first used in commercial telephone-line modems in the mid 1970s.

A convergence analysis similar to that of Section 8.2 can be carried out for fractionally-spaced equalizers. Simulation of QAM in equalizers with $T' = T/2$ over typical voice-grade circuits (Qureshi, 1982) confirms the improvement, predicted by the theory, over symbol-rate-spaced ("synchronous") equalizers. In particular, (a) the $T/2$ equalizer performs almost as well or even better than a synchronous equalizer with the same number of taps, and consequently twice the time span; (b) a receiving filter (other than the anti-aliasing filter) preceding the equalizer is not required with a $T/2$ equalizer; and (c) for channels with severe band-edge distortions the $T/2$ equalizer outperforms the synchronous equalizer regardless of the choice of the sampling instant (Qureshi, 1985).

## 8.5.    Training the equalizer: Cyclic equalization

So far, our analysis of the TDL equalizer performance assumed that the data sequence $(a_n)$ needed to evaluate the error gradient was known at the receiver's front end. A widely used method to render this assumption realistic in practice is now described. In an initial (training, or start-up) period, a particular data sequence, known and available with the right time alignment at the receiver, is sent through the channel. This training sequence, or *preamble*, may consist of isolated pulses, or may be a continuous sequence with a uniform power spectrum (pseudo-noise sequences (Golomb, 1967) with periods significantly greater than $N$ are often used to this purpose). Once the equalized channel quality has become so good that decisions on transmitted symbols can be made with small enough error probability, the gradient is computed by replacing the estimated data symbol sequence $(\hat{a}_n)$ for the transmitted one $(a_n)$ (see the dashed box in Fig. 8.7). Simulations and experimental evidence show that for reasonable error rates this replacement does not alter the convergence of the equalizer.

In some cases, the error probability before equalization is so small that the training period can be avoided. The equalizer is then said to work in a "bootstrap" mode, this name being derived from the saying about pulling oneself up by one's own bootstrap. However, in most situations the equalizer must be trained before it can be switched to a decision-directed mode of operation.

Concerning the selection of the training sequence, a good choice is a periodic sequence whose period is $N$, the number of TDL taps. This choice, which gives rise to *cyclic equalization*, enables us to solve a problem arising in the start-up procedure and concerning the best choice of the delay $D$ to use in the definition (8.27) of the mean-square error to be minimized. In fact, when aligning in time the training sequence generated locally with that sent by the source, $D$ should be chosen in order to best compensate for the delay introduced by the channel. Cyclic equalization provides a rule to choose $D$ such that the minimum MSE might not be achieved, but a relatively simple implementation is obtained, coupled with adequate performance.

Consider a training sequence with period $N$. Assume for a moment that the channel is noiseless and distortionless. Thus, the received samples are just a delayed version of the transmitted symbol sequence. After convergence of the equalizer, only one of the tap weights will have a nonzero value. The corresponding tap position informs us about the time shift between the received sequence and the one generated locally; in particular, any unit time shift in the sequence generated locally will cause the unique nonzero-weight tap to move by one position in the TDL. Let us now return to a channel affected by linear distortion, but without noise. The received sampled sequence is once again periodic with period $N$. One full period is stored in the TDL. After convergence, the start-up procedure will finish with a set of tap-weight values that needs to be only *cyclically* shifted for proper time alignment. As any cyclic shift between the received sample sequence and the data sequence generated locally causes a cyclic shift of the set of tap-weight values, it is not necessary to achieve time alignment before start-up. This can be done *after* start-up by cycling the tap-weight values so that the largest absolute value is found in a reference position (e.g., the center tap).

An equalizer scheme based on this principle is shown in Fig. 8.8. The periodic sequence generator outputs the training word. All tap weights are preset to identical values to reflect that the location of the largest weight is not known a priori. To begin the start-up procedure, the switch at the bottom is set to po-

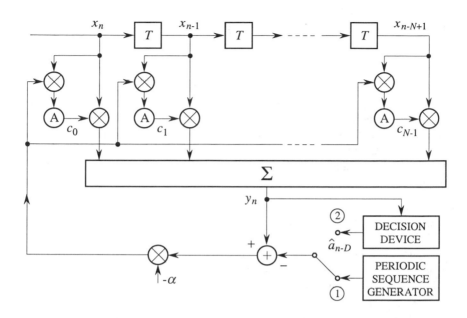

Figure 8.8: *Block diagram of a TDL equalizer with cyclic training.*

sition ①. After training, the tap weight values are aligned by cyclic shifts, as just indicated. The equalizer can now be operated in a decision-directed mode by moving the switch to position ②.

It remains now to analyze the performance of cyclic equalization. Without noise, the sequence $(x_n)$ is periodic; hence, it has a periodic correlation, which makes the matrix $\mathbf{X}$ circulant. As a consequence, the analysis carried out in Section 8.2.1 and based on decomposition (8.35) becomes exact. In particular, from (8.49) we have

$$C_{\text{opt}}\left(\frac{i}{NT}\right) = TQ_{\text{eq}}^{-1}\left(\frac{i}{NT}\right) e^{-j2\pi iD/N}, \qquad i = 0, 1, \ldots, N-1 \qquad (8.88)$$

This shows that equalization of the channel is achieved at a set of $N$ equally spaced frequencies in a frequency interval of width $1/T$. In other words, the inverse channel response is approximated through interpolation at equally spaced points. Thus, equalization after start-up, although nonoptimum in the MSE sense for random transmitted data, can be expected to be reasonably close to the optimum when $N$, the number of taps in the TDL, is sufficiently large.

Concerning the choice of the periodic symbol sequence to be employed in the start-up phase, it has been proved (see Godard, 1981) that the best sequences in the presence of noise are those whose periodic autocorrelation

$$R_k \triangleq \sum_{n=0}^{N-1} a_{(n+k) \bmod N} \, a_n^*$$

is exactly zero for $k \neq iN$, $i = 0, \pm 1, \pm 2, \ldots$. As sequences with this property can be generated for any period $N$ by using constant-amplitude (i.e., purely phase-modulated) signals, they have been called *constant-amplitude, zero-autocorrelation* (CAZAC) sequences. One such sequence with $N = 8$ is obtained by repeating periodically the 4-PSK signals $\pi/4$, $7\pi/4$, $3\pi/4$, $3\pi/4$, $\pi/4$, $3\pi/4$, $3\pi/4$, $7\pi/4$.

## 8.6.  Non-MSE criteria for equalization

The previous sections were devoted to the analysis of equalizers based on a given structure, the TDL, and a given optimization criterion, the minimization of MSE. Although this combination has proved most fruitful in applications, it is by no means the only one, and considerable effort has been spent to devise and analyze different equalization criteria and/or structures. Hereafter we shall review some of the most significant solutions obtained in this framework. In particular, in this section we shall describe two non-MSE criteria for TDL equalization. Other criteria will be described a little later, in the context of blind equalization, and another equalizer structure will be discussed in next section. Notice also that a most sensible optimization criterion would be the minimization of error probability: however, this would result in nonlinear equations, exceedingly more difficult to solve than the linear equations arising from the minimum-MSE criterion.

### 8.6.1.  Zero-forcing equalization

The first approach to automatic equalization assumed a peak-distortion criterion. Peak distortion can be derived from the eye pattern of the received signal, and is closely related to the worst-case bound to error probability (see Section 7.2). Consider for simplicity that a binary ($\pm 1$) stream of independent symbols is transmitted. Further, denote by $(h_\ell)$ the impulse response of the discrete-time system that responds to the source sequence $(a_\ell)$ with the equalized sequence $(y_\ell)$. In other words, $(h_\ell)$ is the response of the discrete-time equalized channel. We define the normalized peak distortion as

$$\mathcal{D}(\mathbf{c}) \triangleq \frac{1}{h_0} \sum_{k \neq 0} |h_k| \tag{8.89}$$

It is assumed that $h_0 = \max_k |h_k|$. In words, $\mathcal{D}(\mathbf{c}) \cdot h_0$ represents the maximum value of the intersymbol interference (ISI) affecting the equalized signal. $\mathcal{D}(\mathbf{c}) = 0$ means that there is no ISI, whereas $\mathcal{D}(\mathbf{c}) \geq 1$ denotes that the eye pattern is completely closed (hence reliable transmission is impossible, irrespective of the noise power level).

If the tap-weight vector $\mathbf{c}$ is chosen so as to minimize $\mathcal{D}(\mathbf{c})$, it can be adjusted by using an iterative algorithm (Lucky, 1965 and 1966) that is guaranteed to converge whenever the unequalized eye is open. This gives the *zero-forcing* algorithm, so called because it forces the ISI to zero. Now, the equalized channel satisfies the zero-ISI condition, or Nyquist's criterion, if the equalizer frequency response is (apart from an unessential linear-phase factor) $TQ_{\text{eq}}^{-1}(f)$, the inverse of the aliased frequency response of the channel seen by the equalizer (a finite-length zero-forcing equalizer simply approximates this inverse). Since this criterion neglects the effects of the noise, it might excessively enhance noise at frequencies where $|Q_{\text{eq}}(f)|$ takes on small values. Nevertheless, due to their simplicity, zero-forcing equalizers were the first incorporated in commercial modems.

## 8.6.2.  Least-squares algorithms

More recently, the expansion of data-transmission systems requiring quick setup and response has created the requirement for equalizers in which a short training time is a premium. This occurs, for instance, in multipoint networks, where the tributary terminals may transmit only when polled by the control modem. The messages from the tributary to a control station are often short, and the control modem must adjust its equalizer whenever a message is received. Quickly convergent equalization algorithms have been sought either by modifying the basic gradient algorithm under an MSE criterion, or by devising other performance criteria. The latter approach can be pursued by introducing a *least squares* (LS) criterion, that is, the sequence of cost functions

$$\mathcal{L}(\mathbf{c}^{(n)}) \stackrel{\triangle}{=} \sum_{i=1}^{n} |\mathbf{c}^{(n)\prime} \mathbf{x}_k - a_{k-D}|^2, \qquad 1 \leq n < \infty \qquad (8.90)$$

to be minimized over the tap-weight vector $\mathbf{c}^{(n)}$. In words, a $\mathbf{c}^{(n)}$ is sought that minimizes the sum of the squared errors that would be obtained if $\mathbf{c}^{(n)}$ were used with all the past received-signal samples. Algorithms matched to the cost function (8.90), called *LS algorithms*, have been proved to provide fast convergence as required.

By taking the gradient of (8.90) and setting it equal to the null vector, the following equation for the optimum tap-weight vector is obtained:

$$\mathbf{X}^{(n)}\mathbf{c}_{\text{opt}}^{(n)} = \mathbf{v}^{(n)} \tag{8.91}$$

where

$$\mathbf{X}^{(n)} \stackrel{\triangle}{=} \sum_{k=1}^{n} \mathbf{x}_k^* \mathbf{x}_k' \tag{8.92}$$

and

$$\mathbf{v}^{(n)} \stackrel{\triangle}{=} \sum_{k=1}^{n} a_{k-D}\mathbf{x}_k^* \tag{8.93}$$

It should be observed that, apart from a factor $1/n$, $\mathbf{X}^{(n)}$ and $\mathbf{v}^{(n)}$ are the time-average counterparts of $\mathbf{X}$ and $\mathbf{v}$, as defined respectively in (8.29) and (8.28). Thus, $\mathbf{X}^{(n)}$ and $\mathbf{v}^{(n)}$ can be viewed as estimates of $\mathbf{X}$, $\mathbf{v}$.

Solution of (8.91) by matrix inversion can be complicated by the fact that $\mathbf{X}^{(n)}$, being only an estimate of a correlation matrix, need not be positive definite. Hence, its inverse may not exist. This problem can be circumvented by simply adding to $\mathbf{X}^{(n)}$ a scalar matrix $\delta\mathbf{I}$, where $\delta$ is a positive constant included to ensure that $\mathbf{X}^{(n)}$ is nonsingular for all $n$. Moreover, the cost function (8.90) can be slightly modified to include a feature desirable when the channel is time varying: by introducing a geometric attenuation factor $0 < \lambda < 1$, that is, by introducing the new cost function

$$\mathcal{L}_\lambda(\mathbf{c}^{(n)}) \stackrel{\triangle}{=} \sum_{i=1}^{n} \lambda^{n-k}|\mathbf{c}^{(n)'}\mathbf{x}_k - a_{k-D}|^2, \qquad 1 \le n < \infty \tag{8.94}$$

the present influences the tap-weight update more than the past. In fact, $\lambda$ weights recent samples more heavily, so that $\mathcal{L}_\lambda(\mathbf{c}^{(n)})$ tends to forget the old samples. Thus, slow channel variations with can be tracked. For a time-invariant channel we may choose $\lambda = 1$. In a time-varying environment $\lambda < 1$, its actual value having no influence on the convergence rate but determining the tracking capabilities of the equalizer (Ling and Proakis, 1984).

For the update of $\mathbf{c}^{(n)}$ several algorithms have been proposed. The *Kalman algorithm* (Godard, 1974) assures rapid start-up, but requires a number of calculations proportional to $N^2$, where $N$ is the number of taps in the TDL. A similar algorithm, usually referred to as the *fast Kalman algorithm* (Falconer and Ljung, 1978), improves the Kalman algorithm as it achieves a lower complexity (linear growth with $N$) without performance degradation because it is mathematically equivalent to the latter. These algorithms have been compared by simulation over several channels by Lim and Mueller (1980). Their convergence properties have been proved to be very similar. They require roughly one-third as many iterations as the stochastic-gradient algorithm. The price for this increase in speed is complexity: the fast Kalman algorithm, which has the lowest complexity, requires about 10 times as many multiplications as the stochastic gradient. Notice also that the fast Kalman algorithm may be unstable when implemented digitally with insufficient accuracy (Lim and Mueller, 1980).

## 8.7.   Non-TDL equalizer structures

It is also possible to use equalizer structures that are not transversal filters. One of these alternative structures is obtained by using the *Kalman filter* as an equalizer (Lawrence and Kaufman, 1971). The Kalman filter, a version of the minimum-MSE linear receiver, has a recursive structure. Comprehensive simulation results (Benedetto and Biglieri, 1974) have shown that the performance of this linear filter is not significantly better than that of a TDL of comparable complexity.

Another equalizer structure is based on *lattice filters* (Lim and Mueller, 1980). Among the properties of lattice filters that make them worthy of special attention are their fast convergence and their high insensitivity to round-off errors deriving from finite-precision digital implementation.

Satorius and Pack (1981) have compared the convergence properties of lattice equalizers based on the minimization of MS or LS error with those of an MSE TDL equalizer. By simulation, the LS lattice equalizer is shown to converge in 40 to 50 iterations where the MSE lattice equalizer needs about 120 iterations. These figures are independent of the eigenvalue spread of the matrix **X**. On the other hand, an MSE TDL working with the stochastic-gradient algorithm requires about 600 iterations for its convergence when the eigenvalue spread $\mu_{\max}/\mu_{\min} = 11$, and about 1000 when $\mu_{\max}/\mu_{\min} = 21$. Thus, not only does the lattice equalizer converge faster, but its convergence properties do not depend, to a certain extent, on the channel. The price paid for this improved performance is increased complexity. In fact, the LS lattice equalizer must perform $12N$ multiplications, $11N$ additions, and $3N$ divisions at each step, while the TDL equalizer needs only $2N$ multiplications and $2N$ additions (Schichor, 1982). The LS lattice equalizer needs even more operations than the fast Kalman algorithm mentioned in Section 8.6 (which requires $10N$ multiplications, $9N$ additions, and $2N$ divisions). On the other hand, the fast Kalman algorithm performs much the same as the LS lattice in terms of convergence speed (see Lim and Mueller, 1980).

## 8.8.   Decision-feedback equalization

In this section we examine a class of *nonlinear* equalizers that are especially useful for channels with severe distortion. The basic idea is the following. Let us

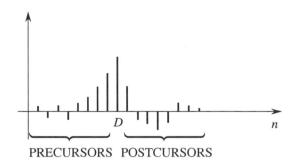

PRECURSORS  POSTCURSORS

Figure 8.9: *Qualitative impulse response of a discrete channel to be equalized.*

observe that, due to propagation delay in the transmission channel, the received signal sample $x_\ell$ is used to make a decision on the symbol that was emitted by the source $D$ discrete-time instants before, say $a_{\ell-D}$. The impulse response $(q_\ell)$ of the discrete channel with input $(a_\ell)$ and output $(x_\ell)$ is sketched qualitatively in Fig. 8.9. The samples $q_\ell$, $\ell < D$, are called the *precursors*, while the samples $q_\ell$, $\ell > D$, are called the *tails*, or *postcursors*, of the impulse response. Assume for a moment that this impulse response and the source symbol $a_{\ell-D}$ are known. Since

$$x_\ell = \sum_{m=-\infty}^{\infty} a_{\ell-m} q_m + n_\ell \tag{8.95}$$

we can subtract the known quantities $a_{\ell-k-D} q_{k+D}$ from the samples $x_{\ell+k}$, $k \neq 0$, thus eliminating all the ISI due to symbol $a_{\ell-D}$. This is the essence of data-aided equalization: if a number of source symbols are correctly detected and the channel impulse response is known, then the ISI can be reconstructed and therefore canceled from the received signal. By implementing this idea when the channel suffers from a large amount of amplitude distortion, we can expect a performance improvement with respect to standard equalization. In fact, this is a situation where an ordinary linear equalizer would considerably enhance the noise, while this data-aided equalizer would not play any role in determining the noise power of the equalizer output. In fact, it will just provide a weighted sum of noise-free symbols to be subtracted from the received symbols. The reader is warned, however, that the assumption of a known transmitted sequence makes the preceding statements only approximately true. Actually, in a real setting there is no hope of canceling completely the ISI without introducing a certain amount of noise enhancement. This is because the minimum distance $d_{\min}$ (see

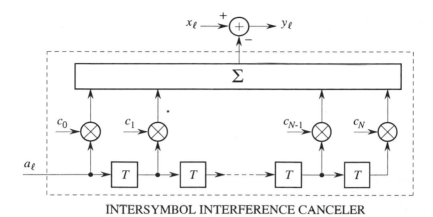

INTERSYMBOL INTERFERENCE CANCELER

Figure 8.10: *Block diagram of an ideal data-aided equalizer.*

Section 7.5), which depends on the source symbol structure and on the channel, imposes a limit to the error performance of *any conceivable receiver*, and hence of any receiver based on data-aided equalization.

The block diagram of an ideal data-aided equalizer is shown in Fig. 8.10. The ISI canceler is a transversal filter with tap weights $\{c_k\}$. Denoting by S the index set $\{0, 1, \ldots, D-1, D+1, \ldots, N\}$, the equalized signal takes the form

$$y_\ell = \sum_{m=-\infty}^{\infty} a_{\ell-m} q_m - \sum_{m \in S} a_{\ell-m} c_m + n_\ell \qquad (8.96)$$

Notice that S does not include the index $D$. This is because we want to restrict the role of the canceler to remove the ISI without altering the useful signal $a_{\ell-D} q_D$. It satisfies intuition, and can be proved, that the canceler weights $\{c_k\}$ that minimize the MSE $\mathcal{E} \triangleq \mathrm{E} |y_\ell - a_{\ell-D}|^2$ are

$$c_k = q_k, \qquad k \in S \qquad (8.97)$$

Consider now the practical implementation of a data-aided equalizer. Since the source symbol sequence $(a_\ell)$ appearing in Fig. 8.10 is not available at the receiver, it must be estimated from the received samples. Thus, it must be assumed that, after a suitable training period, the equalizer yields an error rate so low that near-perfect detection is not an unrealistic assumption. Also, as the channel is now known in advance (and can vary with time), the canceler will have adaptively-varying coefficients.

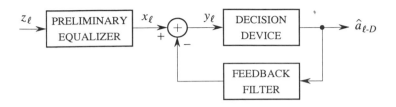

Figure 8.11: *Block diagram of a decision-feedback equalizer.*

The most popular scheme of data-aided equalization is the *decision-feedback equalizer* (DFE). In it, the set S includes only the integers $D + 1, \ldots, N$, so the canceler operates only on the postcursors of the channel impulse response. With this choice for S, we see from (8.96) that at time $\ell$ the source symbols needed for cancellation are $a_{\ell-D-1}, a_{\ell-D-2}, \ldots, a_{\ell-D-N}$. Since at time $\ell$ a decision is taken on $a_{\ell-D}$, the symbols needed can be obtained from the previous decisions.

In conclusion, a DFE is based on a canceler that takes care of the postcursors of the channel impulse response. Now, this cannot be self-sufficient, because the precursors also have to be accounted for. Hence, a preliminary equalizer (also called a *feed-forward filter*) should precede the canceler. Its task may be viewed as the elimination of the precursors, i.e., a function complementary to the canceler. Since the feed-forward filter need not approximate the inverse of the channel transfer function, excessive noise enhancement can be avoided.

The DFE scheme is presented in Fig. 8.11. In it, as is customary, the canceler is referred to as the *feedback filter*. Notice that the inclusion of the decision device into a loop renders this equalizer intrinsically nonlinear (as generally in all data-aided structures).

To analyze the DFE behavior, assume that the preliminary equalizer and the feedback filter are TDL with lengths $N$ and $N'$ and weight vectors c and f, respectively. Thus, the signal sample at the input of the decision device can be written in the form

$$y_\ell = \mathbf{c}'\mathbf{z}_\ell - \mathbf{f}'\widehat{\mathbf{a}}_\ell \qquad (8.98)$$

where $\mathbf{z}_\ell \overset{\triangle}{=} [z_\ell, z_{\ell-1}, \ldots, z_{l-N+1}]'$ and $\widehat{\mathbf{a}}_\ell \overset{\triangle}{=} [\hat{a}_\ell, \hat{a}_{\ell-1}, \ldots, \hat{a}_{l-N'+1}]'$. By defining the two column vectors, with $N + N'$ components each,

$$\mathbf{b} \overset{\triangle}{=} \begin{bmatrix} \mathbf{c} \\ \mathbf{f} \end{bmatrix} \qquad (8.99)$$

and

$$\mathbf{u}_\ell \triangleq \begin{bmatrix} \mathbf{z}_\ell \\ -\hat{\mathbf{a}}_\ell \end{bmatrix} \qquad (8.100)$$

Eq. (8.98) can be rewritten in the more compact form

$$y_\ell = \mathbf{b}'\mathbf{u}_\ell \qquad (8.101)$$

Since (8.101) bears a close resemblance to (8.24), it is not difficult to dupli-cate the arguments of Section 8.2 to find the vector **b**, and hence **c** and **f**, that minimize the MSE, and to devise a gradient algorithm for minimization. (Min-imization of MSE is of course not the only design criterion. For example, a zero-forcing criterion is also applicable (see Price, 1972).) By assuming $\hat{a}_\ell = a_\ell$ for all $\ell$, we obtain

$$\mathrm{E}\{|y_\ell - a_{\ell-D}|^2\} = \mathbf{b}^\dagger \mathbf{U}\mathbf{b} - 2\Re[\mathbf{b}^\dagger \mathbf{w}] + \sigma_a^2 \qquad (8.102)$$

where

$$\mathbf{U} \triangleq \mathrm{E}[u_\ell^* u_\ell'] \qquad (8.103)$$

and

$$\mathbf{w} \triangleq \mathrm{E}[a_{\ell-D} u_\ell^*] \qquad (8.104)$$

By taking the gradient of (8.102) with respect to **b** and setting it equal to the null vector, we obtain the following equation for the optimum tap-weight vector $\mathbf{b}_{\mathrm{opt}}$:

$$\mathbf{U}\,\mathbf{b}_{\mathrm{opt}} = \mathbf{w} \qquad (8.105)$$

Since **U** is not necessarily positive definite, a form for $\mathbf{b}_{\mathrm{opt}}$ similar to (8.32) may not be available. However, a gradient algorithm can be displayed:

$$\mathbf{b}^{(\ell+1)} = \mathbf{b}^{(\ell)} - \alpha \mathrm{E}[(y_\ell - a_{\ell-D})\mathbf{u}_\ell^*] \qquad (8.106)$$

which converges to a tap-weight vector achieving the minimum MSE.

By taking the limit as $N$ and $N'$ both tend to infinity, an expression can be obtained for the minimum achievable MSE (Salz, 1973):

$$[\mathcal{E}_{\min}]_{\mathrm{DFE}} = \exp\left\{ T \int_{-1/2T}^{1/2T} \ln \frac{\sigma_a^2 G_n(f)}{G_n(f) + (\sigma_a^2/T)|Q_{\mathrm{eq}}(f)|^2}\, df \right\} \qquad (8.107)$$

By comparing (8.107) with (8.53), the analogous expression for the infinite-length linear TDL equalizer, one can prove that $[\mathcal{E}_{\min}]_{\mathrm{DFE}}$ is always less than or equal to the value (8.53). The equality holds only when $|Q_{\mathrm{eq}}(f)|$ is a constant and the noise is white: this fact suggests that the DFE equalizer outperforms the linear equalizer especially when the channel is far from flat, i.e., it causes a large distortion. This is true asymptotically, and in the absence of decision errors. Unfortunately, however, there is no definite answer to the question of whether a finite-length DFE achieves a lower MSE than a linear equalizer with the same overall number of taps. In fact, the relative performance of the two equalizers depends on the actual channel characteristics, on the number of taps, and on the choice of the delay $D$ (Qureshi, 1982). Simulation results (Salz, 1973) confirm what is intuitively expected: the DFE is markedly superior to the linear equalizer with the same finite length when operating on channels with spectral nulls in the Nyquist interval. This is the situation where the linear equalizer suffers most from noise enhancement. In addition, the DFE performance is less sensitive to the sampling time (Qureshi, 1982).

Finally, it must be observed that our preceding discussion was based on the assumption that $\hat{a}_\ell = a_\ell$ for all $\ell$, i.e., that the decision process was error-free. Now it is reasonable to ask to what extent the DFE performance is degraded by decision errors. Decision errors tend to propagate

because they produce wrong cancellation of tails. In fact, when the feedback filter is fed by a wrong decision its output reflects this error during the next few time instants. This causes a reduced noise margin for future decisions. In turn, this entails a higher probability of future incorrect decisions, and so on. Simulation results show that this error propagation is not catastrophic: in fact, on typical channels errors tend to cluster in bursts short enough to only slightly degrade performance.

## 8.9.   Blind equalization

The need for an initial training period for the equalizer, which is detrimental for data rate, can be avoided by using *blind* equalization techniques. They aim at providing equalizer convergence without burdening the transmitter with a training overhead. An obvious way of doing this is to use blind identification as described earlier in this chapter, and exploit the channel information thus obtained to set the equalizer to its optimum value. This procedure can be repeated as often as needed in order to track the channel variations.

Another approach to blind equalization is based on a non-MSE criterion: as we shall see, a proper choice of the distortion function to be minimized yields a tap-weight setting algorithm that does not depend on the transmitted data.

The following assumptions on the source symbols are made: the RV $a_n$ are complex and iid. Moreover,

$$\mathrm{E}[a_n^2] = 0 \tag{8.108}$$

(this occurs, for example, when the real and imaginary parts of $a_n$ are uncorrelated and have the same variance). Also, define

$$m_2 \overset{\triangle}{=} E\{|a_n|^2\} \tag{8.109}$$

and

$$m_4 \overset{\triangle}{=} E\{|a_n|^4\} \tag{8.110}$$

Condition (8.108) implies certain symmetries in the signal constellation used by the digital modulator. For example, binary PSK and one-dimensional modulations are excluded.

### 8.9.1.   Constant-modulus algorithm

The idea here is to force the equalizer weights to maintain a constant envelope on the received signal. Under the assumption

$$2m_2^2 > m_4 \tag{8.111}$$

the distortion criterion, *independent of* $a_n$, is then

$$\mathcal{E}(\mathbf{c}) \overset{\triangle}{=} E\left[|y_n|^2 - \frac{m_4}{m_2}\right]^2 \tag{8.112}$$

If we denote by $(h_n)$, as usual in this chapter, the impulse response of the discrete channel extending from the source to the equalizer output, so that

$$y_n = \sum_{k=-\infty}^{\infty} a_{n-k} h_k \tag{8.113}$$

we have, after some algebra,

$$\mathcal{E}(\mathbf{c}) = (m_4 - m_2^2) \sum_{k=-\infty}^{\infty} |h_k|^2$$
$$+ 2m_2^2 \left[\sum_{k=-\infty}^{\infty} |h_k|^2\right]^2 - 2m_4 \sum_{k=-\infty}^{\infty} |h_k|^2 + \frac{m_4^2}{m_2^2} \tag{8.114}$$

Now, (8.114) can be rewritten in the form

$$\mathcal{E}(\mathbf{c}) = m_4(1 - |h_0|^2)^2 + m_4 \sum_{k\neq 0} |h_k|^4$$
$$+ 2m_2^2 \left[\left(\sum_{k\neq 0} |h_k|^2\right)^2 - \sum_{k\neq 0} |h_k|^4\right]$$
$$+ [4m_2^2|h_0|^2 - 2m_4] \sum_{k\neq 0} |h_k|^2 + \frac{m_4^2}{m_2^2} - m_4 \tag{8.115}$$

We have derived (8.115) to clarify the choice of (8.112) for the distortion function. Actually, we prove that this choice leads approximately to the same result as the distortion

$$\mathcal{E}'(\mathbf{c}) \triangleq \mathrm{E}[|y_n|^2 - |a_n|^2]^2, \tag{8.116}$$

which appears at a first glance to be sensible for constant-envelope modulations, i.e., constant $|a_n|^2$. Expanding (8.116), we obtain

$$\mathcal{E}'(\mathbf{c}) = m_4(1 - |h_0|^2)^2 + m_4 \sum_{k \neq 0} |h_k|^4 \tag{8.117}$$

$$+2m_2^2 \left[ \left( \sum_{k \neq 0} |h_k|^2 \right)^2 - \sum_{k \neq 0} |k_k|^4 \right] + [4m_2^2|h_0|^2 - 2m_2^2] \sum_{k \neq 0} |h_k|^2$$

Now, (8.117) has a minimum when $|h_0|^2$ is close to unity and the ISI samples $h_k$, $k \neq 0$, are small in magnitude. Moreover, comparison of (8.115) and (8.117) shows that

$$\mathcal{E}(\mathbf{c}) - \mathcal{E}'(\mathbf{c}) = \frac{m_4}{m_2} - m_4 - 2[m_2^2 - m_4] \sum_{k \neq 0} |h_k|^2 \tag{8.118}$$

The difference (8.118) is almost independent of $(h_n)$, and hence of $\mathbf{c}$, when the distortion term is small enough. So it can be expected that minimizing $\mathcal{E}(\mathbf{c})$ will also provide a minimum for $\mathcal{E}'(\mathbf{c})$. The condition for this to be true is that in (8.115) the term $[4m_2^2|h_0|^2 - 2m_4]$ does not become negative near the minimum; but this is assured, because (8.111) is assumed, and near a minimum $|h_0|^2 \approx 1$.

A stochastic-gradient algorithm can now be exhibited. Since $y_n = \mathbf{c}'\mathbf{x}_n$ as usual, the gradient of $\mathcal{E}(\mathbf{c})$ taken with respect to the tap-weight vector $\mathbf{c}$ is

$$\nabla \mathcal{E}(\mathbf{c}) = 4\mathrm{E}\left[y_n \left(|y_n|^2 - \frac{m_4}{m_2}\right) \mathbf{x}_n^*\right] \tag{8.119}$$

so that the following stochastic-gradient algorithm can be used:

$$\mathbf{c}^{(n+1)} = \mathbf{c}^{(n)} - \alpha y_n \left(|y_n|^2 - \frac{m_4}{m_2}\right) \mathbf{x}_n^* \tag{8.120}$$

Comparison of (8.120) with (8.69) shows that the stochastic-gradient term does not depend on the symbols $a_n$, as required. Also, it can be verified that when $\mathbf{c}'\mathbf{x}_n = a_n$ (i.e., perfect equalization is achieved) the gradient (8.119) is the null vector. Incidentally, this result would not hold if in lieu of $m_4/m_2$ another constant were selected in the brackets of (8.112). As a conclusion, (8.120) offers over (8.69) the significant advantage of not requiring a training sequence. This comes at the cost of a lower convergence speed.

### 8.9.2. Shalvi-Weinstein algorithm

A different distortion function was suggested by Shalvi and Weinstein (1990). Under the assumptions that the kurtosis of the data, i.e.,

$$K(a_n) \triangleq \mathrm{E}\{|a_n|^4\} - 2\,\mathrm{E}^2\{|a_n|^2\} - |\mathrm{E}\{a_n^2\}|^2 = m_4 - 2m_2^2$$

is nonzero, and that the discrete Fourier transform of the sampled channel impulse response has no zeros, i.e.,

$$\sum_n q_n e^{-jn\omega} \neq 0, \qquad 0 < \omega \leq 2\pi,$$

a necessary and sufficient condition for the channel to be perfectly equalized is that $\mathrm{E}\{|y_n|^2\} = m_2$ and $\mathrm{E}\{|y_n|^4\} = m_4$.

The proof goes as follows. From the channel input-output relation

$$y_n = \sum_k a_{n-k} h_k$$

we have, by invoking the iid assumption of the $a_n$:

$$\mathrm{E}\{|y_n|^2\} = \mathrm{E}\{|a_n|^2\} \sum_k |h_k|^2 \quad \text{and} \quad \mathrm{E}\{y_n^2\} = 0 \qquad (8.121)$$

Moreover, by recalling definitions (8.109) and (8.110) and observing that

$$\mathrm{E}\{a_{n-i}a_{n-j}^* a_{n-k} a_{n-\ell}^*\} = \begin{cases} m_4, & i = j = k = \ell \\ m_2^2, & i = j \neq k = \ell,\ i = \ell \neq j = k \\ 0, & \text{otherwise} \end{cases}$$

we have

$$\mathrm{E}\{|y_n|^4\} = m_4 \sum_k |h_k|^4 + 2m_2^2 \left[ \left( \sum_k |h_k|^2 \right)^2 - \sum_k |h_k|^4 \right] \qquad (8.122)$$

Thus we obtain

$$\mathrm{E}\{|y_n|^4\} = m_4 \sum_k |h_k|^4 \qquad (8.123)$$

Now, observe that in general

$$\sum_k |h_k|^4 \leq \left( \sum_k |h_k|^2 \right)^2$$

with equality if and only if there is at most one nonzero term in the summations. Thus, if $\mathrm{E}\{|y_n|^2\} = \mathrm{E}\{|a_n|^2\}$ then from (8.121) we have $\sum_k |h_k|^2 = 1$. Hence,

$\sum_k |h_k|^4 \leq 1$, and $\sum_k |h_k|^4 = 1$ if and only if the impulse response has only one nonzero component of magnitude 1. Thus, we have obtained that if $E\{|y_n|^2\} = m_2$ then $E\{|y_n|^4\} \leq m_4$, with equality if and only if the impulse response of the equalized channel has only one nonzero component of magnitude 1. But this is tantamount to saying that the channel is perfectly equalized.

The above result suggests the following equalization criterion:

$$\text{maximize } E\{|y_n|^4\} \quad \text{subject to} \quad E\{|y_n|^2\} = E\{|a_n|^2\} \qquad (8.124)$$

An interesting feature of this criterion, that we shall not prove here, is that $E\{|y_n|^4\}$ has a single global minimum over $E\{|y_n|^2\} = E\{|a_n|^2\}$, and hence, unlike with the constant-modulus criterion described above, a gradient algorithm is expected to converge to the desired response regardless of initialization.

To apply this criterion, we need an explicit expression for the gradient of $\mathcal{E}(\mathbf{c}) = E\{|y_n|^4\}$. We have

$$\nabla \mathcal{E}(\mathbf{c}) = 4E\{|y_n|^2 y_n \mathbf{x}_n^*\} \qquad (8.125)$$

and hence the stochastic-gradient algorithm

$$\mathbf{c}^{(n+1)} = \mathbf{c}^{(n)} + \alpha |y_n|^2 y_n \mathbf{x}_n^* \qquad (8.126)$$

along with the normalization step

$$\mathbf{c}_0^{(n+1)} = \mathbf{c}^{(n+1)} / \| \mathbf{c}^{(n+1)} \|$$

### 8.9.3.  Stop-and-go algorithm

A different approach to blind equalization leads to the "stop-and-go" algorithm of Picchi and Prati (1987). This algorithm aims at retaining the simplicity of a conventional linear or DFE equalizer working in a decision-directed mode, while endowing it with blind convergence properties. The basic idea here is to stop adaptation whenever the reliability of the error signal is not high enough. More precisely, a binary-valued flag is generated telling the equalizer whether its current decision may be used to generate a reliable error signal: if not, adaptation is stopped for that iteration.

## 8.10.   More on complex equalizers

Throughout this chapter, extensive use has been made of complex notations to denote samples of bandpass signals and channel responses. As a special case,

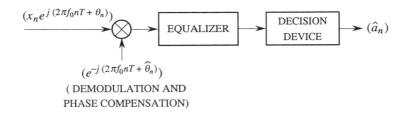

Figure 8.12: *Discrete-time model of a receiver in which equalization is performed after demodulation and carrier-phase compensation.*

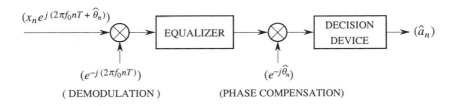

Figure 8.13: *Discrete-time model of a receiver in which equalization is performed after demodulation and before carrier-phase compensation.*

baseband signals and channels can be handled with obvious changes in the equations. In this section we expand briefly on certain features of two-dimensional equalization.

Hitherto we have implicitly assumed that the carrier phase for demodulation has been properly estimated. This estimate can be performed in a decision-directed mode with the receiver arrangement of Fig. 8.12. In it, equalization is performed after coherent demodulation and inside the loop for decision-directed phase compensation (Matyas and McLane, 1974). With this arrangement, as the equalizer itself introduces a many-symbol-interval delay between input and output, the estimated phase sequence $(\hat{\theta}_n)$ is a delayed version of the true phase sequence $(\theta_n)$. This delay prevents the receiver from correctly compensating any time-varying phase shift introduced by the channel. To avoid this impairment source, two different receiver structures can be used (Falconer, 1976). The first one, shown in Fig. 8.13, places phase compensation after the equalizer, while demodulation is performed using a free-running oscillator before the equalizer.

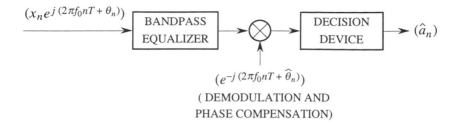

Figure 8.14: *Discrete-time model of a receiver in which equalization is performed before demodulation and carrier-phase compensation.*

The second one (Fig. 8.14) places both demodulation and phase compensation after the equalizer. The latter now operates on bandpass samples. Hence, it is referred to as a *bandpass equalizer*. If equal phase compensation is assumed for both schemes, mathematically these structures are exactly equivalent.

A feature of these two structures worth observing here is the *tap-rotation property* (Gitlin, Ho, and Mazo, 1973). Assume that the equalizer-input samples are rotated by an angle $\varphi$. Then the vector $\mathbf{x}_n$ defined in (8.26) is changed into $\mathbf{x}_n e^{j\varphi}$. Consider the effect of this phase rotation on the operation of a linear equalizer. From definitions (8.28)–(8.29) it can be easily seen that $\mathbf{X}$ is left unchanged, whereas $\mathbf{v}$ is changed into $\mathbf{v}e^{-j\varphi}$. In turn, this implies that the optimum tap-weight vector (8.32) is changed into $\mathbf{c}_{\text{opt}}e^{-j\varphi}$ (i.e., the equalizer taps are rotated by $-\varphi$), the net result being that when the equalizer has settled at its optimum the equalized output samples will not be affected by the phase rotation.

Furthermore, consider the effect on the operation of these two equalizer structures of two transmission impairments typical of telephone lines, phase jitter and frequency offset. Phase jitter acts as a real random sequence $(\varphi_n)$ affecting the phase angle of the channel-output samples. When its time constant is much larger than the equalizer settling time, it can be assumed to cause a constant phase shift. Thus, due to the tap-rotation property, it can be compensated for by the equalizer. Similar considerations hold for a frequency offset (i.e., the perturbation of the carrier frequency $f_0$ by a small amount $\Delta f$). In conclusion, the equalizer, due to its tap-rotation property, can track small amounts of phase jitter and frequency offset so that the phase compensation loop implicit in Figs. 8.13 and 8.14 is not required. When phase jitter or frequency offset are significant, this phase compensation loop is needed.

## 8.11.  Tomlinson-Harashima precoding

In this section we describe an equalization scheme that operates at the transmitter side, and hence avoids the noise enhancement caused by linear equalization or the error propagation caused by decision-feedback equalization. This is called Tomlinson-Harashima (TH) precoding, and is in some ways inspired by decision-feedback equalization. There, intersymbol interference (ISI) due to previously detected symbols is subtracted out in the receiver before detection of the current symbol. In TH precoding, a similar effect *is achieved by operating on the transmitter side*: the source symbols, instead of being directly sent through the channel, are first pre-equalized to counter the distortion that will be introduced there. In general, a linear pre-equalizer would exhibit the effect of boosting the transmitted power, which is highly undesired for channels with a constraint on the value of the latter. The idea in TH precoding is to introduce a nonlinear operation that prevents the transmitted power from increasing. This precoding scheme is easy to implement, and can be used in conjunction with coded modulation (to be described in Chapter 12).

Assume from now on that the impulse response $(q_k)_{k=0}^{\infty}$ of the discrete channel is causal, with $q_0$ having the largest magnitude, and is known at the transmitter. This knowledge can be achieved by sending a training sequence during a startup phase, and subsequently relaying the received sequence to the transmitter. Assume further that $q_0 = 1$.

Let the channel symbols be denoted $a_k$, and the source symbols $b_k$. Since the channel is assumed to be known at the transmitter, then a linear "zero-forcing" transmission filter, i.e., one obeying

$$a_k = b_k - \sum_{i \geq 1} q_i b_{k-i}$$

would cause complete elimination of the ISI (or of its "postcursors," if the channel were not causal). Now, the difficulty with this solution is that the transmitted power may be increased, and can even be unbounded if the zero-forcing filter is unstable. To avoid this increase, a nonlinear operation is introduced so that the values taken by $a_k$ are forced to be in approximately the same range as those of the source output $b_k$.

At each discrete time $k$, given the past channel symbols $a_{k-i}$, $i \geq 1$, and the actual source symbol $b_k$, a symbol $f_k$ is first determined by subtracting from $b_k$ the ISI due to the tail of the channel impulse response, i.e.,

$$f_k = b_k - \sum_{i \geq 1} q_i a_{k-i} = b_k - \sum_{i} q_i a_{k-i} - a_k \qquad (8.127)$$

The new transmitted symbol $a_k$ is then obtained by reducing the coordinates of $f_k$. If the modulation is $M$-ary PAM, with $b_k = \pm 1, \pm 3, \ldots, \pm(M-1)$,

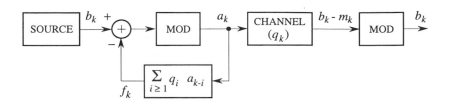

Figure 8.15: *Precoding and decoding with Tomlinson-Harashima equalization.*

then $a_k$ is reduced to the interval $(-M, M]$ by a nonlinear transformation that leaves it invariant if it takes a value in this interval, otherwise it subtracts from it an appropriate $m_k$, an integer multiple of $2M$. Similarly, if the modulation is QAM with real and imaginary parts of $b_k$ taking on $M$ values each, then $a_k$ is reduced to the square $(-M, M]^2$ by subtracting an appropriate $m_k$ in $2M\mathbf{Z}^2$. This is equivalent to a mod-$2M$ reduction of $f_k$, or of its real and imaginary part if $f_k$ is a complex number. Notice that, while the source symbols $b_k$ can take on either $M$ or $M^2$ values, the precoded symbols $a_k$ may take on many more values.

Thus,

$$a_k = f_k - m_k$$

and hence, because of (8.127),

$$a_k = b_k - \sum_i q_i a_{k-i} + a_k - m_k$$

which implies that

$$\sum_i q_i a_{k-i} = b_k - m_k \qquad (8.128)$$

The latter equation can be interpreted by observing that its LHS, i.e., the channel output, consists of $b_k$ (the source symbol) minus a correction term. Since $m_k \in 2M\mathbf{Z}$ (PAM) or $m_k \in 2M\mathbf{Z}^2$ (QAM), the channel outputs take values in an extended signal constellation, whose exact range is channel-dependent. Before decision, the channel outputs are reduced to the source-symbol range by reducing them mod-$2M$. This is equivalent to saying that the source signal estimates can be recovered from the channel output by a memoryless operation that folds the latter into the region $(-M, M]$ (for one-dimensional modulations) or to the square $(-M, M]^2$ (for two-dimensional modulations). If the source symbols $b_k$ form a PAM sequence, the memoryless operation is simply a slicer. Fig. 8.15 summarizes the precoding/decoding scheme resulting from Tomlinson-Harashima equalization.

Finally, observe that if the precoder is forced to choose the all-zero precoding sequence, that is, if $m_k = 0$ for all $k$, then $f_k = a_k$, and the transmitted sequence is simply

$$a_k = \sum_i g_{k-i} b_i$$

where $(g_k)$ is the impulse response of the formal channel inverse: this is still causal, with $g_0 = 1$. The last equation is interpreted by observing that linear zero forcing in the transmitter is equivalent to precoding with $m_k = 0$.

We should also observe a "data-flipping" effect caused in the receiver by the nonlinear operation. Consider, for simplicity's sake, binary transmission with symbols $\pm 1$. Suppose that $a_k = +1$ is transmitted through the channel, and that the noise takes the signal out of the $[-2, 2)$ region, causing for example the value 2.1 to be received. Without slicing, the decision would still be correct, while the mod-4 operation produces the value $2.1 - 4 = -1.9$, and hence an error is made by the decision device.

## Transmitted power with TH precoding

We examine now the average power transmitted with TH precoding. We may confine ourselves to consideration of PAM only: since in QAM the signals are transmitted over orthogonal carriers, the total power is obtained by summing the powers of the individual in-phase and quadrature parts.

We have the following result, that we shall prove later on. The average power of the transmitted symbols $a_k$ is bounded above and below by

$$\frac{M^2 - 1}{3} \leq \mathrm{E}[a_k^2] \leq \frac{M^2 - 1}{3} + 1 \tag{8.129}$$

while, as we know from (5.12),

$$\mathrm{E}[b_k^2] = \frac{M^2 - 1}{3} \tag{8.130}$$

By comparing the last two equations we see that for a binary system the maximum power penalty is 3 dB, which reduces to 0.8 dB for a quaternary system and tends to zero as $M$ increases.

We now prove (8.129). Let us define the function $\mu[\cdot]$ that reduces its argument mod $2M$. Then the average transmitted power can be written in the form (we drop the time index for simplicity):

$$\mathrm{E}[a^2] = \mathrm{E}_b \mathrm{E}_\Theta[\mu^2(b + \Theta)] \tag{8.131}$$

where $\Theta$ is a random variable representing the term added to the source symbol $b$ in (8.127). Since the calculation of the pdf of $\Theta$ is a difficult task, we choose to look for the pdf's that yield a maximum and a minimum value of $\mathrm{E}[a^2]$.

The key to the solution of this problem is the observation that the RV $\Theta$ may be restricted, without loss of generality, to the interval $(-1, +1]$. In fact, with this restriction the RV $\mu(b + \Theta)$ takes on values, with equal probabilities, in

$$-(M - 1) + \Theta, \ -(M - 1) + \Theta + 2, \ \ldots, (M - 1) + \Theta$$

On the other hand, if $\Theta$ takes values in the interval $(-1, +1] + 2$, then $\mu(w + \Theta)$ takes on values, with equal probabilities, in

$$-(M - 1) + \Theta + 2, \ -(M - 1) + \Theta + 4, \ \ldots, \mu[(M - 1) + \Theta + 2]$$

Since
$$\mu[(M - 1) + \Theta + 2] = -(M - 1) + \Theta$$

then $\mu(b + \Theta)$ has the same distribution of $\mu(b + \Theta + 2)$, and, by the same argument, of $\mu(b + \Theta + 2j)$, $j$ any integer.

Now, pick a pair $\pm B$ of values taken by $b$. With the restriction of $\Theta$ to the interval $(-1, 1]$ the contribution of this pair to $E[a_k^2]$ is proportional to

$$
\begin{aligned}
E_\Theta[\mu^2(\Theta + B)] + E_\Theta[\mu^2(\Theta - B)] &= E_\Theta[(\Theta + B)^2] + E_\Theta[(\Theta - B)^2] \\
&= 2E[\Theta^2] + 2B^2
\end{aligned}
$$

Now, for $\Theta$ taking values in $(-1, 1]$, and independently of $B$, the last quantity is minimized by choosing a pdf that assigns to $\Theta$ the value 0 with probability 1, and maximized by choosing a pdf that assigns to $\Theta$ the value 1 with probability 1. Substitution of these pdf's in (8.131) yields the desired bounds (8.129).

## 8.12.  Bibliographical notes

The scheme of Fig. 8.1 was first analyzed by Magee and Proakis (1973) and Proakis (1974). Prior to the mid-1960s, considerable research effort was directed to the specification of digital receivers for channels affected by ISI. The structure of an adaptive receiver based on the TDL, and an iterative optimization technique for adjusting its tap weights, eventually emerged from this work. The history of the TDL filter as an equalizer dates back to Nyquist (1928). The fundamental ideas on which automatic equalization is based were not unknown in the mid-1960s: see, e.g., Goldenberg and Klovsky (1959), a paper that one of the authors claims to be the first to describe time-domain adaptive equalization (Klovsky and Nikolaev, 1978, p. 40); or the papers by Kettel (1961, 1964). However, there is no doubt that it was R. W. Lucky's early work (Lucky, 1965, 1966) that provided the major breakthrough in the problem of equalizing intersymbol-interference channels. The TDL equalizer based on an MSE criterion was first analyzed by

Proakis and Miller (1969) and Gersho (1969). Its convergence properties in the training mode were studied by, among others, Ungerboeck (1972), Mazo (1979), and Gardner (1984) using the "independence assumption." Our treatment of the subject is based on the latter paper, which contains the most comprehensive convergence analysis known to the authors. Convergence analysis of the MSE TDL equalizer working in the tracking mode with the stochastic-gradient algorithm is more complicated. For details, the reader is referred to Macchi and Eweda (1984) and to the references therein.

The interest on developing adaptive algorithms with fast convergence is more recent. Chang (1971) and Gitlin and Magee (1977) approached this problem by transforming the equalizer input sequence in order to decorrelate its samples. Godard (1974) obtained an adaptive algorithm that has the structure of a Kalman filter and exhibits a particularly fast convergence. Gitlin and Magee (1977) showed that Godard's algorithm owes its convergence properties to its capability of decorrelating the equalizer inputs. The "fast Kalman algorithm," having the same convergence properties as Godard's, but requiring a lower complexity, was proposed by Falconer and Ljung (1978). Later, attention was attracted by lattice filters. As lattice filters perform Gram-Schmidt orthogonalization on their input sequence, their application to the fast-converging algorithm problem is natural. An overview of the properties and applications of lattice filters can be found in Makhoul (1978), Friedlander (1982), and in the book by Honig and Messerschmitt (1984). Application of lattice filters to adaptive equalization was first suggested by Satorius and Alexander (1979) and Makhoul (1978). A problem with adaptive lattice filters is that their outputs are uncorrelated only after the adaptation algorithm has reached the steady state. Thus, the equalizer convergence may not be as fast as with Godard's algorithm. An adaptive lattice algorithm whose outputs are uncorrelated at any time was discovered by Morf (1977) (see also Morf *et al.*, 1977) and applied to equalization by Schichor (1982) and Satorius and Pack (1981). Complex adaptive lattice structures are examined in Symons (1979). The Kalman, fast Kalman, and adaptive lattice algorithms are extended to complex fractionally-spaced equalizers by Muller (1981).

The idea of fractionally-spaced TDL equalizers dates back to an unpublished 1969 paper by Gersho, and was rediscovered a few years later (see Guidoux, 1975; Macchi and Guidoux, 1975). Ungerboeck (1976), Qureshi and Forney (1977), and Gitlin and Weinstein (1981) analyze their performance and convergence properties.

The idea of using past decisions to cancel intersymbol interference, and hence the concept of decision-feedback equalization, was introduced by Austin (1967). An overview of the work done in this area before 1978 is contained in Belfiore and Park (1979), where the derivations are based on results of linear prediction and estimation theory. A unified theory of data-aided equalization is provided in the paper by Mueller and Salz (1981). Our treatment of the subject is based on it.

Blind equalization goes back to the pioneering work by Sato (1975). The first approach to blind equalization based on the introduction of a new distortion criterion, different from MSE, was due to Sato (1975) and Godard (1980). The presentation of the constant-modulus algorithm here follows Godard (1980). For an analysis and extension of Sato's method see Benveniste and Goursat (1984), and for an analysis of Godard's method see Foschini (1985). Further work in this area was done by Picchi and Prati (1987), Shalvi and Weinstein (1990) (see also the set of references in this paper), Shalvi and Weinstein (1994), Johnson (1991), Tong, Xu, and Kailath (1994), and Tong, Xu, Hassibi, and Kailath (1995).

Because, in practice, adaptive equalizers are implemented digitally, their parameters, as well as the signal samples, are quantized to a finite number of levels. The effects of digital implementation of the TDL equalizer are examined in Gitlin, Mazo, and Taylor (1973) and Gitlin and Weinstein (1979).

Proakis (1991) provides a survey of adaptive equalization techniques for time-division multiple access digital mobile radio systems. The development of adaptive equalization up to 1984 is summarized in Qureshi (1985), where an extensive list of references can be found.

## 8.13.  Problems

*Problems marked with an asterisk should be solved with the aid of a computer.*

**8.1** Consider mean-square error channel identification when the source symbols $a_\ell$ are not uncorrelated with mean zero. By defining their correlation matrix $\mathbf{A} = \mathrm{E}[a_\ell^* a_\ell']$, and assuming $\mathbf{A}$ to be positive definite (which corresponds to assuming that the $a_\ell$ are linearly independent), show that the minimum achievable mean-square error corresponds to perfect identification.

**8.2** Consider the channel identification problem of Section 8.2. Assume that the TDL used to identify the channel has $\widehat{L}$ delay elements, while the channel memory span is $L > \widehat{L}$. This situation can be dealt with by writing, in lieu of (8.4),

$$x_\ell = \mathbf{q}' \mathbf{a}_\ell + n_\ell$$

and

$$\hat{x}_\ell = \widehat{\mathbf{q}}_0' \mathbf{a}_\ell$$

where

$$\mathbf{q} \triangleq \begin{bmatrix} \mathbf{q}_a \\ \mathbf{q}_b \end{bmatrix}, \qquad \widehat{\mathbf{q}}_0 \triangleq \begin{bmatrix} \widehat{\mathbf{q}} \\ \mathbf{0} \end{bmatrix}$$

and $\widehat{\mathbf{q}}$, $\mathbf{q}_a$ have $\widehat{L} + 1$ components, and $\mathbf{0}$ is the null vector with $L - \widehat{L}$ entries. With these assumptions, derive an expression for the minimum mean-square identification error without the assumption of zero-mean, uncorrelated source symbols. Also, when the source symbols are uncorrelated and have zero mean, show the minimum MSE is achieved for $\widehat{\mathbf{q}} = \mathbf{q}_a$.

**8.3** (*) Consider binary PAM transmission with source symbols $\pm 1$ using a channel bandlimited in the Nyquist interval $(-1/2T, 1/2T)$. The impulse-response samples, taken every T, are 0.833, 1.0, and 0.583. The noise is additive Gaussian, and the receiving filter is an ideal low-pass filter with cutoff frequency $1/2T$. Compute the bit error probability versus $\mathcal{E}_b/N_0$ in the following situations:

(a) Unequalized channel.

(b) Channel equalized by a 5-tap minimum-MSE TDL equalizer (choose the optimal value for the delay $D$).

(c) Same as in (b), with a 7-tap TDL.

(d) Same as in (b), with a 15-tap TDL.

**8.4** Consider a linearly-modulated signal transmitted over a dispersive channel. Assume that an infinitely-long zero-forcing equalizer is used that completely eliminates ISI.

(a) Derive the transfer function of the equalizer.

(b) Derive an expression for the bit error probability of this transmission system that takes into account the noise enhancement introduced by the equalizer.

**8.5** (*) Consider binary PAM transmission with source symbols $\pm 1$ using a channel bandlimited in the Nyquist interval $(-1/2T, 1/2T)$. The channel has a sampled overall impulse response $(h_n)$, the noise is additive Gaussian, and the receiving filter is an ideal low-pass filter with cutoff frequency $1/2T$. Assume that the channel is equalized using a three-tap minimum-MSE equalizer, and compute the bit error probability versus $\mathcal{E}_b/N_0$, with the delay $D$ as a parameter, for these two situations:

(a) $(h_n) = (0.5, 1.0, 0.5)$.

(b) $(h_n) = (1.0, 0.67, 0.45, 0.3, 0.2, 0.135)$.

**8.6** Discuss the Shalvi-Weinstein algorithm in the case where $E\{a_n^2\} = 0$, but the kurtosis is nonzero.

# Carrier and clock synchronization

## 9.1. Introduction

In previous chapters, when computing the performance of a digital communication system, we assumed implicitly that the same clock controlled both the transmitter and the receiver operations. This means that corresponding events in the transmitter and receiver are *synchronous* (i.e., they occur at the same time instants, or at time instants that differ by a fixed and constant delay).

Also, in Chapters 4 and 5, we saw that the most efficient demodulation schemes are *coherent*; they make use of the phase information of the carrier. Optimum demodulation requires then a local carrier at the receiver side whose *frequency* and *phase* are in perfect agreement with that of the transmitted signal. In principle, two pairs of ideal identical oscillators at the transmitter and receiver sides could ensure the synchronization and coherence required for proper operation of the system. In practice, however, the signals emitted by a pair of oscillators with the same nominal frequency will start drifting from each other because of their physical inability to keep the nominal frequency with infinite precision.

A good model, valid for the signals emitted by two independent oscillators with the same nominal frequency $f_0$ synchronized at $t = 0$, is the following:

$$z_1(t) = A_1 \cos[2\pi f_0 t + \theta_1(t)] \tag{9.1}$$

$$z_2(t) = A_2 \cos[2\pi f_0 t + \theta_2(t)] \tag{9.2}$$

where each $\theta_i(t)$ is a Wiener random process with $\theta_i(0) = 0$, zero mean, and variance equal to $t/\tau_i$, $i = 1, 2$. This random process is a nonstationary Gaussian process defined in the interval $(0, \infty)$. Thus, the variance of the random process representing the phase difference between the two oscillators is given by

$$E[\theta_1(t) - \theta_2(t)]^2 = \frac{t}{\tau_1} + \frac{t}{\tau_2} = t\frac{\tau_1 + \tau_2}{\tau_1 \tau_2} \triangleq \frac{t}{\tau_{12}} \qquad (9.3)$$

where we have defined the *joint coherence time* $\tau_{12}$ of the two oscillators as the time required by them to yield a unitary variance of their phase difference. Since the variance (9.3) increases with time, we can conclude that a pair of independent oscillators cannot maintain their synchronization indefinitely. They need to mutually exchange certain informations, that is, to be in some way *locked*. In this chapter, we examine the fundamentals of *clock* and *carrier synchronization* with the aim of clarifying some very basic concepts. In pursuing this scope, we had to make some choices, the most important of which is to treat the subject in the analog (instead than in the digital) domain. The reason is that, in our opinion, the basic synchronization circuits (one for all, the phase-locked loop) are easier to understand in the analog domain. We are perfectly aware that modem implementations (at least for low-to-medium carrier frequencies) tend today to place an analog-to-digital converter, equipped with an anti-aliasing filter, at the input of the receiver, and then to perform all operations, including synchronization, in the digital domain. The readers interested in implementing synchronization circuits in the digital domain are referred to the recent and comprehensive books of Mengali and D'Andrea (1997), and Meyr *et al.* (1997).

## 9.2. Acquisition and training

So far, we have supposed $\theta_1(0) - \theta_2(0) = 0$. This is certainly not true when we switch on the modems to start the transmission. The two oscillators are completely incoherent, and we need an initial period of time to synchronize the oscillators before the transmission of data can be started. This is usually known as *acquisition time* or *acquisition phase*. At the end of the acquisition phase, the two oscillators are locked and data transmission starts. During the data transmission, we also need to keep the phase difference between the two oscillators within certain specified bounds. This operation is known as the *tracking* phase. It is needed only when the transmission time is significantly larger than the joint coherence time of the oscillators. When this is not the case, as in the transmission of characters from a terminal keyboard, we have an *asynchronous* transmission. Different levels of synchronization are often required in the system. As an example, consider a time-division multiplexed pulse-coded modulation system to transmit the voice, employing binary CPSK modulation. We need the four levels of synchronization shown in Fig. 9.1: the frame, word, symbol, and carrier synchronization. Here, we will only be concerned with the last two, carrier and symbol synchronization.

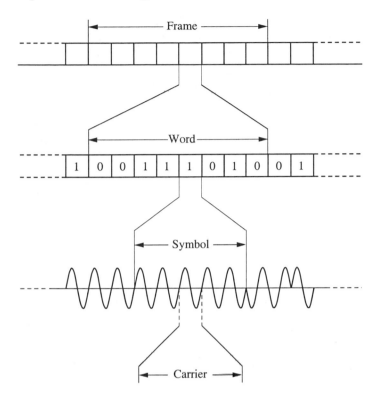

Figure 9.1: *Different synchronization levels in digital transmission.*

The location of carrier and clock synchronizers within a possible receiver structure is shown in Fig. 9.2.

We have seen from the spectral analysis of modulated signals in Chapter 5 that the most efficient digital modulation techniques suppress the carrier completely; all transmitted power resides in the continuous part of the spectrum, and none is "wasted" on a spectral line at the carrier frequency. Also, under the hypothesis that the information-bearing random variables are independent and identically distributed, the spectrum of the digital signal is continuous and does not contain spectral lines at the clock frequency. Thus, any carrier or clock synchronizer will be composed of two conceptually distinct parts: (1) a suitable nonlinear circuit that regenerates a carrier or clock frequency from the data signal that contains neither, and (2) a narrowband device (typically a tuned filter or a *phase-locked loop*, PLL) that separates the regenerated carrier or clock from background disturbances.

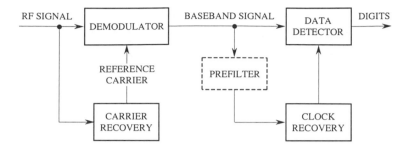

Figure 9.2: *Receiver illustrating locations of synchronizers.*

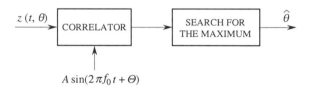

Figure 9.3: *Maximum-likelihood estimator for the carrier phase.*

To give a theoretically sound justification of the structure of the PLL, which is the most widely used circuit for synchronization purposes, let us consider a simplified situation in which the regenerated carrier is only affected by the addition of Gaussian noise; that is,

$$z(t, \theta) = A \sin(2\pi f_0 t + \theta) + n(t) \qquad (9.4)$$

We want to obtain the "best" estimate of the unknown phase $\theta$ based on the observation of $z(t)$ in an interval of length $nT_0$, with $T_0 = 1/f_0$ and $n$ any integer. A straightforward application of detection theory (see Section 2.6) to the continuous case leads to the following expression for the log-likelihood function:

$$\lambda(\Theta) = \int_0^{nT_0} [z(t, \theta) - A \sin(2\pi f_0 t + \Theta)]^2 dt \qquad (9.5)$$

where $\Theta$ is the RV representing the estimate of $\theta$. The optimum unbiased estimate $\Theta = \hat{\theta}$ of $\theta$ under the maximum-likelihood criterion is the one minimizing the RHS of (9.5) or, equivalently, solving the equation

$$\int_0^{nT_0} z(t, \theta) \cos(2\pi f_0 t + \hat{\theta}) dt = 0 \qquad (9.6)$$

A block diagram of the ML estimator of $\theta$ is presented in Fig. 9.3. To obtain an approximate expression of the variance of the estimate $\hat{\theta}$, let us assume that the noise power is low (and consequently the estimate error is low), so that we can write

$$\cos(2\pi f_0 t + \hat{\theta}) \simeq \cos(2\pi f_0 t + \theta) + (\hat{\theta} - \theta)\frac{d}{d\theta}\cos(2\pi f_0 t + \theta) \tag{9.7}$$

Substitution of (9.4) and (9.7) into (9.6) leads to

$$\hat{\theta} - \theta \simeq \frac{\int_0^{nT_0} n(t)\cos(2\pi f_0 t + \theta)\, dt}{A\int_0^{nT_0} \sin^2(2\pi f_0 t + \theta)\, dt} \tag{9.8}$$

Now, accounting for the fact that $n(t)$ is a white Gaussian noise with power spectral density $N_0/2$, we can easily obtain for the variance of the estimate the following expression:

$$\mathrm{E}(\hat{\theta} - \theta)^2 \simeq \frac{N_0}{A^2 n T_0} \tag{9.9}$$

From (9.9), one can conclude that the variance of the estimate is inversely proportional to the signal-to-noise ratio $A^2/N_0$ and to the length $nT_0$ of the observation interval. Figure 9.3 also shows that the optimum estimator has an open-loop structure. However, practical considerations render the solution of Fig. 9.3 impractical in most cases. In fact, an estimate of the unknown phase is available only at the end of the observation interval. Since the phase estimate has to be used for coherent demodulation with the final goal of deciding on the transmitted data, these data should be stored for a time equal to $nT_0$ in order to postpone any decision about them. This procedure should also be repeated periodically in order to follow slow variations of the carrier phase during the tracking period.

A possible way of overcoming these difficulties consists in obtaining the desired estimate using an iterative procedure. Suppose that at the end of the $k$th carrier period we have the estimate $\hat{\theta}_k$, and that we want to modify it on the basis of the observation of the received signal in the subsequent carrier period. Consider the average of the quantity (9.6) in an observation interval of length $T_0$ conditioned on the value $\hat{\theta}_k$ obtained in the previous interval

$$
\begin{aligned}
\mathrm{E}\left\{\frac{d\lambda(\Theta)}{d\Theta}\bigg|\Theta = \hat{\theta}_k\right\} &= \mathrm{E}\left\{\int_{kT_0}^{(k+1)T_0} z(t,\theta)\cos(2\pi f_0 t + \hat{\theta}_k)\, dt\right\} \\
&= \frac{AT_0}{2}\sin(\hat{\theta}_k - \theta) \tag{9.10}
\end{aligned}
$$

In Fig. 9.4 the behavior of this average is shown as a function of $\hat{\theta}_k$. If we have a value of $\hat{\theta}_k$ close to $\theta$ (which is reasonable at the end of the acquisition phase), the average (9.10) is a good error indicator for the estimate at hand. In fact,

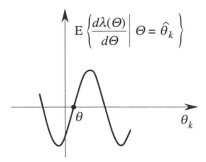

Figure 9.4: *Error indicator function for recursive phase estimation.*

whenever it is positive, we know that $\hat{\theta}_k$ is greater than $\theta$; on the other hand, when it is negative, we know that $\hat{\theta}_k$ is smaller than $\theta$. Moreover, its magnitude tells us how far from $\theta$ our past estimate is. A reasonable recursive algorithm to update our estimate is thus the following:

$$\hat{\theta}_{k+1} = \hat{\theta}_k - \alpha_k \mathrm{E}\{e(\hat{\theta}_k)\} \tag{9.11}$$

where $e(\hat{\theta}_k)$ is defined as

$$e(\hat{\theta}_k) \triangleq \left. \frac{d\lambda(\Theta)}{d\Theta} \right|_{\Theta = \hat{\theta}_k} \tag{9.12}$$

In (9.11) the similarities with the gradient algorithms described in Chapter 8 to recursively update the taps of an adaptive equalizer are evident. Since we are not able to compute the statistical average in the RHS of (9.11), it seems appropriate to modify the recursive algorithm (9.11) by replacing it with a time average

$$\hat{\theta}_{k+1} = \hat{\theta}_k - \frac{1}{k} \sum_{i=0}^{k} \alpha_i e(\hat{\theta}_i) \tag{9.13}$$

where we have extended the influence of the past estimates to the whole time interval $(0, kT_0)$.

## 9.3.  The phase-locked loop

A practical implementation of the recursive algorithm (9.13) is shown in the circuit of Fig. 9.5, which is called a *phase-locked loop* (PLL). In it, the received signal $z(t, \theta)^1$ is multiplied by the output of a *voltage-controlled oscillator* (VCO),

---

[1]Notice that we have multiplied here by $\sqrt{2}$ the magnitude of the received signal in order to simplify the expressions that follow.

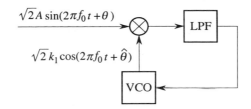

Figure 9.5: *Block diagram of a phase-locked loop.*

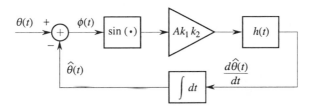

Figure 9.6: *Equivalent block diagram of a phase-locked loop showing the relationship between the phases.*

a sinusoid with magnitude equal to $\sqrt{2}k_1$. This device generates a carrier whose frequency varies linearly with the amplitude of a control signal. The product is low-pass filtered and input to the VCO, whose instantaneous angular frequency is changed according to

$$\frac{d\hat{\theta}(t)}{dt} = k_2 e(t) \tag{9.14}$$

Under reasonable simplifications, it can be shown (see Problem 9.2) that the PLL implements a relationship like (9.13) between successive estimates of $\theta$.

Let us now analyze in some detail the behavior of the PLL, which is the heart of many synchronization circuits. Suppose for the moment that the received signal is noiseless. Let us denote by $h(t)$ the impulse response of the low-pass filter in Fig. 9.5, and suppose that it filters out the high-frequency component at the output of the multiplier. Thus, having defined the phase error

$$\phi(t) \triangleq \theta(t) - \hat{\theta}(t) \tag{9.15}$$

we obtain the following nonlinear equation governing the behavior of the PLL:

$$\frac{d\phi}{dt} = \frac{d\theta}{dt} - Ak_1 k_2 \int_0^t h(t - \tau) \sin \phi(\tau) \, d\tau \tag{9.16}$$

In Fig. 9.6 a block diagram that functionally represents (9.16) is shown. While

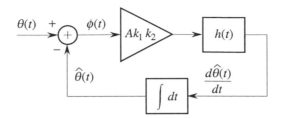

Figure 9.7: *Linearized version of the block diagram of Fig. 9.6.*

in general its behavioral analysis is difficult, it simplifies drastically in the case of a phase error $\phi(t)$ small enough to justify the approximation $\sin\phi \simeq \phi$, which linearizes (9.16) and leads to the circuit of Fig. 9.7. Its analysis is straightforward using Laplace transforms. We obtain the transfer function of interest as

$$H_{eq}(s) \triangleq \frac{\hat{\Theta}(s)}{\Theta(s)} = \frac{Ak_1k_2H(s)}{s + Ak_1k_2H(s)} \tag{9.17}$$

where $\Theta(s)$ is the Laplace transform of $\theta(t)$.

### 9.3.1.  Order of the phase-locked loop

The order of the PLL is defined according to the degree of the denominator of $H_{eq}(s)$, which in turn depends on the loop filter transfer function $H(s)$. Thus we have

**First-order PLL**

$$H(s) = 1 \;\Rightarrow\; H_{eq}(s) = \frac{Ak_1k_2}{s + Ak_1k_2} \tag{9.18}$$

**Second-order PLL**

$$H(s) = \frac{1 + s\tau_2}{s\tau_1} \;\Rightarrow\; H_{eq}(s) = \frac{4\zeta\pi f_n s + (2\pi f_n)^2}{s^2 + 4\zeta\pi f_n s + (2\pi f_n)^2} \tag{9.19}$$

where the two parameters $f_n$ and $\zeta$, called respectively the *natural frequency* and the *damping factor* of the loop, are given by

$$f_n = \frac{1}{2\pi}\sqrt{\frac{Ak_1k_2}{\tau_1}} \tag{9.20}$$

$$\zeta = \frac{\tau_2}{2}\sqrt{\frac{Ak_1k_2}{\tau_1}} \tag{9.21}$$

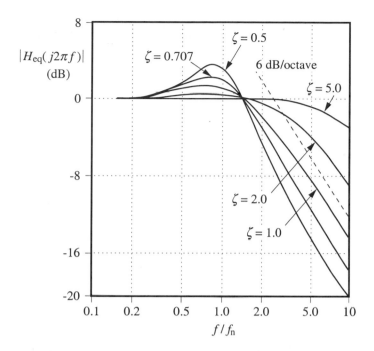

Figure 9.8: *Frequency response of a second-order loop.*

We observe that the order of the PLL corresponds to the number of perfect integrators within the loop. For a first-order PLL, we can control only one parameter, $Ak_1k_2$, which is the 3-dB bandwidth of $H_{eq}(s)$, whereas the second-order loop gives us the two parameters $f_n$ and $\zeta$.

The magnitude of the frequency response $H_{eq}(j2\pi f)$ of a second-order loop for several values of $\zeta$ is plotted in Fig. 9.8. It can be seen that the loop performs a low-pass filtering operation on phase inputs. Using root-locus plot characteristics, it can be shown that first- and second-order loops are always stable, whereas third- and higher-order loops can be stable under certain conditions. Besides the stability considerations, it is important to know the steady-state behavior of the PLL (i.e., the steady-state phase and frequency errors in the presence of particular inputs). We shall examine two different cases of input phases

$$\theta_A(t) = 2\pi t(\Delta f) + \Delta \theta \tag{9.22}$$

$$\theta_B(t) = \pi t^2(\delta f) + 2\pi t(\Delta f) + \Delta \theta \tag{9.23}$$

The first case is the most important in data transmission between fixed points, since it refers to a situation in which the received carrier presents a frequency

| Errors | 1st-order PLL | 2nd-order PLL | 3rd-order PLL |
|--------|---------------|---------------|---------------|
| $\Delta\Phi_A$ | $\dfrac{2\pi\Delta f}{Ak_1k_2}$ | 0 | 0 |
| $\Delta\Phi_B$ | $\infty$ | $\dfrac{2\pi\delta f\tau_1}{Ak_1k_2}$ | 0 |
| $\Delta f_A$ | 0 | 0 | 0 |
| $\Delta f_B$ | $\dfrac{\delta f}{Ak_1k_2}$ | 0 | 0 |

Table 9.1: *Steady-state phase and frequency errors for PLL of various order.*

displacement $\Delta f$ (e.g., due to frequency-division multiplexing) and an initial phase shift $\Delta\theta$. The second case can happen when there is a relative motion between transmitter and receiver, as in a mobile radio communication system. Using the final-value theorem of the Laplace transform (see Problem 9.1), we obtain the steady-state errors of Table 9.1, where $\Delta f_{A,B}$ and $\Delta\Phi_{A,B}$ represent the frequency and phase errors, respectively.

All the preceding results were based on the assumption that the phase error is sufficiently small, thus allowing the loop to be considered linear in its operation. This assumption becomes progressively less useful as error increases until, finally, the loop drops out of lock and the assumption becomes unjustified. Through the analysis of the nonlinear model of the PLL, one can identify important parameters like, for example, the *hold-in range* (i.e., the input frequency range over which the loop will hold lock) or the *acquisition time* (i.e., the time required by the loop to reduce the phase error under a given threshold). A detailed analysis of the behavior of the PLL without the linear assumption can be found in Viterbi (1966) and Lindsey (1972).

When a Gaussian noise is present in additive form at the input of the PLL, an approximate linear analysis is still possible when the signal-to-noise ratio is sufficiently high. This leads (see Viterbi, 1966) to the functional block diagram of Fig. 9.9. In the figure, $n'(t)$ is a Gaussian noise process independent of $n(t)$, with the same spectral properties as the input noise, i.e. with power spectral

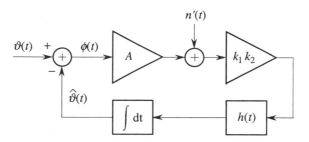

Figure 9.9: *Equivalent block diagram of a linearized PLL including noise.*

density $G_{n'}(f) = N_0$. To evaluate the effect of the noise, we assume that the input phase $\theta(t)$ is constant so that the fluctuations in the phase of the VCO signal can be entirely attributed to the noise. From Fig. 9.9 we can obtain the transfer function between the noise $n'(t)$ and the recovered phase $\hat{\theta}(t)$

$$H_n(s) \triangleq \frac{\hat{\Theta}(s)}{N'(s)} = \frac{k_1 k_2 H(s)}{s + A k_1 k_2 H(s)} = \frac{1}{A} H_{\text{eq}}(s) \qquad (9.24)$$

The noise power contribution to $\theta(t)$ is then

$$\sigma_\theta^2 = \frac{1}{A^2} \int_{-\infty}^{\infty} G_{n'}(f)|H_{\text{eq}}(j2\pi f)|^2 df = \frac{N_0}{A^2} B_{\text{eq}} \qquad (9.25)$$

where $B_{\text{eq}}$ denotes the equivalent noise bandwidth of the PLL. For first- and second-order PLL, it is given by the following expressions:

$$B_{\text{eq}} = \frac{A k_1 k_2}{4} \quad \text{(first order PLL)} \qquad (9.26)$$

$$B_{\text{eq}} = \pi f_n \sqrt{\zeta + \frac{1}{4\zeta}} \quad \text{(second order PLL)} \qquad (9.27)$$

It can be seen from (9.26) and Table 9.1 that, for a first-order PLL, the needs for a small steady-state phase error and a small noise bandwidth are in conflict. For a second-order PLL, a good compromise is achieved with the value $\zeta = 0.707$.

When the linear analysis is valid, the VCO phase error $\phi(t)$ (the so-called phase *jitter*) is a Gaussian random process. In general, this is not true. Nonlinear analysis of a PLL has been concerned with deriving the probability density function (pdf) of the RV $\Phi$ representing the amplitude of the phase error $\phi(t)$. This pdf (see Viterbi, 1966) is found as the steady-state solution of a nonlinear stochastic partial-differential equation known as the Fokker-Planck equation.

Without going into the details, the resulting Tikhonov pdf is

$$f_\Phi(\phi) = \frac{\exp(\rho\cos\phi)}{2\pi I_0(\rho}, \quad |\phi| \le \pi \tag{9.28}$$

where $\rho$ is the signal-to-noise ratio of the loop (the inverse of the variance (9.25)), and $I_0(\cdot)$ is the modified Bessel function of the first kind and order zero. The pdf (9.28) approaches a Gaussian one for large $\rho$.

In its essence, the PLL acts as a narrowband filter whose central frequency tracks the frequency of the received signal (9.4) within a reasonable range without affecting its noise bandwidth. As already stated, it requires the presence of a spectral line, at the frequency to be tracked, contained in the signal at its input. Thus, in addition to the PLL, suitable nonlinear circuits are integral portions of a synchronizer. In the following, we shall examine in some detail some of the most common carrier and clock synchronizers. Only a qualitative description of their behavior will be presented. The reader interested in detailed performance analyses may refer to the Bibliographical Notes at the end of this chapter.

## 9.4.  Carrier synchronization

To simplify, let us consider initially a binary CPSK signal written in its bandpass form

$$v(t) = v_P(t)\cos(2\pi f_0 t + \theta) \tag{9.29}$$

where

$$v_P(t) = \sum_n a_n u_T(t - nT) \tag{9.30}$$

where $u_T(t)$ is the unit step function, equal to 1 for $0 \le t \le T$ and 0 elsewhere, and the information symbols $a_n$ take the values $\pm 1$.

There are three main types of carrier synchronizers, the squaring loop, the remodulator, and the Costas loop. They differ in the position of the nonlinearity, which is entirely separated from the PLL in the squaring loop, whereas it is included in the phase detector for the remodulator and the Costas loop.

The block diagram of the squaring loop is shown in Fig. 9.10. Its nonlinearity is a square-law device, so its output

$$y(t) \triangleq v^2(t) = \frac{1}{2}[1 + \cos(4\pi f_0 t + 2\theta)] \tag{9.31}$$

contains a spectral line at frequency $2f_0$ that can be tracked by a conventional PLL. The VCO output is divided by 2 to provide the desired carrier at frequency $f_0$. In the divide-by-2 operation, there is a phase indeterminacy, which makes it

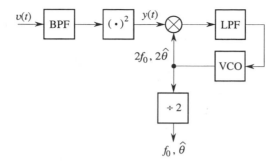

Figure 9.10: *Block diagram of the squaring loop for carrier recovery of a binary PSK signal.*

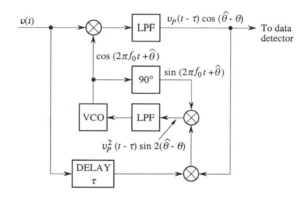

Figure 9.11: *Block diagram of the remodulator for carrier recovery of a binary PSK signal.*

impossible to decide whether the current symbol is 1 or $-1$. This phase ambiguity, inherent in all phase-shift modulation techniques, can be resolved by special encoding, like the differential encoding described in Section 5.8.

A remodulator synchronizer is shown in Fig. 9.11. The received signal is demodulated and the message $v_P(t)$ recovered. It is used to remodulate the received signal so as to remove the modulation. If the baseband waveforms are time-aligned, the output of the balanced modulator has a pure carrier component that can be tracked by the PLL. The delay $\tau$ in Fig. 9.11 is required to compensate for the delay of the low-pass filter following the demodulator. In the figure, the relationships explaining the behavior of the synchronizer in the absence of noise are also given.

A block diagram of the Costas loop is shown in Fig. 9.12. Its behavior

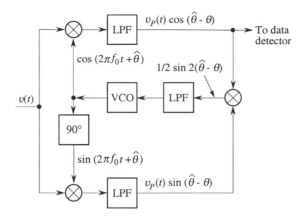

Figure 9.12: *Block diagram of the Costas loop for carrier recovery of a binary PSK signal.*

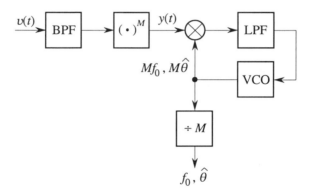

Figure 9.13: *Block diagram of the $M$th power synchronizer for carrier recovery of an $M$-ary PSK signal.*

should be explained by the relationships indicated in the figure, which are valid in the absence of noise.

The carrier recovery circuits described before can be generalized to the situation in which the digital information is transmitted via $M$-ary CPSK modulation. An $M$th power synchronizer is shown in Fig. 9.13. Its operation is easily understood by simple extension of the squaring loop. Because of their wide applications, block diagrams of the 4-PSK remodulator and the Costas loop for 4-PSK are shown in Figs. 9.14 and 9.15. A stable lock can be achieved at any of four different phases. There is an inherent fourfold ambiguity that must be resolved

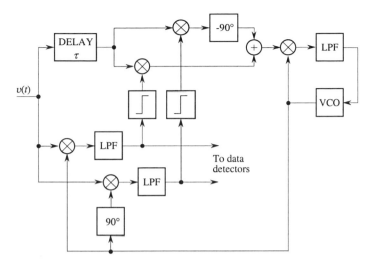

Figure 9.14: *Block diagram of the remodulator for carrier recovery of 4-PSK signals.*

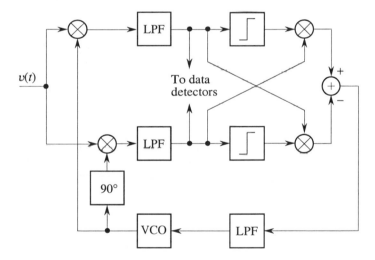

Figure 9.15: *Block diagram of the Costas loop for carrier recovery of 4-PSK signals.*

by other means, such as, for example, the differential co-decoding explained in
Section 5.8. Both the remodulator and the Costas loop perform a multiplication
by the demodulated message in analog form to remove modulation. Better noise
rejection would be possible if the detected digital message were used for modu-

| $M$ | $L_M(\rho_i)$ |
|---|---|
| 1 | 1 |
| 2 | $1+\dfrac{1}{2\rho_i}$ |
| 3 | $1+\dfrac{2}{\rho_i}+\dfrac{2}{3\rho_i^2}$ |
| 4 | $1+\dfrac{3}{\rho_i}+\dfrac{6}{\rho_i^2}+\dfrac{3}{\rho_i^3}$ |

Table 9.2: *Intermodulation losses of $M$th-law regenerators.*

lation removal. This is done in decision-directed synchronizers. Unfortunately, a decision-directed synchronizer cannot acquire the carrier until symbol synchronization has been achieved. Thus, it is not suited for applications requiring fast acquisition. Carrier synchronization circuits based on both the remodulator and Costas loop can be modified to cope with more general two-dimensional modulations such as QAM.

In general, the performance analysis of a synchronizer is very difficult because of the presence of the nonlinear regenerator. Without going into details, we can write the noise-caused VCO phase jitter variance for an $M$-phase synchronizer as

$$\sigma_\theta^2 = M^2 \left( \frac{N_0 B_{\text{eq}}}{A^2} \right) L_M(\rho_i) \tag{9.32}$$

where $M^2$ comes from the $M$-fold phase magnification, $L_M(\rho_i)$ is the loss caused by noise intermodulation in the nonlinearity, and $\rho_i$ is the signal-to-noise ratio at the output of the receiving filter. The quantity within the parentheses represents the jitter variance of an ordinary PLL (see (9.25)). For the special case of an $M$th-power nonlinearity (Butman and Lesh, 1977), some typical losses are given in Table 9.2.

## 9.5.  Clock synchronizers

We assume that carrier and clock are recovered in two distinct steps: first, the phase $\theta$ of the carrier is estimated; then the timing wave is extracted from the demodulated baseband signal. In other words, we shall not consider the approach of the simultaneous estimation of the carrier and clock. This omission does not imply a merit judgment, since in some cases superior performance can be obtained with (admittedly complex) joint estimation methods. Details can be found in the Bibliographical Notes at the end of this chapter.

Consider now the baseband signal obtained from the demodulator:

$$r_D(t) = \sum_n a_n h(t - nT) \tag{9.33}$$

where $(a_n)$ is the message sequence, which is assumed to be a zero-mean stationary discrete random process formed by iid random variables. The objective of the timing synchronization circuit is to extract from $r_D(t)$ a periodic wave with period $T$ (the symbol interval) and a proper phase indicating the sampling instant within each period. Clock synchronizers can be categorized according to the bandwidth of the communication system as wideband or narrowband. We are interested in the most common, and more critical from the timing synchronization point of view, situations where bandwidth occupancy approaches the Nyquist limit of $1/(2T)$ at baseband. More precisely, we assume that the bandwidth does not exceed $1/T$. In this case, the pulse $h(t)$ is spread over many symbol intervals, giving rise to intersymbol interference (ISI). As we have seen in Chapter 7, to avoid ISI the pulses are usually given a Nyquist shape. This yields the elimination of ISI at nominal sampling instants, but it is not sufficient to eliminate the effects of ISI on the clock synchronizer. In general, the recovered clock is affected by a jitter component, called *self-noise* or *data noise*, caused by ISI. In many applications, this self-noise is predominant with respect to the Gaussian noise. For this reason, we have not included the additive Gaussian noise in the RHS of (9.33).

As for carrier acquisition, the available signal $r_D(t)$ has no spectral lines at frequency $1/T$. In fact, $r_D(t)$ is easily recognized as a cyclostationary random process (see Section 2.2.2) with period $T$, zero mean, and mean-square value

$$E\{r_D^2(t)\} = E\{a^2\} \sum_n h^2(t - nT) \tag{9.34}$$

Equation (9.34) shows that the square of $r_D(t)$ does possess a periodic mean value.

Using the "Poisson sum formula" (see (2.109) and Problem 9.3), we can express (9.34) in the more convenient form of a Fourier series whose coefficients

Figure 9.16: *Block diagram of a clock synchronizer.*

are obtained from $H(f)$, the Fourier transform of $h(t)$

$$E\{r_D^2(t)\} = \frac{E\{a^2\}}{T} \sum_\ell \mu_\ell \exp\left(j\frac{2\pi\ell t}{T}\right) \qquad (9.35)$$

with

$$\mu_\ell = \int_{-\infty}^{\infty} H^*\left(f - \frac{\ell}{T}\right) H(f)\, df \qquad (9.36)$$

Due to the assumption of bandwidth limitation for $H(f)$, only the three terms with $\ell = 0, \pm 1$ in the summation of (9.35) are different from zero. The first corresponds to a dc component, whereas the other two give a sinusoidal signal with frequency $1/T$ and amplitude

$$\mu_1 = \int_{-\infty}^{\infty} H^*\left(f - \frac{1}{T}\right) H(f)\, df \qquad (9.37)$$

Note that the sinusoidal component at the clock frequency vanishes when $H(f)$ is strictly bandlimited in the interval $[-1/(2T), 1/(2T)]$. Thus, also in this case, for signals exhibiting some extra bandwidth beyond the Nyquist frequency, we can use a nonlinearity (e.g., a square-law rectifier) to restore the desired spectral line followed by a "tuned" filter (a narrowband filter centered around the timing frequency $1/T$) or a PLL that tracks the restored timing wave. Alternate zero-crossings of the reference waveform $w(t)$ are used by the pulse generator of Fig. 9.16 as indications of the correct sampling instants. Remembering the spectral analysis of cyclostationary processes of Section 2.3, we realize that the spectrum of $y(t) = r_D^2(t)$ presents a continuous part, besides the discrete one giving rise to the desired spectral line. Thus, even in the absence of additive Gaussian noise, we have a self-noise entering the tuned filter (or PLL) of Fig. 9.16 and causing a fluctuation of the zero crossings of $w(t)$ around the nominal sampling instants, the *timing jitter*. To better understand, consider the complex envelope $\tilde{h}(t)$ of $h(t)$ defined with respect to the frequency $1/(2T)$, so that

$$h(t) = \Re\left\{\tilde{h}(t) \exp\left(j\frac{\pi t}{T}\right)\right\} \qquad (9.38)$$

and write $r_D^2(t)$ in terms of $\tilde{h}(t)$

$$r_D^2(t) = y(t) = \frac{1}{2}\Re\left\{\left[\sum_n (-1)^n a_n \tilde{h}^2\right] \exp\left(j\frac{2\pi t}{T}\right)\right\}$$

$$+ \ \frac{1}{2} \left| \sum_n (-1)^n a_n \tilde{h}(t - nT) \right|^2 \qquad (9.39)$$

The second term in the RHS of (9.39) can be disregarded as it is not passed through the tuned filter (or PLL). The first term can be rewritten as

$$A \cos\left(\frac{2\pi t}{T}\right) + b_P(t) \cos\left(\frac{2\pi t}{T}\right) + b_Q(t) \sin\left(\frac{2\pi t}{T}\right) \qquad (9.40)$$

In (9.40) we can recognize the desired periodic component (first term), as well as two in-phase and quadrature disturbances (second and third components, respectively). It is precisely the quadrature component $b_Q(t)$ that produces timing jitter. Using (9.39), this component can be written as (see Problem 9.4)

$$b_Q(t) = \sum_k \sum_m a_k a_m (-1)^{k+m} h_P(t - kT) h_Q(t - mT) \qquad (9.41)$$

where $h_P(t)$ and $h_Q(t)$ are the real and imaginary parts of $\tilde{h}(t)$. It is evident from (9.41) that the timing jitter is strongly dependent on the shape of the date pulse $h(t)$ at the input of the nonlinearity. For this reason, some authors have suggested the insertion of a suitable prefilter before the nonlinearity of Fig. 9.16 in order to eliminate or greatly reduce the data noise (Franks and Bubrouski, 1974; Mengali, 1983). In particular, these authors have shown that using a tuned filter (or a PLL) with a transfer function exhibiting a conjugate symmetry around the symbol rate $1/T$ and a transfer function $H(f)$ limited in bandwidth to the interval $[1/(4T), 3/(4T)]$ with a conjugate symmetry around $1/(2T)$, one can completely eliminate the data noise if the nonlinearity is a square-law rectifier.

## 9.6.    Effect of phase and timing jitter

The first part of Chapter 7 was devoted to the computation of the symbol error probability conditioned on a given value of the phase jitter considered as a RV with a known pdf. Later, we introduced the Tikhonov pdf (see (9.28)), which describes the statistical behavior of the phase error at the output of a PLL. We mentioned that it approaches a Gaussian pdf for large values of the loop signal-to-noise ratio. To give a quantitative idea of the effect of the phase error on the average error probability, we shall consider a binary CPSK system, without ISI, affected by a phase error in the recovered carrier with Gaussian pdf. In this case, it is easily seen that the conditional bit error probability for a given value of the phase error $\phi$ is

$$P_b(e \mid \phi) = \frac{1}{2} \operatorname{erfc}\left(\sqrt{\frac{\mathcal{E}_b \cos^2 \phi}{N_0}}\right) \qquad (9.42)$$

so that the average bit error probability, assuming a Gaussian pdf for the phase jitter $\phi$, becomes

$$P_b(e) \;=\; \mathrm{E}_\phi P_b(e \mid \phi) = \int_{-\infty}^{\infty} P_b(e \mid \phi) f_\Phi(\phi)\, d\phi \qquad (9.43)$$

$$\;=\; \frac{1}{2\sqrt{2\pi\sigma_\phi^2}} \int_{-\infty}^{\infty} \mathrm{erfc}\left( \sqrt{\frac{\mathcal{E}_b \cos^2 \phi}{N_0}} \right) \exp\left( -\frac{\phi^2}{2\sigma_\phi^2} \right) d\phi$$

Figure 9.17 shows the behavior of the average bit error probability as a function of the signal-to-noise ratio $\mathcal{E}_b/N_0$. Different curves are labeled according to a value of the standard deviation of the phase jitter $\sigma_\theta$ in degrees. The curves have been obtained by calculating the conditional error probability and, then, averaging with respect to the phase jitter pdf using standard Gauss quadrature rules (see Problem 9.6). The figures show the typical *error floor* behavior, i.e., the fact that, for a given standard deviation, there is a lower bound to the bit error probability attainable by the system. For an error probability of $10^{-9}$, a standard deviation of 10 degrees induces a penalty in the signal-to-noise ratio close to 0.5 dB. The effect of a symbol synchronization jitter on the error probability is shown in Fig. 9.18. In obtaining the results shown in the figure, we have assumed a simplified situation in which the elementary pulse $h(t)$ in (9.33) is rectangular, so that the only effect of the timing error is to reduce the signal-to-noise ratio at the output of the correlator that implements the optimum receiver for binary CPSK. The timing error is supposed to be a Gaussian RV. The parameter $\sigma_\tau$ labeling the curves of Fig. 9.17 is the standard deviation, multiplied by $\pi$, of the normalized error $\tau \triangleq (t_0 - \hat{t}_0)/T$ in the receiver symbol clock.

When we consider a multilevel signaling scheme employing two quadrature carriers, the effect of the carrier jitter is enhanced, because the phase error also induces, as seen in Chapter 7, a cochannel interference besides the simple attenuation of the binary case. Thus, the accuracy requirements of the carrier recovery circuits become more stringent.

## 9.7.  Bibliographical notes

A huge literature exists in the field of synchronization of digital communication systems. The following books are focused on the phase-locked loop (PLL) theory and applications: Viterbi (1966), Lindsay (1972), Blanchard (1976), and Gardner (1979). In particular, Viterbi's exact analysis of the first-order PLL solving the Fokker-Planck equation has provided much insight for understanding nonlinear operations, whereas Gardner's book is very useful for practicing engineers. The problems related to the design and analysis of digital PLL have

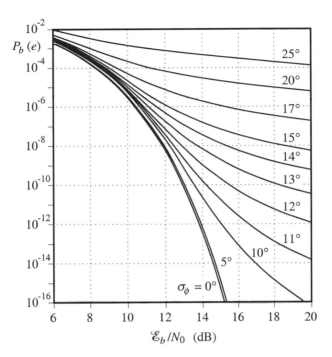

Figure 9.17: *Binary PSK bit error probability with imperfect carrier synchronization as a function of the signal-to-noise ratio $\mathcal{E}_b/N_0$. The parameter labeling each curve represents the standard deviation of the residual phase jitter assumed to have a Gaussian statistics.*

not been considered in this chapter, although they are now very important be- cause of the widespred applications of digital circuitry. A good starting point for the interested reader are the tutorial papers by Gupta (1975) and Lindsey and Chie (1981). A survey of the peculiar methods used in the analysis of digital PLL without noise can be found in D'Andrea and Russo (1983).

The general problem of carrier and clock synchronizers is faced by Stiffler (1971), Lindsey and Simon (1973) and Franks (1983). A comprehensive tuto- rial paper has been written by Franks (1980). The joint recovery of carrier and symbol synchronization has been analyzed by Mengali (1977), Mancianti *et al.* (1979), and Meyers and Franks (1980). The effect of imperfect synchronization on system performance is the subject of certain papers in Stavroulakis (1980). The treatment of this subject is attributable to Franks (1980), Gardner (1979), and Mengali (1979).

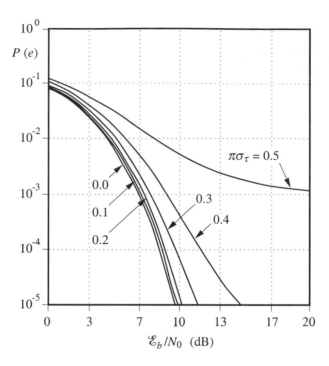

Figure 9.18: *Binary PSK symbol error probability with imperfect timing synchronization as a function of the signal-to-noise ratio $\mathcal{E}_b/N_0$. The parameter labeling each curve represents the standard deviation of the residual, normalized timing jitter assumed to have Gaussian statistics, multiplied by $\pi$.*

Digital synchronization techniques, where timing, phase and frequency synchronization is achieved by operating on signal samples taken at a suitable rate, is the subject of the comprehensive books by Mengali and D'Andrea (1997), and by Meyr *et al.* (1997).

## 9.8. Problems

*Problems marked with an asterisk should be solved with the aid of a computer.*

**9.1** Using the final-value theorem of the Laplace transform, verify the steady-state frequency and phase errors of Table 9.1.

**9.2** Show that the PLL of Fig. 9.5 implements a relationship like (9.13) between successive estimates of $\theta$. *Hint*: Start from the differential equation (9.16) and suppose that the variations of $\hat{\theta}(t)$ are so slow that it is possible to write

$$\frac{d\hat{\theta}(t)}{dt} \simeq \frac{\hat{\theta}(t) - \hat{\theta}(t-T)}{T}$$

**9.3 (a)** Show that (Poisson sum formula)

$$\sum_k h(t - kT) = \frac{1}{T} \sum_m H\left(\frac{m}{T}\right) e^{j2\pi mt/T}$$

*Hint*: Find first the Fourier series expansion of the periodic factor $h(t - kT)$.
**(b)** Using the result in part (a), verify (9.35).

**9.4** Derive the expression (9.41) of the quadrature data noise component $b_Q(t)$.

**9.5** In the absence of noise in the received signal, explain quantitatively the behavior of the remodulator and Costas loop for QPSK modulation, as from the block diagrams of Figs. 9.14 and 9.15.

**9.6 (*)** Using standard quadrature rules, obtain the curves of Fig. 9.17.

**9.7 (*)** Compute the conditional symbol error probability $P(e \mid \phi)$ for the case of QPSK modulation, and then average with respect to $\phi$, assumed to be a Gaussian random variable with standard deviation $\sigma_\phi$, obtaining curves of the symbol error probability versus $\mathcal{E}/N_0$ for various values of $\sigma_\phi$.

**9.8** Using the result of (9.37), evaluate the magnitude of the discrete component at frequency $1/T$ at the output of a square-law device for a raised-cosine impulse response $h(t)$ (see (7.71) for the expression of $h(t)$) as a function of the roll-off factor $\alpha$.

# Improving the transmission reliability: Block codes

Designers of primitive digital communication systems sought to obtain low bit error probabilities by transmitting at high power or by using larger bandwidth than strictly necessary. This approach is adequate if the required error probability is not too low and/or the data rate is not too high: it buys performance with the most precious of resources: spectral bandwidth and power.

The lesson taught by Shannon (see Section 3.3) was that high performance is indeed obtainable by calling a third resource into play, the system complexity. The concurrence of two basic facts, i.e., the sky-rocketing requirements of transmission speed and the affordability, thanks to the modern electronic technology, of extremely sophisticated receivers has made the Shannon dream an every-day reality, so that highly complex co-decoding techniques are now widely used in digital transmission systems to protect the information.

Techniques to control the error probability are based on the addition of redundancy to the information sequence. Traditionally, codes aimed at improving the transmission reliability are called *error correcting codes*. This concept is bound to a particular operating mode of the demodulator and decoder, in which the received signal sequence is *hard-detected* by the demodulator, before being transferred to the decoder. As a consequence, the binary sequence entering the decoder contains errors that the decoder may or may not be able to correct. This mode of operation, however, entails some degree of suboptimality, and is replaced, whenever feasible, by *soft-decoding*, in which the demodulator derives the sufficient statistics in analog or quantized form, and supplies it to the decoder, which, in turn, performs the final task of estimating the transmitted information sequence. When this is the operation mode, talking of "error correcting" codes

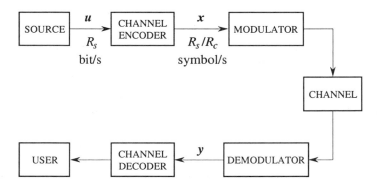

Figure 10.1: *Block diagram of a transmission system employing channel coding.*

does not make sense, since no true correction of error takes place, but, rather, the pair encoder-decoder prevents errors from occurring. In this situation, it would better to talk of *error control* codes (see Blahut, 1983). Most of the algorithms for decoding *tree codes* make use of the soft information in a straightforward manner. The use of soft-decision in block codes is somewhat more involved and generally requires significant changes in the decoding algorithms.

For the reasons previously explained, we will generally speak of codes that improve the transmission reliability, or of *channel codes*, in the sense that these codes aim at protecting the information from impairments occurring during its transmission over the channel.

In this chapter, we will first propose a taxonomy of the codes employed to protect the transmitted information, and then define and analyze *linear* block codes. In the next chapter, we will consider convolutional and concatenated codes.

## 10.1. A taxonomy of channel codes

Consider the simple block diagram of Fig. 10.1. Using the terminology of Blahut (1983), we distinguish a source producing a binary sequence, the *data stream*: it is the binary sequence emitted directly by the source, or by the source encoder. We assume that it is formed by independent identically distributed (iid) binary random variables (RVs). The data stream enters the channel encoder which maps it into a *code stream*. For constructing the code, additional structure may be defined on the data stream by segmenting it into pieces called *data words*. Likewise, the code stream is segmented into pieces called *code words*. For an $(n, k)$ block code, the data words consist of $k$ bits and the code words of $n$ bits. A *channel code* $\mathcal{C}$ is the set of $2^k$ $n$-tuples of bits, the code words $\mathbf{x}$. An *encoder E* is the set of the $2^k$ pairs $(\mathbf{u}, \mathbf{x})$, where $\mathbf{u}$ is a data word, i.e., a $k$-tuple of bits, and $\mathbf{x}$ the corresponding code word. These definitions should clarify the fundamental difference between the notion of a code and the notion of an encoder. The code is a collection of code words and is independent of the way they are obtained. The term encoder refers to the one-to-one correspondence between data words and code words, and also applies to the device that implements this assignment. With respect to how the encoder assigns code words to data words, we say that the $(n, k)$ code is a *block* code when the encoder is memoryless, i.e., when to the same $k$ bits in the data word there correspond the same $n$ code word bits. The block code is an $(n, k)$ code, and the ratio $R_c \triangleq k/n$ is the *rate* of the code. Each data word (block) is encoded independently without interaction with earlier or later data words. When the correspondence between data words and code words has memory, i.e., the $n$ bits of the code word do not depend only on the $k$ bits of the data word, but also on some previous data words, we say that the code is a *tree* code. In this case, it is often convenient to think of infinitely long data streams and code streams, or sequences, which start at time zero and continue indefinitely in the future. A tree code breaks the data stream into segments called *data frames*, each consisting of $k_0$ data bits, $k_0$ normally a small integer. The encoder is a finite-state machine that retains some memory of earlier data frames; in the simplest case, it simply stores the $m$ most recent data frames unchanged. A single code frame consists of $n_0$ bits that are computed from the $mk_0$ data bits of the $m$ data frames stored in the encoder memory, and the $k_0$ bits of the incoming data frame; these $n$ bits are shifted out to the channel as the new $k_0$ data bits enter the encoder. The ratio $R_c \triangleq k_0/n_0$ is still called the code rate. Tree codes with a special memory and linearity structure, to be defined in the next chapter, are called *convolutional* codes.

With respect to the properties of the set of code words, we distinguish between *linear* and *nonlinear* codes. For a linear code, the set of code words (or code streams, for tree codes) is closed under component-wise modulo-2 addition, an operation denoted simply by "+" in this chapter.[1] This property has several important implications that will be made clear in the next sections. According to how the system makes use of the code capabilities, we distinguish between *error detecting* and *error correcting* codes. This does not represent a distinction between the codes themselves, but, rather, between the strategies followed by the system.

Two different strategies can be used in the channel decoder. Conceptually,

---

[1] Modulo-2 addition can also be defined as the addition operation in the Galois field GF(2). Since it is beyond the scope of this book to introduce Galois fields, we will always speak of modulo-2 addition.

these strategies can be related to Fano's inequality (see Chapter 3, (3.67)). In the first strategy, the decoder observes the hardly-demodulated received sequence and detects whether or not errors have occurred. A certain measure of uncertainty is eliminated, which corresponds to the term $H(e)$ in (3.67). Error detection is used to implement one of two possible schemes: error monitoring or automatic repeat request (ARQ). In the case of error monitoring, the decoder supplies to the user a continuous indication regarding the quality of the received sequence, so that, when the reliability becomes too low, the sequence can be discarded. In the case of ARQ, the transmitter is asked to repeat unsuccessful transmissions. To this end, a feedback channel from the receiver to the transmitter must be available.

The second strategy is called *forward error correction* (FEC). The decoder attempts to restore the correct transmitted sequence whenever errors are detected in the received sequence. In this case, an additional quantity of uncertainty must be removed corresponding to the term $P(e) \log(2^k - 1)$ of (3.67). It is intuitive that this strategy requires, for the same code, more complex decoding algorithms. The choice between the two strategies depends on the particular application and on the complexity of the transmission system considered. For example, the ARQ scheme is usually applied in the communication between computers, since a two-way transmission channel is available together with large memory devices that store the incoming information while performing, upon request, the retransmission procedure. On the other hand, FEC is adopted when the information must be protected on a one-way channel, or when real-time, or strictly-controlled delays are required. Examples pertain to deep-space communication and digitized interactive voice transmission.

With respect to the encoder operations, we say that the encoder $E$ is *systematic* when the first $k$ bits of each code word $\mathbf{x}$ coincide with the $k$ bits of the data word $\mathbf{u}$. It is common in textbooks to say that a code, rather that its encoder, is systematic. In the following, we too will sometimes indulge in this imprecision. The reader is warned, though, that the concept of systematicity entails the mapping of data words into code words, and, thus, only pertains to the encoder.

To analyze the benefits due to channel encoding in comparison with the uncoded schemes of Chapter 5, let us consider again the model of Fig. 10.1. The source emits binary digits[2] at a rate of $R_s$ bit/s and the encoder represents each data word of $k$ source bits using $n = k/R_c$ bits. $R_c$ is the code rate. To keep the pace of the source, the transmission speed on the channel must be increased to the value $R_s/R_c$ binary symbols per second, and thus the required bandwidth must also be increased by the same factor $1/R_c$. As a result, the use of chan-

---

[2]Since we make the assumption that the data stream is made of iid binary RVs (0 and 1), we will use indifferently the words "bits" and "binary digits."

nel encoding decreases the bandwidth efficiency with respect to the uncoded transmission by a factor $1/R_c$. The binary symbols produced by the channel encoder are presented to the modulator, and converted into a sequence of waveforms using one of the modulation schemes described in Chapter 5 or 6. For the purposes of this preliminary discussion, we assume that the modulator uses an antipodal binary modulation over an AWGN channel. With this modulation scheme, each binary encoded symbol is mapped by the modulator into a binary waveform of duration $T = T_c = R_c/R_s$ seconds. This duration is shorter than that used in the uncoded case by a factor $R_c$. Denoting with $\mathcal{E}_b$ the energy per information bit, and assuming that the transmitted power is kept constant, we can conclude that coding decreases the energy per channel symbol to the value $\mathcal{E}_b R_c$. As a result, in case of hard decisions, more channel symbols will be incorrectly demodulated than with uncoded transmission. These observations about coding seem rather discouraging. In fact, bandwidth efficiency is decreased and more errors in the demodulated sequence are to be expected. Nevertheless, in a well-designed coded system, the larger number of errors at the demodulator output will be compensated for by the error-correcting capabilities of the decoder. Therefore, a coded transmission should trade bandwidth efficiency for a better overall error performance, using the same transmission power, or, equivalently, for a smaller required power for a given error performance. The decrease in the required power for the coded system is referred to as *coding gain*.

Let us describe the processing that must take place at the channel output to achieve such a result. Consider first the case in which the demodulator is used to make decisions on whether each binary waveform carries a 0 or a 1. To this purpose, the demodulator output is quantized to two levels denoted by 0 and 1 and is said to make *hard* decisions. The sequence of binary digits from the demodulator is fed into the decoder. The decoder attempts to recover the information sequence by using the code word's redundancy for either detecting or correcting the errors that are present at the demodulator output. Such a decoding process is called *hard-decision* decoding. In this model, assuming a binary antipodal coherent modulation and an AWGN channel, the combination of modulator, channel, and demodulator is equivalent to a binary symmetric channel (BSC). Its transition probability is the error probability of a binary antipodal modulation scheme (see Chapter 4). The overall error performance of the coded scheme depends on the implementation of efficient algorithms for error detection and correction.

At the other extreme, consider the case in which the unquantized output of the demodulator, the sufficient statistics, is fed to the decoder. This stores the $n$ outputs corresponding to each sequence of $n$ binary waveforms and builds $2^k$ decision variables. With the optimum decision strategy, the cascaded demodulator and decoder perform the same operation as the optimum coherent demod-

ulator of Chapter 4, i.e., they choose the transmitted sequence corresponding
to the $n$-bit code word which is closest, in the sense of the Euclidean distance,
to the received sequence. Such a decoding process is called *unquantized soft-
decision* decoding. In this model, the combination of modulator, channel, and
demodulator is equivalent to a binary-input, continuous-output channel. It is intuitive that this approach presents a higher reliability than that achieved with the
hard-decision scheme. In fact, the decoder can take advantage of the additional
information contained in the unquantized samples that represent each individual
binary transmitted waveform. An intermediate case, called *soft-decision* decoding, is represented by a demodulator whose output is quantized to $Q$ levels, with
$Q > 2$. In this case, the combination of modulator, channel, and demodulator is
equivalent to a binary input, $Q$-ary output discrete channel. The advantage over
the analog (unquantized) case is that all the processing can be accomplished with
digital circuitry. Therefore, it represents an approximation of the unquantized
soft-decision decoding.

The advantage of a coded transmission scheme over an uncoded one is usually measured by its *coding gain*. This is defined as the difference (in decibels)
in the required value of $\mathcal{E}_b/N_0$ to achieve a given bit error probability between
a binary antipodal uncoded transmission and the encoded one. This concept is
represented qualitatively in Fig. 10.2, where we plot the two curves expressing
the bit error probability $P_b(e)$ versus the signal-to-noise ratio per bit $\mathcal{E}_b/N_0$ for
the uncoded and encoded systems. The typical behavior of the two curves of
Figure 10.2 suggests two considerations:

- The coding gain, which depends on the value of the bit error probability
  (and thus on the signal-to-noise ratio), increases with the signal-to-noise
  ratio and tends (for $\mathcal{E}_b/N_0 \to \infty$ and hence for $P_b(e) \to 0$) to an asymptotic value that will be evaluated later in the chapter.

- For low values of the signal-to-noise ratio, there can be a crossing between
  the uncoded and coded curves, meaning that the coding gain becomes negative. In other words, there is a limit to what a code can do in terms of
  improving a bad channel.

To quantitatively assess the limits of the coding gain, we have plotted in Figure 10.3 the curve of the binary uncoded antipodal scheme (curve A) with the two
channel capacity limits: the first (curve B), which tends to $-1.6$ dB, corresponding to the infinite-bandwidth capacity limit and to soft-decision decoding, and
the second (curve C), which tends to $0.4$ dB, the BSC capacity limit which refers
to a hard-decision demodulator. These limits had been evaluated in Section 3.3.

The regions between the uncoded curve and those of the capacity limits represent the region of potential coding gains. As an example, for a bit error proba-

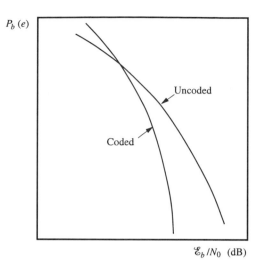

Figure 10.2: *Typical behavior of the bit error probability versus bit signal-to-noise ratio for uncoded and coded systems.*

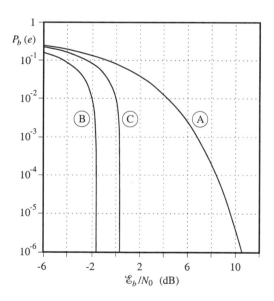

Figure 10.3: *Potential coding gains of coded transmission with respect to binary uncoded antipodal transmission.*

bility of $10^{-5}$, a potential coding gain of 11.2 dB is theoretically available in the case of unquantized soft-decision decoding. Another limit, the *cutoff rate* limit, will be derived when analyzing the performance of coded transmission.

The fifty years that separate us from the channel coding theorem of Shannon has seen a great research effort aiming at filling the channel coding gap through the discovery of codes approaching the capacity limits. Until recently, these efforts had been very successful up to the *cutoff rate* limit (see Section 10.4), a couple of dB short of the capacity limit, but were unable to reach the region between cutoff rate and capacity. As we shall see at the end of next chapter, we now know a way to design codes that can approach to within 0.5 dB of the coding gain promised by the capacity limit at bit error probabilities of $10^{-5}$ to $10^{-7}$.

## 10.2. Block codes

We will consider mainly *binary* codes, i.e., codes for which both the data words and the code words are formed by binary digits 0 and 1. This concept can be extended to $q$-ary codes, and a particularly important case occurs when $q = 2^b$ is a power of 2; in this case, $q$ admits a binary representation with $b$ bits, and the $(n, k)$ code of $q$-ary elements is equivalent to an $(nb, kb)$ binary code.

The basic feature of block codes is that the block of $n$ digits (code word) generated by the encoder depends only on the corresponding block of $k$ digits generated by the source (data word). Therefore, the encoder is memoryless. A great deal of block code theory is an extension of the notion of *parity check*. Take a sequence of $k$ binary digits. Transform it into a sequence of length $n = k + 1$ digits by simply adding in the last position a new binary digit, following the rule that the number of ones in the new sequence must be even. This redundant digit is called a parity-check digit. In this way, any error event on the channel that changes the parity of the sequence from even to odd can be detected by the decoder.

Parity-check codes are a particular class of block codes in which the digits of the code word are a set of $n$ parity checks performed on the $k$ information digits. The code is usually referred to as an $(n, k)$ code. An encoder (or, simply, a code) is called *systematic* when the first $k$ digits in the code word are a replica of the information digits in the data word, and the remaining $(n - k)$ digits are parity checks on the $k$ information digits. Parity checks in binary sequences are formally dealt with using modulo-2 arithmetic, in which the rules of ordinary arithmetic hold true except that the sum $(1 + 1)$ is 0, not 2. Throughout this chapter, modulo-2 arithmetic will be used unless otherwise specified. A functional block diagram of the encoder is shown in Fig. 10.4. It consists of a $k$-stage in-

Input register

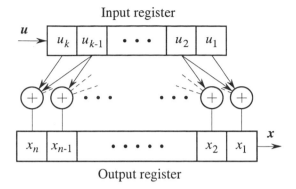

Output register

Figure 10.4: *Block diagram of the encoder for a parity-check code,*

put shift register, $n$ modulo-2 adders, and an $n$-stage output shift register. Each adder is connected to a subset of stages of the input register in order to perform the desired parity checks. The vector $\mathbf{u} = [u_1, u_2, \ldots, u_k]$ of $k$ information digits is fed into the input register. When this register is loaded, the content of each adder is fed in parallel into the corresponding stage of the output register, which shifts out the code word $\mathbf{x} = [x_1, x_2, \ldots, x_n]$. While shifting out one code word, the input register is reloaded and the whole operation repeated. The clocks for the two registers are different, the output rate being higher by a factor $1/R_c$. The following simple examples will clarify these concepts.

**Example 10.1   Repetition code** $(3, 1)$
In this code, each code word of length $n = 3$ is defined by the relations

$$x_1 = u_1, \quad x_2 = u_1, \quad x_3 = u_1 \tag{10.1}$$

The encoder is sketched in Fig.10.5. Obviously, the adders are omitted in this case. The resulting *repetition* encoder is defined by the correspondence

| Data words | Code words |
|:---:|:---:|
| 0 | 000 |
| 1 | 111 |

Notice that the encoder is systematic, and that only two of the eight sequences of length 3 are used in the code.                                                                    □

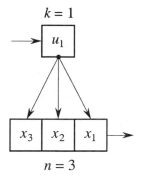

Figure 10.5: *Encoder for the repetition code* $(3, 1)$.

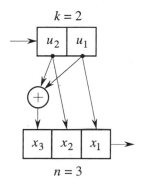

Figure 10.6: *Encoder for the parity-check code* $(3, 2)$.

**Example 10.2  Parity-check code** $(3, 2)$

This is a code in which the third digit is a parity check on the first two digits. The code word is defined by the relations

$$x_1 = u_1, \quad x_2 = u_2, \quad x_3 = u_1 + u_2 \tag{10.2}$$

The systematic encoder is shown in Fig. 10.6. It is defined by the correspondence

| Data words | Code words |
|:---:|:---:|
| 00 | 000 |
| 01 | 011 |
| 10 | 101 |
| 11 | 110 |

Notice that only four of the eight sequences of length 3 are used in the code.　□

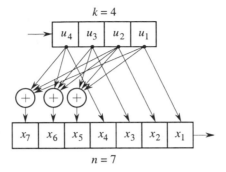

Figure 10.7: *Encoder for the Hamming code* $(7, 4)$.

**Example 10.3   Hamming code** $(7, 4)$
The Hamming code $(7, 4)$ is defined by the relations

$$
\begin{aligned}
x_i &= u_i, & i = 1, 2, 3, 4 \\
x_5 &= u_1 + u_2 + u_3 \\
x_6 &= u_2 + u_3 + u_4 \\
x_7 &= u_1 + u_2 + u_4 .
\end{aligned}
\tag{10.3}
$$

The corresponding systematic encoder is shown in Fig. 10.7. It is defined by the correspondence

| Data words | Code words |
|:---:|:---:|
| 0000 | 0000 000 |
| 0001 | 0001 011 |
| 0010 | 0010 110 |
| 0011 | 0011 101 |
| 0100 | 0100 111 |
| 0101 | 0101 100 |
| 0110 | 0110 001 |
| 0111 | 0111 010 |
| 1000 | 1000 101 |
| 1001 | 1001 110 |
| 1010 | 1010 011 |
| 1011 | 1011 000 |
| 1100 | 1100 010 |
| 1101 | 1101 001 |
| 1110 | 1110 100 |
| 1111 | 1111 111 |

Notice that only 16 of the 128 sequences of length 7 are used in the code. □

These examples show that all the information required to specify the encoder operation is contained in relations of the type of (10.1), (10.2), and (10.3). With reference to Figs. 10.5, 10.6 and 10.7, these relations specify the connections between the input register cells and the adders. If the encoder is systematic, only the $(n - k)$ parity-check equations of the redundant digits must be assigned.

The information that specifies the encoding rule, and thus the structure of the encoder itself, can be concisely represented by the *generator* matrix $\mathbf{G}$ of the code. It is a $k \times n$ matrix whose $(i, j)$ entry is 1 if the $i$-th cell of the input register is connected to the $j$-th adder, and 0 otherwise. Using the row-vector notation for the data word $\mathbf{u}$ and the code word $\mathbf{x}$, the encoding rule is described by the equation

$$\mathbf{x} = \mathbf{u}\mathbf{G} \qquad (10.4)$$

It is easily seen that obtaining a code word $\mathbf{x}$ through (10.4) is equivalent to summing the rows of the matrix $\mathbf{G}$ corresponding to the ones contained in the information sequence $\mathbf{u}$.

**Example 10.4** For the (7, 4) Hamming code of Example 10.3, the generator matrix $\mathbf{G}$ can be found by inspection of the encoder of Fig. 10.7 as follows:

$$\mathbf{G} = \begin{bmatrix} 1 & 0 & 0 & 0 & \vdots & 1 & 0 & 1 \\ 0 & 1 & 0 & 0 & \vdots & 1 & 1 & 1 \\ 0 & 0 & 1 & 0 & \vdots & 1 & 1 & 0 \\ 0 & 0 & 0 & 1 & \vdots & 0 & 1 & 1 \end{bmatrix} \qquad (10.5)$$

If we want the code word corresponding to the data word $\mathbf{u} = [1100]$, we must add the first two rows of $\mathbf{G}$, obtaining

$$\begin{array}{r} 1000101\ + \\ 0100111\ = \\ \hline 1100010 \end{array}$$

and the result agrees with the code table given in Example 10.3. □

For systematic encoders, the first $k$ columns of $\mathbf{G}$ form a $k \times k$ identity matrix, so that $\mathbf{G}$ assumes the form

$$\mathbf{G} = [\mathbf{I}_k \ \vdots \ \mathbf{P}] \qquad (10.6)$$

where $\mathbf{I}_k$ is the $k \times k$ identity matrix and $\mathbf{P}$ is a $k \times (n - k)$ matrix containing the information regarding the parity checks. The knowledge of $\mathbf{P}$ completely defines the encoding rule for a systematic encoder.

The following important properties of parity check codes can be proved.

**Property 1** The block code consists of all possible sums of the rows of the generator matrix.

**Property 2** The sum of two code words is still a code word.

**Property 3** The $n$-tuple of all zeros is always a code word.

Because of these properties, parity-check codes are also called *linear* codes. Linear block codes can be interpreted as being a subspace of the vector space containing all $2^n$ binary $n$-tuples. From this algebraic viewpoint, the rows of the generator matrix $\mathbf{G}$ are a basis of the subspace and consist of $k$ linearly independent code words. In fact, all their $2^k$ linear combinations generate the entire subspace, that is, the code. Note that any $k$ linearly independent code words of an $(n, k)$ linear code can be used to form a generator matrix for the code. For these reasons, it is straightforward to conclude that any generator matrix of an $(n, k)$ block code can be reduced, by means of row operations and column permutations, to the systematic form (10.6), which is also called *reduced-echelon* form. However, while row operations do not alter the code, column permutations may lead to a different set of code words, i.e., to a code that differs from the original one in the arrangement of its binary symbols. Two codes whose generator matrices can be obtained from each other by row operations and column permutations have the same *word* error probability, and, because of that, are said to be *equivalent*. Note, however, that their *bit* error probabilities (it will be defined later in the chapter, together with the word error probability) may be different, because equivalent codes can admit different encoders, and hence different mappings between data words and code words.

Thus, every $(n, k)$ block code is equivalent to a systematic $(n, k)$ block code (see Problem 10.5). Therefore, if the word error probability is the parameter of interest, we can consider only systematic codes without loss of generality.

An important parameter of a code word is its *Hamming weight*, that is, the number of ones that it contains. The set of all distinct weights in a code, together with the number of code words of that weight, is the *weight distribution* of the code. Owing to the previous definition, equivalent codes have the same weight distributions. Given two code words $\mathbf{x}_i$ and $\mathbf{x}_j$, it is useful to define a quantity to measure their difference. This quantity is the *Hamming distance* $d_{ij}$ between the two code words, defined as the number of positions in which the two code words differ. Clearly, $d_{ij}$ satisfies the condition $0 \leq d_{ij} \leq n$. The smallest among the Hamming distances between distinct code words ($i \neq j$) is called the *minimum distance* $d_{\min}$ of the code. The following property allows an easy computation of $d_{\min}$ for linear codes.

**Property 4**   The minimum distance of a linear block code is the minimum weight of its nonzero code words.

In fact, the distance between two binary sequences is equal to the weight of their modulo-2 sum, and the sum of two code words is still a code word (Property 2).

**Example 10.5**   Consider again the (7, 4) Hamming code of Example 10.3. From the code table, we obtain the following weight distribution

| Weight | Number of code words |
|:------:|:--------------------:|
| 0 | 1 |
| 3 | 7 |
| 4 | 7 |
| 7 | 1 |

Using Property 4, we get $d_{\min} = 3$.                                                □

### 10.2.1.   Error-detecting and error-correcting capabilities of a block code

Assume that the demodulator makes hard decisions so that the discrete channel between the channel encoder and decoder can be modeled as a binary symmetric channel (BSC). Each transmitted code word $\mathbf{x}$ is received at the decoder input as a sequence $\mathbf{y}$ of $n$ binary digits (Fig. 10.1). The encoder is systematic. Therefore, the first $k$ digits of $\mathbf{y}$ are the received information digits, while the remaining $(n - k)$ digits are the received parity-check digits. The sequence $\mathbf{y}$ can contain independent random errors caused by the channel noise. Let us define a binary vector e called an *error* vector:

$$\mathbf{e} = [e_1, \ldots, e_n] \tag{10.7}$$

Each component $e_i$ is 1 if the channel has changed the $i$-th transmitted digit; otherwise, it is 0. The received vector is then

$$\mathbf{y} = \mathbf{x} + \mathbf{e} \tag{10.8}$$

where $\mathbf{x}$ is the transmitted code word. The decoder recomputes the $(n - k)$ parity-checks using the first $k$ received bits, and compares them with the $(n - k)$

received parity-checks. If they match, the received sequence is a code word. Otherwise, an error is detected. Therefore, at least for error detection, the decoding rule is very simple: an error pattern is detected whenever at least one of the $(n - k)$ controls on parity checks fails. Let us define a vector s that contains the parity checks performed on the received word $\mathbf{y}$. Its $(n - k)$ binary digits are zeros for all parity checks that are satisfied, and ones for those that are not. The vector s is called the *syndrome* of the received vector $\mathbf{y}$. Recalling the definition (10.6) of the generator matrix $\mathbf{G}$ of a systematic code, it is easily verified that the syndrome can be obtained from the equation

$$s = \mathbf{yH}' \tag{10.9}$$

where the prime means transpose, and where we have introduced the *parity-check* matrix $\mathbf{H}$, defined as

$$\mathbf{H} \triangleq [\mathbf{P}' \vdots \mathbf{I}_{n-k}] \tag{10.10}$$

It is an $(n - k) \times n$ matrix, whose rows represent the parity-check symbols computed by the decoder. A direct calculation using (10.6) shows that

$$\mathbf{GH}' = \mathbf{0} \tag{10.11}$$

where $\mathbf{0}$ is a $k \times (n - k)$ matrix all of whose elements are zero.

**Example 10.6**   Consider again the (7, 4) Hamming code of Example 10.3. The three parity-check symbols computed by the decoder on the received sequence $\mathbf{y}$ can be written by inspection of (10.3) as follows:

$$\begin{array}{rcl}
s_1 & = & (y_1 + y_2 + y_3) + y_5 \\
s_2 & = & (y_2 + y_3 + y_4) + y_6 \\
s_3 & = & (y_1 + y_2 + y_4) + y_7
\end{array} \tag{10.12}$$

The parity check matrix is therefore

$$\mathbf{H} = \begin{bmatrix} 1 & 1 & 1 & 0 & \vdots & 1 & 0 & 0 \\ 0 & 1 & 1 & 1 & \vdots & 0 & 1 & 0 \\ 1 & 1 & 0 & 1 & \vdots & 0 & 0 & 1 \end{bmatrix} \tag{10.13}$$

It can be verified that (10.13) is also obtained from (10.5) using the definition (10.10). The property (10.11) can also be verified.                                                    □

From the definition of the syndrome associated with a received sequence $\mathbf{y}$, the following two properties can be verified:

**Property 5** The syndrome associated with a sequence **y** is a zero vector if and only if **y** is a code word.

**Property 6** The decoder can detect all channel errors represented by vectors **e** that are not code words.

Since the channel can introduce $2^n$ different error vectors, only $2^k$ of them are not detected by the decoder, that is, those corresponding to the set of code words. Finally, since no code word exists with a weight less than $d_{\min}$ (except, of course, the all-zero code word), the following theorem can be proved.

**Theorem 10.1**

A linear block code $(n, k)$ with minimum distance $d_{\min}$ can detect all error vectors of weight not greater than $(d_{\min} - 1)$. $\triangledown$

Until now, we have only explored the error-detection capabilities of a hard-decision decoder. The problem of error correction is more complicated, since the syndrome does not contain sufficient information to locate the errors. Using (10.8), the expression (10.9) for the syndrome can be rewritten as

$$\mathbf{s} = \mathbf{y}\mathbf{H}' = (\mathbf{x} + \mathbf{e})\mathbf{H}' \qquad (10.14)$$

where **x** is a code word. Since $\mathbf{x}\mathbf{H}' = \mathbf{0}$ (Property 5), there are $2^k$ different sequences **y** that generate the same syndrome. They are obtained by summing to a given error vector **e** the $2^k$ code words. Therefore, given a transmitted code word **x**, there are $2^k$ error vectors that give the same syndrome. Which one actually occurred is an uncertainty that cannot be removed by using only the syndrome.

A suitable decoding algorithm must be elaborated. Assume that maximum likelihood (ML) hard decisions are taken by the decoder. This means that it achieves minimum word error probability on the received code words when they are equally likely. If $p$ is the transition probability of the equivalent BSC implied by hard decisions, we have

$$P(\mathbf{y} \mid \mathbf{x}_i) = p^{d_i}(1 - p)^{n-d_i} \qquad (10.15)$$

where $n$ is the block length and $d_i$ is the Hamming distance between the received sequence **y** and the transmitted code word $\mathbf{x}_i$. Assuming, without loss of generality, $p < 1/2$, the probability $P(\mathbf{y} \mid \mathbf{x}_i)$ is a monotonic decreasing function of $d_i$. Therefore, ML decoding is accomplished with minimum Hamming-distance decisions. The "best" decoding algorithm decides for the code word $\mathbf{x}_i$ which is closest to **y**. Recalling the discussion regarding (10.14), we can conclude that

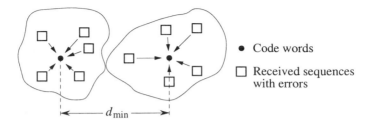

Figure 10.8: *Qualitative representation of the decision regions assigned to code words.*

the minimum-distance decoding algorithm assumes that the error vector **e** that actually occurred is the minimum-weight error vector in the set of the $2^k$ error vectors yielding the syndrome associated with the received sequence **y**. Before considering this decoding rule in detail using the minimum-distance decoding algorithm, let us relate the error-correcting capabilities of a block code to the code parameter $d_{min}$.

**Theorem 10.2**

A linear block code $(n, k)$, with minimum distance $d_{min}$, can correct all error vectors containing no more than $t = \lfloor (d_{min} - 1)/2 \rfloor$ errors, where $\lfloor a \rfloor$ (the "floor" of $a$) denotes the largest integer contained in $a$. The code is then a $t$-error-correcting code, and is often denoted as an $(n, k, t)$ code. $\triangledown$

**Proof of Theorem 10.2**

The decoding algorithm is implemented by assigning to each code word a decision region containing the subset of all the received sequences that are closer to it than to any other (*minimum distance decoding*, see Fig. 10.8). An error vector with no more than $\lceil (d_{min} - 1)/2 \rceil$ errors produces a received sequence lying inside the correct decision region. Error correction is therefore possible. **QED**

The results of Theorems 10.1 and 10.2 are summarized in Table 10.1.

Based on previous Theorems 10.1 and 10.2, a design goal for a block code $(n, k)$ is to use its redundancy to achieve the largest possible minimum distance $d_{min}$. So far, no general solution to this problem is known. Instead, upper and lower bounds to $d_{min}$ are used. Some of them will be described in Section 10.4.2.

| $d_{min}$ | Errors detected | Errors corrected |
|:---:|:---:|:---:|
| 2 | 1 | 0 |
| 3 | 2 | 1 |
| 4 | 3 | 1 |
| 5 | 4 | 2 |
| 6 | 5 | 2 |
| 7 | 6 | 3 |
| 8 | 7 | 3 |
| 9 | 8 | 4 |

Table 10.1: *Error correction and error detection capabilities of linear block codes as a function of $d_{min}$.*

### 10.2.2. Decoding table and standard array of a linear block code

Using the minimum-distance hard decoding algorithm just described, the decoding operation can be performed by looking for the code word nearest to the received sequence. This approach requires the storage of the $2^k$ code words and repeated comparisons with the received sequence. The total storage requirement is on the order of $n \times 2^k$ bits. Hence, the approach becomes rapidly impractical even for moderately-sized codes. Also, the comparison process is unacceptably long when $n$ and $k$ are large.

A more efficient approach is to evaluate the syndrome associated with the received sequence **y** by assuming that the error vector **e** that actually occurred is the minimum-weight vector in the set of the $2^k$ vectors that generate the same syndrome. With this approach, we can build a decoding table by associating with each syndrome the error vector of minimum weight that generated it. The positions of the ones in the error vector indicate the digits that must be corrected in the received sequence **y**. This approach is better clarified by the following example.

**Example 10.7** The $(7, 4)$ Hamming code has minimum distance 3. Thus, it is expected to correct all single errors. There are, of course, 128 possible received words and only 8 different syndromes. All these sequences are included in Table 10.2. They are grouped in rows containing all sequences that share the same syndrome. The syndrome is shown as the first entry in each row. The first column of the table contains all error vectors of minimum weight. It can be verified by inspection that each error vector containing only one error has a different syndrome, and hence it can be corrected. Therefore, the decoding table for this code is the following:

## 10. *Improving the transmission reliability: Block codes*

| Syn-dromes | Coset leaders | Error patterns — Other errors | | | | | | | | | | | | | | |
|---|---|---|---|---|---|---|---|---|---|---|---|---|---|---|---|---|
| 000 | 0000000 | 1000101 | 0100111 | 0010110 | 0001011 | 1100010 | 1010011 | 1001110 | 0110001 | 0101100 | 0011101 | 0111010 | 1011000 | 1101001 | 1110100 | 1111111 |
| 001 | 0000001 | 1000100 | 0100110 | 0010111 | 0001010 | 1100011 | 1010010 | 1001111 | 0110000 | 0101101 | 0011100 | 0111011 | 1011001 | 1101000 | 1110101 | 1111110 |
| 010 | 0000010 | 1000111 | 0100101 | 0010100 | 0001001 | 1100000 | 1010001 | 1001100 | 0110011 | 0101110 | 0011111 | 0111000 | 1011010 | 1101011 | 1110110 | 1111101 |
| 011 | 0001000 | 1001101 | 0101111 | 0011110 | 0000011 | 1101010 | 1011011 | 1000010 | 0111001 | 0100110 | 0010101 | 0110010 | 1010000 | 1100001 | 1111100 | 1110111 |
| 100 | 0000100 | 1000001 | 0100011 | 0010010 | 0001111 | 1100110 | 1010111 | 1001010 | 0110101 | 0101000 | 0011001 | 0111110 | 1011100 | 1101101 | 1110000 | 1111011 |
| 101 | 1000000 | 0000101 | 1100111 | 1010110 | 1001011 | 0100010 | 0010011 | 0001110 | 1110001 | 1101100 | 1011101 | 1111010 | 0011000 | 0101001 | 0110100 | 0111111 |
| 110 | 0010000 | 1010101 | 0110111 | 0000110 | 0011011 | 1110010 | 1000011 | 1011110 | 0100001 | 0111100 | 0001101 | 0101010 | 1001000 | 1111001 | 1100100 | 1101111 |
| 111 | 0100000 | 1100101 | 0000111 | 0110110 | 0101011 | 1000010 | 1110011 | 1101110 | 0010001 | 0001100 | 0111101 | 0011010 | 1111000 | 1001001 | 1010100 | 1011111 |

Table 10.2: *Standard array of the* (7, 4) *Hamming code.*

| Syndrome | Error vector | Digit in error |
|----------|--------------|----------------|
| 000      | 0000000      | None           |
| 001      | 0000001      | 7              |
| 010      | 0000010      | 6              |
| 011      | 0001000      | 4              |
| 100      | 0000100      | 5              |
| 101      | 1000000      | 1              |
| 110      | 0010000      | 3              |
| 111      | 0100000      | 2              |

Obviously, the syndrome 000 corresponds to the set of the 16 code words. The syndrome 111 locates an error in the second position of the received sequence, and so on. Table 10.2 can also be interpreted as follows. Assume that the sequence 1101010 is received. The corresponding syndrome is 011. Therefore, an error in position 4 is assumed and corrected. The code word obtained, which is 1100010, appears at the top of the column containing the received sequence.                                          □

A table such as Table 10.2, containing all the $2^n$ $n$-tuples (the possible received words) of length $n$ organized in that order, is called the *standard array* of the code. It has $2^k$ columns and $2^{n-k}$ rows. The rows are called *cosets*. The first word in each row is nominated a *coset leader*. The top word in a column is a code word, and each coset leader is the minimum-weight word that generates the syndrome common to all words of that coset.

The decoding table is built by simply associating with each syndrome the corresponding coset leader of the standard array. The coset leaders are therefore the correctable error vectors; if the error vector is not a coset leader, then an incorrect decoding will be performed. To minimize the average word error probability, the coset leaders must be the error vectors that are the most likely to occur. For a BSC, the coset leaders are the minimum-weight words associated with a given syndrome. Therefore, the decoding algorithm works as follows:

1. Compute the syndrome for the received sequence.

2. Find the correctable error vector (coset leader) in the decoding table.

3. Get the estimated code word by adding the correctable error vector to the received word.

The decoding table requires the storage of $2^{n-k}$ syndromes of length $(n - k)$ and of $2^{n-k}$ error patterns of length $n$: a total of $2^{n-k} \times (2n - k)$ bits. For high rate codes ($k \simeq n$), the storage requirement is close to $n \times 2^{n-k}$, considerably less as compared to the $n \times 2^k$ bits required by an exhaustive search. In spite of

| $l$ | $(n, k)$ |
|---|---|
| 2 | $(3, 1)$ |
| 3 | $(7, 4)$ |
| 4 | $(15, 11)$ |
| 5 | $(31, 26)$ |
| 6 | $(63, 57)$ |
| 7 | $(127, 120)$ |

Table 10.3: *Parameters of the first Hamming codes.*

this interesting result, the decoding table, too, becomes impractical when $n$ and $k$ are large numbers. In that case a more elaborate algebraic structure must be assigned to the code in order to employ decoding strategies based on computational algorithms, rather than on look-up tables.

### 10.2.3. Hamming codes

Equation (10.14) can be rewritten in the form

$$s = eH'$$  (10.16)

Therefore, the syndrome of a given sequence is the sum of the columns of $H$ corresponding to the position of the ones in the error vector. Consequently, if a column of $H$ is zero, an error in that position cannot be detected. Furthermore, if two columns of $H$ are equal, a single error in one of those two positions cannot be corrected since the two syndromes are not distinct. We can conclude that a block code can correct all single errors if and only if the columns of its parity-check matrix $H$ are nonzero and distinct.

*Hamming codes* are characterized by a matrix $H$ whose columns are all the possible sequences of $(n - k)$ binary digits except the zero sequence. For every $l = 2, 3, 4, \ldots$, there is a $(2^l - 1, 2^l - 1 - l)$ Hamming code. These codes have $d_{min} = 3$ and are thus capable of correcting all single errors. Their rate $R_c = (2^l - 1 - l)/(2^l - 1)$ increases with $l$ and approaches 1 for $l \to \infty$. The parameters of the first six Hamming codes are listed in Table 10.3.

**Example 10.8**   The parity-check matrix of the Hamming code (15,11) is the following:

$$H = \begin{bmatrix} 0 & 0 & 0 & 0 & 0 & 0 & 0 & 1 & 1 & 1 & 1 & 1 & 1 & 1 & 1 \\ 0 & 0 & 0 & 1 & 1 & 1 & 1 & 0 & 0 & 0 & 0 & 1 & 1 & 1 & 1 \\ 0 & 1 & 1 & 0 & 0 & 1 & 1 & 0 & 0 & 1 & 1 & 0 & 0 & 1 & 1 \\ 1 & 0 & 1 & 0 & 1 & 0 & 1 & 0 & 1 & 0 & 1 & 0 & 1 & 0 & 1 \end{bmatrix}$$  (10.17)

Notice that $\mathbf{H}$ is not written in the systematic form of (10.10), its columns being in lexicographical order. It can be reduced to systematic form by a simple rearrangement of columns. The interesting property of (10.17) is that the 4-tuple in each column, as a binary number, identifies the column position. Therefore, an error vector with a single error will generate a syndrome that gives, in binary form, the position of the error in the received sequence. This information can be used for correction. □

Hamming codes have an interesting property that can be verified by inspection of the standard array (see Table 10.2 for the (7, 4) code). All possible received sequences have Hamming distance 1 from one of the code words. Codes of this type are called *perfect codes*. Another property of the Hamming codes is that they are one of the few classes of codes for which the complete weight distribution is known. The weight distribution of a code can be represented in a compact form as a polynomial, called the *weight enumerating function* (WEF) of the code. It is a polynomial in the indeterminate $D$

$$A(D) = \sum_{d=0}^{n} A_d D^d \tag{10.18}$$

where $A_d$ is the number (multiplicity) of code words in the code with weight (or, equivalently, Hamming distance from the all-zero code word) $d$. For the Hamming codes, the WEF can be shown to be

$$A(D) = \frac{1}{n+1}[(1+D)^n + n(1+D)^{(n-1)/2}(1-D)^{(n+1)/2}] \tag{10.19}$$

The result of Example 10.5 can be checked against (10.19).

Each Hamming code can be converted to a new code by adding one parity digit that checks all previous $n$ digits of the code word. This results in a class of $(2^l, 2^l - 1 - l)$ block codes called *extended* Hamming codes. Their parity-check matrix $\mathbf{H}_{\text{ext}}$ is obtained by adding a new row to the Hamming parity-check matrix $\mathbf{H}$ as follows:

$$\mathbf{H}_{\text{ext}} = \begin{bmatrix} & & & \vdots & 0 \\ & & & \vdots & 0 \\ & \mathbf{H} & & \vdots & \vdots \\ & & & \vdots & 0 \\ 1 & 1 & \dots & 1 & \vdots & 1 \end{bmatrix} \tag{10.20}$$

The last row represents the overall parity-check digit. Since, with an overall parity-check, the weight of every code word must be even, the extended Hamming codes have $d_{\min} = 4$. Their particular structure makes it possible to detect all double errors while simultaneously correcting all single errors (as in the original Hamming codes). In fact, the syndromes for double errors form a subset distinct from that of the syndromes for single errors. The decoding algorithm works as follows:

1.  If the last digit of the syndrome is 1, then the number of errors must be odd. Using the minimum-distance algorithm, correction can be performed as for the Hamming codes.

2.  If the last digit of the syndrome is 0, but the syndrome is not all-zero, then no correction is possible since at least two errors must have occurred. A double error is therefore detected.

This property of extending a code by the addition of an overall parity check can be applied to any linear block code other than the Hamming codes. In particular, any linear $(n, k)$ block code with an odd minimum distance can be converted into an extended $(n + 1, k)$ block code with a minimum distance increased by one.

### 10.2.4. Dual codes

The generator matrix $\mathbf{G}$ and the parity-check matrix $\mathbf{H}$ of a linear $(n, k)$ block code are related by (10.11). This relation can be rewritten as

$$\mathbf{HG'} = \mathbf{0} \tag{10.21}$$

Thus, the two matrices can be interchanged and the $\mathbf{H}$ matrix can be the generator matrix of a new $(n, n - k)$ block code. Codes that are so related are said to be *dual* codes. There is a very interesting relationship between the weight distributions of two dual codes. Let $A(D)$ be the weight enumerating function of the $(n, k)$ block code and $A_{\mathrm{dual}}(D)$ the weight enumerating function of its $(n, n - k)$ dual code. Then, the two weight enumerating functions are related by the identity (MacWilliams and Sloane, 1977)

$$A_{\mathrm{dual}}(D) = 2^{-k}(1 + D)^n A\left(\frac{1 - D}{1 + D}\right) \tag{10.22}$$

This relationship is very useful in determining the weight structure of high-rate block codes through an exhaustive computer search performed on their low-rate dual codes.

### 10.2.5. Maximal-length codes

The duals of the Hamming codes are called *maximal-length* codes. Therefore, for every $l = 2, 3, 4, \dots$ there is a $(2^l - 1, l)$ maximal-length code. Its generator matrix is the parity-check matrix of the corresponding Hamming code.

The weight distribution of these codes can be easily determined by introducing (10.19) into (10.22). The weight enumerator $A(D)$ for the maximal-length codes is thus found to be

$$A(D) = 1 + (2^l - 1)D^{2^{l-1}} \tag{10.23}$$

Hence, all nonzero code words have identical weight $2^{l-1}$. Also, this is the minimum distance of the code. These codes are also called *equidistant* or *simplex* codes. Additional insight into the properties of these codes will be obtained later in connection with the description of cyclic codes.

### 10.2.6.    Reed-Muller codes

The *Reed-Muller* codes are a class of linear block codes covering a wide range of rates and minimum distances. They present very interesting properties, among them, the fact that they can be soft-decoded by using a simple trellis (see Forney, 1988b).

For any $m$ and $r < m$, there is a Reed-Muller code with parameters given by

$$n = 2^m, \quad k = \sum_{i=0}^{r} \binom{m}{i}, \quad d_{\min} = 2^{m-r} \tag{10.24}$$

The generator matrix $\mathbf{G}$ of the $r$th-order Reed-Muller code is defined by assigning a set of vectors as follows. Let $\mathbf{v}_0$ be a vector whose $2^m$ elements are all ones, and let $\mathbf{v}_1, \mathbf{v}_2, \ldots, \mathbf{v}_m$ be the rows of a matrix with all possible $m$-tuples as columns. The rows of the $r$th-order generator matrix are the vectors $\mathbf{v}_0, \mathbf{v}_1, \ldots, \mathbf{v}_m$ and all the products of $\mathbf{v}_1, \ldots, \mathbf{v}_m$ two at a time, three at a time, up to $r$ at a time. Here the product vector $\mathbf{v}_i \mathbf{v}_j$ has components given by the products of the corresponding components of $\mathbf{v}_i$ and $\mathbf{v}_j$.

**Example 10.9**    In this example, we show how to generate the Reed-Muller codes with $m = 3$. There are two codes. They have the following parameters:

| $r$ | $n$ | $k$ | $d_{\min}$ |
|---|---|---|---|
| 1 | 8 | 4 | 4 |
| 2 | 8 | 7 | 2 |

The vectors used for building the generator matrices are given in Table 10.4. The first-order code $(r = 1)$ is generated by using the vectors $\mathbf{v}_0, \mathbf{v}_1, \mathbf{v}_2, \mathbf{v}_3$ as rows of the generator matrix. The second-order code $(r = 2)$ is generated by augmenting this matrix with the additional three rows of Table 10.4.                                              □

The first-order Reed-Muller codes are closely related to the maximal-length codes. If a maximal-length code is extended by adding an overall parity check,

| | | | | | | | | |
|---|---|---|---|---|---|---|---|---|
| $v_0$ | 1 | 1 | 1 | 1 | 1 | 1 | 1 | 1 |
| $v_1$ | 0 | 0 | 0 | 0 | 1 | 1 | 1 | 1 |
| $v_2$ | 0 | 0 | 1 | 1 | 0 | 0 | 1 | 1 |
| $v_3$ | 0 | 1 | 0 | 1 | 0 | 1 | 0 | 1 |
| $v_1v_2$ | 0 | 0 | 0 | 0 | 0 | 0 | 1 | 1 |
| $v_1v_3$ | 0 | 0 | 0 | 0 | 0 | 1 | 0 | 1 |
| $v_2v_3$ | 0 | 0 | 0 | 1 | 0 | 0 | 0 | 1 |

Table 10.4: *Vectors for constructing the generator matrix of Reed-Muller codes with $m = 3$.*

we obtain an *orthogonal* code. This code has $2^m$ code words. Each has weight $2^{m-1}$, except for the all-zero code word. Therefore, every code word agrees in $2^{m-1}$ positions and disagrees in $2^{m-1}$ positions with every other code word. If this code is transmitted using an antipodal signaling scheme, each code word is represented by one out of $2^m$ orthogonal signals. This explains the name "orthogonal" code. For the case $m = 3$, the code generator matrix consists of the three rows $v_1, v_2$, and $v_3$ of Table 10.4. In fact, the first column represents the overall parity-check digit, whereas the other columns are all the seven possible triples of binary digits. The first-order Reed-Muller code is obtained from this code (the orthogonal code) by adding to the generator matrix the all-ones vector $v_0$. In terms of transmitted signals, this operation adds to the original orthogonal signal set the opposite of each signal. For this reason, the code is also called a *biorthogonal* code. Finally, notice that the $r$th-order Reed-Muller code is the dual of the Reed-Muller code of order $(m - r - 1)$.

### 10.2.7.  Cyclic codes

The *cyclic codes* are parity-check codes that present a large amount of mathematical structure. These codes share, of course, all the properties previously described for parity-check codes, but, in addition, have peculiar properties that allow easy encoding operations and simple decoding algorithms. Cyclic codes are, for this reason, of great practical interest.

### Definition 10.1

An $(n, k)$ linear block code is a cyclic code if and only if any cyclic shift of a code word produces another code word.

**Example 10.10** It can be verified that the (7, 4) Hamming code of Example 10.3 is a cyclic code. Take, for instance, the code word 0111010. There are six different cyclic shifts of this code word.

1110100  1101001  1010011  0100111  1001110  0011101

They all belong to the set of code words. The same is true for all the code words. □

In dealing with cyclic codes, it is useful to represent a binary sequence of $n$ bits as a polynomial in the indeterminate $Z$ of degree not greater than $(n-1)$ with binary (0 and 1) coefficients. The binary digits of a code word will be numbered in decreasing order from $(n-1)$ to 0, so that each index matches the exponent of $Z$. A code word $\mathbf{x} = [x_{n-1}, x_{n-2}, \dots, x_0]$ is then represented by the code polynomial $x(Z)$ as follows:

$$x(Z) = x_{n-1}Z^{n-1} + x_{n-2}Z^{n-2} + \dots + x_1Z + x_0 \qquad (10.25)$$

The binary coefficients of this polynomial will be manipulated with the rules of modulo-2 arithmetic. In this new notation, the code words of an $(n, k)$ linear block code are in a one-to-one correspondence with code polynomials of degree not greater than $(n-1)$.

By definition of cyclic code, if $x(Z)$ is the code polynomial of a cyclic code, then a cyclic shift of the code word (say to the left) of $i$ positions generates another code polynomial that we denote by $x^{(i)}(Z)$. Theorem 10.3 relates the polynomial representation of a cyclically shifted $n$-tuple to the binomial $Z^n + 1$, which will be shown to play a crucial role for cyclic codes.

**Theorem 10.3**

The code polynomial $x^{(i)}(Z)$ is the remainder resulting from dividing $Z^i x(Z)$ by $(Z^n + 1)$; that is,

$$Z^i x(Z) = q(Z)(Z^n + 1) + x^{(i)}(Z) \qquad (10.26)$$

where $q(Z)$ is the quotient polynomial of degree not greater than $(i-1)$. ▽

**Proof of Theorem 10.3**

Let us write explicitly $Z^i x(Z)$

$$Z^i x(Z) = x_{n-1}Z^{n-1+i} + x_{n-2}Z^{n-2+i} + \dots + x_{n-i}Z^n + \dots + x_0Z^i$$

Sum to this expression twice the terms $x_{n-1}Z^{i-1}, x_{n-2}Z^{i-2}, \ldots, x_{n-i}$; this is possible because $x_{n-j} + x_{n-j} = 0$, $\forall j$. We get

$$\begin{aligned} Z^i x(Z) &= x_{n-1}Z^{i-1}(Z^n + 1) + x_{n-2}Z^{i-2}(Z^n + 1) + \ldots x_{n-i}(Z^n + 1) + \\ &\quad x_{n-i-1}Z^{n-1} + \ldots + x_1 Z^{i+1} + x_0 Z^i + x_{n-1}Z^{i-1} + \ldots + x_{n-i} \end{aligned}$$

and finally,

$$\begin{aligned} Z^i x(Z) &= (x_{n-1}Z^{i-1} + x_{n-2}Z^{i-2} + \ldots x_{n-i})(Z^n + 1) + \\ &\quad x_{n-i-1}Z^{n-1} + \ldots + x_1 Z^{i+1} + x_0 Z^i + x_{n-1}Z^{i-1} + \ldots + x_{n-i} \\ &= q(Z)(Z^n + 1) + x^{(i)}(Z) \end{aligned}$$

**QED**

**Example 10.11**   Let us take again the code word $0111010$ of Example 10.10. The corresponding code polynomial is

$$x(Z) = Z^5 + Z^4 + Z^3 + Z$$

A shift of four positions to the left generates the code polynomial $x^{(4)}(Z)$, which is obtained, according to Theorem 10.3, by dividing $Z^4 x(Z)$ by $(Z^7 + 1)$ as follows:

$$
\begin{array}{r|llllll|l}
Z^7+1 & Z^9 & + & Z^8 & + & Z^7 & + & Z^5 & & & & Z^2 + Z + 1 \\
& Z^9 & + & & & & & & & Z^2 & & \text{quotient} \\
\hline
& & & Z^8 & + & Z^7 & + & Z^5 & + & Z^2 & & \\
& & & Z^8 & + & & & & & & Z & \\
\hline
& & & & & Z^7 & + & Z^5 & + & Z^2 & + & Z \\
& & & & & Z^7 & + & & & & & 1 \\
\hline
& & & & & & & Z^5 & + & Z^2 & + & Z + 1 & \text{remainder}
\end{array}
$$

The remainder is $Z^5 + Z^2 + Z + 1$, that is, the sequence $0100111$. This sequence is obtained from the original one with a four-position shift to the left.     □

Using the polynomial description of cyclic codes, we now want to exploit the algebraic properties of their generator matrices. Let us first proceed through an example that will also enable us to introduce an important theorem.

**Example 10.12**   The $(7, 4)$ Hamming code of Example 10.3 was already claimed to be cyclic in Example 10.10. Let us rewrite its generator matrix (10.5) in polynomial form as follows:

$$\mathbf{G}(Z) = \begin{bmatrix} Z^6+ & & & Z^2+ & & 1 \\ & Z^5+ & & Z^2+ & Z+ & 1 \\ & & Z^4+ & Z^2+ & Z & \\ & & & Z^3+ & & Z+ & 1 \end{bmatrix} \tag{10.27}$$

Consider the last row of this generator matrix, that is, the polynomial

$$g(Z) = Z^3 + Z + 1 \tag{10.28}$$

This code polynomial must have a 1 in the last position (coefficient of $Z^0$); otherwise, six cyclic shifts to the left would generate a code word with $k = 4$ information digits equal to 0 and a parity-check section containing 2 ones, which is impossible. Furthermore, this is the only polynomial of degree $n - k = 3$ in the code. In fact, if there were another, it could be added to $g(Z)$ to generate a code word presenting again an all-zero information section with a nonzero parity section. As a conclusion, there is a unique code polynomial $g(Z)$ of degree $n - k = 3$, and this polynomial has always the form

$$g(Z) = Z^3 + \ldots + 1 \tag{10.29}$$

Let us now derive the remaining rows of the generator matrix (10.27). The third row can be obtained with one cyclic shift to the left of the last row. Should a 1 appear in the fourth position, the last row could be added to cancel it. Each row of $\mathbf{G}(Z)$ can be obtained in a similar way from the row below. The result is

$$\mathbf{G}(Z) = \begin{bmatrix} (Z^3 + Z + 1) & g(Z) \\ (Z^2 + 1) & g(Z) \\ Z & g(Z) \\ & g(Z) \end{bmatrix} \tag{10.30}$$

All the rows of $\mathbf{G}(Z)$ are multiples of the polynomial $g(Z)$. But, since all code words in the code are linear combinations of the rows of $\mathbf{G}(Z)$, we can conclude that all the code polynomials are multiples of the polynomial $g(Z)$.                                              □

The important conclusions drawn from the previous example are stated in general form in Theorem 10.4.

**Theorem 10.4**

Given an $(n, k)$ cyclic code, there is a unique code polynomial of degree $(n - k)$ that has the form

$$g(Z) = Z^{n-k} + \ldots + 1 \tag{10.31}$$

All other $2^k - 1$ code polynomials are multiples of $g(Z)$, and every polynomial of degree $(n - 1)$ or less that is divisible by $g(Z)$ must be a code polynomial. ▽

The proof, involving a generalization of the development in Example 10.12, is left to the reader.

The polynomial $g(Z)$ defined by Theorem 10.4 is called the *generator polynomial* of the cyclic code. Any cyclic code is completely defined by its generator polynomial. The natural question now is whether there exists an $(n, k)$ cyclic code for any $n$ and $k$ and which is the corresponding generator polynomial. Theorem 10.5 provides the answer.

## Theorem 10.5

The generator polynomial $g(Z)$ of an $(n, k)$ cyclic code is a divisor of $(Z^n + 1)$. Conversely, every divisor of $(Z^n + 1)$ of degree $(n - k)$ generates an $(n, k)$ cyclic code. $\triangledown$

## Proof of Theorem 10.5

Consider the polynomial $Z^k g(Z)$ of degree $n$ and divide it by $(Z^n + 1)$. We get

$$Z^k g(Z) = (Z^n + 1) + g^{(k)}(Z) \tag{10.32}$$

where $g^{(k)}(Z)$ is a polynomial of degree not greater than $(n - 1)$. Using (10.26), we can conclude that $g^{(k)}(Z)$ is a code polynomial obtained with $k$ cyclic shifts to the left of $g(Z)$. Therefore, it is also a multiple of $g(Z)$, say $m(Z)g(Z)$. From (10.32), we get

$$Z^n + 1 = [Z^k + m(Z)]g(Z) = h(Z)g(Z) \tag{10.33}$$

and the direct part of the theorem is proved.

Conversely, let $g(Z)$ be a divisor of $(Z^n + 1)$ of degree $(n - k)$ and consider the $k$ polynomials $g(Z), Zg(Z), \ldots, Z^{k-1}g(Z)$ of degree $(n - k), (n - k + 1), \ldots, (n - 1)$. There are $2^k$ linear combinations of these $k$ polynomials. Each of them is a multiple of $g(Z)$, and together they form an $(n, k)$ linear code. Let $x(Z)$ be one of these code polynomials and consider

$$Z^i x(Z) = q(Z)(Z^n + 1) + x^{(i)}(Z) \tag{10.34}$$

where $x^{(i)}(Z)$ is a cyclic shift of $x(Z)$. Since both $x(Z)$ and $(Z^n + 1)$ are multiples of $g(Z)$, then $x^{(i)}(Z)$ is also a multiple of $g(Z)$. Furthermore, it can be expressed as a linear combination of the aforementioned $k$ polynomials. It follows that $x^{(i)}(Z)$ is a code polynomial and that the linear code is cyclic. **QED**

Finally, notice that if $g(Z)$ divides $(Z^m + 1)$ as well as $(Z^n + 1)$, with $m < n$, then $(Z^m + 1)$ is a code word in the cyclic code $(n, k)$ whose minimum distance is therefore 2. To avoid this drawback, $n$ must be taken as the smallest integer such as $(Z^n + 1)$ is a multiple of $g(Z)$.

Considerable algebraic results are available regarding the properties of the polynomials $(Z^n + 1)$. In particular, tables of divisors for different values of $n$

| $n$ | Factors |
|---|---|
| 7 | 6.54.64. |
| 9 | 6.7.444. |
| 15 | 6.7.46.62.76. |
| 17 | 6.471.727. |
| 21 | 6.7.54.64.534.724. |
| 23 | 6.5342.6165. |
| 25 | 6.76.4102041. |
| 27 | 6.7.444.4004004. |
| 31 | 6.45.51.57.67.73.75. |
| 33 | 6.7.4522.6106.7776. |
| 35 | 6.54.64.76.57134.72364. |
| 39 | 6.7.57074.74364.77774. |
| 41 | 6.5747175.6647133. |
| 43 | 6.47771.52225.64213. |
| 45 | 6.7.46.62.76.444.40044.44004. |
| 47 | 6.43073357.75667061. |
| 49 | 6.54.64.40001004.40200004. |
| 51 | 6.7.433.471.637.661.727.763. |
| 55 | 6.76.7776.5551347.7164555. |
| 57 | 6.7.5604164.7565674.7777774. |
| 63 | 6.7.54.64.414.444.534.554.604.634.664.714.724. |
| 127 | 6.406.422.436.442.472.516.526.562.576.602.626.646.652. 712.736.742.756.772. |

Table 10.5: *Factors of the polynomial* $(Z^n + 1)$. *Each polynomial factor is given in octal notation with the lowest-degree terms on the left (MacWilliams and Sloane, 1977).*

can be found. One of these is reproduced in Table 10.5. These tables are very useful because the design of a cyclic code with preassigned properties reduces to an appropriate selection of divisors of $(Z^n + 1)$ as candidates for the code generator $g(Z)$.

Notice that the Table 10.5 considers only odd values of $n$, because in binary algebra we have $(Z^{2m} + 1) = (Z^m + 1)^2$. Furthermore, the values of $n = 3, 5, 11, 13, 19, 29, 37, 53, 59, 61$ are omitted from the table. In fact, for these values the factorization is simply

$$(Z^n + 1) = (Z + 1)(Z^{n-1} + Z^{n-2} + \ldots + Z + 1)$$

Therefore, Table 10.5 gives the factors of $(Z^n + 1)$ for $n \leq 63$ and $n = 127$. The factors are given in octal notation, with the lowest-degree terms on the left. As an example, the second line of the table means that the coefficients of the factors are, respectively, 110, 111, and 100100100, so that we obtain

$$Z^9 + 1 = (Z + 1)(Z^2 + Z + 1)(Z^6 + Z^3 + 1)$$

**Encoding algorithms for cyclic codes**

Given the generator polynomial $g(Z)$ of a cyclic code $(n, k)$, the code polynomial $x(Z)$ corresponding to an information sequence $u(Z)$ can be obtained from Theorem 10.4 as

$$x(Z) = u(Z)g(Z) \tag{10.35}$$

This simple algorithm does not actually represent a systematic encoder, as is verified in the following example.

**Example 10.13**   The $(7, 4)$ Hamming code described in Example 10.12 has the generator $g(Z) = Z^3 + Z + 1$. Let us find the code word corresponding to the information sequence 1101. Since

$$u(Z) = Z^3 + Z^2 + 1$$

we get

$$x(Z) = (Z^3 + Z^2 + 1)(Z^3 + Z + 1) = Z^6 + Z^5 + Z^4 + Z^3 + Z^2 + Z + 1$$

The code word is then 1111111 and the encoder is not systematic. All other code words can be obtained in the same way.    □

The algorithm based on (10.35) can be modified to represent a systematic cyclic encoder $(n, k)$. Given the information sequence $u(Z)$, let us multiply it by $Z^{n-k}$ and divide by the generator polynomial $g(Z)$. We have

$$Z^{n-k}u(Z) = q(Z)g(Z) + r(Z) \tag{10.36}$$

where $q(Z)$ and $r(Z)$ are, respectively, the quotient and the remainder of the division. Notice that $r(Z)$ must have a degree $(n - k - 1)$ or less, since the degree of $g(Z)$ is $(n - k)$. Rearranging (10.36), we get

$$Z^{n-k}u(Z) + r(Z) = q(Z)g(Z) \tag{10.37}$$

This is the key for the desired encoding algorithm. In fact, the LHS of this equation is a polynomial of degree $(n - 1)$ or less, a multiple of $g(Z)$, and hence a code polynomial in the cyclic code generated by $g(Z)$. Let us write it explicitly

$$u_{k-1}Z^{n-1} + u_{k-2}Z^{n-2} + \ldots + u_0 Z^{n-k} + r_{n-k-1}Z^{n-k-1} + \ldots + r_0 \tag{10.38}$$

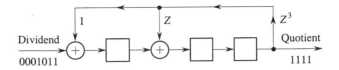

Figure 10.9: *Circuit to divide by* $g(Z) = Z^3 + Z + 1$. *When all seven digits of the dividend have been fed into the shift register, the contents* $r_2 r_1 r_0$ *of the register are the remainder of the division. For the example shown, the remainder will be 001.*

Therefore, the code word consists of the $k$ information digits followed by the $(n-k)$ parity check digits and the encoder is systematic. In conclusion, the parity check section of each code word in a systematic cyclic code $(n, k)$ is obtained as the remainder of the division of $Z^{n-k}u(Z)$ by the generator polynomial $g(Z)$.

**Example 10.14**   Given the $(7, 4)$ cyclic code with the generator polynomial $Z^3 + Z + 1$, let us find the code word for the information sequence $1101$ using the encoding algorithm just described. We have

$$Z^3 u(Z) = Z^6 + Z^5 + Z^3$$

Dividing it by $g(Z)$ yields the remainder

$$r(Z) = 1$$

so that the code word polynomial is $x(Z) = Z^6 + Z^5 + Z^3 + 1$, corresponding to the binary form $1101001$. The encoder is systematic.                                                □

The encoding algorithm based on (10.37) requires the division of $Z^{n-k}u(Z)$ by the generator polynomial $g(Z)$ to get the remainder $r(Z)$. This is the parity-check section of the code polynomial corresponding to $u(Z)$. Therefore, the implementation of the algorithm requires a circuit that performs a division. This task can be accomplished by a shift register having $(n - k)$ stages, the degree of the divisor, and suitable feedback connections that correspond to the coefficients of the divisor. The circuit shown in Figure 10.9 performs the division described in Example 10.14.

At each clock pulse, the digits of the dividend are fed in leftwise starting with the most significant digit, and the quotient is shifted out rightward. The remainder is what remains in the register when all seven digits of the dividend have been fed in. Notice in particular that the feedback connections correspond to the structure of the divisor.

A first possible version of the encoder for the $(7, 4)$ code of Example 10.14 is shown in Figure 10.10. The switches have three positions: First, at Ⓐ, for four

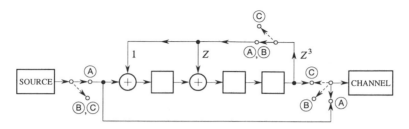

Figure 10.10: *First implementation of the encoder for the* $(7, 4)$ *Hamming code with generator polynomial* $Z^3 + Z + 1$. *The switches are at position* Ⓐ *for 4 clock pulses, at* Ⓑ *for 3 clock pulses, and at* Ⓒ *for 3 clock pulses.*

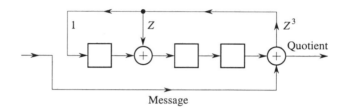

Figure 10.11: *Circuit to divide by* $g(Z) = Z^3 + Z + 1$. *Only four message digits are fed to the shift register at its right end.*

clock pulses, during which the four information digits are fed into the register and sent to the channel; second, at Ⓑ, for three clock pulses, while three zeros enter the register; third, at Ⓒ, for three clock pulses, while the remainder of the division is sent to the channel. The disadvantage of this implementation is that the channel remains idle while the switches are at Ⓑ. To overcome this drawback, the message digits can be fed into the right end of the shift register. This is equivalent to multiplying the symbols by $Z^3$ as they come in. Hence, the divisor circuit of Figure 10.11 is used. The remainder of the division is now available in the register as soon as the last digit had been fed in. The implementation of the encoder based on this concept is shown in Figure 10.12. The switches are at Ⓐ for four clock pulses and at Ⓑ for three clock pulses. The operation of this encoder is described in detail in the following example.

**Example 10.15** We reproduce here the situation described in Example 10.14. The code word for the sequence 1101 is obtained by shifting in the circuit of Figure 10.12 the information seguence and shifting out the remainder of the division. The contents of the shift register, at each step, are shown in the following table

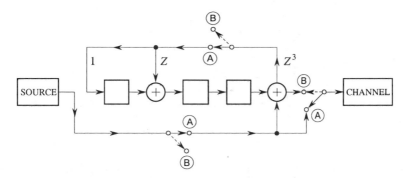

Figure 10.12: *Second implementation of the encoder for the* (7, 4) *Hamming code with generator polynomial* $g(Z) = Z^3 + Z + 1$. *The switches are at position* Ⓐ *for 4 clock pulses, and at* Ⓑ *for 3 clock pulses.*

| $u_i$ | $r_0$ | $r_1$ | $r_2$ |
|---|---|---|---|
| - | 0 | 0 | 0 |
| 1 | 1 | 1 | 0 |
| 1 | 1 | 0 | 1 |
| 0 | 1 | 0 | 0 |
| 1 | 1 | 0 | 0 |

As soon as the four information digits are entered into the register and delivered to the channel, the register contains the sequence 001. This corresponds to the remainder $r(Z) = 1$. □

An encoder similar to that of Figure 10.12 will work for any cyclic code. It requires $(n - k)$ delay elements in the shift register, and the generator polynomial is reflected in the feedback connections structure. For codes with $k < (n - k)$, a simpler circuit with a $k$-stage shift register can be implemented. It is based on the multiplication by the *parity-check* polynomial defined as

$$h(Z) \triangleq \frac{Z^n + 1}{g(Z)} \tag{10.39}$$

The encoder is shown in Figure 10.13. Notice in it the shift register with $k = 4$ delay elements. The connections to the adder are made according to the powers of $h(Z) = Z^4 + Z^2 + Z + 1$. The switch is at position Ⓐ for 4 clock pulses and at Ⓑ for 3 clock pulses. An example of its behavior is deferred to Problem 10.15.

The parity-check polynomial $h(Z)$ is a divisor of $Z^n + 1$ of degree $k$. As such, it can be used as the generator of an $(n, n - k)$ cyclic code. This code is

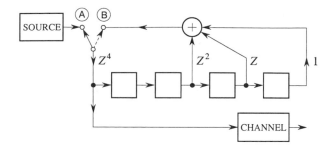

Figure 10.13: *Encoder for the* $(7, 4)$ *Hamming code based on the parity-check polyno-mial* $h(Z) = Z^4 + Z^2 + Z + 1$. *The switches are at position* Ⓐ *for 4 clock pulses, and at* Ⓑ *for 3 clock pulses.*

equivalent to the dual of the code generated by $g(Z)$ (see Problem 10.16). It is customary to refer to the cyclic code generated by $h(Z)$ as the dual code of the cyclic code generated by $g(Z)$, although this is not true according to the formal definition of dual code given in Section 10.2.4.

### Error detection and error correction with cyclic codes

Assume that a code polynomial is transmitted over a noisy channel. In analogy with (10.8), the received sequence can be written in polynomial form as

$$y(Z) = x(Z) + e(Z) \tag{10.40}$$

where $x(Z)$ is the code polynomial and $e(Z)$ is the error polynomial. Let us now divide $y(Z)$ by the generator polynomial of the code. We get

$$y(Z) = m(Z)g(Z) + s(Z) \tag{10.41}$$

where $m(Z)$ is the quotient and $s(Z)$ the remainder of the division. Since only the code polynomials are multiples of the generator polynomial, $y(Z)$ will be a code word if and only if the polynomial $s(Z)$ is zero. This polynomial of degree not greater than $(n - k - 1)$ is the *syndrome* polynomial of $y(Z)$. Since $x(Z) = q(Z)g(Z)$, we can compare (10.40) with (10.41) and obtain

$$e(Z) = [m(Z) + q(Z)]g(Z) + s(Z) \tag{10.42}$$

This equation shows that $s(Z)$ is also the syndrome of $e(Z)$. In conclusion, error detection can be accomplished by simply checking the remainder of the division of the received polynomial $y(Z)$ by the generator $g(Z)$. The detection circuit can be implemented with a circuitry similar to that shown in Figure 10.9. The register is first set to zero and the received sequence is shifted in. The content of the register will represent the syndrome $s(Z)$ as soon as the last digit of $y(Z)$ is entered. One additional property of the syndrome is stated in Theorem 10.6.

### Theorem 10.6

If $s(Z)$ is the syndrome of an error polynomial $e(Z)$, the syndrome of $e^{(i)}(Z)$, that is, of $e(Z)$ shifted cyclically $i$ places to the left, is obtained by shifting $i$ times the syndrome $s(Z)$ inside the division circuit. $\triangledown$

### Proof of Theorem 10.6

First notice that we have

$$e(Z) = q(Z)g(Z) + s(Z) \tag{10.43}$$

and, by Theorem 10.3

$$e^{(i)}(Z) = n(Z)(Z^n + 1) + Z^i e(Z) = n(Z)h(Z)g(Z) + Z^i e(Z) \tag{10.44}$$

Substituting (10.43) into (10.44) yields

$$Z^i s(Z) = [n(Z)h(Z) + Z^i q(Z)]g(Z) + e^{(i)}(Z) \tag{10.45}$$

Expressing $e^{(i)}(Z)$ in terms of $g(Z)$ as

$$e^{(i)}(Z) = m(Z)g(Z) + r(Z)$$

gives

$$Z^i s(Z) = [n(Z)h(Z) + Z^i q(Z) + m(Z)]g(Z) + r(Z) \tag{10.46}$$

which proves that $r(Z)$, the remainder of the division of $e^{(i)}(Z)$ by $g(Z)$, is also the remainder of the division of $Z^i s(Z)$ by $g(Z)$. Remembering the operation of the circuit of Figure 10.9, this operation is precisely obtained by shifting $i$ times the syndrome $s(Z)$ into the division circuit. **QED**

The syndrome $s(Z)$ of a received sequence $y(Z)$ can be obtained using the encoder circuit of the type of Figure 10.10. Also the circuit of Figure 10.11 can be used to the same end. With this circuit, however, the calculated syndrome is that of the sequence $y^{(n-k)}(Z)$. The properties of the syndrome generating circuits are best understood through an example.

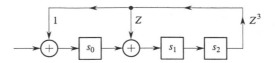

Figure 10.14: *First implementation of the syndrome generator for the (7, 4) code with generator polynomial $g(Z) = Z^3 + Z + 1$.*

**Example 10.17** Let us use again the $(7, 4)$ cyclic code with generator polynomial $g(Z) = Z^3 + Z + 1$ and assume a received sequence $y(Z) = Z^6 + Z^5 + 1$. First, let us derive the syndrome $s(Z)$ from (10.41). Using the polynomial division algorithm shown in Example 10.11 to divide $y(Z)$ by $g(Z)$, we obtain the remainder $s(Z) = Z + 1$, which corresponds to the vector 011. This can also be obtained by using the parity-check matrix (10.13). The circuit of Figure 10.14 can be used to derive the syndrome $s(Z) = Z + 1$. The content of the register at the successive steps is given in the following table:

| Received digit | Register content | | |
|---|---|---|---|
| $y_i$ | $s_0$ | $s_1$ | $s_2$ |
| - | 0 | 0 | 0 |
| 1 | 1 | 0 | 0 |
| 1 | 1 | 1 | 0 |
| 0 | 0 | 1 | 1 |
| 0 | 1 | 1 | 1 |
| 0 | 1 | 0 | 1 |
| 0 | 1 | 0 | 0 |
| 1 | 1 | 1 | 0 |

If, instead, the circuit of Figure 10.15 is used as syndrome generator, we obtain similarly

| Received digit | Register content | | |
|---|---|---|---|
| $y_i$ | $s_0$ | $s_1$ | $s_2$ |
| - | 0 | 0 | 0 |
| 1 | 1 | 1 | 0 |
| 1 | 1 | 0 | 1 |
| 0 | 1 | 0 | 0 |
| 0 | 0 | 1 | 0 |
| 0 | 0 | 0 | 1 |
| 0 | 1 | 1 | 0 |
| 1 | 1 | 0 | 1 |

This syndrome, that is, $s(Z) = Z^2 + 1$, is that of the sequence $y^{(3)}(Z) = Z^3 + Z^2 + Z$. Using Theorem 10.6, the same syndrome can be obtained by shifting three steps

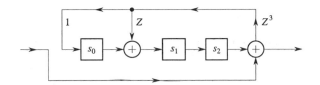

Figure 10.15: *Second implementation of the syndrome generator for the* $(7, 4)$ *code with generator polynomial* $g(Z) = Z^3 + Z + 1.$

the syndrome $s(Z) = Z + 1$ in the register of Figure 10.14. The result is given in the following table:

| Shift | Register | | |
| # | content | | |
| | $s_0$ | $s_1$ | $s_2$ |
| - | 1 | 1 | 0 |
| 1 | 0 | 1 | 1 |
| 2 | 1 | 1 | 1 |
| 3 | 1 | 0 | 1 |

□

To perform error correction, the decoder must find a correctable error pattern $e(Z)$ from the syndrome $s(Z)$. The transmitted code word $x(Z)$ is then obtained by adding $e(Z)$ to the received sequence $y(Z)$. Whether or not this is a practical operation depends on the complexity of the decoder that computes the correctable error pattern $e(Z)$. Special classes of codes have been developed that lead to practical algorithms. But a thorough description of multiple error-correcting schemes is beyond the scope of this book, so that the interested reader should refer to the Bibliographic Section at the end of the chapter. However, one simple technique that is applicable in the case of single-error correction will be described hereafter. It is based on the following general theorem.

**Theorem 10.7**

If the errors of $e(Z)$ are confined to the $(n - k)$ parity-check positions of $y(Z)$, the syndrome $s(Z)$ is identical to $e(Z)$. $\triangledown$

**Proof of Theorem 10.7**

The assumption is equivalent to saying that $e(Z)$ is a polynomial of degree not greater than $(n - k - 1)$. Therefore, the division by $g(Z)$, which has degree

$(n - k)$, gives as a remainder just $e(Z)$. **QED**

Under the conditions of Theorem 10.7, error correction is accomplished by simply adding the syndrome to the $(n - k)$ received parity-check digits. Should the errors be confined to $(n - k)$ consecutive digits different from the parity-check section, then the use of Theorem 10.6 allows again for error correction. In fact, the errors can also in this case be confined in the parity-check section by shifting cyclically the received sequence to the left by $i$ places. The syndrome of this new sequence is that of $y^{(i)}(Z)$. Let us apply these concepts to an example.

**Example 10.18**  Let us use once again the $(7, 4)$ code with generator $g(Z) = Z^3 + Z + 1$. Assume that

$$x(Z) = Z^6 + Z^5 + Z^3 + 1, \quad e(Z) = Z^3$$

Therefore, as in Example 10.17, the received sequence is $y(Z) = Z^6 + Z^5 + 1$. Since $y(Z)$ is shifted into the syndrome generator from the rightmost stage, it corresponds to a preshifted sequence $Z^{n-k}y(Z) = Z^3 y(Z)$. Using the syndrome generator of Figure 10.15, we get from Example 10.17 that $s(Z) = Z^2 + 1$. Therefore, an error located in position $Z^j$ in $y(Z)$ corresponds to an error in position $Z^{n-k+j} = Z^{3+j}$ in the preshifted sequence. When $j = n - 1 = 6$, an error occurs in the first position of $y(Z)$, and appears in position $(n - k + j) = 9$ of the preshifted sequence. Taking into account the end-around shift, this position is the highest of the parity-check section (in fact, $9 - 7 = 2$). Due to Theorem 10.7, the syndrome corresponding to this situation is $s(Z) = Z^2$ (sequence 100). We can now apply Theorem 10.6. The syndrome $Z^2 + 1$ is shifted inside the division circuit. When the syndrome $Z^2$ is identified, this means that the single error is in the first position of the cyclically shifted received sequence.

These concepts are applied to the error-correcting circuit of Figure 10.16. It consists of the syndrome generator of Figure 10.15, a buffer, and an AND gate with $(n - k) = 3$ inputs. The received sequence is shifted into both the buffer and the syndrome generator. While it is read out from the buffer, the syndrome is simultaneously shifted into the register to identify the register contents corresponding to $s(Z) = Z^2$. When this is the case, the error is identified and the digit that will be shifted out will be corrected. The reader is invited to work out the details using Figure 10.16 and the following table, where the line with boldface numbers corresponds to the correction instant.

| Syndrome | | | Shift | | | Output |
|---|---|---|---|---|---|---|
| $s_0$ | $s_1$ | $s_2$ | | $b$ | $c$ | |
| 1 | 0 | 1 | - | 1 | 0 | 1 |
| 1 | 0 | 0 | 1 | 1 | 0 | 1 |
| 0 | 1 | 0 | 2 | 0 | 0 | 0 |
| **0** | **0** | **1** | **3** | **0** | **1** | **1** |
| 1 | 1 | 0 | 4 | 0 | 0 | 0 |
| 0 | 1 | 1 | 5 | 0 | 0 | 0 |
| 1 | 1 | 1 | 6 | 1 | 0 | 1 |

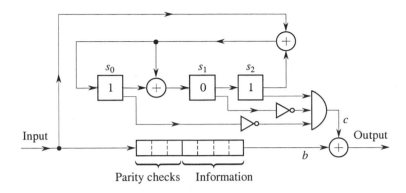

Figure 10.16: *Error trapping correction circuit for the* (7, 4) *cyclic Hamming code generated by* $g(Z) = Z^3 + Z + 1$.

□

The example has shown that one particular syndrome, that is, $s(Z) = Z^2$ corresponding to 100, allows the location and capture of the single error while the received sequence is shifted out from the buffer. This technique for error correction is called *error-trapping* decoding. It can be extended to cases other than single-error correction, like multiple-error and burst correction.

### 10.2.8.   Special classes of cyclic codes

#### Cyclic Hamming codes

The Hamming codes described in Section 10.2.3 can also be shown to be a special class of cyclic codes. To this purpose, let us define an *irreducible polynomial* as a polynomial of degree $l$ that is not divisible by any polynomial of degree less than $l$ and greater than zero. Furthermore, an irreducible polynomial $g(Z)$ of degree $l$ is called *primitive* when the smallest integer $n$, such that $(Z^n + 1)$ is a multiple of $g(Z)$, is $2^l - 1$. Therefore, invoking Theorem 10.5, this primitive polynomial can generate a $(2^l - 1, 2^l - l - 1)$ cyclic code that is a Hamming code. It has been shown that this code can correct all single errors. Let us reconsider the proof using polynomial notation. Let two single-error patterns be $e_i(Z) \triangleq Z^i$ and $e_j(Z) \triangleq Z^j$, where $0 \leq i < j < n$. It will be sufficient to show that the two corresponding syndromes, say $s_i(Z)$ and $s_j(Z)$, are different. Using (10.43), we

| $l$ | Primitive polynomial | $l$ | Primitive polynomial |
|-----|----------------------|-----|----------------------|
| 3   | 64                   | 14  | 60421                |
| 4   | 62                   | 15  | 600004               |
| 5   | 51                   | 16  | 640042               |
| 6   | 604                  | 17  | 440001               |
| 7   | 442                  | 18  | 4020004              |
| 8   | 561                  | 19  | 7100002              |
| 9   | 4204                 | 20  | 4400001              |
| 10  | 4402                 | 21  | 50000004             |
| 11  | 5001                 | 22  | 60000002             |
| 12  | 62404                | 23  | 41000001             |
| 13  | 66002                | 24  | 702000004            |

Table 10.6: *Primitive polynomials of degree l. Each polynomial is expressed in octal notation with the lowest-degree terms on the left.*

have

$$
\begin{aligned}
e_i(Z) &= Z^i = q_i(Z)g(Z) + s_i(Z) \\
e_j(Z) &= Z^j = q_j(Z)g(Z) + s_j(Z)
\end{aligned}
\tag{10.47}
$$

Summing these two equations, we get

$$
Z^i(Z^{j-i} + 1) = [q_i(Z) + q_j(Z)]g(Z) + s_i(Z) + s_j(Z)
\tag{10.48}
$$

Since $j - i < n = 2^l - 1$, and $g(Z)$ is primitive, then $g(Z)$ cannot divide $(Z^{j-i}+1)$, and, consequently, $s_i(Z) \neq s_j(Z)$, for $i \neq j$. Cyclic Hamming codes can be decoded by using the error-trapping algorithm as in Example 10.18. A list of primitive polynomials that generate Hamming codes is given in Table 10.6 for different values of $l$. The polynomials are given in octal notation, with the lowest-degree terms on the left. For example, the first line of the table means

$$
64 \Rightarrow 110100 \Rightarrow g(Z) = 1 + Z + Z^3
$$

**Golay codes**

In searching for perfect codes, Golay discovered a (23,12) code that is a cyclic code with generator polynomial

$$g(Z) = Z^{11} + Z^9 + Z^7 + Z^6 + Z^5 + Z + 1 \qquad (10.49)$$

and with minimum distance $d_{\min} = 7$. Therefore, triple error correction is possible. The important point is that this code is the only possible nontrivial linear binary perfect code with multiple error-correcting capabilities. Besides the Hamming single-error-correcting codes, the repetition codes (with $n$ odd), and the Golay code, no other linear binary perfect codes exist (see MacWilliams and Sloane, 1977, Chapter 6).

**Bose-Chaudhuri-Hocquenghem (BCH) codes**

This class of cyclic codes is one of the most useful for correcting random errors mainly because the decoding algorithms can be implemented with an acceptable amount of complexity. For any pair of positive integers $m$ and $t$, there is a binary BCH code with the following parameters:

$$n = 2^m - 1, \quad n - k \le mt, \quad d_{\min} \ge 2t + 1$$

This code can correct all combinations of $t$ or fewer errors. The generator polynomial for this code can be constructed from the factors of $(Z^{2^m-1}+1)$. Unfortunately, this procedure is not straightforward and is beyond the scope of this book; the interested readers are referred to Chapter 9 of the book by MacWilliams and Sloane (1977). A list of generator polynomials for BCH codes of different parameters is given in Tables 10.7 and 10.8. The polynomials are represented in octal notation, with the highest degree terms on the left[3]. As an example, the third line of the table means

$$721 \Rightarrow 111010001 \Rightarrow g(Z) = Z^8 + Z^7 + Z^6 + Z^4 + 1$$

Notice that this polynomial can be factored as

$$g(Z) = (Z^4 + Z + 1)(Z^4 + Z^3 + Z^2 + Z + 1)$$

It can be verified from Table 10.5 that these two factors are factors of $(Z^{15} + 1)$.

The BCH codes provide a large class of codes. They are useful not only because of the flexibility in the choice of parameters (block length and code

---

[3]The octal notation in this table is different with respect to that of Tables 10.5 and 10.6, in the sense that the highest-degree term is on the left here.

| $n$ | $k$ | $t$ | $g(D)$ |
|---|---|---|---|
| 7 | 4 | 1 | 13 |
| 15 | 11 | 1 | 23 |
| | 7 | 2 | 721 |
| | 5 | 3 | 2467 |
| 31 | 26 | 1 | 45 |
| | 21 | 2 | 3551 |
| | 16 | 3 | 107657 |
| | 11 | 5 | 5423325 |
| | 6 | 7 | 313365047 |
| 63 | 57 | 1 | 103 |
| | 51 | 2 | 12471 |
| | 45 | 3 | 1701317 |
| | 39 | 4 | 166623567 |
| | 36 | 5 | 1033500423 |
| | 30 | 6 | 157464165547 |
| | 24 | 7 | 17323260404441 |
| | 18 | 10 | 1363026512351725 |
| | 16 | 11 | 6331141367235453 |
| | 10 | 13 | 472622305527250155 |
| | 7 | 15 | 5231045543503271737 |
| 127 | 120 | 1 | 211 |
| | 113 | 2 | 41567 |
| | 106 | 3 | 11554743 |
| | 99 | 4 | 3447023271 |
| | 92 | 5 | 624730022327 |
| | 85 | 6 | 130704476322273 |
| | 78 | 7 | 26230002166130115 |
| | 71 | 9 | 6255010713253127753 |
| | 64 | 10 | 1206534025570773100045 |
| | 57 | 11 | 335265252505705053517721 |
| | 50 | 13 | 54446512523314012421501421 |
| | 43 | 14 | 17721772213651227521220574343 |
| | 36 | 15 | 3146074666522075044764574721735 |
| | 29 | 21 | 403114461367670603667530141176155 |
| | 22 | 23 | 123376070404722522435445626637647043 |
| | 15 | 27 | 22057042445604554770523013762217604353 |
| | 8 | 31 | 7047264052751030651476224271567733130217 |
| 255 | 247 | 1 | 435 |
| | 239 | 2 | 267543 |
| | 231 | 3 | 156720665 |
| | 223 | 4 | 75626641375 |
| | 215 | 5 | 23157564726421 |
| | 207 | 6 | 16176560567636227 |
| | 199 | 7 | 7633031270420722341 |
| | 191 | 8 | 2663470176115333714567 |
| | 187 | 9 | 52755313540001322236351 |
| | 179 | 10 | 226247107173404324163004 55 |
| | 171 | 11 | 15416214212342356077061630637 |
| | 163 | 12 | 7500415510075602551574724514601 |

Table 10.7: *List of generator polynomials for BCH codes. Each polynomial is represented in octal notation with the highest-degree terms on the left (Stenbit, 1964).*

| n | k | t | g(D) |
|---|---|---|------|
| 255 | 155 | 13 | 37575130054076650157225064646 77633 |
| | 147 | 14 | 1642130173537165525304165305441011711 |
| | 139 | 15 | 461401732060175561570722730247453567445 |
| | 131 | 18 | 215713331471510151261250277442142024165471 |
| | 123 | 19 | 120614052242066003717210326516141226272506267 |
| | 115 | 21 | 60526665572100247263636404600276352556313472737 |
| | 107 | 22 | 22205772322066256312417300235347420176574750154441 |
| | 99 | 23 | 1065666725347317422274141620157433225241107 6432303431 |
| | 91 | 25 | 675026503032744417272363172473251107555076272 0724344561 |
| | 87 | 26 | 11013676341474323643523163430717204620672254527 33117213 17 |
| | 79 | 27 | 667000356376575000202703442073661746210 1532671176654134 2355 |
| | 71 | 29 | 24024710520644321515554172112331163205444 25036255764322 1706035 |
| | 63 | 30 | 10754475055163544325315217357707003666111 72645526761365 6702543301 |
| | 55 | 31 | 731542520350110013301527530603205432541 4326755010557044 426035473617 |
| | 47 | 42 | 25335420170626465630330413774062331751 23334145446045005 066024552543173 |
| | 45 | 43 | 15202056055234161131101346376423701563 67002447076237303 3202157025051541 |
| | 37 | 45 | 5136330255067007414177447245437530420735 706174323432347 644354737403044003 |
| | 29 | 47 | 302571553667307146552706401236137711534 2242324201174114 060254757410403565037 |
| | 21 | 55 | 12562152570603326560017731536076121032 27341405653074542 52115312161446651 3473725 |
| | 13 | 59 | 46417320050525645444265737142500660043306 77445476561403 174677213570261344 60500547 |
| | 9 | 63 | 15726025217472163201031043255355134614162 36721204407454 5112766115547705561 677516057 |

Table 10.8: *List of generator polynomials for BCH codes. Each polynomial is repre-sented in octal notation with the highest-degree terms on the left (Stenbit, 1964).*

rate), but also because at block lengths of a few hundred or less many of these codes are among the best-known codes of the same length and rate. For the decoding algorithms, see Berlekamp (1968, Chapter 7), Clark and Cain (1981, Chapter 5), and Blahut (1983, Chapters 7 and 9).

**Reed-Solomon codes**

These codes are a subclass of BCH codes generalized to the nonbinary case, that is, to code symbols belonging to a set of cardinality $q = 2^m$. Thus, each symbol can still be represented as a binary $m$-tuple, and the code can be considered as a special type of binary code (see Blahut, 1983, Chapter 7). The parameters of a Reed-Solomon code are the following:

|  |  |
|---|---|
| Symbol | $m$ binary digits |
| Block length $n$ | $=(2^m - 1)$ symbols |
|  | $=m(2^m - 1)$ binary digits |
| Parity checks $(n - k)$ | $= 2t$ symbols |
|  | $= 2mt$ binary digits |

These codes are capable of correcting all combinations of $t$ or fewer symbol errors. Alternatively, interpreted as binary codes, they are well suited for correction of bursts of errors (see Section 10.2.10). In fact, one symbol in error means a number of binary digits in error ranging from 1 to $m$ in adjacent positions within the code word. Perhaps the most important application of these codes is in the concatenated coding scheme described in Chapter 11.

### Shortened cyclic codes

Since the generator polynomial of a cyclic code must be a divisor of $(Z^n + 1)$, it often happens that its possible degree $(n - k)$ does not cover all combinations of $n$ and $k$ that satisfy practical needs. To avoid this difficulty, cyclic codes are sometimes used in a shortened form. To this purpose, the first $i$ information digits are assumed to be always zero and are not transmitted. In this way, a new $(n - i, k - i)$ code is derived whose code words are a subset of the code words of the original code. The code is called *shortened* cyclic code, although it may not be cyclic. The new code has at least the same minimum distance as the code from which it is derived. The encoding and syndrome calculation can be accomplished by the same circuits employed in the original code, since the leading string of zeros does not affect the parity-check computations. Error correction can be accomplished by prefixing to each received vector a string of $i$ zeros, or by modifying accordingly the related circuitry. Therefore, these codes share all the implementation advantages of cyclic codes and are also of practical interest.

### 10.2.9.  Maximal-length (pseudonoise) sequences

The code words of the cyclic $(2^l - 1, l)$ simplex (or maximal-length) code of Section 10.2.5 resemble random sequences of zeros and ones. In fact, we shall see that any nonzero code word of these codes has many of the properties that we would expect from a binary sequence obtained by tossing a coin $2^l - 1$ times.

Maximal-length codes are the duals of the Hamming codes. Remember that a Hamming code of length $2^l - 1$ is generated by a primitive polynomial $g(Z)$ of

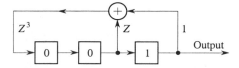

Figure 10.17: *Shift register circuit for encoding the dual code of the Hamming* (7, 4) *code with generator* $g(Z) = Z^3 + Z + 1$. *The circuit generates also a PN sequence of length* $2^3 - 1 = 7$.

degree $l$. The dual code of the same length can be obtained by letting the same $g(Z)$ be its parity-check polynomial. The dual code can therefore be generated by using an $l$-stage encoder of the type of Figure 10.13 with feedback connections reflecting the structure of $g(Z)$. For purposes of clarification, we use the following example.

**Example 10.18**   The dual code of the (7, 4) Hamming code generated by $g(Z) = Z^3 + Z + 1$ is a (7, 3) cyclic code with $g(Z)$ as the parity-check polynomial. A three-stage encoder for the dual code is shown in Figure 10.17. This scheme is a slight modification of the encoder type shown in Figure 10.13. The register is first loaded (from left to right) with the information sequence. Then the register content is shifted out (seven steps) from the right. In the following table, the generation of the code word, corresponding to the sequence 100, is shown, together with the successive states of the register. The last column of the table is the desired code word.

| Register content | | |
|---|---|---|
| 0 | 0 | 1 |
| 1 | 0 | 0 |
| 0 | 1 | 0 |
| 1 | 0 | 1 |
| 1 | 1 | 0 |
| 1 | 1 | 1 |
| 0 | 1 | 1 |
| 0 | 0 | 1 |

In the dual code, all the code words, with the exception of that which is all zero, are different cyclic shifts of a single code word. This property is understood by considering the evolution of the states of the shift register of the encoder of Figure 10.17. When the register is initially loaded and shifted $2^3 - 1$ times, it cycles through all possible $2^3 - 1$ states. Then it returns to the original one. The output sequence, when indefinitely shifted out, is periodic with period $2^3 - 1$. Since there are only $2^3 - 1$ possible states, this period corresponds to the largest possible in this register. This explains the name of *maximal-length* sequence and why the $2^3 - 1$ code words of this cyclic code are different cyclic shifts of one code word.                                                    □

The example can be generalized to show that the encoder of a maximal-length code can be used to generate maximal-length sequences of period $2^l - 1$. Primitive polynomials (see Table 10.6) are suitable for the generation of these sequences. As already stated, these sequences are also called pseudonoise (PN) sequences. They present the following pseudo-randomness properties:

**Property 1.** In any segment of length $2^l - 1$ of the sequence, there are exactly $2^{l-1}$ ones and $2^{l-1} - 1$ zeros. That is, the number of ones and the number of zeros are nearly equal. This property is an immediate consequence of the fact that the considered binary sequence is a code word of the simplex code, whose weight is constant and always equal to $2^{l-1}$ (see (10.23)).

**Property 2.** If we define a *run* to be the maximal string of consecutive identical symbols, then in any segment of the PN sequence of length $2^l - 1$ one-half of the runs have length 1, one-quarter have length 2, one-eighth have length 3, and so on. In each case, the number of runs of zeros is equal to the number of runs of ones.

**Property 3.** The most relevant property is related to the autocorrelation function of the PN sequence. Let us define the autocorrelation function of an infinite real sequence $(a_i)$ of period $n$ as

$$r_m \triangleq \frac{1}{n} \sum_{i=0}^{n-1} a_i a_{i+m}, \quad m = 0, \pm 1, \pm 2, \ldots \qquad (10.50)$$

Notice that $r_m$ is periodic, of period $n$. If the sequence $(a_i)$ is binary, formed by "0" and "1," let us replace it by a sequence $(b_i)$ in which we have substituted the 1's with $-1$'s and the 0's by $+1$'s. Thus, from (10.50), we get

$$r_m \triangleq \frac{1}{n} \sum_{i=0}^{n-1} b_i b_{i+m} = \frac{1}{n} \sum_{i=0}^{n-1} (-1)^{a_i + a_{i+m}} = \frac{A - D}{n} \qquad (10.51)$$

where $A$ and $D$ are the number of places where the sequence $(a_0 a_1 \ldots a_{n-1})$ and its cyclic shift $(a_m a_{m+1} \ldots a_{m+n-1})$ agree and disagree, respectively (so $A + D = n$). Therefore, for a sequence of period $n = 2^l - 1$, we have

$$r_0 = 1 \qquad (10.52)$$

$$r_m = -\frac{1}{n}, \quad \text{for } 1 \leq m \leq 2^l - 2 \qquad (10.53)$$

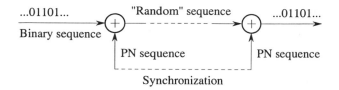

Figure 10.18: *Scrambling and descrambling a binary sequence by adding twice a PN sequence.*

In the sense of minimizing the magnitude of $r_m$, for $m \neq 0$, this is the "best" possible autocorrelation function of any binary sequence of period $n$.

PN sequences are very useful in practice, when it is desired to obtain sequences with random-like properties. To this purpose, the same PN sequence is added modulo-2 to the sequence at hand both at the transmitter and receiver side, as shown in Figure 10.18.   This is possible as the PN sequence is deterministic. The only requirement is that in the two additions the two PN sequences be synchronized. The randomizing operation is known as *scrambling*.

### 10.2.10.   Codes for burst-error detection and correction

In this section, we abandon the model of a channel producing random errors (like an AWGN or its hard-demodulated version BSC) and assume a channel model in which errors tend to be clustered in *bursts*. This is a typical situation in certain communication systems, employing media like magnetic tapes, magnetic disks, magnetic memories and compact disks.  Another situation would be a channel that is basically an AWGN occasionally disturbed by long bursts of noise or radio-frequency interference.  In general, when burst errors dominate, codes designed for correcting random errors may become inefficient.  Nevertheless, cyclic codes again are very useful in this situation.

Let us define a burst of length $b$ as an error pattern in which the errors are confined to $b$ consecutive positions.  Therefore, a burst-error pattern of length $b$ can be represented by the polynomial

$$e(Z) = Z^i e_b(Z) \qquad (10.54)$$

where $Z^i$ locates the burst in the error sequence of length $n$, and $e_b(Z)$ is a polynomial of the type

$$e_b(Z) = Z^{b-1} + \cdots + 1$$

The following theorem holds true.

**Theorem 10.8**

Any cyclic code $(n, k)$ can detect all bursts whose length is not greater than $(n - k)$. $\triangledown$

**Proof of Theorem 10.8**

The syndrome of such bursts is the remainder of the division of $Z^i e_b(Z)$ by the generator polynomial $g(Z)$. But this syndrome is always different from zero, since neither $Z^i$ nor $e_b(Z)$ are multiples of $g(Z)$, provided that $b \leq n - k$. **QED**

When error correction is required, Theorem 10.9 provides a lower bound on the degree of the generator polynomial of the code.

**Theorem 10.9**

A burst-error correcting code can correct all bursts of length $b$ or less provided that the number of check digits satisfies the inequality (*Reiger bound*)

$$n - k \geq 2b \qquad (10.55)$$

$\triangledown$

**Proof of Theorem 10.9**

To correct all bursts of length $b$, the bursts of length $2b$ (or less) must be different from each code word. In fact, if a code word is a burst of length $2b$ (or less), it can be expressed as the sum of two bursts of length $b$ (or less). Consider the standard array of the code. If one of the two bursts (the correctable one) is a coset leader, the other, as a consequence of the assumption made on the code word, must be in the same coset. Therefore, the second burst cannot be corrected. In conclusion, no burst of length $2b$ (or less) can be allowed to be a code word in order to correct all bursts of length $b$. When this condition is met, the number of check digits is at least $2b$. In fact, consider the sequences whose nonzero components are confined to the first $2b$ positions. There are $2^{2b}$ such sequences. These sequences must be in different cosets of the standard array. Otherwise, their sum would be a code word corresponding to a burst of length 2b (or less). Since the cosets are $2^{n-k}$, then the inequality (10.55) follows. **QED**

As a consequence of Theorem 10.9, the ratio

$$z \triangleq \frac{2b}{n - k} \qquad (10.56)$$

| $n - k - 2b$ | Code $(n, k)$ | Burst-correcting ability, $b$ | Generator polynomial |
|---|---|---|---|
| 0 | (7,3) | 2 | 35 |
|   | (15,9) | 3 | 171 |
|   | (19,11) | 4 | 1151 |
|   | (27,17) | 5 | 2671 |
|   | (34,22) | 6 | 15173 |
|   | (38,24) | 7 | 114361 |
|   | (50,34) | 8 | 224531 |
|   | (56,38) | 9 | 1505773 |
|   | (59,39) | 10 | 4003351 |
| 1 | (15,10) | 2 | 65 |
|   | (27,20) | 3 | 311 |
|   | (38,29) | 4 | 1151 |
|   | (48,37) | 5 | 4501 |
|   | (67,54) | 6 | 36365 |
|   | (103,88) | 7 | 114361 |
|   | (96,79) | 8 | 501001 |
| 2 | (31,25) | 2 | 161 |
|   | (63,55) | 3 | 711 |
|   | (85,75) | 4 | 2651 |
|   | (131,119) | 5 | 15163 |
|   | (169,155) | 6 | 55725 |
| 3 | (63,56) | 2 | 355 |
|   | (121,112) | 3 | 1411 |
|   | (164,153) | 4 | 6255 |
|   | (290,277) | 5 | 24711 |
| 4 | (511,499) | 4 | 10451 |
| 5 | (1023,1010) | 4 | 22365 |

Table 10.9: *Efficient cyclic codes and shortened cyclic codes for burst-error correction. The generator polynomial is represented in octal notation with the highest-degree term on the left (Lin, 1970).*

can be assumed as a measure of the burst-correcting efficiency of the code. Some decoding algorithms for burst-error correction are based on error-trapping techniques (see Peterson and Weldon, 1972, Chapters 8 and 11).

A list of efficient cyclic codes and shortened cyclic codes for correcting short bursts is given in Table 10.9. The polynomials are again represented in octal notation, with the highest-degree terms on the left, as in Table 10.7.

### Fire codes

These codes are a versatile class of systematic cyclic codes designed for correcting or detecting a single burst of length $b$ in a block of $n$ digits. Let $p(Z)$ be an irreducible polynomial of degree $m \geq b$, and let $e$ be the smallest positive integer such that $p(Z)$ divides $(Z^e + 1)$. Furthermore, assume that $e$ and $(2b - 1)$ are relatively prime integers. Then the polynomial

$$g(Z) = (Z^{2b-1} + 1)p(Z) \tag{10.57}$$

is the generator of a $b$ burst-error correcting Fire code of length $n = \text{LCM}(e, 2b - 1)$, where LCM means least common multiple. Notice that the number of parity-check digits in these codes is $(m + 2b - 1)$. For the limit case of $m = b$, we obtain a burst-correcting efficiency that cannot exceed 2/3. A proof of the burst-error correcting capabilities of Fire codes, together with a description of error-trapping decoders, can be found in Chapter 9 of Lin and Costello (1983). Under the same conditions as before, given two integers $b$ and $d$, we can generate a Fire code capable of correcting any burst of length $b$ (or less) and simultaneously detecting any burst of length up to $d \geq b$, by using the generator polynomial

$$g(Z) = (Z^c + 1)p(Z) \tag{10.58}$$

with $c$ satisfying the condition $c \geq b + d - 1$ (see Peterson and Weldon, 1972, Chapter 11).

**Example 10.19**   We want to design a Fire code to correct all bursts of length $b \leq 7$ and to detect all bursts of lengths up to 10. We get $c \geq 16$ and $m \geq 7$. Choosing the primitive polynomial of degree 7 in Table 10.6, we obtain the following generator:

$$g(Z) = (Z^{16} + 1)(Z^7 + Z^3 + 1)$$

Since $p(Z)$ is primitive, we have $e = 2^7 - 1 = 127$, and the length of the code is $n = 16 \times 127 = 2032$. Thus, the code is a $(2032, 2009)$ Fire code with a high rate $(R_c = 0.99)$ and a burst-correcting efficiency $z = 0.6$. Notice that the low value of $(n - k)$ makes it easy to implement the encoder. On the other hand, these codes usually have a high length, even for a modest burst-correcting capability. This is a disadvantage, since only one burst per each block length is correctable or detectable. Therefore, a very long guard space between successive bursts is required.                             $\square$

## Interleaved codes

A practical technique to cope with burst errors is that of using random-error-correcting codes in connection with a suitable interleaver/deinterleaver pair. An *interleaver* is a device that rearranges the ordering of a sequence of symbols in a deterministic manner. The *deinterleaver* applies the inverse operation to restore

Figure 10.19: *Block diagram for the application of the interleaver-deinterleaver pair.*

the sequence to its original ordering. Given an $(n, k)$ cyclic code, an $(in, ik)$ interleaved code can be obtained by arranging $i$ code words of the original code into $i$ rows of a rectangular array that will be transmitted by columns. The parameter $i$ is called the *interleaving degree* of the code. If the original code corrects up to $t$ random errors, the interleaved code will have the same random-error-correction capability, but in addition it will be able to correct all bursts of length $i \times t$ (or less). The use of this technique is shown in Figure 10.19 and explained in the following example.

**Example 10.20**    Consider a $(15, 5)$ BCH code, whose generator polynomial is, from Table 10.7 $g(Z) = Z^{10} + Z^8 + Z^5 + Z^4 + Z^2 + Z + 1$. This code corrects all random error configurations with $t = 3$ (or less) errors in sequences of length $n = 15$. Taking $i = 5$, we can derive a $(75, 25)$ interleaved code. The arrangement of the code words is shown in Figure 10.20. An information sequence of 25 digits is divided into five 5-digit message blocks and five code words of length 15 are generated using $g(Z)$. These code words are arranged as five rows of the $5 \times 15$ matrix shown in the figure. The columns of the matrix are transmitted, in the indicated order, as a code word of length 75. Each burst of length 15 (or less) produces no more than three errors in each row of the matrix. A burst from position 18 to position 32 is shown by dashed squares in the figure. Therefore, the decoder can correct the errors by operating on each row. The interleaving process has, in fact, diffused the burst into isolated errors, and all error patterns containing three errors or less in each row of the matrix are correctable.               □

## 10.3.    Performance evaluation of block codes

In Chapter 5, different modulation schemes were compared on the basis of their bit error probability $P_b(e)$. The scope of this section is to provide useful tools for extending those comparisons to coded transmission.

For transmission systems employing block codes, two error probabilities can be introduced:

- The *word error probability* $P_w(e)$, defined as the probability that the decoder output is a wrong code word, i.e., a code word different from that transmitted.

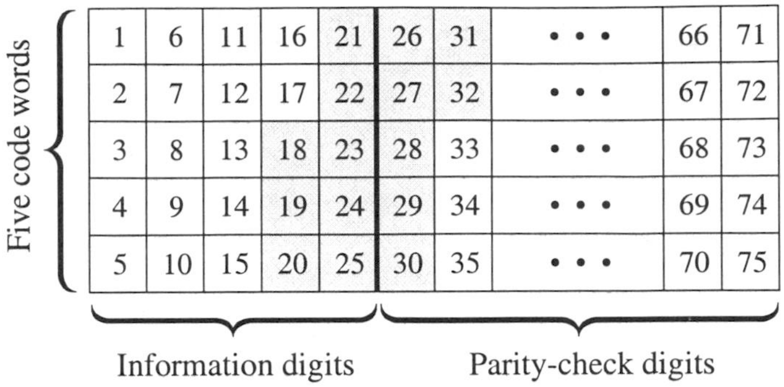

| Five code words | 1 | 6 | 11 | 16 | 21 | 26 | 31 | ••• | 66 | 71 |
|---|---|---|---|---|---|---|---|---|---|---|
| | 2 | 7 | 12 | 17 | 22 | 27 | 32 | ••• | 67 | 72 |
| | 3 | 8 | 13 | 18 | 23 | 28 | 33 | ••• | 68 | 73 |
| | 4 | 9 | 14 | 19 | 24 | 29 | 34 | ••• | 69 | 74 |
| | 5 | 10 | 15 | 20 | 25 | 30 | 35 | ••• | 70 | 75 |

Information digits          Parity-check digits

Figure 10.20: *Scheme for the interpretation of a (75,25) interleaved code derived from a (15,5) BCH code. A burst of length $b = 15$ is spread into $t = 3$ error patterns in each of the five code words of the interleaved code.*

- The *bit error probability* $P_b(e)$ (or symbol error probability for nonbinary codes), defined as the probability that an information bit (symbol) is in error after decoding.

Which of the two probabilities better describes the system performance in a particular situation depends on the system. The significance of the bit error probability comes from the fact that some of the information bits may be correct even if the decoder outputs a wrong code word.

The computation of the word and bit error probabilities depends on the decoding strategies chosen by the system. As an example, when the system employs an ARQ strategy, the decoder will output a wrong code word if and only if the received $n$-tuple is one of the $2^k - 1$ code words different from the transmitted one. This, for linear codes, requires that the channel error vector coincides with one of the nonzero code words. The situation is completely different when an FEC strategy is adopted.

Different decoding strategies are better understood with reference to the standard array of a linear code introduced in Section 10.2.2. We recall that the standard array is an array with $2^k$ columns and $2^{n-k}$ rows that groups all $2^n$ $n$-tuples representing the received words. Each row (a coset) is labeled by a code syndrome, and contains all the $n$tuples that give that syndrome. The first $n$tuple of each row (the coset leader) is the lowest weight word in the row.

Arrange the cosets in order of decreasing weight (i.e., decreasing probability on a BSC) of the coset leader, obtaining the situation of Figure 10.21, and assume that the code has a correction capability of $t$ errors. If the code were a perfect code, such as a Hamming code for $t = 1$, the cosets with a leader of weight up to $t$ would include all $n$-tuples. In general, however, this will not be the case, and

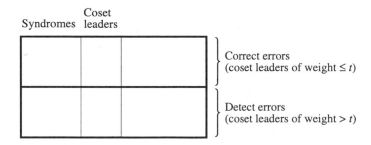

Figure 10.21: *Standard array of a block code with the cosets ordered according to the increasing weight of the coset leaders.*

we will find cosets with leaders having weights beyond $t$. When this happens, we face different decoding strategies:

1. *Complete decoding.* With this strategy, the decoder always outputs a decoded code word. It fully exploits the standard array, so that the top part of it (see Figure 10.21) includes all the received words. Given a received word in the standard array, the decoder assumes that an error vector corresponding to the leader of the coset containing the received word has been added to the transmitted code word on the channel, and decodes accordingly. In doing this, the decoder goes beyond the correction capabilities of the code, so that some error vectors with weight larger than $t$ can lead to wrong decoding.

2. *Bounded t-distance decoding.* This strategy corresponds to the partition of the standard array of Figure 10.21. If the received word **y** lies in the top part of the array, it is maximum-likelihood decoded as the code word found at the top of its column. If **y** lies in the bottom part of the standard array, the decoder just detects that more than $t$ errors have occurred. As a consequence, we have an incomplete decoding, or a mixture of error correction and error detection.

3. *Error detection.* This strategy does not attempt to correct errors; rather, it declares an error whenever the received word does not belong to the code. The upper part of the standard array of Figure 10.21 disappears, and the second includes all received words except code words.

In the following, we shall analyze all decoding strategies. The analysis is highly simplified when the code is linear. In this case, in fact, we can apply, for transmission over the BSC, the *uniform error property*, stating that the error probability conditioned to a given transmitted code word does not depend on that code word. This important property stems from the properties of linear codes (see Problem 10.26). Thus, to compute the average error probability (word or bit), we can choose to transmit every code word; it is customary to choose the all-zero code word, denoted $\mathbf{x}_1$. Code linearity and transmission of $\mathbf{x}_1$ will always be assumed in the following. We start our analysis with the simplest case of hard-decoding and error detection (ARQ) systems.

### 10.3.1. Performance of error detection systems

In ARQ systems, the decoder's function is to answer the binary question of whether a received word is a code word or not. In the negative case, the system asks for the retransmission of the message until a positive answer is obtained from the decoder. This technique has been used for years in computer systems and other applications where a feedback channel is available and retransmission is made possible by the system resources and constraints.

We assume a BSC, and define as $P_w^{(s)}(e)$, $P_w^{(s)}(c)$, $P_w^{(s)}(d)$ the probabilities of decoding error, correct detection, and error detection in a single transmission. A decoding error occurs when the error vector on the channel coincides with a code word different from $\mathbf{x}_1$. We have correct detection when no errors occur over the channel, and, finally, error detection when the error vector is not a code word. These are the only possible events in a single transmission, so that

$$P_w^{(s)}(e) + P_w^{(s)}(c) + P_w^{(s)}(d) = 1 \qquad (10.59)$$

Moreover, the probability of error coincides with the probability that the channel error vector is one of the nonzero code words, given by

$$P_w^{(s)}(e) = \sum_{d=d_{\min}}^{n} A_d p^d (1-p)^{n-d} = (1-p)^n \left[ A\left(\frac{p}{1-p}\right) - 1 \right] \qquad (10.60)$$

where $A_d$ is the number of code words of Hamming weight $d$ and $A(\cdot)$ is the weight enumerating function defined in (10.18).

The probability of correct detection is the probability that no errors occur on the channel, that is

$$P_w^{(s)}(c) = (1-p)^n \qquad (10.61)$$

Noticing then that $A_0 = 1$, we obtain from (10.59), (10.60) and (10.61)

$$P_w^{(s)}(d) = 1 - \sum_{d=0}^{n} A_d p^d (1-p)^{n-d} = 1 - (1-p)^n A\left(\frac{p}{1-p}\right) \qquad (10.62)$$

When the system keeps on retransmitting a message until the received word gets accepted by the decoder, we can have a word error at the $n$th transmission if and only if the previous $n - 1$ transmissions have led to an error detection, and the $n$th to an accepted erroneous code word, so that

$$P_w^{(n)}(e) = P_w^{(s)}(e)[P_w^{(s)}(d)]^{(n-1)} \tag{10.63}$$

The word error probability of this ARQ system is thus given by

$$P_w(e) = \sum_{n=1}^{\infty} P_w^{(n)}(e) = \frac{P_w^{(s)}(e)}{1 - P_w^{(s)}(d)} \tag{10.64}$$

### 10.3.2.    Performance of error correction systems: word error probability

**Hard decoding**

Denote with $P(\mathbf{y}_j \mid \mathbf{x}_i)$ the probability of receiving the $n$-tuple $\mathbf{y}_j$ when the code word $\mathbf{x}_i$ of an $(n, k, t)$ code is transmitted over a BSC with transition probability $p$. Then the word error probability for complete decoding is given by

$$P_w(e) = 1 - \frac{1}{M} \left[ \sum_{i=1}^{M} \sum_{j \in S_i} P(\mathbf{y}_j \mid \mathbf{x}_i) \right] \tag{10.65}$$

where $M = 2^k$ indicates the number of code words assumed to be equally likely, and $S_i$ the set of subscripts $j$ of received words $\mathbf{y}_j$, which are decoded into the code word $\mathbf{x}_i$.

For linear codes and maximum-likelihood decoding, the set $S_i$ identifies the received words lying in the same column of the standard array as $\mathbf{x}_i$. Owing to the uniform error property, we can avoid the average over the transmitted code words in (10.65), obtaining

$$P_w(e) = 1 - \sum_{j \in S_1} P(\mathbf{y}_j \mid \mathbf{x}_1) \tag{10.66}$$

where $\mathbf{x}_1$ is the all-zero code word.

Unfortunately, the evaluation of this apparently simple expression is very hard (Elia, 1983), because exhaustive computations become soon impractical as $n$ increases. Therefore, upper bounds to (10.66) have been sought.

In general, for both complete and bounded $t$-distance decoding, the word error probability is always less than or equal to the probability that more than $t$ errors have occurred on the channel. We obtain then the following upper bound

$$P_w(e) \le \sum_{i=t+1}^{n} \binom{n}{i} p^i (1 - p)^{n-i} = 1 - \sum_{i=0}^{t} \binom{n}{i} p^i (1 - p)^{n-i} \tag{10.67}$$

Notice that the equal sign in (10.67) holds only for perfect codes. When $np \ll 1$, (10.67) can be approximated by its largest term

$$P_w(e) \approx \binom{n}{t+1} p^{t+1}(1-p)^{n-t-1} \tag{10.68}$$

A different approach to obtain a general upper bound to the word error probability stems directly from the union bound explained in Section 4.3. The derivation here will assume the uniform error property. Let us define the pairwise error event $(\mathbf{x}_1 \rightarrow \mathbf{x}_\ell)$ as the set of received words $\mathbf{y}_j$ such that, when the transmitted code word is $\mathbf{x}_1$, the received word $\mathbf{y}_j$ is closer (in the ML sense) to $\mathbf{x}_\ell$ than to $\mathbf{x}_1$. Therefore, for maximum-likelihood decoding

$$(\mathbf{x}_1 \rightarrow \mathbf{x}_\ell) \triangleq \{\mathbf{y}_j : P(\mathbf{y}_j \mid \mathbf{x}_\ell) > P(\mathbf{y}_j \mid \mathbf{x}_1)\} \tag{10.69}$$

Denoting by $S_{1\ell}$ the set of subscripts $j$ for which $(\mathbf{x}_1 \rightarrow \mathbf{x}_\ell)$ occurs, we have, for the pairwise error probability

$$P(\mathbf{x}_1 \rightarrow \mathbf{x}_\ell) = \sum_{j \in S_{1\ell}} P(\mathbf{y}_j \mid \mathbf{x}_1) \tag{10.70}$$

and the union bound (4.50) gives

$$P_w(e) \leq \sum_{l=2}^{M} P(\mathbf{x}_1 \rightarrow \mathbf{x}_\ell) \tag{10.71}$$

Now we derive an upper bound for $P(\mathbf{x}_1 \rightarrow \mathbf{x}_\ell)$ of (10.70). This result, introduced in (10.71), will give us the final answer. Defining the function $f_\ell(\mathbf{y}_j)$ as

$$f_\ell(\mathbf{y}_j) \triangleq \begin{cases} 1, & \text{for } j \in S_{1\ell} \\ 0, & \text{for } j \notin S_{1\ell} \end{cases} \tag{10.72}$$

we can rewrite (10.70) as

$$P(\mathbf{x}_1 \rightarrow \mathbf{x}_\ell) = \sum_{j=1}^{2^n} f_\ell(\mathbf{y}_j) P(\mathbf{y}_j \mid \mathbf{x}_1) \tag{10.73}$$

where the summation has been extended over the whole set of received sequences $\mathbf{y}_j$. Now, we can easily bound $f_\ell(\mathbf{y}_j)$ by

$$f_\ell(\mathbf{y}_j) \leq \sqrt{\frac{P(\mathbf{y}_j \mid \mathbf{x}_\ell)}{P(\mathbf{y}_j \mid \mathbf{x}_1)}} \tag{10.74}$$

Owing to the definition (10.69) of $(\mathbf{x}_1 \to \mathbf{x}_\ell)$, the bound (10.74) is verified for $j \in S_{1\ell}$, whereas for $j \notin S_{1\ell}$ it is trivial. Introducing (10.74) into (10.73), we finally get

$$P(\mathbf{x}_1 \to \mathbf{x}_\ell) \le \sum_{j=1}^{2^n} \sqrt{P(\mathbf{y}_j \mid \mathbf{x}_\ell)P(\mathbf{y}_j \mid \mathbf{x}_1)} \tag{10.75}$$

This expression is called the *Bhattacharyya bound*. Using the memoryless property of the BSC, (10.75) leads to the result (see Problem 10.27)

$$P(\mathbf{x}_1 \to \mathbf{x}_\ell) \le \left[ \sum_{y \in Y} \sqrt{P(y \mid x = 1)P(y \mid x = 0)} \right]^{w_\ell} \tag{10.76}$$

where $Y = \{0, 1\}$, and $w_\ell$ is the weight of the code word $\mathbf{x}_\ell$. Using the transition probability $p$ of the BSC and introducing (10.76) into (10.71), we get

$$P_w(e) \le \sum_{\ell=2}^{M} \left[ \sqrt{4p(1 - p)} \right]^{w_\ell} \tag{10.77}$$

The summation in (10.77) is performed over all $M - 1$ code words different from the all-zero code word. Recalling the meaning of the weight enumerating function $A(D)$ and its expression (10.18), we can transform (10.77) into

$$P_w(e) \le \sum_{d=d_{\min}}^{n} A_d \left[ \sqrt{4p(1 - p)} \right]^{d} = [A(D) - 1]_{D = \sqrt{4p(1-p)}} \tag{10.78}$$

The bound (10.78) requires the knowledge of the weight enumerating function of the code. A simpler, but weaker, bound is obtained if we replace $w_\ell$ with $d_{\min} \; \forall \, \ell$ in (10.77)

$$P_w(e) \le (M - 1) \left[ \sqrt{4p(1 - p)} \right]^{d_{\min}} \tag{10.79}$$

**Unquantized soft-decision decoding**

As described in Section 10.1, unquantized soft-decision decoding entails no quantization of the channel output.

In principle, ML decoding for transmission over an AWGN channel could be performed with the techniques explained in Chapter 4. For an $(n, k)$ code, each code word $\mathbf{x}$ is mapped by the modulator into a waveform $x(t)$. The functions of the demodulator and decoder are integrated within the receiver, which is formed by a bank of $M = 2^k$ parallel filters matched to the waveforms $x(t)$. The sampled output of the $i$th filter yields the correlation between the received signal and the

$i$th modulator signal. The $M$ outputs from the matched filters enter a processor that chooses the largest, thus performing an ML decision. This optimum receiver becomes unrealizable in practice for large values of $k$. Some simplifications are possible, aiming at reducing the number of matched filters, for particular choices of the modulation scheme. Let us assume, as an example, that the $n$ bits of the code word are transmitted using binary antipodal modulation. Each binary waveform is demodulated by the optimum soft demodulator (a single matched filter followed by a sampler) and a code word is represented by a sequence of $n$ random variables. Let $\mathcal{E}$ denote the energy of the modulator waveform. Then, dropping an irrelevant constant, each of the $n$ binary decision variables can be written as

$$z_i = \begin{cases} \sqrt{\mathcal{E}} + \nu_i & \text{if the } i\text{th digit is } 1 \\ -\sqrt{\mathcal{E}} + \nu_i & \text{if the } i\text{th digit is } 0 \end{cases} \qquad (10.80)$$

with $i = 1, 2, \ldots, n$. The random variables $\nu_i$ are samples of the Gaussian noise with zero mean and variance $N_0/2$. From the knowledge of the $M = 2^k$ code words, and upon reception of the sequence $z_1, \ldots, z_n$ from the demodulator, the decoder forms $M$ decision variables as follows:

$$L_i = \sum_{j=1}^{n} (2x_{ij} - 1)z_j, \quad i = 1, \ldots, M \qquad (10.81)$$

where $x_{ij}$ denotes the digit in the $j$th position of the $i$th code word. In this way, the decision variable corresponding to the actual transmitted code word will have a mean value $n\sqrt{\mathcal{E}}$, while the other $(M-1)$ ones will have smaller mean values. Maximum-likelihood decoding is achieved by selecting the largest among the $L_i$'s of (10.81).

Although the computations involved in the previous decoding process are very simple, it may soon become impractical to implement this algorithm because of the exponential growth of the number $M$ of decision variables with $k$. Several different types of soft-decision decoding algorithms have been invented to circumvent this difficulty. Some reference is given to them in the Bibliographical Notes.

The derivation of the exact error probability in the decoding process is not straightforward, as it is complicated by the correlations between the decision variables. Therefore, we resort to a union bound similar to the one employed for the case of hard-decoding. Recalling the uniform error property, we can assume that the all-zero code word is transmitted. Let us define the pairwise error probability $P(\mathbf{x}_1 \to \mathbf{x}_m)$ as

$$P(\mathbf{x}_1 \to \mathbf{x}_m) \triangleq P[L_m > L_1 \mid \mathbf{x}_1] \qquad (10.82)$$

It is the probability that the likelihood of the $m$th code word is higher than that of the transmitted all-zero code word. The union bound (4.50) gives

$$P_w(e) \le \sum_{m=2}^{M} P(\mathbf{x}_1 \to \mathbf{x}_m) \tag{10.83}$$

But the pairwise error probability $P(\mathbf{x}_1 \to \mathbf{x}_m)$ only depends on the Euclidean distance $d_{1m}$ between the two code words. If the $m$th code word has weight $w_m$, then it differs from the all-zero code word in $w_m$ positions. Therefore,

$$d_{1m}^2 = 4w_m \mathcal{E} = 4w_m R_c \mathcal{E}_b \tag{10.84}$$

Introducing this value into (4.52), we obtain $P(\mathbf{x}_1 \to \mathbf{x}_m)$, and, from (10.83), we conclude with

$$P_w(e) \le \frac{1}{2} \sum_{m=2}^{M} \operatorname{erfc}\left(\sqrt{\frac{w_m R_c \mathcal{E}_b}{N_0}}\right) \tag{10.85}$$

A looser bound is obtained by using $d_{\min}$ in (10.85). This yields

$$P_w(e) \le \frac{M-1}{2} \operatorname{erfc}\left(\sqrt{\frac{d_{\min} R_c \mathcal{E}_b}{N_0}}\right) \tag{10.86}$$

Grouping together the $A_d$ code words with the same weight $d$, we can rewrite (10.85) as

$$P_w(e) \le \frac{1}{2} \sum_{d=d_{\min}}^{n} A_d \operatorname{erfc}\left(\sqrt{\frac{d R_c \mathcal{E}_b}{N_0}}\right) \tag{10.87}$$

Using now the inequality (A.5) (see Appendix A) $\frac{1}{2} \operatorname{erfc}(\sqrt{z}) < \frac{1}{2} \exp(-z)$, we transform (10.87) into

$$P_w(e) < \frac{1}{2} \sum_{d=d_{\min}}^{n} A_d e^{-dR_c \mathcal{E}_b/N_0} = \frac{1}{2} [A(D) - 1]_{D=e^{-R_c \mathcal{E}_b/N_0}} \tag{10.88}$$

To estimate the coding gain, we can compare the approximate result (10.86) with (4.36) for binary antipodal transmission. Making use of the exponential approximation (A.5) for the function erfc, we obtain

$$G_c \triangleq \frac{(\mathcal{E}_b/N_0)_{\text{unc}}}{(\mathcal{E}_b/N_0)_{\text{enc}}} = R_c d_{\min} - \frac{k \ln 2}{(\mathcal{E}_b/N_0)_{\text{enc}}} \tag{10.89}$$

Notice that the coding gain $G_c$ depends on both the code parameters and the signal-to-noise ratio. The asymptotic value in dB is $10 \log_{10}(R_c d_{\min})$.

The results obtained in the last two sections will now be illustrated through an example.

**Example 10.21**   Consider the (7, 4) Hamming code that has been used in the examples throughout the chapter. The aim of this example is to assess the bounds introduced in the last two sections. We start with hard decision. To compute the error probability $p$ of the BSC, we assume a binary antipodal transmission, so that

$$p = \frac{1}{2}\text{erfc}\left(\sqrt{\frac{R_c\mathcal{E}_b}{N_0}}\right) \tag{10.90}$$

Since Hamming codes are perfect codes, the expression (10.67) is the exact value of the word error probability

$$P_w(e) = 1 - \sum_{i=0}^{t}\binom{n}{i}p^i(1-p)^{n-i} = 1 - (1-p)^7 - 7p(1-p)^6$$

The exact result for $P_w(e)$ is plotted in Figure 10.22, together with the two bounds (10.78) and (10.79). Recalling the weight enumerating function of Hamming codes (10.19), the bound (10.78) yields

$$P_w(e) \le 7D^3 + 7D^4 + D^7\big|_{D=\sqrt{4p(1-p)}}$$

Similarly, being $d_{\min} = 3$, the bound (10.79) becomes

$$P_w(e) \le 15D^3\big|_{D=\sqrt{4p(1-p)}}$$

From the curves of Figure 10.22, we see that the tighter of the two bounds differs by slightly less than 2 dB from the exact error probability at $P_w(e) = 10^{-6}$, and that the looser bound is worse by a fraction of dB.

In the case of soft decision, we apply the bound (10.87) and the simpler (10.86), obtaining

$$P_w(e) \le \frac{7}{2}\text{erfc}\left(\sqrt{\frac{12\mathcal{E}_b}{7N_0}}\right) + \frac{7}{2}\text{erfc}\left(\sqrt{\frac{16\mathcal{E}_b}{7N_0}}\right) + \frac{1}{2}\text{erfc}\left(\sqrt{\frac{4\mathcal{E}_b}{N_0}}\right)$$

and

$$P_w(e) \le \frac{15}{2}\text{erfc}\left(\sqrt{\frac{12\mathcal{E}_b}{7N_0}}\right)$$

The figure also contains these results pertaining to soft decision.                    □

### 10.3.3.   Performance of error correction systems: bit error probability

The expression of the bit error probability is somewhat more complicated. In this case, in fact, we must enumerate the actual error events and weight the probability of each event by the number of bit errors that occur. Invoking the uniform

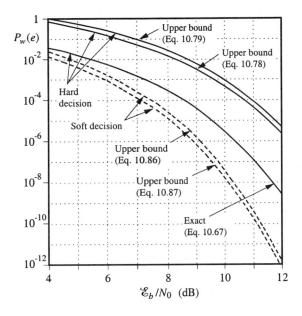

Figure 10.22: *Word error probability for the* (7, 4) *Hamming code: exact value and two upper bounds. Hard decision and soft-decision curves. Binary antipodal transmission.*

error probability, a formal expression for the bit error probability is the following:

$$P_b(e) = \sum_{j=2}^{M} \frac{w_H(\mathbf{u}_j)}{k} P(\mathbf{x}_j \mid \mathbf{x}_1) \tag{10.91}$$

where $P(\mathbf{x}_j \mid \mathbf{x}_1)$ is the probability of decoding a code word $\mathbf{x}_j$ different from the transmitted all-zero code word $\mathbf{x}_1$, and $w_H(\mathbf{u}_j)$ is the Hamming weight of the information word that generates the code word $\mathbf{x}_j$.

Computing (10.91) is a very difficult task, in general. In the following, we will generalize the concept of weight enumerating function so as to obtain a union bound to the bit error probability close to the one already obtained for the word error probability.

The weight enumerating function describes the weight distribution of the code words. It does not provide informations on the data words that generate the code words. In other words, it does not provide informations about the encoder. Let us generalize it by introducing the definition of the *input-output weight enu-*

*merating function* (IOWEF)

$$B(W, D) \triangleq \sum_{w=0}^{k} \sum_{d=0}^{n} B_{w,d} W^w D^d \tag{10.92}$$

where $B_{w,d}$ represents the number of code words with weight $d$ generated by data words of weight $w$. Recalling the definition (10.18) of the weight enumerating function $A(D)$, it is straightforward to derive the relationships

$$A(D) = B(W, D)|_{W=1}, \quad A_d = \sum_{w=0}^{k} B_{w,d} \tag{10.93}$$

As an example, consider the Hamming code $(7, 4)$. From the table defining its encoder (see Example 10.3), we obtain the IOWEF

$$\begin{aligned} B(W, D) &= 1 + W(3D^3 + D^4) + W^2(3D^3 + 3D^4) \\ &\quad + W^3(D^3 + 3D^4) + W^4 D^7 \\ &\triangleq 1 + W B_1(D) + W^2 B_2(D) + W^3 B_3(D) + W^4 B_4(D) \end{aligned} \tag{10.94}$$

where we have defined the *conditional weight enumerating function* (CWEF) $B_w(D)$, as the weight enumerating function of the code words generated by data words of weight $w$. In general, it is obtained through

$$B_w(D) = \sum_{d=0}^{n} B_{w,d} D^d = \frac{1}{w!} \left. \frac{\partial B^w(W, D)}{\partial W^w} \right|_{W=0} \tag{10.95}$$

A union bound to the bit error probability can be obtained by following the derivation of (10.78) and (10.88), and defining a conditional pairwise error event $(\mathbf{x}_1 \rightarrow \mathbf{x}_{w,d})$ as the event that the likelihood of the transmitted all-zero code word is less than that of the code word $\mathbf{x}_{w,d}$ with weight $d$ generated by an information word of weight $w$.

The final upper bounds, for the case of hard and infinitely-soft decoding (see Problem 10.28), are, respectively,

$$P_b(e) \leq \sum_{w=1}^{k} \frac{w}{k} \sum_{d=d_{\min}}^{n} B_{w,d} \left[ \sqrt{4p(1-p)} \right]^d = \sum_{w=1}^{k} \frac{w}{k} \left[ B_w(D) - 1 \right]_{D=\sqrt{4p(1-p)}} \tag{10.96}$$

$$P_b(e) \leq \sum_{w=1}^{k} \frac{w}{2k} \sum_{d=d_{\min}}^{n} B_{w,d} \operatorname{erfc}\left( \sqrt{\frac{dR_c \mathcal{E}_b}{N_0}} \right) \tag{10.97}$$

Making use of the inequality (A.5), (10.97) becomes

$$P_b(e) < \frac{1}{2} \sum_{w=1}^{k} \frac{w}{k} [B_w(D) - 1]_{D = e^{-R_c \mathcal{E}_b / N_0}} \tag{10.98}$$

By exchanging the order of summation, (10.96) (and similarly (10.97)) can also be written in the form

$$P_b(e) \leq \sum_{d=d_{\min}}^{n} \left( \sum_{w=1}^{k} \frac{w}{k} B_{w,d} \right) \left[ \sqrt{4p(1-p)} \right]^d \tag{10.99}$$

$$P_b(e) \leq \frac{1}{2} \sum_{d=d_{\min}}^{n} \left( \sum_{w=1}^{k} \frac{w}{k} B_{w,d} \right) \operatorname{erfc} \left( \sqrt{\frac{d R_c \mathcal{E}_b}{N_0}} \right) \tag{10.100}$$

Comparing (10.99) and (10.100) with the analogous expressions (10.78) and (10.87) for the word error probability, we see that they are formally identical if we interpret the quantity in brackets as the average number of nonzero information bits associated with a code word of weight $d$.

**Example 10.22**  Consider once again the $(7, 4)$ Hamming code. Substituting into (10.96) and (10.98) the CWEFs of this code evaluated in (10.94), we obtain

$$P_b(e) \leq \frac{1}{4} [(3D^3 + D^4) + 2(3D^3 + 3D^4) + 3(D^3 + 3D^4) + 4D^7]$$

from which, replacing $D$ with $\sqrt{4p(1-p)}$ or $\exp(-R_c \mathcal{E}_b / N_0)$, we obtain the bit error probability bounds for the case of hard and soft decision, respectively. □

Computing the previous union bound requires the knowledge of the CWEFs of the code, and this is often a heavy task. A simpler approximation to the bit error probability valid at high signal-to-noise ratios can be obtained as follows. For a code with distance $d_{\min} = 2t + 1$, the most frequent undetected error will occur when the error vector contains exactly $t + 1$ errors. In this case, the decoding algorithm will erroneously assume that the received sequence contains $t$ errors and will choose a code word at distance $d_{\min}$ from the correct one. As a consequence, the decoded word will contain $2t+1$ errors, which can be anywhere in the $n$-digit sequence. The bit error probability can thus be expressed in terms of the word error probability as

$$P_b(e) \approx \frac{2t + 1}{n} P_w(e) \tag{10.101}$$

Approximation (10.101) can be used in conjunction with all expressions of the word error probability previously derived. In the following example, we will show how to estimate the coding gain by comparing uncoded and encoded bit error probabilities curves.

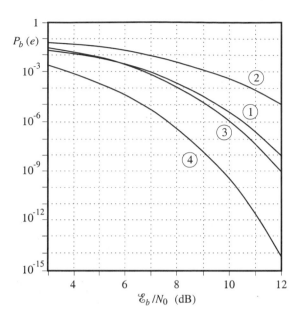

Figure 10.23: *Bit error probability curves for the* $(7, 4)$ *Hamming code with binary antipodal transmission.*

**Example 10.23**   Consider the $(7, 4)$ Hamming code. In Figure 10.23 we have plotted four curves.  Curve ① gives the bit error probability for uncoded transmission with binary antipodal modulation over the AWGN channel

$$P_b(e) = \frac{1}{2}\text{erfc}\left(\sqrt{\frac{\mathcal{E}_b}{N_0}}\right)$$

Curve ② represents the bit error probability of the BSC as seen by the decoder, i.e.

$$P_b(e) = \frac{1}{2}\text{erfc}\left(\sqrt{R_c\frac{\mathcal{E}_b}{N_0}}\right)$$

with $R_c = 4/7$. Comparing these two curves, we better understand the terms of the discussion presented in Section 10.1. In fact, the comparison shows that, when the code is used, the bit error probability on the binary transmitted symbols is higher than with uncoded transmission. The two curves differ by $10\log_{10} R_c \approx 2.4$ dB. It is expected that the error-correcting capabilities of the code are able to eliminate this disadvantage and to achieve a coding gain. The actual result for the Hamming code is shown by curve ③. This curve plots the approximate expression of the bit error probability (10.101), in which $P_w(e)$ is the exact value (10.67). For $P_b(e) = 10^{-6}$, a coding gain of 0.6

dB is obtained. Finally, curve ④ represents the bit error probability when only error detection is used. This curve plots the approximate expression of the bit error probability (10.101), in which $P_w(e)$ is the exact expression (10.64). The apparent dramatic improvement cannot be directly compared with curve ③, because in this case we have a retransmission strategy, so that the effective transmission rate through the channel is decreased: we are gaining performance at the expenses of the bandwidth.                    □

**Example 10.24**   We consider here the performance of some of the BCH codes whose generator polynomials were given in Table 10.7. To this purpose, combining (10.101) with (10.68), we get, for the bit error probability with hard decoding, the following approximation

$$P_b(e) \approx \frac{2t+1}{n} \binom{n}{t+1} p^{t+1}(1-p)^{n-t-1}$$

The BSC channel error probability $p$ is evaluated as before assuming a binary antipodal transmission. Curves of bit error probability versus the bit signal-to-noise ratio $\mathcal{E}_b/N_0$ are plotted in Figure 10.24 for several BCH codes having in common a rate $R_c \approx 0.5$. The curve referring to the uncoded transmission is also plotted. Comparing Figure 10.24 with Figure 10.2, the reader can gain an immediate perception of how much these codes fill the region of potential coding gains. Indeed, it can be verified that substantial coding gains can be obtained, for example, at $P_b(e) = 10^{-5}$. With a code length $n = 511$, about a 4-dB gain is achievable, which is almost half that promised in Figure 10.2. Increasing this coding gain is not easy. Indeed, Figure 10.24 shows that the curves tend to cluster as $n$ increases. It was shown in Wozencraft and Jacobs (1965) that a rather broad maximum of coding gain versus code rate $R_c$ occurs for each block length of BCH codes. This maximum lies in the range from one-third to three-quarters for $R_c$.                    □

**Example 10.25**   We show here in a more complete way the performance of a group of BCH codes and of the Hamming codes. Each code is identified by the triplet $(n, k, t)$. The BCH codes are those of Table 10.7. The results were obtained using for BCH codes the same approximation as in Example 10.24, whereas for Hamming codes the bit error probability is the approximation (10.101) with the exact expression (10.67) of the word error probability. All the details of the computations are omitted. The results are presented in the graph of Figure 10.25. Notice that this presentation is identical to that used for the comparison of different modulation schemes (Chapter 5). The two parameters in the figure are the bandwidth efficiency $R_s/W$ and the bit signal-to-noise ratio $\mathcal{E}_b/N_0$. All the results are given for a fixed bit error probability $P_b(e) = 10^{-6}$. Notice, first, the performance point for uncoded binary antipodal transmission. Comparing with Figure 5.31, it can be seen that the codes fill the region of potential gains in the power-limited region.

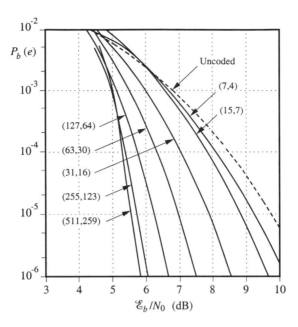

Figure 10.24: *Bit error probability curves for some BCH codes, chosen so as to have a rate $R_c$ of about 0.5.*

The bandwidth expansion of the codes is also evident, leading to bandwidths efficiencies less than 1. Let us briefly comment on the results. To this purpose, three groups of codes are identified by different symbols (stars, black dots, and squares) and connected by dashed lines. These have no other meaning than that of identifying the group.

The black dots refer to the BCH codes of Figure 10.24. Actually, the points are obtained by sectioning the curves of Figure 10.24 at $P_b(e) = 10^{-6}$. The rate of these codes (coinciding with the spectral efficiency) is practically constant and roughly 0.5. These black dots give pictorial evidence to the essence of the channel-coding theorem (see Section 3.3.3). We can improve the performance (in this case we save signal power) by coding with an increasing block length $n$.

A second group of codes, the squares, represents Hamming codes. They are single-error-correcting codes and their coding gain is rather poor. Increasing the block length, we improve both the coding gain and the bandwidth efficiency. The best code at $P_b(e) = 10^{-6}$ is the (63, 57, 1) code. If we keep on increasing the block length, the bandwidth efficiency continues to increase, but the coding gain becomes poorer.

The last group of codes, the stars, are BCH codes of length 255 and different rates, starting from the code (255, 239, 2) down to the code (255, 123, 19). The trend of the stars clearly shows the trade-off between coding gain and bandwidth efficiency when

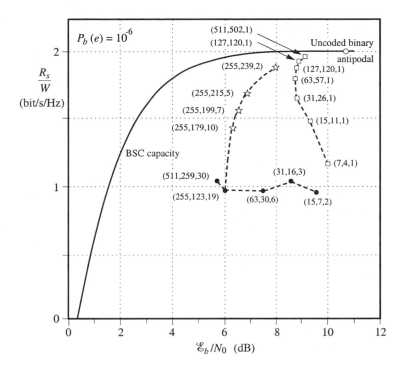

Figure 10.25: *Performance chart of different BCH and Hamming codes. Each code is identified with three numbers: the block length n, the number of information bits k, and the number of corrected errors t.*

the block length is kept constant. □

## 10.4. Coding bounds

When attempting to find new codes, it is very useful to know how close a code is to the best possible code with the same parameters. Coding bounds give quantitative answers to this requirement.

We will present in this section two types of coding bounds. The first is bounds on the minimum distance obtained by examining certain structural aspects of codes. We will consider the four main bounds of this kind, i.e., the *Singleton, Hamming, Plotkin* and *Gilbert* bounds. The first three bounds yield the largest possible minimum distance for given code length and rate, while the fourth bound is an achievable bound, and thus provides a lower bound on the minimum distance of the best code.

The second type of bounds deals with the code performance, based on random coding arguments. The most important result of this approach was already mentioned in Chapter 3 as the Shannon channel coding theorem. Here, we will derive a second bound based on the *cutoff* rate of the channel.

### 10.4.1.  Bounds on the code minimum distance

The design goal when dealing with an $(n, k)$ block code is that of achieving the largest possible $d_{\min}$ with the highest possible code rate $R_c = k/n$. This means that the best use has been made of the available redundancy. Here, we shall present four results obtained in this area. The derivation of the first two bounds is referred to the Problem Section, while that of the last two is omitted. The interested reader can find the details in the book by MacWilliams and Sloane (1977). The first bound is an upper bound known as the *Singleton* bound. The minimum distance $d_{\min}$ of any linear $(n, k)$ code satisfies the inequality

$$d_{\min} \leq n - k + 1 \qquad\qquad (10.102)$$

The second bound is known as the *Hamming*, or *sphere-packing*, bound. The maximum achievable $d_{\min}$ is given implicitly by the expression

$$\sum_{i=0}^{t} \binom{n}{i} \leq 2^{n-k} \qquad\qquad (10.103)$$

where $t$ is the maximum number of correctable errors and $d_{\min} = 2t+1$. The equality sign in (10.103) holds only for perfect codes. This bound is tight for high-rate codes. The proofs of Singleton and Hamming bounds are very simple (see Problem 10.24).

A tight upper bound for low-rate codes is the Plotkin bound, given by

$$d_{\min} \leq \frac{n 2^{k-1}}{2^k - 1} \qquad\qquad (10.104)$$

The derivation of the Plotkin bound is based on the principle that the minimum-weight code word of a group code is not larger than the average weight of all nonzero code words.

An interesting question can be raised at this point. Given that an $(n, k)$ code with an error-correcting ability better than that obtained from (10.103) and (10.104) cannot exist, what can actually be achieved? A partial answer to this question is given by the *Varshamov-Gilbert* bound, which states that it is always possible to find an $(n, k)$ code with minimum distance at least $d_{\min}$, where $n, k,$ and $d_{\min}$ satisfy the inequality

$$\sum_{i=0}^{d_{\min}-2} \binom{n-1}{i} < 2^{n-k} \qquad (10.105)$$

This result represents a lower bound to the achievable $d_{\min}$.

**Example 10.26** We want to find a block code of length $n = 127$ and error-correcting capability $t = 5$ (i.e., $d_{\min} = 11$). Its rate should be the largest possible. The Singleton bound (10.102) yields $k \leq 119$. The Hamming bound (10.103) gives

$$\sum_{i=0}^{5} \binom{127}{i} \leq 2^{127-k}$$

from which we get $k \leq 99$. Instead, the Plotkin bound (10.104) gives

$$2(11 - 1) - \log_2 11 \leq 127 - k$$

so that $k \leq 110$.

On the other side, the Varshamov-Gilbert bound (10.105) gives

$$\sum_{i=0}^{9} \binom{126}{i} < 2^{127-k}$$

from which we argue that codes exist with $k > 82$. Therefore, the maximum value of $k$ should lie between 83 and 99. From Table 10.7, we observe that a (127,92) BCH code exists that provides a satisfactory answer to our problem. $\qquad \square$

### 10.4.2. Bounds on code performance

Bounds on code performance are obtained using random coding techniques, that is, by evaluating the average performance of an ensemble of codes. This implies the existence of specific codes that behave better than the average. The most important result of this approach was already mentioned in Chapter 3 as the *channel-coding theorem*. This theorem states that the word error probability of a coded system can be reduced to any value by simply increasing the code word length $n$, provided only that the code rate does not exceed the channel capacity $C$.

Given the ensemble of binary block codes of length $n$ and rate $R_c$, the minimum attainable error probability over any discrete memoryless channel is bounded by

$$P_w(e) \leq 2^{-nE(R_c)}, \quad R_c \leq C \qquad (10.106)$$

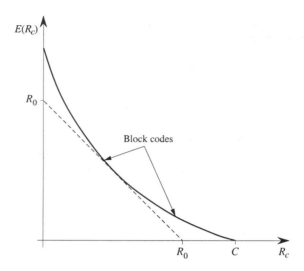

Figure 10.26: *Typical behavior of the reliability function $E(R_c)$ on the AWGN channel.*

The achievable performance is determined by the *reliability function* of the channel $E(\cdot)$, whose typical behavior for a discrete memoryless channel is given in Figure 10.26.   The tangent to $E(R_c)$ with slope $-1$ intercepts the horizontal axis at a value of $R_c$ that we call the *cutoff rate* $R_0$ of the channel. Therefore, we can write a simpler upper bound in the form

$$P_w(e) \le 2^{-n(R_0 - R_c)}, \quad R_c \le R_0 \tag{10.107}$$

The parameter $R_0$ plays an important role in coding theory. Since it is a characteristic of the channel, it allows comparisons of different channels with respect to an ensemble of codes. We shall now derive the expression of $R_0$ for the ensemble of binary block codes using a binary antipodal modulation over an AWGN.

Let us consider the ensemble C of all binary block codes $\mathcal{C}$ of length $n$ and rate $R_c$. Each code $\mathcal{C}$ has $M = 2^k = 2^{nR_c}$ code words, and there is a total of $2^{nM}$ possible codes. This ensemble also includes some very bad codes, such as those having all equal code words. Nevertheless, the bounding technique gives useful results. If we select at random one code $\mathcal{C}$, and denote by $P_w(e \mid \mathcal{C})$ the conditional word error probability, the error probability over the code ensemble is given by

$$P_w(e) = 2^{-nM} \sum_{\mathcal{C} \in C} P_w(e \mid \mathcal{C}) \tag{10.108}$$

Assume now that the code word $\mathbf{x}_i$ of $\mathcal{C}$ is transmitted with probability $P(\mathbf{x}_i)$.

Then

$$P_w(e \mid \mathcal{C}) = \sum_{i=1}^{M} P_w(e \mid \mathcal{C}, \mathbf{x}_i) P(\mathbf{x}_i) \tag{10.109}$$

The conditional error probability $P_w(e \mid \mathcal{C}, \mathbf{x}_i)$ can be upper bounded by using the union bound (see Section 4.3). We get

$$P_w(e \mid \mathcal{C}, \mathbf{x}_i) \leq \sum_{j=1, j \neq i}^{M} P(\mathbf{x}_i \rightarrow \mathbf{x}_j \mid \mathcal{C}) \tag{10.110}$$

where $P(\mathbf{x}_i \rightarrow \mathbf{x}_j \mid \mathcal{C})$ denotes the pairwise error probability between the two code words $\mathbf{x}_i$ and $\mathbf{x}_j$ of the code $\mathcal{C}$.

Introducing (10.110) into (10.109) and going back to (10.108), we obtain

$$P_w(e) \leq 2^{-nM} \sum_{\mathcal{C} \in C} \sum_{i=1}^{M} P(\mathbf{x}_i) \sum_{j=1, j \neq i}^{M} P(\mathbf{x}_i \rightarrow \mathbf{x}_j \mid \mathcal{C}) \tag{10.111}$$

Now comes the crucial step in the derivation, consisting in the interchange of the summations order in (10.111) to get

$$P_w(e) \leq \sum_{i=1}^{M} P(\mathbf{x}_i) \sum_{j=1, j \neq i}^{M} \left[ 2^{-nM} \sum_{\mathcal{C} \in C} P(\mathbf{x}_i \rightarrow \mathbf{x}_j \mid \mathcal{C}) \right] \tag{10.112}$$

The quantity in square brackets is the average of the pairwise error probability $P(\mathbf{x}_i \rightarrow \mathbf{x}_j \mid \mathcal{C})$ over the ensemble of codes and is quite straightforward to compute. Since the code $\mathcal{C}$, and hence the code words, are chosen at random, we perform the average of $\sum_{\mathcal{C} \in C} P(\mathbf{x}_i \rightarrow \mathbf{x}_j \mid \mathcal{C})$ in square brackets of (10.112) by considering the pairs $\mathbf{x}_i, \mathbf{x}_j$ of randomly chosen code words that differ in $h$ symbols (i.e., whose Hamming distance is $d_{ij} = h$) and then averaging with respect to $h$. We have then

$$2^{-nM} \sum_{\mathcal{C} \in C} P(\mathbf{x}_i \rightarrow \mathbf{x}_j \mid \mathcal{C}) = \sum_{h=0}^{n} P(d_{ij} = h) P(\mathbf{x}_i \rightarrow \mathbf{x}_j \mid d_{ij} = h) \tag{10.113}$$

The probability that two code words of length $n$ selected at random differ in $h$ digits is

$$P[d_{ij} = h] = \binom{n}{h} 2^{-n} \tag{10.114}$$

Furthermore, from Chapter 4, (4.29) and using the inequality (A.5), we get

$$P(\mathbf{x}_i \rightarrow \mathbf{x}_j \mid d_{ij} = h) = \frac{1}{2} \mathrm{erfc} \left( \sqrt{\frac{h R_c \mathcal{E}_b}{N_0}} \right) < e^{-h R_c \mathcal{E}_b / N_0} \tag{10.115}$$

Substituting (10.115) and (10.114) into (10.113) and using the binomial expansion, we obtain

$$2^{-nM} \sum_{\mathcal{C} \in C} P(\mathbf{x}_i \to \mathbf{x}_j \mid \mathcal{C}) \leq 2^{-n} \left(1 + e^{-R_c \mathcal{E}_b / N_0}\right)^n \tag{10.116}$$

Introducing (10.116) into (10.112) and observing that the RHS of (10.116) is independent of $i$ and $j$, we get

$$P_w(e) \leq (M - 1)2^{-n} \left(1 + e^{-R_c \mathcal{E}_b / N_0}\right)^n < M 2^{-n} \left(1 + e^{-R_c \mathcal{E}_b / N_0}\right)^n \tag{10.117}$$

Finally, from (10.117) we can obtain the bound (10.107) by letting

$$R_0 = 1 - \log_2 \left(1 + e^{-R_c \mathcal{E}_b / N_0}\right) \tag{10.118}$$

Equation (10.118) represents the cutoff rate of an unquantized AWGN channel with binary antipodal modulation. Similar analyses can be performed to derive the cutoff rate for different types of modulation (see Chapter 12) on the same channel. When hard-decision decoding is used, we have the general model of a discrete memoryless channel with $N_X$ input symbols and $N_Y$ output symbols. The cutoff rate of such a channel, when the input symbols are equally likely, was shown to be (Gallager, 1965)

$$R_0 = -\log_2 \left\{ \sum_{j=1}^{N_Y} \left[ \frac{1}{N_X} \sum_{i=1}^{N_X} \sqrt{P(y_j \mid x_i)} \right]^2 \right\} \tag{10.119}$$

For the BSC, (10.119) specializes to

$$R_0 = 1 - \log_2[1 + 2\sqrt{p(1 - p)}] \tag{10.120}$$

The channel cutoff rate plays an important role in coded transmission. It indicates in fact *both* a region of rates for which arbitrarily small error probability is achievable, *and* an exponent in the bounding expression of the word error probability. For this reason, the cutoff rate was claimed to be the most sensible parameter for comparing different modulation schemes in coded transmission (Massey, 1974). When soft decisions are used, the symbol error probability at the demodulator output is a very poor (practically useless) indication of the quality of the system in the presence of coding. Actually, the demodulator symbol error probability is a straight performance measure in the absence of coding because, in this case, the errors at the demodulator output are immediately reflected in the digits delivered to the user. This approach was first extended to the case of modulation plus coding in the presence of hard-decision decoding, when the purpose of coding is to correct the errors at the demodulator output.

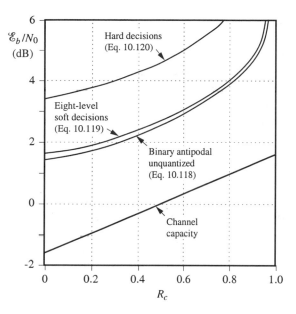

Figure 10.27: *Cutoff rate-based bounds on the required signal-to-noise ratio as a function of the code rate $R_c$ for hard and soft decision and binary antipodal modulation over the AWGN channel.*

When soft decisions are used, the combined process of modulation and coding needs a different approach. From this viewpoint, the purpose of the modulator-demodulator pair is that of presenting the coder-decoder pair with the best possible discrete-input discrete-output (in the case of quantized soft decisions) channel or the best possible discrete-input continuous-output (for unquantized soft decisions) channel for a given available bandwidth and signal-to-noise ratio. The performance measure of this channel is its cutoff rate $R_0$, and the average performance of the coded transmission system is given by (10.107).

To conclude this discussion, we show how the cutoff rate can be used to evaluate bounds on the achievable coding gains for a given channel. Equations (10.118) and (10.120) are plotted in Figure 10.27. They are obtained by letting $R_0 = R_c$ and using for $p$ in (10.120) the expression (10.90). The curves show for a given rate $R_c$ the minimum possible value of $\mathcal{E}_b/N_0$ for which the word error probability (10.107) can still be driven to zero.

As an example, we know from Figure 10.3 that uncoded antipodal transmission requires a value of $\mathcal{E}_b/N_0 = 9.6$ dB to achieve an error probability of $10^{-5}$. From Figure 10.27 we see that, for unquantized soft decisions and code rate

$R_c = 1/2$, we need a value of $\mathcal{E}_b/N_0 = 2.45$ dB. Therefore, a potential coding gain of the order of 7 dB is available at this rate. Moreover, we can observe that there is not much to be gained by increasing the redundancy of the code, since at zero rate the coding gain is only slightly greater than 8 dB.

Another point that results from Figure 10.27 is that hard decisions cause a loss of about 2 dB. In the same figure, we have also plotted the curve of $R_0$ for an eight-level quantized receiver with uniform thresholds on the received binary antipodal signal. This curve plots (10.119). The conclusion is that an eight-level quantization presents a negligible degradation with respect to the unquantized case. Finally, in Figure 10.27 the curve of channel capacity is plotted. It provides an absolute limit on the minimum value of $\mathcal{E}_b/N_0$ required to achieve any desired small error probability. Comparing the cutoff rate and capacity limits, we can see that for codes of rate 1/2 an additional coding gain of 2.4 dB is available over the prediction of the $R_0$ bound. However, decoding algorithms allowing one to operate in this region are usually very complex. A very promising exception are the so-called *turbo* codes, which will be described in the next chapter.

## 10.5.  Bibliographical notes

The reader interested in a deeper approach to the theory of block codes can find some excellent books devoted to these subjects. In particular, recommended reading are the books by Berlekamp (1968), Peterson and Weldon (1972), and MacWilliams and Sloane (1977). A stimulating approach to the decoding of cyclic codes can be found in Blahut (1983). For the reader interested in the applications and implementation problems of coded transmission, the two books by Clark and Cain (1981) and Lin and Costello (1983) are recommended references.

It would be very hard to reference papers on coding theory, besides those motivated by specific details directly in the text. In fact, the literature on the subject is huge. A wide bibliography is included in MacWilliams and Sloane (1977). A selected reading in a historical framework can be found in Berlekamp (1974).

Soft decoding a block code is generally much more complex than using algebraic algorithms for hard decoding. A general approach consists in deriving the *trellis* (see Chapter 11 for its definition) of the code, and then applying maximum likelihood algorithms, such as the Viterbi algorithm described in Appendix F, to decode it. Illuminating references are Wolf (1978), Kschischang and Sorokine (1995), and McEliece (1996).

## 10.6. Problems

**10.1** A $(5, 3)$ block code is defined through the correspondence given in the following table:

| $u_1$ | $u_2$ | $u_3$ | $x_1$ | $x_2$ | $x_3$ | $x_4$ | $x_5$ |
|---|---|---|---|---|---|---|---|
| 1 | 1 | 0 | 1 | 0 | 1 | 0 | 1 |
| 1 | 0 | 1 | 0 | 1 | 0 | 1 | 0 |
| 0 | 1 | 0 | 0 | 1 | 1 | 0 | 0 |

Find the generator matrix of the code.

**10.2** Verify that the Hamming distance between two binary sequences is equal to the weight of their modulo-2 sum.

**10.3** From (10.13) and (10.20) construct the parity check matrix of the extended $(8, 4)$ Hamming code.

1. Show that the last row, if replaced by the modulo-2 sum of all rows, still represents a legitimate parity-check equation.

2. Verify that the matrix obtained is that of a systematic code equivalent to the original one.

**10.4** A $(6, 2)$ linear block code has the following parity-check matrix:

$$\mathbf{H} = \begin{bmatrix} h_1 & 1 & 0 & 0 & 0 & 1 \\ h_2 & 0 & 0 & 0 & 1 & 1 \\ h_3 & 0 & 0 & 1 & 0 & 1 \\ h_4 & 0 & 1 & 1 & 1 & 0 \end{bmatrix}$$

1. Choose the $h$'s in such a way that $d_{min} \geq 3$.

2. Obtain the generator matrix of the equivalent systematic code and list the four code words.

**10.5** Generalize the examples of Problems 10.3 and 10.4 to show that there is always a systematic code equivalent to the one generated by a given parity-check matrix.

**10.6** A systematic $(10, 3)$ linear block code is defined by the following parity-check equations:

$$\begin{aligned} x_4 + x_1 + x_3 &= 0, & x_8 + x_1 + x_2 &= 0, \\ x_5 + x_3 &= 0, & x_9 + x_2 &= 0, \\ x_6 + x_1 + x_3 &= 0, & x_{10} + x_1 + x_2 &= 0, \\ x_7 + x_2 + x_3 &= 0 \end{aligned}$$

$$(10.121)$$

Find the percentage of error patterns with $1, 2, 3, \ldots, 9, 10$ errors that can be detected by the code.

**10.7** A $(5, 2)$ linear block code is defined by the following table:

| $u_1$ | $u_2$ | $x_1$ | $x_2$ | $x_3$ | $x_4$ | $x_5$ |
|-------|-------|-------|-------|-------|-------|-------|
| 0 | 0 | 0 | 0 | 0 | 0 | 0 |
| 0 | 1 | 0 | 1 | 1 | 0 | 1 |
| 1 | 0 | 1 | 0 | 1 | 1 | 1 |
| 1 | 1 | 1 | 1 | 0 | 1 | 0 |

1. Find the generator matrix and the parity-check matrix of the code.

2. Build the standard array and the decoding table to be used on a BSC.

3. What is the probability of making errors in decoding a code word assuming an error detection strategy of the decoder ?

**10.8** Assume that an $(n, k)$ code has minimum distance $d$.

1. Prove that every set of $(d - 1)$ or fewer columns of the parity-check matrix **H** is linearly independent.

2. Prove that there exists at least one set of $d$ columns of **H** that is linearly dependent.

**10.9** Show that the dual of the $(n, 1)$ repetition code is an $(n, n-1)$ code with $d_{\min} = 2$ and with code words always having even weight.

**10.10** Given an $(n, k)$ code, it can be shortened to obtain an $(n - 1, k - 1)$ code by simply taking only the code words that have a 0 in the first position, and deleting this 0. Show that the maximal-length (simplex) code $(2^m - 1, m)$ is obtained by shortening the first-order $(2^m, m + 1)$ Reed-Muller code.

**10.11** Assume that an $(n, k)$ block code with minimum distance $d$ is used on the binary erasure channel of Example 3.14. Show that it is always possible to correctly decode the received sequence provided that no more than $(d - 1)$ erasures have occurred.

**10.12** Consider the following generator matrix of an $(8, 5)$ linear block code:

$$\mathbf{G} = \begin{bmatrix} 1 & 0 & 0 & 0 & 0 & 1 & 1 & 1 \\ 0 & 1 & 0 & 0 & 0 & 1 & 0 & 0 \\ 0 & 0 & 1 & 0 & 0 & 0 & 1 & 0 \\ 0 & 0 & 0 & 1 & 0 & 0 & 0 & 1 \\ 0 & 0 & 0 & 0 & 1 & 1 & 1 & 1 \end{bmatrix}$$

1. Show that the code is cyclic, and find both the generator polynomial $g(Z)$ and the parity-check polynomial $h(Z)$.

   2. Obtain the parity-check matrix **H**.

**10.13** Consider the generator polynomial

$$g(Z) = Z + 1$$

   1. Show that it generates a cyclic code of any length.
   2. Obtain the parity-check polynomial $h(Z)$, the parity-check matrix **H**, and the generator matrix **G**.
   3. What kind of code is obtained?

**10.14** Given the (7, 4) Hamming code generated by the polynomial

$$g(Z) = Z^3 + Z + 1$$

obtain the (7,3) code generated by

$$g(Z) = (Z + 1)(Z^3 + Z + 1)$$

   1. How is it related to the original (7, 4) code?
   2. What is its minimum distance?
   3. Show that the new code can correct all single errors and simultaneously detect all double errors.
   4. Describe an algorithm for correction and detection as in part (3).

**10.15** Illustrate the behavior of the encoder of Figure 10.13 by enumerating the register contents during the encoding of the data word $u(Z) = Z^3 + Z^2 + 1$.

**10.16** Show that the $(n, n - k)$ code generated by the parity-check polynomial $h(Z)$ is equivalent to the dual of the code generated by the generator polynomial $g(Z)$. In particular, show that the dual code of the code generated by $g(Z)$ is generated by $g^{(d)}(Z) = Z^k h(Z^{-1})$.

**10.17** A cyclic code is generated by

$$g(Z) = Z^8 + Z^7 + Z^6 + Z^4 + 1$$

   1. Find the length $n$ of the code.
   2. Sketch the encoding circuits with a $k$ or $(n - k)$ shift register.

**10.18** Discuss the synthesis of a code capable of correcting single errors and adjacent double errors. Develop an example and compare the numbers $n$ and $k$ with those required for the correction of all double and single errors. *Hint*: Count the required syndromes and construct a suitable parity-check matrix.

**10.19** It is desired to build a single-error-correcting $(8, 4)$ linear block code.

1. Define the code by shortening a cyclic code.

2. List the code words and find the minimum distance.

3. Sketch the encoding circuit and verify its behavior with an example.

**10.20** Show that the binary cyclic code of length $n$ generated by $g(Z)$ has minimum distance at least 3, provided that $n$ is the smallest integer for which $g(Z)$ divides $(Z^n + 1)$.

**10.21** Consider a cyclic code generated by the polynomial $g(Z)$ that does not contain $(Z + 1)$ as a factor. Show that the vector of all ones is a code word.

**10.22** Show that the $(7, 4)$ code generated by $g^{(d)}(Z) = Z^3 + Z + 1$ is the dual of the $(7, 3)$ code generated by $g(Z) = Z^4 + Z^3 + Z^2 + 1$.

**10.23** Repeat the computations of Example 10.23 for the $(15, 11)$ Hamming code.

**10.24** Prove the Singleton bound (10.102) and the Hamming bound (10.103).

**10.25** Consider a transmission system that performs error detection over a BSC with transition probability $p$. Using the weight enumerating function $A(D)$, find an exact expression for the probability of undetected errors for the following codes:

1. Hamming codes.

2. Extended Hamming codes.

3. Maximal-length codes.

4. $(n, 1)$ repetition codes.

5. $(n, n - 1)$ parity check codes.

**10.26** Show that for a linear block code the set of Hamming distances from a given code word to the other $(M - 1)$ code words is the same for all code words. *Hint*: Use Property 2 of Section 10.2.

Prove that for any linear code used on a binary-input symmetric channel with ML decoding the *uniform error property* holds, i.e.,

$$P_w(e) = P_w(e|\mathbf{x}_i), \quad i = 1, \ldots, M$$

*Hint*: Write

$$P_w(e|\mathbf{x}_i) = \sum_{j \notin S_i} P(\mathbf{y}_j|\mathbf{x}_i)$$

where $S_i$ is the set of subscripts $j$ of received sequences $\mathbf{y}_j$ that are decoded into the code word $\mathbf{x}_i$, and

$$P(\mathbf{y}|\mathbf{x}_i) = \prod_{k=1}^{n} P(y_k|x_{ik})$$

then use the symmetry of the channel, i.e.,

$$P(y_k = 0|x_{ik} = 1) = P(y_k = 1|x_{ik} = 0)$$

**10.27** Using the memoryless property of the BSC, that is,

$$P(\mathbf{y}_j|\mathbf{x}_i) = \prod_{k=1}^{n} P(y_{jk}|x_{ik})$$

derive (10.76) from (10.75).

**10.28** Following the same steps that led to (10.78), derive the union bound to the bit error probability (10.96) and (10.97).

# Convolutional and concatenated codes

With block codes, the information sequence is segmented into blocks that are encoded independently to form the coded sequence as a succession of fixed-length independent code words. *Convolutional* codes behave differently. The $n_0$ bits that the convolutional encoder generates in correspondence of the $k_0$ information bits depend on the $k_0$ data bits and also on some previous data frames (see Section 10.1): the encoder has *memory*.

Convolutional codes differ deeply from block codes, in terms of their structure, analysis and design tools. Algebraic properties are of great importance in constructing good block codes and in developing efficient decoding algorithms. Good convolutional codes, instead, have been almost invariably found by exhaustive computer search, and the most efficient decoding algorithms (like the Viterbi maximum-likelihood algorithm and the sequential algorithm) stem directly from the sequential-state machine nature of convolutional encoders, rather than from the algebraic properties of the code.

In this chapter, we will start by establishing the connection of binary convolutional codes with linear block codes, and then widen the horizon by assuming a completely different point of view that looks at a convolutional encoder as a finite-state machine and introduces the code *trellis* as the graphic tool describing all possible code sequences.

We will show how to evaluate the distance properties of the code and the error probability performance, and describe in details the application of the Viterbi algorithm to its decoding. A brief introduction to sequential and threshold decoding will also be given.

The second part of the chapter is devoted to *concatenated* codes, a concept

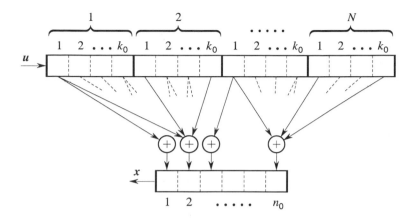

Figure 11.1: *General block diagram of a convolutional encoder in serial form for a* $(n_0, k_0)$ *code with constraint length* $N$.

first introduced by Forney (1966) that has since then found a wide range of applications. After describing the classical concatenation schemes, we will also devote some space to the recently introduced "turbo" codes, a very promising new class of concatenated codes that approach the capacity coding gains at medium-to-low bit error probabilities.

## 11.1. Convolutional codes

A binary convolutional encoder is a finite-memory system that outputs $n_0$ binary digits for every $k_0$ information digits presented at its input. Again, the code rate is defined as $R_c = k_0/n_0$. In contrast with block codes, $k_0$ and $n_0$ are usually small numbers. A scheme that serially implements a *linear, feedforward* binary convolutional encoder is shown in Fig. 11.1. The message digits are introduced $k_0$ at a time into the input shift register, which has $Nk_0$ positions. As a block of $k_0$ digits enters the register, the $n_0$ modulo-2 adders feed the output register with the $n_0$ digits and these are shifted out. Then the input register is fed with a new block of $k_0$ digits, and the old blocks are shifted to the right, the oldest one being lost. And so on. We can conclude that in a convolutional code the $n_0$ digits generated by the encoder depend not only on the corresponding $k_0$ message digits, but also on the previous $(N-1)k_0$ ones, whose number constitutes the *memory* $\nu \triangleq (N-1)k_0$ of the encoder. Such a code is called an $(n_0, k_0, N)$ convolutional code. The parameter $N$, the number of data frames contained in the input register, is called the *constraint length* of the code.[1] With reference to the encoder of Fig. 11.1, a block code can be considered to be the limiting case of a convolutional code, with constraint length $N = 1$.

If we define $\mathbf{u}$ to be the semi-infinite message vector and $\mathbf{x}$ the corresponding encoded vector, we want now to describe how to get $\mathbf{x}$ from $\mathbf{u}$. As for block codes, to describe the encoder we only need to know the connections between the input and output registers of Fig. 11.1. This approach enables us to show both the analogies and the differences with respect to block codes. But, if pursued further, it would lead to complicated notations and tend to emphasize the algebraic structure of convolutional codes. This is less interesting for decoding purposes. Therefore, we shall only sketch this approach briefly. Later, the description of the code will be restated from a different viewpoint.

To describe the encoder of Fig. 11.1, we can use $N$ submatrices $\mathbf{G}_1, \mathbf{G}_2, \mathbf{G}_3, \ldots, \mathbf{G}_N$ containing $k_0$ rows and $n_0$ columns. The submatrix $\mathbf{G}_i$ describes the connections of the $i$-th segment of $k_0$ cells of the input register with the $n_0$ cells of the output register. The $n_0$ entries of the first row of $\mathbf{G}_i$ describe the connections of the first cell of the $i$-th input register segment with the $n_0$ cells of the output register. A "1" in $\mathbf{G}_i$ means a connection, while a "0" means no connection. We can now define the generator matrix of the convolutional code as

$$\mathbf{G}_\infty \triangleq \begin{bmatrix} \mathbf{G}_1 & \mathbf{G}_2 & \ldots & \mathbf{G}_N & & & \\ & \mathbf{G}_1 & \mathbf{G}_2 & \ldots & \mathbf{G}_N & & \\ & & \mathbf{G}_1 & \mathbf{G}_2 & \ldots & \mathbf{G}_N & \\ & & & \mathbf{G}_1 & \mathbf{G}_2 & \ldots & \mathbf{G}_N \\ & & & & \ldots & \ldots & \ldots & \ldots \end{bmatrix} \tag{11.1}$$

All other entries in $\mathbf{G}_\infty$ are zero. This matrix has the same properties as for block codes, except that it is semi-infinite (it extends indefinitely downward and to the right). Therefore, given a semi-infinite message vector $\mathbf{u}$, the corresponding coded vector is

$$\mathbf{x} = \mathbf{u}\mathbf{G}_\infty \tag{11.2}$$

This equation is formally identical to (10.4). A convolutional encoder is said to be *systematic* if, in each segment of $n_0$ digits that it generates, the first $k_0$ are a replica of the corresponding message digits. It can be verified that this condition

---

[1]The reader should be warned that there is no unique definition of constraint length in the convolutional code literature.

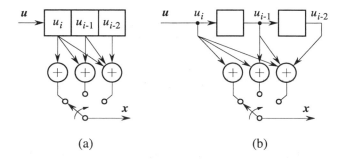

Figure 11.2: *Two equivalent schemes for the convolutional encoder of the (3,1,3) code of Example 11.1.*

is equivalent to have the following $k_0 \times n_0$ submatrices:

$$G_1 = \left[ \begin{array}{ccccc|c} 1 & 0 & 0 & \ldots & 0 & \\ 0 & 1 & 0 & \ldots & 0 & \\ 0 & 0 & 1 & \ldots & 0 & P_1 \\ \ldots & \ldots & \ldots & \ldots & & \\ 0 & 0 & 0 & \ldots & 1 & \end{array} \right] \tag{11.3}$$

and

$$G_i = \left[ \begin{array}{ccccc|c} 0 & 0 & 0 & \ldots & 0 & \\ 0 & 0 & 0 & \ldots & 0 & \\ 0 & 0 & 0 & \ldots & 0 & P_i \\ \ldots & \ldots & \ldots & \ldots & & \\ 0 & 0 & 0 & \ldots & 0 & \end{array} \right] \tag{11.4}$$

for $i = 2, 3, \ldots, N$. All these concepts are better clarified with two examples.

**Example 11.1**    Consider a (3,1,3) convolutional code. Two equivalent schemes for the encoder are shown in Fig. 11.2. The first uses a register with three cells, whereas the second uses two cells, each introducing a unitary delay. The output register is replaced by a commutator that reads sequentially the outputs of the three adders. The encoder is specified by the following three submatrices (actually, three row vectors, since $k_0 = 1$):

$$\begin{array}{rcl} G_1 & = & [1 \ 1 \ 1] \\ G_2 & = & [0 \ 1 \ 1] \\ G_3 & = & [0 \ 0 \ 1] \end{array}$$

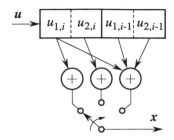

Figure 11.3: *Convolutional encoder for the (3,2,2) code of Example 11.1.*

The generator matrix, from (11.1), becomes

$$
G_\infty = \begin{bmatrix}
111 & 011 & 001 & 000 & \cdots & \cdots & \cdots \\
000 & 111 & 011 & 001 & 000 & \cdots & \cdots \\
000 & 000 & 111 & 011 & 001 & 000 & \cdots \\
\cdots & \cdots & \cdots & \cdots & \cdots & \cdots & \cdots
\end{bmatrix}
$$

It can be verified, from (11.2), that the information sequence $u = (11011\ldots)$ is encoded into the sequence $x = (111100010110100\ldots)$. The encoder is systematic. Notice that the code sequence can be obtained by summing modulo-2 the rows of $G_\infty$ corresponding to the "1" in the information sequence, as for block codes.                     □

**Example 11.2**   Consider a (3,2,2) code. The encoder is shown in Fig. 11.3. The code is now defined by the two submatrices

$$
G_1 = \begin{bmatrix} 1 & 0 & 1 \\ 0 & 1 & 0 \end{bmatrix} \qquad G_2 = \begin{bmatrix} 0 & 0 & 1 \\ 0 & 0 & 1 \end{bmatrix}
$$

The encoder is systematic, since (11.3) and (11.4) are satisfied. The generator matrix is now given by

$$
G_\infty = \begin{bmatrix}
101 & 001 & 000 & \cdots & \cdots \\
010 & 001 & 000 & \cdots & \cdots \\
000 & 101 & 001 & 000 & \cdots \\
000 & 010 & 001 & 000 & \cdots \\
000 & 000 & 101 & 001 & 000 \\
000 & 000 & 010 & 001 & 000 \\
\cdots & \cdots & \cdots & \cdots & \cdots
\end{bmatrix}
$$

The information sequence $u = (11011011\ldots)$ is encoded into the code sequence $x = (111010100110\ldots)$.                     □

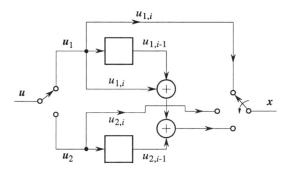

Figure 11.4: *Parallel implementation of the same convolutional encoder of Fig. 11.3.*

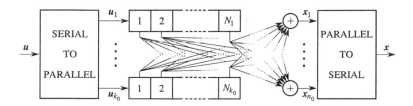

Figure 11.5: *General block diagram of a convolutional encoder in parallel form for an $(n_0, k_0, N)$ code.*

The encoder of Fig. 11.3 requires a serial input. The $k_0 = 2$ input digits can also be presented in parallel, and the corresponding encoder is given in Fig. 11.4.

The parallel representation of the encoder, shown for a general $(n_0, k_0)$ encoder in Fig. 11.5. is more flexible than the serial one of Fig. 11.1, as it allows allocation of a different number of register cells in each parallel section. When $N_i = N$, $\forall i$, we can define the constraint length $N$ as in the case of the serial representation. When the $N_i$'s are different, we define the constraint length $N$ as the largest among the $N_i$'s, i.e., $N \triangleq \max_i N_i$, $i = 1, \ldots, k_0$. The encoder memory is in this case $\nu = \cdot \sum_{i=0}^{k_0} (N_i - 1)$.

If we look for a physical meaning of the constraint length, $N - 1$ represents the maximum number of trellis steps (see Section 11.1.1) that are required to return from any state of the encoder to the zero state. This "remerge" operation, called *trellis termination*, is required to transform $K$ sections of the trellis of a convolutional code into a block code with parameters $k = k_0 \cdot K$, $n = n_0 \cdot (K + N - 1)$. Sometimes, system constraints impose a frame (or "burst") structure on the information stream. For short bursts, terminated convolutional codes are used, and each burst is decoded using the Viterbi algorithm without truncation (see 11.1.3).

Notice that, in the more general case of different $N_i$'s, the structure of the encoder is not identified by the three parameters $(n_0, k_0, N)$; instead, beyond $k_0$ and $n_0$, the whole set $\{N_i\}_{i=1}^{k_0}$ is needed.

From Example 11.1, it can be verified that the operation of the encoder for an $(n_0, 1)$ code is to generate $n_0$ digits of the sequence $\mathbf{x}$ for each digit $u_i$ according to the following expression:

$$
\begin{aligned}
(x_{i1}, x_{i2}, x_{i3}, \dots, x_{in_0}) &= u_i \mathbf{G}_1 + u_{i-1} \mathbf{G}_2 + \dots + u_{i-N+1} \mathbf{G}_N \\
&= \sum_{k=1}^{N} u_{i-k+1} \mathbf{G}_k \qquad\qquad (11.5)
\end{aligned}
$$

This is the discrete convolution of the vectors $\mathbf{G}_1, \mathbf{G}_2, \dots, \mathbf{G}_N$ and the $N$-digit input sequence $(u_i, u_{i-1}, \dots, u_{i-N+1})$. The term *convolutional* code stems from this observation.

Quite often, the number of modulo-2 adders in the encoder is smaller than the constraint length of the code. In fact, code rates of 1/2 or 1/3 are widely used, and in these cases we have only two or three adders, respectively. For this reason, instead of describing the code with the $N$ submatrices $\mathbf{G}_i$ previously introduced, it is more convenient to describe the encoder connections by using the *transfer function* matrix $\mathbf{G}$

$$
\mathbf{G} = \begin{bmatrix} g_{1,1} & \cdots & g_{1,n_0} \\ \vdots & & \\ g_{k_0,1} & \cdots & g_{k_0,n_0} \end{bmatrix} \qquad\qquad (11.6)
$$

where $g_{i,j}$ is a binary row vector with $N$ entries describing the connections from the $i$th input, $i = 1, \dots, k_0$, to the $j$th output, $j = 1, \dots, n_0$. Vectors $g_{i,j}$ are often called *generators* of the encoder.

**Example 11.3**   Let us reconsider the code of Example 11.1. This code has $k_0 = 1$ and $n_0 = 3$. Therefore, it can be described with the following three generators:

$$
\begin{aligned}
g_{1,1} &= (100) \\
g_{1,2} &= (110) \\
g_{1,3} &= (111)
\end{aligned}
$$

In the literature, the binary vectors $g_{i,j}$ are also represented as octal numbers ($110 \rightarrow 6$), or polynomials in the indeterminate $Z$, as was done for cyclic codes. As an example, the previous vector $g_{1,3} = (110)$ would be represented as $g_{1,3}(Z) = Z^2 + Z$. The tables describing the "best" convolutional codes (see Section 11.1.2) shall characterize the codes using the transfer function matrix, in which each generator will be represented as an octal number.

The advantage of this representation is not immediately apparent. As in Example 11.1, we have three vectors. But, for example, in the case of a (3,1,10) code, this second representation always requires three generators of length 10, whereas in the other representation we would need 10 vectors (the submatrices $\mathbf{G}_i$) of length 3. No doubt the first description is more practical. □

### 11.1.1. State diagram representation of convolutional codes

As already noted, there is a powerful and practical alternative to the algebraic description of convolutional codes. This alternative is based on the observation that the convolutional encoder is a finite-memory system, and hence its output sequence depends on the input sequence and on the state of the device. The description we are looking for is called the *state diagram* of the convolutional encoder.

We shall illustrate the concepts involved in this description by taking as an example the encoder of Fig. 11.2. This encoder refers to the (3,1,3) code described in Example 11.1. Notice that each output triplet of digits depends on the input digit and on the content of the shift register that stores the oldest two input digits. The encoder has memory $\nu = N - 1 = 2$. Let us define the state $\sigma_\ell$ of the encoder at discrete time $\ell$ as the content of its memory at the same time. That is,

$$\sigma_\ell \overset{\triangle}{=} (u_{\ell-1}, u_{\ell-2}) \tag{11.7}$$

There are $N_\sigma = 2^\nu = 4$ possible states. That is, 00, 01, 10, and 11. Looking at Fig. 11.2, assume, for example, that the encoder is in state 10. When the input digit is 1, the encoder produces the output digits 100 and moves to the state 11.

This type of behavior is completely described by the state diagram of Fig. 11.6. Each of the four states is represented in a circle. A solid edge represents a transition between two states forced by the input digit "0," whereas a dashed edge represents a transition forced by the input digit "1." The label on each edge represents the output digits corresponding to that transition. Using the state diagram of Fig. 11.6, the computation of the encoded sequence is quite straightforward. Starting from the initial state 00, we jump from one state to the next following a solid edge when the input is 0 or a dashed edge when the input is 1.

If we define the states to be $S_1 = (00), S_2 = (01), S_3 = (10), S_4 = (11)$, we can easily check that the input sequence $\mathbf{u} = (11011\ldots)$, already considered in Example 11.1, assuming $S_1$ as the initial state, corresponds to the *path* $S_1 S_3 S_4 S_2 S_3 S_4 \ldots$ through the state diagram, and the output sequence is $\mathbf{x} = (111\ 100\ 010\ 110\ 100\ \ldots)$, as found by writing down the sequence of edge labels.

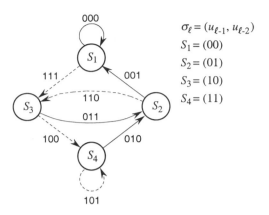

$$\sigma_\ell = (u_{\ell-1}, u_{\ell-2})$$
$$S_1 = (00)$$
$$S_2 = (01)$$
$$S_3 = (10)$$
$$S_4 = (11)$$

Figure 11.6: *State diagram for the (3,1,3) convolutional code of Example 11.1.*

The concept of state diagram can be applied to any $(n_0, k_0, N)$ code with memory $\nu$. The number of states is $N_\sigma = 2^\nu$. There are $2^{k_0}$ edges entering each state and $2^{k_0}$ edges leaving each state. The labels on each edge are sequences of length $n_0$. As $\nu$ increases, the size of the state diagram grows exponentially and becomes very hard to handle. As we are "walking inside" the state diagram following the guidance of the input sequence, it soon becomes difficult to keep track of the past path, because we travel along the same edges many times. Therefore, it is desirable to modify the concept of state diagram by introducing time explicitly. This result is achieved if we replicate the states at each time step, as shown in the diagram of Fig. 11.7. This is called a *trellis diagram*. It refers to the state diagram of Fig. 11.6. In this trellis, the four nodes on the same vertical represent the four states at the same discrete time $\ell$, which is called the *depth* into the trellis. Dashed and solid edges have the same meaning as in the state diagram. The input sequence is now represented by the path $\sigma_0 = S_1, \sigma_1 = S_3, \sigma_2 = S_4, \dots$, and so on. Any encoder output sequence can be found by walking through the appropriate path into the trellis.

Finally, a different representation of the code can be given by expanding the trellis diagram of Fig. 11.7 into the *tree diagram* of Fig. 11.8. In this diagram, the encoding process can be conceived as a walk through a *binary tree*. Each encoded sequence is represented by one particular path into the tree. The encoding process is guided by binary decisions (the input digit) at each *node* of the tree. This tree has an exponential growth. At depth $\ell$, there will be $2^\ell$ possible paths representing all the possible encoded sequences of that length. The path corresponding to the input sequence 11011 is shown as an example in Fig. 11.8. The nodes of the tree are labeled with reference to the states of the state diagram

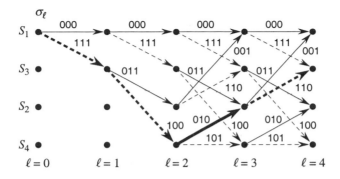

Figure 11.7: *Trellis diagram for the (3,1,3) convolutional code of Example 11.1. The boldface path corresponds to the input sequence 1101.*

shown in Fig. 11.6.

**Distance properties and transfer functions of convolutional codes**

As for block codes, the error-detection and error-correction capabilities of a convolutional code are directly related to the distance properties of the encoded sequences. Due to the uniform error property of linear codes, we assume that the all-zero sequence is transmitted in order to determine the performance of the convolutional code.

Let us start with some definitions. Consider a pair of encoded sequences up to the depth $\ell$ into the code trellis and assume that they disagree at the first branch. We define the $\ell$-th order *column distance* $d_c(\ell)$ as the minimum Hamming distance between all pairs of such sequences. For the computation of $d_c(\ell)$, one of the sequences of the pair can be the all-zero sequence. Therefore, we have to consider all sequences, up to the depth $\ell$ in the code trellis, such that they disagree at the first branch from the all-zero sequence. The column distance $d_c(\ell)$ is the minimum weight of this set of code sequences. The column distance $d_c(\ell)$ is a nondecreasing function of the depth $\ell$. By letting the value of $\ell$ go to infinity, we obtain the so-called *free distance* $d_f$ of the convolutional code, defined as

$$d_f \triangleq \lim_{\ell \to \infty} d_c(\ell) \tag{11.8}$$

From (11.8), we see that the free distance of the code is the minimum Hamming distance between infinitely long encoded sequences.

It can be found on the code trellis by looking for those sequences (paths) that, after diverging from the all-zero sequence, merge again into it. The free distance is the minimum weight of this set of encoded sequences.

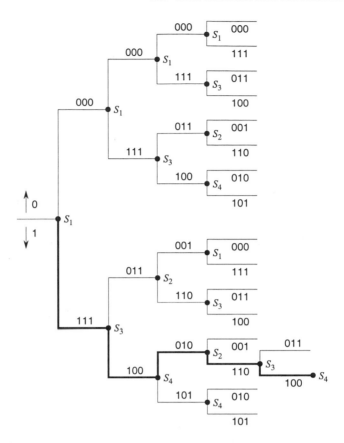

Figure 11.8: *Tree diagram for the (3,1,3) convolutional code of Example 11.1. The solid path corresponds to the input sequence 11011.*

A straightforward algorithm to compute $d_f$ is based on the following steps:

1. Set $\ell = 0$

2. $\ell \rightarrow \ell + 1$

3. Compute $d_c(\ell)$

4. If the sequence giving $d_c(\ell)$ merges into the all-zero sequence, keep its weight as $d_f$ and go to 6.

5. Return to 2

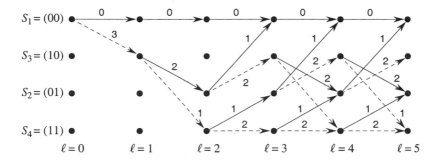

Figure 11.9: *Part of the trellis diagram for the (3,1,3) code of Example 11.4 for the computation of the code distance properties. The trellis is the same as in Fig. 11.7, but the labels represent here the weight of the output sequence of three digits associated with each edge.*

6. Stop.

**Example 11.4**   We want to reconsider the $(3,1,3)$ convolutional code, whose trellis is given in Fig. 11.7, to find the distances just defined. Let us consider Fig. 11.9. Part of the trellis is reproduced in the figure, with the following features. Only the all-zero sequence and the sequences diverging from it at the first branch are reproduced. Furthermore, each edge is labeled with the weight of the encoded sequence. The column distance of the code can be found by inspection. We get

| $\ell$ | $d_c(\ell)$ |
|---|---|
| 1 | 3 |
| 2 | 4 |
| 3 | 5 |
| 4 | $6 \leftarrow d_f$ |

Since the constraint length of this code is $N = 3$, we have the first merge of one sequence into the all-zero sequence for $\ell = 3$. However, the merging sequence has weight 6, and does not give $d_c(3)$, which is instead equal to 5. Thus, we must keep looking for $d_f$. For $\ell = 4$, we have a merging sequence giving $d_c(4)$. Its weight is 6, and therefore we conclude that $d_f = 6$.                                                                              $\square$

The computation of $d_f$, although straightforward, may require the examination of exceedingly long sequences. In practice, the problem is amenable to an algorithmic solution based on the state diagram of the code. We take again the case of Fig. 11.6 as a guiding example. The state diagram is redrawn in Fig. 11.10 with certain modifications made in view of our goal.        First, the edges are labeled with an indeterminate $D$ raised to an exponent that represents the weight

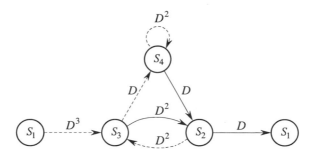

Figure 11.10: *State diagram for the (3,1,3) convolutional code of Fig. 11.6. The labels allow the computation of the weight enumerating function $T(D)$.*

(or, equivalently, the Hamming distance from the all-zero sequence) of the encoded sequence corresponding to that state transition. Furthermore, the self-loop at state $S_1$ has been eliminated, since it does not contribute to the weight of a sequence. Finally, the state $S_1$ has been split into two states, one of which represents the input and the other the output of the state diagram.

Let us now define the label of a path as the product of the labels of all its edges. Therefore, among all the infinitely many paths starting in $S_1$ and merging again into $S_1$, we are looking for the path whose label $D$ is raised to the smallest exponent. This exponent is indeed $d_f$. By inspection of Fig. 11.10, we can verify that the path $S_1 S_3 S_2 S_1$ (see Example 11.4) has label $D^6$, and indeed this code has $d_f = 6$. We can define a *weight enumerating function* [2] $T(D)$ of the output sequence weights as a series that gives all the information about the weights of the paths starting from $S_1$ and merging again into $S_1$. This weight enumerating function can be computed as the transfer function of the signal-flow graph of Fig. 11.10. Using standard techniques for the study of directed graphs (see Appendix D), the transfer function for the graph of Fig. 11.10 is given by

$$T(D) = \frac{2D^6 - D^8}{1 - (D^2 + 2D^4 - D^6)} = 2D^6 + D^8 + 5D^{10} + \ldots \overset{\triangle}{=} \sum_{d=d_f}^{\infty} A_d D^d \quad (11.9)$$

where $A_d$ is the number (multiplicity) of paths with weight $d$ diverging from state $S_1$ and remerging into it later. Thus, we deduce from (11.9) that there are

---

[2]The function $T(D)$ is more often called the *generating* or *transfer function* of the convolutional code, and, sometimes, we will use this denomination, too. The term "weight enumerating function," however, is more appropriate, because, apart from the length of the described code words, which can be infinite for convolutional codes, its meaning is the same as for the function $A(D)$ of block codes defined in Eq. (10.18). The only difference is that $T(D)$ does not contain the all-zero sequence, so in its development as a power series the "1" is missing.

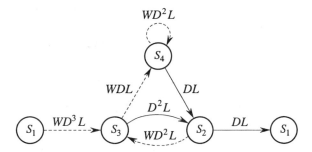

Figure 11.11: *State diagram for the (3,1,3) code of Fig. 11.6. The labels allow the computation of the input-output weight enumerating function* $T_3(W, D, L)$.

two paths of weight 6, one path of weight 8, five paths of weight 10, and so on. Using the terminology of Chapter 4, we can also say that the all-zero path has two *nearest neighbors* at Hamming (instead of Euclidean) distance 6.

Different forms of transfer functions can be used to provide additional information on the code properties. This is done by considering the modified graph of Fig. 11.11. Each edge has now a label containing three indeterminates, $W, D, L$. The exponent of $W$ is the weight of the input data frame (a single digit, in this example) that caused the transition, so that the exponent of $W$ for a given path will represent the Hamming weight of the information sequence that generated it. The indeterminate $D$ has the same meaning as before, and finally, $L$ is present in each edge, so that its exponent will count the length of the paths. According to the expanded labels, we have a new *input-output weight enumerating function* denoted by $T_3(W, D, L)$, where the subscript refers to the number of indeterminates.

For the state diagram of Fig. 11.11 we obtain

$$
\begin{aligned}
T_3(W, D, L) &= \frac{WD^6L^3(1 + WL - WD^2L)}{1 - WD^2L(1 + D^2L + WD^2L^2 - WD^4L^2)} \\
&= WD^6L^3(1 + WL) + W^3D^8L^5 + \ldots \qquad (11.10) \\
&\triangleq \sum_{w=1}^{\infty} \sum_{d=d_f}^{\infty} \sum_{\ell=1}^{\infty} C_{w,d,l} W^w D^d L^\ell
\end{aligned}
$$

where $C_{w,d,\ell}$ is the number of paths diverging from state $S_1$ and remerging into it later generated by an information sequence of weight $w$, having weight $d$, and with length $\ell$. From (11.10) we see that the two paths of weight 6 have lengths 3 and 4, respectively, and that the weights of the input sequences are 1 and 2. The path of weight 8 has length 5, and the corresponding input sequence has weight 3. And so on. These numbers can be checked immediately in Example 11.4.

Comparing (11.9) with (11.10), we realize that $T(D)$ can be obtained from $T_3(W, D, L)$ by setting $W = L = 1$.

Sometimes, the length of the path is not important. In this case, we have a third version of the input-output weight enumerating function, $T_2(W, D)$, which contains only two indeterminates. With obvious notations, it is defined as

$$T_2(W, D) \triangleq \sum_{w=1}^{\infty} \sum_{d=d_f}^{\infty} B_{w,d} W^w D^d \tag{11.11}$$

where $B_{w,d}$ is the number of paths diverging from state $S_1$ and remerging into it later with weight $d$, generated by an information sequence of weight $w$. $T_2(W, D)$ can be obtained from $T_3(W, D, L)$ by setting $L = 1$.

We have thus three distinct weight generating functions. The first, $T(D)$, characterizes the *distance spectrum* of the convolutional code through the pairs $(A_d, d)$ yielding the weights $d$ of the code sequences and their multiplicities $A_d$. The second, the input-output weight enumerating function $T_2(W, D)$, provides information on the encoder mapping between input and code sequences, by keeping distinct code sequences of the same weight generated by input sequences of different weights. The third, $T_3(W, D, L)$, finally, adds to $T_2$ the information about the length of sequences in terms of number of trellis branches.

The multiplicities $A_d, B_{w,d}, C_{w,d,\ell}$ satisfy the following relationships:

$$A_d = \sum_{w=1}^{\infty} B_{w,d} = \sum_{w=1}^{\infty} \sum_{\ell=1}^{\infty} C_{w,d,\ell}, \quad B_{w,d} = \sum_{\ell=1}^{\infty} C_{w,d,\ell} \tag{11.12}$$

We have determined the properties of all code paths with reference to a simple convolutional code. The same techniques can be applied to any code of arbitrary rate and constraint length. We shall see in the next sections how the weight enumerating functions of the code can be used to bound the error probabilities of convolutional codes.

## 11.1.2.   Best known short-constraint-length convolutional codes

When considering the weight enumerating function $T(D)$ of a convolutional code, it was implicitly assumed that $T(D)$ converges. Otherwise, the expansions of (11.9) and (11.10) are not valid. This convergence cannot occur for all values of the indeterminate, because the coefficients are nonnegative. In some cases, certain coefficients are infinite, and the code is called *catastrophic*. An example is given in Problem 11.4. The code is a (2,1,3) code. Its state diagram shows that the self-loop at state $S_4$ does not increase the distance from the all-zero sequence, i.e., its label has an exponent of $D$ equal to zero. Therefore, the path $S_1 S_3 S_4 \ldots S_4 S_2 S_1$ will be at distance 6 from the all-zero path no matter how

| $\nu$ | $d_f$ (Rate 1/2) | | $d_f$ (Rate 1/3) | |
|---|---|---|---|---|
|  | Systematic | Nonsystematic | Systematic | Nonsystematic |
| 1 | 3 | 3 | 5 | 5 |
| 2 | 4 | 5 | 6 | 8 |
| 3 | 4 | 6 | 8 | 10 |
| 4 | 5 | 7 | 9 | 12 |
| 5 | 6 | 8 | 10 | 13 |
| 6 | 6 | 10 | 12 | 15 |
| 7 | 7 | 10 | 12 | 16 |

Table 11.1: *Maximum free distances achievable with systematic codes and nonsystematic noncatastrophic codes with memory $\nu$ and rates 1/2 and 1/3.*

many times it circulates in the self-loop at state $S_4$. We have the unfortunate circumstance where a finite-weight code sequence corresponds to an infinite-weight information sequence. Thus, it is possible to have an arbitrarily large number of decoding errors even for a fixed finite number of channel errors. This explains the name given to these codes.

The presence in the trellis of a self-loop, different from the one in state $S_1$, with zero weight associated, is a sufficient condition for the code to be catastrophic. We may have, however, closed loops (i.e., paths from state $S_i$ to state $S_i$) in the state diagram longer than one trellis branch, and with overall zero weight. In this case, too, the code is catastrophic.

Conditions can be established on the code generators that form the transfer function matrix (11.6) of the code to avoid catastrophic codes. For rate $1/n_0$ codes, the condition is particularly simple, and states that the code generators, in polynomial form, must be relatively prime to avoid catastrophicity (see Problem 11.5). The general conditions can be found in Massey and Sain (1968).

An important consideration here is that systematic convolutional codes cannot be catastrophic. Unfortunately, however, the free distances that are achievable by systematic codes realized with the feed-forward encoder[3] of Fig. 11.1 are usually lower than for nonsystematic codes of the same constraint length $N$. Table 11.1 shows the maximum free distances achievable with systematic (generated by feed-forward encoders) and nonsystematic noncatastrophic codes of rates 1/2 and 1/3 for increasing values of the code memory $\nu$.

Computer search methods have been used to find convolutional codes opti-

---

[3] We insist on the role of the encoder structure, since in Section 11.1.6 we will show that every nonsystematic convolutional encoder admits an equivalent systematic encoder, provided that the encoder is not constrained to be feed-forward.

mum in the sense that, for a given rate and a given constraint length, they have the largest possible free distance. These results were obtained by Odenwalder (1970), Larsen (1973), Paaske (1974), Daut *et al.* (1982), and recently by Chang *et al.* (1997). While the first searches used as selection criterion the maximization of the free distance, the recent search by Chang *et al.* (1997) is aimed at optimizing the input-output weight-enumerating function previously introduced. This criterion, as we will see in Section 11.1.5, is equivalent to minimizing the upper bounds to bit and error event probabilities. The best codes are reproduced in part in Tables 11.2 through 11.7. For the rates and number of states included in the search by Chang *et al.* (1997), the tables reproduce those codes since they have been found using the more complete optimization criterion. The codes are identified by their transfer function matrix defined in (11.6), in which the generators are represented as octal numbers. So, for example, an $(n_0, k_0)$ code will be represented by $k_0 \times n_0$ octal numbers organized in a matrix with $k_0$ rows and $n_0$ columns. The tables also give, when available, upper bounds on $d_f$ derived in Heller (1968) for codes of rate $1/n_0$ and extended to codes of rate $k_0/n_0$ by Daut, Modestino, and Wismer (1982). The Heller bound is described later in this chapter.

**Example 11.5** The rate 1/2 convolutional code of memory $\nu = 3$ of Table 11.2 has generators 15 and 17, which means

$$g_{1,1} = (1101)$$
$$g_{1,2} = (1111)$$

The block diagram of the encoder is shown in Fig. 11.12. For the rate 2/3 code of memory $\nu = 3$ in Table 11.6, the transfer function matrix is

$$G = \begin{bmatrix} 3 & 2 & 1 \\ 4 & 2 & 7 \end{bmatrix}$$

and the block diagram of the encoder is shown in Fig. 11.13.                                    □

**Punctured convolutional codes**

An appropriate measure of the maximum-likelihood decoder complexity for a convolutional code (see next section) is the number of visited edges per decoded bit. Now, a rate $k_0/n_0$ code has $2^{k_0}$ edges leaving and entering each trellis state and a number of states $N_\sigma = 2^\nu$, where $\nu$ is the memory of the encoder. Thus, each trellis section, corresponding to $k_0$ input bits, has a total number of edges

| Memory $\nu$ | Generators in octal notation | | $d_f$ | Upper bound on $d_f$ |
|---|---|---|---|---|
| 1 | 1 | 3 | 3 | 3 |
| 2 | 5 | 7 | 5 | 5 |
| 3 | 15 | 17 | 6 | 6 |
| 4 | 23 | 35 | 7 | 8 |
| 5 | 53 | 75 | 8 | 8 |
| 6 | 133 | 171 | 10 | 10 |
| 7 | 247 | 371 | 10 | 11 |
| 8 | 561 | 753 | 12 | 12 |
| 9 | 1131 | 1537 | 12 | 13 |
| 10 | 2473 | 3217 | 14 | 14 |
| 11 | 4325 | 6747 | 15 | 15 |
| 12 | 10627 | 16765 | 16 | 16 |
| 13 | 27251 | 37363 | 16 | 17 |

Table 11.2: *Feed-forward nonsystematic encoders generating maximum free distance convolutional codes of rate 1/2 and memory $\nu$. (Chang et al., 1997).*

| Memory $\nu$ | Generators in octal notation | | | $d_f$ | Upper bound on $d_f$ |
|---|---|---|---|---|---|
| 1 | 1 | 3 | 3 | 5 | 5 |
| 2 | 5 | 7 | 7 | 8 | 8 |
| 3 | 13 | 15 | 17 | 10 | 10 |
| 4 | 25 | 33 | 37 | 12 | 12 |
| 5 | 47 | 53 | 75 | 13 | 13 |
| 6 | 117 | 127 | 155 | 15 | 15 |
| 7 | 225 | 331 | 367 | 16 | 16 |
| 8 | 575 | 623 | 727 | 18 | 18 |
| 9 | 1167 | 1375 | 1545 | 20 | 20 |
| 10 | 2325 | 2731 | 3747 | 22 | 22 |
| 11 | 5745 | 6471 | 7553 | 24 | 24 |
| 12 | 10533 | 10675 | 17661 | 24 | 24 |
| 13 | 21645 | 35661 | 37133 | 26 | 26 |

Table 11.3: *Feed-forward nonsystematic encoders generating maximum free distance convolutional codes of rate 1/3 and memory $\nu$. (Larsen, 1973, and Chang et al., 1997).*

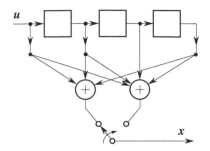

Figure 11.12: *Encoder for the (2,1,4) convolutional code of Example 11.5.*

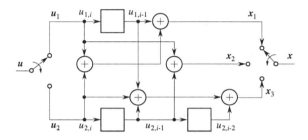

Figure 11.13: *Encoder for the (3,2,2) convolutional code of Example 11.5.*

equal to $2^{k_0+\nu}$. As a consequence, an $(n_0, k_0, N)$ code has a decoding complexity

$$\mathcal{D} = \frac{2^{k_0+\nu}}{k_0} \tag{11.13}$$

The increase of complexity inherent in passing from rate $1/n_0$ to rate $k_0/n_0$ codes can be mitigated using the so-called *punctured* convolutional codes. A rate $k_0/n_0$ punctured convolutional code can be obtained by starting from a rate $1/n_0$ and deleting parity-check symbols. An example will clarify the concept.

**Example 11.6**  Consider the 4-state convolutional encoder of Fig. 11.14 (a). For each input bit entering the encoder, two bits are sent through the channel, so that the code generated has rate 1/2. Its trellis is also shown in Fig. 11.14 (b). Suppose now that for every four parity-check digits generated by the encoder, one (the last) is punctured, i.e., not transmitted. In this case, for every two input bits three bits are generated by the encoder, thus producing a rate 2/3 code. The trellis for the punctured code is shown in Fig. 11.14 (c), and the letter "$x$" denotes a punctured output bit. As an example, the input sequence $\mathbf{u} = 101101\ldots$ would yield $\mathbf{x} = 111000010100$ for the rate 1/2 code, and $\mathbf{x} = 111000010$ for the punctured rate 2/3 code. In a similar way, higher rates can be obtained by increasing the number of punctured parity-check bits.

| Memory $\nu$ | Generators in octal notation | | | | $d_f$ | Upper bound on $d_f$ |
|---|---|---|---|---|---|---|
| 1 | 1 | 1 | 3 | 3 | 6 | 6 |
| 2 | 5 | 5 | 7 | 7 | 10 | 10 |
| 3 | 13 | 13 | 15 | 17 | 13 | 15 |
| 4 | 25 | 27 | 33 | 37 | 16 | 16 |
| 5 | 45 | 53 | 67 | 77 | 18 | 18 |
| 6 | 117 | 127 | 155 | 171 | 20 | 20 |
| 7 | 257 | 311 | 337 | 355 | 22 | 22 |
| 8 | 533 | 575 | 647 | 711 | 24 | 24 |
| 9 | 1173 | 1325 | 1467 | 1751 | 27 | 27 |
| 10 | 2387 | 2353 | 2671 | 3175 | 29 | 29 |
| 11 | 4767 | 5723 | 6265 | 7455 | 32 | 32 |
| 12 | 11145 | 12477 | 15537 | 16727 | 33 | 33 |
| 13 | 21113 | 23175 | 35527 | 35537 | 36 | 36 |

Table 11.4: *Feed-forward nonsystematic encoders generating maximum free distance convolutional codes of rate 1/4 and memory $\nu$. (Larsen, 1973, and Chang et al., 1997).*

| Memory $\nu$ | Generators in octal notation | | | | | $d_f$ | Upper bound on $d_f$ |
|---|---|---|---|---|---|---|---|
| 2 | 7 | 7 | 7 | 5 | 5 | 13 | 13 |
| 3 | 17 | 17 | 13 | 15 | 15 | 16 | 16 |
| 4 | 37 | 27 | 33 | 25 | 35 | 20 | 20 |
| 5 | 75 | 71 | 73 | 65 | 57 | 22 | 22 |
| 6 | 175 | 131 | 135 | 135 | 147 | 25 | 25 |
| 7 | 257 | 233 | 323 | 271 | 357 | 28 | 28 |

Table 11.5: *Feed-forward nonsystematic encoders generating maximum free distance convolutional codes of rate 1/5 and memory $\nu$ (Modestino and Wismer, 1982).*

It is interesting to note that the punctured rate 2/3 code so obtained is equivalent to the unpunctured rate 2/3 code depicted in Fig. 11.15, for which one stage of the trellis corresponds to two stages of the trellis of the punctured code. □

Of course, the way parity-check digits are deleted, or "punctured," should be optimized in order to maximize the free distance of the code (see Problem 11.6). Tables of optimum punctured codes can be found in Cain *et al.* (1979) and Ya-

| Constraint length $N$ | Memory $\nu$ | Transfer function matrix in octal notation | $d_f$ | Upper bound on $d_f$ |
|:---:|:---:|:---:|:---:|:---:|
| 2 | 2 | $\begin{pmatrix} 3 & 1 & 0 \\ 2 & 3 & 3 \end{pmatrix}$ | 3 | 4 |
| 3 | 3 | $\begin{pmatrix} 3 & 2 & 1 \\ 4 & 2 & 7 \end{pmatrix}$ | 4 | - |
| 3 | 4 | $\begin{pmatrix} 6 & 5 & 1 \\ 7 & 2 & 5 \end{pmatrix}$ | 5 | 6 |
| 4 | 5 | $\begin{pmatrix} 07 & 06 & 03 \\ 12 & 01 & 13 \end{pmatrix}$ | 6 | - |
| 4 | 6 | $\begin{pmatrix} 06 & 13 & 13 \\ 13 & 06 & 17 \end{pmatrix}$ | 7 | 7 |
| 5 | 7 | $\begin{pmatrix} 16 & 13 & 03 \\ 25 & 05 & 34 \end{pmatrix}$ | 8 | - |
| 5 | 8 | $\begin{pmatrix} 37 & 31 & 16 \\ 23 & 14 & 35 \end{pmatrix}$ | 8 | - |
| 6 | 9 | $\begin{pmatrix} 27 & 23 & 16 \\ 47 & 17 & 41 \end{pmatrix}$ | 9 | - |
| 6 | 10 | $\begin{pmatrix} 63 & 51 & 34 \\ 52 & 37 & 55 \end{pmatrix}$ | 10 | - |

Table 11.6: *Feed-forward nonsystematic encoders generating maximum free distance convolutional codes of rate 2/3 and constraint length N. (Chang et al., 1997).*

suda *et al.* (1984). They yield rate $k_0/n_0$ codes from a single rate $1/n_0$ "mother" code.

From the previous example, we can derive the conclusion that a rate $k_0/n_0$ convolutional code can be obtained considering $k_0$ trellis sections of a rate 1/2 mother code. Measuring the decoding complexity as done before in (11.13), we obtain for the punctured code

$$\mathcal{D}_{\text{punc}} = \frac{k_0 2^{\nu+1}}{k_o} \tag{11.14}$$

so that the ratio between the case of the unpunctured to the punctured solution yields

$$\frac{\mathcal{D}}{\mathcal{D}_{\text{punc}}} = \frac{2^{k_0}}{2k_o} \tag{11.15}$$

which shows that, for $k_0 > 2$, there is an increasing complexity reduction yielded

| Constraint length $N$ | Memory $\nu$ | Transfer function matrix in octal notation | $d_f$ | Upper bound on $d_f$ |
|---|---|---|---|---|
| 2 | 2 | $\begin{pmatrix} 1 & 1 & 1 & 0 \\ 3 & 0 & 0 & 1 \\ 3 & 2 & 0 & 2 \end{pmatrix}$ | 3 | - |
| 2 | 3 | $\begin{pmatrix} 3 & 2 & 1 & 0 \\ 3 & 1 & 2 & 1 \\ 2 & 2 & 2 & 3 \end{pmatrix}$ | 4 | 4 |
| 3 | 4 | $\begin{pmatrix} 0 & 1 & 2 & 3 \\ 3 & 0 & 1 & 2 \\ 2 & 4 & 1 & 5 \end{pmatrix}$ | 4 | - |
| 3 | 5 | $\begin{pmatrix} 3 & 3 & 2 & 2 \\ 5 & 2 & 7 & 0 \\ 4 & 7 & 0 & 1 \end{pmatrix}$ | 5 | - |
| 3 | 6 | $\begin{pmatrix} 5 & 4 & 3 & 2 \\ 4 & 6 & 5 & 5 \\ 6 & 1 & 4 & 3 \end{pmatrix}$ | 6 | - |
| 4 | 7 | $\begin{pmatrix} 02 & 03 & 04 & 07 \\ 03 & 07 & 03 & 05 \\ 15 & 02 & 02 & 17 \end{pmatrix}$ | 6 | - |
| 4 | 8 | $\begin{pmatrix} 04 & 06 & 07 & 07 \\ 01 & 12 & 05 & 14 \\ 00 & 07 & 14 & 11 \end{pmatrix}$ | 7 | - |
| 4 | 9 | $\begin{pmatrix} 03 & 06 & 10 & 15 \\ 00 & 16 & 03 & 13 \\ 16 & 05 & 02 & 17 \end{pmatrix}$ | 8 | - |

Table 11.7: *Feed-forward nonsystematic encoders generating maximum free distance convolutional codes of rate 3/4 and constraint length $N$. (Chang et al., 1997).*

by the punctured solution. Also, with puncturing, one can obtain several rates from the same mother code, thus simplifying the implementation through a sort of "universal" encoder, and this fact is greatly exploited in VLSI implementations.

There are at least two downsides to the punctured solution. First, punctured codes are normally slightly worse in terms of distance spectrum with respect to unpunctured codes of the same rate (see also Problem 11.6). Second, since the trellis of a punctured $(n_0, k_0)$ code is time-varying with period $k_0$, the decoder needs to acquire frame synchronization.

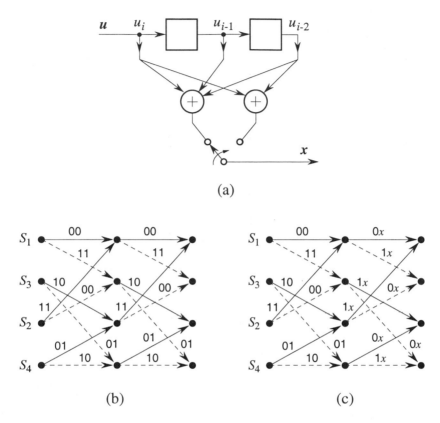

Figure 11.14: *Encoder (a) and trellis (b) for a (2,1,3) convolutional code. The trellis (c) refers to the rate 2/3 punctured code described in Example 11.6.*

### 11.1.3. Maximum-likelihood decoding of convolutional codes and the Viterbi algorithm

We have already seen that ML decoding of block codes is achieved when the decoder selects the code word whose distance from the received sequence is minimum. In the case of hard decoding, the distance considered is the Hamming distance, while for soft decoding it is the Euclidean distance. Unlike a block code, a convolutional code has no fixed block length. But it is intuitive that the same principle works also for convolutional codes. In fact, each possible encoded sequence is a path into the code trellis. Therefore, the optimum decoder must choose that path into the trellis that is *closest* to the received sequence. Also in this case, the distance measure will be the Hamming distance for hard

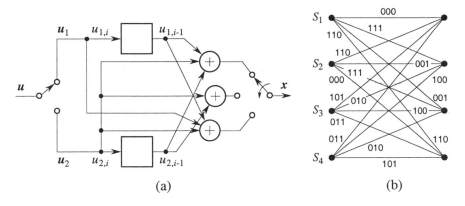

Figure 11.15: *Encoder (a) and trellis (b) for the (3,2,3) convolutional code equivalent to the rate 2/3 punctured code described in Example 11.6.*

decoding and the Euclidean distance for unquantized soft decoding.

Let us start with hard decoding. We assume binary antipodal modulation, and, consequently, the equivalent discrete channel is a BSC with error probability $p$. Denoting with $\mathbf{y}$ and $\mathbf{x}^{(r)}$ the received sequence and the $r$th path in the trellis, respectively, the optimum ML decoder must choose the path $\mathbf{x}^{(r)}$ of the trellis for which the conditional probability $P(\mathbf{y} \mid \mathbf{x}^{(r)})$ is maximum. We may take the logarithm of this probability as well. Therefore, the ML decoder must find the path corresponding to ,

$$U(\sigma_{K-1}) \triangleq \max_r U^{(r)}(\sigma_{K-1}) \triangleq \max_r P(\mathbf{y} \mid \mathbf{x}^{(r)}) \qquad (11.16)$$

$$\equiv \max_r \left[ \ln \prod_{\ell=0}^{K-1} P(\mathbf{y}_\ell \mid \mathbf{x}_\ell^{(r)}) \right] = \max_r \left[ \sum_{\ell=0}^{K-1} \ln P(\mathbf{y}_\ell \mid \mathbf{x}_\ell^{(r)}) \right]$$

where the symbol "$\equiv$" means "equivalent." In (11.16), $K$ indicates the length of the path into the trellis, or, equivalently, $K n_0$ is the length of the binary received sequence, $\mathbf{y}_\ell$ is the sequence of $n_0$ binary digits supplied to the decoder by the demodulator between discrete times $\ell$ and $(\ell + 1)$, and $\mathbf{x}_\ell^{(r)}$ is the $n_0$-digit label of the $r$-th path in the code trellis between states $\sigma_\ell$ and $\sigma_{\ell+1}$.

The maximization of the RHS of (11.16) is already formulated in terms suitable for the application of the Viterbi algorithm and, henceforth, it is assumed that the reader is familiar with the contents of Appendix F. The metric for each branch of the code trellis is defined as

$$V_\ell^{(r)}(\sigma_{\ell-1}, \sigma_\ell) \triangleq \ln P(\mathbf{y}_\ell \mid \mathbf{x}_\ell^{(r)}) \qquad (11.17)$$

and therefore

$$U(\sigma_{K-1}) = \max_r \sum_{\ell=0}^{K-1} V_\ell^{(r)}(\sigma_{\ell-1}, \sigma_\ell) \qquad (11.18)$$

If we denote with $d_\ell^{(r)}$ the Hamming distance between the two sequences $\mathbf{y}_\ell$ and $\mathbf{x}_\ell^{(r)}$, and use (10.15), we can rewrite (11.17) as

$$V_\ell^{(r)}(\sigma_{\ell-1}, \sigma_\ell) = -d_\ell^{(r)} \ln \frac{1-p}{p} + n_0 \ln(1-p) = -\alpha d_\ell^{(r)} - \beta \qquad (11.19)$$

with $\alpha$ and $\beta$ positive constants (if $p < 0.5$).

Using (11.19) into (11.18), and dropping unessential constants, the problem is reduced to finding

$$U'(\sigma_{K-1}) \overset{\triangle}{=} \min_r \sum_{\ell=0}^{K-1} d_\ell^{(r)} \qquad (11.20)$$

As expected intuitively, (11.20) states that ML decoding requires the minimization of the Hamming distance between the received sequence and the path chosen into the code trellis. This conclusion is perfectly consistent with the ML decoding of block codes, provided that the infinitely-long sequences are replaced by $n$-bit code words. The form of (11.20) is such that the minimization can be accomplished with the Viterbi algorithm (described in Appendix F), the metric on each branch being the Hamming distance between binary sequences.

**Example 11.7** We apply the Viterbi decoding algorithm to the code whose trellis is shown in Fig. 11.7, corresponding to the state diagram of Fig. 11.6. We know already (see Example 11.4) that this code has $d_f = 6$. Assume that the transmitted information sequence is 01000000..., whose corresponding encoded sequence is 000 111 011 001 000 000 000 000.... Furthermore, assume that the received sequence is instead 110 111 011 001 000 000 000 000.... It contains two errors in the first triplet of digits, and therefore it does not correspond to any path through the trellis. To apply the Viterbi algorithm, it is more useful to refer to a trellis similar to that of Fig. 11.9, in which, now, the label of each edge corresponds to the Hamming distance between the three digits associated to that edge and the corresponding three received bits. The successive steps of the Viterbi algorithm are shown in Fig. 11.16 and Fig. 11.17. The algorithm, at each step $\ell$ into the trellis, stores for each state the *surviving* path (the minimum distance path from the starting state ($\sigma_0 = S_1$)) and the corresponding accumulated metric. Consider, for example, the situation at step $\ell = 4$. We have

| State $\sigma_4$ | Surviving path | Metric |
|:---:|:---:|:---:|
| $S_1$ | $S_1 S_1 S_3 S_2 S_1$ | 2 |
| $S_3$ | $S_1 S_3 S_2 S_1 S_3$ | 5 |
| $S_2$ | $S_1 S_3 S_2 S_3 S_2$ | 7 |
| $S_4$ | $S_1 S_3 S_4 S_4 S_4$ | 6 |

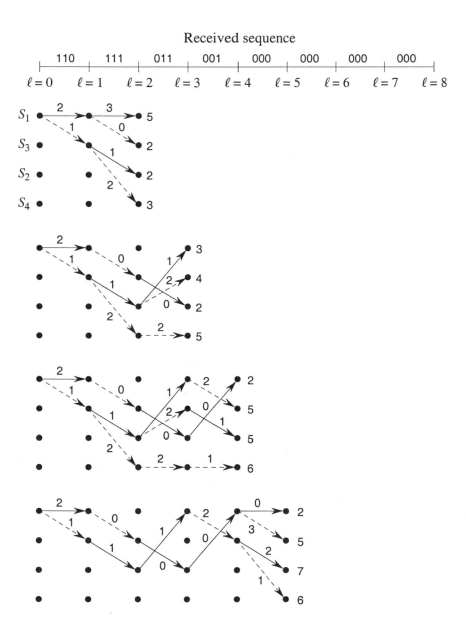

Figure 11.16: *Viterbi decoding algorithm applied to the (3,1,3) convolutional code of Fig. 11.6. The decoded sequence is 01000000.*

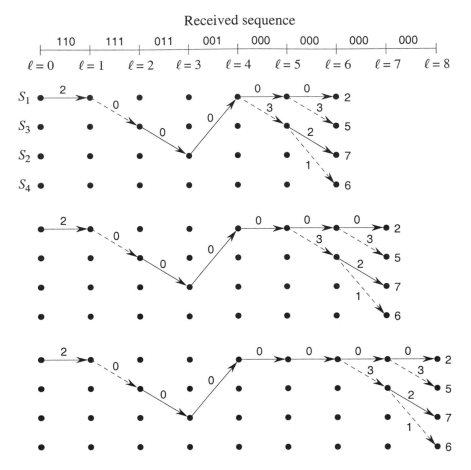

Figure 11.17: *Continuation of Fig. 11.16: Viterbi decoding algorithm applied to the (3,1,3) convolutional code of Fig. 11.6. The decoded sequence is 01000000.*

Therefore, at step $\ell = 4$, the ML path (the one with the smallest distance) is $S_1 S_1 S_3 S_2 S_1$, and the corresponding information sequence is 0100 (follow the dashed and solid edges on the trellis). Consequently, in spite of the two initial channel errors, already at this step the correct information sequence is identified. In case of a tie, i.e., when two or more states exhibit the same lowest path metric, any of the corresponding paths can be chosen.

Let us see in detail how the algorithm proceeds one step farther to compute the situation at $\ell = 5$. Consider first the state $\sigma_5 = S_1$. From the trellis diagram of Fig. 11.7, we can verify that the state $S_1$ can be reached either from the state $S_1$ with a transition corresponding to an encoded triplet 000, or from the state $S_2$ with a transition corresponding to an encoded triplet 001. The received triplet during this transition is 000. Therefore, from (11.19) and (11.18) we have, as potential candidates to $U'(\sigma_5)$

$$U'(\sigma_5)|_{\sigma_5=S_1} = U'(\sigma_4)|_{\sigma_4=S_1} + V_5^{(1)}(S_1, S_1) = 2 + 0 = 2$$
$$U'(\sigma_5)|_{\sigma_5=S_1} = U'(\sigma_4)|_{\sigma_4=S_2} + V_5^{(1)}(S_2, S_1) = 5 + 1 = 6$$

Thus, the minimum-distance path leading to $S_1$ at $\ell = 5$ comes from $S_1$, and the transition from $S_2$ is dropped. The metric $U'(\sigma_5)$ at $S_1$ will be 2, and the surviving path will be that of $\sigma_4 = S_1$ (i.e., 0100) with a new 0 added (i.e., 01000). The interesting feature is that at $\ell = 6$ all surviving paths *merge* at state $\sigma_4 = S_1$. This means that at this step the first four information digits are uniquely decoded in the correct way and the two channel errors are corrected.                                                                                           □

For a general $(n_0, k_0, N)$ convolutional code, there are $2^\nu$ states at each step in the trellis. Consequently, the Viterbi decoding algorithm requires the storage of $2^\nu$ surviving paths and $2^\nu$ metrics. At each step, there are $2^{k_0}$ paths reaching each state, and therefore $2^{k_0}$ metrics must be computed for each state. Only one of the $2^{k_0}$ paths reaching each state does survive, and this is the minimum-distance path from the received sequence up to that transition. The complexity of the Viterbi decoder, measured in terms of number of visited trellis edges per decoded bit, is then

$$\mathcal{D} = \frac{2^{k_0+\nu}}{k_0} \tag{11.21}$$

and grows exponentially[4] with $k_0$ and $\nu$. For this reason, practical applications are confined to the cases for which $k_0 + \nu$ is in the range 2 to 15.[5] The Viterbi algorithm is basically simple, and has properties that yield easy VLSI implementations. Actually, Viterbi decoding has been widely applied and is presently one of the most practical techniques for providing large coding gains.

The trellis structure of the decoding process has the following consequence. If at some point an incorrect path is chosen, it is highly probable that it will merge with the correct path at a later time. Therefore, the typical error sequences of convolutional codes, when decoded by a Viterbi decoder, result in bursts of errors due to the incorrect path diverging from the correct one and soon merging again into it. Typical bursts have a length of a few constraint lengths.

One final consideration concerns the technique used to output the decoded digits. The optimum procedure would be to decode the sequence only at the end

---

[4]The exponential growth with $k_0$ can be avoided using punctured codes, as seen previously.

[5]To our knowledge, the most complex implementation of the Viterbi algorithm concerns a code with $k_0 = 1$ and $\nu = 14$ for deep-space applications (see Dolinar, 1988).

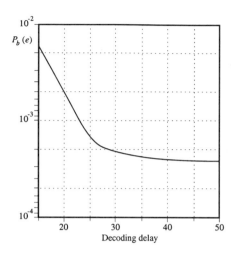

Figure 11.18: *Simulated bit error probability versus the decoding delay for the decoding of the rate 1/2 (2,1,7) convolutional code of Table 11.2. The signal-to-noise ratio $\mathcal{E}/N_0$ is 3 dB.*

of the whole receiving process. However, this would result in unacceptably long decoding delays and excessive memory storage for the surviving sequences. We have seen in the example that all surviving paths tend to merge into one single path when proceeding deeply enough into the trellis. A solution to this problem is thus to use the truncated Viterbi algorithm, described in Appendix F. This forces the decision on the oldest symbol of the minimum distance path after a fixed and sufficiently long delay. Computer simulations show that a delay on the order of $5N$ results in a negligible degradation with respect to the optimum performance. This is shown in Fig. 11.18, where we report the bit error probability evaluated by simulation as a function of the decoding delay for the constraint length 7, rate 1/2 code of Table 11.2.

An important feature of the Viterbi algorithm is that soft-decision decoding, unlike for block codes, requires only a trivial modification of the procedure discussed previously. In fact, it is sufficient to replace the Hamming metric with the Euclidean metric and let all the other decoding operations remain the same. Therefore, the implementation complexity for soft-decision decoding is not significantly different from the hard-decision case.

Let us now derive an expression for the branch metric (11.17) in the case of unquantized soft decisions. In practice, 3-bit quantization of the branch metrics is sufficient to obtain almost ideal performance (Jacobs, 1974).

If $\{y_{j\ell}\}_{j=1}^{n_0}$ is the set of received demodulator outputs in the case of binary

antipodal modulation (with transmitted and received energy $\mathcal{E}$) and assuming
that the $\ell$th branch of the $r$th path has been transmitted, we have

$$y_{j\ell} = \sqrt{\mathcal{E}}(2x_{j\ell}^{(r)} - 1) + \nu_j \qquad (11.22)$$

which is obtained from (10.81). Here, $x_{j\ell}^{(r)}$ is a binary digit and $\nu_j$ is a Gaussian
RV with zero mean and variance $N_0/2$. Therefore, from (11.22) we get

$$
\begin{aligned}
P(\mathbf{y}_\ell \mid \mathbf{x}_\ell^{(r)}) &= \prod_{j=1}^{n_0} P(y_{j\ell} \mid x_{j\ell}^{(r)}) \\
&= \prod_{j=1}^{n_0} \frac{1}{\sqrt{\pi N_0}} \exp\left\{ -\frac{\left[ y_{j\ell} - \sqrt{\mathcal{E}}(2x_{j\ell}^{(r)} - 1) \right]^2}{N_0} \right\}
\end{aligned} \qquad (11.23)
$$

Inserting (11.23) into (11.17) and neglecting all the terms that are common to all
branch metrics, we get

$$V_\ell^{(r)}(\sigma_{\ell-1}, \sigma_\ell) = \sum_{j=1}^{n_0} y_{j\ell}(2x_{j\ell}^{(r)} - 1) \qquad (11.24)$$

This is the branch metric to be used by the soft-decision Viterbi decoder. It is
called, for obvious reasons, *correlation* metric. The best path would correspond
in this case to the highest metric. As an alternative, one can also use the *distance*
metric, which should be minimized.

### 11.1.4.    Other decoding techniques for convolutional codes

The computational effort and the storage size required to implement the Viterbi
algorithm limit its application to convolutional codes with small-medium values
of the memory $\nu$ (typically, $1 \leq \nu \leq 14$). Other decoding techniques can be
applied to convolutional codes. These techniques preceded the Viterbi algorithm
historically, and are quite useful in certain applications. In fact, they can use
longer code constraint lengths than those allowed by practical implementations
of the Viterbi algorithm, and hence yield larger coding gains.

### Sequential decoding techniques

As already pointed out, the operation of a convolutional encoder can be described
as the choice of a path through a binary tree in which each path represents an
encoded sequence. The sequential decoding techniques (in their several variants)
share with the Viterbi algorithm the idea of a probabilistic search of the correct
path, but, unlike the Viterbi algorithm, the search does not extend to all paths

that can potentially be the best. Only some subsets of paths that appear to be the most probable ones are extended. For this reason, sequential decoding is not an optimal (ML) algorithm as the Viterbi algorithm. Nevertheless, sequential decoding is one of the most powerful tools for decoding convolutional codes of long constraint length. Its error performance is not significantly worse than that of Viterbi decoding.

The decoding approach can be conceived as a trial-and-error technique for searching out the correct path into the tree. Let us consider a qualitative example by looking at the code tree of Fig. 11.8. In the absence of noise, the code sequences of length $n_0 = 3$ are received without errors. Consequently, the receiver can start its walk into the tree from the root and then follow a path by simply replicating at each node the encoding process and making its binary decision after comparing the locally generated sequence with the received one. The transmitted message will be recovered directly from the path followed into the tree.

The presence of noise introduces errors, and hence the decoder can find itself in a situation in which the decision entails risk. This happens when the received sequence is different from all the possible alternatives that are locally generated by the receiver. Consider again the code tree of Fig. 11.8, and assume that the transmitted sequence is the one denoted by the heavy line. Let, for instance, the received sequence be 111 100 111. .... Starting from the root, the first two choices are not ambiguous. But, when reaching the second-order node, the decoder must choose between the upward path (sequence 010) and the downward path (sequence 101), having received the sequence 110. The choice that sounds more reasonable is to go downward in the tree, since the Hamming distance between the received and locally generated sequences is only one, instead of two. With this choice, however, the decoder would proceed on a wrong path in the tree, and the continuation would be in error. If the branch metric is the Hamming distance, the decoder can track the cumulative Hamming distance between the received sequence and the path followed into the tree, and eventually notice that this distance grows higher than expected. In this case, the decoder can decide to go back to the node at which an apparent error was made and try the other choice. This process of going forward and backward into the tree is the rationale behind sequential decoding. This movement can be guided by modifying the metric of the Viterbi algorithm (the Hamming distance for hard decisions) with the addition of a negative constant at each branch. The value of this constant is selected such that the metric for the correct path decreases on the average, while that for any incorrect path increases. By comparing the accumulated metric with a moving threshold, the decoder can detect and discard the incorrect paths.

Sequential algorithms trade with the Viterbi algorithm a larger decoding delay with a smaller storage need. Unlike for the Viterbi algorithm, both the decoding delay and the computational complexity are not constant. Instead, they are random variables that depend, among other factors, on the signal-to-noise ratio. When there is little noise, the decoder is usually following the correct path requiring only one computation to advance one node deeper into the code tree. However, when the noise becomes significant, the metric along the correct path may increase and be higher than the metric along an incorrect path. This forces the decoder to follow an incorrect path, so that a large number of steps (and computations) may be required to return to the correct path. To make this statement quantitative, we refer to the *cutoff rate* of the channel $R_0$ already introduced in Section 10.4.2. When the code rate $R_c$ is larger than the channel cutoff rate, the average computational load of sequential decoding, defined as the average number of computations per decoded branch, is unbounded. For this reason, $R_0$ is often called the *computational* cutoff rate, as it indicates a limit on the code rates beyond which sequential decoding becomes impractical (for a proof of these statements, see Lin and Costello, 1983).

### The $M$-algorithm

The idea of the $M$-algorithm is to look at the best $M$ ($M$ less than the number of trellis states $N_\sigma$) paths at each depth of the trellis, and to keep only these paths while proceeding into the trellis (no backtracking allowed). For $M = N_\sigma$, it becomes the Viterbi algorithm. The choice of $M$ trades performance for complexity. Unlike sequential decoding, the $M$-algorithm has the advantage of fixed complexity and decoding delay. For details on the $M$-algorithm, see Anderson and Mohan (1991).

### Syndrome decoding techniques

Unlike sequential decoding, these techniques are deterministic and rely on the algebraic properties of the code. Typically, a syndrome sequence is calculated (as for block codes). It provides a set of linear equations that can be solved to determine the minimum-weight error sequence. The two most widely used among such techniques are *feedback decoding* (Heller, 1975) and *threshold decoding* (Massey, 1963). They have the advantage of simple circuitry and small decoding delays, thus allowing high-speed decoding. However, since the allowable codes presenting the required algebraic properties are rather poor, only moderate coding gain values are achievable with these techniques.

**The maximum-a-posteriori (MAP) symbol decoding algorithm**

The Viterbi algorithm performs the ML estimate of the transmitted sequence. Its output is the code sequence closest in some sense (Hamming or Euclidean distance) to the received one. The Viterbi algorithm thus minimizes the *sequence* error probability when the information sequences are assumed to be equally likely.

In the most general case, the decision rule minimizing the *bit* error probability should be based on the maximization of the a posteriori probabilities (APP) of each individual bit in the sequence

$$P(u_k \mid \mathbf{y}), \quad k = 0, \ldots, K - 1 \tag{11.25}$$

where $u_k$ is the transmitted bit at time $k$, $\mathbf{y}$ is the entire received sequence, and $K$ is the sequence length. We recall that MAP decoding also differs from ML decoding in that it does not assume equally likely information symbols.

The simplest algorithm to compute the a-posteriori probabilities (11.25) was proposed by Bahl *et al.* (1974), but until recently it received very little attention because its complexity exceeds that of the Viterbi algorithm, yet the advantage in bit error rate performance is small. It is described in Appendix F under the name of BCJR algorithm, from the initials of the researchers who proposed it.

The big difference between Viterbi and APP algorithms consists in their outputs. The Viterbi algorithm outputs a *hard* decision on the transmitted digits, whereas the APP algorithm provides the a posteriori probability, which may be interpreted as a *soft* estimate of the transmitted digits reliability, actually the best possible one. When a convolutional code is employed in a *concatenated* coding scheme, like those examined in the last section of this chapter, this difference becomes fundamental, and explains the recent great revival of interest for APP algorithms.

As a final, important comment, we can say that the Viterbi algorithm, and consequently also the other algorithms briefly described previously, is applicable to the decoding of *any* code whose code words can be associated to paths in a trellis. As a consequence, the soft decoding of block codes, which can be represented by time-varying trellises (see, for example, Wolf, 1978), can be performed by using the Viterbi algorithm.

### 11.1.5.  Performance evaluation of convolutional codes with ML decoding

In this section, we will derive upper bounds to the *error event* and *bit* error probabilities of a convolutional code. Since the results are based on the union bound, they provide tight approximations to the actual probabilities for medium-high signal-to-noise ratios. For lower signal-to-noise ratios, these bounds diverge,

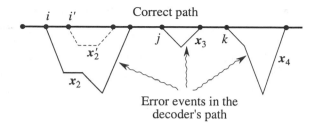

Figure 11.19: *Trellis paths showing possible error events of a Viterbi decoder.*

and one should resort to simulation,[6] or to more sophisticated techniques, like the one described in Poltyrev (1996) and references therein. Moreover, we do not consider the suboptimality of the truncated Viterbi algorithm involved in forcing premature decisions (see Hemmati and Costello, 1977, and Onyszchuk, 1991).

**Error event probability**

Before discussing techniques for bounding the bit error probability $P_b(e)$, it is necessary to analyze in some detail the concept of *error event* in the Viterbi decoding. Since convolutional codes are linear and the uniform error property holds for them, we can assume that the all-zero sequence has been transmitted and evaluate error probabilities under this hypothesis. We denote as the *correct path* the all-zero horizontal path at the top of the trellis diagram (Fig. 11.19).

The decoder bases its decisions on the received noisy version of the transmitted sequence and can choose a path different from the correct one on the basis of the accumulated metric. For a given discrete time $k$, an error event is defined as an incorrect path segment that, after diverging from the correct path at trellis depth $k$, merges again into it at a successive depth. Since the trellis of a convolutional code is time-invariant (with the exception of the initial transient from the zero state to all other states), the performance will not depend on the starting time $k$ of the error event, which can thus be omitted. Fig. 11.19 shows three error events starting at nodes $i, j, k$, and corresponding to the sequences $\mathbf{x}_2, \mathbf{x}_3, \mathbf{x}_4$. Notice also that the dotted path, corresponding to the sequence $\mathbf{x}_2'$ diverging at node $i'$, may have a metric higher than the correct path and yet not be selected, because its accumulated metric is smaller than that of the solid path corresponding to the sequence $\mathbf{x}_2$.

---

[6]Fortunately, simulation is required for low signal-to-noise ratios, where the error probabilities are high, say greater that $10^{-3}$, so that the required computer time is often quite reasonable.

The term "error event" comes from the fact that, when the decoder chooses an incorrect path forming an error event, the errors accumulated during the periods of divergence cannot be corrected, since, after remerging, the correct and error event paths will accumulate the same metrics. We may conclude that a necessary and sufficient condition for an error event to occur at a certain node is that the metric of an incorrect path, diverging from the correct one at that node, accumulates higher metric increments than the correct path over the unmerged path segment.

Thus, we can define an error event at a certain node as the set of all paths diverging from the correct path at that node, and having a path metric larger than the correct one.

If we denote by $x_d$ a path of weight $d$ diverging from the all-zero path, by $x_1$ the correct path (all-zero sequence), and by $P(x_1 \to x_d)$ the *pairwise* error probability between the two sequences[7] $x_1$ and $x_d$, then the probability $P(e)$ of an error event can be upper bounded, using the union bound, as

$$P(e) \leq \sum_{d=d_f}^{\infty} A_d P(x_1 \to x_d) \qquad (11.26)$$

where $A_d$ is the number of paths of weight $d$ diverging from the all-zero path. To proceed further, we must distinguish the cases of hard and soft-decoding.

**Hard decoding**   Using (10.76), we have

$$P(x_1 \to x_d) \leq \left[ \sqrt{4p(1-p)} \right]^d \qquad (11.27)$$

Introducing (11.27) into (11.26) and recalling the definition (11.9) of the weight enumerating function $T(D)$, we finally obtain

$$P(e) \leq T(D)|_{D=\sqrt{4p(1-p)}} \qquad (11.28)$$

This result emphasizes the role of the weight enumerating function $T(D)$ for the computation of the probability $P(e)$ of an error event, paralleling the results obtained for the word error probability of block codes.

**Soft decoding**   The case of soft decoding is the same as the hard decoding one, except that the metric is the Euclidean distance, and, consequently, assuming a

---

[7]The pairwise error probability was defined in Chapter 4 and used in Chapter 10. Here, it represents the conditional probability that the sequence $x_d$ has a larger metric than the correct sequence $x_1$.

serial transmission of the coded bits using a binary antipodal modulation with energy $\mathcal{E}_b$, we can write the pairwise error probability as

$$P(\mathbf{x}_1 \to \mathbf{x}_d) = \frac{1}{2}\text{erfc}\left(\sqrt{\frac{dR_c\mathcal{E}_b}{N_0}}\right) \tag{11.29}$$

where $R_c$ is the code rate. Substituting the right-hand side of (11.29) into (11.26), we get the union bound to the error event probability

$$P(e) \leq \frac{1}{2}\sum_{d=d_f}^{\infty} A_d\,\text{erfc}\left(\sqrt{\frac{dR_c\mathcal{E}_b}{N_0}}\right) \tag{11.30}$$

Using the inequality (A.5) $\frac{1}{2}\text{erfc}(\sqrt{x}) < \frac{1}{2}e^{-x}$, we can express (11.30) as

$$P(e) < \frac{1}{2}T(D)\Bigg|_{D=e^{-R_c(\mathcal{E}_b/N_0)}} \tag{11.31}$$

The difference between the two bounds (11.30) and (11.31), is that the first is tighter. However, the second can be evaluated from the closed-form knowledge of the weight enumerating function, whereas the first requires the *distance spectrum* of the code, i.e., the pairs $\{A_d, d\}_{d=d_f}^{\infty}$, which can be obtained from the power series expansion of $T(D)$. Usually, a small number of pairs are sufficient to obtain a close approximation. A bound in closed form tighter than (11.31) can also be derived (see Problem 11.7).

## Bit error probability

The computation of an upper bound to the bit error probability is more difficult, and, in fact, the result is not always rigorously derived in textbooks.

Consider the trellis section between time $m$ and time $m+1$, and assume that the all-zero sequence has been transmitted. Let $E(w, d, \ell)$ be the event that an error event with input information weight $w$, code sequence weight $d$, and length $\ell$ is active, i.e., has the highest path metric, in the interval $(m, m+1)$. Also, denote with $e(w, d, \ell)$ an error event starting at time $m$ with input information weight $w$, code sequence weight $d$, and length $\ell$.

The probability of the event $E(w, d, \ell)$ is easily bounded as

$$P[E(w, d, \ell)] \leq \ell P[e(w, d, \ell)] = \ell C_{w,d,\ell} P(\mathbf{x}_1 \to \mathbf{x}_d) \tag{11.32}$$

where $P[e(w, d, \ell)]$ is the probability of an error event produced by an incorrect path $\mathbf{x}_d$ of weight $d$, length $\ell$ and input weight $w$, and $C_{w,d,\ell}$ is the multiplicity of such incorrect paths. The first inequality in (11.32) relies on the fact that, to be active in the interval $(m, m+1)$, the error event $e(w, d, \ell)$ of length $\ell$ must have started at a node between $m - \ell + 1$ and $m$.

The bit error probability $P_b(e)$, defined as the probability that a bit is in error in the interval $(m, m+1)$, is given by

$$P_b(e) \leq \sum_{w=1}^{\infty} \sum_{d=d_f}^{\infty} \sum_{\ell=1}^{\infty} P_b[e \mid E(w, d, \ell)] P[E(w, d, \ell)] \tag{11.33}$$

But the probability of a bit error for a rate $k_0/n_0$ code, conditioned to $E(w, d, \ell)$, is simply

$$P_b[e \mid E(w, d, \ell)] = \frac{w}{k_0 \ell} \tag{11.34}$$

In fact, $\ell$ is the number of information digits equal to one (digits in error, since the all-zero sequence was transmitted) in $E(w, d, \ell)$, and $k_0 \ell$ is the overall number of information digits in the error event of length $\ell$. Substituting (11.34) and the upper bound (11.32) into (11.33), we obtain

$$P_b(e) \leq \frac{1}{k_0} \sum_{w=1}^{\infty} \sum_{d=d_f}^{\infty} \sum_{\ell=1}^{\infty} w C_{w,d,\ell} P(\mathbf{x}_1 \rightarrow \mathbf{x}_d) \tag{11.35}$$

or, taking into account (11.12)

$$P_b(e) \leq \sum_{d=d_f}^{\infty} \left( \sum_{w=1}^{\infty} \frac{w}{k_0} B_{w,d} \right) P(\mathbf{x}_1 \rightarrow \mathbf{x}_d) \tag{11.36}$$

Comparing (11.36) with (11.26), we see that the two are formally the same, provided that we substitute for $A_d$ the *bit multiplicity* $A_d^{(b)}$, defined as

$$A_d^{(b)} \triangleq \sum_{w=1}^{\infty} \frac{w}{k_0} B_{w,d} \tag{11.37}$$

The pairs $(A_d^{(b)}, d)$ form the *bit distance spectrum* of the convolutional code. To continue, we must distinguish as before the cases of hard and soft decoding.

**Hard decoding**   Using (11.27) for $P(\mathbf{x}_1 \rightarrow \mathbf{x}_d)$ into (11.36), we obtain

$$\begin{aligned} P_b(e) &\leq \frac{1}{k_0} \sum_{w=1}^{\infty} \sum_{d=d_f}^{\infty} w B_{w,d} \left[ \sqrt{4p(1-p)} \right]^d \\ &= \frac{1}{k_0} \left. \frac{\partial T_2(W, D)}{\partial W} \right|_{D=\sqrt{4p(1-p)}, W=1} \end{aligned} \tag{11.38}$$

which shows the role of the input-output weight enumerating function in the computation of the bit error probability.

When $p$ is very small, we can approximate the general result (11.38) by keeping only the smallest exponent $d = d_f$ in the summation. This yields the asymptotic approximation

$$P_b(e) \approx A_{d_f}^{(b)} 2^{d_f} p^{d_f/2} \tag{11.39}$$

where $A_{d_f}^{(b)}$ is the bit multiplicity of paths with weights $d = d_f$. The importance of the free distance is thus fully evident. A somewhat tighter upper bound can be obtained in (11.35) by using for $P(\mathbf{x}_1 \to \mathbf{x}_d)$ an expression different from the Bhattacharyya bound (see Problem 11.8).

**Soft decoding**  The case of soft decoding is the same as the hard decoding one, except that the metric is the Euclidean distance. Proceeding as for the case of the error event probability, we obtain

$$P(\mathbf{x}_1 \to \mathbf{x}_d) = \frac{1}{2}\mathrm{erfc}\left(\sqrt{\frac{dR_c\mathcal{E}_b}{N_0}}\right) \tag{11.40}$$

Substituting the right-hand side of (11.40) into (11.36), we get the union bound to the bit error probability

$$\begin{aligned}
P_b(e) &\leq \frac{1}{2k_0} \sum_{w=1}^{\infty} \sum_{d=d_f}^{\infty} wB_{w,d}\mathrm{erfc}\left(\sqrt{\frac{dR_c\mathcal{E}_b}{N_0}}\right) \\
&= \frac{1}{2} \sum_{d=d_f}^{\infty} A_d^{(b)}\mathrm{erfc}\left(\sqrt{\frac{dR_c\mathcal{E}_b}{N_0}}\right)
\end{aligned} \tag{11.41}$$

where we have used the definition (11.37) of the *bit multiplicity* $A_d^{(b)}$. To use the bound (11.41) in practice, we need to truncate the summation in $d$ to some finite value. Usually, the first few (say 10) terms are enough to guarantee accurate results.

A looser, but closed-form bound can be obtained through the inequality (A.5), by which (11.41) becomes

$$P_b(e) \leq \frac{1}{2k_0} \left.\frac{\partial T_2(W, D)}{\partial W}\right|_{D=e^{-R_c(\mathcal{E}_b/N_0)}, W=1} \tag{11.42}$$

The bound (11.42) can be further refined and made tighter, although it will remain looser than (11.41) (see Problem 11.7).

For large signal-to-noise ratios $\mathcal{E}_b/N_0$, the general result (11.41) can be approximated as

$$P_b(e) \approx \frac{1}{2} A_{d_f}^{(b)} \operatorname{erfc} \left( \sqrt{\frac{d_f R_c \mathcal{E}_b}{N_0}} \right) \qquad (11.43)$$

where $A_{d_f}^{(b)}$ is the bit multiplicity of error events with weight equal to $d_f$.

We can conclude that, for large signal-to-noise ratios, the behavior of the bit error probability is dominated in all cases by the free distance $d_f$, which plays a role similar to the minimum distance $d_{\min}$ for block codes. Comparing (11.43) with the bit error probability for uncoded transmission (4.36), we can derive the asymptotic ML coding gain as $10 \log_{10} d_f R_c$.

**Example 11.8** In this example we derive the bit error probability $P_b(e)$ for the (3,1,3) convolutional code with state diagram depicted in Fig. 11.11. This code was seen to have a free distance $d_f = 6$.

We plot in Fig. 11.20 five curves. They refer to the upper bounds (11.38) and (11.41) referring to hard and soft ML decoding, to the asymptotic approximations of Eqs. (11.39) and (11.43) that take into account only the first term deriving from the free distance, and, finally, to the results obtained by simulating the Viterbi algorithm with infinitely-soft quantization (curve with "+"). We see from the curves three things: the hard decoding penalty of 2.4 dB, the fact that the asymptotic bounds converge to the bounds using the entire distance spectrum for bit error probabilities below $10^{-3}$, and, finally, that the upper bound (11.41) is very tight, as it almost coincides with the simulation. □

**Example 11.9** To get a more complete idea of the coding gains achievable by ML decoded convolutional codes, some codes of rate 1/2 (Table 11.2) and rate 1/3 (Table 11.3) have been considered. The upper bounds to $P_b(e)$ have been evaluated using the bound (11.41) truncated to the first five nonzero terms of the bit distance spectrum. The results are shown in Figures 11.21 and 11.22.

Notice that the potential coding gains are quite significant and increase as the signal-to-noise ratio increases until they reach the asymptotic coding gain $10 \log_{10} d_f R_c$. As the constraint length $N$ is increased for the codes of the same rate, the coding gain increases by about 0.3 to 0.4 dB for each increase of one unit in $N$. On the other hand, increasing the code rate for the same $N$ entails a loss in coding gain of about 0.4 dB. The curves assume infinitely-soft quantization in soft Viterbi decoding. However, the predicted coding gains are quite real. In fact, if an eight-level quantization is used rather than infinite quantization, the observed degradation in performance is of the order of 0.25 dB. Achievable coding gains with eight-level quantization soft-decision Viterbi decoding are given in Table 11.8 for different values of the bit error probability. The asymptotic

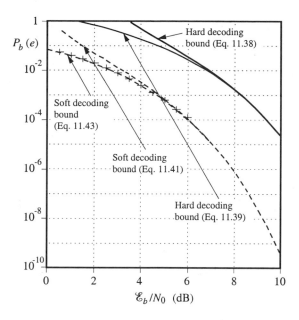

Figure 11.20: *Performance bounds with hard- and soft-decision maximum-likelihood decoding for the (3,1,3) convolutional code of Fig. 11.6 (see Example 11.8). The curve with "+" refers to simulation results obtained with the soft-decision Viterbi algorithm.*

upper bound $10 \log_{10} d_f R_c$ to the coding gain is also indicated. □

### 11.1.6. Systematic recursive convolutional encoders

We have seen previously (see Table 11.1) that systematic convolutional codes generated by feed-forward encoders yield, in general, lower free distances than nonsystematic codes. In this section, we will show how to derive a systematic encoder from every rate $1/n_0$ nonsystematic encoder, which generates a systematic code with the same weight enumerating function as the nonsystematic one. The encoder has feedback connections in it, and we call it a *recursive systematic* encoder. Costello (1969) and Forney (1970) were the first to point out this possibility.

Consider for simplicity a rate 1/2 feed-forward encoder characterized by the two generators (in polynomial form) $g_{1,1}(Z)$ and $g_{1,2}(Z)$. Using the power series $u(Z)$ to denote the input sequence $\mathbf{u}$ and $x_1(Z)$ and $x_2(Z)$ to denote the two

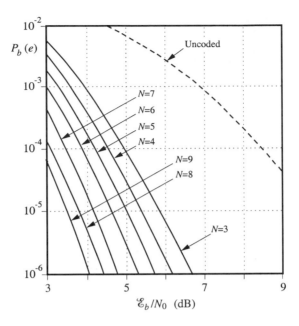

Figure 11.21: *Upper bounds to the soft ML decoding bit error probability of different convolutional codes of rate 1/2. The codes are those listed in Table 11.2 (Clark and Cain, 1981).*

sequences $\mathbf{x}_1$ and $\mathbf{x}_2$ forming the code $\mathbf{x}$, we have the relationships

$$
\begin{aligned}
x_1(Z) &= u(Z)g_{1,1}(Z) \\
x_2(Z) &= u(Z)g_{1,2}(Z)
\end{aligned}
\tag{11.44}
$$

To obtain a systematic code, we need to have either $x_1(Z) = u(Z)$ or $x_2(Z) = u(Z)$. To obtain the first equality, let us divide both equations (11.44) by $g_1(Z)$, so that

$$
\begin{aligned}
\tilde{x}_1(Z) &\triangleq \frac{x_1(Z)}{g_{1,1}(Z)} = u(Z) \\
\tilde{x}_2(Z) &\triangleq \frac{x_2(Z)}{g_{1,1}(Z)} = \frac{u(Z)}{g_{1,1}(Z)}g_{1,2}(Z)
\end{aligned}
\tag{11.45}
$$

Defining now a new input sequence $\tilde{u}(Z)$ as

$$
\tilde{u}(Z) \triangleq \frac{u(Z)}{g_{1,1}(Z)}
\tag{11.46}
$$

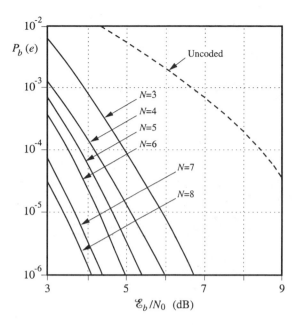

Figure 11.22: *Upper bounds to the soft ML decoding bit error probability of different convolutional codes of rate 1/3. The codes are those listed in Table 11.3 (Clark and Cain, 1981).*

the relations (11.45) become

$$\begin{aligned}
\tilde{x}_1(Z) &= \tilde{u}(Z)g_{1,1}(Z) \\
\tilde{x}_2(Z) &= \tilde{u}(Z)g_{1,2}(Z)
\end{aligned}$$
(11.47)

We notice now that the transformation (11.46) between $u(Z)$ and $\tilde{u}(Z)$ is that of a recursive digital filter with operations in the field $GF(2)$. This transformation operates a simple reordering of the input sequences $u(Z)$, which form the set of all possible binary sequences. Making use of this observation, and comparing (11.47) with (11.44), we can affirm that the set of code sequences $\tilde{x}(Z)$ is the same as the set of $x(Z)$, and thus the two codes have the same weight enumerating functions. As a consequence, the recursive encoder $(\mathbf{u}, \tilde{\mathbf{x}})$ generates a systematic code equivalent, i.e., having the same distance spectrum, to the nonsystematic code generated by the feed-forward encoder $(\mathbf{u}, \mathbf{x})$; for the two encoders, however, the same code sequences correspond to different input sequences. As an example, it is easy to verify that input sequences with weight 1 produce error events in the feed-forward encoder, whereas the recursive encoder

| $\mathcal{E}_b/N_0$ for uncoded transmission (dB) | $P_b(e)$ | $R_c=1/3$ $N$ | | $R_c=1/2$ $N$ | | | $R_c=2/3$ $N$ | | $R_c=3/4$ $N$ | |
|---|---|---|---|---|---|---|---|---|---|---|
| | | 7 | 8 | 5 | 6 | 7 | 6 | 8 | 6 | 9 |
| 6.8 | $10^{-3}$ | 4.2 | 4.4 | 3.3 | 3.5 | 3.8 | 2.8 | 3.1 | 2.6 | 2.6 |
| 9.6 | $10^{-5}$ | 5.7 | 5.9 | 4.3 | 4.6 | 5.1 | 4.2 | 4.6 | 3.6 | 4.2 |
| 11.3 | $10^{-7}$ | 6.2 | 6.5 | 4.9 | 5.3 | 5.8 | 4.7 | 5.2 | 3.9 | 4.8 |
| $\to \infty$ | $\to 0$ | 7.0 | 7.3 | 5.4 | 6.0 | 7.0 | 5.2 | 6.7 | 4.8 | 5.7 |

Table 11.8: *Achievable coding gains with some convolutional codes of rate $R_c$ and constraint length $N$ at different values of the bit error probability. Eight-level quantization soft-decision Viterbi decoding is used. The last line gives the asymptotic upper bound* $10 \log_{10} d_f R_c$ *(Jacobs, 1974).*

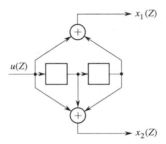

Figure 11.23: *Rate 1/2, 4-state feed-forward encoder generating the nonsystematic code (A) of Example 11.10.*

needs input sequences with weight at least 2 to return to the zero state, and thus to generate an error event. As we shall see in Section 11.3, this difference will prove to be crucial for the performance of turbo codes.

**Example 11.10**　Consider the (2,1,3) feed-forward encoder (A) of Fig. 11.23, with generators $g_1^A(Z) = 1 + Z^2$, $g_2^A(Z) = 1 + Z + Z^2$ and its equivalent recursive encoder of Fig. 11.24 obtained as explained previously with generators $g_1^B(Z) = 1$, $g_2^B(Z) = (1 + Z + Z^2)/(1 + Z^2)$.

Using the state diagram representations for both codes, and computing the input-output weight enumerating functions $T_2^A(W, D), T_2^B(W, D)$, we obtain

$$T_2^A(W, D) = \frac{W D^5}{1 - 2WD}$$

$$T_2^B(W, D) = \frac{(W^2 D^5)(W^2 D - D + 1)}{1 - W^2 D^2 + D^2 - 2D} \tag{11.48}$$

Letting $W = 1$, we obtain the same weight enumerating function for the code sequences

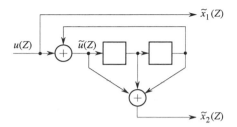

Figure 11.24: *Rate 1/2, 4-state recursive encoder generating the systematic code (B) of Example 11.10.*

of the two codes, i.e.

$$T^A(D) = T^B(D) = \frac{D^5}{1 - 2D} \tag{11.49}$$

As a consequence, the upper bounds to the error event probability of the two codes are exactly the same. On the other hand, taking the derivative of the input-output transfer functions (11.48), we obtain two different results, and thus two different bit error probability bounds. In fact

$$\begin{aligned}
\frac{\partial T_2^A(W, D)}{\partial W}\bigg|_{W=1} &= \frac{D^5}{(1 - 2D)^2} \\
\frac{\partial T_2^B(W, D)}{\partial W}\bigg|_{W=1} &= \frac{2D^5(1 - D - D^2)}{(1 - 2D)^2}
\end{aligned} \tag{11.50}$$

It is instructive to compute the bit distance spectrum $(A_d^{(b)}, d)$ (the multiplicities $A_d^{(b)}$ were defined in (11.37)) of the two codes, whose first terms are reported in the following table (NS stands for "nonsystematic," RS for "recursive systematic"):

| $d$ | 5 | 6 | 7 | 8 | 9 | 10 | 11 | 12 | 13 | 14 | 15 |
|---|---|---|---|---|---|---|---|---|---|---|---|
| $A_d^{(b)}$, NS | 1 | 4 | 12 | 32 | 80 | 192 | 448 | 1024 | 2304 | 5120 | 11264 |
| $A_d^{(b)}$, RS | 2 | 6 | 14 | 32 | 72 | 160 | 352 | 768 | 1664 | 3584 | 7680 |

Comparing the multiplicities $A_d^{(b)}$ for the two codes, we see that for low values of $d$ (and in particular for $d = d_f$) the multiplicities are smaller for the nonsystematic encoder, whereas for large $d$ the situation is reversed. We can thus argue that for medium-to-high signal-to-noise ratios the bit error probability will be lower for the nonrecursive encoder, whereas for low signal-to-noise ratios the recursive encoder will perform better. These observations are confirmed by the curves of Fig. 11.25, where we have plotted the bit error probabilities for the two codes. The curves have been obtained by simulating the MAP decoding algorithm employing the BCJR APP algorithm (see Appendix F). □

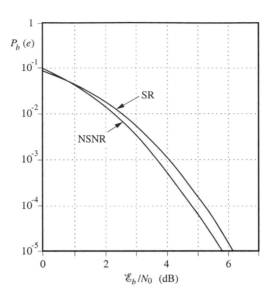

Figure 11.25: *Bit error probability versus signal-to-noise ratio for the nonsystematic nonrecursive (NSNR) and systematic recursive (SR) encoders of Example 11.10.*

The same behavior of the above example holds for codes with different constraint lengths and different rates $1/n_0$. When used in a stand-alone configuration, the nonsystematic codes are the right choice, because of their slightly better behavior at bit error probabilities below $10^{-2}$; this explains the almost universal preference given to them in practice. When used in a concatenated scheme, however, aiming at very large coding gains, the behavior at low signal-to-noise ratios become crucial, and thus recursive systematic encoders should be used.

### 11.1.7. Coding bounds

As explained for block codes in Section 10.3, we can derive also for convolutional codes bounds on the free distance and bounds on code performance based on random coding arguments.

The best known bound on the free distance is the Heller bound, already used in the tables of good codes in Section 11.1.2. It states that the largest achievable free distance by a $(n_0, 1, N)$ convolutional code satisfies the inequality

$$d_f \leq \min_{r>0} \left[ \frac{2^{r-1}}{2^r - 1}(N + r - 1)n_0 \right] \tag{11.51}$$

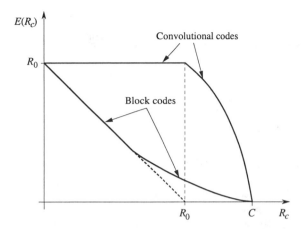

Figure 11.26: *Typical behavior of the reliability function $E(R_c)$ on the AWGN channel for convolutional and block codes.*

where $r$ is an integer. The Heller bound has been extended to codes of any rate by Daut *et al.* (1982).   A version of the channel coding theorem described in Chapter 3 for block codes can be derived also for the class of *time-varying* convolutional codes (see Chapter 5 of Viterbi and Omura, 1979). The final result is that the error event probability for a $(n_0, k_0, N)$ convolutional code is upper bounded by an expression similar to the block code expression given by (10.106), namely

$$P(e) \leq A(R_c)2^{-n_0 N E(R_c)}, \qquad R_c \leq C \qquad (11.52)$$

where $A(R_c)$ is a small constant, and $E(R_c)$ is the *reliability* function for convolutional codes shown in Fig. 11.26, where we have redrawn also the reliability function of block codes of Fig. 9.20, for the sake of comparison.

Therefore, also for convolutional codes we can write a simple upper bound in terms of the cutoff rate $R_0$

$$P(e) \leq A(R_c)2^{-n_0 N R_0} \qquad R_c \leq R_0 \qquad (11.53)$$

Through the inequalities in (10.106) and (11.53) used with $n = n_0 N$, we can compare the relative performance of block and convolutional codes in a homogeneous way.   Choosing $n$ (or $n_0 N$) as a reasonable indicator of the decoder complexity for both block and convolutional codes, we can achieve an exponent in (11.53) for convolutional codes greater than the one in (10.106) for block codes.   Roughly speaking, and with the due caution of comparing ensemble-average bounds, we can say that convolutional codes give a better

Figure 11.27: *Block diagram of a two-level concatenated code.*

performance/complexity tradeoff than block codes, especially for low signal-to-noise ratios.

## 11.2.   Concatenated codes

*Concatenated codes* were introduced by Forney (1966), in his goal to find a class of codes whose probability of error would decrease exponentially at a rate less than capacity, while decoding complexity would increase only algebraically. Initially motivated only by theoretical research interests, concatenated codes have since then evolved as a standard for those applications where high coding gains and affordable complexity are needed.

The basic concept of concatenated codes is illustrated in Fig. 11.27, for the case of a two-level concatenation. Two, or more, encoders are arranged in a cascaded fashion, where the code words of the first encoder become the input words of the second encoder, and so on. In the case of the concatenation of two codes, the first is called the *outer* code, and the second the *inner* code. They are the constituent codes of the concatenated scheme. In the following, we will only consider two-level concatenations.

Let the outer code be an $(n_o, k_o)$ block code, and the inner code be an $(n_i, k_i)$ block code. The parameters $k_i$ and $n_o$ must be multiple (or integer submultiple) of each other. Typically, $n_o$ is larger than $k_i$, so that

$$n_o = mk_i, \quad m \text{ integer} \tag{11.54}$$

Thus, the code word of the outer code contains an integer number of data words of the inner code. Denoting by $R_c^o$, $R_c^i$ the rates of the outer and inner codes, the overall rate of the concatenated code is

$$R_c = \frac{k_o}{mn_i} = \frac{k_o}{n_o}\frac{n_o}{mn_i} = R_c^o R_c^i \tag{11.55}$$

i.e., the product of the rates of the constituent codes. When the constituent codes are convolutional codes, the same previous relationships hold between the code rates.

The most common approach for a long time has been using as the outer code a nonbinary code, typically a long high-rate Reed-Solomon (RS) code. For the inner code, many different solutions have been proposed, like orthogonal codes, convolutional codes, and short block codes.

The crucial advantage of concatenated codes is that they lend themselves to a staged decoding strategy which breaks the burden of decoding the overall $(mn_i, k_o)$ code into a cascade of two simpler $(n_i, k_i)$, $(n_o, k_o)$ codes. In other words, the receiver first decodes the inner code, and then the outer code. There are several ways of performing this cascaded decoding, which depend on the nature of the demodulator and outer decoder operations. The demodulator may output either hard or soft decisions, and the inner decoder, in turn, may provide to the outer decoder hard or soft estimates of the outer code symbols. Forney (1966) has shown that the best way to operate the cascaded decoding requires for the inner decoder to estimate the a posteriori probabilities of the outer code symbols given the received channel sequence. This optimum procedure requires the soft operations of the demodulator and the inner decoder.

In the simpler, suboptimal strategy where the inner decoder produces only hard symbol decisions, the outer decoder sees an equivalent channel formed by the cascade of inner encoder, modulator, channel, demodulator, and inner hard decoder. This equivalent channel is characterized by a symbol error probability $p_s$ that depends on the signal-to-noise ratio over the physical channel, on the modulation scheme, and on the error correction capability of the inner code. For an $(n_o, k_o, t)$ RS outer code using $K$-bit symbols, and assuming a symbol error probability $p_s$ at the RS decoder input, we can estimate the bit error probability as

$$P_b(e) \leq \frac{2^{K-1}}{2^K - 1} \sum_{j=t+1}^{n_o} \frac{j+t}{n_o} \binom{n_o}{j} p_s^j (1 - p_s)^{n_o - j} \qquad (11.56)$$

The result (11.56) has been obtained by starting from the bound (10.67) on the word error probability, and making the pessimistic assumption that a pattern of $j$ channel errors $(j > t)$ will cause the decoded word to differ from the correct one in $j + t$ positions, so that a fraction $(j + t)/n_o$ of the $k_o$ information symbols is decoded erroneously. The factor $[2^{K-1}/(2^K - 1)]$ accounts for the average number of information bit errors per symbol error.

### 11.2.1.  Reed-Solomon codes and orthogonal modulation

An interesting choice for the inner code is that of representing each $K$-digit symbol of the Reed-Solomon code with one of the $2^K$ signals of an orthogonal signal set. The decoder complexity grows with $K$, since a separate correlator is required for each of the $2^K$ signals of the inner code. Therefore, only small values of $K$ are of practical interest. Furthermore, as we know from Chapter 4, a large bandwidth is required to transmit the orthogonal signal set. These codes

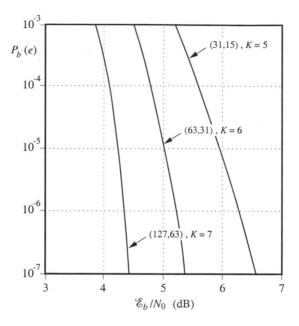

Figure 11.28: *Bit error probability for a concatenated code. The inner code is a set of $2^K$ orthogonal signals with noncoherent demodulation. The outer codes are $(n_o, k_o)$ RS codes with $R_c \simeq 1/2$ and symbol size K. (Clark and Cain, 1981.)*

have been used in deep-space communication systems. The symbol error probability depends on the mo-demodulation scheme, and can be computed using the techniques of Chapter 4. Then, having $p_s$, we can apply (11.56) to evaluate the bit error probability. As an example, we plot in Fig. 11.28 the results concerning the use of orthogonal signals with noncoherent demodulation and RS outer codes with $R_c^o \simeq 1/2$. There is an available coding gain of the order of 9 to 10 dB (at $P_b(e) = 10^{-5}$) with respect to binary orthogonal signaling and noncoherent demodulation.

### 11.2.2. Reed-Solomon and convolutional codes

Perhaps the most used concatenated code configuration employs as inner code a convolutional code. Since the inner code operates at moderately high error probabilities, short constraint length convolutional codes are usually chosen with soft-decision Viterbi decoding.

Since the errors at the Viterbi decoder are bursty, they tend to present to the RS decoder correlated symbol errors, a situation that degrades the performance

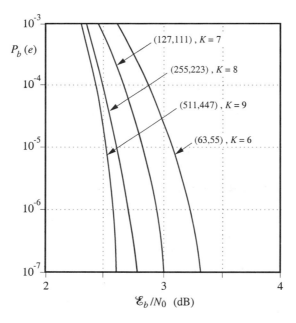

Figure 11.29: *Bit error probability for a concatenated code formed by an inner concatenated code of rate 1/2 and constraint length 7 and an outer Reed-Solomon code. (Clark and Cain, 1981.)*

of the RS decoder. The degree of correlation depends on the constraint length of the convolutional code, and extends up to 6 times the constraint length. To destroy the correlation we can use an interleaver (see Section 10.2.10) between the outer and inner encoders (and correspondingly at the receiver side). The action of the interleaver is such that no two symbols occurring within a decoding depth at the Viterbi decoder output belong to the same RS code word. It should be noted also that (11.56) requires random symbol errors at the RS decoder input, and cannot be applied in the presence of bursts.

In Fig. 11.29 we present results assuming an ideal interleaving, so that (11.56) can be used. The symbol error probability has been estimated by simulation under the assumption of a binary PSK modulation and an AWGN channel. The inner code is a rate 1/2, constraint length 7 convolutional code. The Reed-Solomon code uses symbols of $K$ bits, $K$ ranging from 6 to 9. For each value of $n_o$, the code selected is that with the best performance. Comparing these results with Fig. 10.3, we see that coding gains up to 7 dB can be obtained at $P_b(e) = 10^{-5}$. Moreover, the curves are so steep that a moderate increase in $\mathcal{E}_b/N_0$ results in a

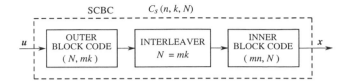

Figure 11.30: *Block diagram of a serially concatenated block code.*

significant decrease of the bit error probability.

## 11.3.    Concatenated codes with interleaver

In a paper by Berrou *et al.* (1993), simulation results concerning a new coding scheme called *turbo code* showed performance close by 0.5 dB to the Shannon capacity limit, at a bit error probability of $10^{-5}$. This astonishing performance gave rise to a large interest in the coding community, and generated several results in a subject that we call *concatenated codes with interleavers*. In this section, we will present these high-performance coding schemes, show how to analyze their average performance, give some design guidelines and, finally, introduce a simple iterative decoding technique.

The main ingredients of a concatenated code with interleaver are two *constituent* codes (CCs) and one interleaver.[8] They can be connected in series, like in Fig. 11.30, or in parallel, as in Fig. 11.31.

In Fig. 11.30 we show the example of a serially concatenated block code (SCBC), composed by two linear cascaded CCs, the *outer* $(N, mk)$ code $\mathcal{C}_o$ with rate $R_c^o = mk/N$ and the *inner* $(mn, N)$ code $\mathcal{C}_i$ with rate $R_c^i = N/(mn)$, linked by an interleaver of size $N$, an integer multiple of the length of code words of the outer code, $N = mp$. The overall SCBC is then a linear $(mn, mk, N)$ code, denoted by $\mathcal{C}_s$, with rate $R_c^s = R_c^o R_c^i = k/n$.

Parallel concatenated codes are obtained as in Fig. 11.31, which refers to the case of a parallel concatenated block code (PCBC). Two linear block codes $\mathcal{C}_1$ with parameters $(n_1, k)$ and rate $R_c^{(1)} = k/n_1$, and $\mathcal{C}_2$ with parameters $(n_2, k)$ and rate $R_c^{(2)} = k/n_2$, the constituent codes, having in common the lenght $k$ of the input information words, are linked through an interleaver of size $N = mk$. The block of $m$ input words to the second encoder is a permuted version of the corresponding input block of the first one. The PCBC code word is formed by concatenating the two code words generated by the first and second encoder. The PCBC, that we denote as $\mathcal{C}_p$, is then an $(n_1 + n_2, k, N)$ linear systematic code

---

[8]We can have more than two CCs and one interleaver, but we will consider only this case for simplicity.

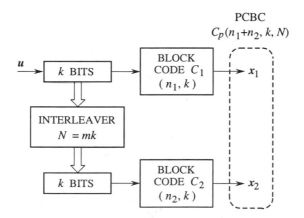

Figure 11.31: *Block diagram of a parallel concatenated block code.*

with rate

$$R_c^p = \frac{R_c^{(1)} R_c^{(2)}}{R_c^{(1)} + R_c^{(2)}} \tag{11.57}$$

Instead of using block codes as CCs, we could employ convolutional codes as well. Indeed, turbo codes are parallel concatenated convolutional codes (PCCC). The nickname "turbo" codes, refers not to the encoder, but to the decoding procedure, a clever suboptimal technique that iterates the individual soft decoding operation on each CC decoder several times, improving the performance at each iteration.

To clarify the behavior of these coding schemes, we will first derive an upper bound to the *average* maximum-likelihood performance, then show some design rules, and, finally, explain the iterative decoding procedure. Part of the analytical details, indeed very heavy, will be avoided; the interested reader can find them in the papers referenced at the end of the chapter, in the bibliographical notes.

### 11.3.1. Performance analysis

Although the case of concatenated convolutional codes is more interesting for the applications, we will first present the analysis for the simpler case of linear block concatenated codes, i.e., SCBC and PCBC, and then outline the extension to convolutional concatenated codes. When the CCs are linear, the resulting serial or parallel concatenated code also is linear (see Problem 11.11), so that the uniform error property (see Section 10.3) applies, and we can assume the transmitted code word to be the all-zero word. In the following, we will limit ourselves to the evaluation of the bit error probability.

**Serially concatenated block codes with interleaver**

To obtain an upper bound to the bit error probability of the SCBC of Fig. 11.30, we capitalize on the result expressed in (10.98), here reported with a slight change in notations

$$P_b(e) \leq \sum_{w=1}^{k} \frac{w}{k} \left[ B_w^{\mathcal{C}_s}(D) \right]_{D=e^{-R_c \mathcal{E}_b/N_0}} \tag{11.58}$$

where $B_w^{\mathcal{C}_s}(D)$ is the conditional weight enumerating function (CWEF) of the SCBC, i.e., the weight enumerating function of the code words of the SCBC generated by information words of weight $w$.

Alternatively, we can proceed as in Section 11.1.5, starting from the definition of the CWEF

$$B_w^{\mathcal{C}_s}(D) = \sum_{d=0}^{n} B_{w,d}^{\mathcal{C}_s} D^d \tag{11.59}$$

and computing the tighter upper bound

$$P_b(e) \leq \frac{1}{2k} \sum_{d=1}^{n} A_d^{(b)} \mathrm{erfc} \left( \sqrt{\frac{d R_c^s \mathcal{E}_b}{N_0}} \right) \tag{11.60}$$

where $A_d^{(b)}$ is the bit multiplicity defined in (11.37) as

$$A_d^{(b)} \triangleq \sum_{w=1}^{k} w B_{w,d}^{\mathcal{C}_s} \tag{11.61}$$

The union bounds (11.58) and (11.60) show that, in order to upper bound the bit error probability of the SCBC, we need to compute its CWEFs. To do this, we should take each data word of weight $w$, encode it by the outer encoder, pass the code word so obtained through the interleaver, and then encode the interleaved code word by the inner encoder. Now, the code word of the inner encoder will depend not only on the weight $w$ of the data word to the outer encoder, but also on the permutation induced by the interleaver. For large interleaver sizes $N$, this exhaustive procedure is exceedingly complicated.

To overcome this difficulty, we introduce the notion of *uniform interleaver*, defined as follows. A uniform interleaver of size $N$ is a probabilistic device that maps a given input word of weight $w$ into all distinct $\binom{N}{w}$ permutations of it with equal probability $1/\binom{N}{w}$. The uniform interleaver transforms a code word of weight $w$ at the output of the outer encoder into all its distinct $\binom{N}{w}$ permutations. As a consequence, all code words of the outer code of weight $j$, through the action of the uniform interleaver, enter the inner encoder generating the *same* set of $\binom{N}{j}$ code words of the inner code.

Thus, assuming for simplicity that the interleaver size equals the length of the outer code words, i.e., $N = k$, the number $B_{w,d}^{\mathcal{C}_s}$ of code words of the SCBC of weight $d$ associated with an input word of weight $w$ is given (with obvious notations) by

$$B_{w,d}^{\mathcal{C}_s} = \sum_{j=0}^{N} \frac{B_{w,j}^{\mathcal{C}_o} \times B_{j,d}^{\mathcal{C}_i}}{\binom{N}{j}} \tag{11.62}$$

From (11.62), we easily derive the expressions of the conditional weight enumerating function of the SCBC

$$B_w^{\mathcal{C}_s}(D) = \sum_{j=0}^{N} \frac{B_{w,j}^{\mathcal{C}_o} \times B_j^{\mathcal{C}_i}(D)}{\binom{N}{j}} \tag{11.63}$$

from which we see that, to obtain the enumerators of the SCBC, we only need to know those of the two constituent codes.

**Example 11.11**   Consider the $(7, 3, 4)$ SCBC code obtained by concatenating the $(4, 3)$ parity-check code with a $(7, 4)$ Hamming code through an interleaver of size $N = 4$. The CWEFs of the outer code are

$$B_0^{\mathcal{C}_o}(D) = 1, \quad B_1^{\mathcal{C}_o}(D) = 3D^2, \quad B_2^{\mathcal{C}_o}(D) = 3D^2, \quad B_3^{\mathcal{C}_o}(D) = D^4$$

so that the only coefficients $B_{w,j}$ different from zero are $B_{0,0} = 1$, $B_{1,2} = 3$, $B_{2,2} = 3$, and $B_{3,4} = 1$. For the inner code, recalling Eq. 10.94, we have

$$\begin{aligned} B_0^{\mathcal{C}_i}(D) &= 1, \quad B_1^{\mathcal{C}_i}(D) = 3D^3 + D^4, \quad B_2^{\mathcal{C}_i}(D) = 3(D^3 + D^4) \\ B_3^{\mathcal{C}_i}(D) &= D^3 + 3D^4, \quad B_4^{\mathcal{C}_i}(D) = D^7 \end{aligned}$$

so that, placing these results into (11.63), we obtain

$$B_0^{\mathcal{C}_s}(D) = 1, \; B_1^{\mathcal{C}_s}(D) = 2.25(D^3 + D^4), \quad B_2^{\mathcal{C}_s}(D) = 1.5(D^3 + D^4), \quad B_3^{\mathcal{C}_s}(D) = D^7$$

and the union bound to the bit error probability (11.58) becomes

$$P_b(e) \leq 1.75 \exp\left(-\frac{9\mathcal{E}_b}{7N_0}\right) + 1.75 \exp\left(-\frac{12\mathcal{E}_b}{7N_0}\right) + \exp\left(-\frac{3\mathcal{E}_b}{N_0}\right)$$

$\square$

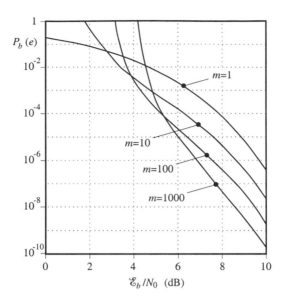

Figure 11.32: *Upper bound to the bit error probability for the SCBC of Example 11.11 using an interleaver of size* $4m$.

Use of the uniform interleaver drastically simplifies the performance evaluation. However, what we obtain is the performance of an *average* SCBC, instead of the *actual* performance of the SCBC with a given interleaver. The significance of the average performance was demonstrated in Benedetto and Montorsi (1996a), where it was proved that, for each value of the signal-to-noise ratio, the performance obtained with the uniform interleaver is achievable by at least one actual interleaver, and that an interleaver chosen at random will behave with high likelihood closely to the average.

Previous result (11.63) can be easily generalized to the case of an interleaver with size $N$ being an integer multiple (by a factor $m$) of the length of the outer code words (see Problem 11.12). As an example, consider again the CCs of Example 11.11, linked by an interleaver of size $N = 4m$. Using the upper bound (11.60) we have obtained the bit error probability curves plotted in Fig. 11.32 for various values of the integer $m$. The curves show the *interleaver gain*, defined as the factor by which the bit error probability is decreased with the interleaver size. A careful examination of the curves shows that the error probability seems to decrease regularly with $m$ as $m^{-1}$. We will explain this behavior later.

**Parallel concatenated block codes with interleaver**

As for the case of SCBCs, to obtain an upper bound to the bit error probability of the PCBC shown in Fig. 11.31, we will use (11.58), here rewritten with the replacement of $C_s$ with $C_p$ to denote that we are dealing with parallel (instead of serial) concatenated codes

$$P_b(e) \leq \sum_{w=1}^{k} \frac{w}{k} \left[ B_w^{C_p}(D) \right]_{D=e^{-R_c \mathcal{E}_b/N_0}} \tag{11.64}$$

where $B_w^{C_p}(D)$ is the conditional weight enumerating function (CWEF) of the PCBC, i.e., the weight enumerating function of the code words of the PCBC generated by information words of weight $w$.

Also in this case we need to compute the CWEF $B_w^{C_p}(D)$. Using, as for SCBCs, a uniform interleaver, a given word of weight $w$ at the input of the interleaver is mapped into all its permutations, which are then encoded by the code $C_2$. As a consequence, all data words of the same weight generate the same set of code words of $C_2$, so that the CWEFs of $C_1$ and $C_2$ become independent, and can be multiplied and suitably normalized to yield the CWEF of the PCBC. In formulas, we have, for $N = k$

$$B_w^{C_p}(D) = \frac{B_w^{C_1}(D) \cdot B_w^{C_2}(D)}{\binom{N}{w}} \tag{11.65}$$

Knowing $B_w^{C_p}(D)$, we can apply the union bound (11.64).

**Example 11.12**   Consider a (10,4,4) PCBC obtained through the use of a (7,4) systematic Hamming code $C_1$ and a (4,3) code $C_2$ whose code words are the three parity-check bits of the systematic (7,4) Hamming code. The CWEFs of the (7,4) Hamming code have already been obtained in Example 11.11. As for those of $C_2$, they can be easily derived, yielding

$$\begin{aligned} B_0^{C_2}(D) &= 1, \quad B_1^{C_2}(D) = 3D^2 + D^3, \quad B_2^{C_2}(D) = 3(D + D^2) \\ B_3^{C_2}(D) &= 1 + 3D \\ B_4^{C_2}(D) &= D^3 \end{aligned}$$

so that, applying (11.65), we obtain the CWEF of the SCBC as

$$\begin{aligned} B_0^{C_p}(D) &= 1 \\ B_1^{C_p}(D) &= 2.25D^5 + 1.5D^6 + 0.25D^7 \\ B_2^{C_p}(D) &= 1.5D^4 + 3D^5 + 1.5D^6 \\ B_3^{C_p}(D) &= 0.25D^3 + 1.5D^4 + 2.25D^5 \\ B_4^{C_p}(D) &= D^{10} \end{aligned}$$

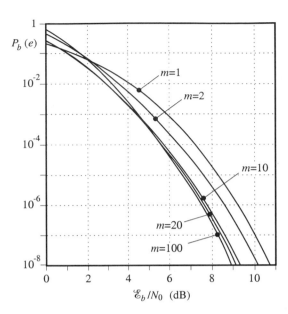

Figure 11.33: *Upper bound to the bit error probability for the PCBC of Example 11.12 using an interleaver of size 4m.*

We then apply the union bound (11.64) to obtain

$$
\begin{aligned}
P_b(e) \;\leq\; & 0.25 \,\exp(-\frac{6\mathcal{E}_b}{5N_0}) + 3\,\exp(-\frac{8\mathcal{E}_b}{5N_0}) + 7.5\exp(-2\frac{\mathcal{E}_b}{N_0}) \\
& +3\,\exp(-\frac{12\mathcal{E}_b}{5N_0}) + 0.25\,\exp(-\frac{14\mathcal{E}_b}{5N_0}) + \exp(-4\frac{\mathcal{E}_b}{N_0})
\end{aligned}
$$

□

Previous result (11.65) can be easily generalized to the case of an interleaver with size $N$ being an integer multiple (by a factor $m$) of the length of the input words (see Problem 11.13). As an example, consider again the PCBC of Example 11.12, and assume an interleaver of size $N = 4m$. We have obtained the bit error probability curves plotted in Fig. 11.33 for various values of the integer $m$. The curves show a gain of roughly 1 dB as $m$ increases. Unlike the case of SCCC, however, the interleaver gain tends to saturate for large $m$. This behavior will be explained later.

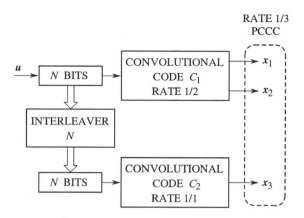

Figure 11.34: *Block diagram of a rate 1/3 parallel concatenated convolutional code.*

### Extension to parallel concatenated convolutional codes

We will now extend the previous analysis starting from the case of parallel concatenated convolutional codes (PCCCs), which were the first to be proposed under the name of turbo codes. The block diagram of a PCCC is shown in Fig. 11.34. This figure represents the case of a rate 1/3 PCCC obtained from the concatenation of a rate 1/2 CC ($\mathcal{C}_1$) with a rate 1/1 CC ($\mathcal{C}_2$). As an example, this rate 1/1 code could be obtained using the same rate 1/2 systematic encoder that generates $\mathcal{C}_1$ and transmitting only the parity-check bit.

We will break the performance analysis of PCCCs into two steps. The first is an exact analysis yielding an upper bound to the bit error probability. The second shows how to obtain an accurate approximation that drastically reduces the computational complexity.

### Exact analysis

Consider a PCCC formed by an interleaver of size $N$ and two convolutional CCs $\mathcal{C}_1$ and $\mathcal{C}_2$ whose trellises have $m_1$ and $m_2$ states, respectively. To examine the whole dynamic of the PCCC, we must consider a super-trellis with $m_1 \cdot m_2$ states, like the one depicted in Fig. 11.35.

The state $S_{ij}$ of the super-trellis corresponds to the pair of states $s_{1i}$ and $s_{2j}$ for the first and second CCs, respectively. Each branch $S_{ij} \rightarrow S_{mn}$ in the super-trellis represents all pairs of paths which start from the pair of states $s_{1i}$, $s_{2j}$ and reach the pair $s_{1m}$, $s_{2n}$ in $N$ steps (see Fig. 11.36).

Thus, when embedded in a PCCC using an interleaver of size $N$, the CC's contributions to the final code word (or code sequence) derive from $N$-truncated versions of their input information sequences, or, equivalently, from trellis paths

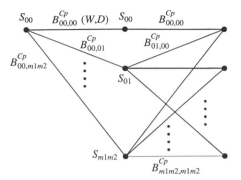

Figure 11.35: *Super-trellis of a PCCC.*

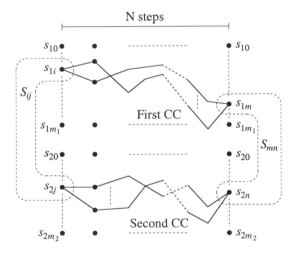

Figure 11.36: *Associations of states and paths of the trellises of constituent codes to states and paths of the PCCC super-trellis.*

of size $N$.

Let $B^{\mathcal{C}_p}_{ij,mn}(W, D)$ be the label of the branch $S_{ij} \to S_{ml}$ of the super-trellis. It represents the *input-output weight enumerating function* (IOWEF), already defined in Section 10.3.3, of the equivalent parallel concatenated block code obtained by enumerating the weights of all $N$-truncated sequences of the PCCC joining the super-states $S_{ij}$ and $S_{mn}$.

The IOWEF of an $(n, k)$ code $\mathcal{C}$ has been defined as

$$B^{\mathcal{C}}(W, D) \triangleq \sum_{l=0}^{k} \sum_{j=0}^{n} B_{w,d}^{\mathcal{C}} W^w D^d \qquad (11.66)$$

where $B_{w,d}^{\mathcal{C}}$ is the number of code words with weight $d$ generated by data words of weight $w$. As a function of the CWEF $B_w^{\mathcal{C}}(D)$ previously introduced in (11.59), the IOWEF can be written as

$$B^{\mathcal{C}}(W, D) = \sum_{w=0}^{k} W^w B_w^{\mathcal{C}}(D) \qquad (11.67)$$

Once we know the labels $B_{ij,ml}^{\mathcal{C}_p}(W, D)$, the performance of the PCCC can be obtained by applying to the super-trellis the standard transfer function approach described in Section 11.1.5 for convolutional codes.

To derive the branch labels of the super-trellis, we can use the same procedure applied previously to parallel concatenated block codes, as we have seen that each label is indeed the IOWEF of a particular equivalent block code[9] with information word size equal to $N$.

We start with the CWEFs $B_{sn,w}^{\mathcal{C}_k}(D)$ of the equivalent block codes associated to the CCs. These functions enumerate the weights of all possible paths generated by data words of weight $w$ and connecting the state $s$ with the state $n$ in $N$ steps for the $k$th constituent encoder, $k = 1, 2$. They can be obtained from the weight enumerating functions of the CCs (see Benedetto and Montorsi, 1996a). From their knowledge, we compute the CWEF $B_{ij,mn,w}^{\mathcal{C}_p}(D)$ of the equivalent block parallel concatenated codes. Owing to the properties of the uniform interleaver, they are simply the normalized product of $B_{im,w}^{\mathcal{C}_1}(D)$ with $D_{jn,w}^{\mathcal{C}_2}(D)$, i.e.

$$B_{ij,mn,w}^{\mathcal{C}_p}(D) = \frac{B_{im,w}^{\mathcal{C}_1}(D) \cdot B_{jn,w}^{\mathcal{C}_2}(D)}{\binom{N}{w}} \qquad (11.68)$$

Successively, we obtain the IOWEF from the corresponding conditional weight enumerating functions through (11.67) and, finally, use the standard transfer function approach to get an upper bound to the bit error probability. An example will clarify the whole procedure.

**Example 11.13**    Consider the rate 1/3 PCCC obtained by concatenating a 2-state rate 1/2 systematic recursive encoder with a 2-state rate 1/1 encoder, obtained from the previous one by neglecting the systematic bit, through an interleaver of size $N = 4$. The resulting encoder structure is depicted in Fig. 11.37. First, we derive the 4 CWEFs

---

[9]Actually, only the label $B_{00,00}^{\mathcal{C}_p}(W, D)$ describes a linear code containing the all "0" word; the other labels refer to cosets of this code. This has no effect on the analysis.

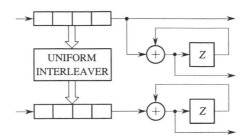

Figure 11.37: *Parallel concatenated convolutional encoder of Example 11.13.*

| | | $B_{ij,w}^{C_1}(D)$ | | | |
|---|---|---|---|---|---|
| $ij$ | $w = 0$ | $w = 1$ | $w = 2$ | $w = 3$ | $w = 4$ |
| 00 | 1 | | $3D^3 + 2D^4 + D^5$ | | $D^6$ |
| 01 | | $D^2 + D^3 + D^4 + D^5$ | | $2D^5 + 2D^6$ | |
| 10 | | $D + D^2 + D^3 + D^4$ | | $2D^4 + 2D^5$ | |
| 11 | $D^4$ | | $D^3 + 2D^4 + 3D^5$ | | $D^6$ |

Table 11.9: *Conditional weight enumerating functions enumerating all paths connecting state $s_i$ with state $s_j$ in 4 steps for the constituent encoder $C_1$ of Example 11.13.*

$B_{ij,w}^{C_1}(D)$ that enumerate all possible paths connecting in 4 steps the state $s_i$ with the state $s_j$ of the CC $C_1$. Then, we evaluate the CWEFs $B_{ij,w}^{C_2}(D)$ for the CC $C_2$. The results for $B_{ij,w}^{C_1}(D)$ are summarized in Table 11.9. Previous results can be used to construct, through (11.68), the CWEFs $B_{ij,mn,w}^{C_P}(D)$ reported in Table 11.10, and, through (11.67), the labeling IOWEFs $B_{ij,mn}^{C_P}(W, D)$ of the super-trellis. From now on, the technique that leads to the computation of the performance of the PCCC is the same as for a standard time-invariant convolutional encoder.                                                □

### An accurate approximation

In the previous example, as the encoder had a very simple structure and the interleaving size was only 4, an analytic approach could be used leading to the exact expression of the average performance of the scheme.

In general, for long interleavers and codes with larger constraint lengths, the super-trellis is completely connected, so that the number of branches increases with the fourth power of the number of states (supposed to be equal) of the CCs. Thus, although the complexity of the analysis is only related to CCs and not to the interleaver size, it may become very heavy. Our experience shows that this is

$$B^{\mathcal{C}_P}_{ij,mn,w}(D)$$

| $ij,mn$ | $w=0$ | $w=1$ | $w=2$ | $w=3$ | $w=4$ |
|---|---|---|---|---|---|
| 00, 00 | 1 | | $\frac{9D^4+12D^5+10D^6+4D^7+D^8}{6}$ | | $D^8$ |
| 00, 01 | | | | | |
| 00, 10 | | | | | |
| 00, 11 | | $\frac{D^3+2D^4+3D^5+4D^6+3D^7+2D^8+D^9}{4}$ | | $\frac{4D^7+8D^8+4D^9}{4}$ | |
| 01, 00 | $D^4$ | | | | |
| 01, 01 | | | $\frac{3D^4+8D^5+14D^6+8D^7+3D^8}{6}$ | | $D^8$ |
| 01, 10 | | $\frac{D^2+2D^3+3D^4+4D^5+3D^6+2D^7+D^8}{4}$ | | $\frac{4D^6+8D^7+4D^8}{4}$ | |
| 01, 11 | | | | | |
| 10, 00 | | | | | |
| 10, 01 | | | $= B^{\mathcal{C}_P}_{01,10}(W,D)$ | | |
| 10, 10 | | | $= B^{\mathcal{C}_P}_{01,01}(W,D)$ | | |
| 10, 11 | | | | | |
| 11, 00 | | $\frac{D+2D^2+3D^3+4D^4+3D^5+2D^6+D^7}{4}$ | | $\frac{4D^5+8D^6+4D^7}{4}$ | |
| 11, 01 | | | | | |
| 11, 10 | | | | | |
| 11, 11 | $D^8$ | | $\frac{D^4+4D^5+10D^6+12D^7+9D^8}{6}$ | | $D^8$ |

Table 11.10: *Conditional weight enumerating functions of the PCCC of Example 11.13.*

the case for CCs with more than 8 states.

To overcome this difficulty, we propose a much simpler analysis. It is based on approximating the complete transfer function of the super-trellis with the IOWEF $B^{\mathcal{C}_P}_{00,00}(W,D)$ that labels the branch joining the zero states of the super-trellis. It describes all pairs of paths which diverge from the zero states of both CCs and remerge into the zero states after $N$ steps.

To check the accuracy of the approximation, we have used the exact and approximate analyses to estimate the performance of the PCCC of Example 11.13 with different interleaver sizes, namely $N = 2, 10, 1000$. The results are reported in Fig. 11.38. For $N = 2$, the approximate and exact curves are significantly different above $10^{-4}$. They merge around $10^{-2}$ for $N = 10$, and are completely indistinguishable for $N = 1000$. Actually, this behavior appears from $N = 20$.

In general, the approximate method gives accurate results when the interleaver size is significantly larger (say 10 times) than the CC memory. For this reason, since the results which follow refer to this situation, we will always use the approximate analysis.

As an example, consider a rate 1/3 PCCC employing as constituent codes the same rate 1/2, 4-state recursive systematic convolutional code with generators

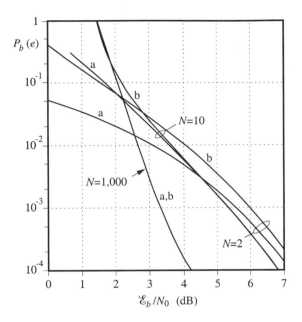

Figure 11.38: *Comparison between exact and approximate techniques to evaluate the performance of a PCCC. The results refer to the 2-state PCCC of Example 11.13, with N = 2, 10, 1000. Curves labeled "a" refer to the approximation, and curves labeled "b" to the exact analysis of the super-trellis.*

$g_1(Z) = 1$, $g_2(Z) = (1 + Z^2)/(1 + Z + Z^2)$ and free distance 5. We have constructed different PCCCs through interleavers of various sizes, and evaluated their performance based on the previously described procedure. The results are reported in Fig. 11.39. Gains beyond 4 dB are achievable. The curves show a decrease as $N^{-1}$ of the bit error probability, due to the interleaver gain.

### Serially concatenated convolutional codes

The block diagram of a serially concatenated convolutional code (SCCC) is shown in Fig. 11.40. It refers to a rate $k/n$ code, obtained by cascading an outer code $\mathcal{C}_o$ with rate $k/p$ and an inner code $\mathcal{C}_i$ with rate $p/n$ through an interleaver of size $N$.

The analysis developed for PCCCs can be easily extended to SCCCs. As an example, consider an SCCC employing as outer code a 4-state, rate 1/2 nonrecursive convolutional encoder, as inner code a 4-state, rate 2/3, recursive convolutional encoder. In Fig. 11.41 we present the upper bounds to the bit error

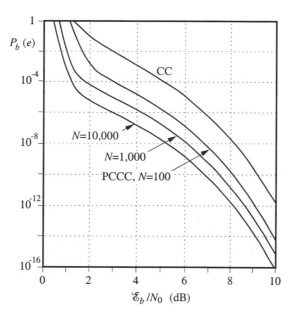

Figure 11.39: *Average upper bounds to the bit error probability for a rate 1/3 PCCC obtained by concatenating two rate 1/2, 4-state CCs, through a uniform interleaver of sizes $N = 100, 1,000, 10,000$.*

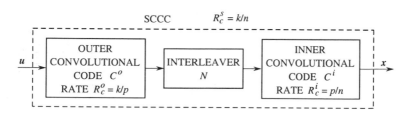

Figure 11.40: *Block diagram of a rate $k/n$ serially concatenated convolutional code.*

probability for different sizes of the uniform interleaver. From the figure, we notice a larger interleaver gain than for PCCCs. In fact, the bit error probability seems to decrease as $N^{-3}$, instead of as $N^{-1}$. This behavior will be clarified in the next section.

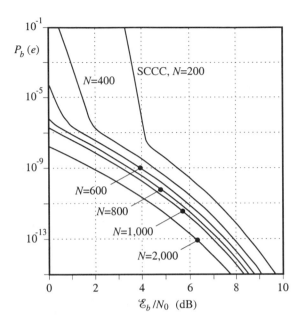

Figure 11.41: *Average upper bound to the bit error probability for a rate 1/3 SCCC using as outer code a 4-state, rate 1/2, nonrecursive convolutional encoder, and as inner code a 4-state, rate 2/3, recursive convolutional encoder with uniform interleavers of various sizes.*

### 11.3.2. Design of concatenated codes with interleaver

The ingredients of parallel and serially concatenated codes with interleavers are two constituent codes and an interleaver with size $N$. For medium-to-large interleavers, the state complexity of the overall code becomes so large as to prevent exhaustive searches for good codes. The only way to get some insight into the design of PCCCs and SCCCs is to split it into the separate designs of the CCs and of the interleaver. The tool to decouple the two problems is still the uniform interleaver defined in Section 11.3.1., which permits identification of the most important performance parameters of the CCs and consequent design of CCs based on a simplified computer search. Using this approach, one first designs the CCs as the optimum ones for the average interleaver, and then, having fixed the CCs, chooses a suitable interleaver.

We have seen that the performance of PCCCs and SCCCs using an interleaver of size $N$ depends on equivalent block codes obtained by considering the $N$-long paths of the CC's trellises starting from and ending at the zero state. The code

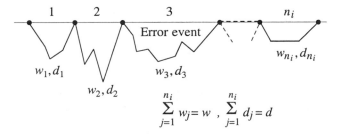

$$\sum_{j=1}^{n_i} w_j = w \ , \ \sum_{j=1}^{n_i} d_j = d$$

Figure 11.42: *The meaning of coefficients* $B^{C_i}_{w,d,n_i}$.

words of the equivalent block code are then concatenations of error events of the CCs (we recall that by error event of a convolutional code we mean a trellis path starting from and ending into the zero state).

Let

$$B^{C_i}(w, D, n_i) = \sum_d B^{C_i}_{w,d,n_i} D^d \tag{11.69}$$

be the weight enumerating function of the code words of the block code equivalent to the $i$th CC formed by the concatenation of $n_i$ error events of the CC, where $B_{w,d,n_i}$ is the number of code words with weight $d$, generated by data words of weight $w$, and number of concatenated error events $n_i$ (see Fig. 11.42 for the meaning of the coefficients $B^{(C_i)}_{w,d,n_i}$). For $N$ large, the CWEF of the equivalent block code can be approximated by

$$B^{C_i}_w(D) \simeq \sum_{n_i=1}^{n_{iM}} \binom{N}{n_i} B^{C_i}(w, D, n_i) \tag{11.70}$$

where $n_{iM}$ is the largest number of error events generated by a weight $w$ information sequence in the $i$th code. In the following, we will use (11.70) to obtain approximate expressions of the bit error probability bounds for PCCCs and SC-CCs valid for large $N$.

**The case of PCCCs**

Using (11.70), assuming a uniform interleaver, and exploiting (11.65), we obtain the CWEF of the PCCC as

$$B^{C_p}_w(D) \simeq \sum_{n_1=1}^{n_{1M}} \sum_{n_2=1}^{n_{2M}} \frac{\binom{N}{n_1}\binom{N}{n_2}}{\binom{N}{w}} B^{C_1}(w, D, n_1) B^{C_2}(w, D, n_2) \tag{11.71}$$

As we are interested in large interleavers, we will use for the binomial coefficient the asymptotic approximation

$$\binom{N}{n} \simeq \frac{N^n}{n!}$$

Comparing (11.71) with (11.59), and making use of the relationship (11.61) between the bit multiplicities $A_d^{(b)}$ and the coefficients $B_{w,d}^{C_p}$, we obtain the following expression for the $A_d^{(b)}$ as a function of the PCCC parameters

$$A_d^{(b)} \simeq \sum_{w=1}^{N} w! w \sum_{n_1=1}^{n_{1M}} \sum_{n_2=1}^{n_{2M}} \frac{N^{n_1+n_2-w-1}}{n_1! n_2!} \sum_{d_1} B_{w,d_1,n_1}^{C_1} B_{w,d-d_1,n_2}^{C_2} \qquad (11.72)$$

From (11.60), we see that the pairs $(A_d^{(b)}, d)$ yield the upper bound to the bit error probability. Moreover, for large $N$, the performance, for each $d$, will be dominated by the terms in (11.72) having the largest exponent of $N$. Define this largest exponent as

$$\alpha(d) \overset{\triangle}{=} \max_{n_1, n_2, w} \{n_1(d) + n_2(d) - w(d) - 1\} \qquad (11.73)$$

Inserting (11.73) into (11.72), and this into (11.60), and retaining only for each $d$ the terms with the exponent $\alpha(d)$, we can write the upper bound to the bit error probability in the form

$$P_b(e) \simeq \sum_{d=d_{fP}}^{\infty} C_d N^{\alpha(d)} \mathrm{erfc}\left( \sqrt{\frac{d R_c E_b}{N_0}} \right) \qquad (11.74)$$

where $d_{fP}$ is the free distance of the PCCC, and $C_d$ is a coefficient that depends on the CCs but not on the interleaver size $N$. Equation (11.74) highlights the contribution of the interleaver size to the performance of the PCCC. Those weights $d$ for which $\alpha(d)$ is larger will dominate the performance when the interleaver size increases. Moreover, if the largest exponent is negative, we shall have an *interleaver gain*, since the bit error probability will decrease with $N$.

We shall consider now two particular values of $\alpha(d)$, for which we give the expressions without the derivations that can be found in Benedetto *et al.* (1998b).

**The exponent $\alpha(d_{fP})$**   The asymptotic (with the signal-to-noise ratio) performance of the PCCC is dominated by the lowest value of $d$, i.e., its free distance $d_{fP}$. Thus, it is then important to evaluate the exponent of $N$ corresponding to this distance. The result is the following:

$$\alpha(d_{fP}) = 1 - w_{fP} \qquad (11.75)$$

where $w_{fP}$ is the minimum input weight among those generating error events of the two CCs yielding the free distance of the PCCC. Equation (11.75) shows that, in order to obtain an interleaver gain (a fact happening if $\alpha < 0$) for medium-large signal-to-noise ratios, we should satisfy the inequality

$$w_{fP} > 1 \tag{11.76}$$

a condition that is always satisfied by recursive constituent encoders.

The design of the CCs consists in the maximization of the parameter $w_{fP}$. The asymptotic expression of the bit error probability is

$$P_b(e) \simeq C_{d_{fP}} N^{1-w_{fP}} \operatorname{erfc}\left(\sqrt{\frac{d_{fP} R_c E_b}{N_0}}\right) \tag{11.77}$$

where the value of $C_{d_{fP}}$ is easily obtained from (11.72).

**The largest exponent $\alpha_M$**  For very large $N$, the performance of the PCCC is dominated by the term in the summation of (11.74) with the largest exponent of $N$, denoted by $\alpha_M$

$$\alpha_M \stackrel{\triangle}{=} \max_{w,n_1,n_2,d} \{n_1(d) + n_2(d) - w(d) - 1\} \tag{11.78}$$

Recalling that $n_{iM}$ is the maximum number of concatenated error events in a code word of the $i$th equivalent block code, we obtain that, for a given $w$, the largest values of $n_1$ and $n_2$ are given by

$$n_{iM} = \left\lfloor \frac{w}{w_{im}} \right\rfloor, \qquad i = 1, 2 \tag{11.79}$$

where $\lfloor x \rfloor$ means "integer part of $x$," and $w_{im}$ is the minimum weight of the input sequence yielding an error event of the $i$-th CC. It is easy to prove (see Problem 11.14) that $w_{im}$ is equal to 1 for nonrecursive convolutional encoders and to 2 for recursive convolutional encoders. As a consequence, from (11.78), we have $\alpha_M = 0$ when at least one CC is a block or nonrecursive convolutional code, and

$$\alpha_M = 2 \left\lfloor \frac{w}{2} \right\rfloor - w - 1 = -1 \tag{11.80}$$

when the CCs are both recursive.

The design consequence is that both CCs must be recursive convolutional encoders, in which case the bit error probability decreases with $N$ as $N^{-1}$ for $N \to \infty$.

For recursive CCs, the asymptotic (in $N$) expression of the bit error probability is

$$P_b(e) \simeq C_{d_{\mathrm{fP,eff}}} N^{-1} \mathrm{erfc}\left(\sqrt{\frac{d_{\mathrm{fP,eff}} R_c E_b}{N_0}}\right)$$

where we have defined the *effective free distance* $d_{\mathrm{fP,eff}}$ of the PCCC as the minimum weight of the sequences of the PCCC generated by input sequences of weight 2, and where $C_{d_{\mathrm{fP,eff}}}$ can be easily obtained from (11.72).

Since $d_{\mathrm{fP,eff}} = d_{\mathrm{f1,eff}} + d_{\mathrm{f2,eff}}$, the choice of the CCs must be aimed at maximizing their $d_{\mathrm{f,eff}}$.

Tables of CCs with various rates optimized with respect to this criterion can be found in Benedetto *et al.* (1998a).

### The case of SCCCs

Using the same approach previously described for PCCCs, which will not be repeated for conciseness, we obtain the following expression for the $A_d^{(b)}$ as functions of the parameters of the SCCC depicted in Fig. 11.40:

$$A_d^{(b)} \simeq \sum_{w=w_m^o}^{N R_o} \sum_{l=d_f^o}^{N} \sum_{n^o=1}^{n_M^o} \sum_{n^i=1}^{n_M^i} N^{n^o+n^i-l-1} \frac{l!}{p^{n^o+n^i} n^o! n^i!} \frac{w}{k} C_{w,l,n^o}^{C_o} B_{l,h,n^i}^{C_i} \qquad (11.81)$$

where the meaning of symbols is the same as for PCCCs, and where the superscripts "o" and "i" refer to the outer and inner codes, respectively.

For large $N$, and for a given $d$, the dominant coefficient of the exponentials in (11.81) is the one for which the exponent of $N$ is maximum. Define this maximum exponent as

$$\alpha(d) \stackrel{\triangle}{=} \max_{w,\ell}\{n^o + n^i - \ell - 1\} \qquad (11.82)$$

Evaluating $\alpha(d)$ in general is not possible without specifying the CCs. Thus, we will consider two important cases, for which only the final results will be reported. For analytical details, see Benedetto *et al.* (1998b).

### The exponent of $N$ for the minimum weight

For large values of $E_b/N_0$, the performance of the SCCC is dominated by the first term of the summation in (11.60) with respect to $d$. We denote by $d_m$ the minimum weight of SCCC; it may be larger than the free distance $d_f^i$ of the inner code, since the input sequences to the inner encoder are not unconstrained binary sequences, but, rather, code words of the outer code. With simple computations, we obtain

$$\alpha(d_{\mathrm{m}}) = 1 - \ell_{\mathrm{m}}(d_{\mathrm{m}}) \leq 1 - d_f^o . \tag{11.83}$$

where $\ell_{\mathrm{m}}(d_{\mathrm{m}})$ is the minimum weight $\ell$ of code words of the outer code yielding a code word of weight $d_{\mathrm{m}}$ of the inner code, and $d_f^o$ is the free distance of the outer code.

The result (11.83) shows that the exponent of $N$ corresponding to the minimum-weight of SCCC code words is always negative for $d_f^o \geq 2$, thus yielding an interleaver gain at high $\mathcal{E}_b/N_0$. Substitution of the exponent $\alpha(d_{\mathrm{m}})$ into (11.74) truncated to the first term of the summation with respect to $d$ yields

$$\lim_{E_b/N_0 \to \infty} P_b(e) \tilde{\leq} C_{d_m} N^{1-d_f^o} \mathrm{erfc}\left(\sqrt{\frac{d_m R_c \mathcal{E}_b}{N_0}}\right) \tag{11.84}$$

where the constant $C_{d_m}$ can be derived from (11.81).

Expression (11.84) suggests the following conclusions:

- For the values of $\mathcal{E}_b/N_0$ and $N$ where the SCCC performance is dominated by its free distance $d_f^{Cs} = d_{\mathrm{m}}$, increasing the interleaver size yields a gain in performance.

- To increase the interleaver gain, one should choose an outer code with large free distance $d_f^o$.

- To improve the performance with $E_b/N_0$, one should choose an inner and outer code combination such that $d_{\mathrm{m}}$ is large.

These conclusions do not depend on the structure of the CCs, and thus they yield for both recursive and nonrecursive encoders.

The second important case regards the largest exponent of $N$, defined as

$$\alpha_M \overset{\triangle}{=} \max_d \{\alpha(d)\} = \max_{w,\ell,d} \{n^o + n^i - \ell - 1\} \tag{11.85}$$

This exponent will permit one to find the dominant contribution to the bit error probability for $N \to \infty$.

**The maximum exponent of $N$** We will give separately the results for nonrecursive and recursive inner encoders.

*Block and nonrecursive convolutional inner encoders*

In this case, it is straightforward to obtain for the value of $\alpha_M$

$$\alpha_M = n_M^o - 1 \geq 0$$

so that interleaving gain is not obtained.

*Recursive convolutional inner encoders*

We have already seen (Problem 11.14) that for recursive convolutional encoders the minimum weight of input sequences generating error events is 2. As a consequence, an input sequence of weight $\ell$ can generate at most $\lfloor \ell/2 \rfloor$ error events. Based on this consideration, and after lengthy computations, we obtain

$$\alpha_M = - \left\lfloor \frac{d_f^o + 1}{2} \right\rfloor \tag{11.86}$$

The value (11.86) of $\alpha_M$ shows that the exponents of $N$ in (11.81) are always negative integers. Thus, for all $d$, the coefficients of the exponents in $d$ decrease with $N$, and we always have an interleaver gain. Moreover, $\alpha_M$ can be sensibly lower for SCCCs than for PCCCs, where it was always equal to $-1$. As an example, for $d_f^o = 5$, $\alpha_M = -3$, so that the interleaver gain goes as $N^{-3}$. This was precisely the behavior of the SCCCs whose performances have been presented in Fig. 11.41.

Denoting by $d_{f,\text{eff}}^i$ the minimum weight of code words of the inner code generated by weight-2 input sequences, we obtain a different weight $h(\alpha_M)$ for even and odd values of $d_f^o$.

## $d_f^o$ even

For $d_f^o$ even, the weight $h(\alpha_M)$ associated with the highest exponent of $N$, is given by

$$d(\alpha_M) = \frac{d_f^o d_{f,\text{eff}}^i}{2}$$

since it is the weight of an inner code word that concatenates $d_f^o/2$ error events with weight $d_{f,\text{eff}}$.

Substituting the exponent $\alpha_M$ into (11.74), approximated only by the term of the summation with respect to $d$ corresponding to $d = d(\alpha_M)$, yields

$$\lim_{N \to \infty} P_b(e) \stackrel{\sim}{\leq} C_{\text{even}} N^{-d_f^o/2} \text{erfc} \left( \sqrt{\frac{d_f^o d_{f,\text{eff}}^i}{2} \cdot \frac{R_c \mathcal{E}_b}{N_0}} \right) \tag{11.87}$$

where $C_{even}$ can be derived from (11.81).

**$d_f^o$ odd**

For $d_f^o$ odd, the value of $d(\alpha_M)$ is given by

$$d(\alpha_M) = \frac{(d_f^o - 3)d_{f,\mathrm{eff}}^i}{2} + d_m^{(3)} \tag{11.88}$$

where $d_m^{(3)}$ is the minimum weight of sequences of the inner code generated by a weight 3 input sequence. In this case, in fact, we have

$$n_M^i = \frac{d_f^o - 1}{2}$$

concatenated error events, of which $n_M^i - 1$ are generated by weight 2 input sequences and one is generated by a weight-3 input sequence.

Thus, substituting the exponent $\alpha_M$ into (11.74) approximated by keeping only the term of the summation with respect to $d$ corresponding to $d = d(\alpha_M)$ yields

$$\lim_{N\to\infty} \tilde{P}_b(e) \lesssim C_{\mathrm{odd}} N^{-\frac{d_f^o+1}{2}} \mathrm{erfc}\left(\sqrt{\left(\frac{(d_f^o - 3)d_{f,\mathrm{eff}}^i}{2} + d_m^{(3)}\right) \cdot \frac{R_c \mathcal{E}_b}{N_0}}\right) \tag{11.89}$$

where the constant $C_{\mathrm{odd}}$ can be derived from (11.81).

In both cases of $d_f^o$ even and odd, we can draw from (11.87) and (11.89) a few important design considerations:

- the use of a recursive convolutional inner encoder always yields an interleaver gain. As a consequence, the first design rule states that the inner encoder must be a convolutional recursive encoder.

- The coefficient that multiplies the signal-to-noise ratio $\mathcal{E}_b/N_0$ in (11.89) increases for increasing values of $d_{f,\mathrm{eff}}^i$. Thus, we deduce that the effective free distance of the inner code must be maximized.

**Comparison between parallel and serially concatenated codes**

In this section, we shall use the bit error probability bounds previously derived to compare the performance of parallel and serially concatenated block and convolutional codes.

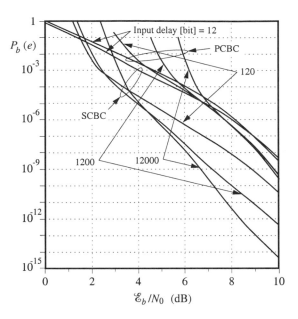

Figure 11.43: *Comparison of an SCBC and PCBC with various interleaver sizes, chosen so as to yield the same input decoding delay for the two codes.*

**Serially and parallel concatenated block codes**   To obtain a fair comparison, we have chosen the following PCBC and SCBC: the PCBC has parameters $(11m, 3m, N)$ and employs two equal $(7,3)$ systematic cyclic codes with generator $g(Z) = (1 + Z)(1 + Z + Z^3)$; the SCBC, instead, is a $(15m, 4m, N)$ SCCC, obtained by the concatenation of the $(7, 4)$ Hamming code with a $(15, 7)$ BCH code.

They have almost the same rate ($R_{\mathcal{C}_S} = 0.266$, $R_{\mathcal{C}_P} = 0.273$), and have been compared by choosing the interleaver size in such a way that the decoding delay due to the interleaver, measured in terms of input information bits, is the same. As an example, to obtain a delay equal to 12 input bits, we must choose an interleaver size $N = 12$ for the PCBC, and $N = 12/R_C^o = 21$ for the SCBC.

The results are reported in Fig. 11.43, where we plot the bit error probability versus the signal-to-noise ratio $\mathcal{E}_b/N_0$ for various input delays. The results show that, for low values of the delay, the performance is almost the same. On the other hand, increasing the delay (and thus the interleaver size $N$), yields a significant interleaver gain for the SCBC, and almost no gain for the PCBC.

**Serially and parallel concatenated convolutional codes**   For a fair comparison, we have chosen the following PCCC and SCCC: the PCCC is a rate 1/3 code obtained by concatenating two equal rate 1/2, 4-state systematic recursive convolutional codes with transfer function matrix in polynomial form

$$G(Z) = \left[1, \frac{1 + Z^2}{1 + Z + Z^2}\right]$$

The SCCC is a rate 1/3 code using as outer code the same rate 1/2, 4-state code as in the PCCC, and, as inner code, a rate 2/3, 4-state systematic recursive convolutional code with generator matrix

$$G(Z) = \begin{bmatrix} 1, & 0, & \dfrac{1 + Z^2}{1 + Z + Z^2} \\[2mm] 0, & ,1, & \dfrac{1 + Z}{1 + Z + Z^2} \end{bmatrix}$$

Also in this case, the interleaver sizes have been chosen so as to yield the same decoding delay, due to the interleaver, in terms of input bits. The results are reported in Fig. 11.44, where we plot the bit error probability versus the signal-to-noise ratio $\mathcal{E}_b/N_0$ for various input delays.

The results show the great difference in the interleaver gain, as anticipated by the discussion made in Section 11.3.2. In particular, the PCCC shows an interleaver gain going as $N^{-1}$, as dictated by (11.80), whereas the interleaver size of the SCCC, as from (11.86), goes as $N^{(d_f^o+1)/2} = N^{-3}$, being the free distance of the outer code equal to 5. This means, for $P_b(e) = 10^{-11}$, a gain of more than 2 dB in favor of the SCCC.

### 11.3.3.   Iterative decoding of concatenated codes with interleavers

Maximum-likelihood decoding algorithms, like Viterbi or MAP algorithms, exhibit a complexity increasing linearly with the number of states of the code trellis. As a consequence, they are not applicable to concatenated codes with interleavers, whose state complexity, although difficult to evaluate exactly (see Benedetto *et al.*, 1997), can be shown to increase exponentially with the interleaver size. In addition to that, the trellis of the overall code is time-varying, and this makes even more difficult the implementation of the Viterbi (or MAP) decoder.

In this section, we will describe a suboptimum iterative decoding algorithm, whose complexity is almost independent from the interleaver size, and increases linearly with the number of states of the CCs. The behavior of the decoding algorithm in terms of convergence conditions (*When does it converge? What*

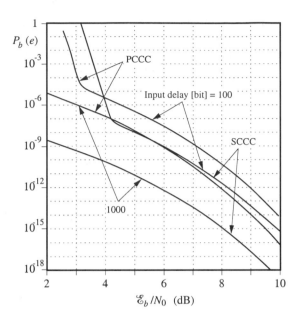

Figure 11.44: *Comparison of an SCCC and PCCC with various interleaver sizes, chosen so as to yield the same input decoding delay for the two codes. The CCs are 4-state convolutional codes.*

*does it converge to?* ) is far from being understood. So far, only explanations based on analogies to similar known algorithms have been provided in the literature (McEliece *et al.*, 1998). However, there is wide simulation evidence that the iterative algorithm works nicely, and yields results very close to the Shannon limits at bit error probabilities in the range $10^{-5}$-$10^{-7}$. Thus, in practice, it can be considered as yielding a very effective trade-off between performance and complexity.

In the following, we will propose a heuristic justification of the iterative algorithm, followed by the input-output relationships of the main decoder block, which implements the evaluation of the a posteriori probabilities (APP) through forward and backward recursions. This APP algorithm is a slight modification and extension of the BCJR algorithm (Bahl *et al.*, 1974) described in Appendix F. Successively, we will present the results obtained by simulating the iterative algorithm applied to a few cases of parallel and serially concatenated codes. The description will be based on PCCCs, and then extended to SCCCs.

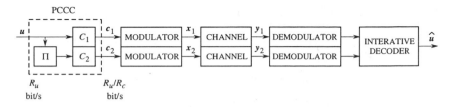

Figure 11.45: *Block diagram of a transmission system employing a parallel concatenated convolutional code.*

### The reference transmission system

The description of the iterative decoding algorithm will be based on the reference transmission system shown in Fig. 11.45, which represents the transmission of sequences encoded by a parallel concatenated convolutional encoder through two parallel channels.[10]

The information sequence **u**, with rate $R_u$ bit/s, is formed by symbols $u$ defined over a finite or infinite index set with cardinality $K$, and drawn from the alphabet

$$U = \{u_1, \ldots, u_{|U|}\}$$

The information symbols are encoded by a parallel concatenated code (PCC) with rate

$$R_c = \frac{R_{c1} R_{c2}}{R_{c1} + R_{c2}}$$

where $R_{c1}$ and $R_{c2}$ are the binary rates of the two constituent codes $C_1$ and $C_2$.

The encoded sequence **c** is formed by concatenating two parallel code sequences $(\mathbf{c}_1, \mathbf{c}_2)$ with rate $R_u/R_{c1}$ and $R_u/R_{c2}$, obtained through the two encoders $C_1$ and $C_2$ that operate on the sequence **u** and on its scrambled replica provided by the interleaver $I$.

The code symbols are drawn from the two alphabets

$$C_i = \{c_{i1}, \ldots, c_{i|C_i|}\} \ , \ i = 1, 2$$

with cardinalities generally different from that of the input alphabet.

The two encoded sequences are then mapped, on a one-to-one basis, into two modulation alphabets producing the signals $\mathbf{x}_1$ and $\mathbf{x}_2$. The modulated signals are transmitted over two parallel memoryless noisy channels, and the received signals $\mathbf{y}_1$ and $\mathbf{y}_2$ are fed to the soft demodulator. Remembering that $K$ is the overall number of information symbols in the whole transmission interval, the number of the transmitted (and received) symbols $x_1$ and $x_2$ will be

---

[10]The same analysis would apply to a system transmitting the two sequences $\mathbf{c}_1$ and $\mathbf{c}_2$ serially.

$N_1 = (K \log_2 |U|)/(R_{c1} \log_2 |C_1|)$, and $N_2 = (K \log_2 |U|)/(R_{c2} \log_2 |C_2|)$, respectively.

The soft demodulator has knowledge of the conditional probability density functions that define the channel

$$p(Y_k \mid X_k) = \mathrm{P}[Y_k = y \mid X_k = x_i] = \mathrm{P}[Y_k = y \mid C_k = c_i], \quad i = 1, 2, \ldots, |C| \tag{11.90}$$

where the subscript $k$ denotes a discrete time. Based on its knowledge of the channel, the soft demodulator provides the values of the conditional probabilities to the decoder for all time instants $k$.

In the following, we shall use the shorthand notation

$$p(y_k \mid c_i) \stackrel{\triangle}{=} p(Y_k = y \mid C_k = c_i)$$

or, equivalently

$$(p(y_{1k} \mid c_i), p(y_{2k} \mid c_i))$$

The task of the decoder is to provide an estimate $\hat{u}$ of the symbols of the transmitted sequence $\mathbf{u}$. In the analysis, we shall assume that the sequence of a priori probability distributions of the information symbols $p_a(\mathbf{u})$ is known. When this probability distribution is not known, it can be assumed to be uniform.

**The iterative decoding algorithm: a heuristic explanation**

The optimal symbol-by-symbol maximum a posteriori (MAP) decision should be based on the maximization of the a posteriori probability (APP) of the information symbols

$$\hat{u}_k \stackrel{\triangle}{=} \arg \max_i [\mathrm{APP}(k, i)] \tag{11.91}$$

where $\arg \max_i [f(i)]$ denotes the argument $i$ of $f(i)$ that maximizes it, and where we have defined the a posteriori probability

$$
\begin{aligned}
\mathrm{APP}(k, i) \stackrel{\triangle}{=} \ & p(u_k = i, \mathbf{y}_1, \mathbf{y}_2) = \\
= \ & \sum_{\mathbf{u}:u_k=i} p(\mathbf{y}_1 \mid c_1(\mathbf{u})) p(\mathbf{y}_2 \mid c_2(\mathbf{u})) p_a(\mathbf{u})
\end{aligned} \tag{11.92}
$$

and

$$p(\mathbf{y}_1 \mid c_1(\mathbf{u})) = \prod_{j=1}^{N_1} p(y_{1j} \mid c_{1j}(\mathbf{u})) \tag{11.93}$$

$$p(\mathbf{y}_2 \mid c_2(\mathbf{u})) = \prod_{m=1}^{N_2} p(y_{2m} \mid c_{2m}(\mathbf{u})) \qquad (11.94)$$

$$p_a(\mathbf{u}) = \prod_{l=1}^{K} p_a(u_l) \qquad (11.95)$$

In order to simplify the mathematical expressions, we will omit in the following the upper limits of the products, limiting ourselves to keep distinct the running variables.

As observed previously, the evaluation of (11.91) for PCCCs has a computational complexity that is linearly proportional to the number of states of the overall code, thus making impossible the evaluation of the true APP when the interleaver size $N$ is large (say, more than 10–15).

Berrou *et al.* (1993, 1996) proposed a technique to iteratively evaluate (11.92) for a PCCC, with a complexity that is almost independent from the interleaver size. Although suboptimal, the iterative decoding procedure has been proved by a large number of simulations to work very well. A heuristic explanation of the iterative algorithm can be based on the *independence* assumption, stating that the probabilities (11.93) and (11.94), which refer to the constituent codes, can be expressed as products of functions defined on the individual information symbols:

$$p(\mathbf{y}_1 \mid c_1(\mathbf{u})) \simeq \prod_l \tilde{P}_{1l}(u_l)$$

$$p(\mathbf{y}_2 \mid c_2(\mathbf{u})) \simeq \prod_l \tilde{P}_{2l}(u_l)$$

Then, after substitution into (11.92), we obtain that the APPs must satisfy the following nonlinear system:

$$\mathrm{APP}(k,i) = \left[ \sum_{\mathbf{u}:u_k=i} p(\mathbf{y}_2 \mid c_2(\mathbf{u})) \prod_{l \neq k} \tilde{P}_{1l}(u_l) p_a(u_l) \right] \times \tilde{P}_{1k}(i) \times p_a(i)$$

$$\mathrm{APP}(k,i) = \left[ \sum_{\mathbf{u}:u_k=i} p(\mathbf{y}_1 \mid c_1(\mathbf{u})) \prod_{l \neq k} \tilde{P}_{2l}(u_l) p_a(u_l) \right] \times \tilde{P}_{2k}(i) \times p_a(i)$$

which admits the following solutions (provided that they exist):

$$\tilde{P}_{1k}(i) = \sum_{\mathbf{u}:u_k=i} p(\mathbf{y}_1 \mid c_1(\mathbf{u})) \prod_{l \neq k} \tilde{P}_{2l}(u_l) p_a(u_l) \qquad (11.96)$$

$$\tilde{P}_{2k}(i) = \sum_{\mathbf{u}:u_k=i} p(\mathbf{y}_2 \mid c_2(\mathbf{u})) \prod_{l \neq k} \tilde{P}_{1l}(u_l) p_a(u_l) \qquad (11.97)$$

Based on these solutions, we can evaluate the APP of the $k$-th information bit as:

$$\text{APP}(k, i) = \tilde{P}_{1k}(i) \times \tilde{P}_{2k}(i) \times p_a(i)$$

The nonlinear system (11.96)–(11.97) is the key point for the iterative evaluation of $\text{APP}(k, i)$. The iterative decoding scheme (Berrou *et al*, 1993, 1996) can indeed be viewed as a way to iteratively solve (11.97) as

$$\tilde{P}_{2k}^{(0)}(i) \;=\; 1, \quad k = 1, \ldots, K$$

$$\vdots$$

$$\tilde{P}_{1k}^{(m)}(i) \;=\; \sum_{\mathbf{u}:u_k=i} p(\mathbf{y}_1 \mid c_1(\mathbf{u})) \prod_{l \neq k} \tilde{P}_{2l}^{(m-1)}(u_l) p_a(u_l), \quad k = 1, \ldots, K$$

$$\tilde{P}_{2k}^{(m)}(i) \;=\; \sum_{\mathbf{u}:u_k=i} p(\mathbf{y}_2 \mid c_2(\mathbf{u})) \prod_{l \neq k} \tilde{P}_{1l}^{(m)}(u_l) p_a(u_l), \quad k = 1, \ldots, K(11.98)$$

### The iterative decoding algorithm employing log-likelihood ratios

For binary ("0" and "1") information and code symbols, we can replace the symbol probabilities with their *log-likelihood ratios* (LLRs). This leads to two important simplifications: first, only one quantity needs to be computed and propagated through the algorithm, leading to considerable savings in computation and memory requirements; second, it permits transformation of the intrinsically multiplicative algorithm into an additive one.

Let us define the following quantities:

$$\lambda_k(\text{APP}) \triangleq \log \frac{\sum_{\mathbf{u}:u_k=0} p(\mathbf{y}_1 \mid c_1(\mathbf{u})) p(\mathbf{y}_2 \mid c_2(\mathbf{u})) p_a(\mathbf{u})}{\sum_{\mathbf{u}:u_k=1} p(\mathbf{y}_1 \mid c_1(\mathbf{u})) p(\mathbf{y}_2 \mid c_2(\mathbf{u})) p_a(\mathbf{u})} \tag{11.99}$$

$$\lambda_k \triangleq \log \frac{p(y_k \mid 0)}{p(y_k \mid 1)}$$

$$\lambda_{1j} \triangleq \log \frac{p(y_{1j} \mid 0)}{p(y_{1j} \mid 1)} \qquad j = 1, \ldots, N_1 \tag{11.100}$$

$$\lambda_{2m} \triangleq \log \frac{p(y_{2m} \mid 0)}{p(y_{2m} \mid 1)} \qquad m = 1, \ldots, N_2 \tag{11.101}$$

$$\lambda_a \triangleq \log \frac{p_a(0)}{p_a(1)} \tag{11.102}$$

$$\pi_{1l} \triangleq \log \frac{\tilde{P}_{1l}(0)}{\tilde{P}_{1l}(1)}$$

$$\pi_{2l} \triangleq \log \frac{\tilde{P}_{2l}(0)}{\tilde{P}_{2l}(1)}$$

Dividing the numerator and denominator of (11.99) by the constant factor $p(\mathbf{y}_1 \mid \mathbf{0})p(\mathbf{y}_2 \mid \mathbf{0})p_a(\mathbf{0})$ does not influence the result. As a consequence, we obtain

$$\lambda_k(\mathrm{APP}) = \log \frac{\sum_{\mathbf{u}:u_k=0} \dfrac{p(\mathbf{y}_1 \mid c_1(\mathbf{u}))}{p(\mathbf{y}_1 \mid \mathbf{0})} \dfrac{p(\mathbf{y}_2 \mid c_2(\mathbf{u}))}{p(\mathbf{y}_2 \mid \mathbf{0})} \dfrac{p_a(\mathbf{u})}{p_a(\mathbf{0})}}{\sum_{\mathbf{u}:u_k=1} \dfrac{p(\mathbf{y}_1 \mid c_1(\mathbf{u}))}{p(\mathbf{y}_1 \mid \mathbf{0})} \dfrac{p(\mathbf{y}_2 \mid c_2(\mathbf{u}))}{p(\mathbf{y}_2 \mid \mathbf{0})} \dfrac{p_a(\mathbf{u})}{p_a(\mathbf{0})}} \qquad (11.103)$$

By definition, we have

$$\frac{p(\mathbf{y}_1 \mid c_1(\mathbf{u}))}{p(\mathbf{y}_1 \mid \mathbf{0})} = \prod_{j=1}^{N_1} \frac{p(y_{1j} \mid c_{1j}(\mathbf{u}))}{p(y_{1j} \mid 0)}$$

and

$$\frac{p(y_{1j} \mid c_{1j}(\mathbf{u}))}{p(y_{1j} \mid 0)} = \left\{ \begin{array}{ll} \exp[-\lambda_{1j}] & \text{if } c_{1j}(\mathbf{u}) = 1 \\ 1 & \text{if } c_{1j}(\mathbf{u}) = 0 \end{array} \right\} = \exp[-c_{1j}(\mathbf{u})\lambda_{1j}]$$

Analogously, we can write

$$\frac{p(y_{2m} \mid c_{2m}(\mathbf{u}))}{p(y_{2m} \mid 0)} = \exp[-c_{2m}(\mathbf{u})\lambda_{2m}]$$

$$\frac{p_a(u_l)}{p_a(0)} = \exp[-u_l\lambda_a]$$

Using then the definitions (11.100), (11.101), (11.102) into (11.103), finally yields

$$\begin{aligned}
\lambda_k(\mathrm{APP}) &= \log \left\{ \sum_{\mathbf{u}:u_k=0} \exp - \left[ \sum_{j=1}^{N_1} c_{1j}(\mathbf{u})\lambda_{1j} + \sum_{m=1}^{N_2} c_{2m}(\mathbf{u})\lambda_{2m} + \sum_{l=1}^{K} u_l\lambda_a \right] \right\} \\
&\quad - \log \left\{ \sum_{\mathbf{u}:u_k=1} \exp - \left[ \sum_{j=1}^{N_1} c_{1j}(\mathbf{u})\lambda_{1j} + \sum_{m=1}^{N_2} c_{2m}(\mathbf{u})\lambda_{2m} + \sum_{l=1}^{K} u_l\lambda_a \right] \right\}
\end{aligned}$$

and the iterative decoding procedure (11.98) can be rewritten as

$$\pi_{2k}^{(0)} = 0$$

$$\vdots$$

$$\pi_{1k}^{(m)} = \log \left\{ \sum_{\mathbf{u}:u_k=0} \exp - \left[ \sum_j c_{1j}(\mathbf{u})\lambda_{1j} + \sum_{l\neq k} u_l(\lambda_a + \pi_{2l}^{(m-1)}) \right] \right\}$$
$$- \log \left\{ \sum_{\mathbf{u}:u_k=1} \exp - \left[ \sum_j c_{1j}(\mathbf{u})\lambda_{1j} + \sum_{l\neq k} u_l(\lambda_a + \pi_{2l}^{(m-1)}) \right] \right\} \quad (11.104)$$

$$\pi_{2k}^{(m)} = \log \left\{ \sum_{\mathbf{u}:u_k=0} \exp - \left[ \sum_m c_{2m}(\mathbf{u})\lambda_{2m} + \sum_{l\neq k} u_l(\lambda_a + \pi_{1l}^{(m)}) \right] \right\}$$
$$- \log \left\{ \sum_{\mathbf{u}:u_k=1} \exp - \left[ \sum_m c_{2m}(\mathbf{u})\lambda_{2m} + \sum_{l\neq k} u_l(\lambda_a + \pi_{1l}^{(m)}) \right] \right\} \quad (11.105)$$

where $k = 1, \ldots, K$, and the final value of the LLR is computed as

$$\lambda_k(\text{APP}) = \pi_{1k} + \pi_{2k} + \lambda_a$$

and the MAP decision is made according to the sign of $\lambda_k$.

The block diagram of the iterative decoding scheme making use of the LLRs is reported in Figure 11.46. In the figure, the blocks labeled by $\Pi$ and $\Pi^{-1}$ represent the interleaver and its inverse.

It remains to see how to perform, at the sequence level, the operations involved in (11.104) and (11.105). They contain implicitly the trellis constraints imposed by the encoders, and can be executed through forward and backward recursions. In Fig. 11.46, these operations are performed by the two blocks SISO 1 (Soft-Input Soft-Output) and SISO 2, whose behavior will be described in the next section.

**Example 11.14** Using a binary antipodal modulation with energy per bit $\mathcal{E}_b$ over an additive white Gaussian noise channel with two-sided noise spectral density $N_0/2$, yields the following probability density functions conditioned on the transmitted bit:

$$p(Y \mid 0) = \frac{1}{\sqrt{\pi N_0}} \exp\left(-\frac{(Y - \sqrt{\mathcal{E}_b})^2}{N_0}\right)$$

$$p(Y \mid 1) = \frac{1}{\sqrt{\pi N_0}} \exp\left(-\frac{(Y + \sqrt{\mathcal{E}_b})^2}{N_0}\right)$$

so that the LLRs become

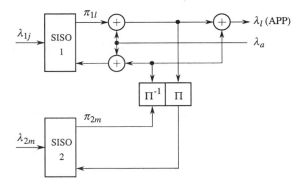

Figure 11.46: *Block diagram of the iterative decoding scheme for binary convolutional codes.*

Figure 11.47: *The trellis encoder*

$$\lambda_k = y_k \frac{4\sqrt{\mathcal{E}_b}}{N_0}$$

$$\lambda_{1k} = y_{1k} \frac{4\sqrt{\mathcal{E}_b}}{N_0}$$

$$\lambda_{2k} = y_{2k} \frac{4\sqrt{\mathcal{E}_b}}{N_0}$$

The computation of the LLR thus requires that the soft demodulator simply multiplies the observed samples at the output of the matched filter (sufficient statistics) by a constant that depends on the channel noise.                                          □

## The soft-input soft-output (SISO) algorithm

In Fig. 11.47 we show a *trellis encoder*, characterized by the input information symbols **u** and the output code symbols **c**.

The trellis encoder takes a group of $n_u$ consecutive input symbols forming a symbol vector called $\underline{u}$, and emits a group of $n_c$ symbols (symbol vector $\underline{c}$) according to the trellis section.

The decoding algorithm underlying the behavior of SISO works for encoders represented in their trellis form. It can be a time-invariant or time-varying trellis, and thus the algorithm can be used for both block and convolutional codes. In the following, for simplicity of the exposition, we will refer to the case of *time-invariant convolutional codes*.

The dynamic of a time-invariant convolutional code is completely specified by a single *trellis section*, which describes the transitions ("edges") between the states of the trellis at two consecutive time instants.

A trellis section is characterized by:

- A set of $N$ states $\mathcal{S} = \{s_1, \ldots, s_N\}$. The state of the trellis at time $k$ is $S_k = s$, with $s \in \mathcal{S}$.

- A set of $N \cdot N_I$ edges obtained by the Cartesian product

$$\mathcal{E} = \mathcal{S} \times \mathcal{U}^{n_u} = \{e_1, \ldots, e_{N \cdot N_I}\}$$

  which represent all possible transitions between the trellis states.

With each edge $e \in \mathcal{E}$ the following functions are associated (see Fig. 11.48):

- The starting state $s^S(e)$ (the projection of $e$ onto $\mathcal{S}$);

- The ending state $s^E(e)$;

- The set of input symbols $\underline{u}(e)$ (the projection of $e$ onto $\mathcal{U}^{n_u}$);

- The set of output symbols $\underline{c}(e)$.

The relationship between these functions depend on the particular encoder. As an example, in the case of systematic encoders $(s^S(e), \underline{c}(e))$ also identifies the edge since $\underline{u}(e)$ is uniquely determined by $\underline{c}(e)$. In the following, we only assume that the pair $(s^S(e), \underline{u}(e))$ uniquely identifies the ending state $s^E(e)$; this assumption is always verified, as it is equivalent to say that, given the initial trellis state, there is a one-to-one correspondence between input sequences and state sequences, a property required for the code to be uniquely decodable.

The soft-input soft-output (SISO) module in its general form is a four-port device (see Fig. 11.49) that accepts at the input the sequences of *likelihood functions*

$$P_i(c; I) \qquad P_i(u; I)$$

and outputs the sequences of *extrinsic* a posteriori likelihood functions required

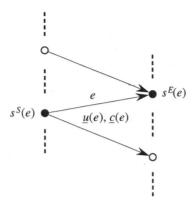

Figure 11.48: *An edge of the trellis section.*

Figure 11.49: *The Soft-Input Soft-Output (SISO) module.*

in the iterative procedure (11.97)

$$P_k(u; O) = \frac{1}{P_k(u; I)} \sum_{u:u_k=u} \prod_i P_i(c_i(\mathbf{u}); I) \prod_j P_j(u_j; I) \quad (11.106)$$

$$P_k(c; O) = \frac{1}{P_k(c; I)} \sum_{u:c_k(u)=c} \prod_i P_i(c_i(\mathbf{u}); I) \prod_j P_j(u_j; I) \quad (11.107)$$

computed according to its inputs and to its knowledge of the trellis section. The name "extrinsic" comes from the original proposal of turbo codes (Berrou *et al.*, 1993): its meaning will be clarified in the following.

The SISO algorithm can be embedded into the iterative decoding procedure (11.98) letting:

$$P_i(u; I) = \tilde{P}_{2i}^{(m-1)}(u)p_a(u)$$
$$P_i(c; I) = p(y_{1i} \mid c)$$

for the first SISO decoder and

$$P_i(u; I) = \tilde{P}_{1l}^{(m)}(u)p_a(u)$$
$$P_i(c; I) = p(y_{2i} \mid c)$$

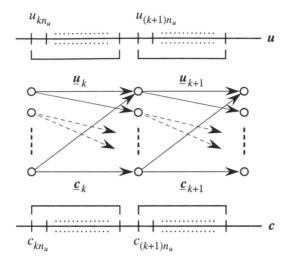

Figure 11.50: *The relationships between the information and code sequences* **u**,**c**, *their constituent symbols* $u_i, c_i$, *and the symbols* $\underline{u}_k$, $\underline{c}$ *that label the trellis edge.*

for the second SISO decoder.

The SISO algorithm allows computation of (11.106) and (11.107), for any code defined by its trellis, with a complexity that grows linearly with the number of states of the trellis and also linearly (and not exponentially as in (11.106) and (11.107)) with the transmission length.

We assume first that the time index set is finite, with cardinality $K$. The algorithm by which the SISO operates will be explained step by step in the following.

- At time $k$, the input likelihood functions relative to the information and code symbols $u$ and $c$ are combined to form the likelihood functions of the symbols $\underline{u}$ and $\underline{c}$ that label a single trellis edge (see Fig.11.50):

$$P_k(\underline{u}; I) = \prod_{i=0}^{n_u-1} P_{kn_u+i}(u; I)$$

$$P_k(\underline{c}; I) = \prod_{i=0}^{n_c-1} P_{kn_c+i}(c; I)$$

- The output likelihood functions are computed according to

$$P_k(\underline{u}; O) =$$

$$\sum_{e:\underline{u}(e)=\underline{u}} A_{k-1}[s^S(e)]P_k[\underline{c}(e);I]P_k[\underline{u}(e);I]B_k[s^E(e)] \quad (11.108)$$

$$P_k(\underline{c};O) =$$

$$\sum_{e:\underline{c}(e)=\underline{c}} A_{k-1}[s^S(e)]P_k[\underline{c}(e);I]P_k[\underline{u}(e);I]B_k[s^E(e)] \quad (11.109)$$

- The quantities $A_k(\cdot)$ and $B_k(\cdot)$ in (11.108) and (11.109) are obtained through the *forward* and *backward* recursions, respectively, as

$$A_k(s) = \sum_{e:s^E(e)=s} A_{k-1}[s^S(e)]P_k[\underline{u}(e);I]P_k[\underline{c}(e);I]$$

$$k = 1,\ldots,K-1 \quad (11.110)$$

$$B_k(s) = \sum_{e:s^S(e)=s} B_{k+1}[s^E(e)]P_{k+1}[\underline{u}(e);I]P_{k+1}[\underline{c}(e);I]$$

$$k = K-1,\ldots,1 \quad (11.111)$$

with initial values:

$$A_0(s) = \begin{cases} 1 & s = S_0 \\ 0 & \text{otherwise} \end{cases} \quad (11.112)$$

$$B_K(s) = \begin{cases} 1 & s = S_K \\ 0 & \text{otherwise} \end{cases} \quad (11.113)$$

- The output likelihood functions at the symbol level are obtained from (11.108) and (11.109) through

$$P_{kn_u+i}(u;O) = \frac{1}{P_{kn_u+i}(u;I)} \sum_{\underline{u}:u_{kn_u+i}=u} P_k(\underline{u};O) \quad (11.114)$$

$$P_{kn_c+i}(c;O) = \frac{1}{P_{kn_c+i}(u;I)} \sum_{\underline{c}:c_{kn_c+i}=c} P_k(\underline{c};O) \quad (11.115)$$

The new likelihood functions $P_{kn_u+i}(u;O), P_{kn_c+i}(c;O)$ represent updated versions of the input distributions $P_{kn_u+i}(u;I), P_{kn_c+i}(c;I)$, based on the code constraints and obtained using the likelihood functions of all symbols of the sequence but the $k$th ones $P_{kn_u+i}(u;I)$ and $P_{kn_c+i}(c;I)$. In the literature of "turbo decoding," $P_{kn_u+i}(u;O), P_{kn_c+i}(c;O)$ are called *extrinsic informations*. They represent the "value added" of the SISO module to the "a priori" distributions $P_{kn_u+i}(u;I), P_{kn_c+i}(c;I)$.

The formulation of the APP algorithm has been presented in a form that generalizes the one given in Appendix F and in other previous formulations, which were not in a form suitable to work with a general trellis code.

**Other forms of the SISO algorithm**

As our previous description should have made clear, the SISO algorithm is a multiplicative algorithm requiring the whole sequence to be received before starting. The reason is due to the backward recursion that starts from the (known) final trellis state $s_K$. As a consequence, its practical application is limited to the case where the duration of the transmission is short ($K$ small), or, for $K$ long, when the received sequence can be segmented into independent consecutive blocks, like for block codes or convolutional codes with *trellis termination*. It cannot be used for continuous decoding of convolutional codes. This constraint imposes a frame rigidity to the system, and also reduces the overall code rate through trellis termination.

A more flexible decoding strategy is offered by modifying the algorithm in such a way that the SISO module operates on a fixed memory span, and outputs the updated likelihood functions after a given delay $D$, like in the Viterbi algorithm. This new version of the algorithm, which is called the *sliding window soft-input soft-output* algorithm (SW-SISO), is described in Benedetto *et al.* (1998c). In essence, it starts the backward recursion assuming that all trellis states are equally likely, and offers almost the same performance as the standard SISO algorithm, provided that the decoding delay $D$ is sufficiently large. The reader will notice here a reminiscence of the practical implementation of the Viterbi algorithm.

The sliding-window SISO algorithm solves the problem of continuously updating the likelihood functions, without requiring trellis terminations. Its computational complexity, however, is still high when compared to the Viterbi algorithm. This is mainly due to the fact that both SISO and SW-SISO are *multiplicative* algorithms, a fact that can be tolerated if the hardware implementation makes use of a DSP, but that becomes computationally heavy for VLSI implementations.

This drawback is overcome by the additive version of the SISO algorithm, that applies as well to SISO and SW-SISO. For a complete description of the sliding window and additive algorithms, together with an analysis of their computational complexity, the reader is referred to Benedetto *et al.* (1998c).

**Example 11.15**  As an example of the performance of the iterative decoding algorithm, consider a rate 1/2 PCCC obtained concatenating a 16-state, rate 2/3 systematic, recursive convolutional encoder with a rate 2/1 encoder obtained from the previous one by eliminating the two systematic bits. The interleaver size is $N = 8920$. The performance is reported in Fig. 11.51 in terms of bit error probability versus $\mathcal{E}_b/N_0$ for several iterations of the decoding algorithm.

The curves show that, at $\mathcal{E}_b/N_0 = 0.8$ dB, an error free transmission is obtained

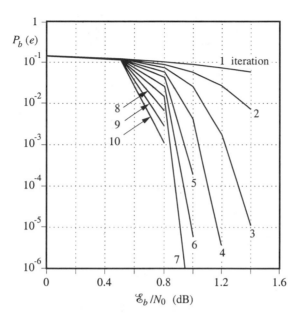

Figure 11.51: *Bit error probability obtained by simulating the iterative decoding algorithm. Rate 1/2 PCCC based on 16-state rate 2/3 and 2/1 CCs, and interleaver with size $N = 8920$.*

based on a 100 million bits simulation. This result is at 0.6 dB from the Shannon limit.
□

**Iterative decoding of serially concatenated convolutional codes**

The iterative decoding algorithm described previously for PCCCs can be extended to the case of SCCCs.

The core of the iterative decoding procedure is still the SISO algorithm. The block SISO is used within the iterative decoding algorithm as shown in Fig. 11.52, where we also exhibit the block diagram of the encoder to clarify the notations.

The symbols $\lambda(\,\cdot\,;I)$ and $\lambda(\,\cdot\,;O)$ at the input and output ports of SISO refer to the logarithmic likelihood ratios (LLRs),[11] binary unconstrained when the sec-

---

[11] When the symbols are binary, only one LLR is needed; when the symbols belong to an $L$-ary alphabet, $L - 1$ LLRs are required.

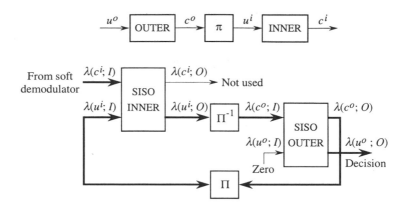

Figure 11.52: *Block diagrams of the encoder and iterative decoder for serially concatenated convolutional codes.*

ond argument is $I$, and modified according to the code constraints when it is $O$. The first argument $u$ refers to the information symbols of the encoder, whereas $c$ refers to code symbols. Finally, the superscript $o$ refers to the outer encoder, and $i$ to the inner encoder. The LLRs are defined as

$$\lambda(x; \cdot) \stackrel{\triangle}{=} \log \left[ \frac{P(x; \cdot)}{P(x_{\text{ref}}; \cdot)} \right] \tag{11.116}$$

When $x$ is a binary symbol, "0" or "1," $x_{\text{ref}}$ is generally assumed to be the "1." When $x$ belongs to an $L$-ary alphabet, we can choose as $x_{\text{ref}}$ each one of the $L$ symbols; a common choice for hardware implementation is the symbol with the highest probability, so that one LLR will be equal to zero and all others will be negative numbers.

Unlike with the iterative decoding algorithm employed for decoding PCCCs, in which only the LLRs of information symbols are updated, we must update here the LLRs of both information and code symbols based on the code constraints.

During the first iteration of the SCCC algorithm,[12] the block "SISO Inner" is fed with the demodulator soft outputs, consisting of the LLRs of symbols received from the channels, i.e., of the code symbols of the inner encoder. The second input $\lambda(u^i; I)$ of the SISO Inner is set to zero during the first iteration, since no a priori information is available on the input symbols $u^i$ of the inner

---

[12]To simplify the description, we assume that the interleaver acts on symbols instead of bits. In the actual decoder, we usually deal with bit LLRs and bit interleavers; the extension of the algorithm to this case can be found in Benedetto *et al.* (1998c).

encoder.

The LLRs $\lambda(c^i; I)$ are processed by the SISO algorithm, which computes the *extrinsic* LLRs of the information symbols of the inner encoder $\lambda(u^i; O)$ conditioned on the inner code constraints. The extrinsic LLRs are passed through the inverse interleaver (block labeled "$\Pi^{-1}$"), whose outputs correspond to the LLRs of the code symbols of the outer code, i.e.

$$\Pi^{-1}[\lambda(u^i; O)] = \lambda(c^o; I)$$

These LLRs are then sent to the block "SISO Outer" in its upper entry, which corresponds to code symbols. The SISO Outer, in turn, processes the LLRs $\lambda(c^o; I)$ of its unconstrained code symbols, and computes the LLRs of both code and information symbols based on the code constraints. The input $\lambda(u^o; I)$ of the SISO Outer is always set to zero, which implies assuming equally likely transmitted source information symbols. The output LLRs of information symbols (which yield the a posteriori LLRs of the SCCC information symbols) will be used in the final iteration to recover the information bits. On the other hand, the LLRs of outer code symbols, after interleaving, are fed back to the lower entry (corresponding to information symbols of the inner code) of the block SISO Inner to start the second iteration. In fact we have

$$\Pi[\lambda(c^o; O)] = \lambda(u^i; I)$$

**Example 11.16**   Consider a rate 1/4 SCCC with a very long interleaver, corresponding to an input decoding delay of 16,384. The constituent codes are 8-state codes: the outer encoder is nonrecursive, and the inner encoder is a recursive encoder. Their generating matrices are

$$G_o(Z) = [1 + Z, 1 + Z + Z^3]$$
$$G_i(Z) = [1, \frac{1 + Z + Z^3}{1 + Z}]$$

respectively.

Its performance in terms of bit error probability versus $\mathcal{E}_b/N_0$ for different number of iterations is presented in Fig. 11.53, They show that the decoding algorithm works at $\mathcal{E}_b/N_0 = -0.05$ dB, at 0.76 dB from the Shannon capacity limit (which is in this case equal to $-0.817$ dB), with very limited complexity (remember that we are using two rate 1/2 codes with 8 states).   □

## Comparison between serially and parallel concatenated codes

The analytical results presented in Section 11.3.2. showed that serial concatenation can yield significantly higher interleaver gains and a steeper asymptotic

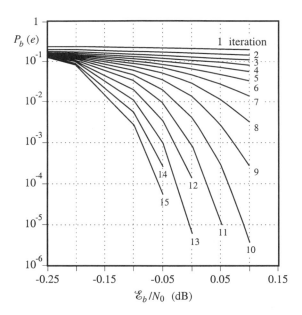

Figure 11.53: *Simulated bit error probability performance of a rate 1/4 serially concate-nated code obtained with two eight-state constituent codes and an interleaver yielding an input decoding delay equal to 16,384.*

slope of the error probability curves. To check if these advantages are retained when the codes are iteratively decoded at very low signal-to-noise ratios, we have simulated the behavior of SCCCs and PCCCs in equal system conditions: the concatenated code rate is 1/3, the CCs are 4-state recursive encoders (rates 1/2 + 1/2 for PCCCs, and rates 1/2 + 2/3 for the SCCCs), and the decoding de-lays in terms of input bits is 16,384. In Fig. 11.54 we report the results, in terms of bit error probability versus $\mathcal{E}_b/N_0$ for six and nine decoding iterations. As can be seen from the curves, the PCCC outperforms the SCCC for high values of the bit error probabilities, down to roughly $10^{-5}$. For lower values of the bit error probability, the SCCC behaves significantly better (the advantage of SCCC at $10^{-6}$ is 0.5 dB with nine iterations), and does not present the phenomenon of "error floor."[13]

---

[13]It is customary in the PCCC literature to call "error floor" what is actually a significant change of slope of the performance curve.

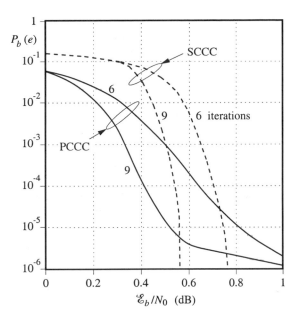

Figure 11.54: *Comparison of rate 1/3 PCCC and SCCC. The PCCC is obtained concatenating two equal rate 1/2 4 states codes, whereas the SCCC concatenates two 4-state rate 1/2 and rate 2/3 codes. The curves refer to six and nine iterations of the decoding algorithm and to an equal input decoding delay of 16,384.*

## 11.4. Bibliographical notes

As for block codes, excellent textbooks exist on the subject of convolutional codes. In particular, the book by Viterbi and Omura (1979) is an invaluable reference book for convolutional codes and for the random coding bounds for all types of codes on different channels. Convolutional (and also block) code applications and implementation problems can be found in the books by Clark and Cain (1981) and Lin and Costello (1983). The three classical papers by Forney (1970, 1974a, and 1974b) provide a complete overview of the structure and decoding techniques of convolutional codes. Forney (1970) and Rosenberg (1971) explore the question of the relative fraction of catastrophic codes in the ensemble of all convolutional codes of a given rate and constraint length. Sequential decoding was first introduced by Wozencraft (1957) and subsequently modified by Fano (1963). An interesting class of sequential decoding algorithms are the stack algorithms, which were proposed independently by Jelinek (1969) and Zigangirov (1966). These algorithms are computationally simpler than Fano's but require more storage. An analysis of the computational problems implied by sequential decoding can be found in Wozencraft and Jacobs (1965), Savage (1966), Forney (1974b), and in the book by Anderson and Mohan (1991).

Concatenated codes were invented by Forney, and are analyzed in his book (1966). Concatenated codes with interleavers are a relatively new subject, started in 1993 with the paper on "turbo" codes by Berrou *et al.* (1993). The excellent performance of turbo codes has been explained in Benedetto and Montorsi (1996a), a paper that contains the analytical details of the analysis presented in Section 11.3 of this chapter, and, independently, by Divsalar and Pollara (1995). A tutorial explanation is contained in Perez *et al.*, (1996). Design techniques and tables of good constituent codes to be employed in parallel concatenated convolutional codes are reported in Benedetto and Montorsi (1996b), Divsalar and McEliece (1996), and Benedetto *et al.* (1998a). The analysis, design and iterative decoding of serially (as opposed to parallel) concatenated codes with interleavers are included in Benedetto *et al.* (1998b).

The BCJR algorithm to compute the a posteriori probability, an essential building block for the iterative decoding of concatenated codes with interleavers, proposed by Bahl *et al.* (1974), is described in Appendix F. The extension to the sliding window and additive algorithms, as well as to other versions of the general SISO algorithm, can be found in Benedetto *et al.* (1998c). The approach of Section 11.3.3 closely follows that paper.

The algorithm that iteratively decodes "turbo" codes has been proposed first by Berrou *et al.* (1993). It is also explained in detail in Hagenauer *et al.* (1996). A general iterative algorithm applicable to all forms of code concatenations has been described in Benedetto *et al.* (1998c).

A great number of papers have appeared on the subject of the "turbo" iterative decoding algorithm, showing that it can be viewed as an istance of previously proposed algorithms (see, for example, McEliece *et al.* (1998) and the extensive references therein). To avoid a huge reference list, the readers are referred to the papers and references in the *European Transactions on Telecommunications* (Biglieri and Hagenauer, 1995), and in the *IEEE Journal on Selected Areas in Communications* (Benedetto *et al.*, 1998d), entirely devoted to concatenated codes and iterative decoding.

## 11.5.  Problems

*Problems marked with an asterisk should be solved with the aid of a computer.*

**11.1** Consider the (3,1,3) convolutional code, defined by the generators

$$g_{1,1} = (111), \quad g_{1,2} = (111), \quad g_{1,3} = (101)$$

1. Draw the state diagram of the code.
2. Obtain the input-output weight enumerating function $T_2(W, D)$.
3. Find the free distance $d_f$ of the code.
4. Evaluate the maximum-likelihood bit error probability $P_b(e)$ over a BSC with $p = 10^{-3}$.

**11.2** Consider the (2,1,5) convolutional code, defined by the generators

$$\mathbf{g}_{1,1} = (11001), \quad \mathbf{g}_{1,2} = (11111)$$

1. Draw the trellis diagram of the code.
2. Find the free distance $d_f$ of the code.
3. (*) Obtain the input-output weight enumerating function implementing in software the matrix technique of Appendix D.

**11.3** Perform a complete study of the (2,1,2) convolutional code with generators

$$\mathbf{g}_{1,1} = (10), \quad \mathbf{g}_{1,2} = (11)$$

**11.4** Consider the (2,1,3) convolutional code with generators

$$\mathbf{g}_{1,1} = (110), \quad \mathbf{g}_{1,2} = (101)$$

1. Draw the state diagram of the code.
2. Verify that the self-loop at state $S_4 = (11)$ does not increase the distance from the all-zero sequence.
3. The code is catastrophic. Verify the effects with an example.
4. Check that the necessary and sufficient condition to avoid catastrophicity given in Section 11.1.2 is not fullfilled.

**11.5** For rate $1/n_0$ codes, prove that the condition to avoid catastrophicity is that the code generators in polynomial form are relatively prime.

**11.6** Using as "mother" code the (2,1,4) convolutional code of Figure 11.12, find the best (in terms of largest free distance) rate 2/3 and 3/4 punctured codes. For the best punctured codes, find the weight enumerating function $T(D)$.

**11.7** (Viterbi and Omura, 1979) The bounds (11.31) and (11.42) can be made tighter through the following steps.

1. Prove, first, the following inequality:

$$\text{erfc}\left(\sqrt{x+y}\right) \le \text{erfc}\left(\sqrt{x}\right) e^{-y}, \quad x \ge 0, \ y \ge 0$$

2. Since $d \geq d_f$, we may bound (11.29) by

$$P(\mathbf{x}_1 \rightarrow \mathbf{x}_d) \leq \frac{1}{2}\mathrm{erfc}\left(\sqrt{\frac{d_f R_c \mathcal{E}_b}{N_0}}\right) \exp\left[-\frac{(d - d_f)R_c \mathcal{E}_b}{N_0}\right]$$

3. Derive the new bound (11.31) in the form

$$P(e) \leq \frac{1}{2}\mathrm{erfc}\left(\sqrt{\frac{d_f R_c \mathcal{E}_b}{N_0}}\right) \exp\left(\frac{d_f R_c \mathcal{E}_b}{N_0}\right) T(D)\Bigg|_{D = e^{-(R_c \mathcal{E}_b / N_0)}}$$

   and similarly for the bound (11.42).

4. (*) Compare the new bound on the bit error probability with (11.42) on a 4-state rate 1/2 code.

**11.8** A bound on $P_b(e)$ tighter than (11.38) can be obtained for convolutional codes if $P(e_{1d})$ is computed exactly instead of being upper bounded as in (11.27). The all-zero path is assumed to be transmitted, and suppose that the path being compared has weight $d$. The incorrect path will be selected if there are more than $(d+1)/2$ errors, with $d$ odd.

1. Show that in this case

$$P(\mathbf{x}_1 \rightarrow \mathbf{x}_d) = \sum_{i=(d+1)/2}^{d} \binom{d}{i} p^i (1-p)^{d-i}$$

   where $p$ is the transition probability of the BSC.

2. When $d$ is even, show that

$$P(\mathbf{x}_1 \rightarrow \mathbf{x}_d) = \frac{1}{2}\binom{d}{d/2} p^{d/2}(1-p)^{d/2} + \sum_{i=d/2+1}^{d} \binom{d}{i} p^i (1-p)^{d-i}$$

3. (*) Use the preceding results in (11.35) and compare the new bound numerically on an example.

**11.9** (*) Evaluate the bit distance spectra $(A_d^{(b)}, d)$ for the codes of Tables 11.2 and 11.3 for $N \leq 8$, and use them to rederive the curves of Figures 11.21 and 11.22.

**11.10** (*) Check the terms of the bit distance spectra of the nonsystematic and recursive systematic codes of Example 11.10.

**11.11** Prove that parallel and serially concatenated block codes with interleaver, whose block diagrams have been shown in Figs. 11.30 and 11.31, are linear codes provided that the two constituent block codes are linear. Is this result also valid for concatenated convolutional codes?

**11.12** Evaluate the conditional weight enumerating function of a serially concatenated block code using a uniform interleaver of size $N = mk$ equal to an integer number of outer code words, as a function of the two conditional weight enumerating functions of the outer and inner code.

**11.13** Evaluate the conditional weight enumerating function of a parallel concatenated block code using a uniform interleaver of size $N = mk$ equal to an integer number of data words words, as a function of the two conditional weight enumerating functions of the constituent codes.

**11.14** Prove that the minimum weight $w_m$ of input sequences to a convolutional encoder generating an error event is equal to 1 for nonrecursive convolutional encoders, and equal to 2 for recursive convolutional encoders.

**11.15** (*) Write a computer program implementing the SISO algorithm described in Section 11.3.3 (in the multiplicative or additive (see Benedetto *et al.*, 1998c) form), and apply it to the MAP decoding of the convolutional code of Problem 11.2.

**11.16** (*) Using the computer program developed in Problem 11.15, implement by software the iterative decoding algorithms for parallel and serially concatenated codes. Test the programs on the PCCC of Example 11.15 and on the SCCC of Example 11.16.

# Coded modulation

In this chapter we introduce coded modulation, with special attention to a version called trellis-coded modulation (TCM). This is used in digital communications for the purpose of gaining noise immunity over uncoded transmission without expanding the signal bandwidth or increasing the transmitted power. The trellis-coded-modulation solution combines the choice of a modulation scheme with that of a convolutional code, while the receiver performs soft demodulation prior to decoding.

## 12.1. The cutoff rate and its role

As we have illustrated in Chapter 4, if coding is not to be included in the design of a modulation/demodulation system, then the sensible design criterion is the minimization of error probability. That is, the best modem is the one which, under the constraints faced by the system designer, generates the discrete channel affected by the lowest error probability.

Now, assume that coding is to be used over the channel. Application of this criterion would lead us to minimize the error probability as seen from the coded-channel input to the decoder output. However, in these terms the design problem would be hardly solvable in practice. In fact, to compare channels used with coding we should pick the best codes for each of them—a prohibitive task. A way out of this impasse comes from a technique that was introduced by Shannon in his proof of the Channel Coding Theorem (see Chapter 3) and was used in Chapter 10 to derive the cutoff rate of a binary-input continuous-output Gaussian channel. This technique, a central one in information theory, is usually referred to as "random coding." It comes from the fact that if we pick a code at random, it is very likely to be good. (Louis Pasteur said, "Chance is the best ally of those who know how to take advantage of it." Think of this as a partial disclaimer of Murphy's law.) Thus, instead of computing the error probability for the optimum code, which we cannot find, we compute the average error probability over the ensemble of all the possible codes to be used on the given channel. Obviously, at least one code will perform at least as well as the ensemble average: hence this ensemble average yields an upper bound to the performance of the optimum code. Moreover, in practice most of the random codes are so good that we cannot beat their performance by using a practical coding scheme. As we have seen in Chapter 3, for a block code of length $n$ and rate $R_c = k/n$ the word error probability is bounded above, for $R_c$ less than the channel capacity, by the quantity $2^{-nE(R_c)}$. A similar result (see, e.g., Viterbi and Omura, 1979) holds for convolutional codes, where the role of $n$ is played by $n_0 N$.

The function $E(R_c)$, which does not depend on the code but only on its rate, tells how good the channel is when we want to use coding on it. In fact, as we increase the complexity of the code by increasing its block length $n$, the higher the value of $E(R_c)$ the lower the error probability bound will be. Loosely speaking, we should expect lower error probabilities for good codes with the same complexity and the same rate over channels with higher $E(R_c)$. For this reason $E(R_c)$ is called the *reliability function* of the channel.

In conclusion, when coding is used, a meaningful comparison among channels should be based on their reliability functions. Unfortunately, the actual computation of this function may be a very demanding task; moreover, it would be far simpler to make comparisons based on a single parameter rather than on a function. How should we select this parameter? It may seem at first that channel capacity is a good choice. However, channel capacity gives only a range of rates where reliable transmission is possible. As discussed in Chapter 10, the sensible parameter here is the *cutoff rate*, $R_0$, of the discrete channel. This is the rate at which the tangent to $E(R_c)$ of slope $-1$ intercepts the $R_c$ axis, so that for $R_c \leq R_0$ we have $E(R_c) \geq R_0 - R$, and hence

$$P(e) \leq 2^{-n(R_0 - R_c)}, \qquad R_c \leq R_0 \qquad (12.1)$$

From the last inequality we see that $R_0$ provides both a range of rates and an exponent to error probability.

### 12.1.1. Computing the cutoff rate: AWGN channel with coherent detection

We now show how to compute the cutoff rate of the discrete channel generated by a given signal constellation transmitted on the additive white Gaussian noise channel with coherent demodulation. We assume, as we did in Chapter 4, a finite signal constellation $\mathcal{S} = \{\mathbf{s}_i\}_{i=1}^{M}$. A *code for the Gaussian channel* is a set of

$\mathcal{M}$ code words $\mathbf{x}$, each one consisting of $n$ signals in $\mathcal{S}$. Transmission of one of these signals corresponds to one "channel use." This code has rate

$$R_c = \frac{\log_2 \mathcal{M}}{n} \quad \text{bit/channel use} \qquad (12.2)$$

For example, for uncoded transmission, $\mathcal{M} = M^n$, and hence $R_c = \log_2 M$. With a binary block code and binary modulation, $\mathcal{M} = 2^k$, and hence $R_c = k/n$; here a channel use corresponds to the transmission of a binary signal.

The average error probability of this code can be evaluated by using the union bound of Section 4.3: $P(e)$ is computed, by averaging over the vectors $\mathbf{x}$ in the code, the pairwise error probability

$$P(e \mid \mathbf{x}) \leq \sum_{\hat{\mathbf{x}} \neq \mathbf{x}} P\{\mathbf{x} \to \hat{\mathbf{x}}\} \qquad (12.3)$$

We recall that $P\{\mathbf{x} \to \hat{\mathbf{x}}\}$ is the probability that the received vector $\mathbf{r}$ be closer (in the sense of Euclidean distance) to $\hat{\mathbf{x}}$ than to $\mathbf{x}$:

$$P\{\mathbf{x} \to \hat{\mathbf{x}}\} = P(|\mathbf{r} - \hat{\mathbf{x}}|^2 < |\mathbf{r} - \mathbf{x}|^2) \triangleq P(X < 0) \qquad (12.4)$$

and

$$\begin{aligned} X &\triangleq |\mathbf{r} - \hat{\mathbf{x}}|^2 - |\mathbf{r} - \mathbf{x}|^2 \\ &= 2(\mathbf{n}, \mathbf{x} - \hat{\mathbf{x}}) + |\mathbf{x} - \hat{\mathbf{x}}|^2 \end{aligned} \qquad (12.5)$$

Here $\mathbf{n}$ is the white Gaussian noise vector affecting the transmission, and $(\cdot, \cdot)$ denotes scalar product. The last equality shows that $X$, being an affine deterministic transformation of a Gaussian random vector, is itself a Gaussian random variable (RV). Its mean and variance can be computed by observing that, with $T$ the duration of the real waveforms corresponding to the signal vectors $\mathbf{s}_i$, we have

$$(\mathbf{n}, \mathbf{x} - \hat{\mathbf{x}}) = \int_0^{nT} n(t)[x(t) - \hat{x}(t)]\, dt \qquad (12.6)$$

Thus, since $\mathrm{E}[n(t)] = 0$ and $\mathrm{E}[n(t)n(\tau)] = (N_0/2)\delta(t - \tau)$, we see immediately that the average of (12.6) is zero, and its variance is equal to

$$\begin{aligned} \mathrm{E}[(\mathbf{n}, \mathbf{x} - \hat{\mathbf{x}})^2] &= \int_0^{nT} \int_0^{nT} \mathrm{E}[n(t)n(\tau)]\,[x(t) - \hat{x}(t)][x(\tau) - \hat{x}(\tau)]\, dt\, d\tau \\ &= \frac{N_0}{2} \int_0^{nT} \int_0^{nT} \delta(t - \tau)[x(t) - \hat{x}(t)][x(\tau) - \hat{x}(\tau)]\, dt\, d\tau \\ &= \frac{N_0}{2} \int_0^{nT} [x(t) - \hat{x}(t)]^2\, dt \\ &= \frac{N_0}{2} |\mathbf{x} - \hat{\mathbf{x}}|^2 \end{aligned}$$

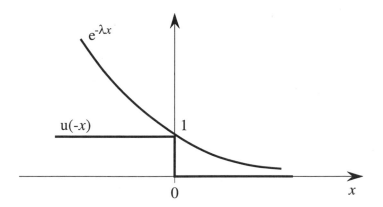

Figure 12.1: *Illustration of the inequality* $\mathrm{u}(-x) \le \exp(-\lambda x)$, $\lambda \ge 0$.

In conclusion, $X$ is a Gaussian RV with mean $|\mathbf{x} - \hat{\mathbf{x}}|^2$ and variance $2N_0|\mathbf{x} - \hat{\mathbf{x}}|^2$.

We shall now derive an upper bound to the pairwise error probability. Observe that, with $\mathrm{u}(x)$ denoting the unit-step function, we have

$$\mathrm{u}(-x) \le e^{-\lambda x}$$

for all real $\lambda \ge 0$ (see Fig. 12.1).

Thus, the following holds for any continuous random variable $\xi$ with probability density function (pdf) $f_\xi(x)$:

$$
\begin{aligned}
P(\xi \le 0) &= \int_{-\infty}^{0} f_\xi(x)\,dx \\
&= \int_{-\infty}^{\infty} f_\xi(x)\mathrm{u}(-x)\,dx \\
&\le \int_{-\infty}^{\infty} f_\xi(x)e^{-\lambda x}\,dx
\end{aligned}
$$

The tightest bound is obtained by choosing the value of $\lambda$ that minimizes the last integral. For $\xi$ a Gaussian RV with mean $\mu$ and variance $\sigma^2$, we have

$$\int_{-\infty}^{\infty} f_\xi(x)e^{-\lambda x}\,dx = e^{-\lambda\mu + \lambda^2\sigma^2/2}$$

which is minimized by choosing $\lambda = \mu/\sigma^2$. This yields the "Chernoff bound"

$$P(\xi \le 0) \le e^{-\mu^2/2\sigma^2}$$

The pairwise error probability (12.3) is then bounded above by

$$P\{\mathbf{x} \to \hat{\mathbf{x}}\} \leq e^{-|\mathbf{x}-\hat{\mathbf{x}}|^2/4N_0} \tag{12.7}$$

Before proceeding further with our derivation of $R_0$ for the AWGN channel, we pause one moment for a comment on the bound just derived. Since the probability that a Gaussian RV takes on a negative value is known (see Appendix A), we could derive an exact expression for the pairwise error probability:

$$P\{\mathbf{x} \to \hat{\mathbf{x}}\} = \frac{1}{2}\text{erfc}\left(\frac{|\mathbf{x}-\hat{\mathbf{x}}|}{2\sqrt{N_0}}\right) \tag{12.8}$$

The reader should recognize that this equality is exactly the same as (4.49), with the role of the signals $\mathbf{s}_i$, $\mathbf{s}_j$ now played by the code words $\mathbf{x}$, $\hat{\mathbf{x}}$. By using the inequality (A.5), valid for $x > 0$

$$\frac{1}{2}\text{erfc}\left(x\right) \leq \frac{1}{2}e^{-x^2} < e^{-x^2}$$

we could obtain (12.7) directly from (12.8). Although the latter derivation is shorter, it obscures the fact that the bounding technique used in the derivation of $R_0$ should be based on the Chernoff bound. Actually, the derivation given before is in a form that lends itself to generalization to other types of channels, an example of which will be provided in Chapter 13.

With an abuse of notation that should not confuse the reader, we now write $\mathbf{x} = (\mathbf{s}_1, \ldots, \mathbf{s}_n)$ and $\hat{\mathbf{x}} = (\hat{\mathbf{s}}_1, \ldots, \hat{\mathbf{s}}_n)$ for the two code words involved in (12.7), which allows us to write the bound in a product form as follows:

$$P\{\mathbf{x} \to \hat{\mathbf{x}}\} \leq \prod_{k=1}^{n} e^{-|\mathbf{s}_k-\hat{\mathbf{s}}_k|^2/4N_0} \tag{12.9}$$

We are now ready for the key step in obtaining a random coding bound: generate a code for the Gaussian channel by randomly selecting, independently and with equal probabilities $1/M$, the $n$ symbols occurring in any code word. The average of the pairwise error probability bound (12.9) over the ensemble of random codes is given by

$$\overline{P\{\mathbf{x} \to \hat{\mathbf{x}}\}} \leq \overline{\prod_{k=1}^{n} e^{-|\mathbf{s}_k-\hat{\mathbf{s}}_k|^2/4N_0}}$$

$$= \prod_{k=1}^{n} \overline{e^{-|\mathbf{s}_k-\hat{\mathbf{s}}_k|^2/4N_0}} \tag{12.10}$$

where the last equality follows from the independence of the code symbols selected in random coding. Since the symbols are equally likely, and since the averaging operation eliminates the dependence on the index $k$, we have explicitly

$$
\begin{aligned}
\overline{e^{-|s_k - \hat{s}_k|^2/4N_0}} &= \frac{1}{M^2} \sum_{s \in \mathcal{S}} \sum_{\hat{s} \in \mathcal{S}} e^{-|s-\hat{s}|^2/4N_0} \\
&= \frac{1}{M} + \frac{1}{M^2} \sum_{s \in \mathcal{S}} \sum_{\hat{s} \neq s} e^{-|s-\hat{s}|^2/4N_0} \\
&= \frac{1}{M} \left( 1 + \frac{1}{M} r(\mathcal{S}, N_0) \right)
\end{aligned}
$$

where

$$
r(\mathcal{S}, N_0) = \sum_{s \in \mathcal{S}} \sum_{\hat{s} \neq s} e^{-|s-\hat{s}|^2/4N_0} \tag{12.11}
$$

Thus, from (12.10) we obtain

$$
\overline{P\{\mathbf{x} \rightarrow \hat{\mathbf{x}}\}} \leq \frac{1}{M^n} \left( 1 + \frac{1}{M} r(\mathcal{S}, N_0) \right)^n \tag{12.12}
$$

This can be put in the form

$$
\overline{P\{\mathbf{x} \rightarrow \hat{\mathbf{x}}\}} \leq 2^{-nR_0} = e^{-\ln 2 \cdot n R_0} \tag{12.13}
$$

by identifying the RHS of (12.12) and (12.13), i.e., by defining

$$
R_0 \overset{\triangle}{=} \log_2 M - \log_2 \left( 1 + \frac{1}{M} r(\mathcal{S}, N_0) \right) \tag{12.14}
$$

From (12.13) we also have, by applying the union bound of Section 4.3 and observing that there are $\mathcal{M}$ equally likely code words $\mathbf{x}$ and $\hat{\mathbf{x}}$

$$
\overline{P(e)} \leq \frac{1}{\mathcal{M}} \sum_{\mathbf{x}} \sum_{\hat{\mathbf{x}} \neq \mathbf{x}} 2^{-nR_0} = (\mathcal{M}-1) 2^{nR_0} \leq \mathcal{M} 2^{-nR_0}
$$

Due to (12.2), we have $\mathcal{M} = 2^{nR_c}$, and consequently

$$
\overline{P(e)} \leq 2^{-n(R_0 - R_c)}
$$

This, which corresponds to (12.1), concludes our derivation of $R_0$.

Before examining some examples of the actual calculation of $R_0$, we hasten to observe that, due to an assumption we made in the derivations, (12.14) might be only a lower bound to the true $R_0$. In fact, we have assumed that in the random codes the channel symbols are equally likely: this may not be the best assumption, because if some symbols were picked with larger frequency than others we might actually decrease the error probability. If our derivation were modified so as not to assume equally likely code symbols, we could easily prove (12.14), but with the new definition

$$r(\mathcal{S}, N_0) = M^2 \max_{\mathbf{Q}} \sum_{\mathbf{s} \in \mathcal{S}} \sum_{\hat{\mathbf{s}} \neq \mathbf{s}} Q(\mathbf{s})Q(\hat{\mathbf{s}})e^{-|\mathbf{s}-\hat{\mathbf{s}}|^2/4N_0}, \qquad (12.15)$$

where $\mathbf{Q}$ denotes the probability distribution of the signals picked from $\mathcal{S}$, and $Q(\cdot)$ their individual probabilities. Since the maximization involved in (12.15) may not be easy to perform, use of the simpler version of $R_0$ (that one should call "symmetric cutoff rate") is often preferred, in spite of the fact that this may not provide the true cutoff rate. In the following, we shall always use this simpler form of $R_0$, without any further comment.

**Example 12.1 (PSK)**   In this example we evaluate the cutoff rate for the discrete channel generated by $M$-ary PSK with coherent detection over the AWGN channel. To evaluate $r(\mathcal{S}, N_0)$ through (12.11), and hence $R_0$, we need only to know the set of Euclidean distances from each vector of the signal set. Here, due to the uniformity of the constellation, the set of distances from any $\mathbf{s}$ does not depend on the selection of $\mathbf{s}$. This allows us to simplify the calculations by picking arbitrarily one $\mathbf{s}$, and writing

$$r(\mathcal{S}, N_0) = M \sum_{\hat{\mathbf{s}} \neq \mathbf{s}} e^{|\mathbf{s}-\hat{\mathbf{s}}|^2/4N_0}$$

For example, for 4PSK with energy $\mathcal{E}$ we have

$$r(\mathcal{S}, N_0) = 4 \left[ e^{-\mathcal{E}/N_0} + 2e^{-\mathcal{E}/2N_0} \right]$$

The values of $R_0$ for binary, quaternary, and octonary PSK are shown in Fig. 12.2.   □

## 12.2.   Introducing TCM

An example will introduce the concept of TCM. Consider a digital communication scheme to transmit data from a source emitting two information bits every $T$. Several solutions are possible (see Fig. 12.3).

(a) Use no coding and 4PSK modulation, with one signal transmitted every $T$. In this situation, every signal carries two information bits.

(b) Use a convolutional code with rate 2/3 and 4PSK. Since every signal carries now 4/3 information bits, it must have a duration of $2T/3$ to match the information rate of the source. This implies that, with respect to the uncoded scheme, the bandwidth increases by a factor 3/2.

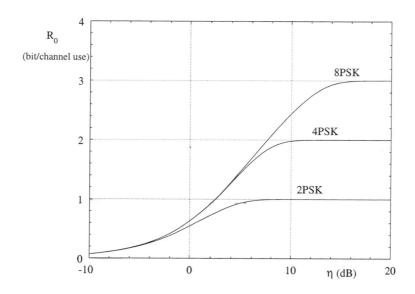

Figure 12.2: *Cutoff rate of the discrete channel generated by M-ary PSK transmitted over the AWGN channel and demodulated coherently. Here $\eta = \mathcal{E}/N_0$.*

**(c)** Use a convolutional code with rate 2/3, and 8PSK to avoid reducing the signal duration. Each signal carries 2 information bits, and hence no bandwidth expansion is incurred because 8PSK and 4PSK occupy the same bandwidth.

We see that with solution (c) we can use coding with no bandwidth expansion. One should expect that the use of a higher-order signal constellation involves a power penalty with respect to 4PSK: thus, the coding gain achieved by the rate 2/3 convolutional code should offset this penalty, the net result being some coding gain at no price in bandwidth.

This idea is indeed not new, since multi-level modulation of convolutionally-encoded symbols was a known concept before the introduction of TCM. The innovative aspect of TCM is the concept that convolutional encoding and modulation should not be treated as separate entities, but rather as a unique operation (this is why we talk about *coded modulation*.) As a consequence, the received signal is processed by a receiver which does "soft demodulation": that is, the detection process will involve *soft* rather than *hard* decisions. Instead of deciding

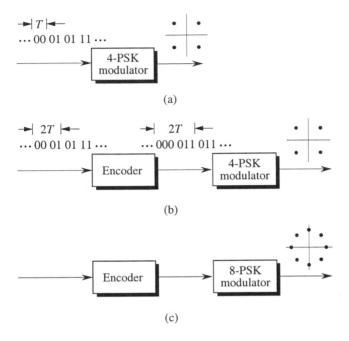

Figure 12.3: *Three digital communication schemes transmitting two bits every T seconds. (a) Uncoded transmission with 4PSK. (b) 4PSK with a rate-2/3 convolutional encoder and bandwidth expansion. (c) 8PSK with a rate-2/3 convolutional encoder and no bandwidth expansion.*

to which transmitted symbols the received signals correspond, the demodulator passes metric information to a soft Viterbi decoder. In conclusion, the parameter governing the performance of the transmission system is not the *free Hamming distance* of the convolutional code, but rather, over the AWGN channel, the *free Euclidean distance* between transmitted signal sequences. Thus, the optimization of the TCM design will be based on Euclidean distances rather than on Hamming distances, and the choice of the code and of the signal constellation will not be performed in separate steps.

### 12.2.1.  Fundamentals of TCM

We assume here transmission over the AWGN channel, and the vector model introduced in Chapter 4. When a signal s is transmitted, the received signal is represented by

$$\mathbf{r} = \mathbf{s} + \mathbf{n}$$

where $\mathbf{n}$ is a noise $N$-tuple whose components are independent Gaussian random variables with mean zero and the same variance $N_0/2$. The signal $\mathbf{s}$ is chosen from a set $\mathcal{S}'$ consisting of $M'$ signals, the *uncoded signal constellation*. Under the assumption that the transmitted signals are all equally likely, the average signal energy is

$$\mathcal{E}' = \frac{1}{M'} \sum_{\mathbf{x} \in \mathcal{S}'} |\mathbf{x}|^2$$

Consider now the transmission of a sequence $(\mathbf{s}_i)_{i=0}^{K-1}$ of $K$ signals, where the subscript $i$ denotes discrete time. The receiver which minimizes the average error probability over the sequence operates by first observing the received sequence $\mathbf{r}_0, \dots, \mathbf{r}_{K-1}$, then deciding that $\hat{\mathbf{x}}_0, \dots, \hat{\mathbf{x}}_{K-1}$ was transmitted if the squared Euclidean distance

$$\delta^2 = \sum_{i=0}^{K-1} |\mathbf{r}_i - \mathbf{x}_i|^2$$

is minimized by taking $\mathbf{x}_i = \hat{\mathbf{x}}_i$, $i = 0, \dots, K-1$, or, in words, if the received sequence is closer to $\hat{\mathbf{x}}_0, \dots, \hat{\mathbf{x}}_{K-1}$ than to any other allowable signal sequence. As we know from Section 4.3, the resulting sequence error probability, as well as the symbol error probability, is upper bounded, at least for high signal-to-noise ratios, by a decreasing function of the ratio $\delta_{\min}^2/N_0$, where $\delta_{\min}^2$ is the minimum squared Euclidean distance between any two allowable signal sequences. With no coding, this is $K$ times the minimum distance among signals in $\mathcal{S}'$.

We can now define the concept of *coding in the signal space*. This consists of restricting the transmitted sequences to a proper subset of $\mathcal{S}'^K$. In other words, we choose a subset of all the possible sequences made of signals of $\mathcal{S}'$, exactly as we generate a binary block or a convolutional code by choosing, among all the possible binary sequences, those that satisfy certain properties. If this is done, the transmission rate will also be reduced because of the decrease in the number of sequences available for transmission. To avoid this unwanted reduction, we choose to increase the size of $\mathcal{S}'$. By substituting for $\mathcal{S}'$ the larger constellation $\mathcal{S} \supset \mathcal{S}'$, and hence increasing $M'$ to $M > M'$, and selecting $M'^K$ sequences as a subset of $\mathcal{S}^K$, we can have sequences which are less tightly packed and hence increase the minimum distance among them.

In conclusion, we obtain a minimum distance $\delta_{\mathrm{free}}$ between any two sequences which turns out to be greater than the minimum distance $\delta_{\min}$ between signals in $\mathcal{S}'$. Hence, use of maximum-likelihood sequence detection will yield a "distance gain" of a factor of $\delta_{\mathrm{free}}^2/\delta_{\min}^2$.

On the other hand, to avoid a reduction of the value of the transmission rate, the constellation was expanded from $\mathcal{S}'$ to $\mathcal{S}$. This may entail an increase in the average energy expenditure from $\mathcal{E}'$ to $\mathcal{E}$, and hence an "energy loss" $\mathcal{E}/\mathcal{E}'$.[1] Thus, we define the *asymptotic coding gain* of a TCM scheme as

$$\gamma = \frac{\delta_{\text{free}}^2/\mathcal{E}}{\delta_{\text{min}}^2/\mathcal{E}'} \tag{12.16}$$

where $\mathcal{E}'$ and $\mathcal{E}$ are the average energies spent to transmit with uncoded and coded transmission, respectively.

This introduction of interdependencies among the signals in a sequence, and the constellation expansion to avoid rate reduction are two of the basic ideas underlying trellis-coded modulation (another one is *set partitioning*, which will be described later.)

### 12.2.2.   Trellis representation

A convenient and fruitful way of describing a set of signal sequences is through a *trellis*, as we learned in Chapter 6 in the context of CPM, and in Chapter 11 in the context of convolutional codes. That is, we choose sequences that correspond to paths in a trellis which (apart from its initial and terminating stages) is periodic.[2] The nodes of the trellis are called the *states* of the encoder. Assume the source emits $M'$-ary symbols. With each of them we associate a branch which stems from each encoder state at time $k$ and reaches a new encoder state at time $k + 1$. The branch is labeled by the corresponding transmitted signal $\mathbf{s} \in \mathcal{S}$.

Thus, with $M'$-ary source symbols, each node must have $M'$ branches stemming from it (one per each source symbol). As we shall see, in some cases two or more branches connect the same pair of nodes. If this occurs, we say that *parallel transitions* take place.[3]

Fig. 12.4 shows an example of a trellis representation. It is assumed that the encoder has four states, the source emits binary symbols, and a constellation with four signals denoted $0, 1, 2, 3$ is used.

### 12.2.3.   Decoding TCM

Due to the one-to-one correspondence between signal sequences and paths traversing the trellis, maximum-likelihood (ML) decoding consists of searching for the

---

[1] Here and in the following we disregard certain losses caused by the demodulation of a constellation of larger size. For example, carrier-phase recovery with 8PSK may entail a loss of performance with respect to 4PSK.

[2] This is by no means the only possible choice. Nonperiodic, i.e., time-varying trellises are also possible, although seldom used in practice.

[3] In this context, we may describe uncoded transmission by a trellis that degenerates to a single state, and all of whose transitions are parallel.

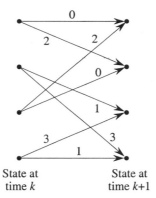

State at                                          State at
time $k$                                          time $k$+1

Figure 12.4: *Example of a trellis describing a TCM scheme with four states and four signals, denoted* $\{0, 1, 2, 3\}$, *used to transmit from a binary source.*

trellis path with the minimum Euclidean distance to the received signal sequence. If a sequence of length $K$ is transmitted, and the sequence $\mathbf{r}_0, \mathbf{r}_1, \ldots, \mathbf{r}_{K-1}$ is observed at the output of the AWGN channel, then the ML receiver looks for the sequence $\mathbf{x}_0, \mathbf{x}_1, \ldots, \mathbf{x}_{K-1}$ that minimizes

$$\sum_{i=0}^{K-1} |\mathbf{r}_i - \mathbf{x}_i|^2$$

This is done by using the Viterbi algorithm (Appendix F). The branch metrics to be used are obtained as follows. If the branch in the trellis used for coding is labeled by signal $\mathbf{x}$, then at discrete time $i$ the metric associated with that branch is $|\mathbf{r}_i - \mathbf{x}|^2$ if there are no parallel transitions. If a pair of nodes is connected by parallel transitions, and the branches have labels $\mathbf{x}', \mathbf{x}'', \ldots$, in the set $\mathcal{X}$, then in the trellis used for decoding the same pair of nodes is connected by a single branch, whose metric is

$$\min_{\mathbf{x} \in \mathcal{X}} |\mathbf{r}_i - \mathbf{x}|^2$$

That is, in the presence of parallel transitions the decoder first selects the signal, among $\mathbf{x}', \mathbf{x}'', \ldots$, at the minimum distance from $\mathbf{r}_i$ (this is a "demodulation" operation), then builds the metric based on the signal selected.

#### 12.2.4.  Free distance of TCM

The distance properties of a TCM scheme can be studied through its trellis diagram in the same way as for convolutional codes. It should be kept in mind, however, that the uniform error property (valid for convolutional codes) does not necessarily hold for TCM schemes, so that certain simplifications may not be possible. Recall that optimum decoding is the search of the most likely path through the trellis once the received sequence has been observed at the channel output. Because of the noise, the path chosen may not coincide with the correct path, i.e., the one traced by the sequence of source symbols, but will occasionally diverge from it (at time $n$, say) and remerge at a later time $n + L$. When this happens, we say that an *error event* of length $L$ has taken place. Thus, the *free distance* of a TCM scheme is the minimum Euclidean distance between two paths forming an error event.

### 12.3.   Some examples of TCM schemes

Here we describe a few examples of TCM schemes and describe their coding gains. We do this before providing tools for performance evaluation, with the aim of motivating the in-depth analysis of TCM that follows.

Consider first the transmission from a source with 2 bits per symbol. With uncoded transmission a channel constellation with $M' = 4$ would be adequate. We shall examine TCM schemes with $M = 2M' = 8$, i.e., such that the redundancy needed to achieve a coding gain is obtained by expanding the constellation size by a factor of 2.

Let us examine PSK signals first. With $M' = 4$ we obtain

$$\frac{\delta_{\min}^2}{\mathcal{E}'} = 2$$

a figure which will be used as a baseline to compute the coding gain of PSK-based TCM. We use TCM schemes based on the octonary PSK constellation whose signals we label $\{0, 1, 2, \ldots, 7\}$ as shown in Fig. 12.5. We have

$$\mathcal{E}' = \frac{\delta'^2}{4 \sin^2 \pi/8}$$

**Two states.**   Consider first a scheme with two states, as shown in Fig. 12.6. If the encoder is in state $S_1$, the subconstellation $\{0, 2, 4, 6\}$ is used. If it is in state $S_2$, constellation $\{1, 3, 5, 7\}$ is used instead. The free distance of this

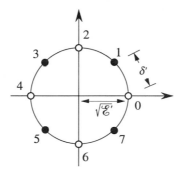

Figure 12.5: *Octonary constellation used in a TCM scheme.*

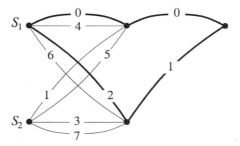

Figure 12.6: *A TCM scheme based on a 2-state trellis, $M' = 4$, and $M = 8$.*

TCM scheme is obtained by choosing the smallest among the distances between signals associated with parallel transitions (error events of length 1) and the distances associated with a pair of paths in the trellis that originate from a common node and merge into a single node at a later time (error events of length greater than 1). The pair of paths yielding the free distance is shown in Fig. 12.6, and, with $\delta(i, j)$ denoting the Euclidean distance between signals $i$ and $j$, we have the following::

$$\frac{\delta_{\text{free}}^2}{\mathcal{E}} = \frac{1}{\mathcal{E}}[\delta^2(0, 2) + \delta^2(0, 1)] = 2 + 4\sin^2\frac{\pi}{8} = 2.586$$

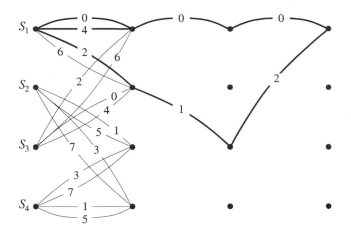

Figure 12.7: *A TCM scheme based on a 4-state trellis, $M' = 4$, and $M = 8$.*

Hence, we obtain an asymptotic coding gain over 4PSK

$$\gamma = \frac{2.586}{2} = 1.293 \Rightarrow 1.1 \text{ dB}$$

**Four states.**    Let us now use a TCM scheme with a more complex structure in the hope of increasing the coding gain. With the same constellation of Fig. 12.5, take a trellis with 4 states as in Fig. 12.7. We associate the constellation $\{0, 2, 4, 6\}$ with states $S_1$ and $S_3$, and $\{1, 3, 5, 7\}$ with $S_2$ and $S_4$. In this case the error event leading to $\delta_{\text{free}}$ has length 1 (a parallel transition), and is shown in Fig. 12.7. We get

$$\frac{\delta_{\text{free}}^2}{\mathcal{E}} = \delta^2(0, 4) = 4$$

and hence

$$\gamma = \frac{4}{2} = 2 \Rightarrow 3 \text{ dB}$$

**Eight states.**    A further step in the road toward higher complexities, and hence higher coding gains, can be taken by choosing a trellis with 8 states as shown in Fig. 12.8. To simplify the figure, the four symbols associated with the branches emanating from each node are used as node labels. The first symbol in each node label is associated with the uppermost transition from the node, the second symbol with the transition immediately below it, etc. The error event leading to

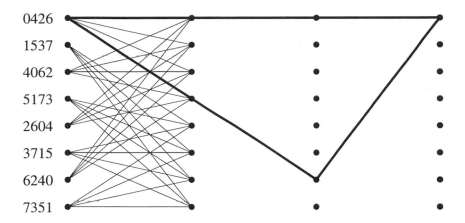

Figure 12.8: *A TCM scheme based on an 8-state trellis, $M' = 4$, and $M = 8$.*

$\delta_{\text{free}}$ is also shown. It yields

$$\frac{\delta^2_{\text{free}}}{\mathcal{E}} = \frac{1}{\mathcal{E}}[\delta^2(0,6) + \delta^2(0,7) + \delta^2(0,6)] = 2 + 4\sin^2\frac{\pi}{8} + 2 = 4.586$$

and hence

$$\gamma = \frac{4.586}{2} = 2.293 \Rightarrow 3.6 \text{ dB}$$

**Consideration of QAM.**    Consider now the transmission of 3 bits per symbol and quadrature amplitude modulation (QAM) schemes. The octonary constellation of Fig. 12.9 (black dots) will be used as the baseline uncoded scheme. It yields

$$\frac{\delta^2_{\text{min}}}{\mathcal{E}'} = 0.8$$

A TCM scheme with 8 states and based on this constellation is shown in Fig. 12.10. The subconstellations used are

$$\{0, 2, 5, 7, 8, 10, 13, 15\}$$

and

$$\{1, 3, 4, 6, 9, 11, 12, 14\}$$

The free distance is obtained from

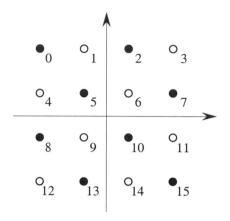

Figure 12.9: *The octonary QAM constellation* $\{0, 2, 5, 7, 8, 10, 13, 15\}$ *and the 16-ary QAM constellation* $\{0, 1, \ldots, 15\}$.

0, 10, 2, 8, 5, 15, 7, 13
1, 11, 3, 9, 4, 14, 6, 12
2, 8, 0, 10, 7, 13, 5, 15
3, 9, 1, 11, 6, 12, 4, 14
5, 15, 7, 13, 0, 10, 2, 8
4, 14, 6, 12, 1, 11, 3, 9
7, 13, 5, 15, 2, 8, 0, 10
6, 12, 4, 14, 3, 9, 1, 11

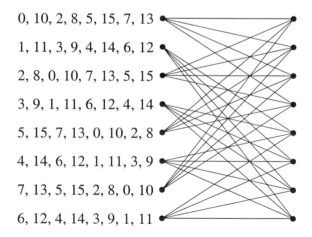

Figure 12.10: *A TCM scheme based on an 8-state trellis,* $M' = 8$, *and* $M = 16$.

$$\frac{\delta^2_{\text{free}}}{\mathcal{E}} = \frac{1}{\mathcal{E}}[\delta^2(10,13) + \delta^2(0,1) + \delta^2(0,5)]$$

$$= \frac{1}{\mathcal{E}}[0.8\mathcal{E} + 0.4\mathcal{E} + 0.8\mathcal{E}]$$

$$= 2$$

so that

$$\gamma = \frac{2}{0.8} = 2.5 \Rightarrow 3.98 \text{ dB}$$

### 12.3.1.   Coding gains achieved by TCM schemes

The values of $\delta_{\text{free}}$ achieved by actual designs based on two-dimensional modulations (PSK and QAM) are shown in Fig. 12.11. Free distances here are expressed in dB relative to the value $\delta^2_{\min} = 2$ of unit-energy uncoded 4PSK. The free distances of various schemes are plotted versus $R_s/W$, the bandwidth efficiency in bit/s/Hz, under the assumption that the signal bandwidth is the Shannon bandwidth $1/T$. Note that significant coding gains can be achieved by TCM schemes having as few as 4, 8, and 16 states. A rule of thumb is that 3 dB can be gained with 4 states, 4 dB with 8 states, 5 dB with 32 states, and up to 6 dB with 128 or more states. With higher numbers of states the returns are diminishing.

### 12.3.2.   Set partitioning

Consider the determination of $\delta_{\text{free}}$. This is the Euclidean distance between the signals associated with a pair of paths that originate from an initial split and, after $L$ (say) time instants, merge into a single node as shown in Fig. 12.12. Assume first that the free distance is determined by parallel transitions, i.e., $L = 1$. Then $\delta_{\text{free}}$ equals the minimum distance between the signals in the set associated with the branches emanating from a given node. Consider next $L > 1$. With A, B, C, D denoting subsets of signals associated with each branch, and $\delta(X,Y)$ denoting the minimum Euclidean distance between one signal in X and one in Y, $\delta^2_{\text{free}}$ will have the expression

$$\delta^2_{\text{free}} = \delta^2(A,B) + \cdots + \delta^2(C,D)$$

This implies that, in a good code, the signal subsets assigned to the same originating state (A and B in Fig. 12.12) or to the same terminating state (C and D in Fig. 12.12) must have the largest possible distance. To put this observation into practice, Ungerboeck (1982) suggested the following technique, called *set partitioning*.

The $M$-ary constellation is successively partitioned into $2, 4, 8, \ldots$, subsets with size $M/2, M/4, M/8, \ldots$, having progressively larger minimum Euclidean distances $\delta^{(1)}_{\min}, \delta^{(2)}_{\min}, \delta^{(3)}_{\min}, \ldots$ (see Fig. 12.13 and Fig. 12.14). Then,

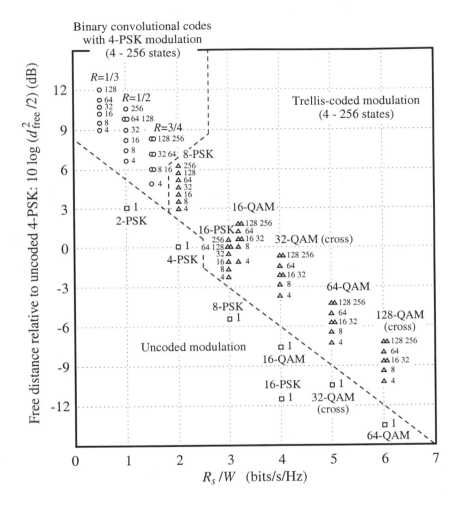

Figure 12.11: *Free distance vs. bandwidth efficiency of selected TCM schemes based on two-dimensional modulations. (Adapted from Ungerboeck, 1987.)*

**U1** Members of the same partition are assigned to parallel transitions.

**U2** Members of the next larger partition are assigned to "adjacent" transitions, i.e., transitions stemming from, or merging into, the same node.

These two rules, in conjunction with the symmetry requirement

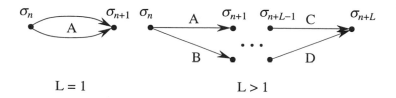

Figure 12.12: *A pair of splitting and remerging paths, for L = 1 (parallel transitions), and L > 1.*

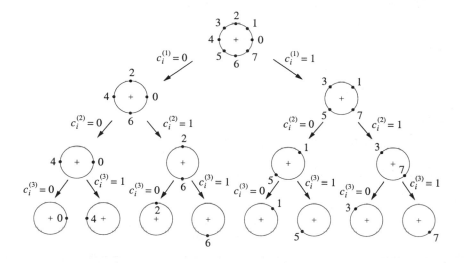

Figure 12.13: *Set partition of an 8PSK constellation.*

**U3** All the signals are used equally often,

are deemed to give rise to the best TCM schemes, and are usually referred to as the three "Ungerboeck's rules."

### 12.3.3.   Representation of TCM

We now examine the design of TCM encoders. In particular, we examine TCM encoders consisting of a convolutional encoder cascaded to a memoryless mapper.

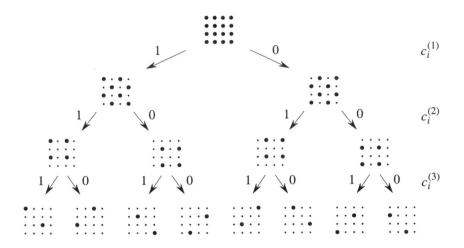

Figure 12.14: *Set partition of a 16-QAM constellation.*

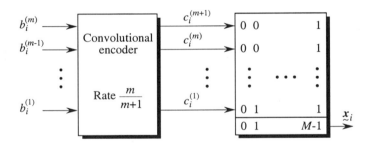

Figure 12.15: *A TCM encoder.*

Consider binary source symbols, grouped into blocks of $m$ bits $b_i^{(1)}, \cdots b_i^{(m)}$ that are presented simultaneously to a convolutional encoder with rate $m/(m + 1)$. The latter determines the trellis structure of the TCM scheme (and, in particular, the number of its states). A memoryless mapper following the encoder generates a one-to-one correspondence between the binary coded $(m+1)$-tuples and a constellation with $M = 2^{m+1}$ signals (see Fig. 12.15).

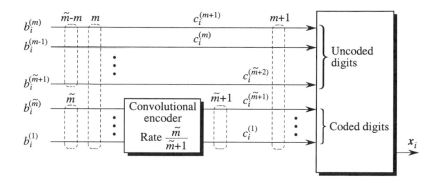

Figure 12.16: *A TCM encoder where the bits that are left uncoded are shown explicitly.*

It is convenient to modify the above representation as described in Fig.12.16, where the fact that some of the source bits are left uncoded is made explicit. The convolutional code appearing here has rate $\tilde{m}/(\tilde{m}+1)$. The presence of uncoded digits causes parallel transitions; a branch in the trellis diagram of the code is now associated with $2^{m-\tilde{m}}$ signals. The correspondence between the encoded digits and the subconstellations obtained from set partitioning is shown in Fig. 12.13 and Fig. 12.14.

**Example 12.2**   Fig. 12.17 shows a TCM encoder and the corresponding trellis. Here $m = 2$ and $\tilde{m} = 1$, so that the trellis nodes are connected by parallel transitions associated with 2 signals each. The trellis has four states, as does the rate-1/2 convolutional encoder, and its structure is determined by the latter. $\quad\Box$

### 12.3.4.   TCM with multidimensional constellations

We have seen that, for a given signal constellation, the performance of TCM can be improved by increasing the number of trellis states. However, as this number exceeds a certain value, the increase of coding gain is progressively diminishing. This suggests that to achieve larger gains the constellation should be changed. A possibility is to move from two-dimensional to multidimensional constellations.

Here we expand briefly on the use of constellations that are generated by time-division of elementary (typically, two-dimensional) constellations. For ex-

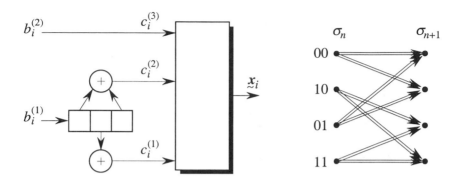

Figure 12.17: *A TCM encoder with* $m = 2$ *and* $\tilde{m} = 1$. *The corresponding trellis is also shown.*

ample, if $N$ two-dimensional signals are sent in a time interval of duration $T$, and each has duration $T/N$, we obtain a $2N$-dimensional constellation.

Use of these multidimensional constellations in TCM offers a number of advantages:

- Spaces with larger dimensionality have more room for the signals, which can consequently be spaced at larger Euclidean distances. Thus, an increased noise margin may come from the constellation itself.

- An inherent cost with one- and two-dimensional constellations is that when the size of the constellation is doubled over that of an uncoded scheme it may occur that $\mathcal{E} > \mathcal{E}'$, that is, the average energy needed for signaling increases. For example, with two-dimensional rectangular constellations (QAM) doubling the constellation size costs roughly 3 dB in energy. Without this increase in energy expenditure, the coding gain of TCM would be greater. Now it happens that if we use a multidimensional rectangular constellation this cost falls from 3 dB for two dimensions to 1.5 dB (four dimensions) or to 0.75 dB (eight dimensions).

- As we shall see, for some applications it might be necessary to design TCM schemes that are transparent to phase rotations. Multidimensional constellations may simplify the design of these "rotationally invariant" schemes.

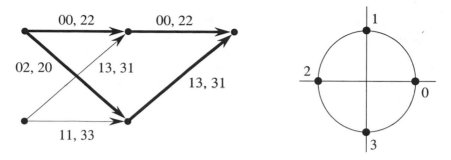

Figure 12.18: *A 2-state TCM scheme based on a 2×4PSK constellation. The error event providing the free Euclidean distance is also shown.*

**Example 12.3**  As an example of a TCM scheme based on a multidimensional signals, consider the 4-dimensional constellation obtained by pairing 4PSK signals. This is denoted 2×4PSK. With the signal labeling of Fig. 12.18, the $4^2 = 16$ four-dimensional signals are

$$\{00, 01, 02, 03, 10, 11, 12, 13, 20, 21, 22, 23, 30, 31, 32, 33\}$$

This constellation achieves the same minimum squared distance as two-dimensional 4PSK, viz.,

$$\delta_{\min}^2 = \delta^2(00, 01) = \delta^2(0, 1) = 2$$

The following subconstellation has 8 signals and a minimum squared distance 4:

$$\mathcal{S} = \{00, 02, 11, 13, 20, 22, 31, 33\}$$

With $\mathcal{S}$ partitioned into the four subsets

$$\{00, 22\} \qquad \{20, 02\} \qquad \{13, 31\} \qquad \{11, 33\}$$

the choice of a two-state trellis provides the TCM scheme shown in Fig. 12.18. This has a squared free distance 8. □

## 12.4. Error probability of TCM

This section is devoted to evaluating the error probability of a TCM scheme. Here we assume the transmission to occur over the additive white Gaussian noise channel, and the detection to be ML. Not surprisingly, we shall find that, asymptotically, the error probability is upper- and lower-bounded by functions that decrease monotonically when $\delta_{\text{free}}$ increases. This fact proves that the free Euclidean distance is the most significant single parameter useful for comparing TCM schemes employed for transmission over the additive white Gaussian noise channel when the signal-to-noise ratio is large enough. This also explains why $\gamma$, the increase from minimum distance to free distance caused by the introduction of TCM, was called "asymptotic coding gain."

Since there appears to be no general technique for choosing an optimum TCM scheme, the selection of any such scheme is typically based on a search among a wide subclass. Thus, it is extremely important that computationally efficient algorithms for the computation of free distance and error probability be available.

### 12.4.1.  Upper bound to error event probability

Recall the scheme of Fig. 12.15. A rate-$m/(m+1)$ convolutional code accepts $m$ binary source symbols $\mathbf{b}_i$ at one time, and transforms them into blocks $\mathbf{c}_i$ of $m+1$ binary symbols, that are fed into a memoryless mapper. This mapper outputs channel symbols $\mathbf{x}_i$. From now on, the binary $(m+1)$-tuple $\mathbf{c}_i$ is called the *label* of signal $\mathbf{x}_i$.

Since there is a one-to-one correspondence between $\mathbf{x}_i$ and its label $\mathbf{c}_i$, two $L$-tuples of signals can be equivalently described by the two $L$-tuples of their labels, namely

$$\mathbf{c}_k, \mathbf{c}_{k+1}, \cdots, \mathbf{c}_{k+L-1}$$

and

$$\mathbf{c}'_k = \mathbf{c}_k \oplus \mathbf{e}_k, \quad \mathbf{c}'_{k+1} = \mathbf{c}_{k+1} \oplus \mathbf{e}_{k+1}, \quad \cdots \quad, \mathbf{c}'_{k+L-1} = \mathbf{c}_{k+L-1} \oplus \mathbf{e}_{k+L-1}$$

where $\mathbf{e}_i$, $i = k, \ldots, k + L - 1$, form a sequence of binary vectors, called from now on *error vectors*, and $\oplus$ denotes modulo-2 addition.

Now, let $\mathbf{X}_L$ and $\widehat{\mathbf{X}}_L$ denote two signal-vector sequences of length $L$. With these notations, an *error event* of length $L$ occurs when the demodulator chooses, instead of the transmitted sequence $\mathbf{X}_L$, a different sequence $\widehat{\mathbf{X}}_L$ corresponding to a trellis path that splits from the correct path at a given time, and remerges exactly $L$ discrete times later. The error probability is then obtained by summing over $L$, $L = 1, 2, \cdots$, the probabilities of error events of length $L$, i.e., the joint probabilities that $\mathbf{X}_L$ is transmitted and $\widehat{\mathbf{X}}_L$ detected.

The union bound of Section 4.3 provides the following inequality for the probability of an error event:

$$P(e) \leq \sum_{L=1}^{\infty} \sum_{\mathbf{X}_L} \sum_{\widehat{\mathbf{X}}_L \neq \mathbf{X}_L} P\{\mathbf{X}_L\} P\{\mathbf{X}_L \to \widehat{\mathbf{X}}_L\} \qquad (12.17)$$

Since we assume a one-to-one correspondence between output symbols and labels, by letting $\mathbf{C}_L$ denote an $L$-sequence of labels $c_i$ and $\mathbf{E}_L$ an $L$-sequence of error vectors $e_i$, we can rewrite (12.17) in the form

$$
\begin{aligned}
P(e) &\leq \sum_{L=1}^{\infty} \sum_{\mathbf{C}_L} P\{\mathbf{C}_L\} \sum_{\widehat{\mathbf{C}}_L \neq \mathbf{C}_L} P\{\mathbf{C}_L \rightarrow \widehat{\mathbf{C}}_L\} \\
&= \sum_{L=1}^{\infty} \sum_{\mathbf{C}_L} P\{\mathbf{C}_L\} \sum_{\mathbf{E}_L \neq 0} P\{\mathbf{C}_L \rightarrow \mathbf{C}_L \oplus \mathbf{E}_L\} \\
&= \sum_{L=1}^{\infty} \sum_{\mathbf{E}_L \neq 0} P\{\mathbf{E}_L\}
\end{aligned}
\tag{12.18}
$$

where we have defined the quantity

$$
P\{\mathbf{E}_L\} \triangleq \sum_{\mathbf{C}_L} P\{\mathbf{C}_L\} P\{\mathbf{C}_L \rightarrow \mathbf{C}_L \oplus \mathbf{E}_L\}
\tag{12.19}
$$

expressing the average pairwise probability of the specific error event of length $L$ caused by the error sequence $\mathbf{E}_L$. The pairwise error probabilities appearing in the last equation can be computed in a closed form. However, we shall not take advantage of this fact, and rather use a bound leading to the Bhattacharyya bound of Section 4.3.

Specifically, denote by $f(\mathbf{c})$ the signal with label $\mathbf{c}$, and, with an abuse of notation, by $f(\mathbf{C}_L)$ the sequence of signals with label sequence $\mathbf{C}_L$. We have, using (A.5)

$$
\begin{aligned}
P\{\mathbf{C}_L \rightarrow \widehat{\mathbf{C}}_L\} &= \frac{1}{2} \mathrm{erfc}\left( \frac{|f(\mathbf{C}_L) - f(\widehat{\mathbf{C}}_L)|}{2\sqrt{N_0}} \right) \\
&\leq \frac{1}{2} \exp\left\{ -\frac{1}{4N_0} |f(\mathbf{C}_L) - f(\widehat{\mathbf{C}}_L)|^2 \right\} \\
&= \frac{1}{2} \exp\left\{ -\frac{1}{4N_0} \sum_{n=1}^{L} |f(\mathbf{c}_n) - f(\widehat{\mathbf{c}}_n)|^2 \right\}
\end{aligned}
\tag{12.20}
$$

Define now the function

$$
W(\mathbf{E}_L) \triangleq \sum_{\mathbf{C}_L} P\{\mathbf{C}_L\} e^{-|f(\mathbf{C}_L) - f(\mathbf{C}_L \oplus \mathbf{E}_L)|^2 / 4N_0}
\tag{12.21}
$$

By observing that $P\{\mathbf{C}_L\} = P\{\mathbf{X}_L\}$, (12.17) can be rewritten in the form

$$
P(e) \leq \frac{1}{2} \sum_{L=1}^{\infty} \sum_{\mathbf{E}_L \neq 0} W(\mathbf{E}_L).
\tag{12.22}
$$

Eq. (12.22), our final result, shows that $P(e)$ is upper-bounded by a sum, over all the possible error-event lengths, of functions of the vectors $\mathbf{E}_L$ causing them. Thus, our next task toward the evaluation of $P(e)$ will be to enumerate these vectors. Before doing this, we pause for a moment to observe that a technique often used (in particular with TCM schemes with a large number of states, or in conjunction with transmission over channels other than AWGN) consists of summing, in the right-hand side of (12.22), a finite number of terms, chosen among the shortest error events. Since these are expected to have the smallest distances, they should contribute most to error-event probability. Needless to say, this technique should be used with the utmost care, since the truncation of a union bound might not result in a bound.

### Enumerating the error events

We enumerate all the error vectors by using the transfer function of an *error state diagram*, i.e., a graph whose branches have labels that are $N_\sigma$ by $N_\sigma$ matrices, where $N_\sigma$ denotes the number of states of the trellis. Specifically, recall that under our assumptions the source symbols have equal probabilities $2^{-m}$, and define the $N_\sigma \times N_\sigma$ "error-weight matrices" $\mathbf{G}(\mathbf{e}_n)$ as follows. The entry $i, j$ of $\mathbf{G}(\mathbf{e}_n)$ is *zero* if no transition from the code trellis state $i$ to the state $j$ is possible. Otherwise, it is given by

$$[\mathbf{G}(\mathbf{e}_n)]_{i,j} = 2^{-m} \sum_{\mathbf{c}_{i \to j}} Z^{|f(\mathbf{c}_{i \to j}) - f(\mathbf{c}_{i \to j} \oplus \mathbf{e}_n)|^2} \qquad (12.23)$$

where $\mathbf{c}_{i \to j}$ are the label vectors generated by the transition from state $i$ to state $j$ (the sum accounts for possible parallel transitions between the two states).

With these notations, to any sequence $\mathbf{E}_L = \mathbf{e}_1, \cdots, \mathbf{e}_L$ of labels in the error state diagram there corresponds a sequence of $L$ error-weight matrices $\mathbf{G}(\mathbf{e}_1), \cdots, \mathbf{G}(\mathbf{e}_L)$, and we have

$$W(\mathbf{E}_L) = \frac{1}{N_\sigma} \mathbf{1}' \left[ \prod_{n=1}^{L} \mathbf{G}(\mathbf{e}_n) \right] \mathbf{1} \Bigg|_{Z = e^{-1/4N_0}} \qquad (12.24)$$

where $\mathbf{1}$ is the column $N_\sigma$-vector all of whose elements are 1 (consequently, for any $N_\sigma \times N_\sigma$ matrix $\mathbf{A}$, $\mathbf{1}'\mathbf{A}\mathbf{1}$ is the sum of all the entries of $\mathbf{A}$.) It should be apparent that the element $i, j$ of the matrix $\prod_{n=1}^{L} \mathbf{G}(\mathbf{e}_n)$ enumerates the Euclidean distances involved in the transitions from state $i$ to state $j$ in exactly $L$ steps. Thus, what we need next to compute $P(e)$ is to sum $W(\mathbf{E}_L)$ over the possible error sequences $\mathbf{E}_L$, according to (12.22).

Figure 12.19: *Trellis diagram for a 2-state, $m = 1$ TCM scheme. The branch labels are the components of* **c**. *The error state diagram is also shown.*

**The error state diagram.**    Due to the linearity of the convolutional code generating the TCM scheme, the set of possible sequences $e_1, \cdots, e_L$ is the same as the set of coded sequences. Thus, the error sequences can be described by using the same trellis associated with the encoder, and can be enumerated by making use of a state diagram which is a copy of the one describing the code. We call it the *error state diagram*. It has a structure determined by the convolutional code, and differs from the code state diagram only in its branch labels, which are now the matrices $\mathbf{G}(e_i)$.

**The transfer function bound.**    From (12.24) and (12.22) we have

$$P(e) \leq \frac{1}{2} T(Z) \Big|_{Z = \exp(-1/4N_0)} \tag{12.25}$$

where

$$T(Z) = \frac{1}{N_\sigma} \mathbf{1}' \mathbf{G} \mathbf{1} \tag{12.26}$$

and the matrix

$$\mathbf{G} \triangleq \sum_{L=1}^{\infty} \sum_{\mathbf{E}_L \neq 0} \prod_{n=1}^{L} \mathbf{G}(e_n) \tag{12.27}$$

is the matrix transfer function of the error state diagram. $T(Z)$ will be called the (scalar) transfer function of the error state diagram.

**Example 12.4**    Consider the TCM scheme one section of whose trellis diagram is shown in Fig. 12.19. Here $m = 1$ and $M = 4$ (binary source, quaternary signals). The error state diagram is also shown. If we denote the error vector as $\mathbf{e} = (e_2 e_1)$ and we let $\bar{e} = 1 \oplus e$ (i.e., $\bar{e}$ denotes the complement of $e$), then we can write the general form of the matrix $\mathbf{G}(\mathbf{e})$ as

$$\mathbf{G}(e_2 e_1) = \frac{1}{2} \begin{bmatrix} Z^{\|f(00)-f(e_2 e_1)\|^2} & Z^{\|f(10)-f(\bar{e}_2 e_1)\|^2} \\ Z^{\|f(01)-f(e_2 \bar{e}_1)\|^2} & Z^{\|f(11)-f(\bar{e}_2 \bar{e}_1)\|^2} \end{bmatrix} \tag{12.28}$$

The transfer function of the error state diagram is

$$\mathbf{G} = \mathbf{G}(10) \left[ \mathbf{I} - \mathbf{G}(11) \right]^{-1} \mathbf{G}(01) \tag{12.29}$$

where $\mathbf{I}$ denotes the $2 \times 2$ identity matrix.

Notice that (12.28) and (12.29) could be written without specifying the signals used in the TCM scheme. Actually, to give the signal constellation corresponds to specifying the four values taken on by the function $f(\cdot)$. These will provide the values of the entries of $\mathbf{G}(e_2 e_1)$ from which the transfer function $T(Z)$ is computed.

Consider first quaternary PAM, with the mapping

$$f(00) = +3 \quad f(01) = 1 \quad f(10) = -1 \quad f(11) = -3$$

In this case we have

$$\mathbf{G}(00) = \frac{1}{2} \begin{bmatrix} 1 & 1 \\ 1 & 1 \end{bmatrix} \tag{12.30}$$

$$\mathbf{G}(01) = \frac{1}{2} \begin{bmatrix} Z^4 & Z^4 \\ Z^4 & Z^4 \end{bmatrix} \tag{12.31}$$

$$\mathbf{G}(10) = \frac{1}{2} \begin{bmatrix} Z^{16} & Z^{16} \\ Z^{16} & Z^{16} \end{bmatrix} \tag{12.32}$$

and

$$\mathbf{G}(11) = \frac{1}{2} \begin{bmatrix} Z^{36} & Z^4 \\ Z^4 & Z^{36} \end{bmatrix} \tag{12.33}$$

so that from (12.29) we obtain

$$\mathbf{G} = \frac{1}{2} \frac{Z^{20}}{1 - \frac{1}{2}(Z^4 + Z^{36})} \begin{bmatrix} 1 & 1 \\ 1 & 1 \end{bmatrix} \tag{12.34}$$

In conclusion, we get the transfer function

$$T(Z) = \frac{1}{2} \mathbf{1}' \mathbf{G} \mathbf{1} = \frac{Z^{20}}{1 - \frac{1}{2}(Z^4 + Z^{36})} \tag{12.35}$$

If we consider instead a unit-energy 4PSK constellation as in Fig. 12.20, we have

$$f(00) = 1 \quad f(01) = j \quad f(10) = -1 \quad f(11) = -j$$

so that

$$\mathbf{G}(00) = \frac{1}{2} \begin{bmatrix} 1 & 1 \\ 1 & 1 \end{bmatrix} \tag{12.36}$$

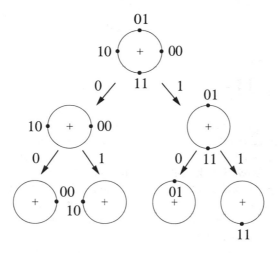

Figure 12.20: *Signal constellation and set partition for 4PSK.*

$$\mathbf{G}(01) = \frac{1}{2}\left[\begin{array}{cc} Z^2 & Z^2 \\ Z^2 & Z^2 \end{array}\right] \qquad (12.37)$$

$$\mathbf{G}(10) = \frac{1}{2}\left[\begin{array}{cc} Z^4 & Z^4 \\ Z^4 & Z^4 \end{array}\right] \qquad (12.38)$$

and

$$\mathbf{G}(11) = \frac{1}{2}\left[\begin{array}{cc} Z^2 & Z^2 \\ Z^2 & Z^2 \end{array}\right] \qquad (12.39)$$

In conclusion,

$$\mathbf{G} = \frac{1}{2}\frac{Z^6}{1-Z^2}\left[\begin{array}{cc} 1 & 1 \\ 1 & 1 \end{array}\right], \qquad (12.40)$$

which yields the transfer function

$$T(Z) = \frac{1}{2}\mathbf{1}'\mathbf{G}\mathbf{1} = \frac{Z^6}{1-Z^2} \qquad (12.41)$$

$\square$

**Interpretation and symmetry considerations**

Let us interpret the meaning of the matrix $\mathbf{G}$ defined in (12.27). We may observe that the entry $i, j$ of $\mathbf{G}$ provides an upper bound to the probability that an error event occurs starting from node $i$ and ending at node $j$. Similarly, $\frac{1}{N_\sigma}\mathbf{G1}$ is a vector whose $i$th entry is a bound to the probability of any error event starting from node $i$, and $\frac{1}{N_\sigma}\mathbf{1}'\mathbf{G}$ is a vector whose $j$-th entry is a bound to the probability of any error event ending at node $j$.

Inspection of the matrix $\mathbf{G}$ leads to the consideration of different degrees of symmetry implied in a TCM scheme. In some cases the matrix $\mathbf{G}$ has equal entries: this is the case of the 4PSK example above. This fact can be interpreted by saying that *all the paths* in the trellis are on an equal footing, i.e., they contribute equally to the error probability[4]. Hence, in the analysis of the TCM scheme we may take any single path as a reference, and compute error probabilities by assuming any given transmitted sequence. A simple sufficient condition for this symmetry to hold is that all the matrices $\mathbf{G}(e)$ have equal entries. However, this condition is not necessary, as proved by examining the 4-PAM example above: it has $\mathbf{G}$ with equal entries although the entries of $\mathbf{G}(11)$ are unequal.

If all the matrices $\mathbf{G}(e)$ have equal entries, then in the computation of the transfer function bound the branches of the error state diagram can be simply labeled by the common entries of these matrices, thus leading to a *scalar* transfer function. However, for the computations to be done in terms of scalars it is not necessary to require such a high degree of symmetry. All is needed is the looser symmetry that arises when *the sum of all the elements in any row (or a column) of $\mathbf{G}$ does not depend on the row (or column) itself.* This symmetry corresponds to having *all the states* on an equal footing, and allows consideration of *a single reference state rather that all the pairs of states* for the computation of error probabilities. More specifically, only the error events leaving from a fixed node (in the case of equal row sums) or reaching a fixed node (in the case of equal column sums) need be considered. This point is discussed and made more precise in the next paragraph.

**Algebraic conditions for scalar transfer functions.**   We now state some simple conditions which, if satisfied, will make it possible to compute a transfer function bound based on *scalar,* rather than matrix, branch labels.

Given a square matrix $\mathbf{A}$, if $\mathbf{1}$ is an eigenvector of its transpose $\mathbf{A}'$, i.e.,

$$\mathbf{1}'\mathbf{A} = \alpha\mathbf{1}'$$

---

[4]More precisely, we should say that they contribute equally to the *upper bound to* error event probability. However, here and in the following we shall avoid making this distinction.

where $\alpha$ is some constant, then the sum of its elements in a column does not depend on the column order. We call $\mathbf{A}$ *column-uniform*. Similarly, if $\mathbf{1}$ is an eigenvector of the square matrix $\mathbf{B}$, i.e.,

$$\mathbf{B}\mathbf{1} = \beta\mathbf{1}$$

where $\beta$ is some constant, then the sum of its elements in a row does not depend on the row order. In this case we call $\mathbf{B}$ *row-uniform*.

Now, the product and the sum of two column- or row-uniform matrices is itself column- or row-uniform. For example, if $\mathbf{B}_1$ and $\mathbf{B}_2$ are row-uniform with eigenvalues $\beta_1$ and $\beta_2$, respectively, and we define $\mathbf{B}_3 = \mathbf{B}_1 + \mathbf{B}_2$ and $\mathbf{B}_4 = \mathbf{B}_1\mathbf{B}_2$, we have

$$\mathbf{B}_3\mathbf{1} = (\beta_1 + \beta_2)\mathbf{1}$$

and

$$\mathbf{B}_4\mathbf{1} = \beta_1\beta_2\mathbf{1}$$

which show that $\mathbf{B}_3$ and $\mathbf{B}_4$ are also row-uniform, with eigenvalues $\beta_1 + \beta_2$ and $\beta_1\beta_2$, respectively. Also, for an $N \times N$ matrix $\mathbf{A}$ which is either row- or column-uniform, we have

$$\mathbf{1}'\mathbf{A}\mathbf{1} = N\alpha$$

From the above it follows that, if all the matrices $\mathbf{G}(\mathbf{e})$ are either row-uniform or column-uniform, then the transfer function (which is a sum of products of error matrices, as seen explicitly in (12.27)), can be computed by using scalar labels on the branches of the error state diagram. These labels are the sums of the elements in a row (column). In this case, we say that the TCM scheme is *uniform*. By recalling the definition of the matrices $\mathbf{G}(\mathbf{e})$, we have that $\mathbf{G}(\mathbf{e})$ is row-uniform if the transitions *stemming from* any node of the trellis carry the same set of labels, irrespective of the order of the transitions. It is column-uniform if the transitions *leading to* any node of the trellis carry the same set of labels.

**Asymptotic considerations**

The entry $i, j$ of the matrix $\mathbf{G}$ is a power series in $Z$. If we denote the general term in the series as $\nu_{ij}(\delta_\ell)Z^{\delta_\ell^2}$, where

$$\nu_{ij}(\delta_\ell) = \frac{1}{M^{L_1}}n_1 + \frac{1}{M^{L_2}}n_2 + \cdots$$

and $n_h$, $h = 1, 2, \cdots$, is the number of error paths starting from node $i$ at time 0 (say), remerging $L_h$ time instants later at node $j$, and whose associated distance is $\delta_\ell$. Since $1/M^{L_h}$ is the probability of a sequence of symbols of length $L_h$,

$\nu_{ij}(\delta_\ell)$ can be interpreted as the average number of competing paths at distance $\delta_\ell$ associated with any path in the code trellis starting at node $i$ and ending at node $j$. Consequently, the quantity

$$\nu_i(\delta_\ell) \triangleq \sum_j \nu_{ij}(\delta_\ell)$$

can be interpreted as the average number of competing paths at distance $\delta_\ell$ associated with any path in the code trellis leaving from node $i$ and ending at any node. Similarly,

$$N(\delta_\ell) \triangleq \frac{1}{N_\sigma^2} \sum_{i,j} \nu_{ij}(\delta_\ell)$$

is the average number of competing paths at distance $\delta_\ell$ associated with a path in the code trellis.

For large signal-to-noise ratios, i.e., $N_0 \to 0$, the only terms in the entries of the matrix $\mathbf{G}$ that provide a contribution to the error probability which is significantly different from zero will be of the type $\nu_{ij}(\delta_{\text{free}}) Z^{\delta_{\text{free}}^2}$. Thus, using (A.5) we have asymptotically

$$P(e) \sim \frac{1}{2} N(\delta_{\text{free}}) e^{-\delta_{\text{free}}^2/4N_0}$$

**An improved upper bound**

An upper bound on $P(e)$ better than (12.25) can be obtained by substituting for the Bhattacharyya bound in (12.20) a tighter expression. Specifically, let us recall that we have, exactly,

$$P\{\mathbf{C}_L \to \mathbf{C}'_L\} = \frac{1}{2} \text{erfc} \left( \frac{|f(\mathbf{C}_L) - f(\mathbf{C}'_L)|}{2\sqrt{N_0}} \right), \qquad (12.42)$$

Since the minimum value taken by $|f(\mathbf{C}_L) - f(\mathbf{C}'_L)|$ is what we call $\delta_{\text{free}}$, use of the inequality (see Problem 11.7)

$$\text{erfc}\left(\sqrt{x+y}\right) \leq \text{erfc}\left(\sqrt{x}\right) e^y, \qquad x \geq 0, \ y \geq 0 \qquad (12.43)$$

leads to the bound

$$P\{\mathbf{C}_L \to \mathbf{C}'_L\} \leq \frac{1}{2} \text{erfc} \left( \frac{\delta_{\text{free}}}{2\sqrt{N_0}} \right) e^{\delta_{\text{free}}/4N_0} \cdot \exp\left\{ -\frac{1}{4N_0} |f(\mathbf{C}_L) - f(\mathbf{C}'_L)|^2 \right\}.$$

$$(12.44)$$

In conclusion, we have the bound on error probability:

$$P(e) \leq \frac{1}{2} \text{erfc} \left( \frac{\delta_{\text{free}}}{2\sqrt{N_0}} \right) e^{\delta_{\text{free}}^2/4N_0} T(Z) \bigg|_{Z=e^{-1/4N_0}} \qquad (12.45)$$

We also have, approximately for high signal-to-noise ratios,

$$P(e) \cong N(\delta_{\text{free}}) \frac{1}{2} \text{erfc} \left( \frac{\delta_{\text{free}}}{2\sqrt{N_0}} \right) \tag{12.46}$$

**Bit error probability**

A bound on bit error probability can also be obtained by following the footsteps of a similar derivation for convolutional codes in the previous chapter. All that is needed here is a change in the error matrices. The entries of the matrix $\mathbf{G}(\mathbf{e})$ associated with the transition from state $i$ to state $j$ of the error state diagram must be multiplied by the factor $W^{\epsilon}$, where $\epsilon$ is the Hamming weight (i.e., the number of ones) of the input vector $\mathbf{b}$ that causes the transition $i \to j$.

With this new definition of the error matrices, the entry $i, j$ of the matrix $\mathbf{G}$ can now be expressed in a power series in the two indeterminates $Z$ and $W$. The general term of the series will be $\mu_{pq}(\delta_\ell, \epsilon_h) Z^{\delta_\ell^2} W^{\epsilon_h}$, where $\mu_{pq}(\delta_\ell, \epsilon_h)$ can be interpreted as the average number of paths having distance $\delta_\ell$ and $\epsilon_h$ bit errors with respect to any path in the trellis starting at node $i$ and ending at node $j$. If we take the derivative of these terms with respect to $W$ and set $W = 1$, each of them will provide the expected number of bit errors per branch generated by the incorrect paths from $i$ to $j$. If we further divide these quantities by $m$, the number of source bits per trellis transition, and we sum the series, we obtain the following upper bound on bit error probability:

$$P_b(e) \leq \frac{1}{2m} \frac{\partial}{\partial W} T_2(Z, W) \Big|_{W=1, Z=e^{-1/4N_0}} \tag{12.47}$$

Another upper bound can be obtained by substituting for the Bhattacharyya bound in (12.20) the tighter inequality derived above. We get

$$P_b(e) \leq \frac{1}{2m} \text{erfc}(\frac{\delta_{\text{free}}}{2\sqrt{N_0}}) \exp(\delta_{\text{free}}^2/4N_0) \frac{\partial}{\partial W} T_2(Z, W) \Big|_{W=1, Z=e^{-1/4N_0}} \tag{12.48}$$

**Convergence considerations and catastrophic codes.**    In our previous analysis we have not considered the issue of the convergence of the series providing the transfer function $T(Z)$ or $T_2(Z, W)$. Actually, we assumed implicitly that $T(Z)$, or the derivative of $T_2(Z, W)$, converges for large enough values of the signal-to-noise ratio. Now, there are situations in which the transfer function does not converge due to the fact that one or more of its coefficients take value infinity. This situation was already examined in the context of convolutional codes in Chapter 11, and may actually occur for certain TCM schemes in which two encoded sequences with a finite Euclidean distance correspond to source symbol sequences with infinite Hamming distance. The scheme is called "catastrophic" in analogy with convolutional codes.

**Lower bound to error probability**

We now derive a lower bound to the probability of an error event. This derivation is based on the fact that the error probability of any real-life receiver is larger than that of a receiver which makes use of side information provided by, say, a genie.

The genie-aided receiver operates as follows. The genie observes a long sequence of transmitted symbols, or, equivalently, the sequence

$$\mathbf{C} = (\mathbf{c}_i)_{i=0}^{K-1}$$

of labels, and tells the receiver that the transmitted sequence was either $\mathbf{C}$ *or the* sequence

$$\mathbf{C'} = (\mathbf{c}_i')_{i=0}^{K-1}$$

where $\mathbf{C'}$ is picked at random from the possibly transmitted sequences having the smallest Euclidean distance from $\mathbf{C}$ (not necessarily $\delta_{\text{free}}$, because $\mathbf{C}$ may not have any sequence $\mathbf{C'}$ at free distance).

The error probability for this genie-aided receiver is that of a binary transmission scheme in which the only transmitted sequences are $\mathbf{C}$ and $\mathbf{C'}$:

$$P_G(e \mid \mathbf{C}) = \frac{1}{2}\text{erfc}\left(\frac{|f(\mathbf{C}) - f(\mathbf{C'})|}{2\sqrt{N_0}}\right) \tag{12.49}$$

Now, consider the unconditional probability $P_G(e)$. We have

$$
\begin{aligned}
P_G(e) &= \sum_{\mathbf{C}} P(\mathbf{C})\frac{1}{2}\text{erfc}\left(\frac{|f(\mathbf{C}) - f(\mathbf{C'})|}{2\sqrt{N_0}}\right) \\
&\geq \sum_{\mathbf{C}} I(\mathbf{C})P(\mathbf{C})\frac{1}{2}\text{erfc}\left(\frac{\delta_{\text{free}}}{2\sqrt{N_0}}\right)
\end{aligned}
\tag{12.50}
$$

where $I(\mathbf{C}) = 1$ if $\mathbf{C}$ admits a sequence at $\delta_{\text{free}}$:

$$\min_{\mathbf{C'}} d[f(\mathbf{C}), f(\mathbf{C'})] = \delta_{\text{free}}$$

and $I(\mathbf{C}) = 0$ otherwise. In conclusion,

$$P(e) \geq \psi \frac{1}{2}\text{erfc}\left(\frac{\delta_{\text{free}}}{2\sqrt{N_0}}\right)$$

where

$$\psi = \sum_{\mathbf{C}} P(\mathbf{C}) I(\mathbf{C}) \tag{12.51}$$

represents the probability that at any given time a code trellis path chosen at random has another path splitting from it at that time, and remerging later, such that the Euclidean distance between them is $\delta_{\text{free}}$. If all the sequences have this property, then we get the lower bound

$$P(e) \geq \frac{1}{2} \text{erfc} \left( \frac{\delta_{\text{free}}}{2\sqrt{N_0}} \right) \tag{12.52}$$

but this is not valid in general. For (12.52) to hold, it is sufficient that all the trellis paths be on an equal footing, so that in particular all of them have a path at $\delta_{\text{free}}$. This is obtained if each one of the error matrices $\mathbf{G}(e)$ has equal entries.

Finally, we may get a lower bound to the *bit* error probability by observing that the average fraction of erroneous information bits in the splitting branch of an error event cannot be lower than $1/m$. Thus

$$P_b \geq \frac{\psi}{m} \frac{1}{2} \text{erfc} \left( \frac{\delta_{\text{free}}}{2\sqrt{N_0}} \right)$$

## 12.4.2.   Examples

In this section we present some examples of computation of error probabilities for TCM schemes. From the theory developed before, it is seen that this computation involves two separate steps. The first one is the evaluation of the transfer function of the error state diagram with formal labels (proper attention should be paid to the noncommutativity of matrix operations if the TCM scheme is nonuniform). This can be accomplished by using the techniques described in Appendix D. Next, the appropriate labels (the error matrices, or the corresponding scalar labels for uniform TCM schemes) are substituted for the formal labels, and the matrix $\mathbf{G}$ computed.

### Four-state code

A four-state code is shown in Fig. 12.21 with the corresponding error state diagram. Denote by $T_\alpha$, $T_\beta$, and $T_\gamma$ the transfer functions of the error state diagram from the starting node to nodes $\alpha$, $\beta$, and $\gamma$, respectively. The relevant equations are

$$\begin{aligned} T_\alpha &= \mathbf{G}(10) + T_\gamma \mathbf{G}(00) \\ T_\beta &= T_\alpha \mathbf{G}(11) + T_\beta \mathbf{G}(01) \end{aligned}$$

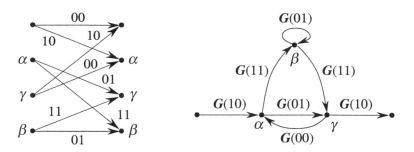

Figure 12.21: *Trellis diagram for a 4-state, $m = 1$ TCM scheme and its corresponding error state diagram.*

$$T_\gamma = T_\alpha \mathbf{G}(01) + T_\beta \mathbf{G}(11)$$
$$T(z) = T_\gamma \mathbf{G}(10)$$

For simplicity we examine here only the case of scalar labels, so that commutativity holds. By defining $g_0 = \mathbf{G}(00)$, $g_1 = \mathbf{G}(01)$, $g_2 = \mathbf{G}(10)$, and $g_3 = \mathbf{G}(11)$, we have the solution

$$T_\beta = \frac{g_2 g_3}{(1 - g_0 g_1)(1 - g_1) - g_0 g_3^2}$$

and finally

$$T(z) = \frac{g_2^2}{1 - g_0 g_1} \left[ \frac{g_3^2}{(1 - g_0 g_1)(1 - g_1) - g_0 g_3^2} + g_1 \right] \qquad (12.53)$$

**4-PAM.** From (12.53) we can obtain an upper bound to the probability of an error event by substituting for the various $g_i$ the values obtained from the calculation of the error matrices $\mathbf{G}(\cdot)$. We do this first for a 4-PAM constellation with

$$f(00) = +3, \qquad f(01) = +1, \qquad f(10) = -1, \qquad f(11) = -3$$

The G-matrices for this constellation were computed in (12.30)–(12.33). Thanks to their row-uniformity, the transfer function can be obtained from (12.53):

$$T(Z) = Z^{36} \frac{4 - 3Z^4 + 2Z^{36} + Z^{68}}{4 - 8Z^8 + 3Z^8 - 2Z^{40} - Z^{72}} = Z^{36} + \frac{5}{4} Z^{40} + \frac{7}{4} Z^{44} + \ldots \quad (12.54)$$

We see that $\delta_{\text{free}}^2 = 36$, obtained with an average energy expenditure $\mathcal{E} = (9 + 1 + 1 + 9)/4 = 5$. An uncoded binary PAM with signals $\pm 1$ would have $\delta_{\text{min}}^2 = 4$ and energy $\mathcal{E}' = 1$, so that the coding gain achieved by this TCM scheme is

$$\gamma = \frac{36}{20} = 1.8 \Rightarrow 2.55 \text{ dB}$$

**4PSK.**   With unit-energy 4PSK and the mapping

$$f(00) = 1 \quad f(01) = j \quad f(10) = -1 \quad f(11) = -j$$

we obtain the G-matrices of (12.36)–(12.39). We have uniformity here, so that the transfer function obtained from (12.53) is

$$T(Z) = \frac{Z^{10}}{1 - 2Z^2} = Z^{10} + 2Z^{12} + 4Z^{14} + \dots \tag{12.55}$$

Here $\delta_{\text{free}}^2 = 10$, obtained with $\mathcal{E} = 1$. An uncoded binary PSK with signals $\pm 1$ would have $\delta_{\text{min}}^2 = 4$ and energy $\mathcal{E}' = 1$, so that the coding gain achieved by this TCM scheme is

$$\gamma = \frac{10}{4} = 2.5 \Rightarrow 4 \text{ dB}$$

Expressing the error probabilities explicitly in terms of $\mathcal{E}_b/N_0$, from (12.55) we have, by observing that $\mathcal{E} = \mathcal{E}_b = 1$,

$$P(e) \leq \frac{1}{2} \frac{e^{-5\mathcal{E}_b/2N_0}}{1 - 2e^{-\mathcal{E}_b/2N_0}}.$$

The improved upper bound (12.45) yields

$$P(e) \leq \frac{1}{2} \text{erfc} \left( \sqrt{\frac{5}{2} \frac{\mathcal{E}_b}{N_0}} \right) \cdot \frac{1}{1 - 2e^{-\mathcal{E}_b/2N_0}}$$

The lower bound (12.52) yields

$$P(e) \geq \frac{1}{2} \text{erfc} \left( \sqrt{\frac{5}{2} \frac{\mathcal{E}_b}{N_0}} \right)$$

These error probabilities should be compared with uncoded binary PSK, for which

$$P(e) = \frac{1}{2} \text{erfc} \left( \sqrt{\frac{\mathcal{E}_b}{N_0}} \right)$$

These four error probabilities are plotted in Fig. 12.22. It can be observed, from the figure and also from their expressions, that the lower bound and the improved

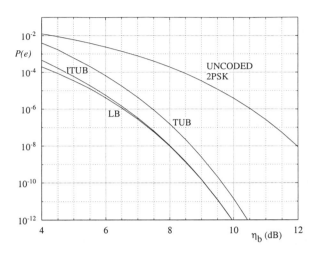

Figure 12.22: *Error probabilities of 4-state TCM based on 4PSK. TUB: Transfer-function upper bound. ITUB: Improved transfer function upper bound. LB: Lower bound. The error probability of uncoded binary PSK is also shown for comparison. Here $\eta_b = \mathcal{E}_b/N_0$.*

upper bound are very close to each other, and hence to the true value of error probability. Unfortunately, this does occur only for TCM schemes based on signal constellations with a small number of points and having a small number of states. Moreover, comparison of $P(e)$ for uncoded binary PSK with the two bounds for PCM shows explicitly that the coding gain is very close to $5/2$, as argued before.

### 12.4.3. Computation of $\delta_{\mathrm{free}}$

The results derived so far for upper and lower bounds on error probability of TCM show that $\delta_{\mathrm{free}}$ plays a central role in determining its performance. Consequently, if a single parameter is to be used to assess the quality of a TCM scheme, the sensible one is $\delta_{\mathrm{free}}$. Thus, it makes sense to look for an algorithm to compute this parameter independently of $P(e)$.

**Using the error state diagram**

The first technique we describe for the computation of $\delta_{\text{free}}$ is based on the error state diagram that was described in the context of the evaluation of an upper bound to error probability. We have already observed that the transfer function $T(Z)$ includes information about $\delta_{\text{free}}$. In the examples of last section we have exploited the fact that the value of $\delta_{\text{free}}^2$ follows immediately from the expansion of that function in a power series: the smallest exponent of $Z$ in that series is $\delta_{\text{free}}^2$. However, in most cases a closed form for $T(Z)$ is not available.

For this reason, we describe here a computational algorithm for evaluating $\delta_{\text{free}}$. This was first described by Saxena (1983) and Mulligan and Wilson (1984), and can be used also to evaluate the minimum distance in intersymbol-interference channels (Chapter 7) and in nonlinear channels (Chapter 14). Consider the trellis describing the TCM scheme. Every pair of branches in a section of the trellis corresponds to one distance between the signals labeling the branches. If there are parallel transitions, every branch will be associated with an entire subconstellation. In this case, only the minimum distance between any two signals extracted from the pair of subconstellations will be used. The squared distance between the signal sequences associated with two paths in the trellis is obtained by summing the individual squared distances. The algorithm is based on the update of the entries of a matrix $\mathbf{D}^{(n)} = (\delta_{ij}^{(n)})$, which are the minimum squared distances between all pairs of paths diverging from any initial state and reaching the states $i$ and $j$ at discrete time $n$. Two such pairs of paths are shown in Fig. 12.23. Notice that the matrix $\mathbf{D}^{(n)}$ is symmetric, and that its elements on the main diagonal are the distances between remerged paths (the "error events"). The algorithm goes as follows.

**Step 1** For each state $i$, find the $2^m$ states (the "predecessors") from which a transition to $i$ is possible, and store them in a table. Set $\delta_{ij} = -1$ for all $i$ and $j \geq i$. If there are parallel transitions, for all $i$ set $\delta_{ii}$ equal to the minimum squared Euclidean distance among all signals associated with the parallel transitions leaving any state.

**Step 2** For each pair of states $(i, j)$, $j \geq i$, find the minimum squared Euclidean distance between pairs of paths diverging from the same initial states (whatever they are) and reaching $i, j$ in one time unit. Two such pairs are shown in Fig. 12.24. This distance is $\delta_{ij}^{(1)}$.

**Step 3** For both states in the pair $(i, j)$, $j > i$, find in the table of Step 1 the $2^m$ predecessors $i_1, \cdots, i_{2^m}$ and $j_1, \cdots, j_{2^m}$ (see Fig. 12.25). In general there are $2^{2m}$ possible paths at time $n - 1$ that pass through $i$ and $j$ at time $n$.

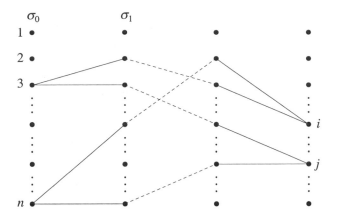

Figure 12.23: *Two pairs of paths diverging at time $n = 0$ and reaching the states $i, j$ at the same time.*

They pass through the pairs

$$(i_1, j_1), (i_1, j_2), \quad \cdots \quad , (i_1, j_{2^m})$$

$$\cdots$$

$$(i_{2^m}, j_1), (i_{2^m}, j_2), \quad \cdots \quad , (i_{2^m}, j_{2^m})$$

The minimum squared distance among all the paths passing through $(i, j)$ at time $n$ is

$$\delta_{ij}^{(n)} = \min \quad \left\{ \delta_{i_1 j_1}^{(n-1)} + \delta^2(i_1 \to i, j_1 \to j), \right.$$
$$\delta_{i_1 j_2}^{(n-1)} + \delta^2(i_1 \to i, j_2 \to j),$$
$$\cdots$$
$$\delta_{i_1 j_{2^m}}^{(n-1)} + \delta^2(i_1 \to i, j_M \to j),$$
$$\cdots \qquad\qquad (12.56)$$
$$\left. \delta_{i_{2^m} j_{2^m}}^{(n-1)} + \delta^2(i_{2^m} \to i, j_{2^m} \to j) \right\}$$

In (12.56), the distances $\delta^{(n-1)}$ come from the calculations of previous Step 2, while for example $\delta(i_1 \to i, j_1 \to j)$ denotes the Euclidean distance between the two signals associated with the transitions $i_1 \to i$ and $j_1 \to j$. These can be computed once for all at the beginning. When one of the previous distances $\delta_{\ell m}^{(n-1)}$ is equal to $-1$, the corresponding term in the

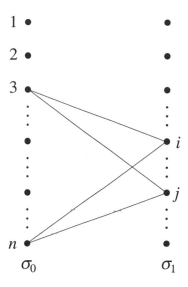

Figure 12.24: *Two pairs of paths starting from different states and reaching the same pair of states in one time instant.*

right-hand side of (12.56) disappears. In fact, $\delta_{\ell m}^{(n-1)} = -1$ means that no pair of paths can pass through the states $\ell$ and $m$ at time $n - 1$. When $i = j$, $\delta_{ii}^{(n)}$ represents the squared distance between two paths remerging at the $n$th step on the state $i$. This is an error event. Thus, if $\delta_{ii}^{(n)} < \delta_{ii}^{(n-1)}$, then $\delta_{ii}^{(n)}$ will take the place of $\delta_{ii}^{(n-1)}$ in matrix $\mathbf{D}^{(n)}$.

**Step 4** If

$$\delta_{ij}^{(n)} < \min_i \delta_{ii}^{(n)} \tag{12.57}$$

for at least one pair $(i, j)$, then set $n = n + 1$ and go back to Step 3. Otherwise, stop iterating and set

$$\delta_{\text{free}}^2 = \min_i \delta_{ii}^{(n)}$$

Condition (12.57) verifies that all the paths still open at time $n$ have distances not less than the minimum distance of an error event, and guarantees that the latter is actually $\delta_{\text{free}}$.

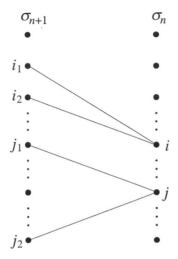

Figure 12.25: *Predecessors of states $i, j$.*

**A word of caution**

While $\delta_{\text{free}}$ provides the best single-parameter description of the quality of a TCM scheme, some caution should be exercised when it is used to compare schemes to be operated at low-to-intermediate signal-to-noise ratios. In fact, besides $\delta_{\text{free}}$, it may be advisable to consider two other parameters as influential over the performance of the TCM scheme. They are:

1. The "error coefficient" $N(\delta_{\text{free}})$. If this is large, it may affect significantly the value of error probability for low-to-intermediate signal-to-noise ratios.

2. The "next distance" $\delta_{\text{next}}$, i.e., the second smallest Euclidean distance between two paths forming an error event. If this is very close to $\delta_{\text{free}}$, the signal-to-noise ratio necessary for a good approximation to $P(e)$ based on (12.46) may become very large.

## 12.5. Power density spectrum

In this section we consider the power density spectrum of the digital signal at the output of the TCM modulator. In particular, we derive simple sufficient conditions for the resulting spectrum to be equal to the spectrum of an uncoded signal.

For simplicity, we shall only consider *linear* and one- or two-dimensional modulations here: i.e., we shall assume that the signal transmitted over the channel has the form

$$y(t) = \sum_{n=-\infty}^{\infty} a_n s(t - nT) \tag{12.58}$$

where $s(t)$ is a waveform defined in the time interval $(0, T)$ with Fourier transform $S(f)$, and $(a_n)$ is a sequence of complex random variables representing the TCM encoder outputs. If we assume that the source symbols are independent and equally likely, from the regular time-invariant structure of the code trellis it follows that the sequence $(a_n)$ is stationary. Under these conditions, from Chapter 2 we have the following result. If

$$\mathbf{E}[a_n] = \mu_a$$

and

$$\mathbf{E}[a_\ell a_m^*] = \sigma_a^2 \rho_{\ell-m} + |\mu_a|^2$$

so that $\rho_0 = 1$ and $\rho_\infty = 0$, then the power density spectrum of $y(t)$ is

$$\mathcal{G}(f) = \mathcal{G}^{(c)}(f) + \mathcal{G}^{(d)}(f) \tag{12.59}$$

where $\mathcal{G}^{(c)}(f)$, the continuous part of the spectrum, is given by

$$\mathcal{G}^{(c)}(f) = \frac{\sigma_a^2}{T} |S(f)|^2 \sum_{\ell=-\infty}^{\infty} \rho_k e^{-j2\pi f\ell T} \tag{12.60}$$

and $\mathcal{G}^{(d)}(f)$, the discrete part of the spectrum (or *line spectrum*) is given by

$$\mathcal{G}^{(d)}(f) = \frac{|\mu_a|^2}{T^2} |S(f)|^2 \sum_{\ell=-\infty}^{\infty} \delta(f - \frac{\ell}{T}) \tag{12.61}$$

When the random variables $a_n$ are uncorrelated (i.e., $\rho_\ell = \delta_{0,\ell}$, $\delta$ the Kronecker symbol) we have the special case

$$\mathcal{G}^{(c)}(f) = \frac{\sigma_a^2}{T} |S(f)|^2 \tag{12.62}$$

This is the power spectral density that we would obtain without TCM from a modulator using the same waveform $s(t)$ for signaling. In the balance of this section we shall investigate the conditions for TCM not to shape (and hence, not to expand) the signal spectrum, i.e., to give the same power density spectrum as for an uncoded signal.

We first assume that $\mu_a = 0$, so that no line spectrum appears. Without loss of generality we also assume $\sigma_a^2 = 1$. Let $\sigma_n$ denote the encoder state when the symbol $x_n$ is transmitted, and $\sigma_{n+1}$ the successive state. The correlations $\rho_\ell$ can be expressed as

$$\rho_k = \sum_{a_k} \sum_{\sigma_k} \sum_{\sigma_1} \sum_{a_0} a_k a_0^* P[a_k, \sigma_k, \sigma_1, a_0]$$

With

$$P[a_k, \sigma_k, \sigma_1, a_0] = P[a_k \mid \sigma_k] \cdot P[\sigma_k, \sigma_1] \cdot P[a_0 \mid \sigma_1]$$

this becomes

$$\rho_k = \sum_{\sigma_k} \sum_{\sigma_1} \mathbf{E}[a_k \mid \sigma_k] \cdot P[\sigma_k, \sigma_1] \cdot \mathbf{E}[a_0^* \mid \sigma_1]$$

Hence, sufficient conditions for $\rho_\ell = 0$, $\ell \neq 0$, are that $\mathbf{E}[a_n \mid \sigma_n] = 0$ or $\mathbf{E}[a_{n-1} \mid \sigma_n] = 0$, for each $\sigma_n$ individually. The first condition is equivalent to stating that for all encoder states the symbols available to the encoder have zero mean. The second condition is equivalent to stating that, for each encoder state, the average of the symbols forcing the encoder to that state is zero. This is the case for a good many TCM schemes.

Consider finally the line spectrum. A sufficient condition for it to be zero is that $\mu_a = 0$, that is, the average of symbols at the encoder output be zero.

## 12.6. Rotationally-invariant TCM

### Channels with a phase offset

We consider here a channel with a phase offset, and how to design a TCM scheme that can cope with it.

Consider, as an example, $M$-PSK transmission and its coherent detection. This detection mode requires that the phase of the carrier be estimated prior to demodulation. Now, most techniques for estimating the carrier phase require that the modulation phase be removed. This was described in Chapter 9. An inherent ambiguity arises: specifically, we say that there is an ambiguity of $2\pi/M$, or, equivalently, that the channel is affected by a phase offset which may take on values $k2\pi/M$, $k = 0, 1, \cdots, M - 1$.

To resolve this ambiguity, i.e., get rid of this phase offset, differential encoding and decoding are often used (see Section 5.8). However, when a TCM scheme is included in the system, we must make sure that it is invariant to phase rotations by multiples of $2\pi/M$. That, is, *any TCM sequence, when rotated by a multiple of $2\pi/M$, must still be a valid TCM sequence.* Otherwise, any phase rotation would cause a long error sequence because, even in the absence of noise, the TCM decoder would not recognize the received sequence as a valid one.

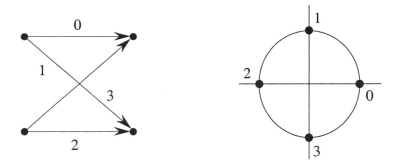

Figure 12.26: *A TCM scheme invariant to rotations of $\pi$, but not to rotations of $\pi/2$.*

**Example 12.5**    An example of this effect is illustrated in Fig. 12.26. Suppose that the all-zero sequence is transmitted. A rotation by $\pi$ causes the all-2 sequence to be received—a valid sequence. However, a rotation by $\pi/2$ generates the all-1 sequence, which is not recognized as valid by the Viterbi algorithm.                                    □

The receiver can handle the phase ambiguity in several ways. One is through a known training sequence sent by the transmitter. A second option is to use a code whose words are not invariant under component-wise rotation. A phase error can then be detected and countered by using a decoder that attempts to distinguish an invalid, rotated sequence from a valid sequence corrupted by channel noise. A third method, which we assume as our model, is to have the receiver arbitrarily assign to the carrier-phase estimate any of the possibilities corresponding to integer multiples of $2\pi/M$. Thus, we should design a TCM scheme that is invariant (or "transparent") to rotations, so that any rotated code sequence is still a coded sequence, and consequently the decoding algorithm will not be affected by a constant phase shift.

What we want to ensure, for a phase offset not to affect a TCM scheme, is the following:

1. The TCM scheme must be transparent to the phase rotations introduced by the carrier-phase recovery subsystem.

2. The encoder must be invariant to the same phase rotations, i.e., all the rotations of a coded TCM sequence must correspond to the same input information sequence.

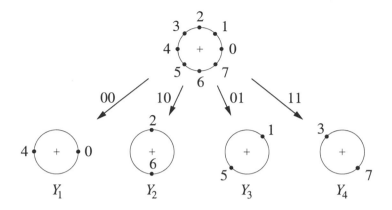

Figure 12.27: *A 4-way rotationally invariant partition of 8PSK.*

Note that the rotational invariance of a TCM scheme is a geometric property: the coded sequences, viewed as a collection of points in an infinite-dimensional Euclidean space, must be invariant under a certain finite set of rotations. The second property is rather a structural property of the encoder, i.e., it has to do with the input-output correspondence established by the latter.

**Rotationally invariant partitions.**   The first fundamental ingredient in the construction of a rotationally invariant TCM scheme is a rotationally invariant partition.

Let $S$ denote a $2N$-dimensional signal constellation, and $\{Y_1, \cdots, Y_K\}$ its partition into $K$ subsets. Consider the rotations about the origin in the two-dimensional Euclidean space; rotations in the $2N$-dimensional space are obtained by rotating separately each 2-dimensional subspace. Consider then the set of rotations that leave $S$ invariant, and denote it by $R(S)$. If $R(S)$ leaves the partition invariant, i.e., if the effect of each element of $R(S)$ on the partition is simply a permutation of its components, then we say that the partition is rotationally invariant.

**Example 12.6**   Consider 8PSK, and the 4-way partition shown in Fig. 12.27. Here $R(S)$ is the set of four distinct rotations by multiples of $\pi/4$. This partition is rotationally invariant. For example, $\rho_{\pi/4}$ corresponds to permutation $(Y_1 Y_3 Y_2 Y_4)$, $\rho_{\pi/2}$ to permutation $(Y_1 Y_2)(Y_3 Y_4)$, $\rho_\pi$ to the identity permutation, etc.                        □

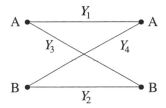

Figure 12.28: *A section of a two-state trellis.*

**Example 12.7**   Take the $2\times4$-PSK four-dimensional signal set of Example 12.3, and its 8-way partition

$$
\begin{aligned}
Y &= \{Y_1, Y_2, Y_3, Y_4, Y_5, Y_6, Y_7, Y_8\} & (12.63)\\
&= \{\{00, 22\}, \{11, 33\}, \{02, 20\}, \{13, 31\}, \\
&\quad \{01, 23\}, \{12, 30\}, \{03, 21\}, \{10, 32\}\} & (12.64)
\end{aligned}
$$

Here the elements of $R(\mathcal{S})$ are pairs of rotations, each by a multiple of $\pi/2$, which we denote $\rho_0$, $\rho_{\pi/2}$, $\rho_\pi$, and $\rho_{3\pi/2}$. It is seen for example that the effect of $\rho_{\pi/2}$ on signal $xy$ is to change it into signal $(x + 1)(y + 1)$, where addition is mod 4. This partition turns out to be rotationally invariant.                                                                                  □

**Rotationally invariant trellis.**   Consider now the effect of a phase rotation on TCM coded sequences. Given a rotationally invariant partition $Y$ of $\mathcal{S}$, the TCM scheme is rotationally invariant if for every $\rho \in R(\mathcal{S})$ the rotated version of any subconstellation sequence compatible with the code is still compatible with the code.

Let us examine one section of the trellis describing the TCM scheme. If we apply one and the same rotation $\rho$ to all the subconstellations labeling the trellis branches, we obtain a new trellis section. Now, for the TCM scheme to be rotationally invariant, this new trellis section must correspond to the original (unrotated) one apart from a permutation of its states.

**Example 12.8**   Consider the trellis segment of the two-state TCM scheme of Fig. 12.28. The partition on which it is based is

$$Y = \{Y_1, Y_2, Y_3, Y_4\}$$

where the notations are the same as in the Example 12.7 above. This partition is rotationally invariant. Denote the trellis segment by the set of its branches $(s_i, Y_j, s_k)$, where $Y_j$ is the subconstellation that labels the branch joining state $s_j$ to state $s_k$. Thus, the trellis is described by the set

$$\mathcal{T} = \{(A, Y_1, A), (A, Y_3, B), (B, Y_4, A), (B, Y_2, B)\}$$

The rotations $\rho_{\pi/2}$ and $\rho_{3\pi/2}$ transform $\mathcal{T}$ into

$$\{(A, Y_2, A), (A, Y_4, B), (B, Y_3, A), (B, Y_1, B)\}$$

which corresponds to the permutation $(A, B)$ of the states of $\mathcal{T}$. Similarly, $\rho_0$ and $\rho_\pi$ correspond to the identity permutation. In conclusion, the TCM scheme is rotationally invariant.                                                                            □

It may happen that a TCM scheme satisfies the property of being rotationally invariant only with respect to a subset of $R(\mathcal{S})$, and not to $R(\mathcal{S})$ itself. In this case we say that $\mathcal{S}$ is *partially* rotationally invariant.

**Example 12.9**   Consider the TCM scheme of Fig. 12.29, with 8 states and subconstellations corresponding to the partition $Y = \{Y_1, Y_2, Y_3, Y_4\}$ of Example 12.6.

The partition, as we know, is rotationally invariant. However, this TCM scheme is *not* rotationally invariant. For example, consider the effect of a rotation of $\pi/4$ summarized in Table 12.1. It is easily seen that $\rho_{\pi/4}$ does not generate a simple permutation of the trellis states: in fact, take the branch $(s_1, Y_3, s_1)$. In the original trellis, there is no state of the type $(s_i, Y_3, s_i)$. Actually, this TCM scheme is *partially* invariant: in fact, rotations by angles multiple of $\pi/2$ leave it invariant. For example, the effect of $\rho_{\pi/2}$ is summarized in Table 12.1: it causes the state permutation $(s_1 s_8)(s_2 s_7)(s_3 s_6)(s_4 s_5)$.   □

**Invariant encoder.**   We finally consider now the invariance of the encoder, i.e., the property that all rotations of a coded TCM sequence correspond to the same input information sequence. If $\mathbf{u}$ denotes a sequence of source symbols and $\mathbf{y}$ the corresponding output sequence of subconstellations, we want any rotation $\rho(\mathbf{y})$ (among those for which the TCM scheme is invariant) to correspond to the same $\mathbf{u}$.

Notice that fulfillment of this condition may require introduction of a differential encoder. We illustrate this point with an example.

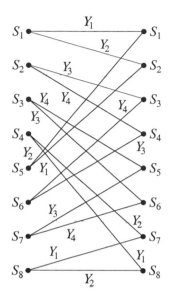

Figure 12.29: *An 8-state TCM scheme which is not rotationally invariant.*

**Example 12.10**  Take once again the eight-state 8PSK TCM scheme of Example 12.6. As we have seen, rotations by $\pi/2$ and $3\pi/2$ cause the permutation $(Y_1 Y_2)(Y_3 Y_4)$. If the encoder associates source symbols with subconstellations according to the rule

$$00 \Rightarrow Y_1 \qquad 10 \Rightarrow Y_2 \qquad 01 \Rightarrow Y_3 \qquad 11 \Rightarrow Y_4$$

then the only effect of $\rho_{\pi/2}$ and of $\rho_{3\pi/2}$ is to change the first bit in the source pair, while $\rho_0$ and $\rho_\pi$ change no bits. Thus, if the first bit is differentially encoded, the TCM encoder is transparent to any rotation multiple of $\pi/2$.  □

### General considerations

It can be said in general that imposing the constraint of rotational invariance may reduce the gain of a TCM scheme based on two-dimensional constellations. Moreover, implementation of a rotationally-invariant decoder may require using a nonlinear convolutional code. On the other hand, multidimensional constellations make it easier to attain rotational invariance, in the sense that they suffer less from this performance loss due to rotational invariance.

| $\rho_0$ | $\rho_{\pi/4}$ | $\rho_{\pi/2}$ |
|---|---|---|
| $(s_1, Y_1, s_1)$ | $(s_1, Y_3, s_1)$ | $(s_1, Y_2, s_1)$ |
| $(s_1, Y_2, s_2)$ | $(s_1, Y_4, s_2)$ | $(s_1, Y_1, s_2)$ |
| $(s_2, Y_3, s_3)$ | $(s_2, Y_2, s_3)$ | $(s_2, Y_4, s_3)$ |
| $(s_2, Y_4, s_4)$ | $(s_2, Y_1, s_4)$ | $(s_2, Y_3, s_4)$ |
| $(s_3, Y_4, s_5)$ | $(s_3, Y_1, s_5)$ | $(s_3, Y_3, s_5)$ |
| $(s_3, Y_3, s_6)$ | $(s_3, Y_2, s_6)$ | $(s_3, Y_4, s_6)$ |
| $(s_4, Y_2, s_7)$ | $(s_4, Y_4, s_7)$ | $(s_4, Y_1, s_7)$ |
| $(s_4, Y_1, s_8)$ | $(s_4, Y_3, s_8)$ | $(s_4, Y_2, s_8)$ |
| $(s_5, Y_2, s_1)$ | $(s_5, Y_4, s_1)$ | $(s_5, Y_1, s_1)$ |
| $(s_5, Y_1, s_2)$ | $(s_5, Y_3, s_2)$ | $(s_5, Y_2, s_2)$ |
| $(s_6, Y_4, s_3)$ | $(s_6, Y_1, s_3)$ | $(s_6, Y_3, s_3)$ |
| $(s_6, Y_3, s_4)$ | $(s_6, Y_2, s_4)$ | $(s_6, Y_4, s_4)$ |
| $(s_7, Y_3, s_5)$ | $(s_7, Y_2, s_5)$ | $(s_7, Y_4, s_5)$ |
| $(s_7, Y_4, s_6)$ | $(s_7, Y_1, s_6)$ | $(s_7, Y_3, s_6)$ |
| $(s_8, Y_1, s_7)$ | $(s_8, Y_3, s_7)$ | $(s_8, Y_2, s_7)$ |
| $(s_8, Y_2, s_8)$ | $(s_8, Y_4, s_8)$ | $(s_8, Y_1, s_8)$ |

Table 12.1: *Effect of rotations by $\pi/4$ and $\pi/2$ on the TCM scheme of Example 12.9.*

## 12.7. Multilevel coded modulation and BCM

Consider again TCM as described through its encoder scheme of Fig. 12.16. We may interpret it by observing that the bits entering the mapper are subject to two levels of protection: specifically, bits $c_i^{(1)}$ to $c_i^{(\tilde{m}+1)}$ are encoded by a convolutional code, while bits $b_i^{(\tilde{m}+2)}$ to $b_i^{(m+1)}$ are left uncoded. The reason why this procedure works is that bits $b_i^{(k)}$ with the highest values of $k$ are more protected than the others by the Euclidean distance of the modulation scheme, and hence need less protection .

**Example 12.11**   In the 8PSK partitioning of Fig. 12.13, it is seen that, if we denote by $\delta_0$, $\delta_1$, and $\delta_2$ the Euclidean distances among the subconstellations at the different partition levels, and we assume unit energy, we have

$$\delta_0^2 = 4\sin^2 \pi/8 \approx 0.5858, \qquad \delta_1^2 = 2, \qquad \delta_2^2 = 4$$

Thus, bit $c_i^{(1)}$ is protected from noise by a Euclidean distance $\delta_0$, bit $c_i^{(2)}$ by a Euclidean distance $\delta_1$, and bit $c_i^{(3)}$ by a Euclidean distance $\delta_2$. To improve transmission reliability, we may decide to add protection to bits $c_i^{(1)}$ and $c_i^{(2)}$ (say) by increasing their Hamming distance through convolutional coding, while $c_i^{(3)}$ is left to the channel's mercy.    □

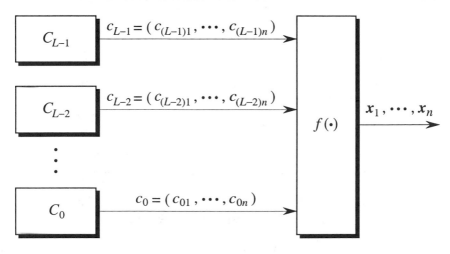

Figure 12.30: *Encoder of an L-level BCM scheme using codes with length $n$.*

Multilevel coded modulation is based on the interpretation above. It generalizes the concept of TCM by combining several codes to provide more than two levels of protection to the bits that label the signals sent to the channel. The resulting scheme provides flexible transmission rates and high asymptotic coding gains, although in many cases it performs worse than TCM at low signal-to-noise ratios, due to its higher number of nearest neighbors.

Here we describe multilevel coded modulation through its most popular version, that is, block-coded modulation (BCM), in which the codes used are block codes. We also assume, for simplicity's sake, that the codes are all binary.

The encoder is shown in Fig. 12.30. We start from an "elementary" signal constellation (8PSK in our previous example) with $M$ signals, and its partition into $L$ levels. At level 0 the elementary constellation is partitioned into 2 subconstellations of $M/2$ signals each. At level 1 each subconstellation is partitioned into 2 sub-subconstellations with $M/4$ signals each, and so forth, until at level $L-1$ we are left with $2^L$ subconstellations with $M/2^L$ signals each. By numbering the subconstellations at partition level $\ell$, we obtain a one-to-one correspondence between the numbers $\{0, 1\}$ and the two subconstellations at that level. The $L$ linear block encoders have the same block length $n$, dimensions $k_0, \cdots, k_{L-1}$, and Hamming distances $d_0, \cdots, d_{L-1}$. Since any $L$-tuple $(c_{0i}, \cdots, c_{(L-1)i})$, for $i = 1, \ldots, n$, defines a unique elementary signal, the mapper $f(\cdot)$ outputs an $n$-tuple of elementary signals.

There are as many different signals as there are $n$-tuples, i.e., code words.

Thus,

$$\mathcal{M} = \prod_{\ell=0}^{L-1} 2^{k_\ell} \tag{12.65}$$

The minimum squared Euclidean distance among two $n$-tuples of elementary signals is bounded below by

$$\delta_{\min}^2 \geq \min\left(\delta_0^2\, d_0, \delta_1^2\, d_1, \cdots, \delta_L^2\, d_L\right) \tag{12.66}$$

where $\delta_\ell$ is the minimum Euclidean distance between any two subconstellations at partition level $\ell$. To prove (12.66), it suffices to observe that at level $\ell$ a minimum Hamming distance $d_\ell$ will cause any two code words of $\mathcal{C}_\ell$ to differ in at least $d_\ell$ positions, which entails the corresponding two $n$-tuples of elementary signals to be at a squared Euclidean distance of at least $\delta_\ell^2 d_\ell$.

If the minimum squared Euclidean distance is specified, then it sounds reasonable to take the minimum Hamming distance of the codes to be

$$d_\ell = \lceil \delta_{\min}^2 / \delta_\ell^2 \rceil$$

where $\lceil x \rceil$ is the smallest integer not less than $x$. However, this is not the best possible choice: for example, Huber and Wachsmann (1994) advocate a rule based on information capacities.

It should also be observed here that this construction may be seen as providing a multidimensional signal with a number of dimensions $nD$, where $D$ is the dimensionality of the elementary signal set. The multidimensional signal thus obtained can be used in conjunction with a TCM scheme.

**Example 12.12**   The concepts and definitions above will now be illustrated and clarified through an example. Choose $L = 3$, $n = 7$, and 8PSK as the elementary constellation. The mapper $f(\cdot)$ associates an 8PSK signal with every triplet of bits at the output of the encoders, and hence it associates $n$ 8PSK signals with each triplet of encoded blocks. Let $\mathcal{C}_2$ be the nonredundant (7,7) code, $\mathcal{C}_1$ the (7,6) single-parity-check code, and $\mathcal{C}_0$ the (7,1) repetition code. From (12.65) the BCM encoder may output $2^{1+6+7} = 2^{14}$ possible signals. The number of dimensions is $nD = 14$, and hence the scheme carries 1 bit per dimension (as 4PSK). Its minimum squared Euclidean distance, normalized with respect to the average signal energy $\mathcal{E}$, is bounded above by (12.66):

$$\frac{d_{\min}^2}{\mathcal{E}} \geq \min\{4 \times 1,\ 2 \times 2,\ 0.586 \times 7\} = 4$$

and direct computation shows it to be exactly equal to 4. Thus, this BCM scheme achieves a 3-dB asymptotic coding gain with respect to 4PSK.                          □

### 12.7.1. Staged decoding of multilevel constructions

An important feature of BCM is that it admits a relatively simple sequential decoding procedure. This is called *staged decoding*. In some cases, namely, for $L = 2$ and $C_1$ a nonredundant code, staged decoding is optimal (see Problem 12.8). Otherwise, it is a suboptimal procedure, but with a complexity lower than for optimum decoding.

The idea underlying staged decoding is the following. An $L$-level BCM scheme based on $[C_0, \cdots, C_{L-1}]$ is decoded by decoding the component codes in sequence. $C_0$, the most powerful code, is decoded first. Then $C_1$ is decoded by assuming that $C_0$ was decoded correctly. Further, $C_2$ is decoded by assuming that the two previous codes were decoded correctly, and so forth.

The block diagram of a staged decoder is shown in Fig. 12.31 for $L = 3$. Let the received signal vector be

$$\mathbf{r} = f(\mathbf{c}_0, \mathbf{c}_1, \mathbf{c}_2) + \mathbf{n}$$

where $\mathbf{c}_0, \mathbf{c}_1, \mathbf{c}_2$ are the code words, and $\mathbf{n}$ the noise vector. The receiver must produce an estimate of the three code words $\mathbf{c}_0, \mathbf{c}_1, \mathbf{c}_2$ in order to decode $\mathbf{r}$. In principle, the metrics of all possible vectors $f(\mathbf{c}_0, \mathbf{c}_1, \mathbf{c}_2)$ should be computed, and the maximum such metric determined. This is an impractical procedure for large signal sets. In staged decoding, decoder $\mathcal{D}_0$ produces an estimate of $\mathbf{c}_0$ for all possible choices of $\mathbf{c}_1, \mathbf{c}_2$. Next, decoder $\mathcal{D}_1$ produces an estimate of $\mathbf{c}_1$ for all the possible choices of $\mathbf{c}_2$, *by assuming that the choices of $\mathbf{c}_0$ were correct*. Finally, decoder $\mathcal{D}_2$ produces an estimate of $\mathbf{c}_2$ *by assuming that the choice of $\mathbf{c}_0$ and of $\mathbf{c}_1$ was correct*. Observe that decoding of $\mathcal{D}_\ell$ will provide us with the estimate of $k_\ell$ source symbols. Thus, at each stage of the procedure we obtain a block of source symbols that are sent to the parallel-to-serial converter.

## 12.8. Bibliographical notes

The idea of using of the cutoff rate to compare channels to be used in conjunction with coding for the AWGN channel was originally advocated by Wozencraft and Kennedy (1966), but sank into oblivion until it was resurrected in the seminal paper by Massey (1974). For a critical view of the role of $R_0$, see Blahut (1987, p. 184). In 1974, Massey formally suggested that the performance of a digital communication system could be improved by looking at modulation and coding as a *combined* entity, rather than as two separate operations (Massey, 1974).

The basic principles of Trellis-Coded Modulation were described by Unger-boeck (1982); see also Ungerboeck (1987), Forney *et al.* (1984), and the book by Biglieri *et al.* (1991). Similar concepts, based on the combination of block

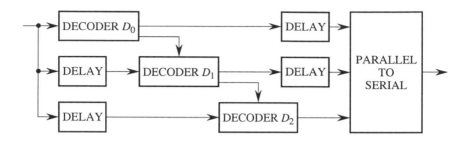

Figure 12.31: *Staged decoder for a 3-level BCM.*

and convolutional codes with modulation, were developed by Imai and Hirakawa (1977). One- or two-dimensional constellations can be used in a TCM scheme in the simple way described in this chapter. Use of higher-dimensional constellations was advocated by Wei (1987). He performed partitioning by an *ad hoc* method, starting from lower-dimension partitions. If lattice constellations are used, powerful algebraic method for partitioning them are available (Calderbank and Sloane, 1987; Forney, 1988a and 1988b).

TCM experienced a fast transition from research to practical applications. In 1984, a generation of modems using TCM became available, achieving reliable transmission at speeds of 14.4 kbit/s on private-line modems and 9.6 kbit/s switched-network modems. The V.32 modem (1984) uses an 8-state two-dimensional TCM scheme (Wei, 1984) based on a nonlinear convolutional encoder to achieve $\pi/2$ rotational invariance. The V.34 modem (1994) specifies three four-dimensional TCM schemes, with 16, 32, and 64 states, respectively. The coding gains at $P(e) \approx 10^{-6}$ are about 3.6 dB for the V.32 modem, and range from 4.2 to 4.7 dB for the V.34 modems (see Forney *et al.*, 1996).

Besides the representation of a TCM encoder based on a convolutional encoder and a memoryless mapper, another description exists, which is analytical in nature. This has been advocated by Calderbank and Mazo (1984). See also Chapter 3 of Biglieri *et al.* (1991).

The discussion on the uniformity properties of TCM schemes, which are useful to evaluate error probabilities in a relatively simple way, was taken from Liu, Oka, and Biglieri (1990), which in turn is an elaboration of Zehavi and Wolf (1987). A comprehensive account of these properties, based on group theory, can be found in Forney (1991). Nonuniform TCM schemes can be dealt with by using the product-trellis method of Biglieri (1984). A tutorial presentation of techniques for evaluating the performance of TCM schemes can be found in (Benedetto *et al.*, 1994).

The theory of rotationally-invariant TCM schemes is fully developed in Trott *et al.* (1996). A generalization of TCM to constellations derived from lattices can be found in Forney (1988a and 1988b).

Ginzburg (1984) develops an algebraic introduction to BCM. A discussion on multilevel codes and multistage decoding is in Calderbank (1989) and Pottie and Taylor (1989). Biglieri (1992) discusses parallel decoding of BCM, and V. Benedetto and Biglieri (1993) show how to evaluate error probabilities. Actual implementations of staged BCM decoders are described in Pellizzoni *et al.* (1997) and Caire *et al.* (1995).

## 12.9.  Problems

*Problems marked with an asterisk should be solved with the aid of a computer.*

**12.1**  Prove that the simplex set of Problem 4.12 maximizes $R_0$ as defined in (12.14) and (12.15) under the constraint of a given average energy. *Hint:* Observe that $R_0$ depends only on the differences between signal vectors, so that for a given average energy an optimal signal set will have its centroid at the origin. Write $R_0$ for a signal set with the centroid at the origin, and use the convexity of the exponential. Find the minimizing probability distribution $\mathbf{Q}$.

**12.2**  Consider the quaternary signal constellation of Fig. 12.32. Design a good 4-state TCM scheme based on this constellation and transmitting 1 bit/signal. Discuss the behavior of the free distance as a function of the angle $\phi$.

**12.3**  Specify the convolutional codes and the mappers that generate trellis-coded modulation schemes with 2 and 4 states, respectively, based on the octonary constellation represented by the black dots of Fig. 12.9 and transmitting 2 bits per signal. Compute their free distances and asymptotic coding gains with respect to 4PSK.

**12.4**  The encoder of a standard TCM scheme for the coded 9.6 kbit/s two-wire full-duplex voiceband modem is shown in Fig. 12.33. The 32-signal constellation used is also shown. Draw one stage of the corresponding trellis diagram.

**12.5**  Consider the TCM scheme shown in Fig. 12.34. The signal constellation used is 16-QAM. Specify the mapper, and compute the resulting free distance.

**12.6**  (*) Write a computer program implementing the algorithm of Section 12.4.3 for the computation of the free distance of a TCM scheme. Use it to verify the result of Problem 12.5.

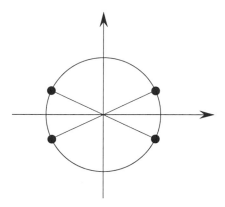

Figure 12.32: *A quaternary signal constellation.*

**12.7** (*) Consider the 32-point cross contellation of Fig. 5.20. Perform set partitioning, and design a 4-state TCM scheme satisfying the Ungerboeck rules and transmitting 4 bits per signal. Compute its free distance and bit error probability.

**12.8** Consider a 2-level BCM scheme with $\mathcal{C}_1$ the nonredundant $(n, n)$ binary code. Show that staged decoding is optimum in this case.

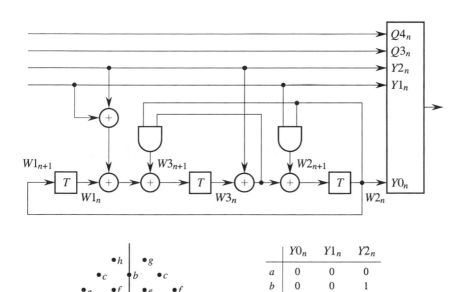

Figure 12.33: *Encoder of a TCM scheme, the corresponding 32-signal constellation, and mapping table.*

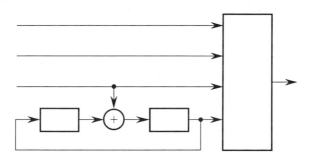

Figure 12.34: *Encoder of a TCM scheme.*

# 13

# Digital transmission over fading channels

In the simple communication channel model considered in Chapter 4 the received signal was affected only by a constant attenuation and a constant delay; Chapter 7 dealt with a more complex form of channel, one in which the signal is distorted by frequency selectivity. In this chapter we describe digital transmission over a channel affected by *fading*. The latter term describes the propagation vagaries of a radio channel which affect the signal strength, and is connected, as we shall see, with a propagation environment referred to as "multipath" and with the relative movement of transmitter and receiver, which causes time variations of the channel. Fading channels generate time-varying attenuations and delays, effects which may significantly degrade the performance of a communication system and hence need analysis in order to take adequate countermeasures against them (as we shall see, diversity and coding are appropriate methods).

Multipath propagation occurs when the electromagnetic energy carrying the modulated signal propagates along more than one "path" connecting the transmitter to the receiver. This simple picture, assuming that the propagation medium includes several paths along which the electromagnetic energy propagates, although not very accurate from a theoretical point of view, is nonetheless useful to understand and to analyze propagation situations that include reflection, refraction, and scattering of radio waves. Examples of such situations occur for example in indoor propagation, where the electromagnetic waves are perturbed by structures inside the building, and in terrestrial mobile radio, where multipath is caused by large fixed or moving objects (buildings, hills, cars, etc.).

**Delay spread.** The signal components arriving from the various paths (direct and indirect) with different delays combine to produce a distorted version of the transmitted signal. Assume for the moment that an ideal impulse $\delta(t)$ is transmitted. The various propagation paths are distinguishable, in the sense that they give rise to various distinguishable copies of the transmitted signal, affected by different delays and attenuations. To characterize by a single constant the various delays incurred by the signal traveling through the channel, we define a *delay spread* as the largest among these delays.

The above is still valid when the signal bandwidth is much larger than the inverse of the time delays. For narrower bandwidths, the received signal copies tend to overlap, and cannot be resolved as distinct pulses. This generates the form of linear distortion that we have previously studied in Chapter 7 as *intersymbol interference*. In the context of the present discussion, we say that this delay spread causes the two effects of *time dispersion* and *frequency-selective fading*.

Let $B_x$ denote the bandwidth of the transmitted signal. If this is narrow enough that the signal is not distorted, there is no frequency selectivity. As $B_x$ increases, the distortion becomes increasingly noticeable. A measure of the signal bandwidth beyond which the distortion becomes relevant is usually given in terms of the so-called *coherence bandwidth* of the channel, denoted by $B_c$ and defined as the inverse of the delay spread. The coherence bandwidth is the frequency separation at which two frequency components of the signal undergo independent attenuations. A signal with $B_x \gg B_c$ is subject to frequency-selective fading. More precisely, the envelope and phase of two unmodulated carriers at different frequencies will be markedly different if their frequency spacing exceeds $B_c$, so that the cross-correlation of the fading fluctuations of the two tones decreases toward zero. The term "frequency-selective fading" expresses this lack of correlation among different frequency components of the transmitted signal.

**Doppler-frequency spread.** When the receiver and the transmitter are in relative motion with constant radial speed, the received signal is subject to a constant frequency shift (the *Doppler shift*) proportional to this speed and to the carrier frequency. Doppler effect, in conjunction with multipath propagation, causes *frequency dispersion* and *time-selective fading*. Frequency dispersion, in the form of an increase of the bandwidth occupancy of a signal, occurs when the channel changes its characteristics during signal propagation. As we shall see in a moment, Doppler-frequency spread is in a sense dual to delay spread.

Assume again an ideal situation: here a single, infinite-duration tone is transmitted. This corresponds to a spectrum made by an ideal impulse. The power spectrum of the signal received from each path is a sum of impulses, each of which has a different frequency shift depending on its path. We have *frequency dispersion.* We define the "Doppler spread" as the largest of the frequency shifts of the various paths.

Let $T_x$ denote the duration of the transmitted signal. If this is long enough, there is no time selectivity. As $T_x$ decreases, the signal spectrum becomes broader, and its copies generated by Doppler effect tend to overlap and cannot be resolved as distinct frequency pulses. The Doppler effect causes a variation of its shape which corresponds to a distortion of the signal waveform. We say that the channel is *time-selective.*

A measure of the signal duration beyond which this distortion becomes relevant is the so-called *coherence time* of the channel, denoted by $T_c$ and defined as the inverse of the Doppler spread. Let $T_x$ denote the duration of a transmitted pulse. If this is so short that during transmission the channel does not change appreciably in its features, then the signal will be received undistorted. Its distortion becomes noticeable when $T_x$ is well above $T_c$, the delay between two time components of the signal beyond which their attenuations become independent.

**Fading-channel classification.** From the previous discussion we have seen that the two quantities $B_c$ and $T_c$ describe how the channel behaves for the transmitted signal. Specifically,

(i) If $B_x \ll B_c$, there is no frequency-selective fading, and hence no time dispersion. The channel transfer function looks constant, and the channel is called *flat* (or *nonselective*) in frequency.

(ii) If $T_x \ll T_c$, there is no time-selective fading, and the channel is called *flat* (or *nonselective*) in time.

Qualitatively, the situation appears as shown in Fig. 13.1. The channel flat in $t$ and $f$ is not subject to fading, neither in time nor in frequency. The channel flat in time and selective in frequency was studied in Chapter 7. The channel flat in frequency is a good model for several terrestrial mobile radio channels, and most of this chapter will be devoted to its analysis. The selective channel, affected by fading both in time and in frequency, is not a good model for terrestrial mobile radio channels, while it can be useful for avionic communications, in which high speeds (and hence short coherence times) combine with long delays (and hence narrow coherence bandwidths) due to earth reflections.

## 13.1. Impulse response and transfer function of a fading channel

As fading channels are typically time-varying, we cannot study them by using the techniques summarized in Chapter 2. Here we briefly review the extension to

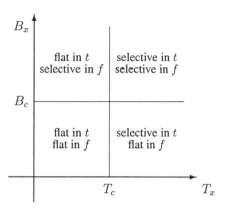

Figure 13.1: *Radio-channel classification.*

time-varying channels of the basic entities (impulse response, transfer function) used in the analysis of time-invariant systems.

Consider the bandpass transmitted signal

$$x(t) = \Re[\tilde{x}(t)e^{j2\pi f_0 t}] \tag{13.1}$$

where as usual $\tilde{x}(t)$ denotes its complex envelope and $f_0$ the carrier frequency. Transmission of $x(t)$ over a multipath time-varying channel yields a received signal

$$y(t) = \sum_n \alpha_n(t)x[t - \tau_n(t)]$$

where $\alpha_n(t)$ and $\tau_n(t)$ denote the attenuation and the delay that affect the copy of $x(t)$ received along the $n$th propagation path. We can also write

$$y(t) = \Re\left\{ \left[\sum_n \alpha_n(t)e^{-j\theta_n(t)}\tilde{x}[t - \tau_n(t)]\right] e^{j2\pi f_0 t}\right\} \tag{13.2}$$

where $\theta_n(t) \triangleq 2\pi f_0 \tau_n(t)$ is the phase shift caused by the delay $\tau_n(t)$. We see that the multipath channel transforms the complex envelope $\tilde{x}(t)$ into the complex envelope

$$\sum_n \alpha_n(t)e^{-j\theta_n(t)}\tilde{x}[t - \tau_n(t)]$$

which is equivalent to saying that the low-pass equivalent channel has a *time-varying impulse response*

$$h(t; \tau) = 2\sum_n \alpha_n(t)e^{-j\theta_n(t)}\delta[\tau - \tau_n(t)] \tag{13.3}$$

This function of the two variables $t$ and $\tau$ describes the attenuations, the phase shifts, and the delays generated by each of the propagation paths. It is the response of the channel at time $t$ to an impulse applied at time $t - \tau$, i.e., $\tau$ seconds before. The channel input-output relationship is expressed by the convolution

$$\tilde{y}(t) = \int_{-\infty}^{\infty} \tilde{x}(t - \tau) h(t; \tau) \, d\tau$$

We obtain a similar characterization of the channel by moving to the frequency domain. The Fourier transform of the impulse response $h(t; \tau)$, taken with respect to the variable $\tau$, provides us with the (time-varying) transfer function of the channel:

$$H(f; t) = \int_{-\infty}^{\infty} h(t; \tau) e^{-j2\pi f \tau} \, d\tau \qquad (13.4)$$

which for our multipath channel takes the form

$$H(f; t) = 2 \sum_n \alpha_n(t) e^{-j\theta_n(t)} e^{-j2\pi f \tau_n(t)}$$

The time-varying transfer function $H(f; t)$ can be interpreted as follows. For a single tone transmitted at frequency $f'$ (relative to the carrier frequency) the received complex envelope is $\exp(j2\pi f't) H(f'; t)$. The latter expression describes the fact that the fading-channel response to a single tone has generally a time-varying envelope and phase. If the transmitted signal bandwidth is much greater than the coherence bandwidth of the channel, then the signal is affected by different gains and phase shifts across its bandwidth, and the channel is frequency-selective. Additional distortion is caused by the time variations in $H(f; t)$: for channel variations related to motions within the propagation medium, which in turn imply Doppler effects, we can interpret the time dependence of $H(f; t)$ in terms of Doppler shifts.

If it is assumed that the propagation medium changes randomly, we may model $h(\tau; t)$ as a random process in the variable $t$, so that the channel behaves as a linear, randomly time-varying system. An assumption often made (the "Rayleigh fading" model, which we shall describe in detail later on) is that $h(\tau; t)$ can be modeled as a zero-mean Gaussian process, whose phase is uniformly distributed in $(0, 2\pi)$ and whose envelope has a Rayleigh distribution. Consequently, also $H(f; t)$ is a Gaussian process.

As for the power density spectrum of this process, we hasten to stress here that it has nothing to do with the fading being selective. Selectivity depends on the transmitted-signal bandwidth, whereas the fading spectrum reflects the relative motion of receiver and transmitter in a multipath environment.

Conditioned on the transmitted signal, the received signal turns out to be a sample function of a Gaussian random process (this has nothing to do with the additive noise, which we are neglecting for the moment). Since the received process and its complex envelope are Gaussian, the requirement that two of its samples be statistically independent is equivalent to the requirement that they be uncorrelated. Consequently, we may define the coherence time as the time beyond which two samples are independent, and the coherence bandwidth as the frequency separation beyond which samples of the Fourier transform of the complex envelope are uncorrelated. For this reason coherence time and coherence bandwidth are sometimes referred to as *correlation time* and *correlation bandwidth*, respectively.

A channel flat in time and frequency has a transfer function that reduces to a constant, albeit a random one. It responds to $x(t)$ with the signal $Re^{j\theta}x(t)$, where $R$ is the envelope and $\theta$ the phase of the fading, respectively. A channel which is flat only in time has a transfer function which depends only on $f$, and hence it behaves like a linear time-invariant system. Its response to an input sinusoid is a sinusoid whose frequency is the same, whose amplitude is multiplied by a constant random gain, and whose phase is shifted by a constant random phase, so that the channel output does not appear to fade in time. A channel which is flat only in frequency has a transfer function which depends only on $t$. The output-signal complex envelope equals the input-signal complex envelope multiplied by a complex function of time. Thus, the channel modulates the transmitted signal in amplitude and phase.

## 13.2. Examples of radio channels

For the purpose of illustration, we provide here some simple models of radio channels giving rise to various types of selectivity.

### 13.2.1. Two-path propagation

Assume that the transmitter and the receiver are fixed, and that two propagation paths exist. This is a useful model for the propagation in terrestrial microwave radio links. The received signal can be written in the form

$$y(t) = x(t) + b\,x(t - \tau) \tag{13.5}$$

where $b$ and $\tau$ denote the relative amplitude and the differential delay of the reflected signal, respectively (in other words, it is assumed that the direct path has attenuation 1 and delay 0). This equation models a static multipath situation in which the propagation paths remain fixed in their characteristics and can be

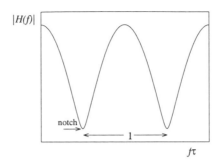

Figure 13.2: *Magnitude of the transfer function of a two-path channel.*

identified individually. The channel is linear and time-invariant. Its transfer
function

$$H(f) = 1 + b\,e^{-j2\pi f\tau}$$

in which the term $b\,\exp(-j2\pi f\tau)$ describes the multipath component, has magnitude

$$\begin{aligned}
|H(f)| &= \sqrt{(1 + b\,\cos 2\pi f\tau)^2 + b^2\,\sin^2 2\pi f\tau} \\
&= \sqrt{1 + b^2 + 2b\,\cos 2\pi f\tau}
\end{aligned}$$

Fig. 13.2 shows a typical behavior of the function $|H(f)|$. For certain delays
and frequencies the two paths are essentially in phase alignment, producing a
large value of $|H(f)|$. For some other values the paths nearly cancel each other,
producing a minimum of $|H(f)|$ usually referred to as a "notch."

If $x(t)$ has a wide enough bandwidth, the two echoes in the RHS of (13.5)
are discernible in time at the receiver (ideally, if $x(t) = \delta(t)$, a pair of pulses
is always observed): the channel is selective in frequency, as the transmitted
signal "sees" a channel with a nonflat transfer function. As the bandwidth $B_x$
of $x(t)$ decreases, and hence its time duration increases, the two echoes become
increasingly indistinguishable, and as $B_x$ becomes small with respect to $1/\tau$ the
received pulse is only affected by a constant attenuation and a constant delay.
The channel becomes nonselective in frequency, and we identify the coherence
bandwidth of the channel with $1/\tau$.

The situation described here does not account for fading, as there is no time
variation of the medium. This may occur in certain situations, due for example
to temperature variations that cause variations in the density of the transmission

Figure 13.3: *Single-path propagation: Effect of movement.*

medium. Thus, both $b$ and $\tau$ vary with time, creating a dynamic multipath situation in which the transfer function of the channel depends on time as well as on frequency.

### 13.2.2. Single-path propagation: Effect of movement

Consider the situation depicted in Fig. 13.3. Here the receiver is in relative motion with respect to the transmitter. The latter transmits an unmodulated carrier with frequency $f_0$. Let $v$ denote the speed of the vehicle (assumed constant), and $\gamma$ the angle between the direction of propagation of the electromagnetic plane wave and the direction of motion. Doppler effect causes the received signal to be a tone whose frequency is displaced (decreased) by an amount

$$f_D = f_0 \frac{v}{c} \cos \gamma \qquad (13.6)$$

(the "Doppler frequency shift"), where $c$ is the speed of propagation of the electromagnetic field in the medium. Notice that the Doppler frequency shift is either greater or lower than 0, depending on whether the transmitter is moving toward the receiver or away from it (this is reflected by the sign of $\cos \gamma$).

By disregarding for the moment the attenuation and the phase shift affecting the received signal, we can write it in the form

$$y(t) = A \exp[j2\pi(f_0 - f_D)t] \qquad (13.7)$$

Notice that we have assumed a constant vehicle speed, and hence a constant $f_D$. Variations of $v$ would cause a time-varying $f_D$ in (13.7).

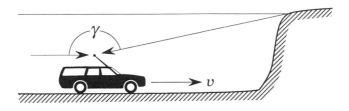

Figure 13.4: *Two-path propagation: Effect of movement.*

More generally, consider now the transmission of a bandpass signal $x(t)$, and take attenuation $\alpha(t)$ and delay $\tau(t)$ into account. The complex envelope of the received signal is

$$\tilde{y}(t) = \alpha(t)e^{-j\theta(t)}\tilde{x}[t - \tau(t)]$$

where

$$\theta(t) = 2\pi\left[(f_0 + f_D)\tau(t) - f_D t\right]$$

This channel can be modeled as a time-varying linear system with low-pass equivalent impulse response

$$h(t;\ \tau) = 2\alpha(t)\,e^{-j\theta(t)}\,\delta[t - \tau(t)]$$

### 13.2.3.  Two-path propagation: Effect of movement

Consider now the more complex situation represented in Fig. 13.4. A vehicle moves at constant speed $v$ along a direction that we take as the reference for angles. The transmitted signal is again an unmodulated carrier at frequency $f_0$. It propagates along two paths, which for simplicity we assume to have the same delay (zero) and the same attenuation. Let the angles under which the two paths are received be 0 and $\gamma$. Due to Doppler effect, the received signal is

$$y(t) = A\exp\left[j2\pi f_0\left(1 - \frac{v}{c}\right)t\right] + A\exp\left[j2\pi f_0\left(1 - \frac{v}{c}\cos\gamma\right)t\right] \quad (13.8)$$

We observe from the above equation that the transmitted sinusoid is received as a pair of tones: this effect can be viewed as a spreading of the transmitted signal frequency, and hence as a special case of frequency dispersion caused by the channel and due to the combined effects of Doppler shift and multipath propagation.

Eq. (13.8) can be rewritten in the form

$$y(t) = A \left[ \exp\left(-j2\pi f_0 \frac{v}{c}t\right) + \exp\left(-j2\pi f_0 \frac{v}{c}\cos\gamma t\right) \right] e^{j2\pi f_0 t} \qquad (13.9)$$

The magnitude of the term in square brackets provides the instantaneous envelope of the received signal:

$$R(t) = 2A \left| \cos\left[2\pi \frac{v}{c} f_0 \frac{1 - \cos\gamma}{2} t\right] \right|$$

The last equation shows an important effect: the envelope of the received signal exhibits a sinusoidal variation with time, occurring with frequency

$$\frac{v}{c} f_0 \frac{1 - \cos\gamma}{2}$$

The resulting channel has a time-varying response. We have time-selective fading, and, as observed before, also frequency dispersion.

### 13.2.4. Multipath propagation: Effect of movement

Assume now that the transmitted signal (an unmodulated carrier as before) is received through $N$ paths. The situation is depicted in Fig. 13.5. Let the receiver be in motion with velocity $v$, and let $A_i$, $\theta_i$, and $\gamma_i$ denote the amplitude, the phase, and the angle of incidence of the ray from the $i$th path, respectively. The received signal contains contributions with a variety of Doppler shifts, say

$$f_i \triangleq f_0 \frac{v}{c} \cos\gamma_i, \qquad i = 1, 2, \ldots, N$$

The (analytic) received signal can be written in the form

$$y(t) = \sum_{i=1}^{N} A_i \exp j[2\pi(f_0 - f_i)t + \theta_i] \qquad (13.10)$$

The complex envelope of the received signal turns out to be

$$R(t)e^{j\Theta(t)} = \sum_{i=1}^{N} A_i e^{-j(2\pi f_i t - \theta_i)}$$

For a large number $N$ of paths, we may assume that the attenuations $A_i$ and the phases $2\pi f_i t - \theta_i$ are random variables (RV), that can be reasonably assumed

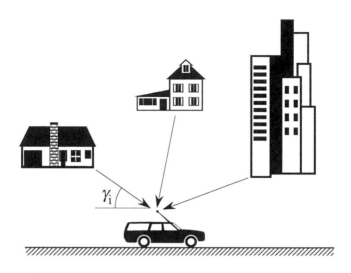

Figure 13.5: *Multipath propagation: Effect of movement.*

to be independent of each other. Then, invoking the central limit theorem, we obtain that at any instant, as the number of contributing paths become large, the resulting sum approaches a Gaussian RV. The complex envelope of the received signal becomes a lowpass Gaussian process whose real and imaginary parts are independent and have mean zero and the same variance $\sigma^2$. In these conditions, $R(t)$ and $\Theta(t)$ turn out to be independent processes, with $\Theta(t)$ being uniformly distributed in $(0, 2\pi)$ and $R(t)$ having a Rayleigh probability density function (pdf), viz.,

$$f_R(r) = \begin{cases} \dfrac{r}{\sigma^2}e^{-r^2/2\sigma^2}, & 0 \le r < \infty \\ 0, & r < 0 \end{cases} \tag{13.11}$$

(see Problem 13.1). Here the average power of the envelope is given by

$$E[R^2] = 2\sigma^2 \tag{13.12}$$

We also have

$$E[R] = \sqrt{\frac{\pi}{2}}\sigma$$

and hence

$$\text{Var}[R] = \left(2 - \frac{\pi}{2}\right)\sigma^2$$

Notice that the Rayleigh pdf is often used in its "normalized" form, obtained by choosing $E[R^2] = 1$:

$$f_R(r) = 2re^{-r^2} \tag{13.13}$$

### 13.2.5.   Multipath propagation with a fixed path

Here we modify the channel model of the previous example by assuming that, as often occurs in practice, the propagation medium has one major strong fixed path in addition to the $N$ weaker paths. Thus, we may write the received-signal complex envelope in the form

$$R(t)e^{j\Theta(t)} = u(t)e^{j\alpha(t)\cdot} + v(t)e^{j\beta(t)}$$

where as before $u(t)$ is Rayleigh-distributed, $\alpha(t)$ is uniform in $(0, 2\pi)$, and $v(t)$ and $\beta(t)$ are deterministic signals. With this model $R(t)$ has the pdf

$$f_R(r) = \frac{r}{\sigma^2} \exp\left\{-\frac{r^2 + v^2}{2\sigma^2}\right\} I_0\left(\frac{rv}{\sigma^2}\right) \tag{13.14}$$

for $r \geq 0$. ($I_0(\cdot)$ denotes the zeroth-order modified Bessel function of the first kind.) Here $R(t)$ and $\Theta(t)$ are not independent.

Let us further assume a certain amount of randomness in the fixed-path signal. Specifically, assume that the phase $\beta$ of the fixed path changes randomly, and that we can model it as a RV uniformly distributed in $(0, 2\pi)$. As a result of this assumption, $R(t)$ and $\Theta(t)$ become independent processes, with $\Theta$ uniformly distributed in $(0, 2\pi)$ and $R(t)$ still having the probability density function (13.14) (see Problem 13.2). The function (13.14) is usually called the "Rice pdf." Its mean-square value is $E[R^2] = v^2 + 2\sigma^2$. This pdf is plotted in Fig. 13.6 for some values of $v$ and $\sigma^2 = 1$.

Notice that in (13.14) $v$ denotes the envelope of the fixed-path component of the received signal, while $2\sigma^2$ is the power of the Rayleigh component (see (13.12) above). Thus, the "Rice factor"

$$K = \frac{v^2}{2\sigma^2}$$

denotes the ratio between the power of the fixed-path component and the power of the Rayleigh component. Sometimes the Rice pdf is written in a slightly

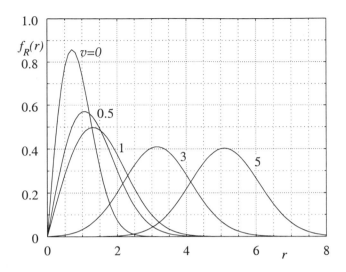

Figure 13.6: *Rice pdf with $\sigma^2 = 1$.*

different form, obtained by assuming $E[R^2] = v^2 + 2\sigma^2 = 1$ and exhibiting the Rice factor explicitly:

$$f_R(r) = 2r(1 + K) \exp\left\{-(1 + K)r^2 - K\right\} I_0\left(2r\sqrt{K(1 + K)}\right) \quad (13.15)$$

for $r \geq 0$.

As $K \to 0$, i.e., as the fixed path reduces its power, since $I_0(0) = 1$ the Rice pdf becomes a Rayleigh pdf. On the other hand, if $K \to \infty$, i.e., the fixed-path power is considerably higher than the power in the random paths, then the Gaussian pdf is a good approximation for the Rice density. In fact, by replacing the Bessel function by its asymptotic expression

$$I_0(x) \sim \frac{e^x}{\sqrt{2\pi x}}, \quad x \to \infty$$

from (13.14) we obtain

$$f_R(r) \sim \frac{\sqrt{r}}{\sqrt{2\pi\sigma}} \exp\left\{-\frac{(r - v)^2}{2\sigma^2}\right\}$$

This shows that when $v$, and hence $K$, becomes large (or $r$ is far out on the tail of the probability density curve) $f_R(r)$ behaves like a Gaussian pdf.

## 13.3.   Frequency-flat, slowly fading channels

The channel models considered in the previous examples assume narrowband transmission. For digital transmission systems this is tantamount to assuming that the duration of a modulated symbol is much greater than the delay spread caused by the multipath propagation. If this occurs, then all frequency components in the transmitted signal are affected by the same random attenuation and phase shift, and the channel is frequency-flat. If in addition the channel varies very slowly with respect to the symbol duration (slow relative motion between transmitter and receiver), then the fading $R(t)\exp[j\Theta(t)]$ remains approximately constant during the transmission of one symbol (if this does not occur the fading process is called *fast*.)

This important special case of this fading model, viz., that of frequency-flat, slow fading, will be subjected to further analysis in the balance of this chapter. The assumption of nonselectivity allows us to model the fading as a process affecting the transmitted signal in a multiplicative form. The assumption of a slow fading leads us to model this process as a constant RV during each symbol interval. From now on we shall also consider the effect of additive Gaussian noise. In conclusion, if $\tilde{x}(t)$ denotes the complex envelope of the modulated signal transmitted during the interval $(0,\,T)$, then the complex envelope of the signal received at the output of a channel affected by slow, flat fading and additive white Gaussian noise can be expressed in the form

$$\tilde{r}(t) = Re^{j\Theta}\tilde{x}(t) + \tilde{n}(t) \qquad (13.16)$$

where $\tilde{n}(t)$ is a complex Gaussian noise, and $Re^{j\Theta}$ is a Gaussian RV, with $\Theta$ uniformly distributed and $R$ having a Rice or Rayleigh pdf.

If we can further assume that the fading is so slow that we can estimate the phase $\Theta$ with sufficient accuracy, and hence compensate for it, then coherent detection is feasible. Thus, model (13.16) can be further simplified to

$$\tilde{r}(t) = R\tilde{x}(t) + \tilde{n}(t) \qquad (13.17)$$

It should be immediately apparent that with this simple model the only difference with respect to an AWGN channel resides in the fact that $R$, instead of being a constant attenuation, is now a RV, whose value affects the amplitude, and hence the power, of the received signal. Assume finally that the value taken by $R$ is known: we describe this situation by saying that we have perfect *channel state information* (CSI). Channel state information can be obtained for example by inserting a pilot tone in a notch of the spectrum of the transmitted signal, and by assuming that the signal is faded exactly in the same way as this tone. Another possible strategy is the insertion of pilot symbols in the transmitted frames.

Detection with perfect CSI can be performed exactly in the same way as for the AWGN channel: in fact, the constellation shape is perfectly known, as is the attenuation incurred by the signal. The optimum decision rule in this case consists of minimizing, as we learned in Section 4.2, the Euclidean distance

$$\int_0^T [r(t) - Rx(t)]^2 \, dt \quad \text{or} \quad |\mathbf{r} - R\mathbf{x}|^2 \qquad (13.18)$$

with respect to the possible transmitted real signals $x(t)$ (or vectors $\mathbf{x}$).

A consequence of this fact is that the error probability with perfect CSI and coherent demodulation of signals affected by frequency-flat, slow fading can be evaluated as follows. We first compute the error probability $P(e \mid R)$ obtained by assuming $R$ constant in (13.17), then we take the expectation of $P(e \mid R)$, with respect to the random variable $R$. The calculation of $P(e \mid R)$ is performed as if the channel were AWGN, but with the energy $\mathcal{E}$ changed into $R^2 \mathcal{E}$. Notice finally that the assumptions of perfect channel-state information and phase-shift estimate make the values of $P(e)$ thus obtained as representing a lower bound to the actual performance.

### 13.3.1.  Coherent detection of binary signals with perfect CSI

A simple case which is amenable to a closed-form expression for the error probability is offered by binary modulation. Let the two transmitted signals have common energy $\mathcal{E}$ and correlation coefficient $\rho$. From (4.34) we know that the error probability, when the received energy is $R^2 \mathcal{E}$ and the detection is coherent, is given by

$$P(e \mid R) = \frac{1}{2} \operatorname{erfc}\left( \sqrt{R^2 \frac{(1-\rho)\mathcal{E}}{2N_0}} \right) \qquad (13.19)$$

Thus, to obtain $P(e)$ we must take the expectation of (13.19) with respect to the pdf of $R$:

$$P(e) = \int_{-\infty}^{\infty} P(e \mid r) \, f_R(r) \, dr \qquad (13.20)$$

**Rayleigh fading**

When $R$ has a Rayleigh pdf, the calculation of $P(e)$ can be done in closed form. By using the definition

$$\operatorname{erfc}(x) \triangleq \frac{2}{\sqrt{\pi}} \int_x^{\infty} e^{-t^2} \, dt$$

in (13.20), and reversing the order of integrations, it is easy to obtain

$$P(e) = \frac{1}{2}\left(1 - \sqrt{\frac{(1-\rho)\bar{\mathcal{E}}/2N_0}{1 + (1-\rho)\bar{\mathcal{E}}/2N_0}}\right) \tag{13.21}$$

where $\bar{\mathcal{E}}$ is the average received energy

$$\bar{\mathcal{E}} \triangleq \mathrm{E}[R^2]\,\mathcal{E} = 2\sigma^2\,\mathcal{E}$$

The special case of antipodal signals ($\rho = -1$) gives

$$P(e) = \frac{1}{2}\left(1 - \sqrt{\frac{\bar{\mathcal{E}}/N_0}{1 + \bar{\mathcal{E}}/N_0}}\right)$$

while for orthogonal signals ($\rho = 0$) we obtain

$$P(e) = \frac{1}{2}\left(1 - \sqrt{\frac{\bar{\mathcal{E}}/2N_0}{1 + \bar{\mathcal{E}}/2N_0}}\right)$$

It can be observed that the 3-dB margin of antipodal signaling over orthogonal signaling, already experienced over the AWGN channel, is retained here. The effect of Rayleigh fading on error probabilities is illustrated in Fig. 13.7.

It is also observed that the error probability over the fading channel decreases with a much lower slope than for AWGN. Hence, the signal-to-noise ratio (SNR) required for reliable transmission over this channel is much higher than with AWGN. By using the asymptotic expression, valid for $x \to \infty$,

$$1 - \sqrt{\frac{x}{1+x}} \sim \frac{1}{2x}$$

we see that, for binary signals and large values of $\bar{\mathcal{E}}/N_0$

$$P(e) \sim \frac{1}{2(1-\rho)}\frac{1}{\bar{\mathcal{E}}/N_0} \tag{13.22}$$

This shows that the error probability decreases only inversely with the signal-to-noise ratio, rather than with an exponential law as in the AWGN channel.

### Rice fading

Fig. 13.7 also shows the error probability for antipodal signals over a Rice channel with different values of the Rice factor $K$, as obtained from numerical integration (see below for a description of this technique). It is seen that, as $K$ increases from 0 to infinity, the channel behavior moves from Rayleigh to AWGN channel, as expected. For $K = 20$ dB the performance is close to AWGN within a fraction of a dB for the signal-to-noise range considered here, while $K = 10$ dB entails a loss of about 7 dB at $P(e) = 10^{-5}$.

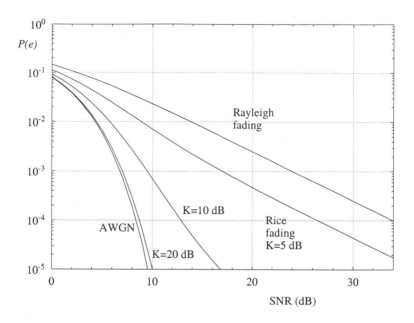

Figure 13.7: *Comparison of error probabilities of antipodal binary modulation over the AWGN, the Rice, and the Rayleigh fading channel. Here SNR=* $\mathrm{E}[R^2]\mathcal{E}/N_0$.

### 13.3.2.  A general technique for computing error probabilities

We now describe a general technique for the evaluation of error probabilities over the fading channel. Although this technique can be used for calculations of $P(e)$ over other types of channels, we describe it here because it finds its most important application for the fading channel.

Recall from Section 4.3.2 that the conditional error probability $P(e \mid \mathbf{x})$ can be bounded above by the *union bound*

$$P(e \mid \mathbf{x}) \leq \sum_{\hat{\mathbf{x}} \neq \mathbf{x}} P\{\mathbf{x} \to \hat{\mathbf{x}}\} \tag{13.23}$$

where $P\{\mathbf{x} \to \hat{\mathbf{x}}\}$ denotes the pairwise error probability, i.e., the probability that the distance of the received signal from $\hat{\mathbf{x}}$ is smaller than that from the transmitted signal $\mathbf{x}$. If $X$ denotes the difference between these two distances when $\mathbf{x}$ is transmitted, then we may write

$$P\{\mathbf{x} \to \hat{\mathbf{x}}\} = P(X < 0) \tag{13.24}$$

This shows how the calculation of error probability in a digital communication scheme can always be reduced to the calculation of the probability that a RV takes on a negative value.

The latter can be evaluated on the basis of the following calculations. Define the (two-sided) Laplace transform of the pdf $f_X(x)$ of the RV $X$

$$\Phi_X(s) \triangleq \mathrm{E}[e^{-sX}] = \int_{-\infty}^{\infty} e^{-sx} f_X(x)\, dx \tag{13.25}$$

where $s = \alpha + j\omega$. From Laplace-transform theory it is known that this integral converges in a vertical strip $\alpha_1 < \alpha < \alpha_2$ of the complex $s$-plane bounded by the closest poles of $\Phi_X(s)$. The restriction of $\Phi_X(s)$ to the real axis is a real positive function, and, since

$$\Phi_X''(\alpha) = \mathrm{E}[X^2 e^{-\alpha X}] \geq 0$$

it is convex. Finally, integrate (13.25) by parts. If $F_X(x)$ denotes the cumulative distribution function of $X$, we obtain

$$\begin{aligned}
\Phi_X(s) &= e^{-sx} F_X(x)|_{-\infty}^{\infty} + s \int_{-\infty}^{\infty} F_X(x) e^{-sx}\, dx \\
&= s \int_{-\infty}^{\infty} F_X(x) e^{-sx}\, dx
\end{aligned}$$

for $\alpha > 0$. Thus, by inverting the Laplace transform, we get

$$F_X(x) = \frac{1}{2\pi j} \int_{c-j\infty}^{c+j\infty} \Phi_X(s) e^{sx} \frac{ds}{s}$$

for $0 < c < \alpha_2$. Consequently, by recognizing that $P(X \leq 0) = F_X(0)$ we have the exact result

$$P(X \leq 0) = \frac{1}{2\pi j} \int_{c-j\infty}^{c+j\infty} \Phi_X(s) \frac{ds}{s} \tag{13.26}$$

where $c$ is chosen so as to obtain convergence.

The latter integral can be evaluated by residues, partial-fraction expansion, or numerical integration. In the following we describe three methods aimed at computing (13.26) exactly or at bounding it.

**Exact calculation through residues.** If $\Phi_X(s)/s = o(s)$ as $|s| \to \infty$, the integral (13.26) can be computed exactly by using either one of the following:

$$P(X \leq 0) = \begin{cases} 1 + \displaystyle\sum_{\text{LH poles}} \mathrm{Residue}[\Phi_X(s)/s] \\[2mm] - \displaystyle\sum_{\text{RH poles}} \mathrm{Residue}[\Phi_X(s)/s] \end{cases} \tag{13.27}$$

where the summations are to be taken over the poles lying in the left-hand (LH) or right-hand (RH) side of the complex $s$-plane, and the residue in $s = 0$ is equal to 1 because $\Phi_X(0) = 1$ by definition.

Calculation of $P(X \leq 0)$ through (13.27) is easy when the poles are simple. If $\Phi_X(s)/s$ contains multiple poles or essential singularities, the actual calculation of (13.27) may become very, if not hopelessly, long and intricate. In fact, in the presence of poles of order $n$ calculation of (13.27) requires determining the derivative of order $n-1$ of a (usually complicated) function. If $\Phi_X(s)/s$ exhibits essential singularities, then derivatives of all orders must be computed.

**Numerical integration.** A different approach to the exact calculation of $P(X \leq 0)$, and one which guarantees arbitrarily high accuracy while not requiring evaluation of the poles or residues of $\Phi_X(s)$, is based on numerical integration. As it can be applied even if $\Phi_X(s)$ is known only numerically, its range of application is exceedingly wide.

Recall (13.26). Since its left-hand side is a real quantity, we are allowed to keep only the real part of its right-hand side. Thus, by observing that

$$\Re\left[\frac{\Phi_X(s)}{s}\right] = \frac{\alpha\Re[\Phi_X(\alpha + j\omega)] + \omega\Im[\Phi_X(\alpha + j\omega)]}{\alpha^2 + \omega^2}$$

we obtain

$$P(X \leq 0) = \frac{1}{2\pi} \int_{-\infty}^{\infty} \frac{c\Re[\Phi_X(c + j\omega)] + \omega\Im[\Phi_X(c + j\omega)]}{c^2 + \omega^2} \, d\omega \qquad (13.28)$$

The change of variable $\omega = c\sqrt{1 - x^2}/x$ transforms (13.28) into

$$P(X < 0) = \frac{1}{2\pi} \int_{-1}^{1} g(x) \frac{dx}{\sqrt{1 - x^2}} \qquad (13.29)$$

where

$$g(x) = c\Re\left[\Phi_X\left(c + jc\frac{\sqrt{1 - x^2}}{x}\right)\right] + \frac{\sqrt{1 - x^2}}{x}\Im\left[\Phi_X\left(c + jc\frac{\sqrt{1 - x^2}}{x}\right)\right]$$

Integrals of the form (13.29) can be evaluated numerically by using Gauss-Chebyshev integration (see, e.g., Abramowitz and Stegun, 1972, p. 889):

$$\int_{-1}^{1} g(x) \frac{dx}{\sqrt{1 - x^2}} = \frac{\pi}{n} \sum_{k=1}^{n} g\left(\cos\frac{(2i - 1)\pi}{2n}\right) + r_n \qquad (13.30)$$

where the remainder term $r_n$ decreases to zero as the number $n$ of terms in (13.30) increases, provided that the derivative of order $2n$ of $g(x)$ remains bounded in

$(-1, 1)$. The actual value of $r_n$ can be bounded from above if the derivatives of $g(x)$ are known, but this is often unnecessary: in practice, to achieve the desired degree of accuracy it suffices to evaluate (13.30) for increasing values of $n$, and accept the result as soon as it does not change significantly.

In our case we obtain explicitly

$$P(X \leq 0) = \frac{1}{\nu} \sum_{k=0}^{\nu/2} \{c\Re[\Phi_X(c + jc\tau_k)] + \tau_k\Im[\Phi_X(c + jc\tau_k)]\} + r_\nu \quad (13.31)$$

where $\tau_k = \tan[(2k - 1)\pi/(2\nu)]$. Notice also that the value of $c$ may affect the value of $\nu$ necessary to achieve a prescribed accuracy. A reasonable choice is usually $c = \alpha_2/2$.

**Chernoff bound.** This provides a very simple, although frequently loose, upper bound to $P(X \leq 0)$. By recalling the definition of $\Phi_X(s)$ and our derivation of the Chernoff bound in Section 12.1, we have, for $\lambda > 0$,

$$P(X \leq 0) \leq \Phi_X(\lambda)$$

The Chernoff bound is thus given by the inequality

$$P(X \leq 0) \leq \min_{0 < \lambda < \alpha_2} \Phi_X(\lambda) \quad (13.32)$$

where, as discussed before, the minimum is unique due to the convexity of the restriction of $\Phi_X(x)$ to the real axis.

It often happens that the minimization involved in (13.32) cannot be obtained in closed form. If this is the case, one may proceed numerically, or choose (by educated guess) a suboptimum value for $\lambda$, which still provides a bound.

**Computing $\Phi_X(s)$**

The former discussion leads to the consequence that once $\Phi_X(s)$ is known the probability $P(X \leq 0)$ can be evaluated in several ways. Thus, it is appropriate to describe a general technique for the evaluation of $\Phi_X(s)$. Before entering into the details of the calculation we list for convenience two equations that will be useful in the sequel. If $v$ is a Gaussian RV with mean $\mu$ and variance $\sigma^2$, we have

$$E[e^{-sv}] = e^{-s\mu + s^2\sigma^2/2} \quad (13.33)$$

If $R$ is a Rice-distributed RV with $E[R^2] = 1$ and Rice parameter $K$, then, for $\Re\{a\} < 1 + K$

$$E[e^{aR^2}] = \frac{1 + K}{1 + K - a} \exp\left\{\frac{Ka}{1 + K - a}\right\} \quad (13.34)$$

The latter equation follows from the observation that the Rice pdf (13.14) integrates to 1, which yields the integral

$$\int_0^\infty re^{-A^2r^2} I_0\left(2A^2rv\right) dr = \frac{1}{2A^2}e^{A^2v^2}$$

As a special case, for a Rayleigh-distributed $R$ we obtain, by taking $K = 0$ in (13.34):

$$E[e^{aR^2}] = \frac{1}{1-a} \tag{13.35}$$

Assume now that, for a given value of the fading random variable $R$, $X$ is conditionally Gaussian with mean $\mu(R)$ and variance $\sigma^2(R)$. From (13.33) we obtain

$$E[e^{-sX} \mid R] = e^{-s\mu(R)+s^2\sigma^2(R)/2} \tag{13.36}$$

The desired function $\Phi_X(s)$ is obtained by averaging the above with respect to the random variable $R$.

### Error probability for coherent detection of general $M$-ary signals with perfect CSI

Here $\mathbf{r} = R\mathbf{x} + \mathbf{n}$, with $\mathbf{n}$ a Gaussian random vector whose components are independent with mean zero and equal variance $N_0/2$. Recalling that

$$P\{\mathbf{x} \to \hat{\mathbf{x}}\} = P(|\mathbf{r} - R\hat{\mathbf{x}}|^2 < |\mathbf{r} - R\mathbf{x}|^2) \tag{13.37}$$

we have

$$X = |\mathbf{r} - R\hat{\mathbf{x}}|^2 - |\mathbf{r} - R\mathbf{x}|^2 = 2R(\mathbf{n}, \mathbf{x} - \hat{\mathbf{x}}) + R^2 d_{x\hat{x}}^2 \tag{13.38}$$

where $(\cdot, \cdot)$ denotes scalar product, and

$$d_{x\hat{x}} \stackrel{\triangle}{=} |\mathbf{x} - \hat{\mathbf{x}}| \tag{13.39}$$

$X$, being a linear transformation of the random vector $\mathbf{n}$, is itself conditionally Gaussian for given $R$. By duplicating calculations performed in Section 12.1, its conditional mean and variance are

$$\mu(R) = R^2 d_{x\hat{x}}^2$$

and

$$\sigma^2(R) = 2R^2 N_0 d_{x\hat{x}}^2$$

respectively. We are now in a position to use (13.36), which yields

$$\Phi_X(s) = E_R[\exp\{R^2 s(N_0 s - 1)d_{x\hat{x}}^2\}] \tag{13.40}$$

If $R$ has a Rice pdf, the integral arising from (13.40) can be given a closed form. In fact, by using (13.34), we obtain

$$\Phi_X(s) = \frac{1+K}{(1+K) - s(N_0 s - 1)d_{x\hat{x}}^2} \exp\left\{ \frac{Ks(N_0 s - 1)d_{x\hat{x}}^2}{(1+K) - s(N_0 s - 1)d_{x\hat{x}}^2} \right\}$$

(13.41)

The special case of Rayleigh fading is obtained by letting $K = 0$ in (13.41):

$$\Phi_X(s) = \frac{1}{1 - s(N_0 s - 1)d_{x\hat{x}}^2}$$

(13.42)

**Method of the residues.**    For Rayleigh fading the function $\Phi_X(s)$ has two simple poles

$$s_{1,2} = \frac{1 \mp \sqrt{1 + 4N_0/d_{x\hat{x}}^2}}{2N_0}$$

where $s_1$ is the pole in the left-hand side of the $s$-plane, and $s_2$ is the pole in its right-hand side. By writing

$$\Phi_X(s) = \frac{s_1 s_2}{(s - s_1)(s - s_2)}$$

we see that the residue of the function $\Phi_X(s)/s$ in $s_2$ is $s_1/(s_2 - s_1)$. After a sign change, from (13.27) we finally obtain the exact result

$$P\{\mathbf{x} \to \hat{\mathbf{x}}\} = \frac{1}{2}\left(1 - \sqrt{\frac{d_{x\hat{x}}^2/4N_0}{1 + d_{x\hat{x}}^2/4N_0}}\right)$$

(13.43)

**Chernoff bound.**    It should be easily recognized that for $s = \lambda$ (a real variable) $\Phi_X(s)$ in (13.41) is an increasing function of $\lambda(N_0\lambda - 1)$, irrespective of $K$. Thus, it achieves its smallest value when $\lambda(N_0\lambda - 1)$ is a minimum, i.e., when

$$\lambda = \frac{1}{2N_0}$$

Hence, the Chernoff bound for Rice fading yields

$$P\{\mathbf{x} \to \hat{\mathbf{x}}\} \le \Phi_X\left(\frac{1}{2N_0}\right)$$

$$= \frac{1+K}{(1+K) + d_{x\hat{x}}^2/4N_0} \exp\left\{ \frac{-Kd_{x\hat{x}}^2/4N_0}{(1+K) + d_{x\hat{x}}^2/4N_0} \right\}$$

(13.44)

For Rayleigh fading we have the Chernoff bound

$$P\{\mathbf{x} \to \hat{\mathbf{x}}\} \le \frac{1}{1 + d_{x\hat{x}}^2/4N_0}$$

(13.45)

Fig. 13.8 compares this Chernoff bound with the exact value (13.43) of pairwise error probability.

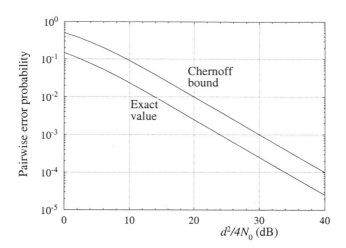

Figure 13.8: *Comparison of the Chernoff bound to the pairwise error probability*
$P\{\mathbf{x} \to \widehat{\mathbf{x}}\}$ *with its exact value on a Rayleigh channel.*

**Example 13.1 (Binary signals)**   For two signals with common energy $\mathcal{E}$ and correlation coefficient $\rho = (\mathbf{x}, \widehat{\mathbf{x}})/\mathcal{E}$ we have $d_{x\hat{x}}^2 = 2\mathcal{E}(1-\rho)$, and hence, for Rayleigh fading, from (13.43)

$$P(e) = \frac{1}{2}\left(1 - \sqrt{\frac{(1-\rho)\mathcal{E}/2N_0}{1 + (1-\rho)\mathcal{E}/2N_0}}\right) \tag{13.46}$$

This coincides (as it should) with (13.21), after observing that $\mathcal{E} = \bar{\mathcal{E}}$ here because we are assuming that $\mathrm{E}[R^2] = 1$.

The Chernoff bound yields

$$P(e) \le \frac{1}{1 + (1-\rho)\mathcal{E}/2N_0} \tag{13.47}$$

or, for large SNRs,

$$P(e) \le \frac{2}{1-\rho}\,\frac{1}{\mathcal{E}/N_0} \tag{13.48}$$

□

### 13.3.3. No channel-state information

So far we have made the assumption of perfect channel-state information in coherent detection. Since an infinitely-accurate CSI is never available, we should
discuss the effect of its imperfect knowledge. For simplicity, we shall limit ourselves to examining the limiting case where no CSI at all is available. In this case,
we choose to demodulate by minimizing with respect to $\mathbf{x}$, in lieu of (13.18), the
Euclidean distance we would use in the Gaussian channel, that is,

$$\int_0^T [r(t) - x(t)]^2 \quad \text{or} \quad |\mathbf{r} - \mathbf{x}|^2 \tag{13.49}$$

The calculation of error probability in this case can be performed in the same
way as with perfect CSI: we first derive $P(e \mid R)$, then we average over the
fading pdf.

Given two competing signals $\mathbf{x}$ and $\hat{\mathbf{x}}$, their pairwise error probability can be
calculated from (13.24), where as usual $X$ denotes the difference between the
squared distances of the received vector $\mathbf{r}$ to $\hat{\mathbf{x}}$ and $\mathbf{x}$:

$$
\begin{aligned}
X &= |\mathbf{r} - \hat{\mathbf{x}}|^2 - |\mathbf{r} - \mathbf{x}|^2 \\
&= 2(\mathbf{n}, \mathbf{x} - \hat{\mathbf{x}}) + |\hat{\mathbf{x}}|^2 - |\mathbf{x}|^2 + 2R(\mathbf{x}, \mathbf{x} - \hat{\mathbf{x}})
\end{aligned} \tag{13.50}
$$

It is recognized, by duplicating calculations performed in Section 12.1, that,
given $R$, $X$ is a conditionally Gaussian RV, with conditional mean and variance

$$\mu(R) = |\hat{\mathbf{x}}|^2 - |\mathbf{x}|^2 + 2R(\mathbf{x}, \mathbf{x} - \hat{\mathbf{x}})$$

and

$$\sigma^2(R) = 2N_0 |\mathbf{x} - \hat{\mathbf{x}}|^2$$

respectively. Thus, from (13.25) we have

$$\Phi_X(s) = \mathrm{E}_R \left[ \exp\left\{ -s \left( |\hat{\mathbf{x}}|^2 - |\mathbf{x}|^2 + 2R(\mathbf{x}, \mathbf{x} - \hat{\mathbf{x}}) \right) + N_0 s^2 |\hat{\mathbf{x}}|^2 - |\mathbf{x}|^2 \right\} \right] \tag{13.51}$$

Consider for simplicity the case of Rayleigh fading. By using the integral

$$\int_0^\infty x e^{-Bx - Ax^2/2} \, dx = \frac{1}{A} \left[ 1 - e^{B^2/2A} \sqrt{\pi} \frac{B}{\sqrt{2A}} \, \mathrm{erfc}\left( \frac{B}{\sqrt{2A}} \right) \right]$$

we obtain for the Rayleigh channel

$$\Phi_X(s) = \exp\left[ -(|\hat{\mathbf{x}}|^2 - |\mathbf{x}|^2)s + N_0|\hat{\mathbf{x}}|^2 - |\mathbf{x}|^2 s^2 \right] \tag{13.52}$$
$$\cdot \left[ 1 - \exp\left( (\mathbf{x}, \mathbf{x} - \hat{\mathbf{x}})^2 s^2 \right) \sqrt{\pi}(\mathbf{x}, \mathbf{x} - \hat{\mathbf{x}})s \, \mathrm{erfc}\left( (\mathbf{x}, \mathbf{x} - \hat{\mathbf{x}})s \right) \right]$$

The above expression can be used in conjunction with numerical integration to
obtain the union bound to error probabilities of any signal constellation.

### 13.3.4.   Differential and noncoherent detection

In a channel for which the fading process is not slow enough to allow accurate estimation of its phase shift, an alternative to coherent detection is differential demodulation. In fact, the latter requires only that the phase remain stable over two adjacent symbols.

The calculation of error probability for binary signaling over a fading channel with differential detection is quite easy. Assume antipodal signaling, and observe that, conditionally on $R$, we have, from Eq. (5.79),

$$P(e \mid R) = \frac{1}{2}e^{-R^2 \mathcal{E}/N_0}$$

For Rayleigh fading, direct integration yields

$$P(e) = \frac{1}{2}\frac{1}{1 + \bar{\mathcal{E}}/N_0}$$

which is twice as large as (13.22). Thus, differentially coherent detection entails a loss, for high SNRs, of 3 dB with respect to coherent detection.

Another possible choice is not to estimate the channel phase shift at all. The error probability for binary orthogonal signals with noncoherent detection is, from (4.95),

$$P(e \mid R) = \frac{1}{2}e^{-R^2 \mathcal{E}/2N_0}$$

For Rayleigh fading, direct integration yields

$$P(e) = \frac{1}{2 + \bar{\mathcal{E}}/N_0}$$

which, for high SNRs, is four time as large as (13.22) when $\rho = -1$ (antipodal signals), and two time as large when orthogonal signals are used with coherent detection.

## 13.4.   Introducing diversity

We have seen that in the presence of fading the transmitter has to deliver a power higher, and in some cases much higher, than for an AWGN channel to achieve the same error probability. For example, passing from AWGN to Rayleigh fading transforms an exponential dependency of error probability on SNR into an inverse linear one. To combat fading, and hence to reduce transmit-power needs, a very effective technique consists of introducing *diversity* in the channel. Based on the observation that on a fading channel the SNR at the receiver is a random variable, the idea is to transmit the same signal through $L$ separate fading channels. They are chosen so as to provide the receiver with $L$ replicas of the same signal, these replicas being affected by fading processes as independent as possible, and hence giving rise to independent SNRs. If $L$ is large enough, at any time instant there is a high probability that at least one of the signals received from the $L$ "diversity branches" will not be affected by a deep fade, and hence its SNR will be above a critical threshold. By suitably combining the received signals, the fading effect will be mitigated.

Many techniques have been advocated for obtaining the independent channels needed by diversity, and several methods are known for combining the signals obtained at their outputs. The most important among them can be categorized as follows.

**Space diversity.** This consists of receiving the transmitted signal through $L$ separate antennas, whose spacing is wide enough with respect to the carrier wavelength so as to obtain sufficient decorrelation. This technique can be easily implemented, and does not require extra spectrum occupancy.

**Polarization diversity.** If a radio channel exhibits decorrelated fading for signals transmitted on orthogonal polarizations, then diversity can be obtained by using a pair of cross-polarized antennas in the receiver. Notice that only two diversity branches are available here, while any value of $L$ can in principle be obtained with space diversity. On the other hand, cross-polarized antennas do not need the large physical separation necessary for space diversity. In scattering environments tending to depolarize a signal, there is no need for separate transmission.

**Frequency diversity.** This is obtained by sending the same signal over different frequency carriers, whose separation must be larger than the coherence bandwidth of the channel. Clearly, frequency diversity is not a bandwidth-efficient solution.

**Time diversity.** If the same signal is transmitted in different time slots separated by an interval longer than the coherence time of the channel, time diversity can be obtained. Since in mobile radio systems slow-moving receivers have a large coherence time, in these conditions time diversity could only be introduced at the price of large delays.

### 13.4.1. Diversity combining techniques

Three main combining techniques, viz., selection, maximal ratio, and equal gain, will be described here. Each of them can be used in conjunction with any of the diversity schemes just listed. Some analyses will follow; however, it should be clear from the onset that the relative advantage of a diversity scheme will be lower as the channel moves away from Rayleigh fading towards Rice fading. In fact, increasing the Rice factor $K$ causes the various diversity branches to exhibit a lower difference in their instantaneous SNRs.

**Selection combining**

Among the forms of diversity combining, this is conceptually the simplest. It consists of selecting at each time, among the $L$ diversity branches, the one with the largest value of signal-to-noise ratio.

We analyze the performance of this combining technique by assuming that each diversity branch is affected by the same Gaussian noise power, so that selecting the branch with the largest instantaneous SNR is tantamount to selecting the branch with the largest instantaneous power. Moreover, let the branches have the same average signal-to-noise ratio

$$\bar{\eta} = \frac{\bar{\mathcal{E}}}{N_0} = \mathrm{E}[R^2]\frac{\mathcal{E}}{N_0} \tag{13.53}$$

and denote by $\eta_i$, $1 \leq i \leq L$, the instantaneous signal-to-noise ratio measured in the $i$th diversity branch during the transmission of a given symbol, that is

$$\eta_i = R_i^2 \frac{\mathcal{E}}{N_0} \tag{13.54}$$

The probability that the SNR in the $i$th branch be lower than a threshold $H$ is given by

$$P(\eta_i \leq H) = \int_0^H f_i(y)\,dy \tag{13.55}$$

where $f_i(y)$ denotes the pdf of $\eta_i$, which we assume to be the same for all the branches. With $L$ independent branches the probability that all of them have an SNR below the same threshold $H$ is given by

$$P(\eta_1 \leq H, \eta_2 \leq H, \ldots, \eta_L \leq H) = [P(\eta_i \leq H)]^L \tag{13.56}$$

and decreases as $L$ increases. This is also the cumulative distribution function of the RV

$$\check{\eta} = \max\{\eta_1, \ldots, \eta_L\}$$

(in fact, $\check{\eta}$ is less than $H$ if and only if $\eta_1, \ldots, \eta_L$ are all less than $H$). Hence, the pdf of the SNR obtained at the output of the selection combiner is obtained by taking the derivative of (13.56) with respect to $H$. This result can be used to obtain the error probability of a digital modulation scheme in the presence of selection combining: once the error probability is known for a given SNR, it suffices to average it over the pdf of $\check{\eta}$.

**Example 13.2** If $R$ has a Rayleigh pdf, then $R^2$ has the same pdf of the sum of the squares of two independent, zero-mean Gaussian random variables with the same variance. Hence, $R^2 \mathcal{E}/N_0$ is a chi-square RV with two degrees of freedom, i.e., an exponentially-distributed RV:

$$f_i(y) = \frac{1}{\bar{\eta}} e^{-y/\bar{\eta}}$$

Thus, the probability that all the $L$ diversity branches have an SNR below the threshold $H$ is given by

$$P(\eta_1 \leq H, \, \eta_2 \leq H, \, \ldots, \, \eta_L \leq H) = \left(1 - e^{-H/\bar{\eta}}\right)^L, \qquad H \geq 0$$

Consequently, this is the cumulative distribution function of the SNR at the output of the selection combiner. The corresponding pdf is obtained by differentiation of the above:

$$f_{\bar{\eta}}(y) = \frac{L}{\bar{\eta}} e^{-y/\bar{\eta}} \left(1 - e^{-y/\bar{\eta}}\right)^{L-1}, \qquad y \geq 0$$

As a special case, let $\bar{\eta} = 20$ dB, and compute the probability $p$ that the instantaneous SNR at the output of the selection combiner be lower than $H = 10$ dB. We have

$$p = \left(1 - e^{-H/\bar{\eta}}\right)^L = \left(1 - e^{-0.1}\right)^L$$

For example, with $L = 1$ (no diversity) we have $p = 0.095$, while for $L = 2$ we have $p = 0.009$ (a decrease of one order of magnitude) and for $L = 4$ we have $p = 8.2 \times 10^{-5}$ (a decrease of three orders of magnitude). □

### Maximal-ratio combining

Selection combining is relatively easy to implement, as it requires only a measure of the powers received from each diversity branch and an antenna switch at the receiver. However, the simple observation that it disregards the information obtained from all branches but one leads us to argue that it is not an optimum combining technique.

In maximal-ratio combining, the signals at the output of the $L$ diversity branches are combined linearly, and the coefficients of the linear combination are selected so as to maximize the ratio between the instantaneous signal energy and the noise power spectral density. Specifically, let us denote by

$$r_i = R_i x + n_i, \qquad 1 \leq i \leq L$$

the complex signals at the output of the diversity branches, and let us assume that the noises $n_i$ are independent and have the same power spectral density

$2N_0$. Moreover, assume that we have perfect channel state information on all the branches, i.e., that the values taken on by the fading random variables $R_i$ are perfectly known. Finally, assume for simplicity that $x$ is an equal-energy signal with $|x|^2 = 2\mathcal{E}$.

Maximal-ratio combining consists of using the linear combination

$$r = \sum_{i=1}^{L} G_i r_i = \sum_{i=1}^{L} G_i R_i x + \sum_{i=1}^{L} G_i n_i \qquad (13.57)$$

as the decision variable for demodulation. The power spectral density of the noise after combining is given by

$$2N_0 \sum_{i=1}^{L} |G_i|^2$$

while the instantaneous signal energy is

$$2\mathcal{E} \left| \sum_{i=1}^{L} G_i R_i \right|^2$$

The ratio between these two quantities

$$\eta = \frac{\mathcal{E}}{N_0} \frac{\left| \sum_{i=1}^{L} G_i R_i \right|^2}{\sum_{i=1}^{L} |G_i|^2}$$

can be maximized as follows. Recall the Cauchy-Schwarz inequality

$$\left| \sum_{i=1}^{L} a_i b_i^* \right|^2 \le \sum_{i=1}^{L} |a_i|^2 \sum_{i=1}^{L} |b_i|^2$$

which holds with equality for $a_i = c b_i$, $c$ any constant. Thus,

$$\eta \le \frac{\mathcal{E}}{N_0} \frac{\sum_{i=1}^{L} |G_i|^2 \sum_{i=1}^{L} R_i^2}{\sum_{i=1}^{L} |G_i|^2} = \frac{\mathcal{E}}{N_0} \sum_{i=1}^{L} R_i^2 \qquad (13.58)$$

The equality in (13.58) is obtained for $G_i = R_i$ for all $i$, which provides the weighting coefficients for maximal-ratio combining. This shows that each diversity branch is weighted proportionally to the fading attenuation that affects it: the branches more faded are counted less, and vice versa. Maximal-ratio combining provides a signal-to-noise ratio

$$\eta_{\mathrm{MR}} = \frac{\mathcal{E}}{N_0} \sum_{i=1}^{L} R_i^2 \qquad (13.59)$$

If we recall that $\mathcal{E} R_i^2 / N_0$ is the SNR per branch, the latter equation shows that with maximal-ratio combining $\eta_{\mathrm{MR}}$ is the sum of the SNRs, and hence can be large even when the individual SNRs are small. The SNR provided by selection combining is simply the largest among the terms in (13.59).

**Example 13.3** Assume that all branches are affected by independent fading, and that this is Rayleigh-distributed. Since the sum of $L$ squares of independent Rayleigh RVs has a chi-square pdf with $2L$ degrees of freedom (Papoulis, 1984, p. 187), we obtain from (13.59) the pdf of the signal-to-noise ratio:

$$f_{\eta_{MR}}(y) = \frac{1}{(L-1)!(\bar{\mathcal{E}}/N_0)^L} y^{L-1} e^{-y/(\bar{\mathcal{E}}/N_0)}, \qquad y \geq 0 \qquad (13.60)$$

$\square$

The result just obtained about the signal-to-noise ratio of the decision variable resulting from maximal-ratio combining can be used to evaluate the error probability of a modulation scheme. For simplicity, here we restrict our attention to binary antipodal signals. From (13.19) with $\rho = -1$ we obtain

$$P(e \mid \eta_{MR}) = \frac{1}{2} \operatorname{erfc}\left(\sqrt{\eta_{MR}}\right) \qquad (13.61)$$

where $\eta_{MR}$ is the SNR (13.59) obtained from maximal-ratio combining. Thus,

$$P(e) = \mathrm{E}_{\eta_{MR}}[P(e \mid \eta_{MR})]$$

The latter expectation is computed by multiplying $P(e \mid \eta_{MR})$ by the pdf of $\eta_{MR}$ and integrating.

**Example 13.3 (continued)** In the conditions of this example, $P(e)$ can be computed in closed form. We obtain

$$P(e) = \left(\frac{1-\mu}{2}\right)^L \sum_{k=0}^{L-1} \binom{L-1+k}{k} \left(\frac{1+\mu}{2}\right)^k \qquad (13.62)$$

where we have defined

$$\mu = \sqrt{\frac{\bar{\mathcal{E}}/N_0}{1 + \bar{\mathcal{E}}/N_0}}$$

The values of $P(e)$ for $L = 1$ to $L = 4$ are shown in Fig. 13.9. When the signal-to-noise ratio $\bar{\mathcal{E}}/N_0$ is large enough, from the same asymptotic approximation used in Section 13.3.1. we obtain $(1+\mu)/2 \sim 1$ and $(1-\mu)/2 \sim 1/(4\bar{\mathcal{E}}/N_0)$. Furthermore,

$$\sum_{k=0}^{L-1} \binom{L-1+k}{k} = \binom{2L-1}{L}$$

so that the probability of error $P(e)$ can be asymptotically approximated by

$$P(e) \sim \left(\frac{1}{4\bar{\mathcal{E}}/N_0}\right)^L \binom{2L-1}{L} \qquad (13.63)$$

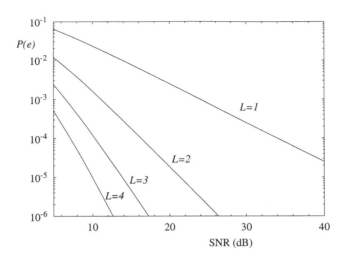

Figure 13.9: *Error probability for binary antipodal transmission with coherent detection and L-branch maximal-ratio diversity combining. Here SNR= $\mathrm{E}[R^2]\mathcal{E}/N_0$.*

The latter equation shows an important fact: the probability of error, with diversity order $L$, decreases inversely with the $L$th power of the signal-to-noise ratio. □

### Equal-gain combining

Should the derivation of channel-state information be too complex, one may use a simplified version of maximal-ratio combining, in which all the coefficients $G_i$ in (13.57) are set equal to 1.

## 13.5. Coding for the Rayleigh fading channel

In this section we provide an analysis of coding for fading channels, with emphasis on the Rayleigh fading channel. As we shall see, the concept of *code diversity* plays a crucial role here: in fact, coding will be seen as a way of introducing diversity in the transmission system, and all channel codes can be interpreted as having a degree of built-in diversity.

Assume transmission of a coded sequence $\mathbf{x} = (\mathbf{s}_1, \mathbf{s}_2, \ldots, \mathbf{s}_n)$ where the components of $\mathbf{x}$ are signal vectors selected from a constellation $\mathcal{S}$. We do not distinguish here among block or convolutional codes (with soft decoding), or block- or trellis-coded modulation. We also assume that, thanks to perfect (i.e., infinite-depth) interleaving, the fading RVs affecting the various symbols $\mathbf{s}_k$ are independent. Hence we write, for the components of the received sequence $(\mathbf{r}_1, \mathbf{r}_2, \ldots, \mathbf{r}_n)$

$$\mathbf{r}_k = R_k \mathbf{x}_k + \mathbf{n}_k \tag{13.64}$$

where the $R_k$ are independent, and, under the assumption that the noise is white, all the components of the random vectors $\mathbf{n}_k$ are also independent.

Coherent detection of the coded sequence, with the assumption of perfect channel-state information, is based upon the search for the coded sequence $\mathbf{x}$ that minimizes the distance

$$\sum_{k=1}^{n} |\mathbf{r}_k - R_k \mathbf{s}_k|^2 \tag{13.65}$$

Thus, the pairwise error probability can be expressed in this case as we did in 13.3.2.

$$P\{\mathbf{x} \to \hat{\mathbf{x}}\} = P(X < 0) \tag{13.66}$$

where now

$$
\begin{aligned}
X &= \sum_{k=1}^{n} \left[ |\mathbf{r}_k - R_k \hat{\mathbf{s}}_k|^2 - |\mathbf{r}_k - R_k \mathbf{s}_k|^2 \right] \\
&= \sum_{k=1}^{n} \left[ |R_k(\mathbf{s}_k - \hat{\mathbf{s}}_k) + \mathbf{n}_k|^2 - |\mathbf{n}_k|^2 \right] \\
&= \sum_{k=1}^{n} \left[ R_k^2 |\mathbf{s}_k - \hat{\mathbf{s}}_k|^2 + 2R_k(\mathbf{n}_k, \mathbf{s}_k - \hat{\mathbf{s}}_k) \right]
\end{aligned}
$$

Under our assumptions, all the terms in the last summation are independent. Thus, from (13.40) and (13.41), by observing that the RVs $R_k$ are independent and equally distributed we obtain

$$
\begin{aligned}
\Phi_X(s) &= \prod_{k=1}^{n} E_{R_k} [\exp s(N_0 s - 1) R_k^2 |\mathbf{s}_k - \hat{\mathbf{s}}_k|^2] \tag{13.67} \\
&= \prod_{k \in \mathcal{K}} E_{R_k} [\exp s(N_0 s - 1) R_k^2 |\mathbf{s}_k - \hat{\mathbf{s}}_k|^2] \tag{13.68}
\end{aligned}
$$

where the last equality differs from the previous one in that the index set is reduced from $\{1, \ldots, n\}$ to $\mathcal{K}$, the set of $k$ such that $\mathbf{s}_k \neq \hat{\mathbf{s}}_k$. This can be done because for the values of $k$ such that $s_k = \hat{s}_k$ the exponential in (13.68) takes value 1, and hence carries no contribution to $\Phi_X(s)$.

Since $\mathcal{K}$ has as many elements as there are components of the code sequence $\mathbf{x}$ that differ from the corresponding components of $\hat{\mathbf{x}}$, the cardinality of the set $\mathcal{K}$ turns out to be the Hamming distance between $\mathbf{x}$ and $\hat{\mathbf{x}}$, that is, the number of components in which $\mathbf{x}$ and $\hat{\mathbf{x}}$ differ. We denote this Hamming distance $d_H(\mathbf{x}, \hat{\mathbf{x}})$.

By using the result (13.41), $\Phi_X(s)$ can be given a closed form in the case of Rice fading, and a fortiori for Rayleigh fading, obtained by letting $K = 0$. In the latter case

$$\Phi_X(s) = \prod_{k \in \mathcal{K}} \frac{1}{1 - s(N_0 s - 1)|s_k - \hat{s}_k|^2} \tag{13.69}$$

Computation of the pairwise error probability through the Chernoff bound is especially simple in this case, because the choice $s = 1/2N_0$ minimizes each term of the product, and hence $\Phi_X(s)$, for real $s$. We obtain

$$P\{\mathbf{x} \to \hat{\mathbf{x}}\} \leq \prod_{k \in \mathcal{K}} \frac{1}{1 + |s_k - \hat{s}_k|^2/4N_0} \tag{13.70}$$

**Example 13.4**  For illustration purposes, let us compute the Chernoff upper bound to the word error probability of a block code with rate $R_c$. Assume that binary antipodal modulation is used, with waveforms of energies $\mathcal{E}$, and that the demodulation is coherent with perfect CSI. We use (13.70) by observing that for $\hat{x}_k \neq x_k$ we have

$$|s_k - \hat{s}_k|^2 = 4\bar{\mathcal{E}} = 4R_c\bar{\mathcal{E}}_b,$$

where $\bar{\mathcal{E}}_b$ denotes the average energy per bit. For two code words $\mathbf{x}$, $\hat{\mathbf{x}}$ at Hamming distance $d_H(\mathbf{x}, \hat{\mathbf{x}})$ we have

$$P\{\mathbf{x} \to \hat{\mathbf{x}}\} \leq \left( \frac{1}{1 + R_c\bar{\mathcal{E}}_b/N_0} \right)^{d_H(\mathbf{x}, \hat{\mathbf{x}})}$$

and hence for a linear code

$$P(e) = P(e \mid \mathbf{x}) \leq \sum_d A_d \left( \frac{1}{1 + R_c\bar{\mathcal{E}}_b/N_0} \right)^d$$

where the sum runs over the set of nonzero Hamming weights of the code, and $A_d$ is the number of words with Hamming weight $d$. It can be seen that for high enough signal-to-noise ratios the dominant term in the expression of $P(e)$ is the one with exponent $d_{\min}$, the minimum Hamming distance of the code.

By recalling Example 13.3 and (13.63), the fact that the probability of error decreases inversely with the signal-to-noise ratio raised to power $d_{\min}$ can be expressed by saying that we have introduced a *code diversity* $d_{\min}$. In this context, the various diversity schemes discussed in the previous section may be seen as implementations of the simplest among the coding schemes, the repetition code, which provides a diversity equal to the number of diversity branches.                                                                ☐

### 13.5.1.    Guidelines of code design for the Rayleigh fading channel

We may further upper bound the RHS of (13.70) by writing

$$P\{\mathbf{x} \to \widehat{\mathbf{x}}\} \leq \prod_{k \in \mathcal{K}} \frac{1}{|s_k - \widehat{s}_k|^2/4N_0} = \frac{1}{[\delta^2(\mathbf{x}, \widehat{\mathbf{x}})/4N_0]^{d_H(\mathbf{x},\widehat{\mathbf{x}})}} \qquad (13.71)$$

(which is close to the true Chernoff bound for small enough $N_0$). Here

$$\delta^2(\mathbf{x}, \widehat{\mathbf{x}}) \triangleq \left[ \prod_{k \in \mathcal{K}} |s_k - \widehat{s}_k|^2 \right]^{1/d_H(\mathbf{x},\widehat{\mathbf{x}})}$$

is the geometric mean of the nonzero squared Euclidean distances between the components of $\mathbf{x}, \widehat{\mathbf{x}}$. The latter result shows the important fact that the error probability is (approximately) inversely proportional to the *product* of the squared Euclidean distances between the components of $\mathbf{x}, \widehat{\mathbf{x}}$ that differ, and to a power of the signal-to-noise ratio whose exponent is the Hamming distance between $\mathbf{x}$ and $\widehat{\mathbf{x}}$.

We hasten to observe that the expression obtained here for the pairwise error probability is an *upper bound*, rather than an exact expression. Thus, the results obtained should be interpreted with some care.

Now, we know from the results referring to block codes, convolutional codes, and coded modulation that the union bound to error probability for a coded system can be obtained by summing up the pairwise error probabilities associated with all the different "error events." For small noise spectral density $N_0$, i.e., for high signal-to-noise ratios, a few equal terms will dominate the union bound. In our framework, these correspond to error events with the smallest value of the Hamming distance $d_H(\mathbf{x}, \widehat{\mathbf{x}})$. We denote this quantity by $L_c$ to stress the fact that it reflects a diversity residing in the code. We have

$$P\{\mathbf{x} \to \widehat{\mathbf{x}}\} \overset{\sim}{\leq} \frac{\nu}{[\delta^2(\mathbf{x}, \widehat{\mathbf{x}})/4N_0]^{L_c}} \qquad (13.72)$$

where $\nu$ is the number of dominant error events. For error events with the same Hamming distance, the values taken by $\delta^2(\mathbf{x}, \widehat{\mathbf{x}})$ and by $\nu$ are also of importance.

This observation may be used to design coding schemes for the Rayleigh fading channel with high SNR: the Euclidean distance, which is the central parameter used in the design of coding schemes for the AWGN channel, plays a minor role here.

From the discussion above, we have learned that over the high-SNR Rayleigh fading channel with perfect interleaving the choice of a coding scheme should be based on the maximization of the code diversity, i.e., the minimum Hamming distance among pairs of error events. Since for the Gaussian channel code diversity does not play the same central role, coding schemes optimized for the Gaussian channel are likely to be suboptimum for the Rayleigh fading channel.

One should also observe that for "conventional" systems, i.e., those separating modulation and coding with binary modulation, Hamming distance is proportional to Euclidean distance, and hence a system optimized for the additive white Gaussian channel is also optimum for the Rayleigh fading channel. This solution has the advantage of being robust, i.e., of providing good performance with a fading channel as well as with an AWGN channel (and, consequently, with a Rice fading channel, which can be seen as intermediate between the latter two). This observation has prompted some authors to advocate "pragmatic" schemes in which coded modulation is generated by pairing an $M$-ary signal set with a binary convolutional code with the largest free Hamming distance. Decoding is achieved by designing a suitable metric to be used in conjunction with a standard, off-the-shelf Viterbi decoder for the convolutional code. For a thorough discussion of this point, see Caire, Taricco, and Biglieri, 1998.

### 13.5.2.   Cutoff rate of the fading channel

We supplement our discussion of the effect of coding on the fading channel by providing the computation of the cutoff rate of the discrete channel generated by a modulation scheme used on the Rayleigh fading channel with coherent detection, perfect channel-state information, and infinite-depth interleaving. We also assume that the fading process is ergodic, i.e., that every realization is representative of the whole process.

Our calculation parallels that of Section 12.1. Pick two code words $\mathbf{x} = (s_1, \ldots, s_n)$ and $\hat{\mathbf{x}} = (\hat{s}_1, \ldots, \hat{s}_n)$, whose components are selected randomly, independently, and with equal probabilities. From (13.67) we have the Chernoff bound

$$P\{\mathbf{x} \to \hat{\mathbf{x}}\} \leq \min_{\lambda > 0} \prod_{k=1}^{n} \mathrm{E}_{R_k} \left[ e^{\lambda(N_0\lambda - 1)R_k^2 |s_k - \hat{s}_k|^2} \right]$$

By observing that the arguments of the exponentials are minimized for $\lambda =$

$1/2N_0$ (irrespective of $k$), we have for the average pairwise error probability

$$
\begin{aligned}
\overline{P\{\mathbf{x} \to \widehat{\mathbf{x}}\}} &\leq \overline{\prod_{k=1}^{n} \mathrm{E}_{R_k} \left[ e^{-R_k^2 |s_k - \widehat{s}_k|^2 / 4N_0} \right]} \\
&= \prod_{k=1}^{n} \overline{\mathrm{E}_{R_k} \left[ e^{-R_k^2 |s_k - \widehat{s}_k|^2 / 4N_0} \right]} \\
&= \prod_{k=1}^{n} \left[ \frac{1}{1 + |s_k - \widehat{s}_k|^2 / 4N_0} \right] \\
&= \frac{1}{M^n} \left( 1 + \frac{1}{M} r(\mathcal{S}, N_0) \right)^n \\
&= 2^{-nR_0}
\end{aligned}
$$

where, by following the footsteps of the derivation of (12.12), we have defined

$$ R_0 = \log_2 M - \log_2 \left( 1 + \frac{1}{M} r(\mathcal{S}, N_0) \right) \qquad (13.73) $$

and

$$ r(\mathcal{S}, N_0) = \sum_{s \in \mathcal{S}} \sum_{\widehat{s} \neq s} \frac{1}{1 + |s - \widehat{s}|^2 / 4N_0} \qquad (13.74) $$

**Example 13.5 (PSK)** By exploiting the symmetry of the constellation as indicated in Example 12.1, we have, for quaternary PSK:

$$ r(\mathcal{S}, N_0) = 4[1/(1 + \mathcal{E}/N_0) + 2/(1 + \mathcal{E}/2N_0)] $$

and

$$ R_0 = \log_2 4 - \log_2[1 + r(\mathcal{S}, N_0)/4] $$

The values of $R_0$, in bits per channel use for binary, quaternary, and octonary PSK, are shown in Fig. 13.10. By comparing these results with those of Fig. 12.2 we can see that the gap between the values of the cutoff rate of AWGN and Rayleigh fading channel is not as wide as the gap in error probability for uncoded systems. This suggests that coding is highly beneficial for this fading channel. □

## 13.6. Bibliographical notes

The books on mobile, or wireless, or cellular radio usually have one or more chapters devoted to fading channels, their characterization, and measurement. The interested reader may see the classic Jakes (1974), or the recent books by

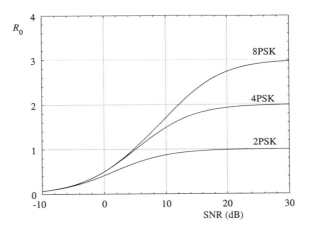

Figure 13.10: *Cutoff rate $R_0$ (in bits per channel use) of the discrete channel generated by M-ary PSK transmitted over the Rayleigh fading channel with coherent detection, perfect channel-state information and infinite-depth interleaving. Here SNR= $\mathrm{E}[R^2]\mathcal{E}/N_0$.*

Pahlavan and Levesque (1995), Rappaport (1996), Stüber (1996), or the collection of articles reprinted in Rappaport (1995). The book by Lee (1989) is devoted to the engineering aspects of mobile radio communications. Chapter 14 of Proakis (1995) contains a variety of topics on fading channels that were not covered here. The paper by Biglieri, Proakis, and Shamai (1998) includes a thorough discussion of the information-theoretic aspects of fading channels.

Further details on the calculation of error probabilities for transmission on a fading channel can be found for example in Lindsey (1964), Ho and Fung (1992), Cavers and Ho (1992), Simon and Alouini (1998), and Biglieri, Caire, Taricco, and Ventura (1998).

A good recent review of diversity techniques can be found in Chapter 12 of Gibson (1996). The paper by Brennan (1959) contains a thorough discussion of linear diversity combining techniques. See also Jakes (1974).

The book by Biglieri *et al.* (1991) contains two chapters on trellis-coded modulation for the fading channel, while Jamali and Le-Ngoc (1994) is entirely devoted to this subject. Our discussion on coding for the Rayleigh channel follows closely the review paper by Seshadri and Sundberg (1993), to which the interested reader is referred for further details and an extensive set of references on this topic.

The synergy of coding and diversity over fading channels is thoroughly studied by Ventura, Caire, Biglieri, and Taricco (1997a, 1997b, and 1997c). Coding strategies that perform well on both the AWGN and the fading channels are the subject of two papers by Zehavi (1992) and Caire, Taricco, and Biglieri (1998). Block fading channels are studied, among others, by Knopp and Humblet (1997 and 1998).

## 13.7.   Problems

*Problems marked with an asterisk should be solved with the aid of a computer.*

**13.1**  Consider two independent Gaussian random variables $X, Y$, with mean zero and common variance $\sigma^2$. Let $R$ denote the magnitude and $\Theta$ the phase of the complex random variable $X + jY$. Prove that $R$ and $\Theta$ are independent, that $R$ has a Rayleigh pdf, and $\Theta$ is uniformly distributed in $(0, 2\pi)$.

**13.2**  Consider two independent Gaussian random variables $X, Y$, with mean values $\mu_X$ and $\mu_Y$, respectively, and common variance $\sigma^2$. Let $R$ denote the magnitude and $\Theta$ the phase of the complex random variable $X + jY$.

   (a)  Prove that $R$ and $\Theta$ are generally not independent, and that $R$ has a Rice pdf,

   (b)  Assume next that the phase of the complex number $\mu_X + j\mu_Y$ is a RV uniformly distributed in $(0, 2\pi)$ , and prove that in these conditions $R$ has still a Rice pdf, $\Theta$ is uniformly distributed in $(0, 2\pi)$, and $R$ and $\Theta$ are independent.

**13.3**  We introduce here the "block Rayleigh fading" channel model. This model is motivated by the fact that, in many mobile radio situations, the channel coherence time is much longer than one symbol interval, and hence several transmitted symbols are affected by the same fading value. This model assumes that a code word of length $n = \mu\nu$ spans $\mu$ blocks of length $\nu$. An interleaver spreads the code symbols over the $\mu$ blocks. The value of the fading in each block is a Rayleigh-distributed RV, constant over a block and independent from block to block.

   (a)  Prove that for this channel the pairwise error probability decreases exponentially with exponent $d_H(\mu)$, the Hamming distance between code words on a block basis (this is the number of *blocks* in which two code words differ, rather than the number of different *symbols*).

   (b)  Derive a Singleton upper bound on $d_H(\mu)$, depending on the code rate, the value of $\mu$, and the number of signals used by the modulator.

**13.4** Prove that, with constant envelope signals ($|\mathbf{x}|$ constant), the error probabilities obtained with (13.18) and (13.49) coincide.

**13.5** (*) Consider the $(8, 4)$ extended Hamming code with $d_{\min} = 4$ and soft decoding. Compare the coding gain obtained for an error probability $P(e) = 10^{-5}$ by using coherently-demodulated binary antipodal modulation over the AWGN channel and over the Rayleigh fading channel.

**13.6** (*) Consider binary antipodal modulation with coherent detection and a Rayleigh fading channel with $L$-branch diversity and selection combining. Compute and plot the resulting error probability for $L = 1, 2, 4$, and 8.

**13.7** Compute the average signal-to-noise ratio obtained at the output of a selection combiner. Assume an $L$-branch diversity receiver.

# Digital transmission over nonlinear channels

In Chapters 7 and 13 we considered some of the major impairments affecting digital transmission besides additive Gaussian noise, such as intersymbol interference (ISI), interchannel interference (ICI), and fading. They are due to the nonideal characteristics of the linear devices present in the system and to the transmission medium. We have seen in Chapter 7 that ISI and ICI can be modeled as terms that sum to the useful signal. This is a direct consequence of the linear assumption made for the components of the system.

There are many cases, however, where this assumption is not true, as nonlinear devices significantly contribute to system degradation. One example is encountered in high-speed digital transmission over telephone channels, where nonlinear signal distortion arises principally from inaccuracies in signal companding (compressing-expanding) in telephone transmission. A second example is the digital satellite link, in which both the earth station and the satellite repeater are equipped with amplifiers operated in a nonlinear region of the input-output characteristics for a better exploitation of the power of the device. The earth station amplifier nonlinearity is often mild, either because it operates some decibels below the saturation point (it is "backed off"), in a nearly linear region, or because a predistortion linearizer is inserted in the transmitter chain. To the contrary, the satellite amplifier (a traveling wave tube (TWT) or a solid-state device) is driven near to the saturation point and exhibits highly nonlinear characteristics, which must be included in the analysis of the system performance. A third example is offered by power amplifiers in hand-held terminals of mobile radio communications, whose power effieciency is at a premium.

In most engineering fields the gap between the tractability of linear and non-

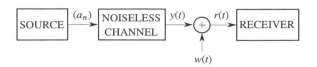

Figure 14.1: *Block diagram of a nonlinear transmission system.*

linear problems is wide. Whereas refined mathematical tools and comprehensive theories are normally available in the linear case, only very special categories of problems can be analyzed in nonlinear situations. Digital transmission theory is no exception to this rule. No well-established theories exist for the analysis and/or design of a nonlinear digital transmission system. Because of this, the structure of this chapter will be in a sense peculiar with respect to the rest of the book. Most of the topics analyzed in detail before in a linear context will be discussed concisely here, aiming at extending the results, wherever possible, to the nonlinear situations.

Much of the technical literature shows the effort that has been devoted to smoothing out the differences among models and analyses used to attack nonlinear problems. However, this chapter is idiosyncratic, in the sense that it reflects, in its choice of topics and analytical tools, the preferences of the authors and their research experience. The particular nature of the chapter renders more important the Bibliographical Notes at the end, where reference is also made to different approaches to the same problems.

Section 14.1 is devoted to the modeling of a finite-memory nonlinear channel. Then this model is used as a starting point to perform spectral analysis of nonlinear signals (Section 14.2), to derive the optimum symbol-by-symbol linear receiver (Section 14.4), and the optimum maximum likelihood (ML) sequence receiver (Section 14.5). A more specialized model of the nonlinear channel, based on the discrete Volterra series, is introduced in Section 14.3. This model permits the computation of an explicit expression for the received signal in which all the significant contributions (i.e., useful signal, linear and nonlinear interferences, and noise) appear separately. In the same section, this result is used in evaluating the performance of a nonlinear system employing QPSK modulation. In Section 14.6, the Volterra model is employed to derive the structure of a nonlinear equalizer and of a linearizer. Finally, in Section 14.7, precompensation of nonlinearities is examined.

## 14.1.  A model for the nonlinear channel

The model applied to the system analyzed is shown in Fig. 14.1. The source

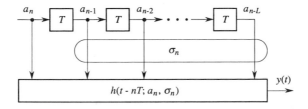

Figure 14.2: *Model of the noiseless part of a nonlinear channel with finite memory.*

emits a stationary sequence $(a_n)$ of discrete independent random variables (RVs) of known statistics, one every $T$; the block labeled "channel" represents the noiseless part of the real channel and includes all devices between the source and the receiver (modulator, filters, physical channel, nonlinear devices, etc.). It transforms the discrete-time sequence at its input into a continuous waveform $y(t)$. After the addition of Gaussian noise, the signal enters the receiver, in which it is processed to estimate the transmitted sequence $(a_n)$.

We assume that the channel has a finite memory; that is, at any time instant $t$ the channel output $y(t)$ depends only on a finite number, say $L$, of past source symbols besides the one emitted at time $t$. Using the definition of shift-register state sequence given in Section 2.2.1, we can define the state $\sigma_n$ of the channel during the $n$th interval $[(n-1)T, nT]$ as the $n$-tuple of the $L$ consecutive symbols $a_i$

$$\sigma_n \triangleq (a_{n-1}, a_{n-2}, \ldots, a_{n-L}) \tag{14.1}$$

The state $\sigma_n$ then represents the memory of the channel. The emission of symbol $a_n$ from the source forces a transition of the channel state from $\sigma_n$ to $\sigma_{n+1}$. A sequence of states forms the special case of the Markov chain studied in Section 2.2.1 under the name of shift-register state sequence.

With the finite-memory assumption, we can model the noiseless channel of Fig. 14.1 as in Fig. 14.2, where $h(t - nT); a_n, \sigma_n)$ is the waveform generated in the $n$th time interval of duration $T$. The channel output $y(t)$ will then consist of a sum of nonoverlapping waveforms, each defined in an interval of duration $T$

$$y(t) \triangleq \sum_n h(t - nT; a_n, \sigma_n) \tag{14.2}$$

In (14.2), the waveforms $h(t; a_n, \sigma_n)$ are zero outside the interval $(0, T)$ and can assume no more than $M^{L+1}$ different shapes, where $M$ is the cardinality of the set of values assumed by $a_n$. The waveforms $h(t; \cdot, \cdot)$ will be called *chips*. The signal $y(t)$ at the output of the channel is then obtained as a sequence of chips chosen according to the values of the sequence $(a_n)$ or, equivalently, $(\sigma_n)$.

Figure 14.3: *Noiseless part of a bandpass nonlinear transmission system using 2-PSK modulation.*

A final observation is pertinent. This system model can also accommodate a channel (block or trellis) encoder with finite memory.

## 14.2.   Spectral analysis of nonlinear signals

A general method to evaluate the power spectrum of a random digital signal was presented in Section 2.3. The signal $y(t)$ of (14.2) has the form (2.104), which is the starting point for the derivation of the power spectrum. We can use then formulas (2.149)–(2.151) to compute the spectrum, taking into account the simplification that the state sequence $(\sigma_n)$ is a shift-register sequence (Section 2.2.1). This reduces the infinite summation in (2.151) to a finite summation ranging from 1 to $L$, $L$ the channel memory as defined above. The following example shows some results obtained through the application of this method.

**Example 14.1**   The aim of this example is to pictorially describe the effect of the non-linearity on the power spectrum of a digital signal. Consider a channel formed by cascading a linear filter with a memoryless nonlinear device like that in Fig. 14.3, used by a system employing a 2-PSK modulation. The whole channel can be considered as a nonlinear system with memory, in which the memory is due to the linear filter and the nonlinearity to the memoryless nonlinear device. The filter, represented by its impulse response $s(t)$, is a sixth-order Butterworth filter with normalized equivalent noise bandwidth $B_{eq}T = 1.2$. The nonlinear device is a traveling-wave tube whose AM/AM and AM/PM characteristics are shown in Fig. 14.4 (see Section 14.3 for a detailed explanation of the meaning of these characteristics). Inspection of the impulse response of the filter shows that a value of $L = 3$ is sufficient to account for the memory of the system. We have then $2^4 = 16$ different chips $h(t; a, \sigma)$ for this system. Using the technique previously outlined, we may compute the power spectra of the signals $x(t)$, the output of an ideal 2-PSK modulator, $z(t)$, its filtered version, and $y(t)$, the filtered signal passed through the nonlinearity of Fig. 14.4 for different values of the *input backoff*, defined as the difference in dB between the actual input power and that corresponding to the maximum output power. Looking at the results shown in Fig. 14.5, the phenomenon of the sidelobes restoration due to the TWT appears evident: the sidelobes of the CPSK spectrum are first attenuated by the filter, then restored by the TWT. Also evident is the role played by the input backoff and the consequent trade-off between power efficiency and sidelobes enhancement. The sidelobes restoration gives rise to interchannel interference (ICI) in a system employing frequency-division multiplexing.

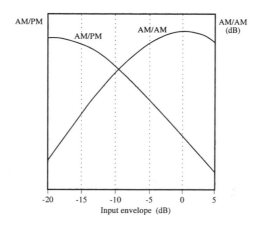

Figure 14.4: *Typical AM/AM and AM/PM characteristics of a TWT.*

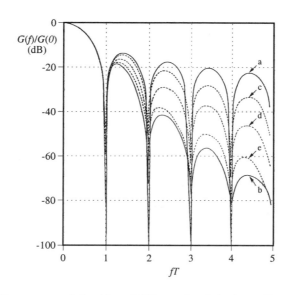

Figure 14.5: *Power spectral densities of different signals in the nonlinear transmission system of Example 14.1. Curve a refers to the modulator, curve b to the filter, curve c to the TWT at saturation, and curves d and e to the TWT with an input backoff of 6 and 12 dB, respectively.*

□

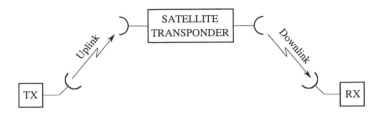

Figure 14.6: *Satellite link.*

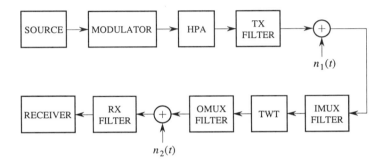

Figure 14.7: *Block diagram of a satellite link.*

## 14.3.  Volterra model for bandpass nonlinear channels

As previously mentioned, satellite digital transmission represents one of the most important cases of digital communication systems employing a nonlinear channel. It consists of two earth stations (TX and RX in Fig. 14.6), usually far from each other, connected by a repeater traveling in the sky (the satellite) through two radio links (uplink and downlink in the figure). A block diagram of the system of Fig. 14.6 is shown in Fig. 14.7. The block labeled HPA (high-power amplifier) represents the earth station power amplifier; its input-output power characteristics are nonlinear, of a saturating type like that presented in Fig. 14.4. Although the highest power efficiency is obtained by letting the HPA operate at (or near) the saturation point, it is common practice to operate the HPA a few decibels below saturation in a nearly linear region. This backing-off the operation point facilitates the attenuation of the effects caused by nonlinearity (e.g., the spreading of the spectrum of the input signal, which, as seen in Example 14.1, gives rise to ICI).

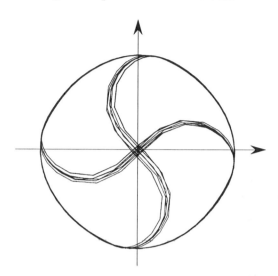

Figure 14.8: *Simulated phasor diagram of a QPSK signal at the output of a bandpass nonlinear channel.*

The TX filter limits the bandwidth of each channel in an FDMA system, whereas the IMUX (input multiplexing) filter limits the amount of uplink noise entering the satellite transponder. The block labeled TWT represents the satellite's on-board amplifier. Owing to the satellite's limited power resources, this amplifier is usually operated at saturation to obtain the maximum efficiency. Typical input-output characteristics of TWTs are of the kind shown in Fig. 14.4.

Due to bandwidth limitations, the modulated signal at the TWT input does not have a constant envelope. In particular, when transitions between opposite points in the signal space occur (as in CPSK), the envelope may pass through zero. This phenomenon is represented in Fig. 14.8, where the set of envelope values is drawn in the signal space at the input of the TWT in a typical system situation. These input envelope fluctuations are translated at the TWT output in phase shifts that deteriorate the system performance. The functions of the OMUX (output demultiplexing) filter and RX filter are, respectively, similar to those of TX and IMUX. The receiver is assumed to be a conventional symbol-by-symbol receiver. In this section the receiver optimization is not considered. It will be dealt with later. Our objective is to determine the performance of the system in terms of error probability.

Two main tracks have been followed to assess the error performance of a nonlinear satellite link. The first uses simulation tools to analyze a complete model of the system. The second is based on some simplifications of the model

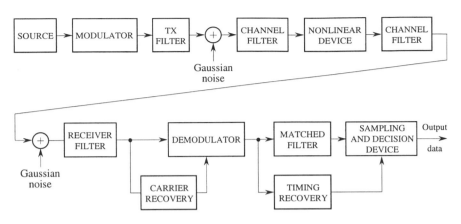

Figure 14.9: *Block diagram of a digital satellite transmission link.*

such that an analytical approach is feasible. Based on the channel model of Section 14.1, an intermediate approach could be used when the uplink noise is omitted from the system model. This often occurs owing to the higher signal-to-noise ratio in the uplink. The approach consists in evaluating the error probability conditioned on the channel state $\sigma_n$, and then averaging over all possible states by generating the whole set of chips $h(t; \cdot, \cdot)$ in (14.2). This approach stems directly from the direct-enumeration method for computing the error probability in the presence of the ISI described in Section 7.2.1.

In this section we present a method for the performance analysis based on Volterra series. It renders explicit the dependence of the received signal on the sequence of source symbols $(a_n)$ and allows the application to the nonlinear case of the methods for computing the error probability with ISI.

It is felt that the model based on Volterra series represents a balance between two often conflicting requirements (i.e., faithful representation of the physical system and analytical tractability of the model).

Consider the block diagram of Fig. 14.9. It represents a slightly more detailed version of Fig. 14.7, as the block "receiver" has been expanded. This block diagram will be the basis of our analysis. Note that it does not include some types of disturbances that can affect the two radio links of Fig. 14.6, (e.g., fading and interferences from other users of the transmission medium, cochannel and interchannel interference, etc.). Furthermore, carrier and timing recovery are supposed to be ideal. The nonlinear device is a bandpass memoryless nonlinear system whose input-output characteristics are described by the relationship (see (2.218)):

$$y(t) = F[A_x(t)]e^{j\{\phi_x(t) + \Phi[A_x(t)]\}} \tag{14.3}$$

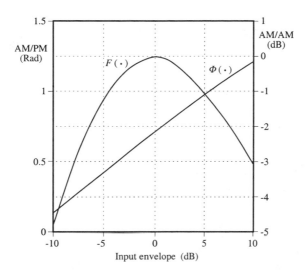

Figure 14.10: *Measured input-output AM/AM and AM/PM characteristics of a commercial TWT.*

where $y(t)$ is the complex envelope of the output signal and

$$x(t) = A_x(t)e^{j\phi_x(t)} \tag{14.4}$$

is the complex envelope of the input signal. The nonlinearity is then character-ized by two real-valued functions, $F(\cdot)$ and $\Phi(\cdot)$, which describe the AM/AM and AM/PM conversion effects, respectively. A typical example related to satel-lite links is shown in Fig. 14.10, where $F(\cdot)$ and $\Phi(\cdot)$ represent measured charac-teristics of a commercial TWT. Since the whole system represented in Fig. 14.9 is bandpass, we can use the low-pass representation of systems and signals (Sec-tion 7.1), which leads to the block diagram of Fig. 14.11. Here

- $(a_n)$ is a sequence of (generally complex) discrete iid generally complex RVs.

- $x(t)$ is the modulated signal

$$x(t) = \sum_n a_n\delta(t - nT) \tag{14.5}$$

- $s(t)$ is the overall impulse response of the filters preceding the nonlinearity.

- $c(\cdot)$ is a complex function that represents the input-output relationship of the nonlinearity

$$c(\cdot) = F(\cdot)e^{j\Phi(\cdot)} \tag{14.6}$$

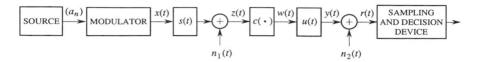

- $u(t)$ is the impulse response of a filter that represents the cascade of all linear devices following the nonlinearity.

- $n_1(t)$ and $n_2(t)$ are generally complex baseband Gaussian processes, with zero mean and variances $\sigma_1^2$ and $\sigma_2^2$ (in a satellite link, they represent the uplink and downlink noises, respectively).

The analysis of the system in Fig. 14.11 will follow two steps. First, we assume that $n_1(t) = 0$ (i.e., no uplink noise is present in the system). Although motivated by the desire for simplicity, this assumption is reasonable for satellite systems in which the greater power of the earth transmitter keeps the signal-to-noise ratio in the uplink higher than in the downlink. Successively, we will include uplink noise in the analysis.

Consider a Volterra system, as defined in Section 2.1.2, whose input-output relationship has the form

$$y(t) = \sum_{k=1}^{\infty} y_k(t) \tag{14.7}$$

where

$$y_k(t) \triangleq \int_{-\infty}^{\infty} \cdots \int_{-\infty}^{\infty} h_k(\tau_1, \ldots, \tau_k) \left[ \prod_{i=1}^{k} x(t - \tau_i) d\tau_i \right] \tag{14.8}$$

and $h_k(\tau_1, \ldots, \tau_k)$ are the Volterra *kernels* of the system. We want to obtain input-output expressions like (14.7) and (14.8) in the particular case of the *bandpass* Volterra system, for which the input and output signals are the complex envelopes of the actual signals, and the system is characterized by its *equivalent low-pass* Volterra kernels. Extending the development of Example 2.18, which concerned memoryless polynomial-law nonlinear devices, to the case of the bandpass nonlinear system comprised between signals $x(t)$ and $y(t)$ in Fig. 14.11, we obtain the following result (see Problem 14.2):

$$
\begin{aligned}
y(t) = &\sum_{m=0}^{\infty} \int_{-\infty}^{\infty} \cdots \int_{-\infty}^{\infty} k_{2m+1}(\tau_1, \ldots, \tau_{2m+1}) \\
&\cdot \prod_{i=1}^{m+1} x(t - \tau_i) \prod_{j=m+2}^{2m+1} x^*(t - \tau_j) d\tau_1 \ldots d\tau_{2m+1}
\end{aligned} \tag{14.9}
$$

where $*$ means complex conjugate.

Substituting now the expression (14.5) of $x(t)$ into (14.9), we have

$$
y(t) = \sum_{m=0}^{\infty} \sum_{n_1=-\infty}^{\infty} \cdots \sum_{n_{2m+1}=-\infty}^{\infty} a_{n_1} \cdots a_{n_{m+1}} a_{n_{m+2}}^* \cdots a_{n_{2m+1}}^*
$$
$$
\cdot k_{2m+1}(t - n_1 T, \ldots, t - n_{2m+1} T) \tag{14.10}
$$

where $k_{2m+1}$ are the low-pass Volterra kernels of the system. Owing to the particular structure of the nonlinear system of Fig. 14.11 that confines the memory into linear components, the low-pass Volterra kernels can be derived in a relatively easy way.

Consider first the following Taylor series expansion of the nonlinear function $c(\cdot)$ defined in (14.6):

$$
c(A) = \sum_{n=0}^{\infty} \gamma_{2m+1} A^{2m+1} \tag{14.11}
$$

in which the presence of only odd powers of the argument $A$ is a consequence of the bandpass nature of the nonlinearity (see Example 2.18). Using (14.11), the output of the nonlinearity (Fig. 14.11) can be expressed as

$$
w(t) = \sum_{m=0}^{\infty} \gamma_{2m+1} z^{m+1} z^{*m}(t) \tag{14.12}
$$

Since now

$$
z(t) = \int_{-\infty}^{\infty} s(\tau) x(t - \tau) d\tau \tag{14.13}
$$

and

$$
y(t) = \int_{-\infty}^{\infty} u(\tau) w(t - \tau) d\tau \tag{14.14}
$$

we get, through some easy algebra

$$
y(t) = \sum_{m=0}^{\infty} \gamma_{2m+1} \int_{-\infty}^{\infty} \cdots \int_{-\infty}^{\infty} u(\tau) \prod_{r=1}^{m+1} s(\tau_r - \tau) \prod_{s=m+2}^{2m+1} s^*(\tau_s - \tau)
$$
$$
\cdot \prod_{i=1}^{m+1} x(t - \tau_i) \prod_{j=m+2}^{2m+1} x^*(t - \tau_j) d\tau d\tau_1 \ldots d\tau_{2m+1} \tag{14.15}
$$

Comparing (14.15) with (14.9), one finds the desired expression for the low-pass equivalent kernels

$$
k_{2m+1}(\tau_1, \ldots, \tau_{2m+1}) = \gamma_{2m+1} \int_{-\infty}^{\infty} u(\tau) \prod_{r=1}^{m+1} s(\tau_r - \tau) \prod_{s=m+2}^{2m+1} s^*(\tau_s - \tau) d\tau
$$
$$
\tag{14.16}
$$

We now have all the ingredients to find an explicit expression for the sampled received signal entering the decision device in Fig. 14.11. The received signal $r(t)$ is given by

$$r(t) = y(t) + n_2(t) \qquad (14.17)$$

where $y(t)$ has been defined in (14.10). Sampling now at $t = t_0$ and defining

$$K_{2m+1}(n_1, \ldots, n_{2m+1}) \overset{\triangle}{=} k_{2m+1}(t_0 - n_1 T, \ldots, t_0 - n_{2m+1} T) \qquad (14.18)$$

$$N_2 \overset{\triangle}{=} n_2(t_0) \qquad (14.19)$$

$$R \overset{\triangle}{=} r(t_0) \qquad (14.20)$$

leads to

$$R = \sum_{m=0}^{\infty} \sum_{n_1=-\infty}^{\infty} \cdots \sum_{n_{2m+1}=-\infty}^{\infty} a_{n_1} \cdots a_{n_{m+1}} a^*_{n_{m+2}} \cdots a^*_{n_{2m+1}}$$
$$\cdot K_{2m+1}(n_1, \ldots, n_{2m+1}) + N_2 \qquad (14.21)$$

This result is the starting point for the analysis of a synchronous digital communication system using nonlinear devices. The computation of the discrete kernels $K_{2m+1}(\cdot)$ only involves a convolution integral (see (14.16)), which can be solved using standard numerical techniques.

To apply (14.21), we also have to deal with two kinds of infinite summations. The first one, index $m$, comes from the power series expansion (14.11) of the nonlinearity characteristics. It is usually truncated to some value $m_M$ large enough to accurately represent the function $c(\cdot)$. As an example, in the case of the TWT characteristics shown in Fig. 14.10, $m_M = 3$ allows a reasonable approximation of the curves in the range from zero to a few decibels beyond the saturation point. The second kind of infinite series, indexes $n_i$, depends on the memory of the linear components of the communication system, that is, $s(t)$ and $u(t)$. As is usual for linear systems, we shall suppose that both $s(t)$ and $u(t)$ have a finite duration. Thus, we can say that each $n_i$ takes value in a finite set of integers. In the following, summations limits will often be omitted. It is intended that the set of values taken by the indexes have finite cardinality.

As seen in Chapter 7, the decision device operates on samples of the inphase and quadrature demodulated signals, which correspond to the real and imaginary parts of $R$ (i.e., $R_P$ and $R_Q$). From (14.21), we can extract all the terms containing only the transmitted symbol $a_0$, which contribute to form what we call the "useful sample" $R_0$:

$$R_0 \overset{\triangle}{=} R_{0P} + j R_{0Q} = a_0 \sum_{m=0}^{m_M} |a_0|^{2m} K_{2m+1}(0, \ldots, 0) \qquad (14.22)$$

This allows one to rewrite (14.21) in the form

$$R = a_0 \sum_{m=0}^{m_M} |a_0|^{2m} K_{2m+1}(0,\ldots,0) + \sum_{n_1 \neq 0} a_{n_1} K_1(n_1)$$
$$+ \sum_{m=0}^{m_M} \sum_{n_1} \cdots \sum_{n_{2m+1}} a_{n_1} \cdots a_{n_{m+1}} a_{n_{m+2}}^* \cdots a_{n_{2m+1}}^*$$
$$\cdot K_{2m+1}(n_1,\ldots,n_{2m+1}) + N_2 \qquad (14.23)$$

Let us define

$$R_{0P} \triangleq \Re\{a_0 \sum_{m=0}^{m_M} |a_0|^{2m} K_{2m+1}(0,\ldots,0)\} \qquad (14.24)$$

$$R_{0Q} \triangleq \Im\{a_0 \sum_{m=0}^{m_M} |a_0|^{2m} K_{2m+1}(0,\ldots,0)\} \qquad (14.25)$$

$$R_P \triangleq \Re\{\sum_{n_1 \neq 0} a_{n_1} K_1(n_1) + \sum_{m=0}^{m_M} \sum_{n_1} \cdots \sum_{n_{2m+1}} a_{n_1} \cdots a_{n_{m+1}} a_{n_{m+2}}^* \cdots a_{n_{2m+1}}^*$$
$$\cdot K_{2m+1}(n_1,\ldots,n_{2m+1})\} \qquad (14.26)$$

$$R_Q \triangleq \Im\{\sum_{n_1 \neq 0} a_{n_1} K_1(n_1) + \sum_{m=0}^{m_M} \sum_{n_1} \cdots \sum_{n_{2m+1}} a_{n_1} \cdots a_{n_{m+1}} a_{n_{m+2}}^* \cdots a_{n_{2m+1}}^*$$
$$\cdot K_{2m+1}(n_1,\ldots,n_{2m+1})\} \qquad (14.27)$$

(where the summation indexes $n_1,\ldots,n_{2m+1}$ in the RHS of (14.26) and (14.27) cannot be all zeros) and

$$N_{2P} \triangleq \Re\{N_2\} \qquad (14.28)$$

$$N_{2Q} \triangleq \Im\{N_2\} \qquad (14.29)$$

so that finally (14.23) becomes

$$R = (R_{0P} + R_P + N_{2P}) + j(R_{0Q} + R_Q + N_{2Q}) \qquad (14.30)$$

The structure of (14.23) is quite similar to that of (7.18) and can be viewed as an extension of it. In fact, we recognize in (14.23) the useful signal (first term in RHS), the linear ISI (second term), the nonlinear contribution to ISI (third term), and, finally, the additive Gaussian noise (last term).

As in Chapter 7 for the linear case, we can now apply the methods there described (in particular, the Gauss quadrature rule approach explained in detail in Appendix E) to compute the error probability of the system. Those methods are based on the knowledge of a certain number of moments involving the RVs that represent ISI (i.e., RP and RQ in our case). With respect to the linear situation, two new factors are present. They give rise to considerable complications. First, it is necessary to compute the discrete Volterra kernels $K_{2m+1}(\cdot)$. This can be achieved with the aforementioned numerical algorithms. Second, the RVs $R_P$ and $R_Q$ cannot be written (as in the linear case) as a sum of independent RVs. The next section is almost entirely devoted to solve this problem. Although our treatment will deal with $M$-ary CPSK, it requires only minor changes to extend it to any coherent modulation scheme using a two-dimensional signal constellation.

### 14.3.1.  Error probability evaluation for $M$-ary CPSK

Let $a_n = \exp\{j\phi_n\}$ in (14.5), with $\phi_n$ assuming equally likely values in the set

$$\left\{\frac{(2k+1)}{M}\right\}_{k=0}^{M-1} \tag{14.31}$$

which corresponds to $M$-ary PSK. The decision device determines the received phase angle $\Phi_R$

$$\Phi_R = \tan^{-1}\frac{R_{0Q} + R_Q + N_{2Q}}{R_{0P} + R_P + N_{2P}} \tag{14.32}$$

and decides according to the phase thresholds $2k\pi/M + \Theta$, $k = 0, \ldots, M - 1$, where $\Theta$ is a constant phase offset taking into account the value of AM/PM conversion of the TWT at the nominal operating point. This phase conversion has the effect of rotating the signal space by a constant value. This is compensated for by shifting the phase thresholds.

Following the procedure described in Section 7.2.2 for the linear case, we obtain the following bounds for the error probability, conditioned on the transmitted symbol $a_0 = \exp\{j\pi/M\}$:

$$\max(I_1, I_2) \le P(e \mid a_0) \le I_1 + I_2 \tag{14.33}$$

where

$$I_1 \overset{\triangle}{=} \frac{1}{2}\int_{\mathcal{L}} \operatorname{erfc}\left\{\frac{\lambda(\Theta + 2\pi/M) + \lambda_0(\Theta + 2\pi/M)}{\sqrt{2}\sigma_2}\right\} f_\Lambda[\lambda(\Theta + 2\pi/M)]d\lambda \tag{14.34}$$

$$I_2 \overset{\triangle}{=} \frac{1}{2}\int_{\mathcal{L}} \operatorname{erfc}\left\{\frac{-\lambda(\Theta) - \lambda_0(\Theta)}{\sqrt{2}\sigma_2}\right\} f_\Lambda[\lambda(\Theta)]d\lambda \tag{14.35}$$

In (14.34) and (14.35), $\sigma_2^2$ is the variance of the Gaussian RVs $N_{2P}$, $N_{2Q}$; $\mathcal{L}$ and $f_\Lambda(\lambda)$ are the range and pdf of the RV $\Lambda$, defined as

$$\Lambda(\beta) \overset{\triangle}{=} R_P \sin\beta - R_Q \cos\beta \tag{14.36}$$

The useful sample in (14.34) and (14.35) is present in $\lambda_0$, defined as

$$\lambda_0(\beta) \triangleq R_{0P} \sin \beta - R_{0Q} \cos \beta \qquad (14.37)$$

As in the linear case, the symmetry of the received signals implies that the average error probability coincides with the conditional one. As a matter of fact, the transformations operated by the nonlinear device on the input signal depend on the envelope of the signal itself, which in turn is independent of the transmitted phase.

To evaluate (14.34) and (14.35) using the approaches outlined in Appendix E, we need a few conditional moments of the RV $\Lambda(\beta)$. In the following, we shall derive a recurrent relationship that permits their computation. From (14.36), the conditional moments of $\Lambda$ can be written as

$$\begin{aligned}
\mathrm{E}\{\Lambda^n(\beta) \mid a_0\} &= \mathrm{E}\{(R_P \sin \beta - R_Q \cos \beta)^n \mid a_0\} \qquad (14.38)\\
&= \sum_{k=0}^{n} \binom{n}{k} \mathrm{E}\{R_P^k R_Q^{n-k} \mid a_0\}(-1)^{n-k} \sin^k \beta \cos^{n-k} \beta
\end{aligned}$$

and the problem of computing the moments of $\Lambda(\beta)$ is reduced to the computation of the joint conditional moments

$$\mu_{kn} \triangleq \mathrm{E}\{R_P^k R_Q^{n-k} \mid a_0\} \qquad (14.39)$$

Define the complex random variable $\xi$ as

$$\xi \triangleq R_P + jR_Q \qquad (14.40)$$

so that

$$R_P = \frac{1}{2}(\xi + \xi^*) \qquad (14.41)$$

$$R_Q = \frac{1}{2j}(\xi - \xi^*) \qquad (14.42)$$

Thus, we can write (14.39) as

$$\mu_{kn} = \frac{1}{2^n j^{n-k}} \sum_{i=0}^{k} \sum_{\ell=0}^{n-k} \binom{k}{i} \binom{n-k}{\ell} (-1)^{n-k-\ell} \mathrm{E}\{\xi^{i+\ell} \xi^{*n-i-\ell}\} \qquad (14.43)$$

Let us get a deeper insight into the structure of the powers of $\xi$ and $\xi^*$. From (14.26) and (14.27), $\xi$ is given by

$$\begin{aligned}
\xi &= \sum_{m=0}^{m_M} \sum_{n_1} \cdots \sum_{n_{2m+1}} a_{n_1} \cdots a_{n_{m+1}} a_{n_{m+2}}^* \cdots a_{n_{2m+1}}^* \\
&\quad \cdot K_{2m+1}(n_1, \ldots, n_{2m+1})
\end{aligned} \qquad (14.44)$$

where the multiple summation excludes the term all of whose indexes are zero. Observe that any power of $\xi$ (and $\xi^*$) can be considered as output of a Volterra system having $\xi$ (or $\xi^*$) as input. We shall relate the new Volterra kernels defining the output to the ones that define the input. Let us start with $\xi^2$. When computing it, the terms $a_{n_1} a_{n_2}$, $a_{n_1} a_{n_2} a_{n_3} a_{n_4}^*, \ldots$, are multiplied by coefficients obtained as sums of products of Volterra kernels $K_{2m}(\cdot)$ as follows:

$$a_{n_1} a_{n_2} \qquad\qquad \Rightarrow \quad K_1(n_1) K_1(n_2) \triangleq K_2^{(2)}(n_1, n_2)$$

$$a_{n_1} a_{n_2} a_{n_3} a_{n_4}^* \qquad \Rightarrow \quad K_1(n_1) K_3(n_2, n_3, n_4)$$
$$+ K_3(n_1, n_2, n_4) K_1(n_3) \triangleq K_4^{(2)}(n_1, n_2, n_3, n_4)$$

$$a_{n_1} a_{n_2} a_{n_3} a_{n_4} a_{n_5}^* a_{n_6}^* \quad \Rightarrow \quad K_1(n_1) K_5(n_2, n_3, n_4, n_5, n_6)$$
$$+ K_3(n_1, n_2, n_5) K_3(n_3, n_4, n_6)$$
$$+ K_5(n_1, n_2, n_3, n_5, n_6) K_1(n_4)$$
$$\triangleq K_6^{(2)}(n_1, n_2, n_3, n_4, n_5, n_6)$$

where the symbol $K_\ell^{(2)}$ refers to the Volterra coefficient of the $\ell$-th order relative to the second power of $\xi$.

Based on (14.44) and on previous definitions of $K_\ell^{(2)}$, we can write $\xi^2$ as

$$\xi^2 = \sum_{m=0}^{m_M} \sum_{n_1} \cdots \sum_{n_{2m+1}} a_{n_1} \cdots a_{n_{2+m}} a_{n_{2+m+1}}^* \cdots a_{n_{2+2m}}^* \cdot K_{2+2m}^{(2)}(n_1, \ldots, n_{2+2m})$$
$$(14.45)$$

where $K_{2+2m}^{(2)}(\cdot)$ satisfies the recurrence

$$K_{2+2m}^{(2)}(n_1, \ldots, n_{2+2m})$$
$$= \sum_{i=0}^{m} K_{2i+1}(n_1, \ldots, n_{i+1}, n_{2+m+1}, \ldots, n_{2+m+i}) \qquad (14.46)$$
$$\cdot K_{2(m-1)+1}(n_{i+2}, \ldots, n_{2+m}, n_{2+m+i+1}, \ldots, n_{2+2m})$$

The previous procedure can be extended (see Problem 14.4) to derive the general formula

$$\xi^l = \sum_{m=0}^{m_M} \sum_{n_1} \cdots \sum_{n_{\ell+2m}} a_{n_1} \cdots a_{n_{\ell+m}} a_{n_{\ell+m+1}}^* \cdots a_{n_{\ell+2m}}^* K_{\ell+2m}^\ell(n_1, \ldots, n_{\ell+2m})$$
$$(14.47)$$

with

$$K_{\ell+2m}^{(\ell)}(n_1, \ldots, n_{\ell+2m})$$

$$= \sum_{i=0}^{m} K_{2i+1}(n_1, \ldots, n_{i+1}, n_{\ell+m+1}, \ldots, n_{\ell+m+i}) \qquad (14.48)$$

$$\cdot K_{2(m-1)+\ell-1}^{(\ell-1)}(n_{i+2}, \ldots, n_{\ell+m}, n_{\ell+m+i+1}, \ldots, n_{\ell+2m})$$

and $K_m^{(10)}(\cdot) \equiv K_m(\cdot)$. The reader is invited to challenge his patience in deriving similar relationships for the powers of $\xi^*$ (see Problem 14.5).

Using (14.47)–(14.48), and the corresponding formulas for $(\xi)^*$, the averages in the RHS of (14.43) can be given the form

$$E\{\xi^\ell \xi^{*m} \mid a_0\} = \qquad (14.49)$$

$$\sum_{m=0}^{m_M} \sum_{n_1} \cdots \sum_{n_{\ell+m+2k+2i}} E\{a_{n_1} \cdots a_{n_{\ell+k+i}} a^*_{n_{\ell+k+i+1}} \cdots a^*_{n_{\ell+2k+2i}} \mid a_0\}$$

$$\cdot K_{\ell+2k}^{(\ell)}(n_1, \ldots, n_{\ell+k}, n_{\ell+k+i+1}, \ldots, n_{\ell+2k+i})$$

$$\cdot K_{m+2k}^{(m)*}(n_{\ell+k+1}, \ldots, n_{\ell+k+i}, n_{\ell+2k+i+1}, \ldots, n_{\ell+2k+2i})$$

and the final step toward the knowledge of the moments of $\Lambda$ (see (14.36)) is the computation of the conditional averages in the RHS of (14.49).

To this end, define

$$A_k \triangleq E\{a_n^k\} \qquad (14.50)$$

and remember that the indexes $n_i$ range in a finite set, say $\{0, \ldots, N\}$. Denote by $\nu_m$ (and $\nu_m^*$) the number of indexes of the $a_{n_i}$ (and $a^*_{n_i}$) that take on the values $m$ ($m = 0, 1, \ldots, N$). Taking into account that

$$a_n a_n^* = \exp\{j\phi_n\} \exp\{-j\phi_n\} = 1 \qquad (14.51)$$

we have

$$E\{a_{n_1} \cdots a_{n_{\ell+k+i}} a^*_{n_{\ell+k+i+1}} \cdots a^*_{n_{\ell+m+2k+2i}} \mid a_0\} = a_0^{\nu_0 - \nu_0^*} \prod_{i=1}^{N} A_{\nu_i - \nu_i^*} \qquad (14.52)$$

and

$$A_k = \begin{cases} 1, & \text{if } k = \pm 2iM, & i = 0, 1, \ldots \\ -1, & \text{if } k = \pm(2i+1)M, & i = 0, 1, \ldots \\ 0, & \text{otherwise} \end{cases} \qquad (14.53)$$

As an example, assuming $M = 4$ (quaternary PSK), we have

$$E\{a_5 a_4 a_0 a_3 a_5 a_0 a_4^* a_4^* \mid a_0\} = a_0^2 A_{-1} A_1 A_2 = 0$$

To end this section, a step-by-step summary of the whole procedure needed to obtain the error probability may prove useful for applicative purposes.

**Step 1** Compute the Volterra coefficients $K_{2i+1}(\cdot)$ by using, for example, their definition (14.18) and (14.16).

**Step 2** Compute the Volterra coefficients of higher order using recurrent relationships like (14.48).

**Step 3** Compute the moments $\mu_{kn}$ defined in (14.43) through (14.49)–(14.53).

**Step 4** For a given $\beta$, compute the moments of the RV $\Lambda(\beta)$ through (14.38), taking into account (14.39).

**Step 5** Compute $I_1$ and $I_2$ (and thus the bounds on error probability) defined in (14.34) and (14.35) using the methods described in Appendix E.

### 14.3.2.   Including uplink noise in the analysis

In general, a wide-sense stationary Gaussian process $n_1(t)$ can be represented in the following form (see Masry *et al.*, 1968):

$$n_1(t) = \sum_{i=1}^{\infty} \beta_i b_i(t) \qquad (14.54)$$

where $b_i(t)$ are appropriately chosen deterministic functions and $(\beta_i)$ is a sequence of zero-mean, unit-variance independent Gaussian RVs (possibly complex).

The general representation (14.54) reduces to the following form:

$$n_1(t) = \sum_{i=1}^{\infty} \beta_i \delta(t - iT) * s(t) \stackrel{\triangle}{=} n_1'(t) * s(t) \qquad (14.55)$$

when (see Problem 14.7) the lowpass equivalent linear systems that limit the uplink noise power can be approximated by an ideal lowpass filter with bandwidth $1/2T$ .

If the representation (14.55) is valid, we can use the model of 14.12, where $n_0(t)$ is added to the modulated signal before the baseband shaping filter. In this way, the uplink noise is accounted for by simply considering as an input signal $x(t)$, instead of the one given by (14.5), the new signal

$$x(t) = \sum_{i=-\infty}^{\infty} (a_i + \beta_i)\delta(t - iT) \qquad (14.56)$$

The received signal can now be given the same form as (14.21), with the substitution of $(a_n + \beta_n)$ in place of $a_n$, and the same previous procedure can be applied

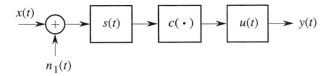

Figure 14.12: *Block diagram of a nonlinear transmission system including uplink noise.*

to the computation of the moments (see Problem 14.8), since the averages to be performed as the final step of the moment computation, that is,

$$\begin{aligned} \text{E}\{(a_{n_1} + \beta_{n_1}) \ldots (a_{n_{\ell+k+i}} + \beta_{n_{\ell+k+i}})(a^*_{n_{\ell+k+i+1}} + \beta^*_{n_{\ell+k+i+1}}) \\ \ldots (a_{n_{\ell+2k+2i}} + \beta_{n_{\ell+2k+2i}}) \mid a_0\} \end{aligned} \tag{14.57}$$

involve products of independent RVs whose moments are known, as in the previous analysis. The only significant change in the procedure worth mentioning here is that the presence of the uplink noise also affects the "useful" signal. In fact, the expression for $R_0$ in (14.22) now becomes

$$(a_0 + \beta_0) \sum_{m=0}^{m_M} |a_0 + \beta_0|^{2m} K_{2m+1}(0, \ldots, 0) \tag{14.58}$$

To obtain the error probability, a final average must be computed with respect to $\beta_0$. This can be done by using, for instance, a standard Gauss-Hermite numerical quadrature formula (Appendix E).

**Example 14.2** In this example, some results obtained by applying the Volterra series method of analysis will be presented. The system model is that shown in Fig. 14.11. The TWT characteristics were given in Fig. 14.10. The filters $s(t)$ and $u(t)$ are fourth- and second-order Butterworth, respectively. The curves of error probability are given as a function of the parameter

$$\eta_2 \triangleq 10 \log_{10} \frac{(\mathcal{E}_b)_{\text{sat}}}{N_0}$$

where $(\mathcal{E}_b)_{\text{sat}}$ is the energy per bit at the input of the receiver in correspondence with the TWT operated at its saturation point. The sampling instant $t_0$ was chosen as the one in which the convolution of the impulse responses $s(t)$ and $u(t)$ is maximum, whereas the chosen phase offset was that minimizing the error probability (a good starting point for the optimization is the value corresponding to the AM/PM conversion at saturation).

In Figs. 14.13 to 14.15 the error probability for a QPSK system is plotted versus the downlink signal-to-noise ratio $\eta_2$ for different values of the uplink parameter $\eta_1 \triangleq 10 \log_{10}(1/\sigma_1^2)$. It can be seen that, due to the combined effects of uplink noise and nonlinearity, a "floor" effect takes place in the error probability curves. For low values

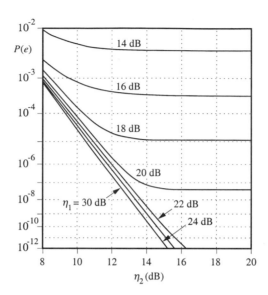

Figure 14.13: *Error probability of QPSK versus downlink signal-to-noise ratio at satu-ration $\eta_2$, with uplink $\eta_1$ as a parameter. Transmission filter: fourth-order Butterworth with equivalent noise bandwidth $B_{eq} = 1.8/T$. Receiving filter: second-order Butter-worth with $B_{eq} = 1.1/T$.*

of $\eta_2$, the effect of downlink noise dominates. For large $\eta_2$, the uplink noise dominates and the error probability does not depend on downlink-noise power.

By comparing Figs. 14.13 and 14.15, another relevant feature is observed. As a result of an increase of the transmitting filter bandwidth from 1.8 to 2.5, the error prob-ability decreases, due to the smaller amount of ISI introduced by the filtering. This is explained by the lack of neighboring channels in the models. Actually, in a multichannel environment, for any increase in the transmitting filter bandwidth, a corresponding in-crease in interference power occurs. By comparing Figs. 14.13 and 14.14, the effects of the nonlinearity can be observed. The most dramatic feature is the increased sensitivity to the uplink noise. In fact, when signal plus uplink noise enter a nonlinearity with a sat-urating characteristic, a signal-suppression effect takes place. In the linear case this does not occur because uplink and downlink noise simply sum up without further corrupting the signal.

Finally, in Fig. 14.16, the sensitivity of the system to the offset in timing recov-ery is shown. The error probability is plotted versus the normalized deviation from the "optimum" sampling time $t'_0$. It is apparent that the presence of the nonlinear device (continuous lines) renders the system behavior more critical to the choice of the sam-pling instant.        □

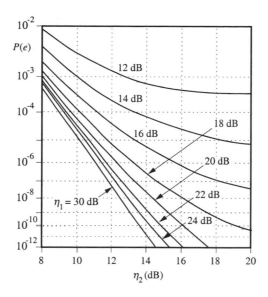

Figure 14.14: *Same as Fig. 14.13, without nonlinearity.*

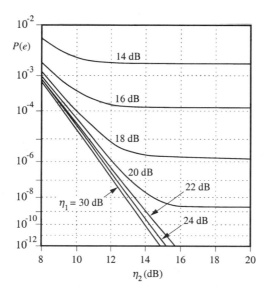

Figure 14.15: *Same as Fig. 14.13, with transmission bandwidth $B_{\text{eq}} = 2.5/T$.*

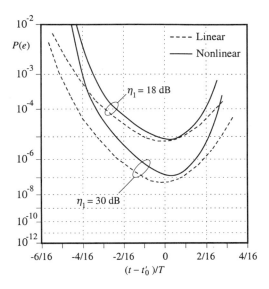

Figure 14.16: *Error probability of QPSK versus normalized deviation from the "optimum" sampling instant $t'_0$, with uplink $\eta_1$ as a parameter and for $\eta_2 = 12.5$ dB. Same channel as in Fig. 14.13.*

## 14.4.   Optimum linear receiving filter

In the preceding sections, we presented models for the description of nonlinear channels and demonstrated their use to compute certain performance parameters like power spectrum or error probability. So far, nonlinearity has been seen as an unwanted source of performance degradation imposed by some particular system demands. Also, the receiver structure was assumed to be the same as in the linear case. In this section and the following, we shall consider the problem of modifying the receiver according to some optimality criterion in order to take into account the channel nonlinearity. The analysis is based on the general model introduced in Section 14.1.

Let us consider the symbol-by-symbol receiver shown in Fig. 14.17, in the form of a linear filter followed by a symbol-rate sampler and a memoryless decision device. Here we want to choose the filter impulse response $u(t)$ in such a way that the sample of the received signal $r(t)$ taken at time $t_n = t_0 + nT$ is as close as possible (in the mean-square sense) to the transmitted symbol $a_n$. In

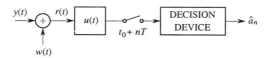

Figure 14.17: *Block diagram of a linear symbol-by-symbol receiver.*

formulas, we aim at finding the $u(t)$ that minimizes the quantity

$$\mathcal{E} \triangleq \mathrm{E}\{|r(t_n) * u(t_n) - a_n|^2\} \tag{14.59}$$

This optimization has been already considered for the linear case in Section 7.4. Here, we shall extend those results to a nonlinear environment. Minimization of $\mathcal{E}$ in (14.59) will be performed in the frequency domain. The received signal (see Fig. 14.1) is given by

$$r(t) = y(t) + w(t) \tag{14.60}$$

where $y(t)$ has been defined in (14.2) and $w(t)$ is a noise process independent of $y(t)$ with power spectral density $G_w(f) > 0$ for all $f$.

Let us denote with $U(f)$ the transfer function of the receiving filter and with $Y(f)$ the Fourier transform of the signal $y(t)$. We have

$$Y(f) = \sum_n H(f; a_n, \sigma_n) e^{-j2\pi n fT} \tag{14.61}$$

where $H(f; a_n, \sigma_n)$ is the Fourier transform of the waveform $h(t; a_n, \sigma_n)$. It can assume $N = M^{L+1}$ shapes $\{H(f; i)\}_{i=1}^N$ not necessarily distinct, according to the finite-memory assumption made for the channel.

In the frequency domain, (14.59) can be rewritten as

$$\begin{aligned}
\mathcal{E} = {}& \sigma_a^2 + \int_{-\infty}^{\infty} \int_{-\infty}^{\infty} U^*(f_1)\Gamma_y(f_1, f_2)U(f_2)e^{j2\pi(f_1 - f_2)t_n}\, df_1 df_2 \\
& -2\Re\left\{\int_{-\infty}^{\infty} V(f)U^*(f)df\right\} + \int_{-\infty}^{\infty} G_w(f)|U(f)|^2 df
\end{aligned} \tag{14.62}$$

where the following definitions have been used:

$$\Gamma_y(f_1, f_2) \triangleq \mathrm{E}\{Y(f_1)Y^*(f_2)\} \tag{14.63}$$

$$V(f) \triangleq e^{-j2\pi f t_n} \mathrm{E}\{a_n Y^*(f)\} \tag{14.64}$$

$$\sigma_a^2 \triangleq \mathrm{E}\{|a_n|^2\} \tag{14.65}$$

Using the standard variational calculus technique summarized in Appendix C, it can be shown that a necessary and sufficient condition for $U(f)$ to minimize $\mathcal{E}$ is that it be the solution to the integral equation

$$\int_{-\infty}^{\infty} \Gamma_y^*(f, f') U(f') e^{-j2\pi(f-f')t_n} df' + G_w(f) U(f) = V(f) \qquad (14.66)$$

Thus far, we have not yet exploited our knowledge of the structure of $Y(f)$, as provided by (14.61). Substituting (14.61) into (14.63) and exploiting the fact that the state sequence is a shift-register sequence, we obtain (see the derivation of (2.110))

$$\Gamma_y^*(f_1, f_2) = \frac{1}{T} \sum_{k=-\infty}^{\infty} G_k(f_1) \delta(f_1 - f_2 - k/T) \qquad (14.67)$$

where $G_k(f)$ has been defined as

$$G_k(f) \triangleq \sum_{\ell=-\infty}^{\infty} E\{H(f; a_n, \sigma_n) H^*(f - k/T; a_{n+\ell}, \sigma_{n+\ell})\} e^{-j2\pi\ell fT} \qquad (14.68)$$

so that the integral equation (14.66) takes the form

$$\frac{1}{T} \sum_{k=-\infty}^{\infty} G_k^*(f) U(f - k/T) e^{-j2\pi kt_n/T} + G_w(f) U(f) = V(f) \qquad (14.69)$$

Let us now compute the averages involved in the definitions of $V(f)$ and $G_k(f)$

$$E\{a_n Y^*(f)\} = \sum_{k=-\infty}^{\infty} e^{j2\pi kfT} E\{a_n H^*(f; a_k, \sigma_k)\} \qquad (14.70)$$

The average in RHS of (14.70) can be computed using a method similar to that followed in Section 2.3, leading to the result of (2.146). With the same notations, except for the replacement of $s(t; a_n, \sigma_n)$ with $h(t; a_n, \sigma_n)$, we obtain

$$E\{a_n H^*(f; a_k, \sigma_k)\} = \begin{cases} \mathbf{c}_2^*(f) \mathbf{P}^{k-n-1} \mathbf{c}_3', & n < k \\ \mathbf{c}_2^*(f) [\mathbf{P}^{1-k+n}]' \mathbf{c}_3', & n \geq k \end{cases} \qquad (14.71)$$

where $\mathbf{c}_2(f)$ has been defined in (2.135), $\mathbf{P}$ (the transition probability matrix of the state sequence $(\sigma_n)$) in (2.133), and

$$\mathbf{c}_3 \triangleq \sum_{h=1}^{M} p_h a_h \mathbf{w} \mathbf{E}_k \qquad (14.72)$$

where $p_h = P(a_h)$[1] and $\mathbf{w}$, $\mathbf{E}_h$ were defined in Section 2.3. Note that $\mathbf{c}_3$ is a vector whose $j$th entry is the average of the source symbols $a_h$ that lead the channel to the state $S_j$, being $\{S_j\}_{j=1}^M$ the set of values assumed by the RV $\sigma_n$.

Substitution of (14.71) into (14.70) and of this into (14.64) yields

$$V(f) = e^{-j2\pi f t_0}\mathbf{c}_2^*(f)\mathbf{\Lambda}(f)\mathbf{c}_3' + e^{-j2\pi f t_0}\mathbf{c}_2^*(f)\mathbf{P}^\infty \mathbf{c}_3' \sum_\ell e^{-j2\pi f\ell T} \qquad (14.73)$$

where

$$\mathbf{\Lambda}(f) \triangleq \sum_{\ell=1}^\infty [\mathbf{P}^{\ell-1} - \mathbf{P}^\infty]e^{-j2\pi f\ell T} + \sum_{\ell=0}^\infty [\mathbf{P}^{1+\ell} - \mathbf{P}^\infty]'e^{j2\pi f\ell T} \qquad (14.74)$$

The second term in the RHS of (14.73) contains spectral lines at dc and multiples of the symbol rate $1/T$. It disappears when either $\mathbf{P}^\infty \mathbf{c}_3'$ or $\mathbf{c}_2(f)\mathbf{P}^\infty$ is equal to zero. This means that the average value of symbols $a_j$'s or of waveforms $H(f; a_k, \sigma_k)$ is zero. We shall make this assumption in the following; thus we have

$$V(f) = e^{-j2\pi f\ell t_0}[\mathbf{c}_2^*(f)\mathbf{\Lambda}(f)\mathbf{c}_3'] \qquad (14.75)$$

Turning our attention to the average in the RHS of (14.68), which defines $G_k(f)$, using a straightforward replica of the algebra leading to (2.146), we obtain

$$E\{H(f; a_n, \sigma_n)H^*(f - k/T; a_{n+\ell}, \sigma_{n+\ell})\} \qquad (14.76)$$
$$= \begin{cases} \mathbf{c}_2^*(f)[\mathbf{P}^{\ell-1}]'\mathbf{c}_1^\dagger(f - k/T), & \ell > 0 \\ \mathbf{c}_2^*(f)[\mathbf{P}^{1-\ell}]\mathbf{c}_1^\dagger(f - k/T), & \ell \leq 0 \end{cases}$$

where $\mathbf{c}_1(f)$ was defined in (2.138).

Substitution of (14.76) into (14.76) yields

$$G_k(f) = \mathbf{c}_2(f)\mathbf{\Lambda}(f)\mathbf{c}_1^\dagger(f - k/T) + \mathbf{c}_2(f)\mathbf{P}^\infty \mathbf{c}_1^\dagger(f) \sum_{\ell=0}^\infty \cos(2\pi\ell fT) \qquad (14.77)$$

where $\mathbf{\Lambda}(f)$ was defined in (14.74). With the hypothesis

$$\mathbf{c}_2(f)\mathbf{P}^\infty \mathbf{c}_1^\dagger(f) = 0 \qquad (14.78)$$

we can write finally

$$G_k(f) = \mathbf{c}_2(f)\mathbf{\Lambda}(f)\mathbf{c}_1^\dagger(f - k/T) \qquad (14.79)$$

---

[1]Here and in (14.72) the subscript $h$ runs over the set of $M$ values assumed by the RV $a_n$. It might have been more appropriate to employ two different notations (one for the RV and the other for its values). However, we opted for just one to avoid notational aggravations.

Following the procedure used in Section 7.4 for the linear case, we shall prove that the equation (14.69) admits the solution

$$U_{opt} = \frac{c_2(f)\gamma(f)}{G_w(f)} e^{-j2\pi f t_0} \tag{14.80}$$

where $\gamma(f)$ is a column $M$-vector of frequency functions periodic with period $1/T$.

Recalling through definition (2.135) that $c_2(f)$ is a vector whose $i$th component is the average amplitude spectrum of the waveforms available to the modulator when it is in the $i$th state $S_i$, the similarity of (14.80) with the result (7.87) obtained in the linear case becomes apparent. As a matter of fact, we have that the optimum receiving filter may be thought of as being composed of a bank of filters matched to the average transmitted waveforms, each one followed by an infinite-length transversal filter.

Let us define

$$\mathbf{b}(f) \triangleq \mathbf{\Lambda}(f)\mathbf{c}_3' \tag{14.81}$$

so that (14.75) can be rewritten as

$$V(f) = e^{-j2\pi f t_0} \mathbf{c}_2^*(f)\mathbf{b}(f) \tag{14.82}$$

Substitution of (14.79), (14.80), and (14.82) into (14.69) yields

$$\mathbf{c}_2^*(f)\left[\frac{1}{T}\mathbf{\Lambda}^*(f)\sum_k \frac{\mathbf{c}_1'(f-k/T)\mathbf{c}_2^*(f-k/T)}{G_w(f-k/T)} + \mathbf{I}\right]\gamma(f) = \mathbf{c}_2^*(f)\mathbf{b}(f) \tag{14.83}$$

Equivalently, (14.83) can be rewritten in the form

$$\mathbf{c}_2^*(f)\mathbf{d}(f) = 0 \tag{14.84}$$

where $\mathbf{d}(f)$ is the $M$-vector, periodic with period $1/T$, defined by

$$\mathbf{d}(f) \triangleq \mathbf{A}(f)\gamma(f) - \mathbf{b}(f) \tag{14.85}$$

and

$$\mathbf{A}(f) \triangleq \frac{1}{T}\mathbf{\Lambda}^*(f)\sum_k \frac{\mathbf{c}_1'(f-k/T)\mathbf{c}_2^*(f-k/T)}{G_w(f-k/T)} + \mathbf{I} \tag{14.86}$$

The vector $\mathbf{d}(f)$ is now periodic with period $1/T$, so the inverse Fourier transforms of its components, $d_j(t)$, $i = 1, \ldots, M$, have the following form:

$$d_j(t) = \sum_{j=-\infty}^{\infty} \rho_{ij}\delta(t-jT) \tag{14.87}$$

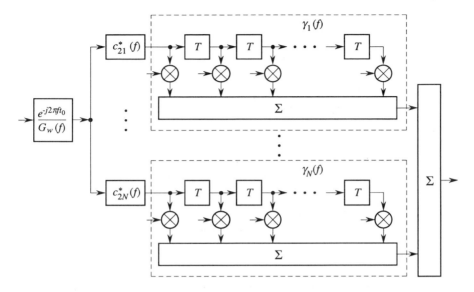

Figure 14.18: *Structure of the optimum receiving filter.*

Hence, the time-domain version of (14.84) is

$$\sum_{i=1}^{M} \sum_{j=-\infty}^{\infty} \rho_{ij} c_2^*(t;i) \delta(t - jT) = \sum_{i=1}^{M} \sum_{j=-\infty}^{\infty} \rho_{ij} c_2^*(t - jT;i) = 0 \qquad (14.88)$$

where $c_2(t;i)$ is the inverse Fourier transform of the $i$th component of $\mathbf{c}_2(f)$.

Thus, if the waveforms $c_2(t;i)$ are linearly independent, (14.88), and hence (14.84), can hold if and only if all the $\rho_{ij}$ are equal to zero. This means that (14.84) admits the only solution $\mathbf{d}(f) = 0$, that is, assuming that $\mathbf{A}(f)$ is nonsingular:

$$\boldsymbol{\gamma}(f) = \mathbf{A}^{-1}(f)\mathbf{b}(f) \qquad (14.89)$$

Equation (14.89) shows that $\boldsymbol{\gamma}(f)$ is periodic with period $1/T$. This proof can be extended to the case when the waveforms $c_2(t;i)$ are not linearly independent (see Biglieri *et al.* (1984)).

As already noted, the structure of the optimum filter $U_{\text{opt}}(f)$ in (14.80) is that of a bank of matched filters followed by infinite-length transversal filters. This is shown in Fig. 14.18.

**Example 14.3**    As an example of application, consider binary CPSK signal transmitted over a nonlinear channel consisting of a fourth-order Butterworth filter cascaded to a nonlinear amplifier (a TWT exhibiting both AM/AM and AM/PM conversion, as in

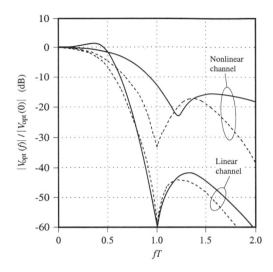

Figure 14.19: *Transfer function of the optimum receiving filter for binary CPSK (solid line) and power spectral density (dashed line) of the received signal; forth-order Butterworth TX filter with $B_{eq} = 1.0128/T$; $N_0 = 5 \cdot 10^{-3}$ W/Hz. Linear and nonlinear channels are shown for comparison.*

Fig. 14.10. For comparison's sake, Fig. 14.19 shows the transfer function of the optimum filter for a channel with and without the nonlinearity, for the sake of comparison. The power spectral densities of the received signal are also shown for comparison. The performance of the optimum filter is shown in Fig. 14.20. For comparison, the mean-square error resulting from a second-order Butterworth receiving filter with optimum 3-dB bandwidth is also shown.                                                                 □

## 14.5.  Maximum-likelihood sequence receiver

In this section we derive the structure of the maximum-likelihood (ML) sequence receiver for a system using a nonlinear channel. The channel model is still that of Section 14.1 and Figs. 14.1 and 14.2. We suppose that the transmission lasts from time 0 to time $KT$. This duration is taken large enough to disregard end effects due to the channel memory (in practice, we assume $K \gg L$). The signal at the output of the noiseless part of the channel, represented by (14.2), can thus

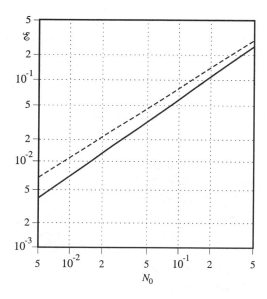

Figure 14.20: *Mean-square error with optimum linear filter (solid line) and with bandwidth-optimized second-order Butterworth filter (dashed line). Same nonlinear system as in Fig. 14.19.*

be written as

$$y_{\mathbf{a}}(t) \triangleq \sum_{n=0}^{K-1} h(t - nT; a_n, \sigma_n) \tag{14.90}$$

where the subscript a denotes the dependence of $y_{\mathbf{a}}(t)$ on the sequence $\mathbf{a} = (a_0, \ldots, a_{K-1})$ of source symbols that must be estimated from the receiver.

From (14.90) we can see that the entire waveform $y_{\mathbf{a}}(t)$, $t \in (0, KT)$, is defined by the sequence of states $(\sigma_L, \ldots, \sigma_K)$, or, equivalently, by the sequence of symbols $(a_0, \ldots, a_{K-1})$. Thus, we have no more than $M^K$ possible received waveforms in the observation interval, and the ML reception is equivalent to the detection of one out of a finite set of waveforms in additive Gaussian noise. Thus, in principle, the optimum receiver will be made up of a bank of $M^K$ matched filters, one for each possible waveform. The filters' outputs are then sampled at the end of the transmission, and the largest sample is used to select the most likely symbol sequence.

In Chapter 7, solutions were found in the linear case for two major problems arising in the implementation of an ML receiver, (1) the number of matched filters required, and (2) the number of comparisons needed to select the largest output. In particular, it was shown that just one matched filter was required to obtain the sufficient statistics for the ML receiver and, also, that the Viterbi algorithm can be used to select the most likely sequence with a complexity that grows only linearly with the message length. In the nonlinear case the results are different. The second problem will be given a satisfactory solution, still invoking the Viterbi algorithm. With respect to the number of matched filters needed, we shall show that it grows exponentially with the channel memory $L$. This is better than the exponential growth with the sequence length $K$ arising from a brute-force approach, but it still makes a practical application of this theory confined to those situations in which the channel memory is very short and/or $M$ is small.

The key to the ML receiver design is the expression of the log-likelihood ratio for the detection of the finite sequence of symbols $\mathbf{a} = (a_0, a_1, \ldots, a_{K-1})$ based on the observation of the waveform

$$r(t) = \sum_{n=0}^{K-1} h(t - nT; a_n, \sigma_n) + w(t) \tag{14.91}$$

where $w(t)$ is a white Gaussian noise process. The log-likelihood ratio for $\mathbf{a}$ has the expression (see Section 2.6)

$$\lambda_{\mathbf{a}} = \frac{2}{N_0} \Re \left\{ \int_0^{KT} y_{\mathbf{a}}^*(t) r(t) \right\} - \frac{1}{N_0} \int_0^{KT} |y_{\mathbf{a}}(t)|^2 dt \tag{14.92}$$

where $y_{\mathbf{a}}(t)$ is the noiseless waveform defined in (14.90). Using (14.91), (14.92) can be rewritten in the form

$$\lambda_{\mathbf{a}} = \frac{2}{N_0} \Re \left\{ \sum_{n=0}^{K-1} \int_0^{KT} h^*(t - nT; a_n, \sigma_n) r(t) dt \right\} \tag{14.93}$$

$$- \frac{1}{N_0} \sum_{n=0}^{K-1} \sum_{\ell=0}^{K-1} \int_0^{KT} h(t - nT; a_n, \sigma_n) h^*(t - \ell T; a_\ell, \sigma_\ell) dt$$

Notice now that $h(t; \cdot, \cdot)$ has a finite duration $T$. Thus, we have

$$\int_0^{KT} h^*(t - nT; a_n, \sigma_n) r(t) dt \tag{14.94}$$

$$= \int_{nT}^{(n+1)T} h^*(t - nT; a_n, \sigma_n) dt, \quad n = 0, \ldots, K-1$$

and

$$\int_0^{KT} h(t - nT; a_n, \sigma_n) h^*(t - \ell T; a_\ell, \sigma_\ell) dt \tag{14.95}$$

$$= \begin{cases} 0, & \ell \neq n \\ \int_{nT}^{(n+1)T} |h(t - nT; a_n, \sigma_n)|^2 dt, & \ell = n \end{cases}$$

So, defining

$$Z_n(a_n, \sigma_n) \triangleq \int_{nT}^{(n+1)T} h^*(t - nT; a_n, \sigma_n) r(t) dt \qquad (14.96)$$

and

$$\mathcal{E}_n(a_n, \sigma_n) \triangleq \int_0^T |h(t; a_n, \sigma_n)|^2 dt \qquad (14.97)$$

we finally get

$$\lambda_{\mathbf{a}} = \frac{2}{N_0} \Re \left\{ \sum_{n=0}^{K-1} Z_n(a_n, \sigma_n) \right\} - \frac{1}{N_0} \sum_{n=0}^{K-1} \mathcal{E}_n(a_n, \sigma_n) \qquad (14.98)$$

We can observe that:

(i) $\mathcal{E}_n(a_n; \sigma_n)$ is the energy of the waveforms $h(t; a_n, \sigma_n)$;

(ii) $Z_n(a_n; \sigma_n)$ can be obtained as the response, sampled at time $(n+1)T$, of a filter matched to $h(t; a_n, \sigma_n)$, to a segment of the input $r(t)$ in the interval $[nT; (n + 1)T]$. The number of matched filters required is then equal to $M^{L+1}$, that is, the number of different values of the pair $(a_n, \sigma_n)$.

The ML sequence decoding rule requires $\lambda_{\mathbf{a}}$ to be maximized over the set of possible sequences **a**. Equivalently, multiplying (14.98) by the constant factor $N_0$ and changing signs, we can say that the ML sequence **a** is the one that minimizes the quantity

$$\ell_{\mathbf{a}} \triangleq -2\Re \left\{ \sum_{n=0}^{K-1} Z_n(a_n, \sigma_n) \right\} + \sum_{n=0}^{K-1} \mathcal{E}_n(a_n, \sigma_n) \qquad (14.99)$$

The ML receiver is formed by a bank of filters matched to $h(t; a_n, \sigma_n)$ followed by one sampler per branch and by a processor, the ML sequence detector, which determines the most likely transmitted data sequence as the one minimizing $\ell_{\mathbf{a}}$.

Define now the transition between states as

$$\tau_{n+1} \triangleq (\sigma_n, \sigma_{n+1}) \qquad (14.100)$$

and observe that there is a one-to-one correspondence between each pair $(a_n, \sigma_n)$ and $\tau_n$. Thus, we can write $Z_n(\tau_n)$ and $\mathcal{E}_n(\tau_n)$, so that, defining

$$\ell_{\mathbf{a}}^{(n)} \triangleq -2\Re\{Z_n(\tau_n)\} + \mathcal{E}_n(\tau_n) \qquad (14.101)$$

we can rewrite (14.99) as

$$\ell_{\mathbf{a}} = \sum_{n=0}^{K-1} \ell_{\mathbf{a}}^{(n)}(\tau_n) \qquad (14.102)$$

Decomposition (14.102), together with the fact that the sequence $(\tau_n)$ originates from a shift-register sequence, ensures that the Viterbi algorithm (see Appendix E) can be applied to the minimization problem at hand. As in the linear case described in Chapter 7, the ML detection problem reduces to the selection of a path through a trellis whose branches have been labeled with the values taken by the function $\ell_a^{(n)}(\tau_n)$, referred to as the *branch metrics*. The same steps illustrated in Chapter 7 to describe the algorithm can be applied with straightforward modifications.

### 14.5.1. Error performance

The problem of evaluating the error probability of an ML sequence receiver can be conceptually reduced to the general problem examined in Chapter 4 and solved in several instances throughout this book. We need only to think of data sequences as points in a signal space: here, error events happen whenever noise and interferences cause the receiver to decide for a point different from the transmitted one. Of course, the error probability depends on the distribution of the Euclidean distances between all possible pairs of points, and, roughly speaking, they are mainly related to the minimum value $d_{\min}$ of these distances.

Usually, finding $d_{\min}$ requires exhaustive comparisons of the received signal waveforms corresponding to all possible pairs of symbol sequences. This can be avoided when the structure of points representing the received sequences is completely symmetric, in such a way that comparisons of all sequences with respect to a particular one are representative of the larger set of comparisons that would be needed. In this case, we say that the uniform error property applies. A significant example is represented by linear codes, where the error performance was computed using as a reference the all-zero word (block codes) or sequence (convolutional codes).

In the present case the uniform error property does not hold, so all comparisons are needed. In some cases, like the typical satellite channel with a short memory, the number of sequence pairs that needs to be considered is not very large. Hence, a brute-force computation is feasible. A method to do that can be found in Herrmann (1978). It consists of an algorithm to compute systematically the so-called "chip functions" and "chip distances." It uses these to compute the distances between pairs of received waveforms and then to estimate the symbol error probability.

Here we use a different approach. We first derive an upper bound to the sequence error probability, defined as the probability of choosing as true a transmitted sequence different from the actual one. Then, we shall see that the most relevant contribution to the error probability, for medium-to-large values of signal-to-noise ratios, depends on the minimum Euclidean distance between all pairs of possible received waveforms. At that point, the general and efficient algorithm described in Chapter 12 to compute the minimum Euclidean distance can be used to assess the asymptotic performance.

Suppose that the sequence $\mathbf{a}_j$ has been transmitted. The probability that the estimated sequence $\hat{\mathbf{a}}$ is different from $\mathbf{a}_j$ can be written as

$$P(e \mid \mathbf{a}_j) \triangleq \mathrm{P}\{\hat{\mathbf{a}} \neq \mathbf{a}_j \mid \mathbf{a}_j\} \tag{14.103}$$

Application of the union bound (see Section 4.3.2) allows us to write

$$P(e \mid \mathbf{a}_j) \leq \sum_{\mathbf{a}_i \neq \mathbf{a}_j} \mathrm{P}\{\ell_{\mathbf{a}_i} \leq \ell_{\mathbf{a}_j} \mid \mathbf{a}_j\} = \frac{1}{2} \sum_{\mathbf{a}_i \neq \mathbf{a}_j} \mathrm{erfc}\left(\frac{d(\mathbf{a}_i, \mathbf{a}_j)}{2\sqrt{N_0}}\right) \tag{14.104}$$

where $d(\mathbf{a}_i, \mathbf{a}_j)$ is the Euclidean distance between the two signal sequences obtained at the output of the noiseless part of the channel in correspondence with the symbol sequences $\mathbf{a}_i$ and $\mathbf{a}_j$

$$d^2(\mathbf{a}_i, \mathbf{a}_j) \triangleq \int_0^{KT} |y_{\mathbf{a}_i}(t) - y_{\mathbf{a}_j}(t)|^2 dt \tag{14.105}$$

The last equality in (14.104) becomes evident if one thinks of the noiseless received waveforms of duration $KT$ as points in a signal space and then applies the standard formula for the binary error probability between two points at distance $d^2(\mathbf{a}_i, \mathbf{a}_j)$.

To obtain the average sequence error probability, we need only to average $P(e \mid \mathbf{a}_j)$ over all the possible symbol sequences $\mathbf{a}_j$ assumed to be equally likely:

$$P(e) = \sum_{\mathbf{a}_j} P(\mathbf{a}_j) P(e \mid \mathbf{a}_j) \leq \frac{1}{2} \frac{1}{M^K} \sum_{\mathbf{a}_j} \sum_{\mathbf{a}_i \neq \mathbf{a}_j} \mathrm{erfc}\left(\frac{d(\mathbf{a}_i, \mathbf{a}_j)}{2\sqrt{N_0}}\right) \tag{14.106}$$

The final step is to consider the discrete finite set of all Euclidean distances $d(\mathbf{a}_i, \mathbf{a}_j)$, $\mathbf{a}_i \neq \mathbf{a}_j$, denoted by $\mathrm{D} = \{d_\ell\}$. Denoting by $N(d_\ell)$ the number of pairs of sequences $(\mathbf{a}_i, \mathbf{a}_j)$ giving rise to noiseless received waveforms at distance $d_\ell$, we can rewrite (14.106) as

$$P(e) \leq \frac{1}{2} \frac{1}{M^K} \sum_{d_\ell \in \mathrm{D}} N(d_\ell) \mathrm{erfc}\left(\frac{d_\ell}{2\sqrt{N_0}}\right) \tag{14.107}$$

For large values of signal-to-noise ratio, the dominant term of the summation in the RHS of (14.107) is the one containing $d_{\min} \triangleq \min_{d_\ell \in \mathrm{D}} d_\ell$, so that, asymptotically, we can approximate $P(e)$ as

$$P(e) \overset{\sim}{<} \frac{1}{2} \frac{1}{M^K} N(d_{\min}) \mathrm{erfc}\left(\frac{d_{\min}}{2\sqrt{N_0}}\right) \tag{14.108}$$

To compute the sequence error probability (14.107), an extension of the method of the transfer function of directed graphs, used to evaluate the performance of convolutional codes, could be applied (see Viterbi and Omura, 1979, Problem 5.14). However, that approach presents a computational complexity that grows with the fourth power of the number of the states (14.1), and is thus often impractical. A related technique is described in (Liu *et al.*, 1990). If one contents himself with the approximation given in (14.108), which only requires knowledge of the minimum Euclidean distance $d_{min}$, then an efficient algorithm to obtain the latter is easily obtained as a straightforward modification of the one described in Section 12.4.

## 14.6.  Identification and equalization of nonlinear channels

In the preceding sections we have derived the form of the optimum (in the mean-square sense) linear symbol-by-symbol receiver and of the optimum unconstrained (maximum-likelihood) sequence receiver. Both receivers require knowledge of the channel. When the channel is not completely known a priori, a preliminary phase, before the transmission of information, must be devoted to channel identification. Moreover, the procedure has to be occasionally repeated as the channel characteristics vary with time. It is the same phenomenon encountered when dealing with adaptive equalization (see Chapter 8), but now it is complicated by the nonlinear behavior of the channel.

The conventional linear receiver, possibly equipped with an adaptive equalizer, does not attain the required performance in some applications. An example is represented by the satellite channel previously described: its behavior is driven by a strong nonlinearity in conjunction with a short memory span. The problem of using an ML sequence receiver in this channel has more to do with the speed of digital integrated circuits, which must cope with the high symbol rate (up to hundreds of Mbit/s), than with the complexity related to the channel memory.

We shall describe in this section a receiver that is intermediate (in terms of complexity and optimality) between the conventional linear receiver and the ML sequence receiver. It can be used indifferently for channel identification purposes and/or as a nonlinear equalizer. Its structure can easily be made adaptive. The form of the receiver is a nonlinear extension of the tapped-delay-line (TDL) mean-square-error (MSE) equalizer described in Chapter 8. Its nonlinear structure is suggested by the Volterra model of Section 14.3. As such, it can be used to estimate the parameters of a discrete Volterra model of the channel. The equalizer, which can be seen as a structure-constrained optimum (in the mean-square sense) embedded in a symbol-by-symbol receiver, will be described in its simplest version. It is intended that some of the refinements to the basic TDL equalizer presented in Chapter 8 (such as decision feedback) can be fruitfully applied also to this case. The description of the equalizer, as well as the numerical results, refers to a system employing CPSK.

Recalling the expression (14.10) of the signal $y(t)$ (see Fig. 14.11), we can write the samples $r_n$ that form the received sampled sequence $(r_n)$ as

$$
r_n = \sum_{m=0}^{\infty} \sum_{n_1} \cdots \sum_{n_{2m+1}} a_{n_1} \cdots a_{n_{m+1}} a^*_{n_{m+2}} \cdots a^*_{n_{2m+1}}
$$
$$
\cdot K_{2m+1}(n - n_1, \ldots, n - n_{2m+1}) + \nu_n \qquad (14.109)
$$

where $\nu_n$ is the noise sample.

The structure of (14.109) reflects how the channel output depends both on the channel (through the Volterra coefficients) and on the information symbols. In particular, the symbol structure of PSK modulation $(a_n = e^{j\phi_n})$ results in insensitivity to certain kinds of nonlinearities. In fact, since $a_n a_n^* = 1$, it is apparent from (14.109) that certain Volterra coefficients $K_{2m+1}$ will contribute to nonlinearities of order less than $2m + 1$. To be more specific, consider first the third-order Volterra coefficients $K_3(n - n_1, n - n_2, n - n_3)$; for $n_1 = n_3$ or $n_2 = n_3$, the channel nonlinearities reflected by these coefficients will not affect a PSK signal, because

$$
a_{n_1} a_{n_2} a^*_{n_3} K_3(n - n_1, n - n_2, n - n_3) \qquad (14.110)
$$
$$
= \begin{cases} a_{n_2} K_3(n - n_1, n - n_2, n - n_3), & \text{if } n_1 = n_2 \\ a_{n_1} K_3(n - n_1, n - n_2, n - n_3), & \text{if } n_2 = n_3 \end{cases} \qquad (14.111)
$$

and the only contribution is to the linear part of the channel. Similar considerations on the higher-order coefficients show that some of them contribute to the linear part, others only to the third-order nonlinearity, and so on. These considerations can be further pursued if we observe that, for an $M$-ary CPSK, $a_n^M$ is a constant, which results in a further reduction of sensitivity of PSK to certain nonlinearities. This leads to the noteworthy conclusion that, for CPSK, certain nonlinearities need only a linear compensation, while others affect the signal to a lower degree than other modulation schemes. The overall effect is a further reduction of the number of Volterra coefficients to be taken into account in the channel model.

The Volterra series representation (14.109) provides us with a basis for representing a general signal processor in the same form. In fact, it seems quite natural to choose, for the general discrete-time processor, the structure suggested by (14.109) after truncating the infinite sums. This processor can be implemented using a TDL, a nonlinear combiner, a number of complex multipliers, and a summing bus, as shown in Fig. 14.21 for the special case $N_1 = N_2 =$

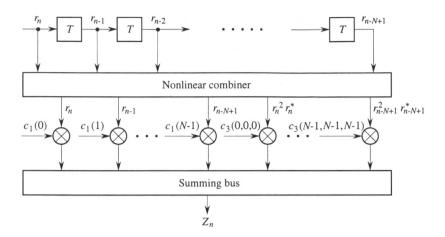

Figure 14.21: *Third-order nonlinear TDL equalizer.*

$N_3 = N$. ($N_i$, $i = 1, 2, 3$, to be defined shortly.) Thus, assuming a $(2k + 1)$th order equalizer (i.e., a processor with nonlinearities up to the order $2k + 1$), its output sequence $(z_n)$ is related to the input (received) sequence $(r_n)$ by the relationship

$$z_n = \sum_{n_1=0}^{N_1-1} r_{n-n_1} c_1(n_1) \tag{14.112}$$

$$+ \sum_{n_1=0}^{N_1-1} \sum_{n_2=0}^{N_2-1} \sum_{n_3=0}^{N_3-1} r_{n-n_1} r_{n-n_2} r^*_{n-n_3} c_3(n_1, n_2, n_3) + \dots$$

$$+ \sum_{n_1=0}^{N_1-1} \cdots \sum_{n_{2k+1}=0}^{N_{2k+1}-1} r_{n-n_1} \cdots r_{n-n_{k+1}} r^*_{n-n_{k+2}} \cdots r^*_{n-n_{2k+1}} c_{2k+1}(n_1, \dots, n_{2k+1})$$

where $N_i$ represents the number of values assumed by the summation index $n_i$.

Equation (14.112) shows that the output is related to the input by a finite set of constants $c_1(n_1)$, $c_3(n_1, n_2, n_3)$, ..., $c_{2k+1}(n_1, \dots, n_{2k+1})$, which will be referred to hereafter as the *tap weights* of the equalizer, in analogy with the linear case described in Chapter 8. Thus, the design of the equalizer is equivalent to the choice of its tap weights. If we arrange them in the column vector

$$\mathbf{c} \triangleq [c_1(0), \dots, c_1(N_1 - 1), c_3(0, 0, 0), \dots, c_3(N_1 - 1, N_2 - 1, N_3 - 1),$$
$$\dots, c_{2k+1}(0, \dots, 0), \dots, c_{2k+1}(N_1 - 1, \dots, N_{2k+1} - 1)]' \tag{14.113}$$

and define

$$\mathbf{r}_n \triangleq [r_n, r_{n-1}, \dots, r_{n-N_1-1}, \dots, r_{n-N_1+1} \cdots r^*_{n-N_{2k+1}+1}]' \tag{14.114}$$

the input-output relationship for the equalizer can be written in vector form as

$$z_n = \mathbf{c}'\mathbf{r}_n \tag{14.115}$$

The input-output relationship (14.115) governing the behavior of the equalizer is linear in the tap-weight vector $\mathbf{c}$; thus, the methods used in the linear case can also be applied here. We want to design the processor so as to get, at its output, a sequence of samples approximating, under an MSE criterion, the sequence of channel input symbols $a_{n-D}$, where $D$ denotes the allowed delay. Thus, we need to minimize the MSE

$$\mathcal{E} \overset{\triangle}{=} \mathrm{E}\{|\mathbf{c}'\mathbf{r}_n - a_{n-D}|^2\} = \mathbf{c}^\dagger \mathbf{R}\mathbf{c} - 2\Re\{\mathbf{c}'\mathbf{g}\} + \mathrm{E}\{|a_{n-D}|^2\} \tag{14.116}$$

with respect to $\mathbf{c}$, where

$$\mathbf{R} \overset{\triangle}{=} \mathrm{E}\{\mathbf{r}_n^* \mathbf{r}_n'\} \tag{14.117}$$

$$\mathbf{g} \overset{\triangle}{=} \mathrm{E}\{a_{n-D}\mathbf{r}_n^*\} \tag{14.118}$$

The optimum $\mathbf{c}$ can thus be obtained by solving the set of linear equations

$$\mathbf{R}\mathbf{c} = \mathbf{g} \tag{14.119}$$

which admits the solution

$$\mathbf{c}_{\mathrm{opt}} = \mathbf{R}^{-1}\mathbf{g} \tag{14.120}$$

provided that $\mathbf{R}$ is positive definite. It is seen from the definition of $\mathbf{R}$ that this condition is fulfilled if, for any arbitrary complex vector $\mathbf{b}$, we have

$$\mathbf{b}^\dagger \mathbf{R}\mathbf{b} = \mathrm{E}\{|\mathbf{b}^\dagger \mathbf{r}_n|^2\} > 0 \tag{14.121}$$

The RHS of (14.121) can be thought of as the average power of the output of an equalizer with tap weights $\mathbf{b}^*$. This power cannot be zero, due to the presence of the noise added to the samples entering the equalizer. In the absence of noise, (14.121) can only be zero if $\mathbf{b}^\dagger \mathbf{r}_n = 0$ (i.e., the entries of $\mathbf{r}_n$ are linearly dependent).

The solution (14.120) for the optimum tap-weight vector requires knowledge of $\mathbf{R}$ and $\mathbf{g}$. Computation of the averages involved in their definitions (14.117) and (14.118) can be done using the method described in Section 14.3 to evaluate the moments of the RVs representing ISI. In particular, it is possible to evaluate by exhaustive enumeration the part of the averages depending on information symbols and, analytically, the part depending on Gaussian noise. These procedures, of course, allow only off-line computation of $\mathbf{R}$ and $\mathbf{g}$, and are not suitable for the equalizer working in an adaptive mode. However, the algorithms described for adaptive linear equalizers and, in particular, the stochastic-gradient

| Linear part |
| --- |
| $K_1(0) = 1.22 + j0.646$ |
| $K_1(1) = 0.063 - j0.001$ |
| $K_1(2) = -0.024 - j0.014$ |
| $K_1(3) = 0.036 + j0.031$ |

| Third-order nonlinearities |
| --- |
| $K_3(0,0,2) = 0.039 - j0.022$ |
| $K_3(3,3,0) = 0.018 - j0.018$ |
| $K_3(0,0,1) = -0.035 + j0.035$ |
| $K_3(0,0,3) = -0.040 - j0.009$ |
| $K_3(1,1,0) = -0.001 - j0.017$ |

| Fifth-order nonlinearities |
| --- |
| $K_5(0,0,0,1,1) = 0.039 - j0.022$ |

Table 14.1: *Reduced Volterra coefficients for the nonlinear QPSK system described in Example 14.4.*

algorithm of Chapter 8 can also be applied fruitfully in this case. Its application leads to the following recursion for tap updating:

$$\mathbf{c}^{(n+1)} = \mathbf{c}^{(n)} - \alpha(z_n - a_{n-D})\mathbf{r}_n^* \qquad (14.122)$$

**Example 14.4**    Consider the scheme of Fig. 14.11. The transmission filter $s(t)$ includes a rectangular shaping filter and a fourth-order Butterworth with 3-dB bandwidth $1.7/T$. The TWT is described by the characteristics of Fig. 14.10, and is driven at saturation by the PSK symbols. The receiving filter $u(t)$ is a second-order Butterworth with 3-dB bandwidth $1.1/T$. The computed Volterra coefficients for this channel, after neglecting the linear Volterra coefficients whose value lies below 0.001 and the nonlinear coefficients lying below 0.005, and after the further reduction that takes into account the structure of PSK symbols, as previously mentioned, are shown in Table 14.1. We want to study the effect of a nonlinear equalizer cascaded to the channel. An important question is worth mentioning at this point. For a given complexity (i.e., a given number of first, third,..., $n$th-order coefficients in the TDL), what is the best allocation of those coefficients, i.e., how can we choose the range of indexes $n_1, n_2, n_3, \ldots$ in (14.112)?

Simulation experience shows that often a good use of the allowable complexity consists in allocating the TDL taps so that they match the most significant Volterra coefficients of the channel. In other words, it is convenient to introduce a tap, say $c_3(i, j, k)$, if the corresponding Volterra coefficient $K_3(i, j, k)$ of the channel model has a relevant

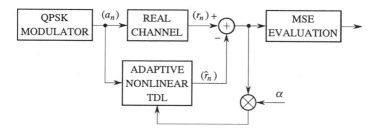

Figure 14.22: *Block diagram for the simulation of the channel identifier.*

magnitude. This way of matching the structure of the nonlinear TDL to that of the channel requires previous knowledge of the channel structure. Thus, the first step consists of channel identification using the model of Fig. 14.21. The system behavior has been simulated as in Fig. 14.22. A stochastic-gradient algorithm was used to iteratively modify the coefficients of the nonlinear TDL, with the goal of minimizing $E\{|r_n - \hat{r}_n|^2\}$, the mean-square difference between the true sample at the channel output and the sample generated by the channel estimator.

The convergence of the identification process is described through a sequence $(\xi_k)$ of running averages of $|r_n - \hat{r}_n|^2$ evaluated over successive blocks of $K$ symbols. If the parameter $\alpha$ in (14.122) is chosen in the field of values allowing the convergence of the algorithm, the sequence $(\xi_k)$ will decrease to the minimum value achievable with the complexity chosen for the channel model.

For the sake of comparison, $(\xi_k)$ is plotted in Fig. 14.23 in the linear case (i.e., when the TWT is not present in the channel and the channel estimator is a linear TDL with 10 taps). In the nonlinear case, we have considered a nonlinear TDL with Volterra coefficients of first, third, and fifth order, allowing a memory of 10 for the linear part, 4 for the third-order part, and 3 for the fifth order. In Fig. 14.24 the behavior of $(\xi_k)$ is shown for $K = 100$. For curve A, the initial choice of the coefficient $\alpha$ was taken outside the convergence interval of the algorithm. Thus, the curve shows an initial divergence that ends when the value of $\alpha$ is suitably reduced. Curves B and C differ in the choice of the coefficient $\alpha$ governing the updating of the nonlinear coefficients. Comparison with the linear case of Fig. 14.23 shows that the MSE settles to a value sensibly lower in the linear case and that the convergence is faster. This happens because of the following:

(i) In the linear case the complexity of the TDL ($N = 10$) is lower. This allows the use of larger $\alpha$ and accelerates the convergence.

(ii) The structure of the estimator in the linear case is much closer to the true channel than in the nonlinear case. In the latter, other nonlinear coefficients should be added to the model in order to decrease the steady-state MSE.

Let us now proceed to the design of the equalizer. Having chosen the structure of the nonlinear TDL on the basis of the estimate of the channel, the optimum values of the

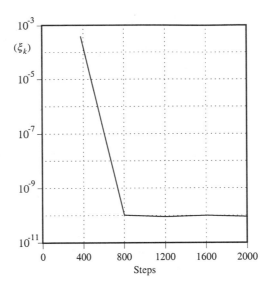

Figure 14.23: *Identification of the channel of Example 14.4 in the absence of the TWT (linear channel). The curves present the behavior of the MSE estimated through a running time average made over blocks of $K = 400$ symbols. The linear equalizer has $N = 10$ taps, and $\alpha = 0.1$.*

tap weights $c_i(\cdot)$ of the equalizer can be found through the stochastic gradient algorithm (14.122) aiming at minimizing the MSE $\mathcal{E}$ defined in (14.116). Also in this case we have used a running average $(\mathcal{E}_k)$ over $K$ symbols, which represents a time average of $|z_n - a_{n-D}|^2$.

The results in terms of $\mathcal{E}_k$ in the linear case are shown in Fig. 14.25, for the sake of comparison, for various numbers of the TDL taps. The two sets of curves refer to different values of $\alpha$. The continuous curves have been obtained with $\alpha = 0.1$, whereas the dashed ones have $\alpha = 0.05$. The value of $K$ is equal to 100. Turning now to the nonlinear case, we can examine the results presented in Fig. 14.26. The set of curves refers to equalizers of increasing complexity cascaded with the nonlinear channel. The results suggest the following observations:

(i) Linear taps cannot compensate for the distortion, since after addition of a certain number of them the MSE will not show any significant reduction.

(ii) The addition of a few nonlinear taps allows a significant reduction of the MSE.

□

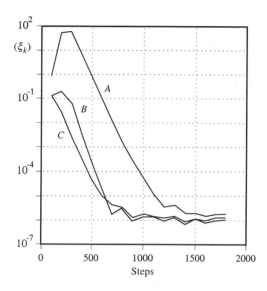

Figure 14.24: *Identification of the channel of Example 14.4 in the presence of the TWT (nonlinear channel). The curves present the behavior of the MSE estimated through a running time average made over blocks of K = 100 symbols. Curve A refers to a case in which the initial choice of the parameter controlling the gradient algorithm is made outside the convergence interval and then reduced so as to have convergence. Curve B has a value of the parameter controlling the updating of the linear part of the equalizer equal to 0.1, and a value of the parameter controlling the nonlinear part equal to 0.04. Curve C has both parameters equal to 0.01.*

## 14.7. Compensation of nonlinear channels

A fundamental limit to the performance of optimum receivers and equalization schemes (and, more generally, of any conceivable receiver, either linear or nonlinear, as we have shown in Section 7.5) depends on the minimum Euclidean distance between the signals observed at the output of the noiseless part of the channel. Stated in words, this limitation is due to the fact that the signal to be processed by the receiver is affected by noise, and hence any attempt to compensate for the channel distortion by introducing a sort of "inverse distortion" will enhance the noise. For this reason, it appears logical to investigate solutions based on the compensation of the nonlinear channel *before* noise addition. With this procedure, the channel should be made to look as similar as possible to a Gaussian channel. This idea bears obvious similarities to Tomlinson-Harashima precoding, that we examined in Chapter 8.

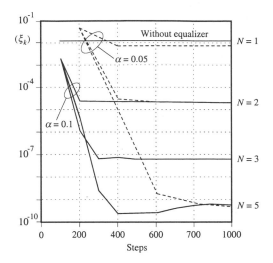

Figure 14.25: *Equalization of the channel of Example 14.4 in the absence of the TWT (linear channel). The curves present the behavior of the MSE estimated through a running-time average made over blocks of $K = 100$ symbols. The number $N$ of taps of the linear equalizer is a parameter. The solid curves refer to a value of the parameter $\alpha = 0.1$, whereas the dashed curves have $\alpha = 0.05$.*

If this approach is chosen, several factors and constraints should be taken into account. One of them is, of course, the ultimate performance that the nonlinearity-compensating scheme can achieve. The second one is the implementation complexity. The third one is the fact that the compensator itself may expand the signal bandwidth. Finally, if the channel is time-varying, provision must be made for *adaptive* compensation.

In this section we examine digital predistorters, i.e., devices to be inserted at the input of the digital channel, and whose aim is to compensate for the unwanted effects of the nonlinear channel. Their design will be based on the concept of $p$th-order inverse of a nonlinear system.

### 14.7.1.  $p$th-order compensation

Assume first, for simplicity, that the channel has no memory, i.e., that no bandwidth-limiting components exist, and that the channel's input-output relationship is invertible. In this situation a memoryless predistorter may act by skewing the transmitted signal constellation in such a way that, when passed through the nonlinear channel, it will resume its original shape. Here the compensator's

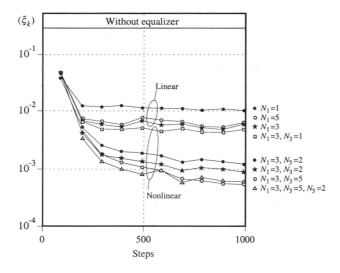

Figure 14.26: *Equalization of the channel of Example 14.4 in the presence of the TWT (nonlinear channel). The curves prsent the behavior of the MSE estimated through a running-time average made over blocks of $K = 100$ symbols. The curves marked with "linear" refer to a linear equalizer, whereas the ones marked with "nonlinear" refer to a nonlinear equalizer. The numbers of linear and nonlinear taps are parameters of the curves. The value of $\alpha$ is equal to 0.1 for the linear taps, and equal to 0.01 for the nonlinear taps.*

task is to invert the nonlinearity. The channel symbols $a_n$ are changed into the symbols $b_n = g(a_n)$, where $g(\cdot)$ is a suitable complex function, and hence the design of a predistorter for a channel without memory can be viewed as the selection of a new modulation scheme.

Consider next the more general assumption of a channel with memory. Inversion of a nonlinear system with memory may not be possible: not all nonlinear systems possess an inverse, and many systems can be inverted only for a restricted range of input amplitudes. However, it is always possible to define a *pth-order inverse*, for which the amplitude range is not restricted.

Given a nonlinear system H, its *p*th-order inverse is a system which, when cascaded to H, results in a system in which the first-order Volterra kernel is a unit impulse, and the second- through the *p*th-order Volterra kernels are zero (Schetzen, 1980). In other words, if the *p*th-order nonlinear inverse channel is synthesized at the transmitter's front end, the compensated transmission channel will exhibit no linear distortion, and no nonlinear distortion up to order *p*. Obviously, the performance of the *p*th-order compensated channel will depend on the effects of the residual distortion.

To describe $p$th-order compensation, it is convenient to use tensor notations. These imply that any subscript occurring twice in the same term is to be summed over the appropriate range of discrete time. Thus, for example, we write $x_i y_i$ to denote $x_1 y_1 + x_2 y_2 + \cdots$.

These notations allow us to write the input-output relationship of a noiseless system with memory, whose input is the sequence $(a_n)$ and whose output is the sequence $(y_n)$, in the form

$$y_n = k^{(1)}_{n;i} a_i + k^{(3)}_{n;i,j,k} a_i a_j a_k^* + k^{(5)}_{n;i,j,k,l,m} a_i a_j a_k a_l^* a_m^* + \cdots \qquad (14.123)$$

which corresponds to the noiseless part of (14.109).

Consider now two bandpass nonlinear systems. Let the first one (the compensator) be characterized by Volterra coefficients $f$, and the second (the channel) by Volterra coefficients $g$. Denote by $h$ the coefficients of the system resulting from the cascade of the two. The first- and third-order $h$-coefficients are explicitly given by

$$h^{(1)}_{n;i} = g^{(1)}_{n;v} f^{(1)}_{v;i} \qquad (14.124)$$

$$h^{(3)}_{n;i,j,k} = g^{(1)}_{n;v} f^{(3)}_{v;i,j,k} + g^{(3)}_{n;v,w,z} f^{(1)}_{v;i} f^{(1)}_{w;j} f^{(1)*}_{z;k} \qquad (14.125)$$

Higher-order coefficients can be expressed in a similar form, although with more terms appearing in the right-hand side: for example, 5th-order coefficients have five terms.

We are now ready to provide equations for the $p$th-order compensator. Under the assumption that the linear part of system $f$, i.e., the linear functional determined by the first-order coefficient of $f$, is invertible, it is possible to find a system $g$ such that its cascade with $f$ yields a system with no linear distortion, i.e.,

$$g^{(1)}_{n;v} f^{(1)}_{v;i} = f^{(1)}_{n;v} g^{(1)}_{v;i} = \delta_{n;i} \overset{\triangle}{=} \begin{cases} 1 & n = i \\ 0 & \text{elsewhere} \end{cases} \qquad (14.126)$$

This choice generates the first-order compensator. Equation (14.126) expresses nothing but the Nyquist criterion for the absence of intersymbol interference in the overall channel. Since we are interested in finite-complexity compensators, it is useful to examine an approximate solution to (14.126) including a finite number of coefficients. This problem is equivalent to the design of a zero-forcing equalizer with finite length, as studied in Section 8.6. The third-order compensator is obtained by choosing $g^{(3)}$ so as to have $h^{(3)} = 0$. By taking the discrete convolution of both sides of (14.125) with $g^{(1)} g^{(1)} g^{(1)*}$, and recalling (14.126), we obtain

$$g^{(3)}_{n;i,j,k} = -g^{(1)}_{n;y} f^{(3)}_{y;u,v,z} g^{(1)}_{u;i} g^{(1)}_{v;j} g^{(1)*}_{z;k} \qquad (14.127)$$

**Example 14.5** Assume that the channel nonlinearity can be modeled as the cascade of a linear system L and a memoryless device D. The $p$th-order compensator for this channel turns out to be the cascade of a linear filter, the inverse $L^{-1}$ of L, preceded by a nonlinear memoryless device, the $p$th-order inverse of D. Notice that the cascade of $L^{-1}$ and L gives rise to a Nyquist filter. This result shows that in this case compensation of the channel nonlinearity consists of removing the channel memory, then compensating for the resulting memoryless nonlinearity by memoryless predistortion. □

**Example 14.6** Consider a nonlinear channel whose linear part has already been compensated by a suitable combination of channel filtering and linear equalization at the receiver's front-end. In this situation some simplifications of the compensator are possible. In particular we obtain, for the first- and third-order compensators,

$$g_{n;i}^{(1)} = \delta_{n;i}$$
$$g_{n;i,j,k}^{(3)} = -f_{n;i,j,k}^{(3)}$$

The first equation above shows that no linear compensation should be added, while the third-order compensator coefficient is simply obtained by changing the sign of the third-order channel coefficient. □

## Spectrum-shaping effects

Under our assumption of linear modulation, the effect of the compensator on the transmitted signal is to transform the symbol sequence $(a_n)$ into a new sequence $(b_n)$, while keeping the basic modulator waveform unchanged. Consequently, the power density spectrum of the transmitted signal can be analyzed by examining the autocorrelation function of the sequence $(b_n)$.

**Example 14.7** Consider a linear compensator responding to the symbol sequence $(a_n)$, $E|a_n|^2 = 1$, with the sequence $(b_n) = (a_n + A a_{n-1})$, $A$ a real constant. Computation of the autocorrelation function of $(b_n)$ shows that this compensator causes a spectral shaping $(1 + A^2) + 2A\cos(2\pi fT)$, $T$ the symbol period. In some cases the nonlinear terms of the compensator may be irrelevant in shaping the spectrum. Consider, for instance, the compensator-output sequence $(b_n) = (a_n + A a_{n-1} + B a_n a_{n-1} a_{n-2}^*)$, $A$ and $B$ two real numbers. Direct calculation shows that $(r_k)$, the autocorrelation sequence of $(b_n)$, takes values

$$r_k = \begin{cases} 1 + A^2 + B^2 & k = 0, \\ A & k = \pm 1, \\ 0 & |k| > 1 \end{cases}$$

Thus, for $A^2 \gg B^2$ the third-order nonlinearity has very little effect on the spectrum. □

## 14.8. Bibliographical notes

Our treatment of the modeling and performance evaluation of bandpass nonlinear digital systems follows closely Benedetto *et al.* (1976, 1979). Discrete Volterra series applied to sampled-data systems are analyzed in Barker and Ambati (1972). Inverses of nonlinear systems represented by continuous and discrete Volterra series are investigated in Schetzen (1976) and Wakamatsu (1981).

The problem of evaluating the performance of a nonlinear digital transmission system has received considerable attention, particularly as applied to satellite links. The interested reader is invited to scan the extensive reference list in Benedetto *et al.* (1979).

The derivation of the optimum receiving filter follows that presented in Biglieri *et al.* (1984). Different approaches have been pursued by Fredricsson (1975) and Mesiya *et al.* (1978).

The model and the approach followed in the derivation of the ML sequence receiver are original. Based on an analytical model of the bandpass nonlinearity, Mesiya *et al.* (1977) have derived the ML sequence receiver for binary PSK. An ML receiver taking into account also the uplink noise in satellite links has been described by Benedetto *et al.* (1981). The Volterra series technique has been applied by Falconer (1978) and, previously, by Thomas (1971) to the design of adaptive nonlinear receivers. The use of orthogonal Volterra series to achieve rapid adaptation in connection with a Volterra series approach is proposed in Biglieri *et al.* (1984).

Our treatment of nonlinear equalization follows closely that of Benedetto and Biglieri (1983), while that of $p$th-order compensation is taken from Biglieri *et al.* (1988), where applications to satellite channels are also described. Use of Volterra series in the compensation of nonlinearities in magnetic recording channels is described in Biglieri *et al.* (1994). Other contributions to the analysis and compensation of nonlinear distortion can be found in Karam and Sari (1989, 1990, 1991a, and 1991b).

## 14.9. Problems

*Problems marked with an asterisk should be solved with the aid of a computer.*

**14.1** (*) Good analytical approximations for typical AM/AM and AM/PM characteristics of TWTs are

$$F(r) = \frac{\alpha_a r}{1 + \beta_a r^2}$$

$$\phi(r) = \frac{\alpha_\phi r^2}{1 + \beta_\phi r^2}$$

(see Saleh, 1981), where the coefficients $\alpha_a$, $\beta_a$, and $\alpha_\phi, \beta_\phi$ are found by fitting these equations to the experimental data through a minimum mean-square-error procedure. Apply this procedure to approximate the curves of Fig. 14.10 and draw the resulting $F(r)$ and $\phi(r)$.

**14.2** Following the calculations of Example 2.18, extend the results obtained there for memoryless nonlinear devices to the case of the bandpass nonlinear system of Fig. 14.11 by obtaining the result (14.9).

**14.3** (*) Consider a channel formed by cascading a 2nd-order Butterworth filter with normalized equivalent noise bandwidth $B_{eq}T = 2.0$ with the TWT of Fig. 14.10. Using a polynomial approximation of the complex TWT characteristics, that is, $F(r)\exp[j\phi(r)]$, with powers up to the seventh, compute the discrete Volterra coefficients (14.18) discarding those smaller than $10^{-3} \cdot k_1(t_0)$.

**14.4** Derive the recurrent formulas (14.47) for the powers $\xi^\ell$ of discrete Volterra coefficients.

**14.5** Derive the recurrent formulas for the powers $(\xi^*)^m$ of discrete Volterra coefficients using the derivation of Problem 14.4.

**14.6** Prove the result (14.52).

**14.7** A wide-sense stationary Gaussian process $n(t)$ can be represented as

$$n(t) = \sum_i \beta_i b_i(t)$$

where $\beta_i$ are unit-variance independent Gaussian RVs and $b_i(t)$ are appropriately chosen deterministic functions. Evaluate the functions $b_i(t)$ in the case of a power spectral density of $n(t)$ equal to $G_0$ in the interval $(-F, F)$, and zero outside. *Hint*: If $G_n(f)$ is the power spectral density of the process, the functions $b_i(t)$ are obtained as inverse Fourier transforms of $\gamma_i(f)G_n(f)$, where $\gamma_i(f)$ form a complete sequence of orthonormal functions in the Hilbert space with norm

$$\|\gamma\|^2 = \int_{-\infty}^{\infty} |\gamma(f)|^2 G_n(f)\, df$$

**14.8** Repeat the computations of the moments (14.52) in the case of uplink noise represented as in (14.55).

**14.9** Using variational calculus techniques (Appendix C), derive the integral equation (14.66) that minimizes the mean-square error (14.59).

**14.10** Compute the average (14.71).

**14.11** Compute the average (14.76).

**14.12** (*) Modify the algorithm described in Section 12.4.3 and write a computer program implementing it. Use it to compute $d_{min}$ for the nonlinear system of Example 14.1.

**14.13** (*) Repeat the simulation of Example 14.4 by setting the 3-dB bandwidth of the transmission filter equal to $2/T$.

# Useful formulas and approximations

## A.1. Error function and complementary error function

Definitions

$$\text{erf}(x) \triangleq \frac{2}{\sqrt{\pi}} \int_0^x e^{-t^2} dt \tag{A.1}$$

$$\text{erfc}(x) \triangleq \frac{2}{\sqrt{\pi}} \int_x^\infty e^{-t^2} dt = 1 - \text{erf}(x) \tag{A.2}$$

Relation with the Gaussian probability density function with mean $\mu$ and variance $\sigma^2$

$$\frac{1}{\sqrt{2\pi}\sigma} \int_{-\infty}^x e^{-(t-\mu)^2/(2\sigma^2)} dt = \frac{1}{2}\left\{ 1 + \text{erf}\left(\frac{x-\mu}{\sqrt{2}\sigma}\right) \right\} \tag{A.3}$$

The function $\text{erfc}(x)$ admits the following asymptotic $(x \to \infty)$ expansion:

$$\text{erfc}(x) \simeq \frac{e^{-x^2}}{\sqrt{\pi}x}\left\{ 1 + \sum_{i=1}^\infty (-1)^i \frac{1 \cdot 3 \cdot 5 \cdots (2i-1)}{(2x^2)^i} \right\} \tag{A.4}$$

Table A.1. shows a comparison, for values of $x$ between 3 and 5, of the approximation (A.4) truncated to its first term with the exact value of $\text{erfc}(x)$. Relative errors are also indicated. To obtain closed-form upper bounds to the error probabilities for block and convolutional codes, we will use sometimes the following upper bound:

$$\frac{1}{2}\text{erfc}(\sqrt{x}) < \frac{1}{2}e^{-x} \tag{A.5}$$

In Figure A.1 the two functions $\frac{1}{2}\text{erfc}(\sqrt{x})$ and $\frac{1}{2}\exp(-x)$ are plotted.

| $x$ | $A \overset{\triangle}{=} \text{erfc}(x)$ | $B \overset{\triangle}{=} \frac{\exp(-x^2)}{\sqrt{\pi}x}$ | $\frac{B-A}{A} \cdot 100$ |
|---|---|---|---|
| 3.00 | $0.221\text{E} - 04$ | $0.232\text{E} - 04$ | $0.506\text{E} + 01$ |
| 3.25 | $0.430\text{E} - 05$ | $0.449\text{E} - 05$ | $0.437\text{E} + 01$ |
| 3.50 | $0.743\text{E} - 06$ | $0.771\text{E} - 06$ | $0.380\text{E} + 01$ |
| 3.75 | $0.114\text{E} - 06$ | $0.118\text{E} - 06$ | $0.334\text{E} + 01$ |
| 4.00 | $0.154\text{E} - 07$ | $0.159\text{E} - 07$ | $0.295\text{E} + 01$ |
| 4.25 | $0.185\text{E} - 08$ | $0.190\text{E} - 08$ | $0.263\text{E} + 01$ |
| 4.50 | $0.197\text{E} - 09$ | $0.201\text{E} - 09$ | $0.236\text{E} + 01$ |
| 4.75 | $0.185\text{E} - 10$ | $0.189\text{E} - 10$ | $0.213\text{E} + 01$ |
| 5.00 | $0.154\text{E} - 11$ | $0.157\text{E} - 11$ | $0.193\text{E} + 01$ |

Table A.1: *Comparison between the values of* erfc(x) *and its approximation.*

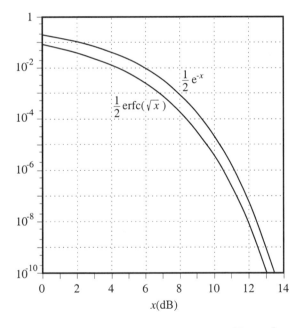

Figure A.1: *Graph of the functions* $1/2\text{erfc}(\sqrt{x})$ *and* $\frac{1}{2}e^{-x}$.

Using relation (A.3) and definitions (A.1) and (A.2), it is straightforward to compute the following probabilities related to the Gaussian random variable

(RV) $\xi$:

$$P\{\xi < x\} = \frac{1}{2}\left\{1 + \mathrm{erf}\left(\frac{x-\mu}{\sqrt{2}\sigma}\right)\right\} = \frac{1}{2}\mathrm{erfc}\left(\frac{\mu-x}{\sqrt{2}\sigma}\right) \qquad (A.6)$$

$$P\{\xi > x\} = \frac{1}{2}\left\{\mathrm{erfc}\left(\frac{x-\mu}{\sqrt{2}\sigma}\right)\right\} \qquad (A.7)$$

$$P\{|\xi| < x\} = \frac{1}{2}\left\{\mathrm{erf}\left(\frac{x-\mu}{\sqrt{2}\sigma}\right) + \mathrm{erf}\left(\frac{x+\mu}{\sqrt{2}\sigma}\right)\right\} \qquad (A.8)$$

$$P\{|\xi| > x\} = \frac{1}{2}\left\{\mathrm{erfc}\left(\frac{x-\mu}{\sqrt{2}\sigma}\right) + \mathrm{erfc}\left(\frac{x+\mu}{\sqrt{2}\sigma}\right)\right\} \qquad (A.9)$$

## A.2. The modified Bessel function $I_0$

Definition

$$I_0(x) \triangleq \frac{1}{\pi}\int_0^\pi e^{\pm x\cos\theta}\, d\theta = \frac{1}{\pi}\int_0^\pi \cosh(x\cos\theta)\, d\theta \qquad (A.10)$$

It admits the following asymptotic $(x \to \infty)$ expansion:

$$I_0(x) \sim \frac{e^x}{\sqrt{2\pi x}}\left\{1 + \frac{1}{8x} + \frac{9}{2!(8x)^2} + \frac{9\cdot 25}{3!(8x)^3} + \cdots\right\} \qquad (A.11)$$

## A.3. Marcum Q-function and related integrals

Definitions and basic properties

$$Q(a,b) \triangleq \int_b^\infty \exp\left(-\frac{a^2+b^2}{2}\right)I_0(ax)x\, dx \qquad (A.12)$$

$$Q(0,b) = \exp\left(-\frac{b^2}{2}\right) \qquad (A.13)$$

$$Q(a,0) = 1 \qquad (A.14)$$

Asymptotic expansion, valid for $b \gg 1$ and $b \gg b - a$

$$Q(a,b) \sim \frac{1}{2}\mathrm{erfc}\left(\frac{b-a}{\sqrt{2}}\right) \qquad (A.15)$$

Symmetry and antisymmetry relations

$$Q(a, b) + Q(b, a) = 1 + \exp\left(-\frac{a^2 + b^2}{2}\right) I_0(ab) \tag{A.16}$$

$$Q(a, a) = \frac{1}{2}\left[1 + e^{-a^2} I_0(a^2)\right] \tag{A.17}$$

$$1 + Q(a, b) - Q(b, a) = \frac{b^2 - a^2}{b^2 + a^2} \int_{(a^2+b^2)/2}^{\infty} e^{-y} I_0\left(\frac{2aby}{a^2 + b^2}\right) dy, \quad b > a > 0 \tag{A.18}$$

Asymptotic expansions, valid for $b \gg 1$, $a \gg 1$, and $b \gg b - a > 0$

$$1 + Q(a, b) - Q(b, a) \sim \text{erfc}\left(\frac{b - a}{\sqrt{2}}\right) \tag{A.19}$$

$$Q(a, b) + Q(b, a) - 1 \sim \frac{\exp[-(b - a)^2/2]}{\sqrt{2\pi ab}} \tag{A.20}$$

$$Q(a, b) \sim \frac{1}{2}\left\{\text{erfc}\left(\frac{b - a}{\sqrt{2}}\right) + \frac{\exp[-(b - a)^2/2]}{\sqrt{2\pi ab}}\right\} \tag{A.21}$$

## A.4.    Probability that one Rice-distributed RV exceeds another one

Given a pair $(\xi_1, \xi_2)$ of independent Rice-distributed RVs with pdf

$$f_{\xi_k}(x_k) = \frac{x_k}{\sigma_k^2} \exp\left(-\frac{a_k^2 + x_k^2}{2\sigma_k^2}\right) I_0\left(\frac{a_k x_k}{\sigma_k^2}\right), \quad a_k > 0, \quad x_k > 0, \quad k = 1, 2 \tag{A.22}$$

the probability $P\{\xi_2 > \xi_1\}$ that one exceeds the other can be expressed by one of the following equivalent forms:

$$P\{\xi_2 > \xi_1\} = Q(\sqrt{a}, \sqrt{b}) - \frac{\nu^2}{1 + \nu^2} \exp\left(-\frac{a + b}{2}\right) I_0(\sqrt{ab}) \tag{A.23}$$

$$P\{\xi_2 > \xi_1\} = \frac{\nu^2}{1 + \nu^2}[1 - Q(\sqrt{b}, \sqrt{a})] + \frac{1}{1 + \nu^2} Q(\sqrt{a}, \sqrt{b}) \tag{A.24}$$

$$P\{\xi_2 > \xi_1\} = \frac{1}{2}[1 - Q(\sqrt{b}, \sqrt{a}) + Q(\sqrt{a}, \sqrt{b})] - \tag{A.25}$$
$$\frac{1}{2}\frac{\nu^2 - 1}{\nu^2 + 1} \exp\left(-\frac{a + b}{2}\right) I_0(\sqrt{ab})$$

where

$$a \triangleq \frac{a_2^2}{\sigma_1^2 + \sigma_2^2}, \quad b \triangleq \frac{a_1^2}{\sigma_1^2 + \sigma_2^2}, \quad \nu \triangleq \frac{\sigma_1}{\sigma_2}$$

# Some facts from matrix theory

In this appendix we collect together, for ease of reference, some basic results from matrix theory that are needed throughout the book, and in particular in Chapter 8, where extensive use of matrix notations is made. We assume that the reader has had previous exposure to matrix calculus. Thus, we focus our attention mainly on the results that are specifically needed in the text, whereas more elementary material is either skipped or included for reference's sake only.

## B.1. Basic matrix operations

A real (complex) $N \times M$ matrix is a rectangular array of $NM$ real (complex) numbers, called its *entries*, or *elements*, arranged in $N$ rows and $M$ columns and indexed as follows:

$$\begin{bmatrix} a_{11} & a_{12} & \cdots & a_{1M} \\ a_{21} & a_{22} & \cdots & a_{2M} \\ & & \cdots & \\ a_{N1} & a_{N2} & \cdots & a_{NM} \end{bmatrix} \tag{B.1}$$

We write $\mathbf{A} = (a_{ij})$ as shorthand for the matrix (B.1). If $N = M$, $\mathbf{A}$ is called a *square matrix*; if $M = 1$, $\mathbf{A}$ is called a *column vector*, and if $N = 1$, $\mathbf{A}$ is called a *row vector*. We denote column vectors by using boldface lowercase letters, such as $\mathbf{x}$, $\mathbf{y}$, . . ..

Standard operations for matrices are the following:

**(a)** Multiplication of $\mathbf{A}$ by the real or complex number $c$. The result, denoted by $c\mathbf{A}$, is the $N \times M$ matrix with entries $ca_{ij}$.

**(b)** Sum of two $N \times M$ matrices $\mathbf{A} = (a_{ij})$ and $\mathbf{B} = (b_{ij})$. The result is the

$N \times M$ matrix $\mathbf{C}$ whose entries are $a_{ij} + b_{ij}$. The sum is commutative (i.e., $\mathbf{A} + \mathbf{B} = \mathbf{B} + \mathbf{A}$).

**(c)** Product of the $N \times K$ matrix $\mathbf{A}$ by the $K \times M$ matrix $\mathbf{B}$. The result is the $N \times M$ matrix $\mathbf{C}$ with entries

$$c_{ij} \stackrel{\triangle}{=} \sum_{k=1}^{K} a_{ik} b_{kj}, \qquad i = 1, \ldots, N, \quad j = 1, \ldots, M$$

The matrix product is not commutative (i.e., in general $\mathbf{AB} \neq \mathbf{BA}$), but it is associative (i.e., $\mathbf{A}(\mathbf{BC}) = (\mathbf{AB})\mathbf{C}$) and distributive with respect to the sum (i.e., $\mathbf{A}(\mathbf{B} + \mathbf{C}) = \mathbf{AB} + \mathbf{AC}$ and $(\mathbf{A} + \mathbf{B})\mathbf{C} = \mathbf{AC} + \mathbf{BC}$).

When $\mathbf{AB} = \mathbf{BA}$, the two matrices $\mathbf{A}$ and $\mathbf{B}$ are said to *commute*. The notation $\mathbf{A}^n$ is used to denote the $n$th power of a square matrix $\mathbf{A}$ (i.e., the product of $\mathbf{A}$ by itself performed $n - 1$ times); that is,

$$\mathbf{A}^n \stackrel{\triangle}{=} \underbrace{\mathbf{A} \cdot \mathbf{A} \cdots \mathbf{A}}_{n}$$

If we define the *identity matrix* $\mathbf{I} \stackrel{\triangle}{=} (\delta_{ij})$ as the square matrix all of whose elements are 0 unless $i = j$, in which case they are equal to 1, multiplication of any square matrix $\mathbf{A}$ by $\mathbf{I}$ gives $\mathbf{A}$ itself, and we can set

$$\mathbf{A}^0 = \mathbf{I}$$

**(d)** Given a square matrix $\mathbf{A}$, there may exist a matrix, which we denote by $\mathbf{A}^{-1}$, such that $\mathbf{A}\mathbf{A}^{-1} = \mathbf{A}^{-1}\mathbf{A} = \mathbf{I}$. If $\mathbf{A}^{-1}$ exists, it is called the *inverse* of $\mathbf{A}$, and $\mathbf{A}$ is said to be nonsingular. For $n$ a positive integer, we define a negative power of a nonsingular square matrix as follows:

$$\mathbf{A}^{-n} \stackrel{\triangle}{=} (\mathbf{A}^{-1})^n$$

We have:

$$(\mathbf{AB})^{-1} = \mathbf{B}^{-1}\mathbf{A}^{-1}$$

**(e)** The *transpose* of the $N \times M$ matrix $\mathbf{A}$ with entries $a_{ij}$ is the matrix with entries $a_{ji}$, which we denote by $\mathbf{A}'$. If $\mathbf{A}$ is a complex matrix, its *conjugate* $\mathbf{A}^*$ is the matrix with elements $a_{ij}^*$, and its conjugate transpose $\mathbf{A}^\dagger \stackrel{\triangle}{=} (\mathbf{A}')^*$ has entries $a_{ji}^*$. The following properties hold:

$$(\mathbf{AB})' = \mathbf{B}'\mathbf{A}', \qquad (\mathbf{AB})^\dagger = \mathbf{B}^\dagger\mathbf{A}^\dagger$$

The *scalar product* of two real column vectors $\mathbf{x}$, $\mathbf{y}$ is

$$\mathbf{x}'\mathbf{y} \triangleq \sum_{i=1}^{N} x_i y_i = \mathbf{y}'\mathbf{x} \tag{B.2}$$

If $\mathbf{x}$ and $\mathbf{y}$ are complex, their scalar product is defined as

$$\mathbf{x}^\dagger\mathbf{y} \triangleq \sum_{i=1}^{N} x_i^* y_i = (\mathbf{y}^\dagger\mathbf{x})^* \tag{B.3}$$

Two vectors are called *orthogonal* if their scalar product is zero.

## B.2.   Numbers associated with a matrix

**(a) The trace.** Given a square $N \times N$ matrix $A$, its *trace* (or *spur*) is the sum of the elements of the *main diagonal* of $\mathbf{A}$

$$\operatorname{tr}(\mathbf{A}) \triangleq \sum_{i=1}^{N} a_{ii} \tag{B.4}$$

The trace operation is linear; that is, for any two given numbers $\alpha$, $\beta$, and two square $N \times N$ matrices $\mathbf{A}$, $\mathbf{B}$, we have

$$\operatorname{tr}(\alpha\mathbf{A} + \beta\mathbf{B}) = \alpha\operatorname{tr}(\mathbf{A}) + \beta\operatorname{tr}(\mathbf{B}) \tag{B.5}$$

In general, $\operatorname{tr}(\mathbf{AB}) = \operatorname{tr}(\mathbf{BA})$ even if $\mathbf{AB} \neq \mathbf{BA}$. In particular, the following properties hold:

$$\operatorname{tr}(\mathbf{A}^{-1}\mathbf{BA}) = \operatorname{tr}(\mathbf{B}) \tag{B.6}$$

and, for any $N \times M$ matrix $\mathbf{A}$

$$\operatorname{tr}(\mathbf{A}\mathbf{A}^\dagger) = \sum_{i=1}^{N}\sum_{j=1}^{N} |a_{ij}|^2 \tag{B.7}$$

Notice also that for two column vectors $\mathbf{x}$, $\mathbf{y}$

$$\mathbf{x}^\dagger\mathbf{y} = \operatorname{tr}(\mathbf{x}\mathbf{y}^\dagger) \tag{B.8}$$

**(b) The determinant.** Given an $N \times N$ square matrix $\mathbf{A}$, its *determinant* is the number defined as the sum of the products of the elements in any row of $\mathbf{A}$ with their respective *cofactors* $\gamma_{ij}$

$$\det \mathbf{A} \triangleq \sum_{j=1}^{N} a_{ij}\gamma_{ij}, \qquad \text{for any } i = 1, 2, \ldots, N \tag{B.9}$$

The cofactor of $a_{ij}$ is defined as $\gamma_{ij} \triangleq (-1)^{i+j}m_{ij}$, where the *minor* $m_{ij}$ is the determinant of the $(N-1) \times (N-1)$ submatrix obtained from $\mathbf{A}$ by removing its $i$th row and $j$th column. The determinant has the following properties:

$$\det \mathbf{A} \;=\; 0 \text{ if one row of } \mathbf{A} \text{ is zero}$$
$$\text{or } \mathbf{A} \text{ has two equal rows} \qquad (\text{B.10})$$
$$\det \mathbf{A} \;=\; \det \mathbf{A}' \qquad (\text{B.11})$$
$$\det (\mathbf{AB}) \;=\; \det \mathbf{A} \cdot \det \mathbf{B} \qquad (\text{B.12})$$
$$\det (\mathbf{A}^{-1}) \;=\; (\det \mathbf{A})^{-1} \qquad (\text{B.13})$$
$$\det(c\mathbf{A}) \;=\; c^N \cdot \det \mathbf{A} \quad \text{for any number } c \qquad (\text{B.14})$$

A matrix is nonsingular if and only if its determinant is nonzero.

**(c) The eigenvalues.** Given an $N \times N$ square matrix $\mathbf{A}$ and a column vector $\underline{u}$ with $N$ entries, consider the set of $N$ linear equations

$$\mathbf{Au} = \lambda\mathbf{u} \qquad (\text{B.15})$$

where $\lambda$ is a constant and the entries of $\mathbf{u}$ are the unknown. There are only $N$ values of $\lambda$ (not necessarily distinct) such that (B.15) has a nonzero solution. These numbers are called the *eigenvalues* of $\mathbf{A}$, and the corresponding vectors $\mathbf{u}$ the *eigenvectors* associated with them. Note that if $\mathbf{u}$ is an eigenvector associated with the eigenvalue $\lambda$ then, for any complex number $c$, $c\mathbf{u}$ is also an eigenvector.

The polynomial $a(\lambda) \triangleq \det(\lambda\mathbf{I} - \mathbf{A})$ in the indeterminate $\lambda$ is called the *characteristic polynomial* of $\mathbf{A}$. The equation

$$\det (\lambda\mathbf{I} - \mathbf{A}) = 0 \qquad (\text{B.16})$$

is the *characteristic equation* of $\mathbf{A}$, and its roots are the eigenvalues of $\mathbf{A}$. The Cayley-Hamilton theorem states that every square $N \times N$ matrix $\mathbf{A}$ satisfies its characteristic equation. That is, if the characteristic polynomial of $\mathbf{A}$ is $a(\lambda) = \lambda^N + \alpha_1\lambda^{N-1} + \cdots + \alpha_N$, then

$$a(\mathbf{A}) \triangleq \mathbf{A}^N + \alpha_1\mathbf{A}^{N-1} + \cdots + \alpha_N\mathbf{I} = 0 \qquad (\text{B.17})$$

where $\mathbf{0}$ is the null matrix (i.e., the matrix all of whose elements are zero). The monic polynomial $\mu(\lambda)$ of lowest degree such that $\mu(\mathbf{A}) = 0$ is called the *minimal polynomial* of $\mathbf{A}$. If $f(x)$ is a polynomial in the indeterminate $x$, and $\mathbf{u}$ is an eigenvector of $\mathbf{A}$ associated with the eigenvalue $\lambda$, then

$$f(\mathbf{A})\mathbf{u} = f(\lambda)\mathbf{u} \qquad (\text{B.18})$$

That is, $f(\lambda)$ is an eigenvalue of $f(\mathbf{A})$ and $\mathbf{u}$ is the corresponding eigenvector. The eigenvalues $\lambda_1, \ldots, \lambda_N$ of the $N \times N$ matrix $\mathbf{A}$ have the properties

$$\det(\mathbf{A}) = \prod_{i=1}^{N} \lambda_i \tag{B.19}$$

and

$$\mathrm{tr}(\mathbf{A}) = \sum_{i=1}^{N} \lambda_i \tag{B.20}$$

From (B.19), it is immediately seen that $\mathbf{A}$ is nonsingular if and only if none of its eigenvalues is zero.

**(d) The spectral norm and the spectral radius.** Given an $N \times N$ matrix $\mathbf{A}$, its *spectral norm* $\|\mathbf{A}\|$ is the nonnegative number

$$\|\mathbf{A}\| \stackrel{\triangle}{=} \sup_{\mathbf{x} \neq 0} \frac{\|\mathbf{A}\mathbf{x}\|}{\|\mathbf{x}\|} \tag{B.21}$$

where $\mathbf{x}$ is an $N$-component column vector, and $\|\mathbf{u}\|$ denotes the Euclidean norm of the vector $\mathbf{u}$

$$\|\mathbf{u}\| \stackrel{\triangle}{=} \sqrt{\sum_{i=1}^{N} |u_i|^2} = \sqrt{\mathbf{u}^\dagger \mathbf{u}} \tag{B.22}$$

We have

$$\|\mathbf{A}\mathbf{B}\| \leq \|\mathbf{A}\| \cdot \|\mathbf{B}\| \tag{B.23}$$
$$\|\mathbf{A}\mathbf{x}\| \leq \|\mathbf{A}\| \cdot \|\mathbf{x}\| \tag{B.24}$$

for any matrix $\mathbf{B}$ and vector $\mathbf{x}$. If $\lambda_i$, $i = 1, \ldots, N$, denote the eigenvalues of $\mathbf{A}$, the radius $\rho(\mathbf{A})$ of the smallest disk centered at the origin of the complex plane that includes all these eigenvalues is called the *spectral radius* of $\mathbf{A}$

$$\rho(\mathbf{A}) \stackrel{\triangle}{=} \max_{1 \leq j \leq N} |\lambda_i| \tag{B.25}$$

In general, for an arbitrary complex $N \times N$ matrix $\mathbf{A}$, we have

$$\rho(\mathbf{A}) \leq \|\mathbf{A}\| \tag{B.26}$$

and

$$\|\mathbf{A}\| = \sqrt{\rho(\mathbf{A}^\dagger \mathbf{A})} \tag{B.27}$$

If $\mathbf{A} = \mathbf{A}^\dagger$, then

$$\rho(\mathbf{A}) = \|\mathbf{A}\| \tag{B.28}$$

**(e) Quadratic forms.** Given an $N \times N$ square matrix $\mathbf{A}$ and a column vector $\mathbf{x}$ with $N$ entries, we call a *quadratic form* the quantity

$$\mathbf{x}^\dagger \mathbf{A} \mathbf{x} = \sum_{i=1}^{N} \sum_{j=1}^{N} x_i^* a_{ij} x_j \tag{B.29}$$

## B.3.  Some classes of matrices

Let $\mathbf{A}$ be an $N \times N$ square matrix.

(a)  $\mathbf{A}$ is called *symmetric* if $\mathbf{A}' = \mathbf{A}$.

(b)  $\mathbf{A}$ is called *Hermitian* if $\mathbf{A}^\dagger = \mathbf{A}$.

(c)  $\mathbf{A}$ is called *orthogonal* if $\mathbf{A}^{-1} = \mathbf{A}'$.

(d)  $\mathbf{A}$ is called *unitary* if $\mathbf{A}^{-1} = \mathbf{A}^\dagger$.

(e)  $\mathbf{A}$ is called *diagonal* if its entries $a_{ij}$ are zero unless $i = j$. A useful notation for a diagonal matrix is

$$\mathbf{A} = \operatorname{diag}(a_{11}, a_{22}, \ldots, a_{NN})$$

(f)  $\mathbf{A}$ is called *scalar* if $\mathbf{A} = c\mathbf{I}$ for some constant $c$; that is, $\mathbf{A}$ is diagonal with equal entries on the main diagonal.

(g)  $\mathbf{A}$ is called a *Toeplitz* matrix if its entries $a_{ij}$ satisfy the condition

$$a_{ij} = a_{i-j} \tag{B.30}$$

That is, its elements on the same diagonal are equal.

(h)  $\mathbf{A}$ is called *circulant* if its rows are all the cyclic shifts of the first one

$$a_{ij} = a_{(i-j) \bmod N} \tag{B.31}$$

(i)  A symmetric real matrix $\mathbf{A}$ is called *positive (nonnegative) definite* if all its eigenvalues are positive (nonnegative). Equivalently, $\mathbf{A}$ is positive (nonnegative) definite if and only if for any nonzero column vector $\mathbf{x}$ the quadratic form $\mathbf{x}^\dagger \mathbf{A} \mathbf{x}$ is positive (nonnegative).

**Example B.1**    Let $A$ be Hermitian. Then the quadratic form $f \stackrel{\triangle}{=} x^\dagger A x$ is real. In fact

$$f^* = (x^\dagger A x)^* = x' A^* x^* = (A^* x^*)' x = x^\dagger A^\dagger x \qquad (B.32)$$

Since $A^\dagger = A$ this is equal to $x^\dagger A x = f$, which shows that $f$ is real.                                    □

**Example B.2**    Consider the random column vector $x = [x_1, x_2, \ldots, x_N]$, and its *correlation matrix*

$$R \stackrel{\triangle}{=} E[x x^\dagger] \qquad (B.33)$$

It is easily seen that $R$ is Hermitian. Also, $R$ is nonnegative definite; in fact, for any nonzero deterministic column vector $a$,

$$a^\dagger R a = a^\dagger E[x x^\dagger] a = E[a^\dagger x x^\dagger a] = E[|a^\dagger x|^2] \geq 0 \qquad (B.34)$$

with equality only if $a^\dagger x = 0$ almost surely; that is, the components of $x$ are *linearly dependent*.

If $x_1, \ldots, x_N$ are random variables taken from a wide-sense stationary discrete-time random process, and we define

$$r_{i-j} \stackrel{\triangle}{=} E[x_{n+i} x_{n+j}^*] \qquad (B.35)$$

it is seen that the entry of $R$ in the $i$th row and the $j$th column is precisely $r_{|i-j|}$. This shows in particular that $R$ is a Toeplitz matrix.

If $\mathcal{G}(f)$ denotes the discrete Fourier transform of the autocorrelation sequence $(r_n)$, that is $\mathcal{G}(f)$ is the power spectrum of the random process $(x_n)$ (see (2.81)), the following can be shown:

(a)  The eigenvalues $\lambda_1, \ldots, \lambda_N$ of $R$ are samples (not necessarily equidistant) of the function $\mathcal{G}(f)$.

(b)  For any function $\gamma(\cdot)$, we have the *Toeplitz distribution theorem* (Grenander and Szegö, 1958):

$$\lim_{N \to \infty} \frac{1}{N} \sum_{i=1}^{N} \gamma(\lambda_i) = \int_{-1/2}^{1/2} \gamma[\mathcal{G}(f)] \, df \qquad (B.36)$$

□

**Example B3**    Let $\mathbf{C}$ be a circulant $N \times N$ matrix of the form

$$\mathbf{C} = \begin{bmatrix} c_0 & c_1 & c_2 & \cdots & c_{N-1} \\ c_{N-1} & c_0 & c_1 & \cdots & c_{N-2} \\ & & \cdots & & \\ c_1 & c_2 & c_3 & \cdots & c_0 \end{bmatrix} \tag{B.37}$$

Let also $w \overset{\triangle}{=} e^{j2\pi/N}$, so that $w^N = 1$. Then the eigenvector associated with the eigenvalue $\lambda_i$ is

$$\mathbf{u}_i = [w^0 w^i w^{2i} \cdot w^{(N-1)i}]', \qquad i = 0, 1, \ldots, N - 1 \tag{B.38}$$

The eigenvalues of $\mathbf{C}$ are

$$\lambda_i = \sum_{m=0}^{N-1} c_m w^{mi}, \qquad i = 0, 1, \ldots, N - 1 \tag{B.39}$$

and $\lambda_i^*$ can be interpreted as the value of the Fourier transform of the sequence $c_0$, $c_1$, ..., $c_{N-1}$, taken at frequency $i/N$. $\qquad\square$

**Example B.4**    If $\mathbf{U}$ is a unitary $N \times N$ matrix, and $\mathbf{A}$ is an $N \times N$ arbitrary complex matrix, pre- or postmultiplication of $\mathbf{A}$ by $\mathbf{U}$ does not alter its spectral norm; that is,

$$\|\mathbf{AU}\| = \|\mathbf{UA}\| = \|\mathbf{A}\| \tag{B.40}$$

$\qquad\square$

## B.4.    Convergence of matrix sequences

Consider the sequence $(\mathbf{A}^n)_{n=0}^\infty$ of powers of the square matrix $\mathbf{A}$. As $n \to \infty$, for $\mathbf{A}^n$ to tend to the null matrix $\mathbf{0}$ it is necessary and sufficient that the spectral radius of $\mathbf{A}$ be less than 1. Also, as the spectral radius of $\mathbf{A}$ does not exceed its spectral norm, for $\mathbf{A}^n \to \mathbf{0}$ it is sufficient that $\|\mathbf{A}\| < 1$.

Consider now the matrix series

$$\mathbf{I} + \mathbf{A} + \mathbf{A}^2 + \cdots \mathbf{A}^n + \cdots \tag{B.41}$$

For this series to converge, it is necessary and sufficient that $\mathbf{A}^n \to \mathbf{0}$ as $n \to \infty$. If this holds, the sum of the series equals $(\mathbf{I} - \mathbf{A})^{-1}$.

## B.5.   The gradient vector

Let $f(\mathbf{x}) = f(x_1, \ldots, x_N)$ be a differentiable real function of $N$ real arguments. Its gradient vector, denoted by $\nabla f$, is the column vector whose $N$ entries are the derivatives $\partial f / \partial x_i$, $i = 1, \ldots, N$. If $x_1, \ldots, x_N$ are complex, that is

$$x_i = x_i' + j x_i'', \qquad i = 1, \ldots, N \qquad (\text{B.42})$$

the gradient of $f(\mathbf{x})$ is the vector whose components are

$$\frac{\partial f}{\partial x_i'} + j \frac{\partial f}{\partial x_i''}, \qquad i = 1, \ldots, N \qquad (\text{B.43})$$

**Example B.5**   If $\mathbf{a}$ denotes a complex column vector, and $f(\mathbf{x}) \triangleq \Re[\mathbf{a}^\dagger \mathbf{x}]$, we have

$$\nabla f(\mathbf{x}) = \mathbf{a} \qquad (\text{B.44})$$

$\square$

**Example B.6**   If $\mathbf{A}$ is a Hermitian $N \times N$ matrix, and $f(\mathbf{x}) \triangleq \mathbf{x}^\dagger \mathbf{A} \mathbf{x}$, we have

$$\nabla f(\mathbf{x}) = 2 \mathbf{A} \mathbf{x} \qquad (\text{B.45})$$

$\square$

## B.6.   The diagonal decomposition

Let $\mathbf{A}$ be a Hermitian $N \times N$ matrix with eigenvalues $\lambda_1, \ldots, \lambda_N$. Then $\mathbf{A}$ can be given the following representation:

$$\mathbf{A} = \mathbf{U} \boldsymbol{\Lambda} \mathbf{U}^{-1} \qquad (\text{B.46})$$

where $\boldsymbol{\Lambda} \triangleq \operatorname{diag}(\lambda_1, \ldots, \lambda_N)$, and $\mathbf{U}$ is a unitary matrix, so that $\mathbf{U}^{-1} = \mathbf{U}^\dagger$. From (B.46) it follows that

$$\mathbf{A} \mathbf{U} = \mathbf{U} \boldsymbol{\Lambda} \qquad (\text{B.47})$$

which shows that the $i$th column of $\mathbf{U}$ is the eigenvector of $\mathbf{A}$ corresponding to the eigenvalue $\lambda_i$. For any column vector $\mathbf{x}$, the following can be derived from (B.46):

$$\mathbf{x}^\dagger \mathbf{A} \mathbf{x} = \sum_{i=1}^{N} \lambda_i |y_i|^2 \qquad (\text{B.48})$$

where $y_1, \ldots, y_N$ are the components of the vector $\mathbf{y} \triangleq \mathbf{U}^\dagger \mathbf{x}$.

## B.7.    Bibliographical notes

There are many excellent books on matrix theory, and some of them are certainly well known to the reader. The books by Bellman (1968) and Gantmacher (1959) are encyclopedic treatments in which details can be found about any topic one may wish to study in more depth. A modern treatment of matrix theory, with emphasis on numerical computations, is provided by Golub and Van Loan (1996). The most complete reference about Toeplitz matrices is the book by Grenander and Szegö (1958). For a tutorial introductory treatment of Toeplitz matrices and a simple proof of the distribution theorem (B.36), the reader is referred to Gray (1971 and 1972). In Athans (1968) one can find a number of formulas about gradient vectors.

# Variational techniques and constrained optimization

In this appendix, we briefly list some of the optimization theory results used in the book. Our treatment is far from rigorous, because our aim is to describe a technique for constrained optimization rather than to provide a comprehensive development of the underlying theory. The reader interested in more details is referred to Luenberger (1969, pp. 171–190) from which our treatment is derived; alternatively, to Gelfand and Fomin (1963).

Let R be a function space (technically, it must be a *normed linear* space). Assume that a rule is provided assigning to each function $f \in$ R a complex number $\varphi[f]$. Then $\varphi$ is called a *functional* on R.

**Example C.1**  Let $f(x)$ be a continuous function defined on the interval $(a, b)$. We write $f \in C(a, b)$. Then

$$\varphi[f] \triangleq f(x_0), \qquad a \leq x_0 \leq b$$

$$\varphi[f] \triangleq \int_a^b w(x) f(x) \, dx, \qquad w \in C(a, b)$$

and

$$\varphi[f] \triangleq \int_a^b f^2(x) \, dx$$

are functionals on the space $C(a, b)$. ☐

If $\varphi$ is a functional on R, and $f, h \in$ R, the functional

$$\delta\varphi[f; h] \triangleq \frac{d}{d\alpha} \varphi[f + \alpha h]_{\alpha=0} \tag{C.1}$$

is called the Fréchet differential of $\varphi$. The concept of Fréchet differential provides a technique to find the maxima and minima of a functional. We have the following result:

A necessary condition for $\varphi[f]$ to achieve a maximum or minimum value for $f = f_0$ is that $\delta\varphi(f_0; h) = 0$ for all $h \in R$.

In many optimization problems the optimal function is required to satisfy certain constraints. We consider in particular the situation in which a functional $\varphi$ on R must be optimized under $n$ constraints given in the implicit form $\psi_1[f] = C_1, \psi_2[f] = C_2, \ldots, \psi_n[f] = C_n$, where $\psi_1, \ldots, \psi_n$ are functionals on R, and $C_1, \ldots, C_n$ are constants. We have the following result:

If $f_0 \in R$ gives a maximum or a minimum of $\varphi$ subject to the constraints $\psi_i[f] = C_i$, $1 \leq i \leq n$, and the $n$ functionals $\delta\psi_i[f_0; h]$ are linearly independent, then there are $n$ scalars $\lambda_1, \ldots, \lambda_n$ which make the Fréchet differential of

$$\varphi[f] + \sum_{i=1}^{n} \lambda_i \psi_i[f] \tag{C.2}$$

vanish at $f_0$.

This result provides a rule for finding constrained maxima and minima. The procedure is to form the functional (C.2), to compute its Fréchet differential, and to find the functions $f$ that make it vanish for every $h$. The values of the "Lagrange multipliers" $\lambda_1, \ldots, \lambda_n$ can be computed using the constraint equations. Whether the solutions correspond to maxima, or to minima, or neither, can be best determined by a close analysis of the problem at hand.

**Example C.2** We want to find the real function $f \in C(0,1)$ which minimizes the functional

$$\varphi[f] \overset{\triangle}{=} \int_0^1 \frac{1}{f^2(x)} w(x)\, dx$$

(where $w(x) \in C(0,1)$) subject to the constraint $\psi[f] = 1$, where

$$\psi[f] \overset{\triangle}{=} \int_0^1 f(x)\, dx$$

The Fréchet differential of $\varphi[f] + \lambda\psi[f]$ is

$$\frac{d}{d\alpha} \int_0^1 \left\{ \frac{w(x)}{[f(x) + \alpha h(x)]^2} + \lambda[f(x) + \alpha h(x)] \right\} dx \bigg|_{\alpha=0}$$

$$= \int_0^1 \left[ \frac{-2w(x)}{f^3(x)} + \lambda \right] h(x)\, dx \tag{C.3}$$

For this functional to be zero for any $h(x)$, we must have

$$f(x) = \left(\frac{2w(x)}{\lambda}\right)^{1/3}$$

If this result is inserted in the constraint expression, the value of $\lambda$ can be determined. $\square$

**Example C.3**   Find the complex function $f(x)$ that minimizes

$$\int_{-\infty}^{\infty} w(x)|f(x)|^2 \, dx$$

(where $w(x) > 0$ for $-\infty < x < \infty$) subject to the constraint

$$\int_{-\infty}^{\infty} \frac{1}{|f(x)|^2} \, dx = 1$$

The relevant Fréchet differential is

$$\frac{d}{d\alpha} \int_{-\infty}^{\infty} \left\{ w(x)[f(x) + \alpha h(x)]f^*(x) + \frac{\lambda}{[f(x) + \alpha h(x)]f^*(x)} \right\} dx \Bigg|_{\alpha=0}$$

$$= \int_{-\infty}^{\infty} \left\{ w(x)f^*(x) - \frac{\lambda}{f(x)|f(x)|^2} \right\} h(x) \, dx$$

and this is zero for any $h(x)$ provided that

$$|f(x)|^4 = \frac{\lambda}{w(x)}$$

$\square$

# D

# Transfer functions of directed graphs

The state diagram of a convolutional code such as that of Fig. 11.6 is a *directed graph*. To define a directed graph, we give a set V= $\{v_1, v_2, \dots, \}$ of vertices (the *states* of the encoder, when dealing with convolutional codes) and a subset E of ordered pairs of vertices from V, called *edges*. A graph can be represented by drawing a set of points corresponding to the vertices and a set of arrows corresponding to each edge of E, and connecting pairs of vertices or one vertex to itself. A *path* in a graph is a sequence of edges, and can be identified by the sequence of subsequent vertices included into it. In the study of convolutional codes, we are interested in the enumeration of all paths of a directed graph.

A simple directed graph is shown in Fig. D.1. There are three vertices, $v_1$, $v_2$ and $v_3$, and and four edges $(v_1, v_2), (v_1, v_3), (v_2, v_4)$, and $(v_2, v_2)$. In this graph there are infinitely many paths between $v_1$ and $v_3$, because of the loop at the

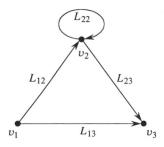

Figure D.1: *A directed graph.*

791

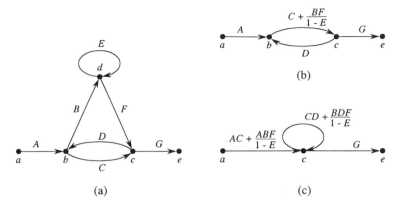

Figure D.2: *(a) Directed graph with five vertices and seven edges. (b) Reduced graph by removing vertex "d." (c) Reduced graph by further removing vertex "b."*

vertex $v_2$. One path of length 4, is, for instance, $v_1v_2v_2v_2v_3$. Each edge of a directed graph can be assigned a label. An important quantity, called the *transfer function* between a pair of vertices, is defined as the sum of the labels of all paths of any length connecting the two vertices. The label of a path is defined as the product of the labels of its edges. For example, considering again the directed graph of Fig. D.1, the label of the path $v_1v_2v_2v_2v_3$ is $L_{12}L_{22}^2L_{23}$. The transfer function $T$ between $v_1$ and $v_3$ is then given by

$$\begin{aligned} T(v_1, v_3) &= L_{13} + L_{12}L_{23} + L_{12}L_{22}L_{23} + L_{12}L_{22}^2L_{23} + \dots \\ &= L_{13} + L_{12}L_{23}(1 + L_{22} + L_{22}^2 + \dots) \qquad\qquad \text{(D.1)} \\ &= L_{13} + \frac{L_{12}L_{23}}{1 - L_{22}} \end{aligned}$$

Therefore, the graph of Fig. D.1 can be replaced by a new graph with a single equivalent edge joining the vertices $v_1$ and $v_3$ with label $L'_{13}$ given by (D.1).

Given a directed graph, it is thus possible to compute the transfer function between a pair of vertices by removing one by one the intermediate vertices on the graph and by redefining the new labels. As an example, consider the graph of Fig. D.2(a), with vertices $a, b, c, d, e$ and labels $A, B, C, ..., G$. Using the result (D.1), this graph can be replaced with that of Fig. D.2 (b), in which the vertex $d$ has been removed. By removing also the vertex $b$, we get the graph of Fig. D.2 (c), and finally the transfer function between $a$ and $e$

$$T(a, e) = \frac{ACG(1 - E) + ABFG}{1 - E - CD + CDE - BDF} \qquad\qquad \text{(D.2)}$$

When the number of vertices is large (say larger than 8), the reduction technique previously explained becomes too complex, and we must resort to a different technique for evaluating the transfer function. This technique is based on a matrix description of the graph, and can be very useful in computations since it lends itself to a software implementation. We will explain this technique with reference to the graph of Fig. D.2.

Let us define by $x_i$ the value of the accumulated path labels from the initial vertex $a$ to the vertex $i$, as influenced by all other vertices. Therefore, the state equations for the graph of Fig. D.2 (a) are

$$\begin{cases} x_b = A + Dx_c \\ x_c = Cx_b + Fx_d \\ x_d = Bx_b + Ex_d \\ x_e = Gx_c \end{cases} \tag{D.3}$$

With this approach, we obviously have $T(a, e) = x_e$, and therefore we can solve the system (D.3) and verify that $x_e$ is given again by (D.2).

The system of equations (D.3) can be given a more general and formal expression. Define the two column vectors

$$\mathbf{x}' \overset{\triangle}{=} (x_b, x_c, x_d, x_e) \qquad \mathbf{x}_0' \overset{\triangle}{=} (A, 0, 0, 0) \tag{D.4}$$

and the *state transition* matrix $\mathbf{T}$

$$\mathbf{T} \overset{\triangle}{=} \begin{bmatrix} 0 & D & 0 & 0 \\ C & 0 & F & 0 \\ B & 0 & E & 0 \\ 0 & G & 0 & 0 \end{bmatrix} \tag{D.5}$$

Using (D.4) and (D.5), the system (D.3) can be rewritten in matrix form as

$$\mathbf{x} = \mathbf{T}\mathbf{x} + \mathbf{x}_0 \tag{D.6}$$

The formal solution to this equation can be written as

$$\mathbf{x} = (\mathbf{I} - \mathbf{T})^{-1}\mathbf{x}_0 \tag{D.7}$$

or as the matrix power series

$$\mathbf{x} = (\mathbf{I} + \mathbf{T} + \mathbf{T}^2 + \cdots)\mathbf{x}_0 \tag{D.8}$$

Notice that the solution in terms of a power series is very useful when considering the state diagram as being described by a walk into a trellis (see Chapters 11 and 12). Each successive multiplication by $\mathbf{T}$ corresponds to one further step into the trellis.

When the number of states is small, the matrix inversion (D.7) is also useful to get directly the result in a closed form similar to (D.2).

# Approximate computation of averages

The aim of this appendix is to describe some techniques for the evaluation of bounds or numerical approximations to the average $E[g(\xi)]$, where $g(\cdot)$ is an explicitly known deterministic function, and $\xi$ is some random variable (RV) whose probability density function is not known explicitly, or is highly complex and hence very difficult to compute exactly. It is assumed that a certain amount of knowledge is available about $\xi$, expressed by a finite and usually small set of its moments. An efficient recursive algorithm to compute those moments for the important special case of $\xi$ being a sum of independent RVs will be presented in the next section. Also, we shall assume that the range of $\xi$ lies in the interval $[a, b]$, where both $a$ and $b$ are finite unless otherwise stated. The techniques described hereafter are not intended to exhaust the set of possible methods for solving this problem. However, they are general enough to handle a large class of situations and are computationally efficient in terms of speed and accuracy. Also, instead of providing a single technique, we describe several, as we advocate that the specific problem handled should determine the technique best suited to it from the viewpoint of computational effort required, accuracy, and applicability.

## E.1. Computation of the moments of a RV

As this is the most relevant case for the applications considered in this book (see Section 7.2), we assume that the RV $\xi$ is the sum of $N$ independent RVs $\xi_n$

$$\xi = \sum_{n=1}^{N} \xi_n \tag{E.1}$$

whise first moments are known. For example, in the case of $M$-PAM modulation, discussed in Section 7.2, $\xi_n = h_n a_n$, where $h_n$ is one of the $N$ significant samples of the overall system impulse response, and $a_n$ are the information-bearing RVs, assuming $M$ equally likely and equally spaced values as in Table 7.1.

A recursive method to compute the moments of $\xi$, which is particularly suited for computer implementation is described step by step in the following.

**Step 1** Compute the moments of the individual RVs $\xi_n$. We have

$$
\begin{aligned}
\gamma_k^{(n)} &\triangleq \mathrm{E}\{\xi_n^k\} = h_n^k \mathrm{E}\{a_n^k\} \\
&= h_n^k \frac{d^k}{2^k M} \sum_{i=1}^{M} (2i - M - 1)^k, \quad n = 1, \ldots, N, \quad k = 0, \ldots, K
\end{aligned}
$$

**Step m** Define the partial sums

$$
\eta_m \triangleq \sum_{n=1}^{m} \xi_n, \qquad m = 2, \ldots, N - 1
$$

with $\eta_N = \xi$, and compute recursively, for each $k$, the moments

$$
\lambda_k^{(m)} \triangleq \mathrm{E}\{\eta_m^k\}
$$

through

$$
\begin{aligned}
\lambda_k^{(m+1)} &= \mathrm{E}\{\eta_{m+1}^k\} = \mathrm{E}\{(\eta_m + \xi_{m+1})^k\} \\
&= \sum_{j=0}^{k} \binom{k}{j} \mathrm{E}\{\eta_m^j\} \mathrm{E}\{\xi_{m+1}^{k-j}\} = \sum_{j=0}^{k} \binom{k}{j} \lambda_j^{(m)} \gamma_{k-j}^{(m+1)}
\end{aligned}
$$

where the independence between $\eta_m$ and $\xi_{m+1}$ has been exploited.

**Step N** When $m = N$ stop the procedure, since

$$
\lambda_k^{(N)} = \mathrm{E}\{\eta_N^k\} = \mathrm{E}\{\xi^k\}
$$

The method for computing the moments, although described with reference to $M$-PAM, is fairly general, since it only requires $\xi$ to be written as a sum of independent RVs. When this is not possible, as for instance in systems employing codes that correlate the symbols in the sequence $(a_n)$, a more elaborate approach must be used to compute the error probability (see Cariolaro and Pupolin, 1975).

## E.2.   Series expansion technique

In this section, we shall assume that the function $g(x)$ is analytic at point $x = x_0$. Hence, we can represent it in a neighborhood of $x_0$ by using the Taylor's series expansion

$$g(x) = g(x_0) + (x - x_0)g'(x_0) + (x - x_0)^2 \frac{g''(x_0)}{2!} + \cdots + (x - x_0)^n \frac{g^{(n)}(x_0)}{n!} + \cdots$$

$$\text{(E.2)}$$

If the radius of convergence of (E.2) is large enough to include the range of the RV $\xi$, and we define

$$c_n \stackrel{\triangle}{=} E[\xi - x_0]^n \tag{E.3}$$

then, averaging termwise the Taylor's series expansion of $g(\xi)$, we get from (E.2) and (E.3)

$$E[g(\xi)] = g(x_0) + c_1 g'(x_0) + c_2 \frac{g''(x_0)}{2!} + \cdots + c_n \frac{g^{(n)}(x_0)}{n!} + \cdots \tag{E.4}$$

It can be seen from (E.4) that $E[g(\xi)]$ can be evaluated on the basis of the knowledge of the sequence of moments $(c_n)_{n=1}^{\infty}$, provided that the series converges. In particular, an approximate value of $E[g(\xi)]$ based on a finite number of moments can be obtained by truncating (E.4)

$$E[g(\xi)] \simeq E_N[g(\xi)] \stackrel{\triangle}{=} \sum_{k=0}^{N} c_k \frac{g^{(k)}(x_0)}{k!} \tag{E.5}$$

The error of this approximation is

$$E[g(\xi)] - E_N[g(\xi)] = \sum_{k=N+1}^{\infty} c_k \frac{g^{(k)}(x_0)}{k!} \tag{E.6}$$

$$= \frac{1}{(N + 1)!} E\{(\xi - x_0)^{N+1} f^{(N+1)}[x_0 + \theta(\xi - x_0)]\}$$

where $0 \le \theta \le 1$. Depending on the specific application characteristics, either one of the second and third terms in (E.6) can be used to obtain a bound on the truncation error. In any case, bounds on the values of the derivatives of $g(\cdot)$ and of the moments (E.3) must be available.

As an example of application of this technique, let us consider the function

$$g(x) = \mathrm{erfc}\left(\frac{h + x}{\sqrt{2}\sigma}\right) \tag{E.7}$$

where $h$ and $\sigma$ are known parameters. This function can be expanded in a Taylor's series in the neighborhood of the origin by observing that (Abramowitz and Stegun, 1972, p. 298)

$$\frac{d^k}{dz^k}\mathrm{erfc}(z) = (-1)^k \frac{2}{\sqrt{\pi}} H_{k-1}(z) e^{-z^2} \tag{E.8}$$

where $H_{k-1}(\cdot)$ is the Hermite polynomial of degree $(k-1)$ (Abramowitz and Stegun, 1972, pp. 773–787). Thus,

$$g^{(k)}(0) = (-1)^k \frac{2}{\sqrt{\pi}(\sqrt{2}\sigma)^k} H_{k-1}\left(\frac{h}{\sqrt{2}\sigma}\right) \exp\left(-\frac{h^2}{2\sigma^2}\right) \tag{E.9}$$

and we finally get

$$\mathrm{E}[g(\xi)] = \mathrm{erfc}\left(\frac{h}{\sqrt{2}\sigma}\right) + \frac{2}{\sqrt{\pi}} \exp\left(-\frac{h^2}{2\sigma^2}\right) \cdot \sum_{k=1}^{\infty} \frac{(-1)^k \mu_k}{(\sqrt{2}\sigma)^k k!} H_{k-1}\left(\frac{h}{\sqrt{2}\sigma}\right) \tag{E.10}$$

where

$$\mu_k \triangleq \mathrm{E}[\xi^k], \qquad k = 1, 2, \ldots \tag{E.11}$$

are the central moments of the RV $\xi$.

The proof that the series (E.10) is convergent, as well as an upper bound on the truncation error, can be obtained by using the following bound on the value of Hermite polynomials (Abramowitz and Stegun, 1972, p. 787):

$$|H_n(z)| \le \beta 2^{n/2} \sqrt{n!} \exp\left(\frac{z^2}{2}\right), \qquad \beta \simeq 1.086435, \quad n = 1, 2, \ldots \tag{E.12}$$

and the following bound on central moments:

$$|\mu_{k+s}| \le \mathrm{E}[|\xi|^k] \chi^s, \quad k \ge 0, \ s \ge 0 \tag{E.13}$$

where $\chi$ denotes the maximum value taken by $|\xi|$. The bound (E.13) can be easily derived under the assumption that $\xi$ is bounded. Using (E.12) and (E.13), we get the inequality (Prabhu, 1971)

$$|R_N[g]| \triangleq |\mathrm{E}[g(\xi)] - \mathrm{E}_N[g(\xi)]| \tag{E.14}$$

$$\le \sqrt{\frac{\sqrt{2}}{\pi} \frac{\beta \mathrm{E}|\xi^N|}{\sigma^{N+1}} \frac{\chi \exp[-h^2/(4\sigma^2)]}{(N+1)\sqrt{N!}} \left[1 - \frac{\chi}{\sigma\sqrt{N+1}}\right]^{-1}}$$

which holds provided that

$$\left(\frac{\chi}{\sigma}\right)^2 < N + 1 \tag{E.15}$$

The condition (E.15) can always be met for finite values of $\chi$, provided that $N$ is sufficiently large.

A case of special interest arises when $\xi$ is a symmetric RV, so that its odd-order central moments are zero

$$\mu_{2k-1} = 0, \qquad k = 1, 2, \ldots$$

In this case, (E.10) specializes to

$$E[g(\xi)] = \text{erfc}\left(\frac{h}{\sqrt{2}\sigma}\right) + \frac{2}{\sqrt{\pi}} \exp\left(-\frac{h^2}{2\sigma^2}\right) \cdot \sum_{k=1}^{\infty} \frac{\mu_{2k}}{(\sqrt{2}\sigma)^{2k}(2k)!} H_{2k-1}\left(\frac{h}{\sqrt{2}\sigma}\right)$$

(E.16)

and, using the inequality (Abramowitz and Stegun, 1972, p. 787)

$$|H_{2n+1}(z)| \leq |z| \exp\left(\frac{z^2}{2}\right) \frac{(2n+2)!}{(n+1)!}, \qquad n = 0, 1, \ldots$$

(E.17)

the following bound can be derived (Prabhu, 1971):

$$|R_{2N}[g]| \triangleq |E[g(\xi)] - E_{2N}[g(\xi)]|$$

(E.18)

$$\leq \frac{|h|}{2\sqrt{2\pi}\sigma} \exp\left(-\frac{h^2}{4\sigma^2}\right) \frac{\mu_{2N}}{\sigma^{2N+2}} \frac{\chi^2}{(N+1)!} \cdot \left[1 - \frac{\chi^2}{\sigma^2(N+2)}\right]^{-1}$$

under the constraint

$$\left(\frac{\chi}{\sigma}\right)^2 < N + 2$$

(E.19)

It can be seen from (E.14) and (E.18) that the truncation error can be made vanishingly small by taking $N$, the number of terms retained for the computation, sufficiently large. Thus, at first it may seem that the average $E[g(\xi)]$ can be approximated with as great an accuracy as desired. But, in practice, roundoff errors may make it impossible to add up too many terms in the series and still retain a satisfactory accuracy. Notice in particular that in (E.10), and even in (E.16), the terms of the series do not have the same sign. Also, in practice, it is virtually impossible to compute a very large set of moments. The inaccuracies in the computations (or in measurements) of the moments increase with their order, so the process of adding more and more terms to the series cannot be extended very far as the computed values of the series become unreliable.

The Taylor's series technique presented so far can be modified by considering a different series expansion for the function $g(\cdot)$. One may want to consider, in lieu of (E.2), a series of the form

$$g(x) = \sum_{n=0}^{\infty} \alpha_n P_n(x), \qquad x \in I$$

(E.20)

where $P_0(x), P_1(x), \ldots$ form a sequence of polynomials orthonormal in the interval I. Consequently, the coefficients of the series expansion are given by

$$\alpha_n = \int_I g(x) P_n(x) dx \tag{E.21}$$

If I includes the range of the RV $\xi$, by averaging (E.20) termwise we obtain

$$E[g(\xi)] = \sum_{n=0}^{\infty} \alpha_n E[P_n(\xi)] \tag{E.22}$$

which can in turn be approximated by a finite sum. The computation of (E.22) requires the knowledge of the "generalized moments" $E[P_n(\xi)]$, which can be obtained, for example, as finite linear combinations of the central moments (E.11). This variant of the first technique may lead to a better convergence of the series.

## E.3.   Quadrature approximations

In this section we shall describe an approximation technique for $E[g(\xi)]$ based on the observation that this average can be formally expressed as the integral

$$E[g(\xi)] = \int_a^b g(x) f_\xi(x) \, dx \tag{E.23}$$

where $f_\xi(\cdot)$ denotes the probability density function (pdf) of the RV $\xi$. Having ascertained that the problem of evaluating $E[g(\xi)]$ is indeed equivalent to the computation of an integral, we can resort to the numerical techniques developed to compute approximate values of integrals of the form (E.23). The most widely investigated techniques for approximating a definite integral lead to the formula

$$\int_a^b g(x) f_\xi(x) \, dx \simeq \sum_{i=1}^{N} w_i g(x_i) \tag{E.24}$$

i.e., a linear combination of values of the function $g(\cdot)$. The $x_i$, $i = 1, 2, \ldots, N$, are called the *abscissas* (or *points* or *nodes*) of the formula, and the $w_i$, $i = 1, 2, \ldots, N$, are called its *weights* (or *coefficients*). The set of abscissas and weights is usually referred to as a *quadrature rule*. A systematic introduction to the theory of quadrature rules of the form (E.24) is given in Krylov (1962).

   The quadrature rule is chosen to render (E.24) as accurate as possible. A first difficulty with this theory arises when one wants to define how to measure the accuracy of a quadrature rule. Since we want the abscissas and weights to be independent of $g(\cdot)$, and hence be the same for all possible such functions, the definition of what is meant by "accuracy" must be made independent of the particular choice of $g(\cdot)$. The classical approach here is to select a number of *probe functions* and constrain the quadrature rule to be exact for these functions. By choosing $g(\cdot)$ to be a polynomial, it is said that the quadrature rule (E.24) has *degree of precision* $\nu$ if it is exact whenever $g(\cdot)$ is a polynomial of degree $\leq \nu$ (or, equivalently, whenever $g(x) = 1, x, \ldots, x^\nu$) and it is not exact for $g(x) = x^{\nu+1}$.

Once a criterion of goodness for quadrature rules has been defined, the next step is to investigate which are the best quadrature rules and how they can be computed. The answer is provided by the following result from numerical analysis, slightly reformulated to fit our framework (see Krylov, 1962, for more details and a proof):

Given a random variable $\xi$ with range $[a, b]$ and all of whose moments exist, it is always possible to define a sequence of polynomials $P_0(x), P_1(x), \ldots$, with deg $P_i(x) = i$, that are orthonormal with respect to $\xi$; that is,

$$\mathrm{E}[P_n(\xi)P_m(\xi)] = \delta_{mn}, \qquad m, n = 0, 1, \ldots \qquad (\text{E.25})$$

Denote by $x_1 < x_2 < \cdots < x_N$ the $N$ roots of the polynomial $P_N(x)$ (they are all real, and lie inside $[a, b]$), and by $k_n$ the coefficient of $x_n$ in the polynomial $P_n(X)$, $n = 0, 1, \ldots$. By defining

$$w_i \triangleq -\frac{k_{N+1}}{k_N} \frac{1}{P_{N+1}(x_i)P'_N(x_i)}, \quad i = 1, 2, \ldots, N \qquad (\text{E.26})$$

the set $\{x_i, w_i\}_{i=1}^N$ is a quadrature rule with degree of precision $2N - 1$. This is the highest degree of precision that can be attained by any quadrature rule with $N$ weights and abscissas.

These quadrature rules are usually called *Gauss quadrature rules* because they were first studied by Gauss. He considered the special case $f_\xi(x) = $ constant.

If $\{x_i, w_i\}_{i=1}^N$ is a Gauss quadrature rule, the error involved in the approximate integration of the function $g(\cdot)$, that is, the difference between the two sides of (E.24), is given by

$$R_N[g] = \frac{1}{(2N)!k_N^2} g^{(2N)}(\eta) \qquad (\text{E.27})$$

provided that $g(\cdot)$ has in $[a, b]$ a continuous derivative of order $2N$; $\eta$ is a point in $[a, b]$.

**Example E.1**  For a number of probability density functions $f_\xi(\cdot)$, results concerning Gauss quadrature rules are available in tabular form. For example, if $\xi$ is a Gaussian RV with mean $\mu$ and variance $\sigma^2$, the corresponding Gauss quadrature rule is

$$E[g(\xi)] \simeq \frac{1}{\sqrt{\pi}} \sum_{i=1}^{N} w_i g(\sqrt{2}\sigma x_i + \mu) \qquad \text{(E.28)}$$

where the abscissas $x_i$, $i = 1, \ldots, N$ are the zeros of the $N$th-degree Hermite polynomial. The actual values of $w_i$ and $x_i$, for various values of $N$, can be found for instance in Abramowitz and Stegun (1972, p. 924). The error is given by

$$R_N[g] = \frac{N! \sigma^{2N}}{(2N)!} g^{(2N)}(\eta), \qquad -\infty < \eta < \infty \qquad \text{(E.29)}$$

It can be seen from this example that the range of the random variable $\xi$ need not be finite for the application of a Gauss quadrature rule to the approximation of $E[g(\xi)]$. □

### E.3.1. Computation of Gauss quadrature rules

In the example just shown, the orthogonal polynomials relevant to the computation of the Gauss quadrature rules were well-known classical polynomials. In most instances, however, the pdf of the RV $\xi$ does not give rise to a polynomial set available in tabular form. In these situations, the Gauss quadrature rule must be computed. The relevant fact here is that the set $\{x_i, w_i\}_{i=1}^{N}$ can be evaluated on the basis of the moments $\mu_1, \mu_2, \ldots, \mu_{2N-1}$ of $\xi$. In other words, only the first $2N - 1$ moments of $\xi$ are needed to determine explicitly a Gauss quadrature rule with $N$ weights and abscissas.[1]

To see how to undertake this, we use the fact that a Gauss quadrature rule with $N$ weights and abscissas has degree of precision $2N - 1$. Since

$$\mu_k = \int_a^b x^k f_\xi(x) \, dx \qquad \text{(E.30)}$$

for any $0 \le k \le 2N - 1$ the Gauss quadrature rule is exact, that is,

$$\mu_k = \sum_{i=1}^{N} w_i x_i^k, \qquad k = 0, 1, \ldots, 2N - 1 \qquad \text{(E.31)}$$

This system of $2N$ nonlinear equations has the weights and abscissas as unknowns; by solving it, the Gauss quadrature rule can be found.

---

[1] It is often claimed that the moment $\mu_{2N}$ is also needed to perform this task; see for instance Golub and Welsch (1969) and Benedetto *et al.* (1973). Actually, the role of $\mu_{2N}$ is just that of normalizing the polynomial $P_N(\cdot)$, and its values affect neither the abscissas nor the weights of the quadrature rule (Gautschi, 1970).

In general, it is not convenient to solve directly equations (E.31), except for a very few simple cases. A computationally effective technique to determine Gauss quadrature rules based on the moments of $\xi$ has been proposed in Golub and Welsch (1969). The reader is referred to the original paper for details about the computational algorithms. Here it suffices to say that Golub and Welsch's technique consists essentially of two steps:

**Step 1** Evaluation of the coefficients $\{\alpha_n\}_{n=1}^N$, $\{\beta_n\}_{n=1}^N$ of the three-term recurrence relationship satisfied by the polynomials $P_0(x), \ldots, P_N(x)$ orthogonal with respect to the random variable $\xi$

$$\beta_n P_n(x) = (x - \alpha_n) P_{n-1}(x) - \beta_{n-1} P_{n-2}(x), \quad n = 1, 2, \ldots, N \quad \text{(E.32)}$$

with the initial conditions

$$P_{-1}(x) \equiv 0, \quad P_0(x) \equiv 1 \quad\quad\quad \text{(E.33)}$$

**Step 2** Generation of a symmetric tridiagonal matrix whose entries depend on $\{\alpha_n\}_{n=1}^N$ and $\{\beta_n\}_{n=1}^N$. The weights $\{w_n\}_{n=1}^N$ are found as the first components of the normalized eigenvectors of this matrix, whereas the abscissas $\{x_n\}_{n=1}^N$ are the corresponding eigenvalues. The computations can be performed on the basis of the knowledge of the moments $\mu_1, \mu_2, \ldots, \mu_{2N-1}$. First, the Gram matrix $\mathbf{M}$ of the moments is formed. It has entries

$$(\mathbf{M})_{ij} = \mu_{i+j}, \quad i, j = 0, 1, \ldots, N$$

Notice that $\mathbf{M}$ also includes the moment $\mu_{2N}$ as an element. Anyhow, as discussed before, the exact value of $\mu_{2N}$ is irrelevant. Thus, if $\mu_{2N}$ is unknown, any value for $\mu_{2N}$ will suffice, provided that $\mathbf{M}$ is positive definite. Let $\mathbf{M} = \mathbf{\Gamma}\mathbf{\Gamma}'$ be the Cholesky decomposition of $\mathbf{M}$, where $\mathbf{\Gamma}$ is a lower triangular matrix with positive diagonal entries (see, e.g., Golub and Van Loan, 1983, p. 88). The elements of $\mathbf{\Gamma}$, which can be computed using standard recursive formulas, provide the coefficients of the recursion (E.32).

### E.3.2.    Round-off errors in Gauss quadrature rules

In principle, if a sufficiently large number of moments of $\xi$ is available, the error term (E.27) can be made as small as desired by increasing the number $N$ of abscissas and weights of the quadrature rule. In fact, as $N \to \infty$, the RHS of (E.24) converges to the value of the LHS "for almost any conceivable function" $f_\xi(\cdot)$ "which one meets in practice" (Stroud and Secrest, 1966, p. 13). However, this is not true in practice, essentially because the moments of $\xi$ needed in the computation are not known with infinite accuracy. Computational experience shows that the Cholesky decomposition of the moment matrix $\mathbf{M}$ is the crucial step of the algorithm for the computation of Gauss quadrature rules, since $\mathbf{M}$ gets increasingly ill-conditioned with increasing $N$. Roundoff errors may cause the computed $\mathbf{M}$ to be no longer positive definite. Thus its Cholesky decomposition cannot be performed because it implies taking the square root of negative numbers (Luvison and Pirani, 1979). In practice, values of $N$ greater than 10 can rarely be achieved; the accuracy thus obtained is, however, satisfactory in most situations.

## E.4. Moment bounds

We have seen in the preceding section that the quadrature rule approach allows $\mathrm{E}[g(\xi)]$ to be approximated in the form of a linear combination of values of $g(\cdot)$. This is equivalent to substituting, for the actual probability density function $f_\xi(x)$, a discrete density in the form

$$f_\xi(x) \simeq \sum_{i=1}^{N} w_i \delta(x - x_i) \tag{E.34}$$

where $\{x_i, w_i\}_{i=1}^{N}$ are chosen so as to match the first $2N - 1$ moments of $\xi$ according to (E.31).

A more refined approach can be taken by looking for upper and lower bounds to $\mathrm{E}[g(\xi)]$, still based on the moments of $\xi$. In particular, we can set the goal of finding bounds to $\mathrm{E}[g(\xi)]$ that are in some sense *optimum* (i.e., they cannot be further tightened with the available informations on $\xi$). The problem can be formulated as follows: given a random variable $\xi$ with range in the finite interval $[a, b]$, whose first $M$ moments $\mu_1, \ldots, \mu_M$ are known, we want to find the sharpest upper and lower bounds to the integral

$$\mathrm{E}[g(\xi)] = \int_a^b g(x) f_\xi(x) dx \tag{E.35}$$

where $g(\cdot)$ is a known function and $f_\xi(\cdot)$ is the (unknown) pdf of the RV $\xi$. To solve this problem, we look at the set of all possible $f_\xi(\cdot)$ whose range is $[a, b]$ and whose first $M$ moments are $\mu_1, \ldots, \mu_M$. Then we compute the maximum and minimum value of (E.35) as $f_\xi(\cdot)$ runs through that set. The bounds obtained are optimum, because it is certain that a pair of random variables exists, say $\xi'$ and $\xi''$, with range in $[a, b]$ and meeting the lower and the upper bound, respectively, with the equality sign.

This extremal problem can be solved by using a set of results due essentially to the Russian mathematician M. G. Krein (see Krein and Nudel'man, 1977). These results can be summarized as follows.

**(a)** If the function $g(\cdot)$ has a continuous $(M+3)$th derivative, and $g^{(M+3)}(\cdot)$ is everywhere nonnegative in $[a, b]$, then the optimum bounds to $E[g(\xi)]$ are in the form

$$\sum_{i=1}^{N'} w_i' g(x_i') \leq E[g(\xi)] \leq \sum_{i=1}^{N''} w_i'' g(x_i'') \tag{E.36}$$

This is equivalent to saying that the two "extremal" pdfs are discrete, which allows the upper and lower bounds to be written in the form of quadrature rules. If $g^{(M+3)}(\cdot)$ is nonpositive, instead of being nonnegative, it suffices to consider $-g(\cdot)$ instead of $g(\cdot)$.

**(b)** *If $M$ is odd*, then $N' = (M+1)/2$ and $N'' = (M+3)/2$. Also, $\{x_i', w_i'\}_{i=1}^{N'}$ is a Gauss quadrature rule, and $\{x_i'', w_i''\}_{i=1}^{N''}$ is the quadrature rule having the maximum degree of precision (i.e., $2N''+1$) under the constraints $x_1'' = a$, $x_{N''}'' = b$. *If $M$ is even*, then $N' = N'' = (M+2)/2$. Also, $\{x_i', w_i'\}_{i=1}^{N'}$ (and, respectively, $\{x_i'', w_i''\}_{i=1}^{N''}$) is the quadrature rule having the maximum achievable degree of precision (i.e., $2N'$) under the constraint $x_i' = a$ (and, respectively, $x_{N''}'' = b$).

A technical condition involved with the derivation of these results requires that the Gram matrix of the moments $\mu_1, \ldots, \mu_M$ be positive definite. For our purpose, a simple sufficient condition is that the cumulative distribution function of $\xi$ has more than $M+1$ points of increase. If $\xi$ is a continuous RV, this condition is immediately satisfied. If $\xi$ is a discrete RV, it means that $\xi$ must take on more than $M + 1$ values. As $M + 1$ is generally a small number, the latter requirement is always satisfied in practice; otherwise, the value of $E[g(\xi)]$ can be evaluated explicitly, with no need to bound it.

### E.4.1.   Computation of moment bounds

Once the moments $\mu_1, \ldots, \mu_M$ have been computed, in order to use Krein's results explicitly, the quadrature rules $\{w_i', x_i'\}_{i=1}^{N'}$ and $\{w_i'', x_i''\}_{i=1}^{N''}$ must be evaluated. From the preceding discussion, it will not be surprising that the algorithms for computing the moment bounds (E.36) bear a close resemblance to those developed for computing Gauss quadrature rules. Indeed, the task is still to find abscissas and weights of a quadrature rule achieving the maximum degree of precision, possibly under constraints about the location of one or two abscissas. Several algorithms are available for this computation (Yao and Biglieri, 1980; Omura and Simon, 1980).

The first approach (Yao and Biglieri, 1980), is based on the assumption that Golub and Welsch's algorithm described in Section E.2 has been implemented. In particular, the known moments of $\xi$ must be modified before being used as inputs to that algorithm, and, also, the weights and abscissas obtained as the outcomes of the algorithm must be modified. These modifications can be found in Yao and Biglieri (1980), and the interested reader is referred to this reference.

## E.5. Approximating the averages depending on two random variables

Before ending this appendix, we shall briefly discuss the problem of evaluating approximations of the average $E[g(\xi, \eta)]$, where $g(\cdot, \cdot)$ is a known deterministic function, and $\xi, \eta$ are two (possibly correlated) RVs with range in a region $\mathcal{R}$ of the plane. Exact computation of this average requires knowledge of the joint pdf $f_{\xi,\eta}(x, y)$ of the pair of RVs $\xi, \eta$. This may not be available, or the evaluation of the double integral

$$E[g(\xi, \eta)] = \int \int_{\mathcal{R}} g(x, y) f_{\xi,\eta}(x, y) \, dx \, dy \tag{E.37}$$

may be unfeasible. In practice, it is often exceedingly easier to compute a small number of joint moments

$$\mu_{\ell,m} \triangleq E[\xi^\ell \eta^m], \qquad \ell, m = 0, 1, \ldots, M \tag{E.38}$$

and use this information to obtain $E[g(\xi, \eta)]$.

The first technique that can be used to this purpose is based on the expansion of $g(\xi, \eta)$ in a Taylor's series. The terms of this series will involve products like $\xi^\ell \eta^m$, so that truncating the series and averaging it termwise will provide the desired approximation.

Another possible technique uses *cubature rules*, a two-dimensional generalization of quadrature rules discussed in Section E.3. With this approach, the approximation of $E[g(\xi, \eta)]$ takes the form

$$E[g(\xi, \eta)] = \sum_{i=1}^{N} w_i g(x_i, y_i) \tag{E.39}$$

As a generalization of the one-dimensional case, we say that the cubature rule $\{w_i, x_i, y_i\}_{i=1}^{N}$ has degree of precision $\nu$ if (E.39) holds with the equality sign whenever $g(x, y)$ is a polynomial in $x$ and $y$ of degree $\leq \nu$, but not for all polynomials of degree $\nu + 1$. Unfortunately, construction of cubature rules with a maximum degree of precision is, generally, an unsolved problem, and solutions are only available in some special cases. For example, in Mysovskih (1968) a cubature rule with degree of precision 4 and $N = 6$ was derived. This is valid when the region $\mathcal{R}$ and the function $f_{\xi,\eta}(\cdot,\cdot)$ are symmetric relative to both coordinate axes. Thus, we must have

$$f_{\xi,\eta}(x,y) = f_{\xi,\eta}(-x,y) = f_{\xi,\eta}(x,-y), \qquad (x,y) \in \mathcal{R} \qquad \text{(E.40)}$$

With these assumptions, $\mu_{i,k} = 0$ if at least one of the numbers $i$ and $k$ is odd. The moments needed for the computation of the cubature rule are then $\mu_{2,0}$, $\mu_{0,2}$, $\mu_{4,0}$, $\mu_{0,4}$, and $\mu_{2,2}$. Under.the same symmetry assumptions, a cubature rule with $N = 19$ and degree of precision 9 can be obtained by using the moments $\mu_{2,0}$, $\mu_{0,2}$, $\mu_{4,0}$, $\mu_{0,4}$, $\mu_{2,2}$, $\mu_{2,4}$, $\mu_{4,2}$, $\mu_{0,6}$, $\mu_{6,0}$, $\mu_{2,6}$. $\mu_{6,2}$, $\mu_{8,0}$, $\mu_{0,8}$, and $\mu_{4,4}$.

If a higher degree of precision is sought or the symmetry requirements are not satisfied, one can resort to "good" cubature rules that can be computed through the joint moments (E.38). Formulas of the type

$$E[g(\xi,\eta)] = \sum_{i=1}^{N_x} \sum_{j=1}^{N_y} w_{ij} g(x_i, y_j) \qquad \text{(E.41)}$$

with degree of precision $\nu = \min (N_x - 1, N_y - 1)$ can be found by using the moments $\mu_{h,0}$, $\mu_{0,k}$, $h = 1,\ldots,2N_x$, $k = 1,\ldots,2N_y$, and $\mu_{h,k}$, $h = 1,\ldots,N_x - 1$, $k = 1,\ldots,N_y - 1$. Equivalent algorithms for the computation of weights and abscissas in (E.41) were derived in Luvison and Navino (1976) and Omura and Simon (1980).

We conclude by commenting briefly on the important special case in which the two random variables $\xi$ and $\eta$ are independent. In this situation, by using moments of $\xi$ one can construct the Gauss quadrature rule $\{w_i, x_i\}_{i=1}^{N_x}$, and by using moments of $\eta$ one can similarly obtain the Gauss quadrature rule $\{u_j, y_j\}_{j=1}^{N_y}$. Then it is a simple matter to show that the following cubature rule can be obtained:

$$E[g(\xi,\eta)] = \sum_{i=1}^{N_x} \sum_{j=1}^{N_y} w_i u_j g(x_i, y_j) \qquad \text{(E.42)}$$

and this has degree of precision $\nu = \min (2N_x - 1, 2N_y - 1)$.

# Viterbi algorithm

## F.1. Introduction

The Viterbi algorithm (VA) was originally proposed in 1967 for decoding convolutional codes. Shortly after its discovery, it was observed that the VA was based on the principles of *dynamic programming*, a general technique for solving extremum (that is, maximization or minimization) problems.

Our application of the VA consists of finding, among the paths traversing a *trellis* from left to right, the one with the maximum or minimum *metric*. Specifically, we define a trellis as a diagram representing all the allowable trajectories of a Markov chain with $N_\sigma$ states $\{S_i\}_{i=1}^{N_\sigma}$ from time $k = 0$ to time $k = K$. The trellis begin and ends at two known states, and there is a one-to-one correspondence between the sequences of $K + 1$ states and the paths through the trellis. Fig. F.1 shows an example of a four-state trellis with $K = 6$. A *branch metric* is associated with each branch (or edge) of the trellis, in the form of a label. The branch metrics are additive, i.e., the metric associated with a pair of adjoining branches is the sum of the two metrics. Consequently, the total metric associated with a path traversing the whole trellis from left to right is the sum of the labels of the branches forming the path. The problem here is to find the path traversing the trellis with the maximum (or minimum) total metric (the choice between maximum or minimum depends of course on the problem being solved). Formally, if $\sigma_k$ denotes the state at time $k$, taking values $\{S_i\}_{i=1}^{N_\sigma}$, and $m(\sigma_k, \sigma_{k+1})$ denotes the metric associated with the branch emanating from node $\sigma_k$ and joining node $\sigma_{k+1}$, we want to maximize (or minimize) the function

$$\lambda(\sigma_0, \sigma_1, \ldots, \sigma_K) \triangleq \sum_{k=0}^{K-1} m(\sigma_k, \sigma_{k+1}) \tag{F.1}$$

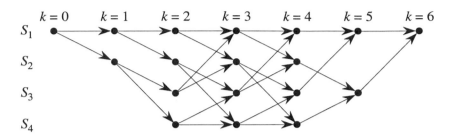

Figure F.1: Trellis of a four-state Markov process.

over all the possible choices of the state sequences $(\sigma_0, \sigma_1, \ldots, \sigma_K)$ compatible with the trellis structure.

Clearly, the problem above could be solved in principle by a brute-force approach, consisting of evaluating all the possible values of the function $\lambda(\,\cdot\,)$ in (F.1), and choosing the largest (or the smallest). However, this algorithm would suffer from two main drawbacks, viz.,

**Complexity:** The number of computations required, and the storage needed, grow exponentially with the length $K$ of the sequence.

**Delay:** If the branch labels are computed sequentially from time $k = 0$ to time $k = K$ (as it occurs in the applications considered in this book), then the decision on the best path must be deferred until the whole sequence of labels is computed, which entails a delay $K$.

As we shall see, the Viterbi algorithm solves the maximization problem without suffering from exponential complexity: actually, its computational complexity (and storage requirements) grow only *linearly* with $K$. Moreover, the *truncated* version of the VA has a delay which may be much smaller than $K$, at the price of a minor loss of optimality.

We start our description of the VA with the illustration of its key step, commonly called ACS (for Add, Compare, and Select). Consider Fig. F.2, where (and from now on) we shall consider a *maximum* problem. It shows the trellis states at time $k$ (denoted $\sigma_k$) and at time $k + 1$ (denoted $\sigma_{k+1}$). The branches joining pairs of paths are labeled by the corresponding branch metrics, while the states $\sigma_k$ are labeled by the *accumulated state metrics*, to be defined soon. The ACS step consists of the following: For each state $\sigma_{k+1}$, examine the branches

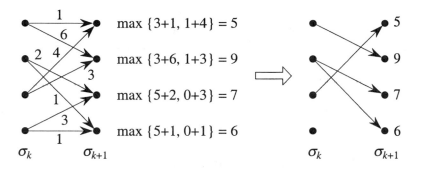

Figure F.2: The ACS step of Viterbi algorithm.

stemming from states $\sigma_k$ and leading to it (there are two such branches in Fig. F.2). For these branches, ADD the metric accumulated at the state from which it stems to the metric of the branch itself. Then COMPARE the results of these sums, and SELECT the branch associated with the maximum value (and consequently discard, for each state, all the other branches entering it; and if two or more of the quantities being compared are equal, choose one at random). The maximum value is associated with the state, and forms its accumulated metric. This value is retained only for the next ACS step and is then discarded.

The VA consists of repeating the ACS step from the starting state to the ending state of the trellis. After each ACS step the VA retains, for each state, one value of accumulated metric and one path, usually called the *survivor* corresponding to the state. Thus, at any time $k$ we are left, for each $\sigma_k$, with a single survivor path traversing the trellis from the initial state to $\sigma_k$, and with one value of accumulated metric. This survivor path is the maximum-metric path to the corresponding state. After $K$ ACS steps, at the termination of the trellis we obtain a single $K$-branch path and a single accumulated metric. These are the maximum-metric path and the maximum-metric value, respectively. Fig. F.3 illustrates the determination of a maximum-metric path through a four-state trellis via the VA.

To prove the optimality of the VA it suffices to observe the following. Assume that the optimum path passes through state $S_i$ (say) at time $k$. Then, *its first $k$ branches must be the same as for the survivor corresponding to $S_i$*. In fact, if they did not, the optimum path would begin with a path passing through $S_i$ and having a metric lower than the survivor of $S_i$, which is a contradiction. In other words, no path discarded in favor of a survivor can provide a contribution to the total metric larger than the survivors.

The computational complexity of the VA is the same at each time instant (we disregard initial and final transients). Hence, it grows only linearly with $K$. More specifically, the VA requires $N_\sigma$ storage locations, one for each state, with each location storing an accumulated metric and a surviving path. In terms of the number of computations, at each time instant the VA must make $Q$ additions, where $Q$ is the number of transitions in a trellis section (for example, $Q = 8$ in Fig. F.1), and $N_\sigma$ comparisons. Thus, the amount of storage is proportional to the number of states, and the amount of computation per time unit is proportional to the number of transitions.

### F.1.1.  The truncated Viterbi algorithm

The VA as described above leaves the delay problem unsolved. In fact, the algorithm cannot reach a decision about the maximum-metric sequence before time $K$. On the other hand, it is obvious that a decision about the best sequence cannot be reached before scanning all the states from $k = 0$ to $k = K$, so that reducing the delay would necessarily entail a loss of optimality of the algorithm.

When the delay of $K$ time instants cannot be tolerated, the *truncated* Viterbi algorithm may be used. This consists of forcing decisions at stage $k$ on all paths prior to stage $k - D$, for some *truncation depth* (or *decision depth*) $D$. The approach consists of comparing the partial path metrics for the paths at stage $k$, and noting which one is the largest. The branch chosen at this time is the one belonging to this path at time $k - D$. Thus, after a latency of $D$ time instants, the truncated VA outputs one branch at a time. Doing this entails also a reduction in the storage needed, since only the last $D$ branches of the survivor paths must be kept in memory. The loss of optimality is reduced when $D$ is increased, because when $D$ is large there is a high probability that all the surviving paths leading to any node have an initial part in common: so this initial path will be a part of the optimum one, and we say that a *merge* has taken place.

### F.1.2.  An example of application

Here we describe a simple example of application of the Viterbi algorithm to a decoding problem. Assume that a symbol sequence x, consisting of $K$ equally likely binary symbols taking values 0 or 1, is transmitted over a memoryless channel. Assume also that the symbols, rather than being independent, are correlated with each other, and that their correlation can be described in the form of a trellis as defined before. Specifically, all the admissible symbol sequences are in one-to-one correspondence with the the paths traversing the trellis from $k = 0$ to $k = K$, with one symbol associated with each branch. This occurs for example when the symbol sequence can be thought of as the output of a finite-state machine driven by an independent, identically distributed sequence of random variables, so that the sequence of states forms a Markov chain.

Let $y_k$ denote the components of the received signal sequence $\mathbf{y}$, and $p(y_k \mid x_k)$ the probability density function of the received samples given that $x_k$ was transmitted. Maximum-likelihood detection of the transmitted sequence consists of maximizing the conditional pdf

$$p(\mathbf{y} \mid \mathbf{x}) = \prod_{k=1}^{K} p(y_k \mid x_k) \qquad (F.2)$$

over all the admissible sequences $\mathbf{x}$. Here the assumption of a memoryless channel has been used to factorize the pdf.

By taking the logarithm of (F.2), we obtain the additive form

$$\ln p(\mathbf{y} \mid \mathbf{x}) = \sum_{k=1}^{K} \ln p(y_k \mid x_k) \qquad (F.3)$$

We can then use $\ln p(y_k \mid x_k)$ as the metric that labels the trellis branches associated at time $k$ with the symbol $x_k$ when the observed channel output is $y_k$. Maximization of the sum (F.3) leads to choosing the most likely sequence of transmitted symbols. As a special case, for the additive white Gaussian noise channel the above leads to a problem equivalent to the minimization of a Euclidean distance, or, for equal-energy signals, to the maximization of a scalar product.

## F.2.    Maximum a posteriori detection. The BCJR algorithm

It is known (Section 2.6) that maximum-likelihood detection minimizes the probability that the whole detected sequence be in error. Assume instead that in the example of Section F.1.2 we are interested in minimizing the symbol error probability for the detected symbols (motivation for this choice is provided in Section 11.3). To do this, for each $k$ we should choose the value of $x_k$ that leads to the greater between the two quantities (a posteriori probabilities) $P(x_k = 0 \mid \mathbf{y})$ and $P(x_k = 1 \mid \mathbf{y})$. This is tantamount to comparing with a unit threshold the a posteriori probability ratio

$$\Lambda_k = \frac{P(x_k = 1 \mid \mathbf{y})}{P(x_k = 0 \mid \mathbf{y})} \qquad (F.4)$$

Now, observe that the transmitted symbol $x_k$ is associated with one or more branches of the trellis stage at time $k$, and that each one of these branches can be characterized by the pair of states, say $(\sigma_k, \sigma_{k+1})$, that it joins. Thus, we can write

$$\Lambda_k = \frac{\displaystyle\sum_{(\sigma_k,\sigma_{k+1}):x_k=1} p(\mathbf{y}, \sigma_k, \sigma_{k+1})}{\displaystyle\sum_{(\sigma_k,\sigma_{k+1}):x_k=0} p(\mathbf{y}, \sigma_k, \sigma_{k+1})} \tag{F.5}$$

where the two summations are over those pairs of states for which $x_k = 1$ and $x_k = 0$, respectively, and the conditional probabilities of (F.4) are replaced by joint probabilities after using Bayes' rule and cancelling out the pdf of $\mathbf{y}$, common to numerator and denominator.

We proceed now to the computation of the pdf $p(\mathbf{y}, \sigma_k, \sigma_{k+1})$. By defining $\mathbf{y}_k^-$, the components of the received vector before time $k$, and $\mathbf{y}_k^+$, the components of the received vector after time $k$, we can write

$$\mathbf{y} = (\mathbf{y}_k^-, y_k, \mathbf{y}_k^+)$$

and consequently

$$
\begin{aligned}
p(\mathbf{y}, \sigma_k, \sigma_{k+1}) &= p(\mathbf{y}_k^-, y_k, \mathbf{y}_k^+, \sigma_k, \sigma_{k+1}) \\
&= p(\mathbf{y}_k^-, y_k, \sigma_k, \sigma_{k+1})\, p(\mathbf{y}_k^+ \mid \mathbf{y}_k^-, y_k, \sigma_k, \sigma_{k+1}) \\
&= p(\mathbf{y}_k^-, \sigma_k)\, p(y_k, \sigma_{k+1} \mid \mathbf{y}_k^-, \sigma_k) p(\mathbf{y}_k^+ \mid \mathbf{y}_k^-, y_k, \sigma_k, \sigma_{k+1})
\end{aligned}
$$

Now, observe that due to the dependences among observed variables and trellis states, reflected by the trellis structure or, equivalently, by the Markov-chain property of the trellis states, $\mathbf{y}_k^+$ depends on $\sigma_k, \sigma_{k+1}, \mathbf{y}_k^-$, and $y_k$ only through $\sigma_{k+1}$, and, similarly, the pair $y_k, \sigma_{k+1}$ depends on $\sigma_k, \mathbf{y}_k^-$ only through $\sigma_k$. Thus, by defining the functions

$$\alpha_k(\sigma_k) \triangleq p(\mathbf{y}_k^-, \sigma_k) \tag{F.6}$$

$$\beta_{k+1}(\sigma_{k+1}) \triangleq p(\mathbf{y}_k^+ \mid \sigma_{k+1}) \tag{F.7}$$

$$\gamma_{k,k+1}(\sigma_k, \sigma_{k+1}) \triangleq p(y_k, \sigma_{k+1} \mid \sigma_k) = p(y_k \mid \sigma_k, \sigma_{k+1})\, p(\sigma_{k+1} \mid \sigma_k) \tag{F.8}$$

we may write

$$p(\mathbf{y}, \sigma_k, \sigma_{k+1}) = \alpha_k(\sigma_k)\, \gamma_{k,k+1}(\sigma_k, \sigma_{k+1})\, \beta_{k+1}(\sigma_{k+1}) \tag{F.9}$$

In conclusion, the a posteriori probability ratio (F.5) can be rewritten in the form

$$\Lambda_k = \frac{\displaystyle\sum_{\sigma_k,\sigma_{k+1}:x_k=1} \alpha_k(\sigma_k)\gamma_{k,k+1}(\sigma_k, \sigma_{k+1})\beta_{k+1}(\sigma_{k+1})}{\displaystyle\sum_{\sigma_k,\sigma_{k+1}:x_k=0} \alpha_k(\sigma_k)\, \gamma_{k,k+1}(\sigma_k, \sigma_{k+1})\, \beta_{k+1}(\sigma_{k+1})} \tag{F.10}$$

To complete our calculations, we now describe how the functions $\alpha_k(\sigma_k)$ and $\beta_{k+1}(\sigma_{k+1})$ can be evaluated recursively. We note the forward recursion

$$
\begin{aligned}
\alpha_{k+1}(\sigma_{k+1}) &= p(\mathbf{y}_{k+1}^-, \sigma_{k+1}) \\
&= p(\mathbf{y}_k^-, y_k, \sigma_{k+1}) \\
&= \sum_{\sigma_k} p(\mathbf{y}_k^-, y_k, \sigma_k, \sigma_{k+1}) \\
&= \sum_{\sigma_k} p(\mathbf{y}_k^-, \sigma_k) p(y_k, \sigma_{k+1} \mid \sigma_k) \\
&= \sum_{\sigma_k} \alpha_k(\sigma_k)\, \gamma_{k,k+1}(\sigma_k, \sigma_{k+1})
\end{aligned}
$$

with the initial condition $\alpha_0(s_1) = 1$ ($s_1$ denotes the initial state of the trellis). Similarly, we note the backward recursion

$$
\begin{aligned}
\beta_k(\sigma_k) &= p(\mathbf{y}_{k-1}^+ \mid \sigma_k) \\
&= \sum_{\sigma_{k+1}} p(y_k, \mathbf{y}_k^+, \sigma_{k+1} \mid \sigma_k) \\
&= \sum_{\sigma_{k+1}} p(y_k, \sigma_{k+1} \mid \sigma_k) p(\mathbf{y}_k^+ \mid \sigma_{k+1}) \\
&= \sum_{\sigma_{k+1}} \gamma_{k,k+1}(\sigma_k, \sigma_{k+1})\, \beta_{k+1}(\sigma_{k+1})
\end{aligned}
$$

with the final value $\beta_K(s_K) = 1$.

The combination of the latter two recursions with (F.10) forms the BCJR algorithm, named after the authors who first derived it (Bahl, Cocke, Jelinek, and Raviv, 1974). This algorithm, that was derived here with the aim of maximizing the a posteriori probabilities of the symbols, is used in Section 11.3 to the purpose of computing a posteriori probabilities.

Roughly speaking, we can state that the complexity of the BCJR algorithm is about three times that of Viterbi algorithm. A truncated version of this algorithm (the "sliding-window" algorithm) is described in Section 11.3.

## F.3.   Bibliographical notes

The Viterbi algorithm was proposed by Viterbi (1967) as a method for decoding convolutional codes (see Chapter 10). Since then, it has been applied to a variety of maximization problems arising in demodulation of digital signals generated by a modulator with memory or in sequence estimation for channels with intersymbol interference. A survey of applications of the Viterbi algorithm, as well as a number of details regarding its implementation, can be found in Forney (1973).

The connections between the Viterbi algorithm and dynamic programming techniques were first recognized by Omura (1969).

The maximum a posteriori decoding algorithm, now commonly referred to as the BCJR algorithm, was proposed in (Bahl, Cocke, Jelinek, and Raviv, 1974). Until recently it received limited attention because of its increase in complexity over the Viterbi algorithm, which yields very close results in terms of bit error probability. Recently the interest in the BCJR algorithm was rekindled by the discovery of the class of "turbo" codes (Berrou, Glavieux, and Thitimajshima, 1993) described in Chapter 11.

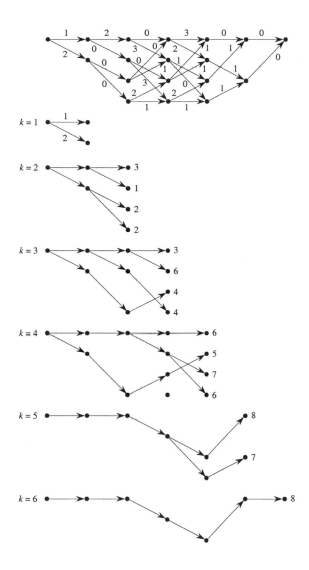

Figure F.3: Determination of the maximum-metric path through a trellis with $K = 6$ and four states via the Viterbi algorithm.

# References

M. R. Aaron and D. W. Tufts, 1966. "Intersymbol interference and error probability," *IEEE Trans. Inform. Theory*, Vol. IT-12, pp. 24–36.

K. Abend and B. D. Fritchman, 1970. "Statistical detection for communication channels with intersymbol interference," *IEEE Proceedings*, Vol. 58, pp. 779–785.

M. Abramowitz and I. L. Stegun, 1972. *Handbook of Mathematical Functions*. New York: Dover.

F. Amoroso, 1980. "The bandwidth of digital signals," *IEEE Communications Magazine*, Vol. 18, pp. 13–24.

F. Amoroso and J. A. Kivett, 1977. "Simplified MSK signaling technique," *IEEE Trans. Commun.*, Vol. 25, pp. 433–441.

J. B. Anderson, T. Aulin, and C.-E. W. Sundberg, 1986. *Digital Phase Modulation*. New York: Plenum Press.

J. B. Anderson and S. Mohan, 1991. *Source and Channel Coding: An Algorithmic Approach*. Boston, MA: Kluwer Academic Press.

R. R. Anderson and G. J. Foschini, 1975. "The minimum distance for MLSE digital data systems of limited complexity," *IEEE Trans. Inform. Theory*, Vol. IT-21, pp. 544–551.

R. Arens, 1957. "Complex processes for envelopes of normal noise," *IEEE Trans. Inform. Theory*, Vol. IT-3, pp. 203–207.

J. Arsac, 1966. *Fourier Transforms and the Theory of Distributions,* translated by A. Nussbaum and G. C. Heim. Englewood Cliffs, NJ: Prentice-Hall.

E. Arthurs and H. Dym, 1962. "On the optimum detection of digital signals in the presence of white Gaussian noise. A geometric interpretation and a study of three basic data transmission systems," *IRE Trans. Communication Systems*, Vol. CS-10, pp. 336–372.

R. B. Ash, 1967. *Information Theory*. New York: Wiley-Interscience.

M. Athans, 1968. "The matrix minimum principle," *Information and Control*, Vol. 11, pp. 592–606.

T. Aulin and C.-E. W. Sundberg, 1981. "Continuous-phase modulation—Part I: Full response signaling," *IEEE Trans. Commun.*, Vol. COM-29, pp. 196–209.

M. Austin, 1967. *Decision-Feedback Equalization for Digital Communication Over Dispersive Channels*, MIT Lincoln Lab., Lexington, MA, Technical Report, August.

L. R. Bahl, J. Cocke, F. Jelinek, and J. Raviv, 1974. "Optimal decoding of linear codes for minimizing symbol error rate," *IEEE Trans. Inform. Theory*, Vol. IT-20, pp. 284–287.

H. A. Barker and S. Ambati, 1972. "Nonlinear sampled-data system analysis by multidimensional $z$-transforms," *IEE Proceedings*, Vol. 119, pp. 1407–1413.

C. T. Beare, 1978. "The choice of the desired impulse response in combined linear Viterbi algorithm equalizers," *IEEE Trans. Commun.*, Vol. COM-26, pp. 1301–1307.

E. Bedrosian, 1962. "The analytic signal representation of modulated waveforms," *IRE Proceedings*, Vol. 50, pp. 2071–2076.

E. Bedrosian and S. O. Rice, 1971. "The output properties of Volterra systems driven by harmonic and Gaussian inputs," *IEEE Proceedings*, Vol. 59, pp. 1688–1707.

C. Belfiore and J. H. Park, Jr., 1979. "Decision feedback equalization," *IEEE Proceedings*, Vol. 67, pp. 1143–1156.

T. C. Bell, J. G. Cleary, and I. H. Whitten, 1990. *Text Compression*. Englewood Cliffs, NJ: Prentice-Hall.

R. E. Bellman, 1968. *Matrix Analysis*, 2nd ed. New York: McGraw-Hill.

S. Benedetto and E. Biglieri, 1974. "On linear receivers for digital transmission systems," *IEEE Trans. Commun.*, Vol. COM-22, pp. 1205–1215.

S. Benedetto, E. Biglieri, and R. Daffara, 1976. "Performance of multilevel baseband digital systems in a nonlinear environment," *IEEE Trans. Commun.*, Vol. COM-24, pp. 1166–1175.

S. Benedetto, E. Biglieri, and R. Daffara, 1979. "Modeling and performance evaluation of nonlinear satellite link: A Volterra series approach," *IEEE on Aerospace and Elec. Sys.*, Vol. AES-15, pp. 494–507.

S. Benedetto, E. Biglieri, and J. K. Omura, 1981. "Optimum receivers for nonlinear satellite channels," *5th International Conference on Digital Satellite Communications*, Genova, Italy.

S. Benedetto and E. Biglieri, 1983. "Nonlinear equalization of digital satellite channels," *IEEE Journal on Selected Areas in Commun.*, Vol. SAC-1, pp. 57–62.

S. Benedetto, G. De Vincentiis, and A. Luvison, 1973. "Error probability in the presence of intersymbol interference and additive noise for multilevel digital signals," *IEEE Trans. Commun.*, Vol. COM-21, pp. 181–188.

S. Benedetto, D. Divsalar, G. Montorsi, and F. Pollara, 1998b. "Serial concatenation of interleaved codes: Performance analysis, design, and iterative decoding," *IEEE Trans. Inform. Theory*, Vol. 44, pp. 909–926, May.

S. Benedetto, D. Divsalar, G. Montorsi, and F. Pollara, 1998c. "Soft-input soft-output modules for the construction and distributed iterative decoding of code networks," *European Trans. Telecommun.*, Vol. 9, pp. 155–172.

S. Benedetto, D. Divsalar and J. Hagenauer (eds.), 1998d. "Concatenated coding techniques and iterative decoding: Sailing toward channel capacity," *IEEE Journal on Selected Areas in Communications*, Vol. 16, no. 2, the whole issue.

S. Benedetto, R. Garello, and G. Montorsi, 1997. "The trellis complexity of turbo codes," *IEEE Communication Theory Mini Conference - Globecom '97*, Phoenix, AZ.

S. Benedetto, R. Garello, and G. Montorsi, 1998a. "A search for good convolutional codes to be used in the construction of turbo codes," *IEEE Trans. Commun.*, to be published.

S. Benedetto, M. Mondin, and G. Montorsi, 1994. "Performance evaluation of trellis-coded modulation schemes," *IEEE Proceedings*, Vol. 82, pp. 833–855.

S. Benedetto and G. Montorsi, 1996a. "Unveiling turbo codes: Some results on parallel concatenated coding schemes," *IEEE Trans. Inform. Theory*, Vol. COM-44, pp. 591–600.

S. Benedetto and G. Montorsi, 1996b. "Design of parallel concatenated convolutional codes," *IEEE Trans. Commun*, Vol. IT-43, pp. 409–428.

V. Benedetto and E. Biglieri, 1993. "Computing error probabilities of multi-level modulation codes," *Annales des Télécommunications*, Vol. 48, pp. 378–383.

A. Benveniste and M. Goursat, 1984. "Blind equalizers," *IEEE Trans. Commun.*, Vol. COM-32, pp. 871–883.

T. Berger, 1971. *Rate Distortion Theory*. Englewood Cliffs, NJ: Prentice-Hall.

T. Berger and D. W. Tufts, 1967. "Optimum pulse amplitude modulation. Part I: Transmitter-receiver design and bounds from information theory," *IEEE Trans. Inform. Theory*, Vol. IT-13, pp. 196–208.

E. R. Berlekamp, 1968. *Algebraic Coding Theory*. New York: McGraw-Hill.

E. R. Berlekamp, ed., 1974. *Key Papers in the Development of Coding Theory*. New York: IEEE Press.

C. Berrou and A. Glavieux, 1996. "Near optimum error-correcting coding and decdoding: Turbo codes," *IEEE Trans. Commun.*, Vol. COM-44, pp. 1261–1271.

C. Berrou, A. Glavieux, and P. Thitimajshima, 1993. "Near Shannon limit error-correcting codes: Turbo codes," *IEEE International Conference on Communications (ICC'93)*, Geneva, Switzerland, pp. 1064–1070.

E. Biglieri, 1984. "High-level modulation and coding for nonlinear satellite channels," *IEEE Trans. Commun.*, Vol. COM-32, pp. 616–626.

E. Biglieri, 1992. "Parallel demodulation of multidimensional signals," *IEEE Trans. Commun.*, Vol. 40, No. 10, pp. 1581–1587.

E. Biglieri, S. Barberis, and M. Catena, 1988. "Analysis and compensation of nonlinearities in digital transmission systems," *IEEE Journal Select. Areas Commun.*, Vol. 6, No. 1, pp. 42–51, January.

E. Biglieri, G. Caire, G. Taricco, and J. Ventura-Traveset, 1998. "Computing error probabilities over fading channels: A unified approach," *European Trans. Telecommun.*, Vol. 9, pp. 15–25.

E. Biglieri, E. Chiaberto, G. P. Maccone, and E. Viterbo, 1994. "Compensation of nonlinearities in high-density magnetic recording channels," *IEEE Trans. Magnetics*, Vol. 30, No. 6, pp. 5079–5086, November.

E. Biglieri, D. Divsalar, P. J. McLane, and M. K. Simon, 1991. *Introduction to Trellis-Coded Modulation with Applications*. New York: Macmillan.

E. Biglieri, M. Elia, and L. Lo Presti, 1984. "Optimal linear receiving filter for digital transmission over nonlinear channels," *GLOBECOM '84*, Atlanta, Georgia.

E. Biglieri, A. Gersho, R. D. Gitlin, and T. L. Lim, 1984. "Adaptive cancellation of nonlinear intersymbol interference for voiceband data transmission," *IEEE Journal on Selected Areas in Commun.*, Vol. SAC-2, pp. 765-777.

E. Biglieri and J. Hagenauer (eds.), 1995. *European Trans. Telecommun.*, Vol. 6, September, the whole issue.

E. Biglieri, J. Proakis, and S. Shamai (Shitz), 1998. "Fading channels: Information-theoretic and communications aspects," *IEEE Trans. Inform. Theory*, Vol. 44, pp. 2619–2692, October.

E. Biglieri and M. Visintin, 1990. "A simple derivation of the power spectrum of full-response CPM and some of its properties," *IEEE Trans. Commun.*, pp. 267–269.

J. A. Bingham, 1990. "Multicarrier modulation for data transmission: An idea whose time has come," *IEEE Commun. Mag.*, Vol. 28, pp. 5–14, May.

N. M. Blachman, 1971. "Detectors, bandpass nonlinearities and their optimization: Inversion of the Chebyshev transform," *IEEE Trans. Inform. Theory*, Vol. IT-17, pp. 398–404.

N. M. Blachman, 1982. *Noise and Its Effects on Communication*, 2nd ed. Malabar, FL: R. E. Krieger Publ. Co.

R. E. Blahut, 1983. *Theory and Practice of Error Control Codes*. Reading, MA: Addison-Wesley.

R. E. Blahut, 1987. *Principles and Practice of Information Theory*. Reading, MA: Addison-Wesley.

A. Blanchard, 1976. *Phase-Locked Loops*. New York: J. Wiley & Sons.

A. Blanc-Lapierre and R. Fortet, 1968. *Theory of Random Functions,* Vol. 2. New York: Gordon and Breach.

N. S. Boutin, S. Morissette, and C. Porlier, 1982. "Extension of Mueller's theory on optimum pulse shaping for data transmission," *IEE Proceedings*, Vol. 129, pp. 255–260.

R. N. Bracewell, 1978. *The Fourier Transform and Its Applications,* 2nd ed. New York: McGraw-Hill.

D. G. Brennan, 1959. "Linear diversity combining techniques," *Proc. IRE*, Vol. 47, pp. 1075–1102.

L. E. Brennan and I. S. Reed, 1965. "A recursive method of computing the Q function," *IEEE Trans. Inform. Theory*, Vol. IT-11, pp. 312–313.

J. Buisán and E. Biglieri, 1996. "Algorithms for blind identification of digital communication channels," *IEEE Trans. Signal Processing*, Vol. 44, pp. 3154–3156.

S. A. Butman and J. R. Lesh, 1977. "The effects of bandpass limiters on n-phase tracking systems," *IEEE Trans. Commun.*, Vol. COM-25, pp. 569–576.

J. B. Cain, G. C. Clark, and J. M. Geist, 1979. "Punctured convolutional codes of rate $(n-1)/n$ and simplified maximum likelihood decoding," *IEEE Trans. Inform. Theory*, Vol. IT-25, pp. 97-100.

G. Caire, G. Taricco, and E. Biglieri, 1998. "Bit-interleaved coded modulation," *IEEE Trans. Inform. Theory*, Vol. 44, pp. 927–946, May.

G. Caire, J. Ventura, M. Hollreiser, and E. Biglieri, 1995. "Systolic architecture for the VLSI implementation of high-speed staged decoders/quantizers," *Journal of VLSI Signal Processing,* Vol. 10, pp. 153–168.

A. R. Calderbank, 1989. "Multilevel codes and multistage decoding," *IEEE Trans. Commun.*, Vol. COM-37, pp. 222–229.

A. R. Calderbank and J. E. Mazo, 1984. "A new description of trellis codes," *IEEE Trans. Inform. Theory,* Vol. IT-30, pp. 784–791.

A. R. Calderbank and N. J. A. Sloane, 1987. "New trellis codes based on lattices and cosets," *IEEE Trans. Inform. Theory*, Vol. IT-33, pp. 177–195.

S. Cambanis and B. Liu, 1970. "On harmonizable stochastic processes," *Inform. and Control*, Vol. 17, pp. 183–202.

S. Cambanis and E. Masry, 1971. "On the representation of weakly continuous stochastic processes," *Information Sciences*, Vol. 3, pp. 277–290.

L. L. Campbell, 1969. "Series expansions for random processes," in *Proc. International Symposium on Probability and Information Theory*, Lecture Notes in Mathematics, No. 89, pp. 77–95. New York: Springer Verlag.

L. L. Campbell and P. H. Wittke, 1997. "Mathematical problems in error calculations for interference,"*IEEE Trans. Commun.*, Vol. COM-45, pp. 1527–1528.

G. L. Cariolaro, G. L. Pierobon, and G. P. Tronca, 1983. "Analysis of codes and spectra calculations," *International Journal of Electronics*, Vol. 55, pp. 35–79.

G. L. Cariolaro and S. Pupolin, 1975. "Moments of correlated digital signals for error probability evaluation," *IEEE Trans. Inform. Theory,* Vol. IT-21, pp. 558–568.

G. L. Cariolaro and G. P. Tronca, 1974. "Spectra of block coded digital signals," *IEEE Trans. Commun.*, Vol. COM-22, pp. 1555–1564.

J. K. Cavers and P. Ho, 1992. "Analysis of the error performance of trellis-coded modulations in Rayleigh-fading channels," *IEEE Trans. Commun.*, Vol. 40, No. 1, pp. 74–83.

J. H. H. Chalk, 1950. "The optimum pulse shape for pulse communication," *IEE Proceedings*, Vol. 97, pp. 88–92.

J.-J. Chang, D.-J. Hwang, and M.-C. Lin, 1997. "Some extended results on the search for good convolutional codes," *IEEE Trans. Inform. Theory*, Vol. IT-43, pp. 1682–1697.

R. W. Chang, 1971. "A new equalizer structure for fast start-up digital communication," *Bell Syst. Tech. J.*, Vol. 50, pp. 1969–2001.

R. W. Chang and J. C. Hancock, 1966. "On receiver structures for channels having memory," *IEEE Trans. Inform. Theory*, Vol. IT-12, pp. 463–468.

G. C. Clark and J. B. Cain, 1981. *Error-Correction Coding for Digital Communications*. New York: Plenum Press.

J. H. Conway and N. J. A. Sloane, 1988. *Sphere Packings, Lattices and Groups*. New York: Springer Verlag.

D. J. Costello, 1969. "Construction of convolutional codes for sequential decoding," *Tech. Report EE-692*, University of Notre Dame, Indiana.

T. Cover and J. Thomas, 1991. *Elements of Information Theory*. New York: Wiley-Interscience.

N. A. D'Andrea and F. Russo, 1983. "First-order DPLL's: A survey of peculiar methodology and some new applications," *Alta Frequenza*, Vol. 52, pp. 495–505.

D. G. Daut, J. W. Modestino and L. D. Wismer, 1982. "New short constraint length convolutional code construction for selected rational rates," *IEEE Trans. Inform. Theory*, Vol. IT-28, pp. 794–800.

W. B. Davenport, Jr., 1970. *Probability and Random Processes. An Introduction for Applied Scientists and Engineers*. New York: McGraw-Hill.

R. de Buda, 1972. "A coherent demodulation of frequency-shift keying with low deviation ratio," *IEEE Trans. Commun.*, Vol. 20, pp. 429–435.

D. Divsalar, 1978. *Performance of Mismatched Receivers*. Ph.D. Dissertation, University of California, Los Angeles.

D. Divsalar and R. J. McEliece, 1996. "Effective free distance of turbo codes," *Electronics Letters*, Vol. 32, no. 5.

D. Divsalar and F. Pollara, 1995. "Turbo codes for deep-space communications," *TDA Progress Report 42-120*, pp. 29–39, Jet Propulsion Laboratory, Pasadena, California.

M. L. Doelz and E. H. Heald, 1961. "Minimum shift data communication system," Collins Radio Co., U.S. Patent 2 977 417, March 28.

S. Dolinar, 1988. "A new code for Galileo," *TDA Progress Report 42-93*, pp. 83–96, Jet Propulsion Laboratory, Pasadena, California.

Dj. Dugundji, 1958. "Envelopes and pre-envelopes of real wave-forms," *IRE Trans. Inform. Theory*, Vol. IT-4, pp. 53–57.

H. Dym and H. P. McKean, 1972. *Fourier Series and Integrals*. New York: Academic Press.

M. Elia, 1983. "Symbol error rate for binary block codes," *Proceedings of the 9th Prague Conference on Information Theory, Statistical Decision Functions, Random Processes*, Prague, pp. 223–227.

T. Ericson, 1971. "Structure of optimum receiving filters in data transmission systems," *IEEE Trans. Inform. Theory*, Vol. IT-17, pp. 352–353.

T. Ericson, 1973. "Optimum PAM filters are always band-limited," *IEEE Trans. Inform. Theory*, Vol. IT-19, pp. 570–573.

M. V. Eyuboglu and G. D. Forney, Jr., 1992. "Trellis precoding: Combined coding, precoding and shaping for intersymbol interference channels," *IEEE Trans. Inform. Theory*, Vol. 38, No. 2, pp. 301–314.

D. D. Falconer, 1976. "Jointly adaptive equalization and carrier recovery in two-dimensional digital communication systems," *Bell Syst. Tech. J.*, Vol. 55, pp. 317–334.

D. D. Falconer, 1978. "Adaptive equalization of channel nonlinearities in QAM data transmission systems," *Bell Syst. Tech. J.*, Vol. 57, pp. 2589–2611.

D. D. Falconer and L. Ljung, 1978. "Application of fast Kalman estimation to adaptive equalization," *IEEE Trans. Commun.*, Vol. COM-26, 1439–1446.

D. D. Falconer and F. R. Magee, Jr., 1973. "Adaptive channel memory truncation for maximum likelihood sequence estimation," *Bell Syst. Tech. J.*, Vol. 52, pp. 1541–1562.

R. M. Fano, 1961. *Transmission of Information.* Cambridge, MA: MIT Press.

R. M. Fano, 1963. "A heuristic discussion of probabilistic decoding," *IEEE Trans. Inform. Theory*, Vol. IT-9, pp. 64–74.

W. Feller, 1968. *An Introduction to Probability Theory and Its Applications,* 3rd ed. New York: J. Wiley & Sons.

R. H. Flake, 1963. "Volterra series representation of nonlinear systems," *AIEE Transactions*, Vol. 81, pp. 330–335.

G. D. Forney, Jr., 1966. *Concatenated Codes.* Cambridge, MA: MIT Press.

G. D. Forney, Jr., 1970. "Convolutional Codes I: Algebraic structure," *IEEE Trans. Inform. Theory*, Vol. IT-16, pp. 720–738.

G. D. Forney, Jr., 1972. "Maximum likelihood sequence estimation of digital sequences in the presence of intersymbol intreference," *IEEE Trans. Inform. Theory*, Vol. IT-18, pp. 363–378.

G. D. Forney, Jr., 1973. "The Viterbi algorithm," *IEEE Proceedings*, Vol. 61, pp. 268–278.

G. D. Forney, Jr., 1974a. "Convolutional Codes II: Maximum-likelihood decoding," *Information and Control*, Vol. 25, pp. 223–265.

G. D. Forney, Jr., 1974b. "Convolutional Codes III: Sequential decoding," *Information and Control*, Vol. 25, pp. 267–297.

G. D. Forney, Jr., 1988a. "Coset codes — Part I: Introduction and geometrical classification," *IEEE Trans. Inform. Theory*, Vol. 34, pp. 1123–1151.

G. D. Forney, Jr., 1988b. "Coset codes — Part II: Binary lattices and related codes," *IEEE Trans. Inform. Theory*, Vol. 34, pp. 1152–1187.

G. D. Forney, Jr., 1989. "Multidimensional constellation — Part II: Voronoi constellations," *IEEE J. Select. Areas Commun.*, Vol. 7, pp. 941–958, August.

G. D. Forney, Jr., 1991. "Geometrically uniform codes," *IEEE Trans. Inform. Theory*, Vol. 37, pp. 1241-1260.

G. D. Forney, Jr., R. G. Gallager, G. R. Lang, F. M. Longstaff, and S. H. Qureshi, 1984. "Efficient modulation for band-limited channels," *IEEE J. Sel. Areas Commun.*, Vol. SAC-2, pp. 632–647, September.

G. D. Forney, Jr., L. Brown, M. V. Eyuboglu, and J. L. Moran III, 1996. "The V.34 high-speed modem standard," *IEEE Commun. Mag.*, Vol. 34, pp. 28–33, December.

G. D. Forney, Jr., and G. Ungerboeck, 1998. "Modulation and coding for linear Gaussian channels," *IEEE Trans. Inform. Theory*, pp. 2384–2415, October.

G. D. Forney, Jr., and L.-F. Wei, 1989. "Multidimensional constellation — Part I: Introduction, figures of merit, and generalized cross constellations," *IEEE J. Select. Areas Commun.*, Vol. 7, pp. 877–892.

G. J. Foschini, 1975. "Performance bound for maximum-likelihood reception of digital data," *IEEE Trans. Inform. Theory*, Vol. IT-21, pp. 47-50.

G. J. Foschini, 1977. "A reduced state variant of maximum likelihood sequence detection attaining optimum performance for high signal-to-noise ratios," *IEEE Trans. Inform. Theory*, Vol. IT-23, pp. 605–609.

G. J. Foschini, 1985. "Equalizing without altering or detecting data," *AT&T Technical Journal*, Vol. 64, pp. 1885–1911.

L. E. Franks, 1968. "Further results on Nyquist's problem in pulse transmission," *IEEE Trans. Commun.*, Vol. COM-16, pp. 337-340.

L. E. Franks, 1969. *Signal Theory*. Englewood Cliffs, NJ: Prentice-Hall.

L. E. Franks, 1980. "Carrier and bit synchronization in data communications: A tutorial review," *IEEE Trans. Commun.*, Vol. COM-28, pp. 1107–1120.

L. E. Franks, 1983. "Synchronization subsystems: Analysis and design," in K. Feher, *Digital Communications: Satellite/Earth Station Engineering*. Englewood Cliffs, NJ: Prentice-Hall.

L. E. Franks and J. P. Bubrouski, 1974. "Statistical properties of timing jitter in a PAM recovery scheme," *IEEE Trans. Commun.*, Vol. COM-22, pp. 913–920.

S. Fredricsson, 1974. "Optimum transmitting filter in digital PAM systems with a Viterbi detector," *IEEE Trans. Inform. Theory*, Vol. IT-20, pp. 479–489,

S. Fredricsson, 1975. "Optimum receiver filters in digital quadrature phase-shift-keyed systems with a nonlinear repeater," *IEEE Trans. Commun.*, Vol. COM-23, pp. 1389–1400.

B. Friedlander, 1982. "Lattice filters in adaptive processing," *IEEE Proceedings*, Vol. 70, pp. 829–867.

P. Galko and S. Pasupathy, 1981. "The mean power spectral density of Markov chain driven signals," *IEEE Trans. Inform. Theory*, Vol. IT-27, pp. 746–754.

R. G. Gallager, 1965. "A simple derivation of the coding theorem and some applications," *IEEE Trans. Inform. Theory*, Vol. IT-11, pp. 3–18.

R. G. Gallager, 1968. *Information Theory and Reliable Communication*. New York: J. Wiley & Sons.

R. G. Gallager, 1995. *Discrete Stochastic Processes*. Boston, MA: Kluwer Academic Press.

F. R. Gantmacher, 1959. *The Theory of Matrices*, Vols. I and II. New York: Chelsea Publ. Co.

F. M. Gardner, 1979. *Phaselock Techniques*, 2nd Edition. New York: J. Wiley & Sons.

W. A. Gardner, 1978. "Stationarizable random processes," *IEEE Trans. Inform. Theory*, Vol. IT-24, pp. 8–22.

W. A. Gardner, 1984. "Learning characteristics of stochastic-gradient descent algorithms: A general study, analysis, and critique," *Signal Processing*, Vol. 6, pp. 113–133.

W. A. Gardner (ed.), 1994. *Cyclostationarity in Communications and Signal Processing*. Piscataway, NJ: IEEE Press.

W. Gardner and L. E. Franks, 1975. "Characterization of cyclostationary random signal processes," *IEEE Trans. Inform. Theory*, Vol. IT-21, pp. 4–14.

G. J. Garrison, 1975. "A power spectral density analysis for digital FM," *IEEE Trans. Commun.*, Vol. 23, pp. 1228–1243, November.

W. Gautschi, 1970. "On the construction of Gaussian quadrature rules from modified moments," *Mathematics of Computation*, Vol. 24, pp. 245–260.

I. M. Gelfand and S. V. Fomin, 1963. *Calculus of Variations*. Englewood Cliffs, NJ: Prentice-Hall.

D. A. George, 1965. "Matched filters for interfering signals," *IEEE Trans. Inform. Theory*, Vol. IT-11, pp. 153-154.

A. Gersho, 1969. "Adaptive equalization of highly dispersive channels," *Bell Syst. Tech. J.*, Vol. 48, pp. 55–70.

G. Giannakis, Y. Inouie, and J. Mendel, 1989. "Cumulant-based identification of multichannel moving average models," *IEEE Trans. Automatic Control*, Vol.. 34, pp. 783–787.

G. Giannakis and J. Mendel, 1989. "Identification of non-minimum phase systems using higher-order statistics," *IEEE Trans. ASSP*, Vol. 37, pp. 360–377.

R. A. Gibby and J. W. Smith, 1965. "Some extensions of Nyquist's telegraph transmission theory," *Bell Syst. Tech. J.*, Vol. 44, pp. 1487–1510.

J. D. Gibson (ed.), 1996. *The Mobile Communications Handbook*. Boca Raton, FL: CRC Press.

V. V. Ginzburg, 1984. "Mnogomerniye signaly dlya nepreryvnogo kanala," (in Russian), *Problemy Peredachi Informacii*, pp. 28–46. English translation: "Multidimensional signals for a continuous channel," *Problems of Information Transmission*, Vol. 23, No. 4, pp. 20–34.

R. D. Gitlin, E. Y. Ho, and J. E. Mazo, 1973. "Passband equalization of differentially phase-modulated data signals," *Bell Syst. Tech. J.*, Vol. 52, pp. 219–238.

R. D. Gitlin and F. R. Magee, 1977. "Self-orthogonalizing adaptive equalization algorithms," *IEEE Trans. Commun.* Vol. COM-25, pp. 666–672.

R. D. Gitlin, J. E. Mazo, and M. G. Taylor, 1973. "On the design of gradient algorithms for digitally implemented adaptive filters," *IEEE Trans. Circuit Theory*, Vol. CT-20, pp. 125–136.

R. D. Gitlin and S. B. Weinstein, 1979. "On the required tap-weight precision for digitally implemented, adaptive, mean-square equalizers," *Bell Syst. Tech. J.*, Vol. 58, pp. 301–321.

R. D. Gitlin and S. B. Weinstein, 1981. "Fractionally-spaced equalization: An improved digital transversal equalizer," *Bell Syst. Tech. J.*, Vol. 60, pp. 275–296.

F. E. Glave, 1972. "An upper bound on the probability of error due to intersymbol interference for correlated digital signals," *IEEE Trans. Inform. Theory*, Vol. IT-18, pp. 356–363.

D. Godard 1974. "Channel equalization using a Kalman filter for fast data transmission," *IBM J. Res. Develop.*, Vol. 18, pp. 267–273.

D. Godard, 1980. "Self-recovering equalization and carrier-tracking in two-dimensional data communication systems," *IEEE Trans. Commun.*, Vol. COM-28, pp. 1867–1875.

D. Godard, 1981. "A 9600 bit/s modem for multipoint communication systems," *Nat. Telecomm. Conf. Rec.*, New Orleans, LA, pp. B3.3.1–B3.3.5.

L. M. Goldenberg and D. D. Klovsky, 1959. "Computer-aided detection of pulse signals," *Trudy LEIS*, Vol. 44, pp. 17–26.

S. W. Golomb, 1967. *Shift Register Sequences*. San Francisco, CA: Holden-Day.

G. H. Golub and C. F. Van Loan, 1996. *Matrix Computations*, 3rd ed. Baltimore, MD: Johns Hopkins University Press.

G. H. Golub and J. H. Welsch, 1969. "Calculation of Gauss quadrature rules," *Mathematics of Computation*, Vol. 23, pp. 221–230.

R. M. Gray, 1971. *Toeplitz and Circulant matrices: A Review*. Stanford Electron. Lab. Tech. Rep. 6501-2, June 1971.

R. M. Gray, 1972. "On the asymptotic eigenvalue distribution of Toeplitz matrices," *IEEE Trans. Inform. Theory*, Vol. IT-18, pp. 725–730.

U. Grenander and G. Szegö, 1958. *Toeplitz Forms and Their Applications*. Berkeley, CA: Univ. of California Press.

T. L. Grettenberg, 1975. "A representation theorem for complex normal processes," *IEEE Trans. Inform. Theory*, Vol. IT-11, pp. 305–306.

S. A. Gronemeyer and A. L. McBride, 1976. "MSK and offset QPSK modulation," *IEEE Trans. Commun.*, Vol. COM-24, pp. 809–819.

M. L. Guidoux, 1975. "Egaliseur autoadaptif à double échantillonnage appliqué à la transmission de donnés à 9600 bit/seconde" (in French), *L'Onde Electrique*, Vol. 55, pp. 9–13.

S. C. Gupta, 1975. "Status of digital phase-locked loops," *IEEE Proceedings*, Vol. 63, pp. 291–306.

J. Hagenauer, E. Offer, and L. Papke, 1996. "Iterative decoding of binary block and convolutional codes," *IEEE Trans. Inform. Theory,* Vol. IT-42, pp. 429–445.

E. Hansler, 1971. "Some properties of transmission systems with minimum mean-square error," *IEEE Trans. Commun.,* Vol. COM-19, pp. 576–579.

H. Harashima and H. Miyakawa, 1972. "Matched-transmission technique for channels with intersymbol interference," *IEEE Trans. Commun.,* Vol. COM-20, No. 4, pp. 774–780.

H. Hashemi, 1993. "The indoor radio propagation channel," *IEEE Proceedings,* Vol. 81, No. 7, pp. 943–968.

J. F. Hayes, 1975. "The Viterbi algorithm applied to digital data transmission," *IEEE Communications Magazine,* Vol. 13, pp. 15–20.

J. A. Heller, 1968. "Short constraint-length convolutional codes," *JPL Pasadena Space Program Summary 37-54,* Vol. 3, pp. 171–174.

J. A. Heller, 1975. "Feedback decoding of convolutional codes," in A. J. Viterbi (ed.) *Advances in Communication Systems,* Vol. 4. New York: Academic Press.

C. W. Helstrom, 1958. "The resolution of signals in white Gaussian noise," *IRE Proceedings,* Vol. 46, pp. 1603–1619.

C. W. Helstrom, 1968. *Statistical Theory of Signal Detection,* 2nd ed. Elmsford, NY: Pergamon Press.

F. Hemmati and D. J. Costello, Jr., 1977. "Truncation error probability in Viterbi decoding," *IEEE Trans. Commun.,* Vol. COM-25, pp. 530–532.

G. F. Herrmann, 1978. "Performance of maximum-likelihood receiver in the nonlinear satellite channel," *IEEE Trans. Commun.,* vol. COM-26, pp. 373–378.

E. H. Ho and Y. S. Yeh, 1970. "A new approach for evaluating the error probability in the presence of intersymbol interference," *Bell Syst. Tech. J.,* Vol. 49, pp. 2249–2265.

E. H. Ho and Y. S. Yeh, 1971. "Error probability of a multilevel digital system with intersymbol interference and Gaussian noise," *Bell Syst. Tech. J.,* Vol. 50, pp. 1017–1023.

P. Ho and D. Fung, 1992. "Error performance of interleaved trellis-coded PSK modulation in correlated Rayleigh fading channels," *IEEE Trans. Commun.,* Vol. 40, pp. 1800–1809.

M. L. Honig and D. G. Messerschmitt, 1984. *Adaptive Filters: Structures, Algorithms, and Applications.* Boston, MA: Kluwer Academic Publisher.

J. Huber and U. Wachsmann, 1994. "Capacities of Equivalent Channels in Multilevel Coding Schemes," *Electronics Letters*, Vol. 30, pp. 557–558.

W. H. Huggins, 1957. "Signal-flow graphs and random signals," *IRE Proceedings*, Vol. 45, pp. 74–86.

P. H. Humblet and M. G. Troulis, 1996. "The information driveway," *IEEE Comm. Magazine*, Vol. 34, pp. 64–68.

H. Hurd, 1969. *An Investigation of Periodically Correlated Stochastic Processes*. Ph.D. Dissertation, Duke University, Durham, NC.

H. Imai and S. Hirakawa, 1977. "A new multilevel coding method using error-correcting codes," *IEEE Trans. Inform. Theory*, Vol. IT-23, No. 3, pp. 371-377.

I. M. Jacobs, 1974. "Practical applications of coding," *IEEE Trans. Inform. Theory*, Vol. IT-20, pp. 305–310.

F. De Jager and C. B. Dekker, 1978. "Tamed frequency modulation, a novel method to achieve spectrum economy in digital transmission," *IEEE Trans. Commun.*, Vol. 26, pp. 534–542.

W. C. Jakes, 1974. *Microwave Mobile Communications*. New York: John Wiley & Sons.

S. H. Jamali and T. Le-Ngoc, 1994. *Coded-Modulation Techniques for Fading Channels*. Boston, MA: Kluwer Academic Publishers.

F. Jelinek, 1969. "Fast sequential decoding algorithm using a stack," *IBM Journal of Research and Development*, Vol. 13, pp. 675–685.

M. C. Jeruchim, P. Balaban, and K. S. Shanmugan, 1992. *Simulation of Communication Systems*. New York: Plenum Press.

C. R. Johnson, 1991. "Admissibility in blind adaptive channel equalization," *IEEE Control Syst. Mag.*, pp. 3–15.

D. S. Jones, 1966. *Generalised Functions*. New York: McGraw-Hill.

S. K. Jones, R. K. Kavin, III, and W. M. Reed, 1982. "Analysis of error-gradient adaptive linear estimators for a class of stationary dependent processes," *IEEE Trans. Inform. Theory*, Vol. IT-28, pp. 318–329.

T. Kailath, 1971. "RKHS approach to detection and estimation problems—Part I: Deterministic signals in Gaussian noise," *IEEE Trans. Inform. Theory*, Vol. IT-17, pp. 530–549.

G. Karam and H. Sari, 1989. "Analysis of predistortion, equalization, and ISI cancellation techniques in digital radio systems with nonlinear transmit amplifiers," *IEEE Trans. Commun.*, Vol. 37, pp. 1245–1253, December.

G. Karam and H. Sari, 1990. "Data predistortion techniques using intersymbol interpolation," *IEEE Trans. Commun.*, Vol. 38, pp. 1716–1723, October.

G. Karam and H. Sari, 1991a. "A data predistortion technique with memory for QAM radio systems," *IEEE Trans. Commun.*, Vol. 39, pp. 336–344, February.

G. Karam and H. Sari, 1991b. "Generalized data predistortion using intersymbol interpolation," *Philips Journal of Research*, Vol. 46, pp. 1–22.

J. G. Kemeny and J. L. Snell, 1960. *Finite Markov Chains*. New York: Van Nostrand Reinhold.

E. Kettel, 1961. "Übertragungssysteme mit idealer Impulsfunktion" (in German), *A.E.Ü.*, Vol. 15, pp. 207–214.

E. Kettel, 1964. "Ein automatisches Optimisator für den Abgleich des Impulsentzerrers in einer Datenübertragung" (in German), *A.E.Ü.*, Vol. 18, pp. 271–278.

D. Klovsky and B. Nikolaev, 1978. *Sequential Transmission of Digital Information in the Presence of Intersymbol Interference* (in Russian). Moscow, USSR: MIR Publishers.

R. Knopp, P. A. Humblet, 1997. "Maximizing diversity on block fading channels," *Proceedings of ICC'97*, Montréal, Canada.

R. Knopp and P. A. Humblet, 1998. "On coding for block fading channels," *Submitted for publication*.

H. Kobayashi, 1971. "Correlative level coding and maximum-likelihood decoding," *IEEE Trans. Inform. Theory*, Vol. IT-17, pp. 586–594.

M. G. Krein and A. A. Nudel'man, 1977. *The Markov Moment Problem and Extremal Problems*, Transl. Math. Monographs, vol. 50. Providence, R.I.: American Mathematical Society.

V. J. Krylov, 1962. *Approximate Evaluation of Integrals*. New York: Macmillan.

F. R. Kschischang and V. Sorokine, 1995. "On the trellis structure of block codes," *IEEE Trans. Inform. Theory*, Vol. IT-41, pp. 1924–1937.

R. Laroia, 1996. "Coding for intersymbol interference channels—Combined coding and precoding," *IEEE Trans. Inform. Theory*, Vol. 42, No. 4, pp. 1053–1061.

R. Laroia, N. Farvardin, and S. A. Tretter, 1994. "On optimal shaping of multidimensional constellations—High-rate, high-dimensional constellations," *IEEE Trans. Inform. Theory*, Vol. IT-40, No. 4, pp. 1044–1056.

R. Laroia, S. A. Tretter, and N. Farvardin, 1993. "A simple and effective precoding scheme for noise whitening on intersymbol interference channels," *IEEE Trans. Commun.*, Vol. 41, No. 10, pp. 1460–1463.

K. J. Larsen, 1973. "Short convolutional codes with maximal free distance for rates 1/2, 1/3 and 1/4," *IEEE Trans. Inform. Theory*, Vol. IT-19, pp. 371–372.

R. E. Lawrence and H. Kaufman, 1971. "The Kalman filter for the equalization of a digital communication channel," *IEEE Trans. Communication Technology*, Vol. COM-19, pp. 1137–1141.

W. C. Y. Lee, 1989. *Mobile Cellular Telecommunications Systems.* New York: McGraw-Hill.

W. U. Lee and F. S. Hill, Jr., 1977. "A maximum likelihood sequence estimator with decision feedback equalization," *IEEE Trans. Commun.*, Vol. COM-25, pp. 971–980.

H. Leib and S. Pasupathy, 1993. "Error-control properties of Minimum Shift Keying," *IEEE Communications Magazine*, pp.52–61, January.

T. L. Lim and M. S. Mueller, 1980. "Rapid equalizer start-up using least-squares algorithms," *IEEE International Conference on Communications (ICC'80)*, Seattle, WA, June.

S. Lin, 1970. *An Introduction to Error Correcting Codes.* Englewood Cliffs, NJ: Prentice-Hall.

S. Lin and D. J. Costello, 1983. *Error Control Coding: Fundamentals and Applications.* Englewood Cliffs, NJ: Prentice-Hall.

W. C. Lindsey, 1964. "Error probabilities for Ricean fading multichannel reception of binary and $N$-ary signals," *IEEE Trans. Inform. Theory*, Vol. IT-10, pp. 339–350.

W. C. Lindsey, 1972. *Synchronous Systems in Communication and Control.* Englewood Cliffs, NJ: Prentice-Hall.

W. C. Lindsey and C. M. Chie, 1981. "A survey of digital phase locked loops," *IEEE Proceedings*, Vol. 69, pp. 410–432.

W. C. Lindsey and M. K. Simon, 1973. *Telecommunication Systems Engineering.* Englewood Cliffs, NJ: Prentice-Hall.

F. Ling and J. G. Proakis, 1984. "A generalized multichannel least squares lattice algorithm based on sequential processing stages," *IEEE Trans. Acoust., Speech, Signal Process.*, Vol. ASSP-32, pp. 381–389.

Y. J. Liu, I. Oka, and E. Biglieri, 1990. "Error probability for digital transmission over nonlinear channels with applications to TCM," *IEEE Trans. Inform. Theory,* Vol. IT-36, pp. 1101–1110.

M. Loève, 1963. *Probability Theory.* Princeton, NJ: Van Nostrand Reinhold.

R. W. Lucky, 1965. "Automatic equalization for digital communication," *Bell Syst. Tech. J.*, Vol. 44, pp. 547–588.

R. W. Lucky, 1966. "Techniques for adaptive equalization of digital communication systems," *Bell Syst. Tech. J.*, Vol. 45, pp. 255–286.

R. W. Lucky, 1975. "Modulation and detection for data transmission on the telephone channel," in J. K. Skwirzynski (ed.), *New Directions in Signal Processing in Communication and Control.* Leiden, Holland: Noordhoff.

R. W. Lucky, J. Salz and E. J. Weldon, 1968. *Principles of Data Communications.* New York: McGraw-Hill.

D. G. Luenberger, 1969. *Optimization by Vector Space Methods.* New York: J. Wiley & Sons.

R. Lugannani, 1969. "Intersymbol interference and probability of error in digital systems," *IEEE Trans. Inform. Theory*, Vol. IT-15, pp. 682–688.

A. Luvison and V. Navino, 1976. "Theoretical and experimental development of two-dimensional quadrature rules," *ICCAD International Conference on Numerical Methods in Electrical and Magnetic Field Problems*, Santa Margherita Ligure, Italy.

A. Luvison and G. Pirani, 1979. "Calculation of error rates in digital communication systems," *Fifth International Symposium on Information Theory*, Tbilisi, USSR.

O. Macchi and E. Eweda, 1984. "Convergence analysis of self-adaptive equalizers," *IEEE Trans. Inform. Theory*, Vol. IT-30, pp. 161–176.

O. Macchi and L. Guidoux, 1975. "A new equalizer and double sampling equalizer," *Annales des Télécommunications*, Vol. 30, pp. 331–338.

F. J. MacWilliams and N. J. A. Sloane, 1977. *The Theory of Error Correcting Codes.* Amsterdam: North Holland.

F. R. Magee, Jr., and J. G. Proakis, 1973. "Adaptive maximum-likelihood sequence estimation for digital signaling in the presence of intersymbol interference," *IEEE Trans. Inform. Theory*, Vol. IT-19, pp. 120–124.

J. Makhoul, 1978. "A class of all-zero lattice digital filters: Properties and applications," *IEEE Trans. Acoust., Speech, Signal Process.*, Vol. ASSP-26, pp. 304–314.

M. Mancianti, U. Mengali and R. Reggiannini, 1979. "A fast start-up algorithm for channel parameter acquisition in SSB-AM data transmission," *IEEE International Conference on Communications (ICC'79)*, Boston, MA.

E. B. Masry, B. Liu, and K. Steiglitz, 1968. "Series expansion of wide-sense stationary random processes," *IEEE Trans. Inform. Theory*, Vol. IT-14, pp. 792–796.

J. L. Massey, 1963. *Threshold Decoding*. Cambridge, MA: MIT Press.

J. L. Massey, 1974. "Coding and modulation in digital communications," *1974 International Zurich Seminar on Digital Communications*, Zurich, Switzerland.

J. L. Massey, 1980. "A generalized formulation of minimum shift keying modulation," *Proc. of ICC'80*, Seattle, WA, pp. 26.5.1–26.5.4.

J. L. Massey, 1995. "Towards an information theory of spread-spectrum systems," in: S. G. Glisic and P. Leppänen (eds.), *Code Division Multiple Access Communications*, pp. 29–46. Boston, MA: Kluwer Academic Publishers.

J. L. Massey and M. K. Sain, 1968. "Inverses of linear sequential circuits," *IEEE Trans. Computers*, Vol. C-17, pp. 330–337.

R. Matyas and P. J. McLane, 1974. "Decision-aided tracking loops for channels with phase jitter and intersymbol interference," *IEEE Trans. Commun.*, Vol. COM-22, pp. 1014–1023.

J. W. Matthews, 1973. "Sharp error bounds for intersymbol interference," *IEEE Trans. Inform. Theory*, Vol. IT-19, pp. 440–447.

J. E. Mazo, 1979. "On the independence theory of equalizer convergence," *Bell Syst. Tech. J.*, Vol. 58, pp. 963–993.

J. E. Mazo and J. Salz, 1976. "On the transmitted power in generalized partial response," *IEEE Trans. Commun.*, Vol. COM-24, No. 3, pp. 348–352.

R. J. McEliece, 1977. *The Theory of Information and Coding*. Reading, MA: Addison-Wesley.

R. J. McEliece, 1996. "On the BCJR Trellis for linear block codes," *IEEE Trans. Inform. Theory*, Vol. IT-42, pp. 1072-1091.

R. J. McEliece, D. J. C. MacKay, and J. F. Cheng, 1998. "Turbo decoding as an istance of Pearl's 'Belief Propagation' algorithm," *IEEE Journal Sel. Areas Commun.*, Vol. 16, pp. 140–152.

R. G. McKay, P. J. McLane, and E. Biglieri, 1991. "Error bounds for trellis-coded MPSK on a fading mobile satellite channel," *IEEE Trans. Commun.*, Vol. 39, No. 12, pp. 1750–1761.

P. J. McLane, 1980. "A residual intersymbol interference error bound for truncated-state Viterbi detector," *IEEE Trans. Inform. Theory*, Vol. IT-26, pp. 548–553.

J. Mendel, 1991. "Tutorial on higher-order statistics (spectra) in signal processing and system theory: Theoretical results and some applications," *IEEE Proceedings*, Vol. 79, pp. 278–305.

U. Mengali, 1977. "Joint phase and timing acquisition in data transmission," *IEEE Trans. Commun.*, Vol. COM-25, pp. 1174-1185.

U. Mengali, 1979. *Teoria dei Sistemi di Comunicazione* (in Italian). Pisa: ETS Università.

U. Mengali, 1983. "A new look at the pulse shaping problem in timing recovery," *1983 International Tirrenia Workshop on Digital Communications*, Tirrenia, Italy.

U. Mengali and A. N. D'Andrea, 1997. *Synchronization Techniques for Digital Receivers*. New York: Plenum Press.

M. F. Mesiya, P. J. McLane, and L. Lorne Campbell, 1977. "Maximum-likelihood sequence estimation of digital sequences in the presence of intersymbol interference," *IEEE Trans. Inform. Theory*, Vol. IT-18, pp. 363-378.

M. F. Mesiya, P. J. McLane, and L. Lorne Campbell, 1978. "Optimal receiver filters for BPSK transmission over a bandlimited nonlinear channel," *IEEE Trans. Commun.*, Vol. COM-26, pp. 12–22.

D. G. Messerschmitt, 1973. "A geometric theory of intersymbol interference. Part II: Performance of the maximum likelihood detector," *Bell Syst. Tech. J.*, Vol. 52, pp. 1521–1539, October.

D. G. Messerschmitt, 1974. "Design of a finite impulse response for the Viterbi algorithm and decision-feedback equalizer," *IEEE Conference on Communications (ICC'74)*, Minneapolis.

M. H. Meyers and L. E. Franks, 1980. "Joint carrier phase and symbol timing recovery for PAM systems," *IEEE Trans. Commun.*, Vol. COM-28, pp. 1121–1129.

H. Meyr, M. Moeneclaey, and S. A. Fechtel, 1997. *Digital Communication Receivers: Synchronization, Channel Estimation, and Signal Processing*. New York: J. Wiley & Sons.

K. S. Miller, 1974. *Complex Stochastic Processes*. Reading, MA: Addison-Wesley.

F. Morales-Moreno and S. Pasupathy, 1984. "Convolutional codes and MSK modulation: A combined optimization," *Proc. 12th Biennial Symp. Communications*, Queen's University, Kingston, ON, pp. c.1.4–c.1.7.

M. Morf, 1977. "Ladder forms in estimation and system identification," *IEEE 11th Annual Asilomar Conference on Circuits, Systems, and Computers*, Pacific Grove, CA, November.

M. Morf, A. Vieira, and D. T. Lee, 1977. "Ladder forms for identification and speech processing," *Proc. 1977 IEEE Conf. Decision Contr.*, New Orleans, LA, pp. 1074–1078.

E. Moulines, P. Duhamel, J.-F. Cardoso, and S. Mayrargue, 1995. "Subspace methods for the blind identification of multichannel FIR filters," *IEEE Trans. Signal Processing*, Vol. 43, pp. 516–526.

K. H. Mueller, 1973. "A new approach to optimum pulse shaping in sampled systems using time-domain filtering," *Bell Syst. Tech. J.*, Vol. 52, pp. 723–729.

M. S. Mueller, 1981. "Least-squares algorithms for adaptive equalizers," *Bell Syst. Tech. J.*, Vol. 60, pp. 1905–1925.

M. S. Mueller and J. Salz, 1981. "A unified theory of data-aided equalization," *Bell Syst. Tech. J.*, Vol. 60, pp. 2023–2038.

M. G. Mulligan and S. G. Wilson, 1984. "An improved algorithm for evaluating trellis phase codes," *IEEE Trans. Inform. Theory*, Vol. IT-30, pp. 846–851.

K. Murota and K. Hirade, 1981. "GMSK modulation for digital mobile radio telephony," *IEEE Trans. Commun.*, Vol. 29, pp. 1044–1050.

I. P. Mysovskih, 1968. "On the construction of cubature formulas with the smallest number of nodes," (in Russian), *Doklady Akademii Nauk USSR*, Vol. 178, pp. 1252–1254; english translation in *Soviet Mathematics Doklady*, Vol. 9, pp. 277-280.

H. Nyquist, 1928. "Certain topics in telegraph transmission theory," *AIEE Transactions*, Vol. 47, pp. 617–644.

J. P. Odenwalder, 1970. *Optimal Decoding of Convolutional Codes*, Ph.D. Dissertation, University of California, Los Angeles.

J. K. Omura, 1969. "On the Viterbi decoding algorithm," *IEEE Trans. Inform. Theory*, Vol. IT-15, pp. 177–179.

J. K. Omura, 1971. "Optimal receiver design for convolutional codes and channel with memory via control theoretical concepts," *Information Sciences*, Vol. 3, pp. 243–266.

J. K. Omura and M. K. Simon, 1980. "Satellite communication performance evaluation: Computational techniques based on moments," *JPL Publication 80-71.*

I. M. Onyszchuk, 1991. "Truncation length for Viterbi decoding," *IEEE Trans. Commun.*, Vol. 39, pp. 1023–1026.

A. V. Oppenheim and R. W. Schafer, 1989. *Discrete-Time Signal Processing.* Englewood Cliffs, NJ: Prentice-Hall.

A. V. Oppenheim, A. S. Willsky, and I. T. Young, 1983. *Signals and Systems.* Englewood Cliffs, NJ: Prentice-Hall.

W. P. Osborne and M. B. Luntz, 1974. "Coherent and noncoherent detection of CPFSK," *IEEE Trans. Commun.*, Vol. COM-22, pp. 1023–1036.

E. Paaske, 1974. "Short binary convolutional codes with maximal free distance for rates 2/3 and 3/4," *IEEE Trans. Inform. Theory*, Vol. IT-20, pp. 683-689.

K. Pahlavan and A. H. Levesque, 1995. *Wireless Information Networks.* New York: John Wiley & Sons.

A. Papoulis, 1962. *The Fourier Integral and Its Applications.* New York: McGraw-Hill.

A. Papoulis, 1965. *Probability, Random Variables, and Stochastic Processes.* New York: McGraw-Hill.

A. Papoulis, 1977. *Signal Analysis.* New York: McGraw-Hill.

A. Papoulis, 1984. *Probability, Random Variables, and Stochastic Processes*, 2nd ed. New York: McGraw-Hill.

E. Parzen, 1962. *Stochastic Processes.* San Francisco, CA: Holden-Day.

S. Pasupathy, 1979. "Minimum Shift Keying: A spectrally efficient modulation," *IEEE Communications Magazine*, pp.14–22, July.

P. Z. Peebles, Jr., 1987. *Digital Communication Systems.* Englewood Cliffs, NJ: Prentice-Hall.

R. Pellizzoni, A. Sandri, A. Spalvieri, and E. Biglieri, 1997. "Analysis and implementation of an adjustable-rate multilevel coded modulation system," *IEE Proc.-Commun.*, Vol. 144, pp. 1–5.

L. C. Perez, J. Seghers, and D. J. Costello, 1996. A distance spectrum interpretation of turbo codes," *IEEE Trans. Inform. Theory*, Vol. IT-42, pp. 1698–1709.

W. W. Peterson and E. J. Weldon, 1972. *Error Correcting Codes*, 2nd Edition. Cambridge, MA: MIT Press.

G. Picchi and G. Prati, 1987. "Blind equalization and carrier recovery using a 'stop-and-go' decision-directed algorithm," *IEEE Trans. Commun.*, Vol. COM-35, pp. 877–887.

R. Piessens and A. Haegemans, 1976. "Cubature formulas of degree nine for symmetric planar regions," *Mathematics of Computation*, Vol. 29, pp. 810–815.

G. Poltyrev, 1996. "Bounds on the decoding error probability of binary linear codes via their spectra," *IEEE Trans. Inform. Theory*, Vol. IT-40, pp. 1261–1271.

G. J. Pottie and D. P. Taylor, 1989. "Multilevel codes based on partitioning," *IEEE Trans. Inform. Theory*, Vol. IT-35, pp. 87–98.

V. K. Prabhu, 1971. "Some considerations of error bounds in digital systems," *Bell Syst. Tech. J.*, Vol. 50, pp. 3127–3151.

R. Price, 1972. "Nonlinearly feedback-equalized PAM vs. capacity for noisy filter channels," *IEEE International Conf. Commun.*, Philadelphia, PA, pp. 22-12 to 22-17.

J. G. Proakis, 1974. "Channel identification for high speed digital communications," *IEEE Trans. Automatic Control*, Vol. AC-19, pp. 916–922.

J. G. Proakis, 1991. "Adaptive equalization for TDMA digital mobile radio," *IEEE Trans. Vehic. Technology*, Vol. 40, No. 2, pp. 333–341.

J. G. Proakis, 1995. *Digital Communications*, 3rd ed. New York: McGraw-Hill.

J. G. Proakis and D. G. Manolakis, 1992. *Introduction to Digital Signal Processing*, 2nd ed. New York: Macmillan.

J. G. Proakis and J. H. Miller, 1969. "An adaptive receiver for digital signaling through channels with intersymbol interference," *IEEE Trans. Inform. Theory*, Vol. IT-15, pp. 484–497.

S. U. H. Qureshi, 1973. "An adaptive decision-feedback receiver using maximum-likelihood sequence estimation," *IEEE International Conference on Communications (ICC'73)*, Seattle, WA.

S. U. H. Qureshi, 1982. "Adaptive equalization," *IEEE Communications Magazine,* Vol. 20, pp. 9–16.

S. U. H. Qureshi, 1985. "Adaptive equalization," *IEEE Proceedings*, Vol. 73, No. 9, pp. 1340–1387.

S. U. H. Qureshi and G. D. Forney, Jr., 1977. "Performance and properties of a $T/2$ equalizer," *National Telecommunication Conference (NTC'77)*, Los Angeles, CA.

S. U. H. Qureshi and E. E. Newhall, 1973. "An adaptive receiver for data transmission over time-dispersive channels," *IEEE Trans. Inform. Theory*, Vol. IT-19, pp. 448-457.

L. R. Rabiner and B. Gold, 1975. *Theory and Applications of Digital Signal Processing*. Englewood Cliffs, NJ: Prentice-Hall.

T. S. Rappaport (ed.), 1995. *Cellular Radio and Personal Communications. A Book of Selected Readings*. Piscataway, NJ: IEEE Press.

T. S. Rappaport, 1996. *Wireless Communications–Principles and Practice*. Englewood Cliffs, NJ: Prentice-Hall.

S. O. Rice, 1982. "Envelopes of narrow-band signals," *IEEE Proceedings*, Vol. 70, pp. 692–699.

B. Rimoldi, 1988. "A decomposition approach to CPM," *IEEE Trans. Inform. Theory*, Vol. 24, pp. 260–270.

B. Rimoldi, 1994. "Five views of differential MSK: A unified approach," in: R. E. Blahut, D. J. Costello Jr., U. Maurer, and T. Mittelholzer (eds.): *Communications and Cryptography: Two Sides of One Tapestry*. Boston, MA: Kluwer Academic Publishers, pp. 333–342.

G. H. Robertson, 1969. "Computation of the noncentral chi-square distribution," *Bell Syst. Tech. J.*, Vol. 48, pp. 201–207.

W. J. Rosenberg, 1971. *Structural Properties of Convolutional Codes*, Ph.D. Dissertation, University of California, Los Angeles.

W. J. Rugh, 1981. *Nonlinear System Theory. The Volterra-Wiener Approach*. Baltimore, MD: Johns Hopkins University Press.

A. A. M. Saleh, 1981. "Frequency-independent and frequency-dependent nonlinear models of TWT amplifiers," *IEEE Trans. Commun.*, Vol. COM-29, pp. 1715–1720.

J. Salz, 1973. "Optimum mean-square decision-feedback equalization," *Bell Syst. Tech. J.*, Vol. 52, pp. 1341–1373.

B. R. Saltzberg, 1968. "Intersymbol interference error bounds with application to ideal bandlimited signaling," *IEEE Trans. Inform. Theory*, Vol. IT-14, pp. 563–568.

Y. Sato, 1975. "A method of self-recovering equalization for multilevel amplitude modulation," *IEEE Trans. Commun.*, Vol. COM-23, pp. 679–682.

E. Satorius and S. T. Alexander, 1979. "Channel equalization using adaptive lattice algorithms," *IEEE Trans. Commun.*, Vol. COM-27, pp. 899–905.

E. H. Satorius and J. D. Pack, 1981. "Application of least squares lattice algorithms to adaptive equalization," *IEEE Trans. Commun.*, Vol. COM-29, pp. 136–142.

J. E. Savage, 1966. "Sequential decoding. The computational problem," *Bell Syst. Tech. J.*, Vol. 45, pp. 149-176.

R. C. P. Saxena, 1983. *Optimum Encoding in Finite State Coded Modulation.* Report TR83-2, Dept. Electrical, Computer, and Systems Engineering, Rensselaer Polytechnic Institute, Troy, NY.

M. Schetzen, 1976. "Theory of $p$th order inverses of nonlinear systems," *IEEE Trans. Circuit and Sys.*, Vol. CAS-23, pp. 285–291.

M. Schetzen, 1980. *The Volterra and Wiener Theories of Nonlinear Systems.* New York: J. Wiley & Sons.

M. Schetzen, 1981. "Nonlinear system modeling based on the Wiener structure," *IEEE Proceedings*, Vol. 69, pp. 1557–1573.

E. Schichor, 1982. "Fast recursive estimation using the lattice structure," *Bell Syst. Tech. J.*, Vol. 61, pp. 97–115.

T. A. Schonhoff, 1976. "Symbol error probabilities for $M$-ary CPFSK: Coherent and noncoherent detection," *IEEE Trans. Commun.*, Vol. COM-24, pp. 644-652.

M. Schwartz, W. R. Bennett, and S. Stein, 1966. *Communication Systems and Techniques.* New York: McGraw-Hill.

M. Schwartz and L. Shaw, 1975. *Signal Processing. Discrete Spectral Analysis, Detection, and Estimation.* New York: McGraw-Hill.

N. Seshadri and C.-E. W. Sundberg, 1993. "Coded modulations for fading channels—An overview," *European Trans. Telecomm.*, Vol. ET-4, pp. 309–324.

O. Shalvi and E. Weinstein, 1990. "New criteria for blind deconvolution of nonminimum phase systems (channels)," *IEEE Trans. Inform. Theory*, Vol. 36, No. 2, pp. 312–321.

O. Shalvi and E. Weinstein, 1994. "Maximum likelihood and lower bounds in system identification with non-Gaussian inputs," *IEEE Trans. Inform. Theory*, Vol. IT-40, pp. 328–339.

C. E. Shannon, 1948. "A mathematical theory of communication," *Bell Syst. Tech. J.*, Vol. 27, pt. I, pp. 379–423; pt. II, pp. 623–656.

O. Shimbo and M. I. Celebiler, 1971. "The probability of error due to intersymbol interference and Gaussian noise in digital communication systems," *IEEE Trans. Commun.*, Vol. COM-19, pp. 113–119.

D. A. Shnidman, 1967. "A generalyzed Nyquist criterion and an optimum linear receiver for a pulse modulation system," *Bell Syst. Tech. J.*, Vol. 46, pp. 2163–2177.

M. K. Simon and M.-S. Alouini, 1998. "A unified approach to the performance analysis of digital communication over generalized fading channels," *IEEE Proceedings*, Vol. 86, pp. 1860–1877, September.

M. K. Simon, S. M. Hinedi, and W. C. Lindsey, 1995. *Digital Communication Techniques – Signal Design and Detection*. Englewood Cliffs, NJ: Prentice-Hall.

B. Sklar, 1983. "A structural overview of digital communications—A tutorial review," Part I, *IEEE Communications Magazine*, Vol. 21, pp. 4–17, August; Part II, *IEEE Communications Magazine*, Vol. 21, pp. 4–17, October;

D. Slepian, 1976. "On bandwidth," *IEEE Proceedings*, Vol. 64, pp. 292–300.

N. J. A. Sloane and A. D. Wyner, eds., 1993. *Collected Papers of C. E. Shannon*. New York: IEEE Press.

J. W. Smith, 1968. "A unified view of synchronous data transmission system design," *Bell Syst. Tech. J.*, Vol. 47, pp. 273-300.

J. M. Smith, 1977. *Mathematical Modeling and Digital Simulation for Engineers and Scientist*, New York: J. Wiley & Sons.

D. A. Spaulding, 1969. "Synthesis of pulse-shaping networks in the time domain," *Bell Syst. Tech. J.*, Vol. 48, pp. 2425–2444.

P. Stavroulakis, ed., 1980. *Interference Analysis of Communication Systems*, New York: IEEE Press.

J. P. Stenbit, 1964. "Table of generators for BCH codes," *IEEE Trans. Inform. Theory*, Vol. IT-10, pp. 390–391.

J. J. Stiffler, 1971. *Theory of Synchronous Communications*. Englewood Cliffs, NJ: Prentice-Hall.

A. H. Stroud and D. Secrest, 1966. *Gaussian Quadrature Formulas*. Englewood Cliffs, NJ: Prentice-Hall.

G. L. Stüber, 1996. *Principles of Mobile Communication*. Boston, MA: Kluwer Academic Publishers.

C.-E. W. Sundberg, 1986. "Continuos phase modulation," *IEEE Commun. Mag.*, Vol. 24, pp. 25–38, April.

C.-E. W. Sundberg and N. Seshadri, 1993. "Coded modulation for fading channels: An overview," *European Trans. Telecommun.*, Vol. 4, No. 3, pp. 309–324.

F. W. Symons, Jr., 1979. "The complex adaptive lattice structure," *IEEE Trans. Acoust., Speech, Signal Process.*, Vol. ASSP-27, pp. 292–295.

E. J. Thomas, 1971. "Some considerations on the application of the Volterra representation of nonlinear networks to adaptive echo cancellers," *Bell Syst. Tech. J.*, Vol. 50, pp. 2797–2805.

M. Tomlinson, 1971. "New automatic equaliser employing modulo arithmetics," *Electronics Letters*, Vol. 7, Nos. 5–6, pp. 138–139.

L. Tong, G. Xu, B. Hassibi, and T. Kailath, 1995. "Blind channel identification based on second-order statistics: A frequency-domain approach," *IEEE Trans. Inform. Theory*, Vol. 41, pp. 329–334.

L. Tong, G. Xu, and T. Kailath, 1994. "Blind identification and equalization based on second-order statistics: A time domain approach," *IEEE Trans. Inform. Theory*, Vol. 40, pp. 340–349.

M. D. Trott, S. Benedetto, R. Garello, and M. Mondin, 1996. "Rotational invariance of trellis codes—Part I: Encoders and precoders," *IEEE Trans. Inform. Theory*, Vol. 42, pp. 751–765.

J. K. Tugnait, "On blind identifiability of multipath channels using fractional sampling and second-order cyclostationary statistics," *IEEE Trans. Inform. Theory*, Vol. 41, pp. 308–311.

G. L. Turin, 1969. *Notes on Digital Communications.* New York: Van Nostrand Reinhold.

G. Ungerboeck, 1972. "Theory on the speed of convergence in adaptive equalizers for digital communication," *IBM Journal of Research and Development*, Vol. 16, pp. 546–555.

G. Ungerboeck, 1974. "Adaptive maximum-likelihood receiver for carrier-modulated data transmission systems," *IEEE Trans. Commun.*, Vol. COM-22, pp. 624-636.

G. Ungerboeck, 1976. "Fractional tap-spacing equalizer and consequences for clock recovery in data modems," *IEEE Trans. Commun.*, Vol. COM-24, pp. 856–864.

G. Ungerboeck, 1982. "Channel coding with multilevel/phase signals," *IEEE Trans. Inform. Theory*, Vol. IT-28, pp. 56-67.

G. Ungerboeck, 1987. "Trellis-coded modulation with redundant signal sets – Part I: Introduction," *IEEE Communications Magazine,* Vol. 25, pp. 5–11, February. "Trellis-coded modulation with redundant signal sets – Part II: State of the art," *Ibidem*, pp. 12-21.

H. L. Van Trees, 1968. *Detection, Estimation, and Modulation Theory.* Vol. 1. New York: J. Wiley & Sons.

J. Ventura-Traveset, G. Caire, E. Biglieri, and G. Taricco, 1997a. "Impact of diversity reception on fading channels with coded modulation—Part I: Coherent detection," *IEEE Trans. Commun.*, Vol. 45, pp. 563–572.

J. Ventura-Traveset, G. Caire, E. Biglieri, and G. Taricco, 1997b. "Impact of diversity reception on fading channels with coded modulation—Part II: Differential block detection," *IEEE Trans. Commun.*, Vol. 45, pp. 676–686.

J. Ventura-Traveset, G. Caire, E. Biglieri, and G. Taricco, 1997c. "Impact of diversity reception on fading channels with coded modulation—Part III: Co-channel interference," *IEEE Trans. Commun.*, Vol. 45, pp. 809–818.

F. L. Vermeulen and M. E. Hellman, 1974. "Reduced state Viterbi decoders for channels with intersymbol interference," *IEEE International Conference on Communications (ICC'74)*, Minneapolis, MN.

A. J. Viterbi, 1965. "Optimum detection and signal selection for partially coherent binary communication," *IEEE Trans. Inform. Theory*, Vol. IT-11, pp. 239–246.

A. J. Viterbi, 1966. *Principles of Coherent Communications.* New York: McGraw-Hill.

A. J. Viterbi, 1967. "Error bounds for convolutional codes and an asymptotically optimum decoding algorithm," *IEEE Trans. Inform. Theory*, Vol. IT-13, pp. 260–269.

A. J. Viterbi and J. K. Omura, 1979. *Principles of Digital Communication and Coding.* New York: McGraw-Hill.

V. Volterra, 1959. *Theory of Functionals and of Integro-Differential Equations.* New York: Dover.

U. Wachsmann and J. Huber, 1995. "Power and bandwidth efficient digital communication using turbo codes in multilevel codes," *European Trans. Telecommun.*, Vol. 6, pp. 557–567.

H. Wakamatsu, 1981. "Inverse systems of nonlinear plant represented by discrete Volterra functional series," *VIII World IFAC Symposium*, Tokyo.

L.-F. Wei, 1984. "Rotationally invariant convolutional channel coding with expanded signal space. Part II: Nonlinear coding," *IEEE Journal on Selected Areas in Communications,* Vol. SAC-2, pp. 672-686.

L.-F. Wei, 1987. "Trellis-coded modulation with multidimensional constellations," *IEEE Trans. Inform. Theory*, Vol. IT-33, pp. 483–501.

D. D. Weiner and J. F. Spina, 1980. *Sinusoidal Analysis and Modeling of Weakly Nonlinear Circuits with Application to Nonlinear Interference Effects.* New York: Van Nostrand Reinhold.

B. Widrow, J. M. McCool, M. G. Larimore, and C. R. Johnson, Jr., 1976. "Stationary and nonstationary learning characteristics of the LMS adaptive filter," *IEEE Proceedings*, Vol. 64, pp. 1151–1162.

J. K. Wolf, 1978. "Efficient maximum-likelihood decoding of linear block codes using a trellis," *IEEE Trans. Inform. Theory*, Vol. IT-24, pp. 76–80.

J. M. Wozencraft, 1957. "Sequential decoding for reliable communications," *Tech. Report 325*, RLE MIT, Cambridge, MA.

J. M. Wozencraft and I. M. Jacobs, 1965. *Principles of Communication Engineering.* New York: J. Wiley & Sons.

J. M. Wozencraft and R. S. Kennedy, 1966. "Modulation and demodulation for probabilistic decoding," *IEEE Trans. Inform. Theory*, Vol. IT-12, pp. 291–297.

K. Yao, 1972. "On minimum average probability of error expression for a binary pulse communication system with intersymbol interference," *IEEE Trans. Inform. Theory*, Vol. IT-18, pp. 528–531.

K. Yao and E. Biglieri, 1980. "Multidimensional moment error bounds for digital communication systems," *IEEE Trans. Inform. Theory*, Vol. IT-26, pp. 454–464.

Y. Yasuda, K. Kashiki, and Y. Hirata, 1984. "High-rate punctured convolutional codes for soft-decision Viterbi decoding," *IEEE Trans. Commun.*, Vol. COM-32, pp. 315–319.

L. A. Zadeh, 1957. "Signal-flow graphs and random signals," *IEEE Proceedings*, Vol. 45, pp. 1413–1414.

E. Zehavi, 1992. "8-PSK trellis codes for a Rayleigh channel," *IEEE Trans. Commun.*, Vol. 40, pp. 873-884.

E. Zehavi and J. K. Wolf, 1987. "On the performance evaluation of trellis codes," *IEEE Trans. Inform. Theory*, Vol. IT-33, No. 2, pp. 196–201.

K. S. Zigangirov, 1966. "Some sequential decoding procedures," *Problemy Peredachi Informacii*, Vol. 2, pp. 13–25.

J. Ziv and A. Lempel, 1977. "A universal algorithm for sequential data-compression," *IEEE Trans. Inform. Theory*, Vol. IT-23, pp. 337–343.

# Index

DEMCO